Lithography

Edited by

Michael Wang

Lithography
Edited by Michael Wang

Republished by InTech
Janeza Trdine 9, 51000 Rijeka, Croatia

Lithography, Edited by Michael Wang
p. cm.
ISBN 978-953-307-064-3

For EU product safety concerns:
IN TECH d.o.o., Prolaz Marije Krucifikse Kozulić 3, 51000 Rijeka, Croatia,
info@intechopen.com or visit our website at intechopen.com.

Preface

Lithography, the fundamental fabrication process of semiconductor devices, is playing a critical role in micro- and nano-fabrications and the revolution in high density integrated circuits. Traditional optical lithography (photolithography) including contact and project photolithography has contributed significantly to the semiconductor device advancements. As the resolution requirement increases for fabrication of finer and smaller components and devices, the technological dependence on photolithography becomes a serious problem since the photolithography resolution is restricted by the diffraction limitation of optics. Reducing the light wavelength from blue to near ultraviolet (UV) and to deep UV is expected to improve the photolithography resolution, but it is far not enough to catch up with the pace in resolution demand in integrated circuit fabrication.

To overcome the limitation of photolithography, several advanced lithography techniques have emerged including maskless laser direct write lithography to eliminate the needs of photo masks, gray-scale lithography to increase the aspect ratio of the lithographic features, immersion lithography to increase the numerical aperture of the focusing optics and thus the resolution, and lithographic techniques based on further reducing wavelengths for better resolution such as extreme ultraviolet (EUV) lithography and X-ray lithography. The uses of particle waves like electrons and ion beams have resulted in high resolution electron beam (E-beam) lithography and ion beam lithography. Besides these wave and beam based lithographic techniques, there are also direct contact lithography such as soft lithography using soft molding and nanoimprint lithography with extremely high resolution. The plasmonic lithography is now in the horizon.

This book is the result of inspirations and contributions from many researchers worldwide. Although the inclusion of the book chapters may not be a complete representation of all lithographic arts, it does represent a good collection of contributions in this field. We hope readers will enjoy reading the book as much as we have enjoyed bringing the book together. We would like to thank all contributors and authors of book chapters who entrusted us with their best work. We also acknowledge the great effort of people who had invested their time in reviewing manuscripts and revision updates.

Each lithographic technique has its advantages and limitations. The conventional photolithography is not discussed here due to its technological maturity. This book begins with maskless, gray-scale, and immersion lithography since they are most close to the conventional photolithography in terms of resolution. Laser nonlinear lithography with capability of non-flat surface fabrication and character projection lithography are also included in this chapter. Further chapters present discussion on objective lens, resist materials and processing, and optical proximity correction.

Part II collects a number of development efforts related to EUV and X-ray lithography. Because of the use of extreme UV light beam of short wavelength, the lithographic resolution is greatly enhanced. The selected chapters discuss EUV light sources, mirrors, and plasma modeling, followed by a chapter discussing X-ray lithography.

Part III of the book is devoted to E-beam lithography which is widely used for high resolution patterning. Because of very short electron particle wavelength, the E-beam lithography is one of the most attractive lithographic techniques at this time. Its resolution limitation comes from unwanted E-beam resist exposure by electron scattering. Great discussions have been presented in this part of the book including design, exposure control, and nano-scale fabrication. Simulations on E-beam lithography and ion beam lithography are also presented.

Apart from above wave and beam lithography, contact lithography is promising for large area fast patterning of fine features. One of the representing lithographic techniques is the soft lithography. It uses a soft mold to reproduce needed features through the techniques of replica molding, micromolding in capillaries, microcontact printing, microtransfer molding, and microfluidic fabrication. This part IV collects chapters related to micro optical device fabrication, SiC microstructure fabrication, and soft lithography for single-object level study of cells and molecules.

Because of appearing nano scale resolution, nanoimprint lithography as a kind of contact lithography has attracted a great deal of attention in recent years. This part V is devoted to the nanoimprint lithography. The selected chapters discuss nanoimprint lithography techniques including hot embossing, UV-nanoimprint lithography, and micro contact printing. The effect of ultrasonic vibration and molecular dynamics are also discussed. There are further discussions on three-dimensional patterning and various applications.

Other emerging lithographic technologies are collected in Chapter VI including plasmonic lithography, nanosphere lithography, self assembly of nanoparticles, and high resolution polymer patterning.

Editor

Michael Wang
University of Miami
U.S.A.

Contents

II EUV and X-Ray Lithography

III E-Beam and Ion Beam Lithography

Advanced Photolithography

1

Direct Laser Lithography and Its Applications

Hyug-Gyo Rhee
Korea Research Institute of Standards and Science
Republic of Korea

1. Introduction

Computer Generated Hologram (CGH) is widely used for testing of large aspheric surfaces. It, however, is difficult to fabricate by the well-known E-beam methods because most of the CGH requires a large diameter (ø75 to 1,000 mm) and a tough precision (position accuracy of each line in the CGH should be less than 50 nm.). In this case, the direct laser lithography can be a proper choice to fabricate the CGH becuse it can easily extend the size (patterned area) with high precision.

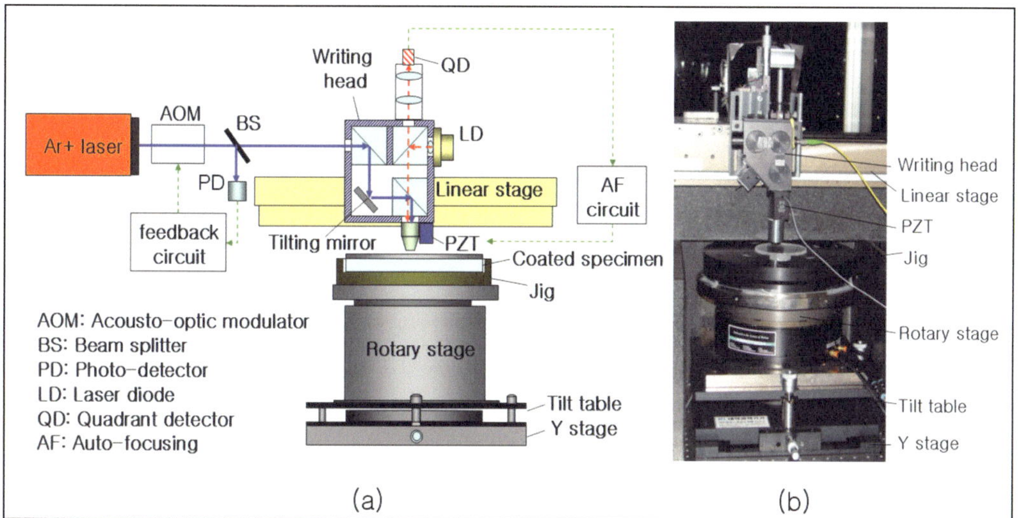

Fig. 1. (a) Configuration and (b) photographic view of the direct laser lithographic system.

Figure 1 (a) shows the configuration of typical direct laser lithographic system, which includes (1) the intensity stabilization and control part, (2) the writing head with autofocusing mechanism, and (3) the moving part. The photographic view of the assembled lithographic system is shown in Fig. 1 (b). The blue light in this figure is the lithographic beam whose wavelength is 457.9 nm. The laser lithographic system requires a high stability of the intensity of the source. In the fluctuating spectrum of a gaseous laser, large variations may be found in the low frequency range, from dc to several hundred Hz, and considerably smaller variations in the frequency band to several hundred kHz. The first is attributed to such main factors as thermal variations of the resonant cavity, mechanical vibrations, dust particles and air currents, instability, and the hum of the power supply. The second is mainly due to the oscillations in the plasma of the discharge column, especially in the region of the space charge at the cathode. The Ar+ laser that is used as a microfabrication source in our system also shows the above-mentioned beam fluctuations. For the source stabilization, we have introduced an Acousto-Optic Modulator (AOM), a photodetector, and produced a servo controller for controlling the AOM modulation depth.

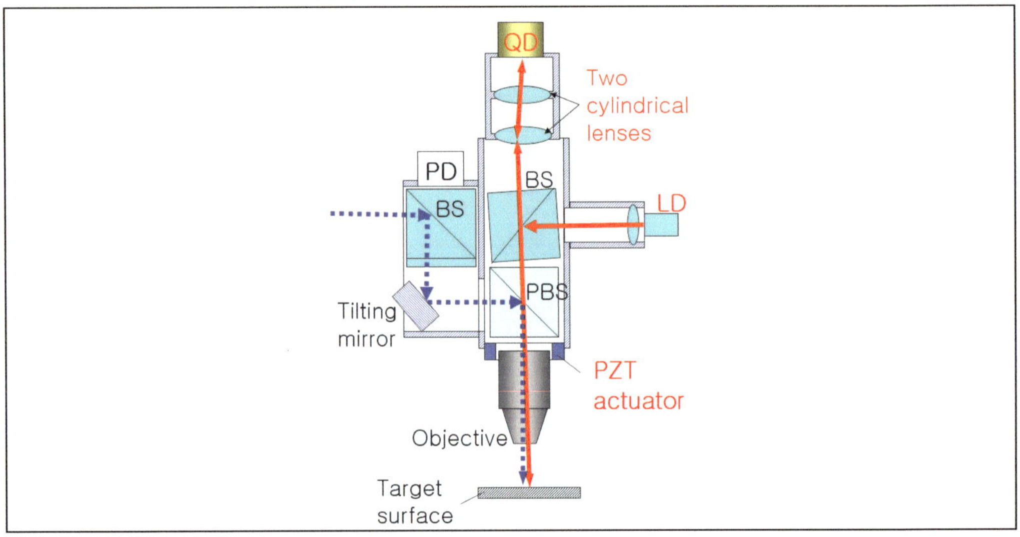

Fig. 2. Configuration of the writing head.

The stabilized lithographic beam from the AOM comes into the writing head as shown in Fig. 2. The dotted line and the solid line represent the lithographic and the autofocusing beam, respectively. The tilting mirror permits a direction change of the lithographic beam with a 0.02° resolution in order to compensate for the run-out error of the rotary motor. One of important functions of the writing head is autofocusing. Furthermore, 20X, 50X, and 100X objectives are available in our system to alter the lithographic spot size. Each objective requires a different set of astigmatic lenses for the best autofocusing performance. Table 1 shows the specifications of the moving part. A laser interferometer was also placed in the system to check the exact position of the writing head.

Specification	Rotary stage	Linear stage
Range	360°	200 mm
Control type	Closed loop	Closed loop
Feedback sensor	Rotary encoder	Linear encoder
Resolution	0.0547 sec.	10 nm
Max. Speed	600 rpm	100 mm/s
Axis loads	Wafer chuck + 1 kg	7 kg

Table 1. Specification of the rotary and the linear stage.

Figure 3 shows the main page of the operating software we developed. A pattern for fabrication is displayed in Part 1. Part 2 is a stage motion test panel. Part 3 shows the signals

from the motor encoders, the laser interferometer, the PD of the intensity control part, and the status of the shutter. The overall fabrication state is displayed in Part 4.
The details of the intensity stabilization and the autofocusing are presented in Section 2. In Section 3, some applications are described.

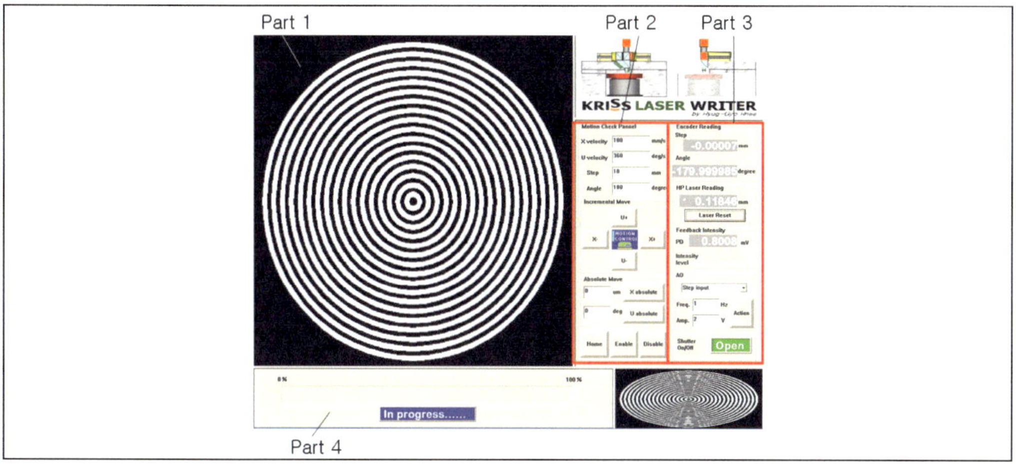

Fig. 3. Operating software.

2. Key technologies

2.1 Laser power stabilization

The source beam is the Ar+ laser with and output power of 1.5 W at 514.5 nm and of 300 mW at 457.9 nm wavelength. Its beam property is linearly polarized owing to plasma tube ends with Brewster angle cut. This is an important factor for power stabilization with our system. The stabilization system consists of the AOM, a cube beam splitter (BS), the photo-detector (PD), and the control servo circuit. The description of each part is as follow: The AOM is installed in the direction of propagation of the beam to conduct active power control, so that it is independent of the laser system used. For this reason, this setup will be able to apply to different kind of laser. The power control scheme using the AOM is that if its modulation depth is changed to minimize the intensity fluctuation in real time, the constant output power can be obtained. Hear the allowable modulation voltage limit corresponding to the modulation depth is up to 1 V. Therefore, to be well operated the AOM, the voltage between 0 V and 1 V have to be introduced to the AOM-RF driver. If the input voltage exceeds the limitation, the AOM loss their function as the active power controller. The first order beam passing through the AOM is split into two parts by BS. A small portion of reflected beam (about 8 %) is measured by the PD for stabilization itself. A linear polarizer (LP) is employed in front of the PD to prevent bring about serious damage to the PD. The used PD has the damage threshold of 100 mW for continuous wave and 0.5 J/cm^2 for 10 ns pulse, respectively. We carried out our experiment with the power of 65 mW. The servo controller was designed with an upper unity gain at 10 kHz to achieve high gain at low frequencies. The gain at frequencies below 100 Hz was at least 60 dB, which was sufficient gain to reduce the main fluctuation noise of around 100 Hz.
The mechanism of the stabilization part is that the detected photocurrent by PD is converted to voltage and then is compared with extremely low noise voltage reference (Analogue

Devices AD581JH) in the servo controller. An error signal out of the controller is introduced into the AOM-RF driver to change the diffractive ratio of the AOM instantaneously. Also owing to time taken to circulate through the loop is very short (about 1 μs), continuous stabilization is achieved instantly. In this way, the first order beam power is maintained constantly. In our experiment, the AOM and the controller have 20 MHz and 5 MHz bandwidth, respectively. These are enough values that cover the noise frequency band of the used Ar+ laser.

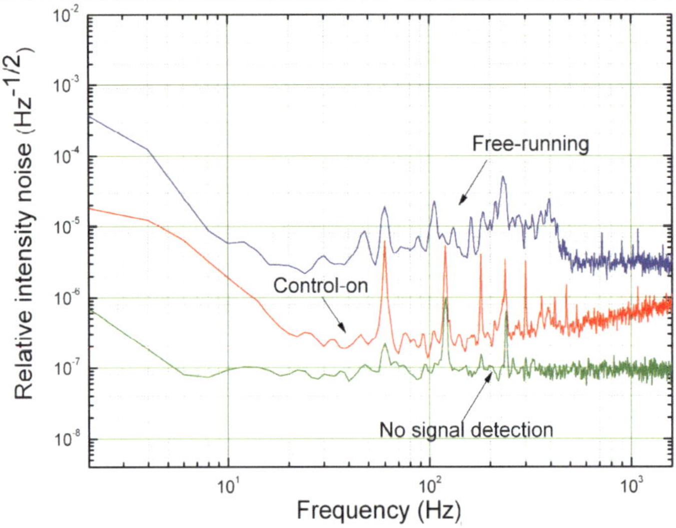

Fig. 4. Relative intensity noise of the Ar+ laser. The free-running mode shows the main noise frequency band of the laser.

Figure 4 shows relative intensity noise as measured by a FFT spectrum analyzer with DC to 102.4 kHz bandwidth (Stanford Research SR785) according to three different conditions which are free-running, control-on, and no signal detection. The free-running mode shows the large fluctuating noise at the low frequency range, from 20 Hz to 500 Hz. Once the control loop is on, we can see the noise level is dramatically reduced. If we look around the relative intensity noise level at relatively large fluctuation frequency of 100 Hz, we can see the value of 1.1×10^{-5} Hz$^{-1/2}$ is reduced to 2.1×10^{-7} Hz$^{-1/2}$ by about two orders. This result shows our stabilization system is able to carry out a function as a power controller. In the Fig. 4, there are regular bursts with 60 Hz spacing at the control-on or the no signal detection node. The reason is due to the florescent light incident on the PD. Later we could see that these bursts disappeared on the screen of the FFT spectrum analyzer as soon as the light was off. And if we expand the frequency band in the FFT spectrum analyzer, we can confirm our stabilization part has the noise control band up to 4 kHz. This range is sufficient to control the output power of various gaseous lasers including the Ar+ laser.
The performance of the stabilization part is shown in Fig. 5. It shows clearly the difference of the power stabilities before and after operating the control loop. Long term stability obtained by the control loop is ± 0.20 % for 12 minutes. This is a considerably reduced quantity comparable with the free-running mode of ± 0.77 %. Normally, it takes 11 or 12 hours to complete a piece of CGH. So we observed the long term stability for over 10 hours but the stability shown in the above result was not nearly changed.

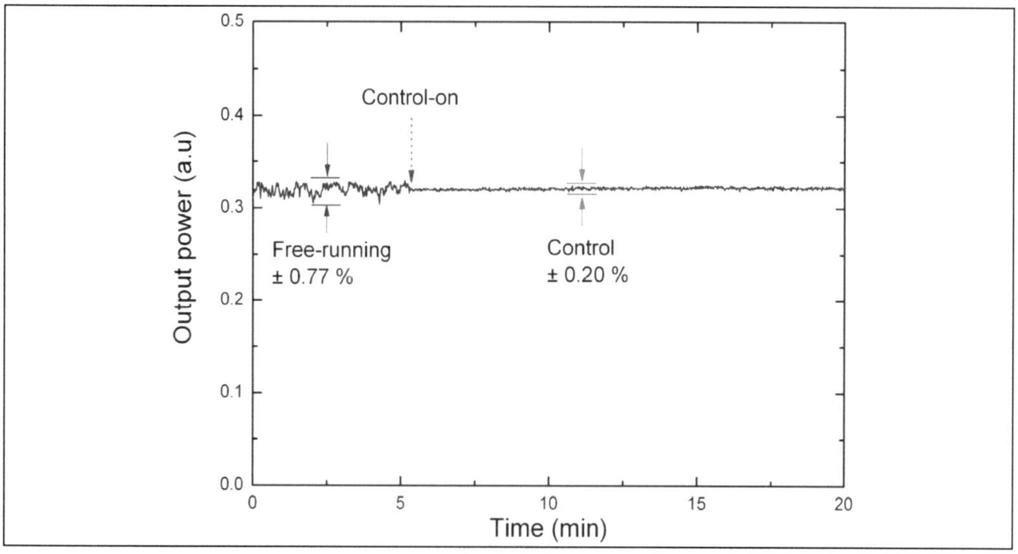

Fig. 5. Long term stability of the Ar+ laser before and after operating the control loop. The a.u means the arbitrary unit.

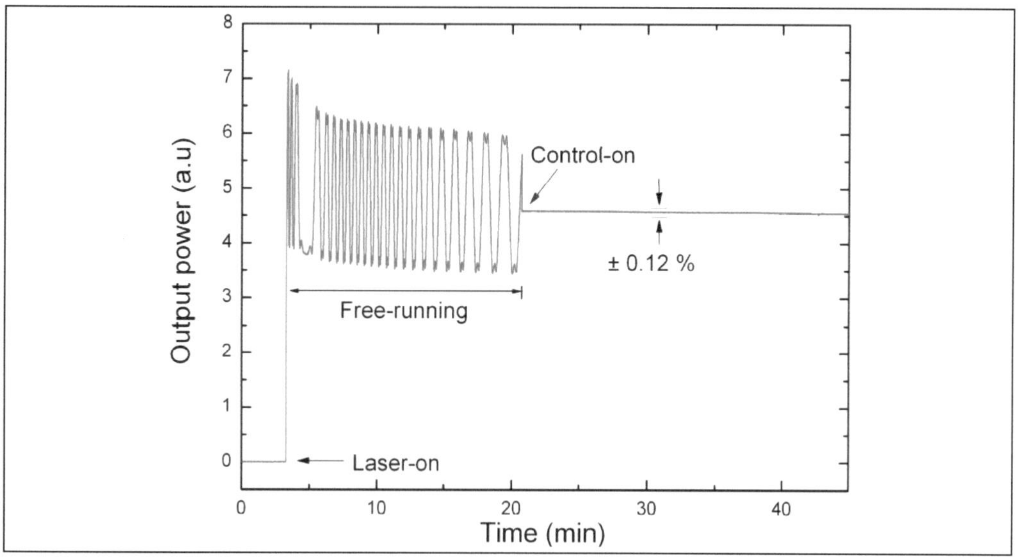

Fig. 6. Long term stability of the internal mirror He-Ne laser.

In addition, we have applied the stabilization part to a He-ne laser, and obtained the ± 0.12 % stability as shown in Fig. 6. An important thing is that the laser has always to be linearly polarized. The reason is as follow. First, if non-polarized beam is used, the first order beam power is considerably decreased because the AOM efficiency is maximized when the direction of polarization of the beam has to be perpendicular to the direction of propagation of acoustic wave passing through the AOM crystal. The second, if we look into a laser gain profile with respect to multi longitudinal mode, each mode ha P or S-polarization, so that the oscillating beam is mixed with P and S-polarization each other. These non-polarization beams were not controlled properly. In case of Ar+ laser, the plasma tube ends were cut with Brewster angle. Therefore the extracted beam was linearly polarized, so that we need not use any additional linear polarizer.

2.2 Beam focusing using the autofocusing technique

It is important to maintain constantly writing beam focus on the all of a surface during a fabrication for producing good quality of CGH and DOE. A principal factor affecting the variation of the focus is the tilted surface of a substrate. Because of the inclination, the size of the focal point is varied during the fabrication. In the result the line width is broader and the depth is shallower. In order to correct the defocusing amount which arises from the above reason, we have introduced an astigmatic strategy but enhanced new one, one of the active autofocusing methods, and produced an independent autofocusing controller to overcome speed limitation. The mechanism of the autofocusing system is as follows: an auxiliary reflection beam (LD in Fig. 1) from the surface of a substrate goes through a set of cylindrical lenses, and makes various intensity shapes on the focal plane depending on the distance variation between the surface and the cylindrical lens set. Here, the perpendicular cylindrical lens set plays an important role because it change sharply the intensity shape in x or y-direction according to the distance, so it maximizes astigmatism that it can increase defocusing amount and sensitivity. To this end, it is possible to make large scale optical fabrication maintaining uniform precision in comparison with the previous astigmatic method which is only applicable to small scale cases such as CD/DVD pickup. The intensity shape variation is accepted by a quadrant detector (EG&G UV140BQ-4) and then introduced to a computer by four different cables. The four signals from the quadrant detector (QD) make an error signal in the computer and feedback the error signal to the PZT actuator (PI P-721.0LQ) supported to the micro-objective lens to maintain the constant focal point on the substrate. By doing so, the constant focus can be formed on the surface. However, this method has a speed limitation of 9 Hz - if a spindle rotates one revolution in one second, then the autofocusing operates one time every 40 degrees - so that it is impossible to control the autofocusing on the high speed rotation with more than the angular velocity of 360 ^{o}/s. To improve the limitation, we have made an autofocusing controller built-in an electronic circuitry independent of the computer. The details of our system are explained as two following subsections.

Depending on the LD beam shape reflected from a target surface through an optical system, QD makes different photocurrents at four detecting areas for the autofocusing. Figure 7 shows the optical configuration of the autofocusing part. Firstly, we attached a band-pass filter, a linear polarizer, a biconvex lens, and a cube beam splitter to the writing head. The role of each part is as follows: the band-pass filter passes only 635 nm LD beam except for the 488 nm and 457.9 nm writing beam. The linear polarizer plays a role of reducing the residue reflection beam in the writing head that acts as a noise source on QD. The biconvex lens adjusts suitably beam scaling on QD and the optical distance from the target surface to QD, and the cube beam splitter reflects off most LD beam (about 90 %) coming from the surface to the cylindrical lens assembly. The other beam from the beam splitter (about 10 %) is monitored by a Charge Coupled Device (CCD). Secondly, we installed the cylindrical lens assembly composed of two cylindrical lenses right angle to each other on the linear stage. Lastly, QD mounted on a XYZ translator was also set up on the stage. In Fig. 7, the beam reflected from the material surface goes though the series of optics and makes a particular spot pattern near the focal planes. The QD is placed between two astigmatic foci that two cylindrical lenses originally have, namely at the location where the intensity pattern looks to be a perfect circle as shown in Fig. 7 (position 2). The orientation of QD lines should be 45 ^{o} or 135 ^{o} with respect to the tangential plane of the cylindrical lenses. If the distance between the material surface and the objective is changed, the intensity patterns on QD will also be changed, thereby bringing about the error signal change. The radius of curvature of each cylindrical lens was 15.5 mm, the focal length was 30 mm, the space between two lenses was 5.7 mm, and the effective focal length of two lenses was 15.7 mm.

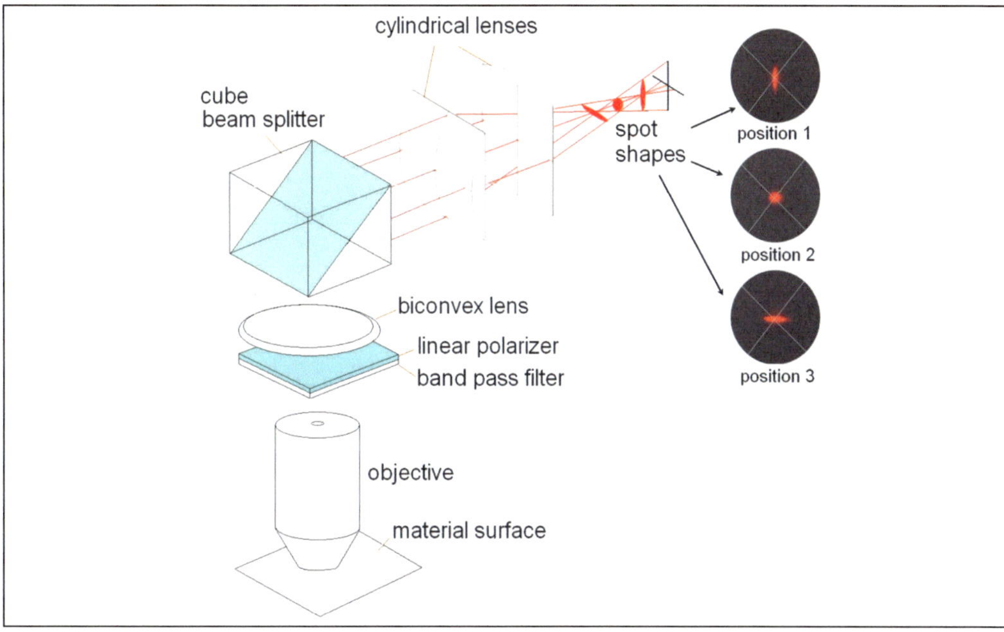

Fig. 7. Optical components to generate the auto-focusing error signal. The three different kinds of spot shape are formed on the QD according to the distance between the objective and the material surface. The position 2 indicates the exact focal point.

Our autofocusing controller is possible to achieve the high speed control up to 150 Hz (PZT modulation limit). Each signal received by QD is guided into the controller through four different BNC cables. In the controller, first, the guided current signals are converted into the voltage signals by current to voltage converters. In this process, the capacitance of a condenser affect considerably to the response time of the PZT actuator. Figure 8 shows Bode diagram according to the capacitance change in which the reducer the capacitance, the faster the actuator response. The optimal capacitance we found here is 1 nF as shown in Fig. 8(c). When the capacitance was less than this, the actuator was overshot even to the minimum gain in our experiment. Especially if no condenser was used like in Fig. 8(d), a lot of noise was occurred in the process of current to voltage conversion. Under the larger capacitance such as Fig. 8(a), the autofocusing speed could not catch up with the rotation speed of 360 ^{o}/s. Next, the converted voltage signals undergo a series of calculation (addition and subtraction), then makes a normalized error signal $e_N(I)$ as shown in the following Eq. (1).

$$e_N(I) = \frac{I_a + I_d - (I_b + I_c)}{I_a + I_d + (I_b + I_c)} \tag{1}$$

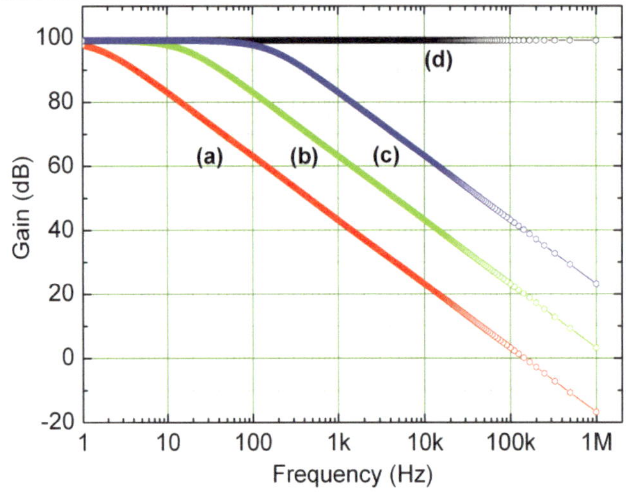

Fig. 8. Bode diagram about current to voltage conversion. (a) and (b) are the capacitance of 100 nF and 10 nF. (c) is in the case of optimal capacitance of 1 nF for fast response of the PZT actuator. (d) shows no condenser case.

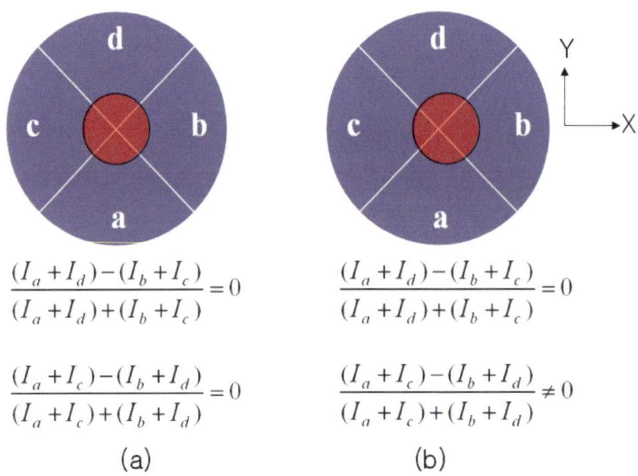

Fig. 9. Alignment of QD to find out center at the exact focal point. (a) The exact QD center, (b) a case of off-center.

where a, b, c, and d indicate the four sections of QD as shown in the Fig. 9(b), respectively. This normalization is able to reduce noise arising from the sudden light intensity fluctuation caused by dusts and/or stains on the surface of the substrate. In other words, this sudden intensity variation degrades the autofocusing function because of acting as a noise source. The normalized error $e_N(I)$ is divided into three parts and applied PID (Proportional Integral Derivative) control to each part. Multiplying suitable gains and summing each part, then a final focusing error signal (FES) is made. It can be written as

$$FES = K_p \times e_N(I) + K_i \times \int e_N(I)dI + K_d \times \frac{de_N(I)}{dI}, \tag{2}$$

where K_p, K_i, and K_d are the proportional, integral, and derivative gain, respectively. We have gotten better result than the case in which the proportional gain was only given. By doing so, residual error after the autofocusing was reduced more (less than 1 μm) and besides the control range was extended twice from ± 25 μm to ± 50 μm. Now, the remainder of work is to locate the laser beam shape to the center of QD. One problem is that when we look for the zero point for the autofocusing, the error signal would be zero even in the case of Fig. 9(b). To correct this we added a circuit which calculates position error signal (PES) to the controller. PES is given by

$$e_{NP}(I) = \frac{I_a + I_c - (I_b + I_d)}{I_a + I_c + (I_b + I_d)}, \tag{3}$$

$$PES = K_p \times e_{NP}(I), \tag{4}$$

where $e_{NP}(I)$ represents a normalized error signal for the estimation of position error deviated from QD center and K_p is the proportional gain. By setting FES and PES to be zero, respectively, we can improve the accuracy of our autofocusing control. To find the zero point, QD position with respect to X and Y axes in Fig. 9 was tuned FES and PES to be 0.000 ± 0.001 V at the initial focal point, and then we turned the control switch on with suitable gains (K_p, K_i, and K_d, respectively). The autofocusing controller gives the error signal to PZT actuator to keep the focusing position during the fabrication.

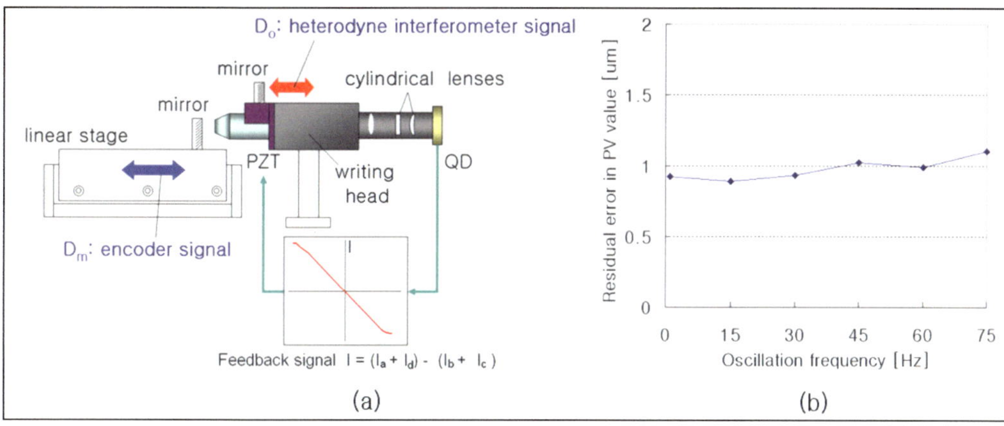

Fig. 10. (a) The experimental setup to check the performance of the autofocusing system, and (b) the residual error after the autofocusing. The higher frequency (larger than 75 Hz) test is meaningless because it is over the Nyquist sampling limitation.

Figure 10 (a) shows a setup to measure the autofocusing error. The mirror attached on the linear stage oscillated, and we measured D_m (the mirror movement) and D_o (the objective movement) at the same time. After whole test, the autofocusing error D_m-D_o was less than 1.1 μm in peak-to-valley (PV) value as shown in Fig. 10 (b). When the focusing point moved 1 μm, the line width change of the pattern was approximately 3.6 % with 100X objective (NA: 0.7, depth of focus: 0.6 μm). Therefore we suppose that the 1-μm focusing error is allowable.

Figure 11 shows CCD (charge coupled device) snap shots showing the focusing variation on the rotating surface. As shown in Fig. 11(a), the variation of the focus, due to surface tilting (the greatest contribution to the focus variation), according to the rotation angle can be observed. The focus size at 0 ° is increasing as the substrate rotates, and then back to normal after one rotation. On the contrary, however, once we apply the autofocusing control to it, the focus size is nearly invariant for one rotation. To confirm system performance in detail, we carried out writing tests divided into two parts, that is to say, with and without the autofocusing control. The target glass wafer is coated with chromium of 100 nm thickness. The test writing results given the surface radius change from 4.6 mm to 14 mm and 10 μm line to line spacing is shown in Fig. 12. As in the inner area of Fig. 12(a), when turn the autofocusing off, the fabricated pattern in the region out of the focus is entirely missed, and what is more, the written areas are also dimming even if the focusing region is little deviated from the focus. Applying the autofocusing control to the whole surface, on the

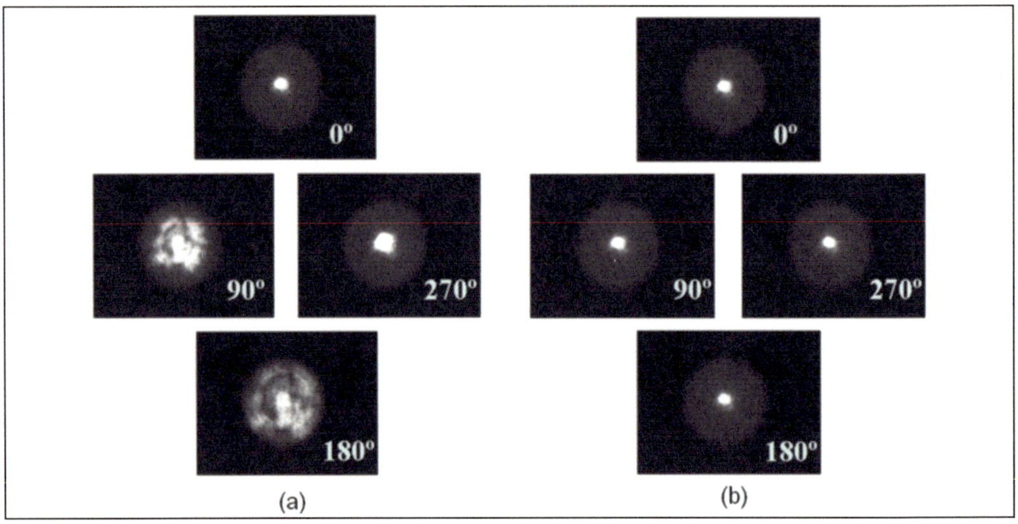

Fig. 11. CCD snap shots of a focal point variation on the rotating substrate (a) before and (b) after the auto-focusing control.

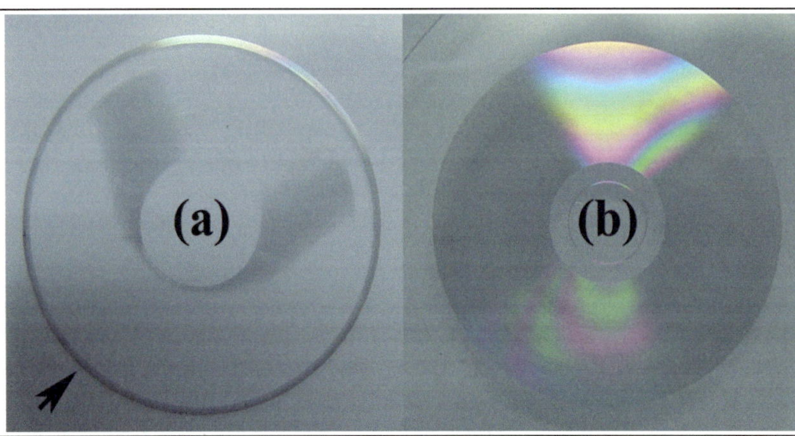

Fig. 12. Fabrication results: comparison of (a) before with (b) after the auto-focusing control. The arrowed circles in (a) is a result with the auto-focusing control to show the significance of it.

other hand, uniform patterns are well written as shown in the outermost arrowed circle in Fig. 12(a). Even in a small area there is a big difference whether the autofocusing is present or not, to say nothing of a large one. From the comparative results, the significance of autofocusing is emphasized here. Figure 12(b) shows that the patterns on the surface are successfully fabricated to the whole area by means of the autofocusing control. We could also confirm the uniform linewidth of 2.0 μm as shown in Fig. 13 using a commercial white-light scanning interferometer. In addition, we have accomplished the linewidth of 0.6 μm by means of controlling the source laser power. In the light of these facts, we can assure that our autofocusing system is well operated.

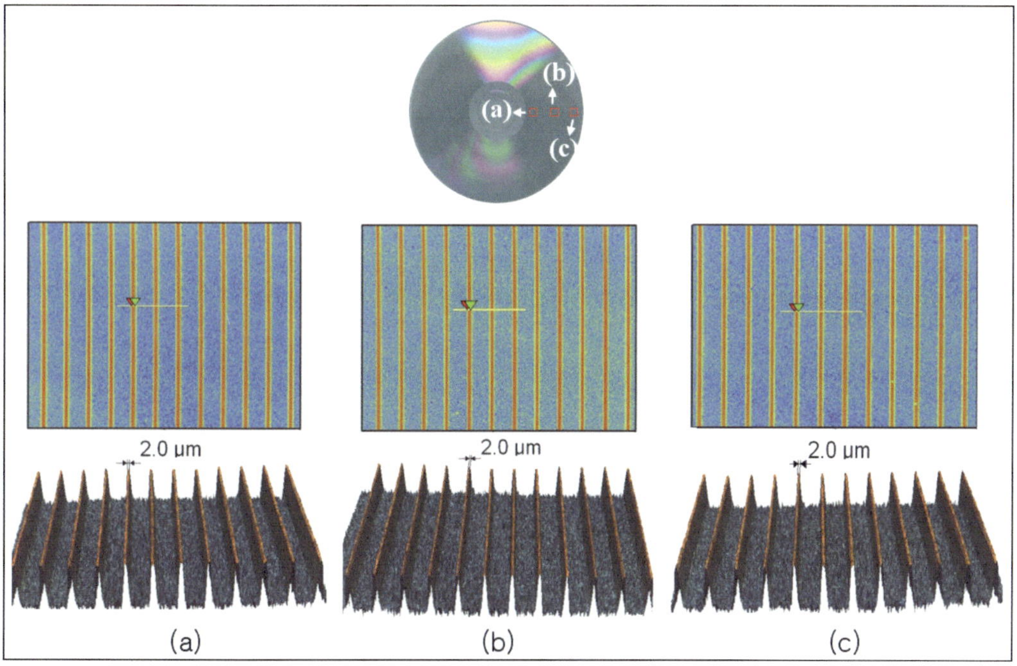

Fig. 13. Fabrication results obtained by a commercial white-light scanning interferometer (field of view: 124 μm × 93 μm, magnification: 50X). Region (a), (b), and (c) show uniform linewidth of 2.0 μm, respectively.

3. Applications

3.1 Computer Generated Hologram (CGH)

Figure 14 (a) show a typical CGH fabricated by the direct laser lithographic system. The root-mean-square wavefront error of the CGH was 0.03 λ (λ means the wavelength of the He-Ne laser, 632.8 nm.) as shown in Fig. 15.

When we fabricate a CGHm the center alignment of the writing head is an important error source. To align the center (origin) precisely, we used the tilt table and the Y stage as shown in Fig. 1.

We used a new alignment method by using four spirals. The procedure is here: (a) Fabricating the first spiral on the sample. (b) Rotating the sample 90° using the rotary stage, and then fabricating the second spiral. (c) Rotating the sample 90° again, and fabricating the third spiral. (d) Rotating the sample 90°, and fabricating the last spiral. If the center alignment is good enough, the spiral pattern looks like Fig. 16 (a). In this figure, the amount of misalignment was 168 nm. This number was calculated by the least square line fitting.

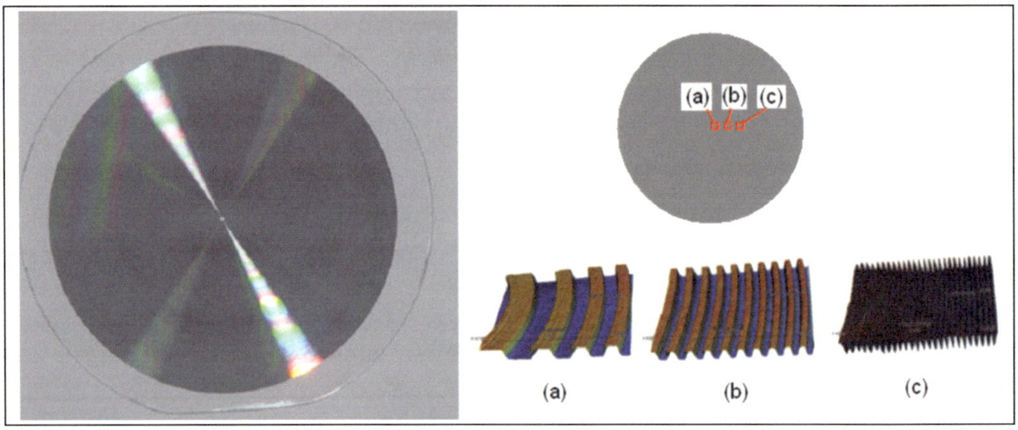

Fig. 14. Typical CGH we fabricated. The fabrication results measured by the white-light scanning interferometer (field of view: 124 μm × 93 μm, magnification: 50X).

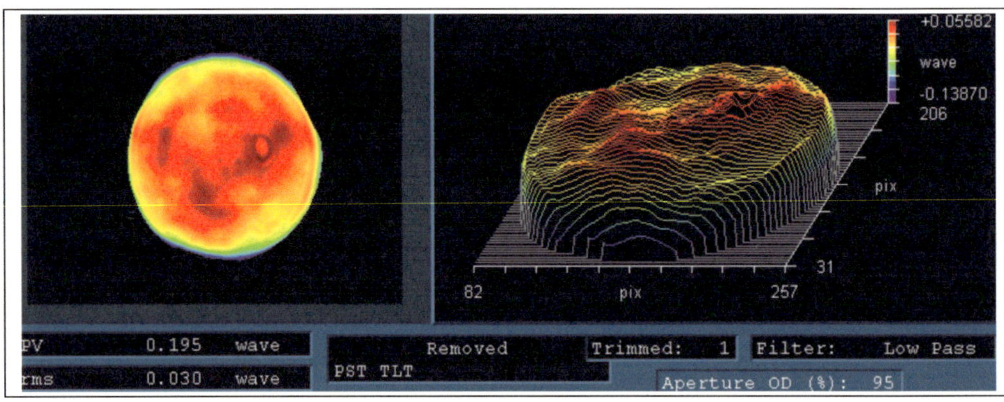

Fig. 15. Wavefront error of the CGH measured by a commercial Fizeau interferometer.

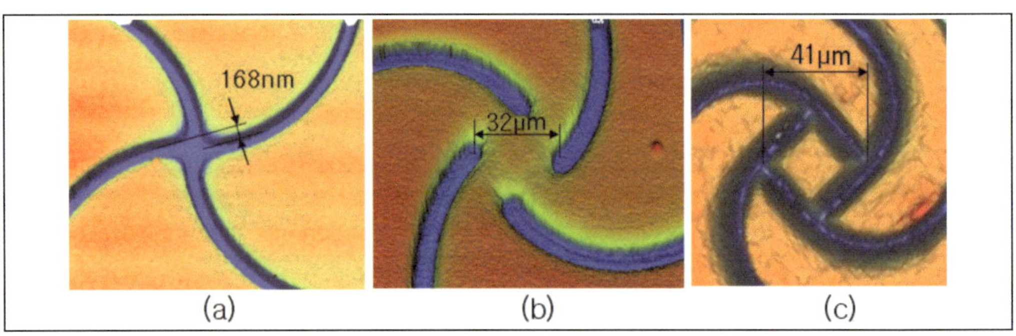

Fig. 16. Four spiral patterns: (a) good alignment case, (b) plus-direction misalignment case, and (c) minus-direction misalignment case.

3.2 Reference chromium patterns on a silicon wafer

In this section, we describe the second application of the direct laser lithographic system. In recent years, the semi-conductor industry has required a new inspection method for internal defects of the silicon wafer. To effectively find these small defects, some companies are developing new inspection equipment using the infrared light source. The infrared beam usually penetrates the silicon, and is scattered on the defect. Using this phenomenon the inspection equipment is able to find out the precise position and the size of each defect in the silicon wafer. At this time the precision of the equipment mainly depends on the coordinate system of the equipment itself. Therefore the equipment should be calibrated before inspection by using a well-made reference specimen that has two-dimensional array patterns whose xy-coordinates are already known. A typical fabrication procedure of the reference wafer is as follows: (a) preparing a wafer that has no internal defect, (b) polishing both sides of the wafer, (c) coating the chromium on the top side of the wafer, (d) patterning on the coated side, and (e) etching the wafer. In this procedure, chromium is preferable to other materials because it can effectually block the infrared beam with a relatively thin thickness. The other advantage of chromium is its easiness for fabrication.

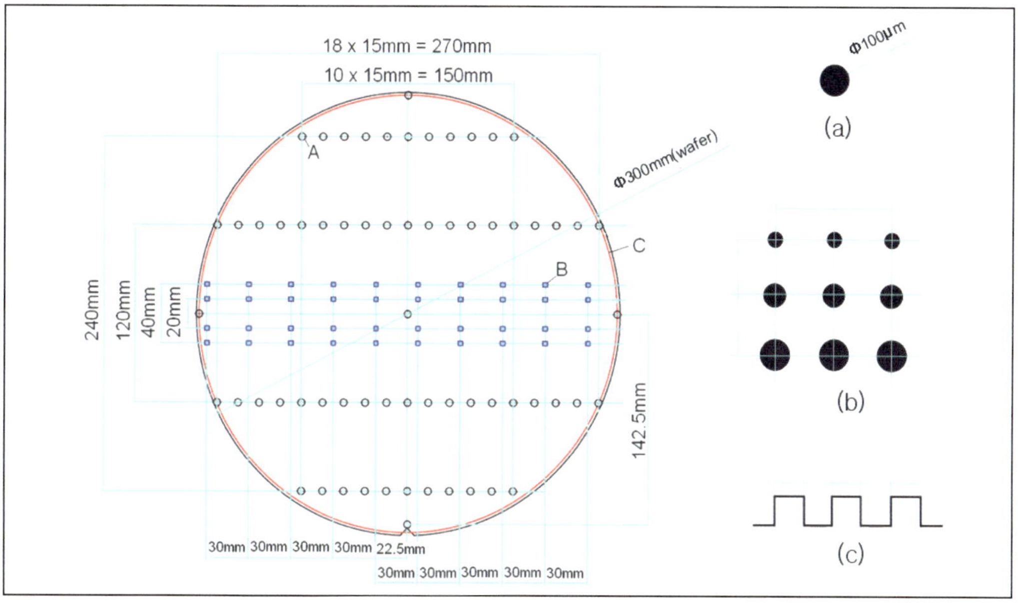

Fig. 17. 300-mm-diameter reference wafer. Using pattern C, the equipment can automatically level the reference wafer.

In spite of the advantages of the chromium, there are two problems. First, the semi-conductor industry requires a maximum 300-mm wafer as illustrated in Fig. 17. With the well-known E-beam method, however, it is hard to achieve this size. Therefore we applied the direct laser lithography technique instead of the e-beam. The second problem is that the most effective chromium etchant (we used the etchant consisting of six parts of 25% solution of $K_3Fe(CN)_6$ and one part of 25% solution of NaOH.) seriously erodes the silicon. To prevent this, we propose a new method using a SiO_2 layer whose thickness is from 100 nm to 200 nm. This layer can protect the silicon wafer from the etchant, and does not disturb the measurement since the infrared beam penetrates the SiO_2 laser. The details are described in Fig. 18.

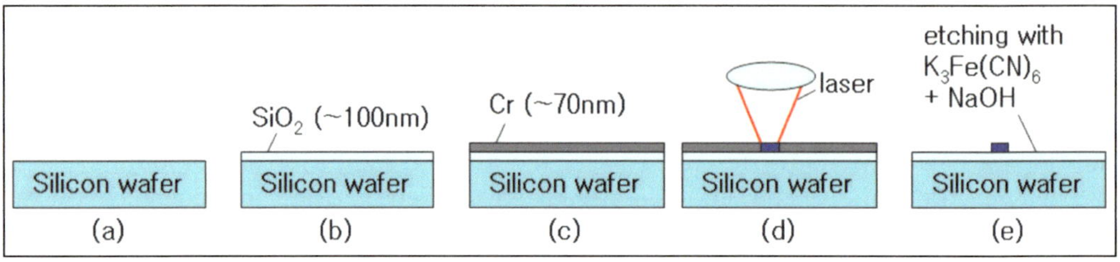

Fig. 18. Proposed procedure to fabricate the reference wafer

Figures 19, 20 and 21 show the fabrication result of each pattern. The thicknesses of the patterns are nearly 70 nm, which is enough to block the infrared light as shown in Fig. 19 (a) and 20 (a).

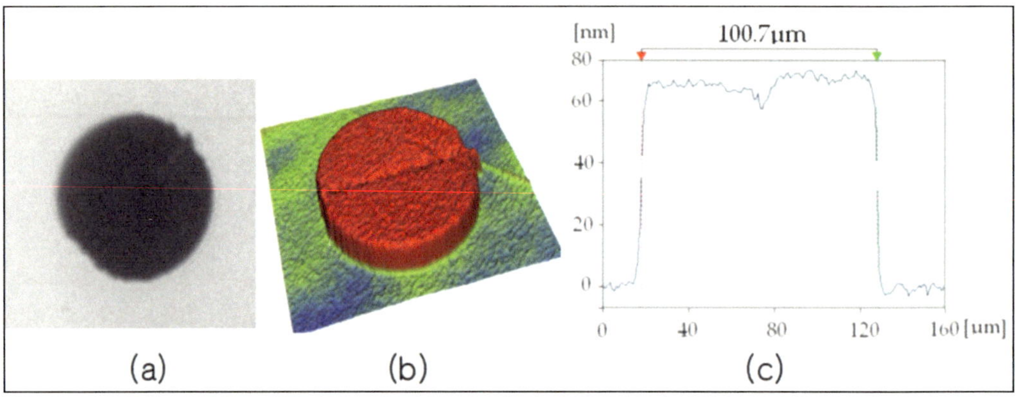

Fig. 19. Pattern A: (a) infrared microscopic view, (b) three-dimensional shape measured by a white-light scanning interferometer, and (c) a section profile. The measured diameter of the pattern was 100.7 μm instead of 100.0 μm. We supposed that this deviation is mainly caused by the fluctuation of the intensity and the focus point. It is also affected by the etching time.

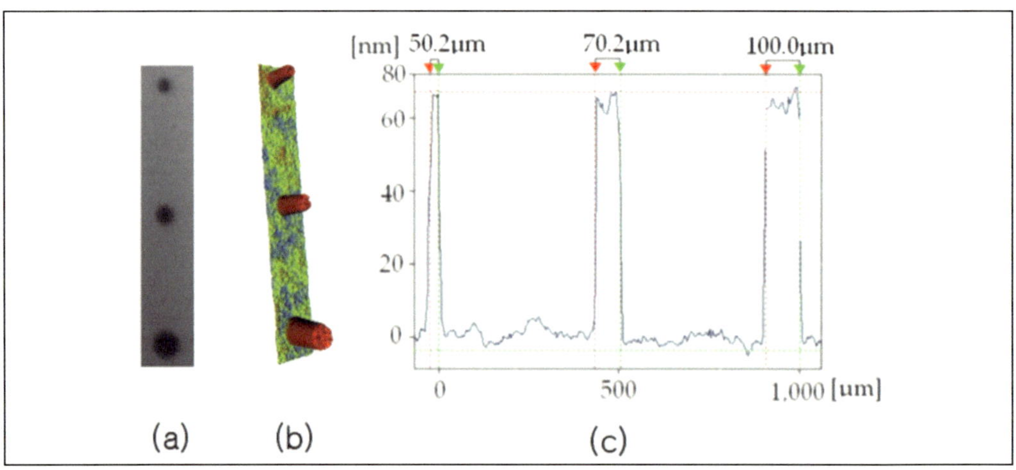

Fig. 20. A column of pattern B: (a) infrared microscopic view, (b) three-dimensional shape measured by a commercial white-light scanning interferometer, and (c) a section profile.

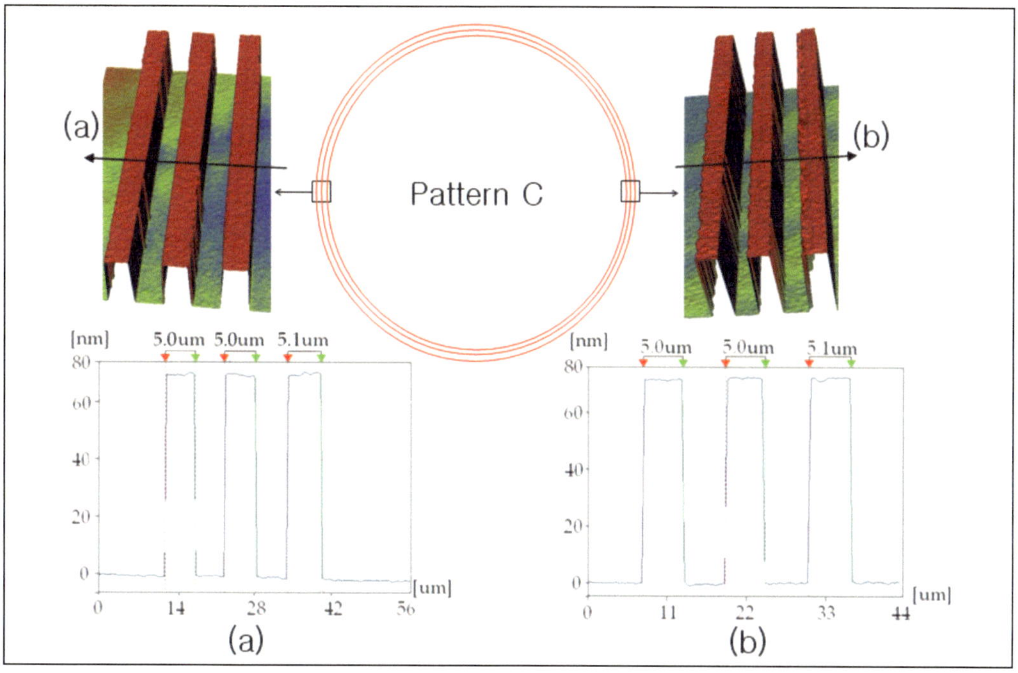

Fig. 21. Pattern C.

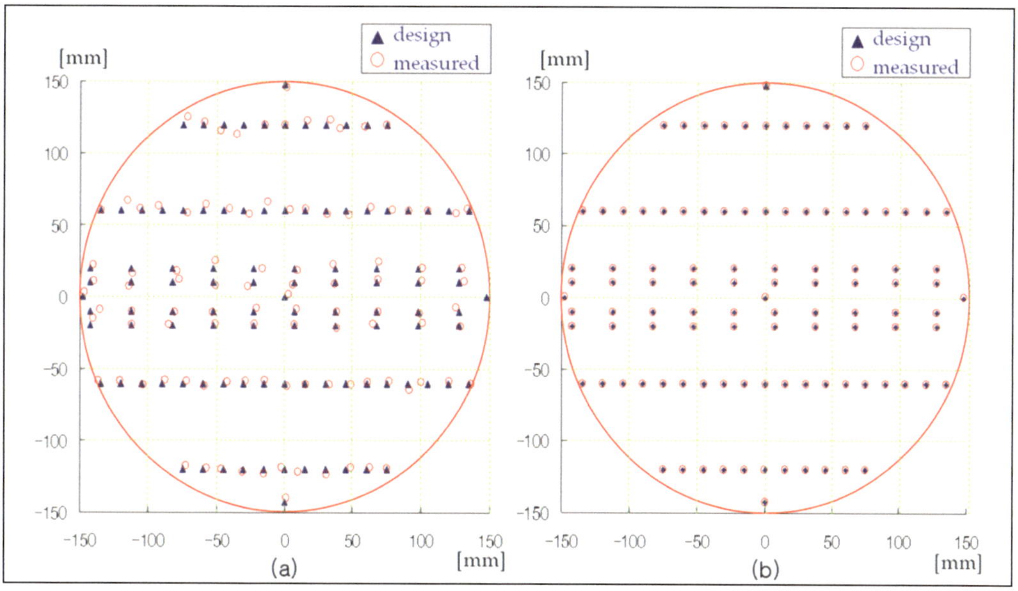

Fig. 22. Systematic error of the infrared inspection equipment (a) before and (b) after calibration. The triangular mark represents the designed position (see Fig. 17), while the circular mark means the measuring result. The positions of the circular marks are intentionally exaggerated.

Using the reference wafer, we finally tested the inspection equipment. The circular marks shown in Fig. 22 (a) represent the systematic error of the inspection equipment, in which the absolute values of the maximum error were measured as 8.9 μm for horizontal- and 12.6 μm for vertical-direction. We calibrated the equipment with the reference wafer, and successfully reduced the systematic error of the equipment as shown in Fig. 22 (b). After calibration, the maximum errors were 0.7 nm for horizontal- and 0.9 nm for vertical-direction in absolute value.

4. Conclusion

The direct laser lithography is a useful technique to fabricate a large precision patterns such as CGHs, DOEs, and reference wafers. The typycal lithograpic system we have built can write up to 360-mm diameter substrate coated with chromium or photoresist film. The writing source were stabilized by using the AOM, the PD, and the servo controller. We also achieve the high speed and large range autofocusing system using two cylindrical lenses. Then we fabricated various CGH, zone plates, and the 300-reference wafer . 150-mm, 200-mm, reference wafers were also successfully fabricated using our system.

5. References

Offner, A. & Malacara, D. (1992). *Optical Shop Testing 2nd edition*, Wiley, 0-471-52232-5, New York

Poleshchuk, A. G.; Churin, E. G.; Koronkevich, V. P.; Korolkov, V. P.; Kharussov, A. A.; Cherkashin, V. V.; Kiryanov, V. P.; Kiryanov, A. V.; Kokarev, S. A. & Verhoglyad, A. G. (1999). Polar coordinate laser pattern generator for fabrication of diffractive optical elements with arbitrary structure, *Appl. Opt.*, Vol. 38, No. 8, 1295-1301, 0003-6935

Asfour, J. & Poleshchuk, A. G. (2006). Asphere testing with a Fizeau interferometer based on a combined computer-generated hologram, *J. Opt. Soc. Am. A*, Vol. 23, No. 1, 172-178, 1084-7529

Ogata, S.; Tada, M. & Yoneda, M. (1994). Electron-beam writing system and its application to large and high-density diffractive optic elements, *Appl. Opt.*, Vol. 33, No. 10, 2032-2038, 0003-6935

Kim, D.; Rhee H.; Song, J. & Lee, Y. (2007). Laser output stabilization for direct laser writing system by using an acousto-optic modulator, *Rev. Sci. Instrum.*, Vol. 78, No. 10, 103110/1-103110/4, 0034-6748

Rhee H.; Kim, D. & Lee, Y. (2008). 300-mm reference wafer fabrication by using direct laser lithography, *Rev. Sci. Instrum.*, Vol. 79, No. 10, 103103/1-103103/5, 0034-6748

Rhee H.; Kim, D. & Lee, Y. (2009). Realization and performance evaluation of high speed autofocusing for direct laser lithography, *Rev. Sci. Instrum.*, Vol. 80, No. 7, 073103/1-073103/5, 0034-6748

Goodman, J. W. (1996). *Introduction to Fourier optics 2nd edition*, McGraw-Hill, 0-07-114257-6, Singapore

Deck L. & deGroot P. (1994). High-speed noncontact profiler based on scanning white-light interferometer, *Appl. Opt.*, Vol. 33, No. 31, 7334-7338, 0003-6935

Ye, J.; Takac, M.; Berglund, C. N.; Owen, G. & Pease, R. F. (1997). An exact algorithm for self-calibration of two-dimensional precision metrology stages, *Prec. Eng.*, Vol. 20, No. 1, 16-32, 0141-6359

2

High Aspect Ratio Sloping and Curved Structures Fabricated by Proximity and UV-LED Backside Exposure

Yoshinonori Matsumoto
Keio University,
Japan

1. Introduction

Fabrication technique for micro three-dimensional structures has been in wide demand in various devices in MEMS, micro optics, and the biomedical field. However, high aspect ratio sloping and curved structures of a few hundred micrometers height are generally difficult to fabricate by using the conventional method. The injection molding method is not suitable for small-lot production because the die assembly is expensive. The thermal reflow method does not provide high controllability of profile because the size depends on the surface tension. In the case of the gray scale mask method, the mask cost is high and it does not provide a high aspect ratio in the size of a few hundred micrometers. In this study, a new fabrication technique has been developed using a novel lithography process. This process provides structures with high uniformity over a large area and also high flexibility of design by only one exposure process. In this paper, details of the process and experimental results are described.

2. High aspect ratio sloping structures

2.1 Exposure system

Micro three-dimensional structures have been fabricated by a layer-by-layer process[1], a UV-LIGA process[2][3], moving mask UV lithography[4], and gray-scale lithography[5]. Another fabrication method has been proposed that utilizes chemically amplified epoxy-based negative resist(SU-8 etc.) and backside exposure[6]. The backside exposure is effective in creating a negative resist structure. It prevents resist peeling off from the substrate because the exposure and the photochemical reaction are occuring at the interface of the glass substrate.

However, the high aspect ratio sloping structure is still difficult to fabricate using these lithography techniques. In this chapter, a fabrication method for this type of structure is proposed based on use of the proximity exposure technique[7]. The method exposes the SU8 resist from the backside of the glass substrate to change the amount of exposure using a conventional mask. By using backside exposure, the exposure amount can be transformed into a resist structure. The top layer of the resist is removed in the developing process so that the uniformity and roughness of the resist during the coating process do not affect the final resist structure.

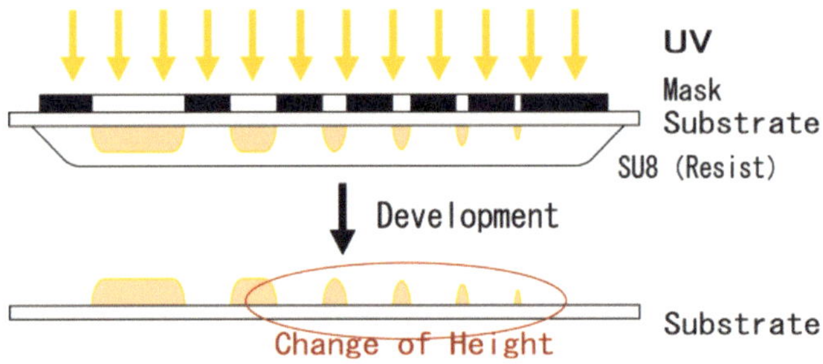

Fig. 1. Diagram of proximity back-side exposure.

2.2 Simulation of and results for the L&S structure

Diffraction phenomena can be positively used to change the amount of exposure at the boundary of the mask. In this study, a glass substrate 150 μm thick was used for the three-dimensional fabrication. The gap between the mask and the resist makes the lithography process in the proximity region dominated by Fresnel diffraction. The light intensity on the resist surface could be calculated based on the model proposed by Dill et al. (the "Dill model") [8]. The Dill model calculates the light intensity on the resist surface through numerical calculations of all elements of the processes of resist exposure and development. Grindle et al. have proposed a method (the "Grindle model") in which the light intensity distribution is obtained by numerical calculations, but the simulation of development is performed based on the actually measured dissolution rate values [9]. The lithography simulator ProxSim-2 (Lithotech Japan) is based on the Grindle model and designed for proximity lithography using a mask aligner.

The light intensity through the slit pattern from 20 μm to 100 μm in width was calculated by the lithography simulator. The simulation was preformed for the exposure dose of 90mJ/cm^2 at wave length of 365 nm, with a collimation angle of ±3°. The light intensity in the resist changed according to the glass thickness and mask opening. The light beams passing through the mask opening were diffracted on the Hopkins equation[10]. Figure 2 (a) shows the calculated UV intensity in SU8 resist. The UV intensity and actual exposure amount is changed depending on the slit width. The final resist structure was determined by the chemical reaction occurring during post-exposure bake and the resist development process. The simulator calculates the final strucuture based on the calculated light intensity distribution and the actually measured dissolution rate values for each intensities of SU-8 3000 resist. Figure 2 (b) shows simulated results for the final resist profile. The simulation predicts that the height changes depending on the slit width and that a slope is formed in the resist sidewall.

As described above, the SU-8 resist structures were fabricated with a mask in which the slit width changed between 7 μm and 35μm. SU-8 10 resist was dip-coated with a thickness of around 500 μm on a 150 μm thick glass substrate. After the prebake process, the glass substrate was backside-exposed for 5 s by the mask aligner LA310k(Nanometric Technology Inc., Japan). The light intensity of the wave length at 365 nm was 18mW/cm^2 with a collimation angle of ±3°. The glass substrate was post-exposure baked for 5min at 100°C. An SEM photograph of the fabricated SU-8 resist structures is shown in Fig. 3. The UV dose and the photochemical reaction changed based on the slit width under the proximity condition. The height of the structures is measured in Fig. 4. The height changes nearly constantly if the slit width is less than 30μm.

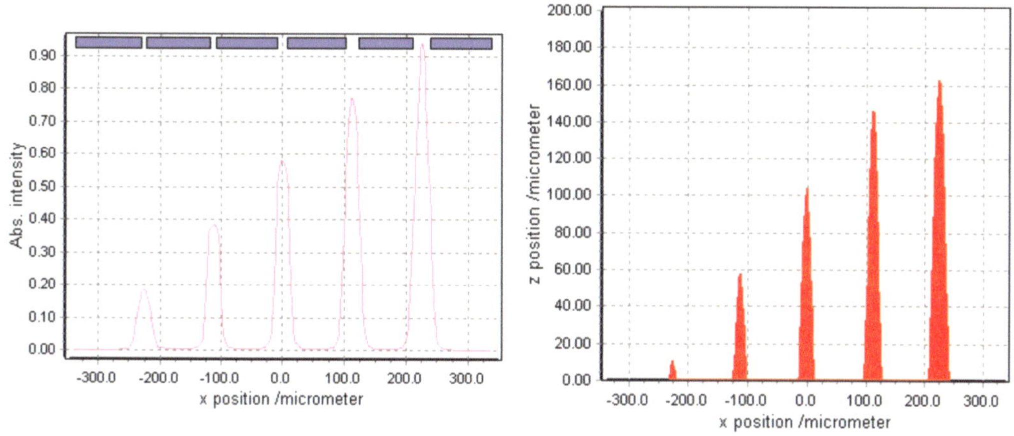

(a) UV intensity in SU-8 resist.

(b) Final resist structure after development process.

Fig. 2. Simulation results for different L&S pattern masks.

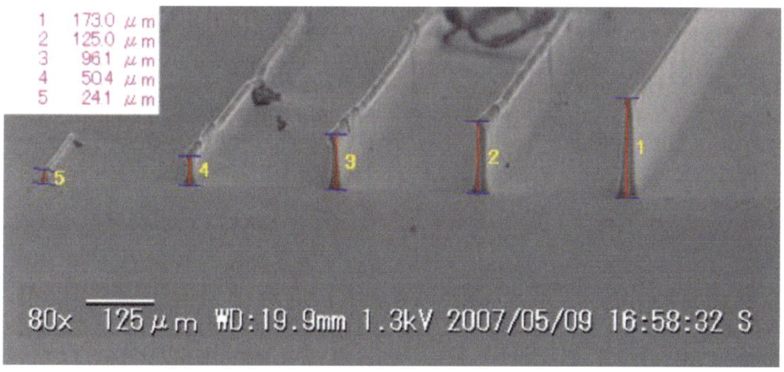

Fig. 3. SEM photograph of different height resist structures for different L&S pattern masks.

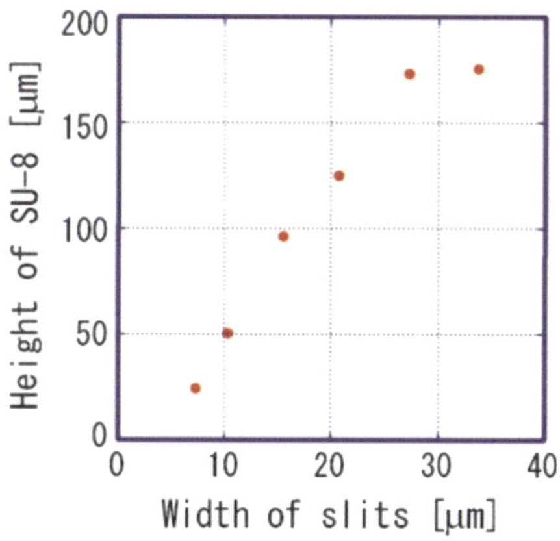

Fig. 4. Relation between slit width and the height of the SU8 resist.

A smooth sloping structure was fabricated by changing the exposure amount using the triangle mask pattern. The mask pattern design is shown in Fig. 5.

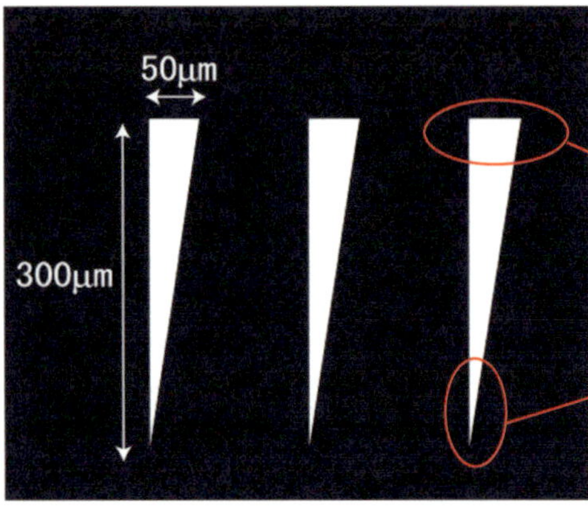

Mask pattern

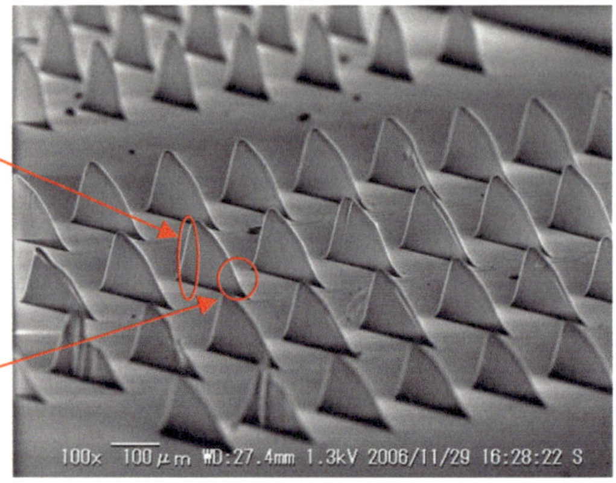

Exposure time 7.5 s

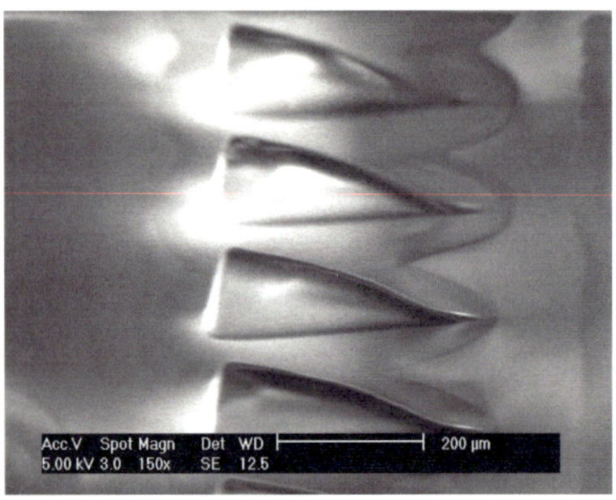

Exposure time 5 s

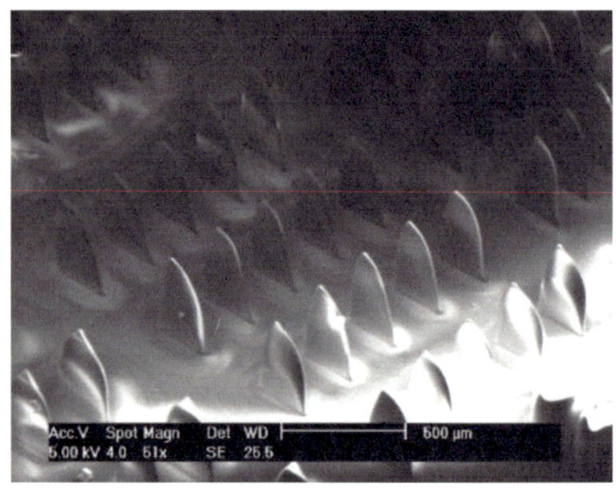

Exposure time 15 s

Fig. 5. Mask pattern and SEM photograph of micro sloping structures

The exposure times are 5 s, 7.5 s, and 15 s by the mask aligner LA310k(Nanometric Technology Inc., Japan). The fabrication results are shown in Fig. 5. These structures had a smooth sloping formation because the width was changed continuously in a triangular mask pattern. A micro sloping structure up to 200 μm was realized. As shown in Fig. 5, the structure height was controlled by adjusting the exposure time. Structures with a height of 250 μm were obtained when the exposure time was 15 sec. The SU-8 structures were transferred to metal by a molding technique for application in MEMS and biomedical engineering.

2.3 Simulation of and results for cylinder structure

Taper cylinder structures with a diameter of several tens of μm were obtained by using the diffraction phenomenon that occurred at the edge of the mask pattern. Figure 6 provides an example of the mask pattern.

The effect of the diffraction was calculated by ProxSim-2. Precise calculations for the mask pattern of Fig. 6 should be performed with a three-dimensional simulator. Because the ProxSim-2 is a two-dimensional simulator, the diffraction effect for the L&S pattern is calculated as shown in Fig. 7. The simulation results predict that taper structures will be formed by the diffraction phenomenon at the side of structure.

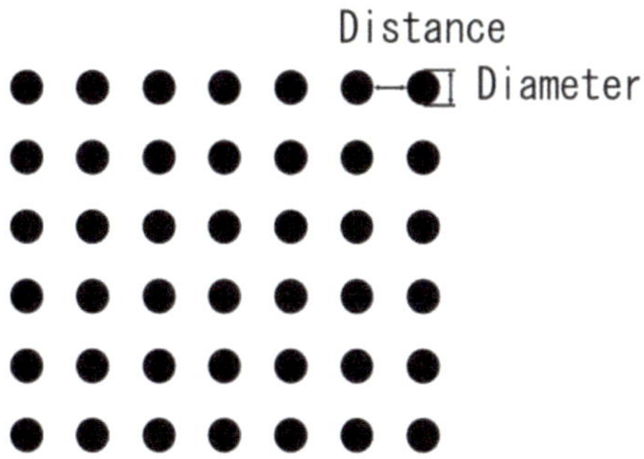

Fig. 6. Mask pattern of square array.

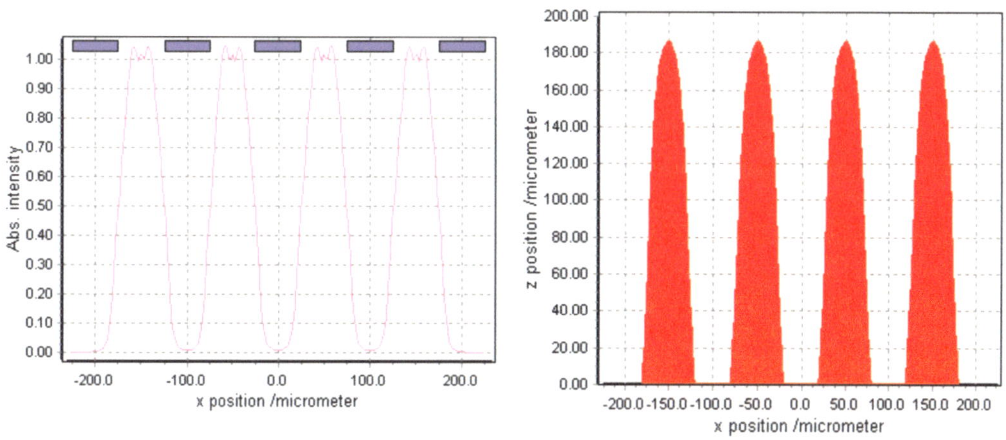

(a) 50 μm L&S pattern.

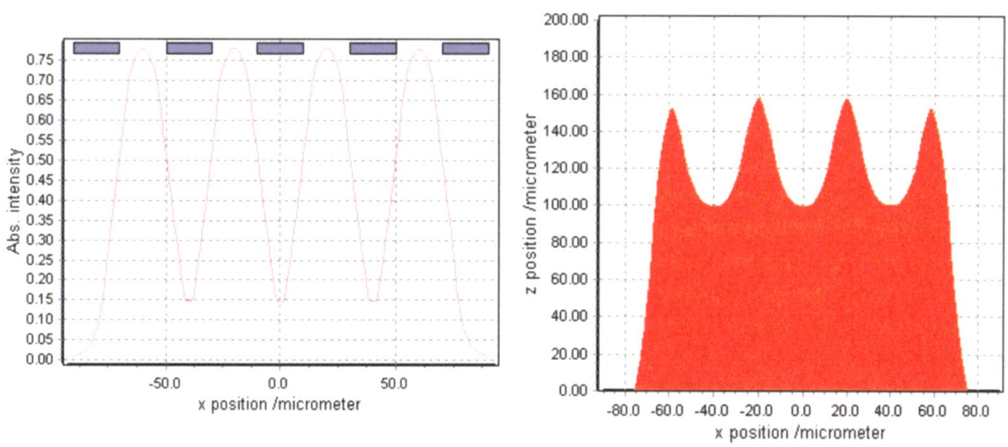

(b) 20 μm L&S pattern.

Fig. 7. Simulations result for exposure dose and resist structure

The SEM photograph obtained after backside exposure using the mask is shown in Fig. 7. Sloping hollows were formed under the diffraction in proximity condition.

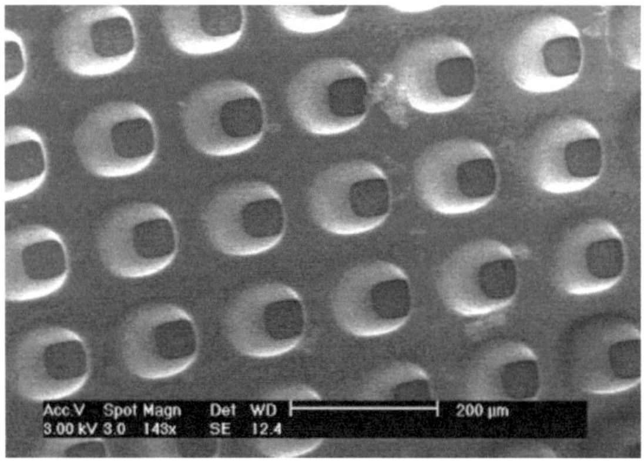

Fig. 8. SEM photograph of resist structure .

The structure shown in Fig. 8 can be used as a mold component. PDMS (Polydimethylsiloxane) was poured onto the structure and polymerized, and the resulting PDMS structures were easy peeled away from the SU-8 resist structures because of the tapered shape. An SEM photograph of the structures that were transferred to PDMS is shown in Fig. 9.
However, the height of Fig. 9 is approximately half of the calculated result of the L&S pattern in Fig. 7. The axisymmetrical diffraction effect should be large at the edge of the circle mask pattern in the experiment. The diffraction UV light forms a resist layer on the masking region. The resist layer reduces the hole depth of the resist and the height of the PDMS cylinder structure.

2.4 Application for super-hydrophobic surface

Super-hydrophobic surfaces have attracted much attention due to their practical application potential such as in micro TAS applications. The PDMS structures form a super hydrophobic surface that depend on the micro relief structures on the PDMS surface. The contact angle of the PDMS surface is measured by using a contact angle measurement apparatus. The normal PDMS surface of the contact angle was 105 degrees, and the super hydrophobic PDMS surface was achieved at over 160 degrees[7]. The size of the fabrication structure was changed in this study, and the relation between the size and the contact angle was measured, as shown in Fig. 10. The contact angle increases as the fabrication structure becomes smaller until reaching a size of 20 μm. However, a taper structure is not fabricated by the diffraction phenomenon when the size of the structure is less than 20 μm. Therefore, the contact angle becomes small. The repellency was highest at a diameter of 20 μm, and the maximum contact angle was 172° and, on average, 165°.

3. High aspect ratio curved structures

3.1 Exposure system

Another fabrication technique for high aspect ratio curved structures has been developed using a UV-LED array and a rotary stage[11]. The high aspect ratio curved surface structures are needed for optical devices such as micro lenses, light-guiding devices, and so on. Smooth surface structures can be fabricated with this technique because the UV-LED makes the difference of the UV dose with its wide directivity characteristics. In addition, the

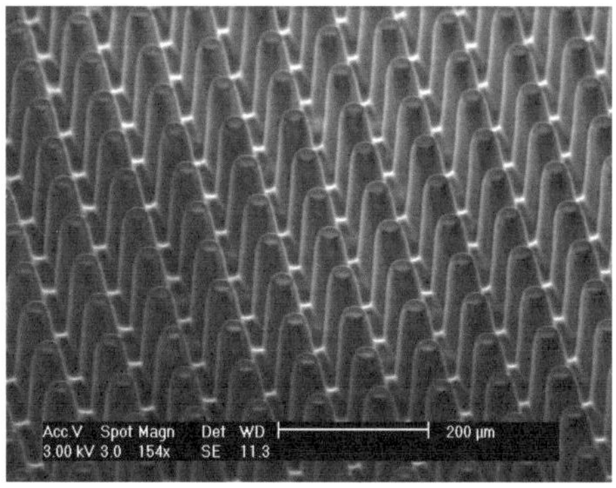

(a) Diameter of 50 μm

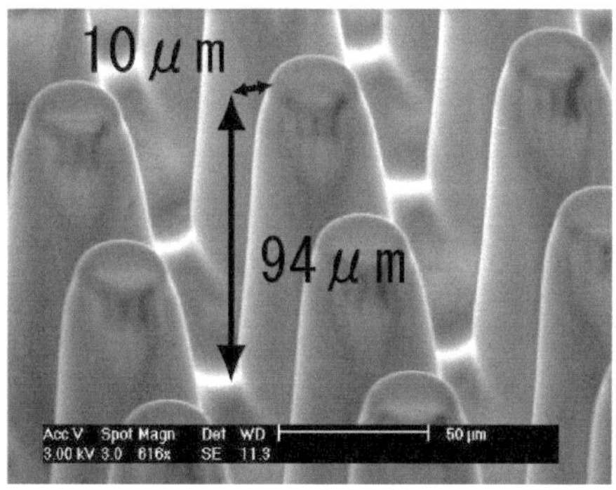

(b) Scale of the 50 μm diameter structure

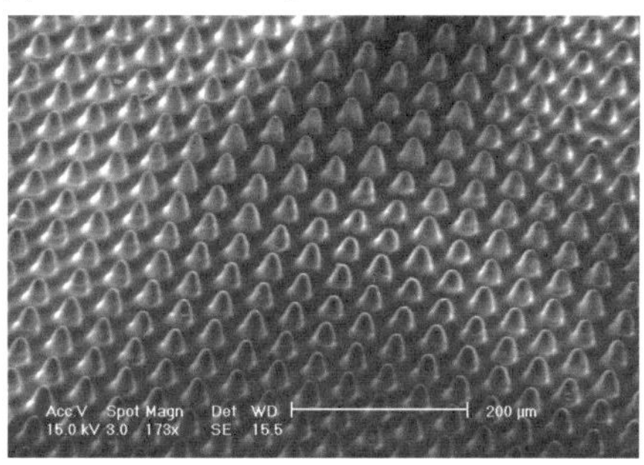

(c) Diameter of 20 μm

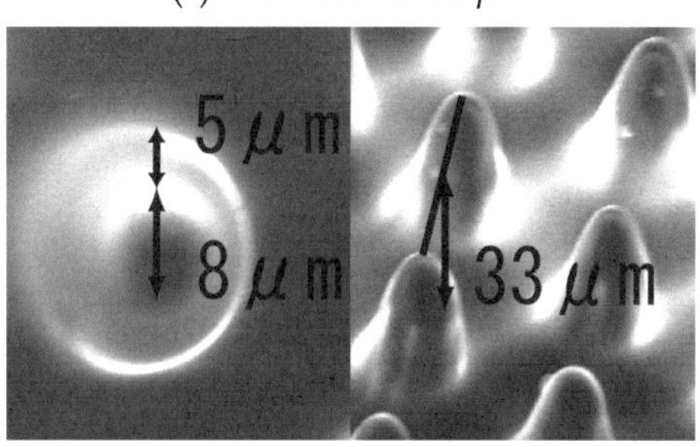

(d) Scale of the 20 μm diameter structure

Fig. 9. SEM photographs of PDMS structures.

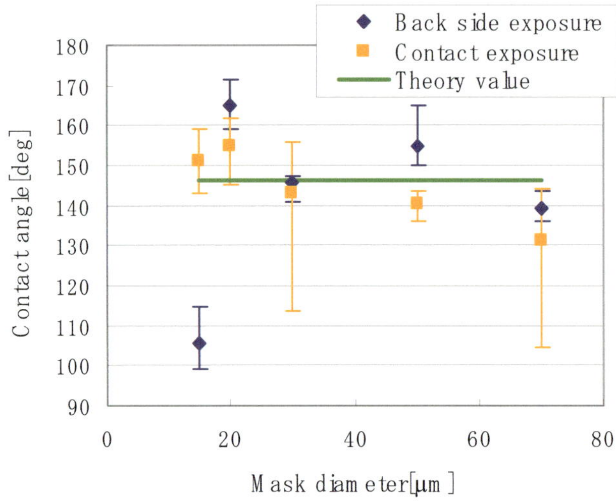

Fig. 10. Mask diameter versus contact angle.

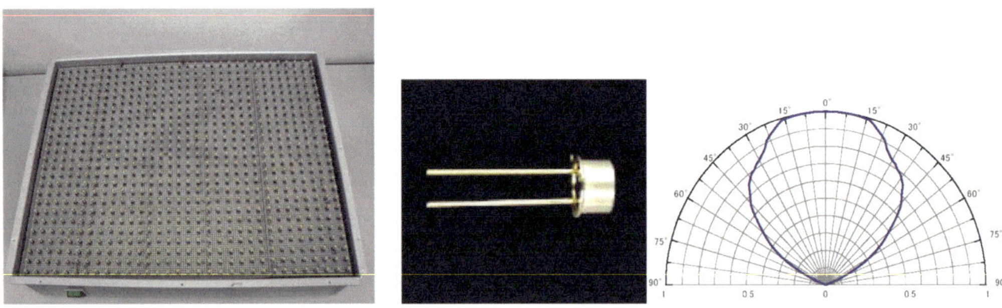

(a) NS370L-5CFA.

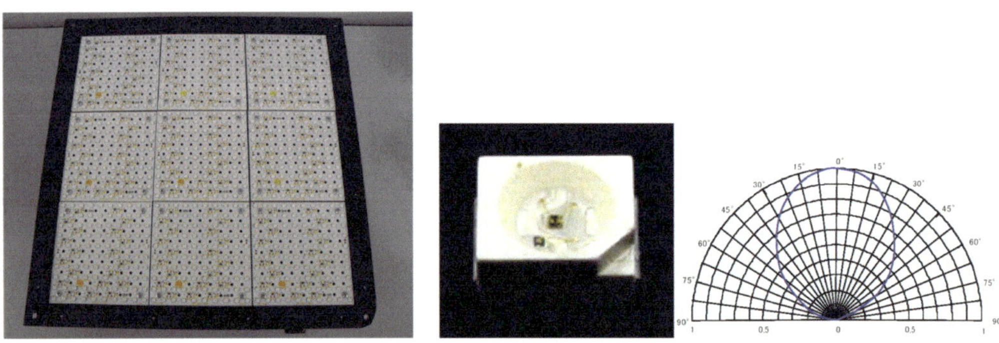

(b) NS370L-7SFF.

Fig. 11. Photograph of the UV-LED array light source.

structures can be formed with high uniformity by only one exposure process because the rotation reduces the unevenness of exposure. This technique can control the height and curvature of structures by changing the exposure time. Large devices can be fabricated by increasing the number of UV-LED array. Furthermore, this technique also provides high flexibility of design; for instance, the semi-cylinder shape is fabricated with the mask patterned rectangular aperture. Because the incidence range of UV rays depends on the aperture size, the use of apertures of various sizes allows for structures of various heights to be fabricated by only one exposure process.

Two UV-LED arrays shown in Fig. 11 were used in this experiment. The UV-LED NS370L-5CFA (Nitride semiconductors Co. Ltd.) is the round type without a collecting lens. The center wavelength is 370 nm, and the directional characteristics angle is 50 deg. The light intensity is 1.8 mW, and the light intensity tolerance is 10%. The pitch of the UV-LED array is 12.5 mm, and the array is arranged in 32 lines and 25 rows. The light intensity of the UV-LED array was 0.315 mW/cm^2. The UV-LED NS370L-7SFF (Nitride semiconductors Co. Ltd.) is a surface mount device type. The center wavelength is 370 nm, and the directional characteristics angle is 50 deg. The light intensity is 2.0 mW, and the light intensity tolerance is 10%. The pitch of the UV-LED array is 10 mm, and the array is arranged in 30 lines and 30 rows. The light intensity of the UV-LED array was 0.830 mW/cm^2.

The rotary stage was located in parallel to the UV-LED array. The distance between the rotary stage and the UV-LED array was determined within a few cm by measuring the uniformity of the UV intensity. Exposure was performed while the stage was rotating. The exposure system is shown in Fig. 12.

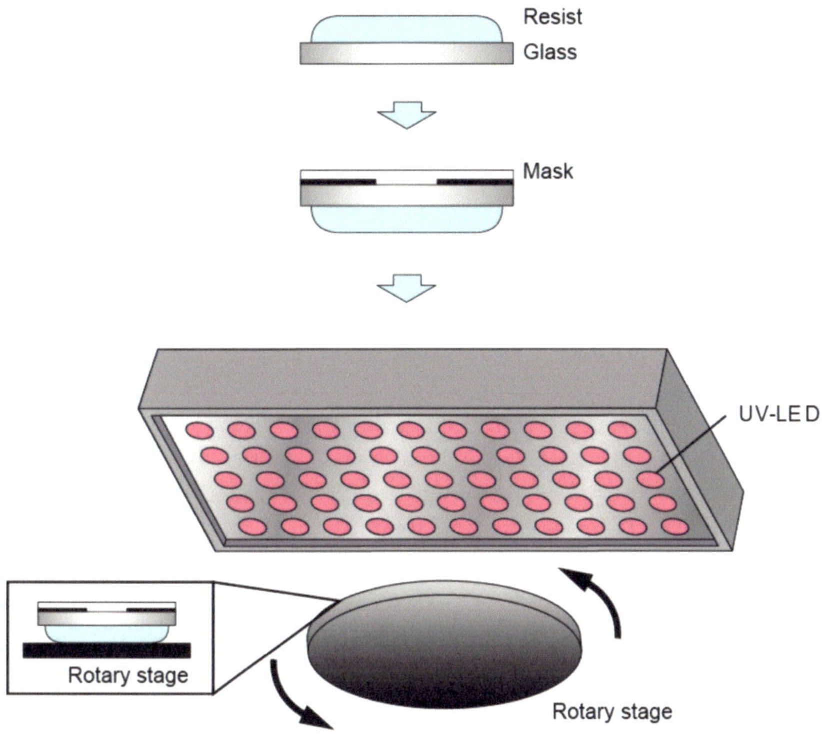

Fig. 12. Exposure system for fabrication of a curved surface structure.

3.2 Fabrication process

A negative photoresist of SU-8 10 was utilized in this technique. SU-8 resist was dip-coated with a thickness of around 500 μm on a 150-μm thick glass substrate. After the prebake

process, the lithography process was performed with a UV-LED array. The mask was set on the glass substrate coated with SU-8 resist, and they were fixed on the rotary stage. They were set at a 5 cm radius of rotation. The rotation speed was 100 rpm. SU-8 was exposed by UV light through the glass and apertures of the mask by the UV-LED array. The amount of UV light depends on the mask pattern and the directional characteristics of the UV-LED. The distributions of light arriving at the resist were simulated by the lithography simulator ProxSim-2 (Lithotech Japan). Figure 13 shows the calculated results for the UV intensity and the resist structure at wave length of 365 nm, with a collimation angle of ±50°. The width of the opening windows is 300μm. If the opening windows is circle, the axisymmetrical diffraction effect should be large at the edge of the mask pattern in the experiment.
The structure shape is determined by the UV dose; therefore the curved surface shape can be formed after the development process. The shape can be controlled by the mask pattern, exposure time, intensity, and directivity of the light source.

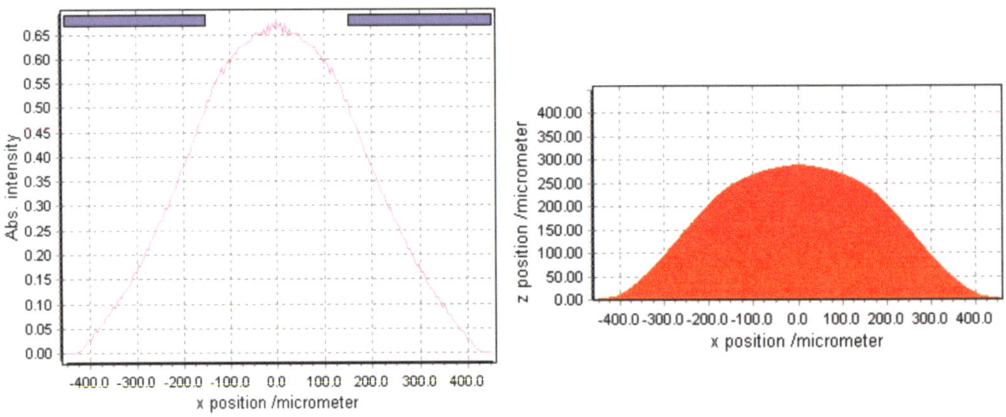

Fig. 13. Simulation results for the exposure dose and structure

3.3 Fabrication results

Semi-cylindrical structures were fabricated using this technique. In this technique, the structure height depends on the distance from the edge of the aperture. A semi-cylindrical structure with a smooth surface can therefore be fabricated using the mask with a rectangular pattern. The mask pattern and the fabrication results for the semi-cylindrical structures are shown in Fig. 14 (a). A semi-cylindrical structure bent at 90 degree pattern was fabricated. The mask pattern and fabrication results are shown in Fig. 14 (b). By using a gray-scale method, a semi-cone pattern was also fabricated. The gray-scale mask of the rectangular pattern and the fabrication results are shown in Fig. 14 (c). A narrow, pointed structure was obtained.
Hemisphere structures with smooth surface were fabricated as shown in Fig. 15(a). The diameters of the structures were 400 μm, and the maximum heights were 200 μm. Hemisphere structures of various sizes were fabricated by only one-step exposure. The exposure time was 18 min. The various diameters of the mask pattern and the fabricated structures are shown in Fig. 15(b). The structure diameters were 270, 380, 480, and 600 μm, in increasing order. The maximum heights were 110, 170, 220, and 250 μm.

(a) Line pattern and semi-cylindrical structure.

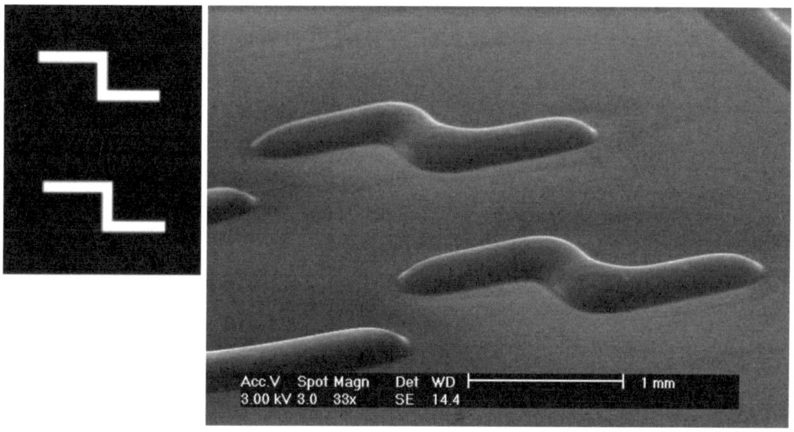

(b) Pattern and semi-cylindrical structure bent at 90 degree.

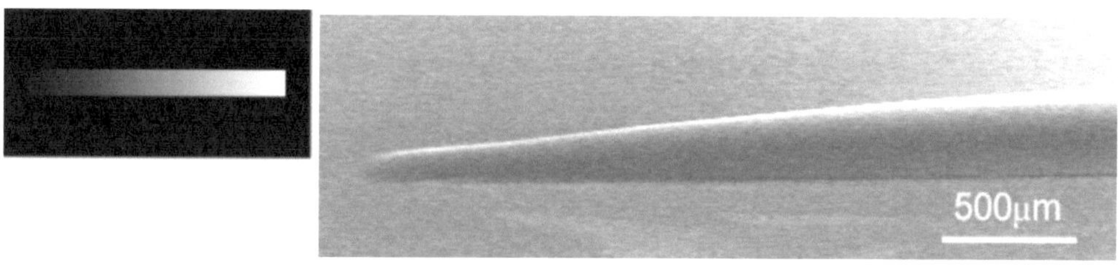

(c) Gray-scale mask and narrow, pointed structure.

Fig. 14. Mask pattern and SEM images of SU-8 structures.

Hemisphere structures of various sizes were fabricated by 16 min exposure. They are shown in Fig. 16. The mask pattern was used the same one shown in Fig. 15(b). The heights of the structures were lower and the curvatures were smaller than the results shown in Fig. 15(b). The structure diameters were 210, 350, 450, and 570 μm. The maximum heights were 50, 100, 130, and 150 μm.

Figure 17 shows the schematic diagram of the shape change versus exposure time.

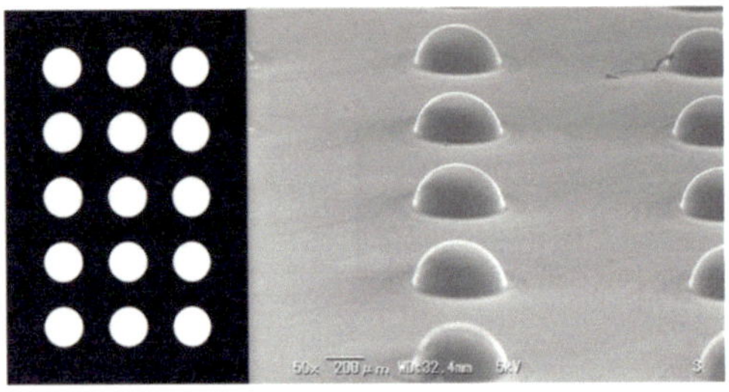

(a) Hemisphere array pattern

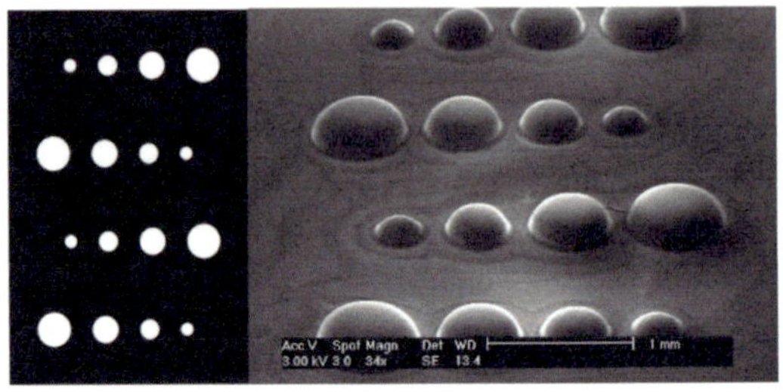

(b) Hemisphere pattern of various sizes

(c) Side view.

Fig. 15. Mask pattern and SEM image of SU-8 structure

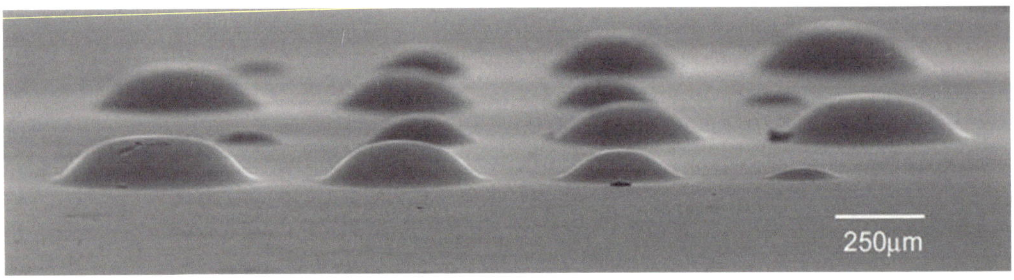

Fig. 16. SEM image of SU-8 structure fabricated with a 16-min exposure time.

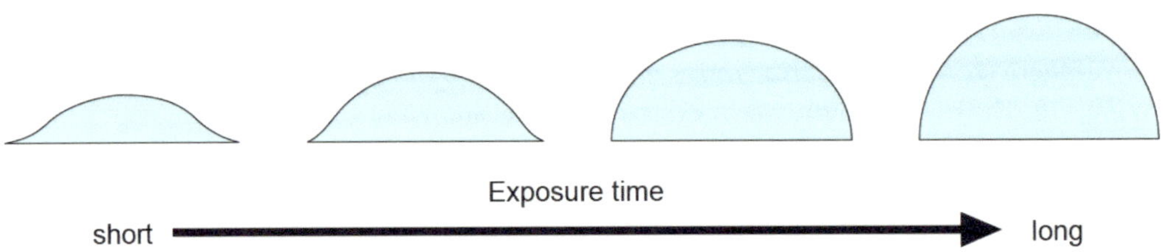

Fig. 17. Relationship between exposure time and shape of resist.

3.4 Micro lens fabrication on another substrate

Micro lenses of UV curable resins were fabricated on another substrate such as a silicon wafer using the above-described hemisphere resist structures. The fabrication process is shown in Fig. 18. Fabricated structures were put into a Petri dish, and PDMS was poured. After defoaming in the vacuum desiccators, the PDMS was cured at room temperature for 6 h and peeled off from the glass plate. UV curable resin (Opster PJ3001, JSR Co.) was then dropped on the silicon wafer, and the PDMS-formed micro lens pattern was pressed on it. The exposure was performed with a mask aligner to penetrate the PDMS. The surface tension was 150 mN/m for Si, over 300 mN/m for SiO_2, and 20 mN/m for PDMS so that PDMS was easily peeled away from the silicon wafer. As a result, micro lenses of UV curable resin were obtained. The fabrication result is shown in Fig. 19.

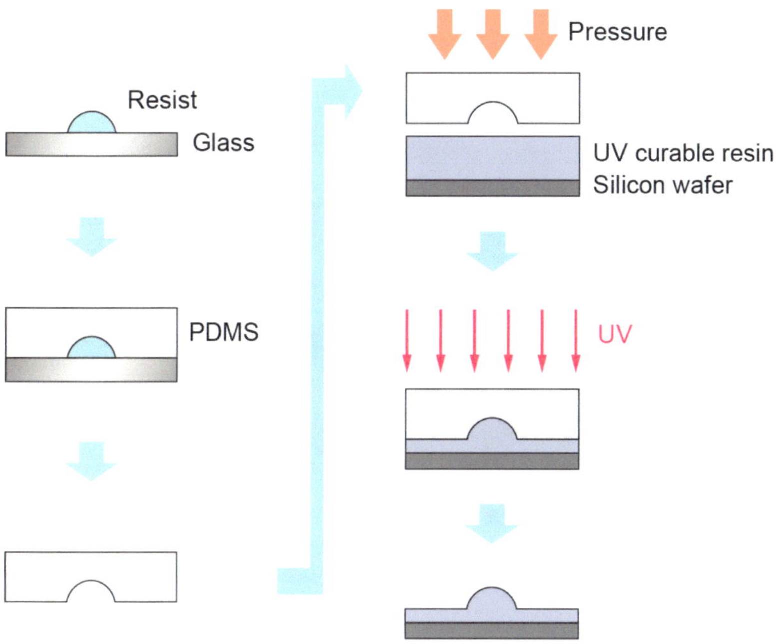

Fig. 18. Fabrication process of a microlens structure using UV curable resin.

Fig. 19. Micro lenses of UV curable resin.

The focusing property of the fabricated micro lenses was simulated by the lay trace simulator Light Tools (Optical Research Associates). Figure 20 shows the calculated results for the focusing property for 240 μm and 440 μm diameter lenses. The lenses effectively collected a laser beam of 550 nm wavelength under the glass substrate.

The focusing property of the fabricated micro lenses was evaluated by irradiating laser beam. The lenses collect laser beam to 12.9-68.5 μm spot diameter and 70.0-380 μm collective distance from the 150 μm thick glass substrate. These results show the same tendency as the ray trace simulation and indicate that the focal length of the fabricated micro lens was shorter than that of the conventional micro lens for optical communication. Therefore, the size of the optical device can be reduced by using fabricated micro lenses.

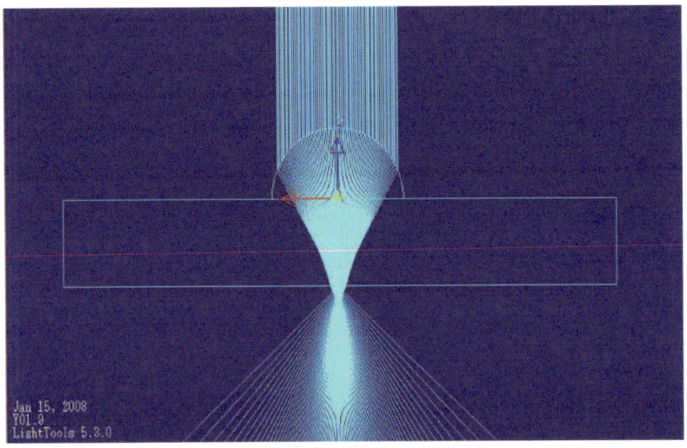

Parameters
Lens diameter: 240 [μm]
Lens curvature radius: 120 [μm]
Thickness of lens: 120 [μm]
Thickness of glass substrate: 150 [μm]
Refractive index of lens: 1.51
Refractive index of substrate: 1.52
Number of lays: 100

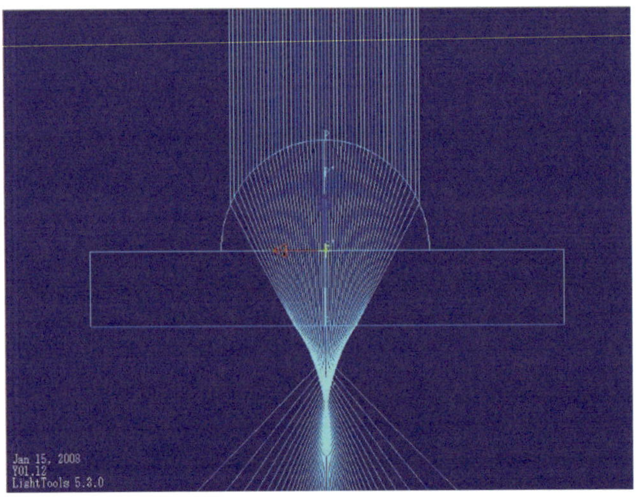

Parameters
Lens diameter: 440[μm]
Lens curvature radius: 220 [μm]
Thickness of lens: 220[μm]
Thickness of glass substrate: 150 [μm]
Refractive index of lens : 1.51
Refractive index of substrate: 1.52
Number of lays: 100

Fig. 20. Lay trace simulation results for Microlenses.

Furthermore, a micro fly eye lens was fabricated by reducing the pitch distance of mask circular apertures. Micro lenses with diameters of 200 μm, 300 μm, 400 μm, and 500 μm were linked to each other in the pitch at distances less than 50 μm, 450 μm, 570 μm, and 690 μm.

Fig. 21. SEM photograph of micro fly-eye lens.

4. Acknowledgments

This research was partially supported by the Ministry of Education, Science, Sports and Culture, Grant-in-Aid for Exploratory Research, 19651064, 2007.

5. References

[1] S. Maruo, O. Nakamura, and S. Kawata,"Three dimensional microfabrication with two-photon absorbed photopolymerization," Optics letters, 122-2, pp.132-134, 1997.

[2] H. Huang, Y. Wang, T. Chang and C. Fu, "3D high aspect ratio micro structures fabricated by one step UV lithography," Journal of Micromechanics and micro engineering, 17, pp.291-296, 2007.

[3] I. Song, K. Kang, Y. Jin, D. S. Park, P. K. Ajmera," Microlens array fabrication by backside exposure using Fraunhofer diffraction", J Micromech Microeng14, pp. 1285-1290, 2008.

[4] W.Däschner, P. Long, R. Stein, C. Wu, and S. H. Lee, "Cost effective mass fabrication of multilevel diffractive optical elements by use of a single optical exposure with a gray-scale mask on high-energy beam-sensitive glass," Appl. Optics, 36, No.20, pp.4675-4680, 1997.

[5] Y. Hirai, Y. Inamoto, K. Sugano,T. Tsuchiya and O.Tabata, "Moving mask UV lithography for three-dimensional structuring", J. Micromech. Microeng. 17, pp.199–206, 2007

[6] Peterman MC, Huie P, Bloom DM, Fishman HA," Building thick photoresist structures from the bottom up", J Micromech Microeng 13, pp.380–382, 2003.

[7] J. Onishi, K. Makabe, Y. Matsumoto, "Fabrication of micro sloping structures of SU-8 by substrate penetration lithography", Microsystem Technologies, Vol.14, No.9-11,pp.1305-1310, (2008).

[8] Dill F. H., Neureuther A. R., Ruttle J. A. and Walker E. J.:"Modeling projection Printing of Positive Photoresists" IEEE Trans. Electron Devices, ED-22, 7, pp. 456-464 (July 1975).

[9] Yeung M. S.: "Modeling high numerical aperture optical lithography" , SPIE, 922, Optical/Laser Microlithography, pp. 149-167 (1988).

[10] Y. Sensu, M. Isono, A. Sekiguchi, M. Kadoi and T. Matsuzawa,"Study of proximity lithography simulations using measurements of dissolution rate and calculation of

the light intensity distributions in the photoresist", Proc. SPIE 5376 pp.1040–52, 2004.
[11] Shinya Suzuki, Yoshinori Matsumoto, "Lithography with UV-LED array for curved surface structure", Microsystem Technologies, Vol.14, No.9-11,pp.1291-1297, (2008).

3

Influence of Immersion Lithography on Wafer Edge Defectivity

K. Jami, I. Pollentier, S. Vedula and G. Blumenstock
[1]KLA-Tencor
[2]IMEC
[1]USA
[2]Belgium

1. Introduction

In semiconductor manufacturing, the control of defects at the edge of the wafer is a key factor to keep the number of yielding die as high as possible per wafer. Using dry lithography, this control is typically done by an edge bead removal (EBR) process, which is understood well. The recent production introduction of immersion lithography however changes this situation significantly. During immersion exposure, the wafer edge is locally in contact with water from the immersion hood, and particles can then be transported back and forth from the wafer edge area to the scanner wafer stage. Material in the EBR region can also potentially be damaged by the dynamic force of the immersion hood movement. In this chapter, we have investigated the impact of immersion lithography on wafer edge defectivity. In the past, such work has been limited to the inspection of the flat top part of the wafer edge, due to the inspection challenges at the curved wafer edge and lack of a comprehensive defect inspection solution. The development of edge inspection & metrology tools now allows us to probe this area of the wafer.

This study used KLA-Tencor's VisEdge CV300-R, an automated edge inspection system that provides full wafer edge inspection (top, side, and bottom) using laser-illumination and multi-sensor detection, and where defects can be classified with Automated Defect Classification (ADC) software. In addition to defectivity capture, the tool performs simultaneous, multi-layer, film edge and EBR metrology, indicating the distance from the wafer edge or wafer top depending on the location of the film/EBR edge. It also can review defects using an integrated, hi-resolution microscope.

In this paper the impact from immersion lithography towards wafer edge defectivity is investigated. The work revealed several key factors related to wafer edge related defectivity control: choice of resist, optimization of EBR recipes, scanner and immersion-fluid contamination, wafer handling and device processing procedures. Understanding the mechanisms of wafer edge related immersion defects is believed to be critical to the successful integration of the immersion process into semiconductor manufacturing.

2. Process control at the wafer edge

In semiconductor manufacturing, control of the process at the wafer edge is a key factor in determining the total number of yielding die on a wafer. The removal of photoresist from the wafer backside and edges is especially important to avoid contact between the resist and the scanner stage or wafer handling hardware. Typically, a solvent EBR rinse is the last step in the coating recipe: the combination of a solvent stream from a static nozzle toward the wafer back side and a dynamic nozzle toward the wafer front side dissolves the resist up to a few millimeters from the wafer's outer edge. The desired position of the EBR material edge at top side (the so-called EBR-width) can depend on the coated material (e.g., antireflective topcoat vs. photoresist material) and/or on the layer within the device: for example, a contact hole lithography process might use a slightly different EBR width than the gate process. To increase die yield, it is desirable to have EBR widths that are as small as possible.

Immersion lithography [1-4] has changed the way we view defectivity issues at the wafer edge significantly. During the immersion exposure sequence, the wafer edge is in contact with the water from the immersion hood (IH), introducing additional concerns beyond direct contact of resist with the scanner. First, when the IH is scanning in the EBR region, its movement can damage material edges (Fig. 1a). IMEC's program on immersion lithography found that, for example, photoresist material can partially peel off during the IH pass (Fig. 1b).

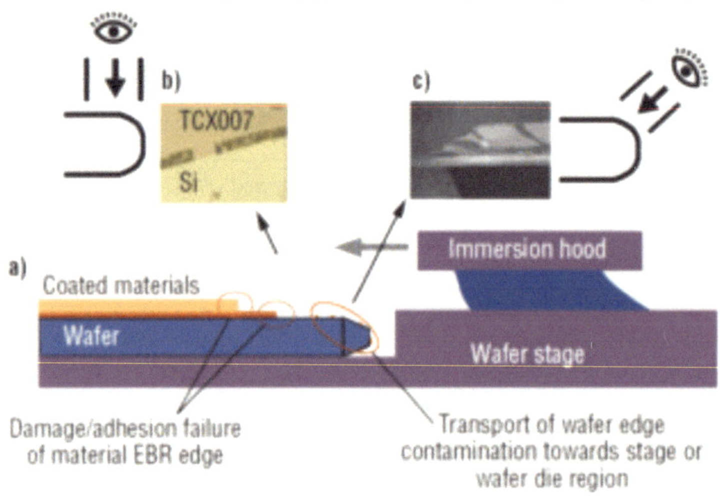

Fig. 1. a) Schematic representation of possible defect issues with immersion lithography; b) example of damaged photoresist material inspected with top-down optical microscope; and c) example of damaged residues at the wafer edge, inspected by tilted SEM (45°).

A second concern involves the cleanliness of the wafer edge outside the EBR edges. The IH pass wets not only the near-edge top surface, but also the curved wafer edge and even part of the bottom surface. Defects can be released from this area and re-deposited either on the wafer or on the wafer stage. In the first case, there will be a direct impact on the wafer defectivity. In the latter case, defects present on the wafer stage can be transported onto wafers in subsequent wafer processing. IMEC's program on immersion lithography found that resist residues left on the curved wafer part by an incomplete EBR step can be damaged by the IH pass, releasing fragments into the system (Fig. 1c).

Traditional defect inspection techniques have serious limitations when monitoring these new issues. Conventional dark field or bright field inspection tools cannot access the wafer edge since these systems typically have an edge exclusion of ~3mm. While microscopy tools can inspect the edge area, they can only give qualitative information with limited sampling.

3. Technology for wafer edge defect inspection

The new measurement system for wafer edge defect inspection (Fig. 2) is based on a laser source directed to the wafer edge surface. Three detectors simultaneously collect the scattered light, the specular or reflected light and the phase difference between different polarization states. As the laser scans the wafer edge surface, each signal can be converted into an image. Each type of defect produces a specific combination of signals from the three detection channels, allowing automated defect classification.

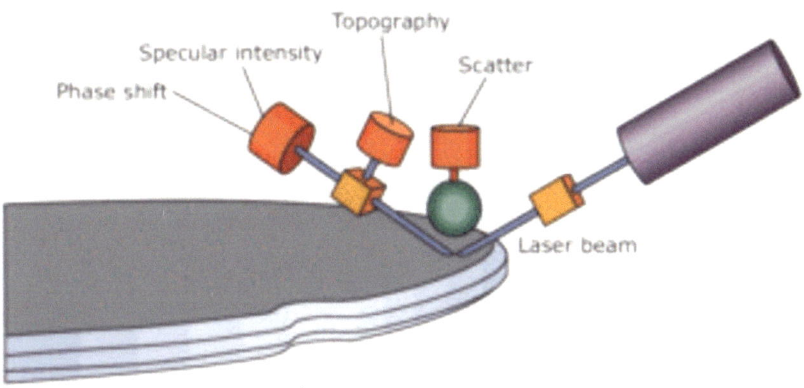

Fig. 2. Schematic representation of the VisEdge measurement principle.

3. Imaging of the wafer edge

Imaging covers the entire edge region including the following areas: ~5mm bottom near-edge, bottom bevel, apex, top bevel, and ~5mm top near-edge. Scanning generates a continuous high-resolution image for the entire wafer edge, which can be represented as a Mercator projection: an unfolding of the wafer edge surface into a flat plane. Excursions in eccentricity and/or in EBR width, which might result in a layer's edge ending on the wrong underlying substrate, can be easily monitored using this kind of inspection.
For wafer edge cleanliness, a high-resolution view of the wafer edge is typically more useful. Here, the images view only a few millimeter of the edge. Figure 3 uses this representation to show resist flakes observed along the apex-bevel regions in the specular channel.

4. Immersion defect process characterization and optimization

As indicated in the introduction, immersion-related defects at the wafer edge can be due to edge damage to the coated material in the EBR area, from the IH passing over this region. On the other hand, defects can be caused by transport of particles present on the bevel. These might be released by forces of the immersion hood, transported by the water in the hood, and re-deposited on the wafer and/or stage. This work focuses on the latter, and in particular on the flake defects observed in past work [5].

5. Edge region flake defects

Flake defects are related to material residues that are present on the wafer edge after coating. Typically, these residues are only present on the apex of the bevel, and therefore are

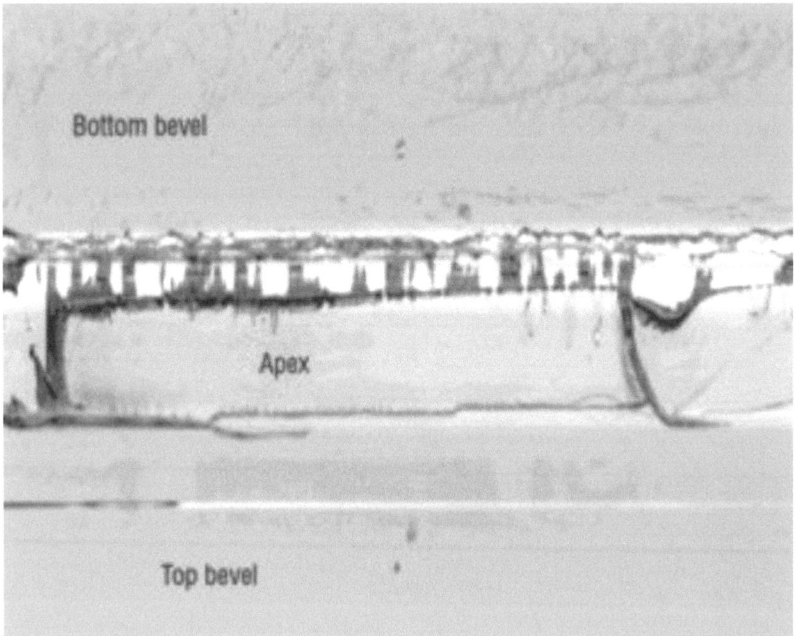

Fig. 3. Example of a specular image using the new technology showing part of the bevel/ apex region. This kind of representation is valuable for the evaluation of wafer edge quality.

difficult to detect by conventional top-down inspection methods. The residues result from a non-optimized EBR process: since the coated material on the wafer edge can be significantly thicker than on the flat top region, an insufficient solvent supply can leave edge residues while the top surface is clean. This phenomenon is more commonly observed with photoresist materials, rather than with BARC and topcoat materials.

The morphology of edge residues can depend on the resist. For some resists, the residue can be quite uniform along the apex. For other resists, large areas of thick residues are combined with areas with thin residues.

Once detected, the problem can be solved by adjusting the EBR recipe. Because making the EBR recipe longer limits the throughput of the immersion cluster, however, fabs try to avoid this adjustment if possible, there is a clear risk of processing wafers with resist residues over to the immersion scanner.

When wafers with resist residues are exposed on an immersion scanner, it is difficult to predict if the IH pass over the edge of the wafer will damage the resist residue, and if (part of) the residues will redeposit on the wafer top-side or on the scanner wafer stage. Tilted SEM review suggested qualitatively that such damage can happen with certain resists.

6. Experimental conditions of edge flake characterization

We experimented with three resists with different chemistries (**Fig. 4**). Sensitivity to edge damage was expected to vary across the three resist types. A dedicated exposure job spatially separated the areas where flakes are expected and where they are not expected. One section, consisting of two rows of 11 fields, was exposed close to the wafer edge at the opposite side of the notch. A similar area of two rows of 11 fields was exposed in the region of the notch. During the exposure of these 2 × 11 sections at both locations near the wafer edge (Region II), the IH makes continuous up- and down-scans over the wafer edge area, increasing the probability of defect generation. The exposure job was also designed so that on another part of the wafer (Region I, on the right hand side), the immersion hood did not pass over the wafer edge. In Region I, no flake-like defects were expected.

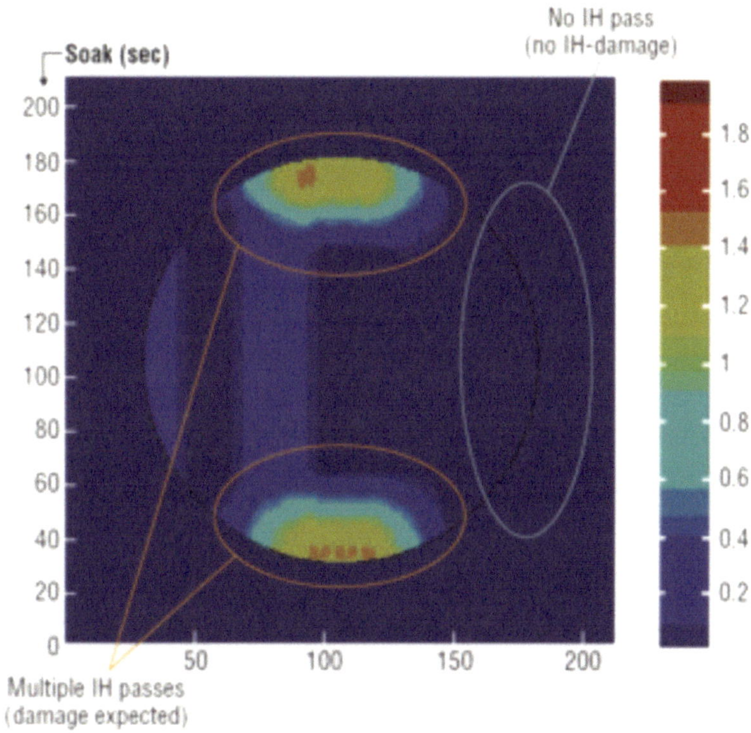

Fig. 4. IH exposure sequence for edge flake characterization.

7. Qualification

The specular images of regions with resist residues clearly showed differences in reflected intensity: dark areas in the resist residues refer to thick layers, while light areas indicate much thinner layers. The results obtained from Resist Type A are detailed below.

We compared the SideScan images of areas where the IH did and did not pass. Figure 5a is a typical SideScan specular image for region I (where the IH did not pass). Differences in thick and thin resist residues are visible, but no fragments of the resist residues are evident. In contrast, in Fig. 5b, taken from Region II, indicated that parts from the thick residue at the bottom of the apex were released. The close-up in the image indicates that some of these edge flakes were re-deposited on the apex closer to the top.

To determine whether any of these edge flakes end up on the top region (where edge die can be damaged), we analyzed the TopScan image of the corresponding areas of Fig. 5a and 5b using the scatter channel, as shown in Figs. 5c and 5d. In Region II, a lot of particles were detected, while in Region I, no particles were observed in the images. This observation was encouraging for further ADC work.

Classification of the edge region flakes was found to vary by the defect location. Re-deposited edge flakes on the apex side were best classified using their signal in the specular

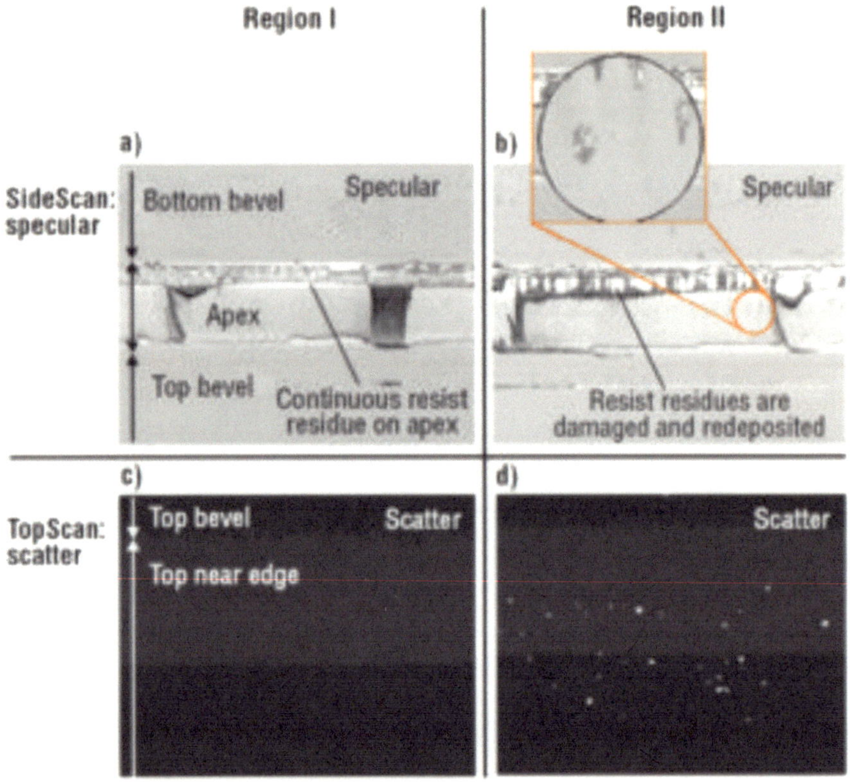

Fig. 5. Specular and scatter images from Region I (a and c) and Region II (b and d) of a test wafer. IH damage is more likely in Region II.

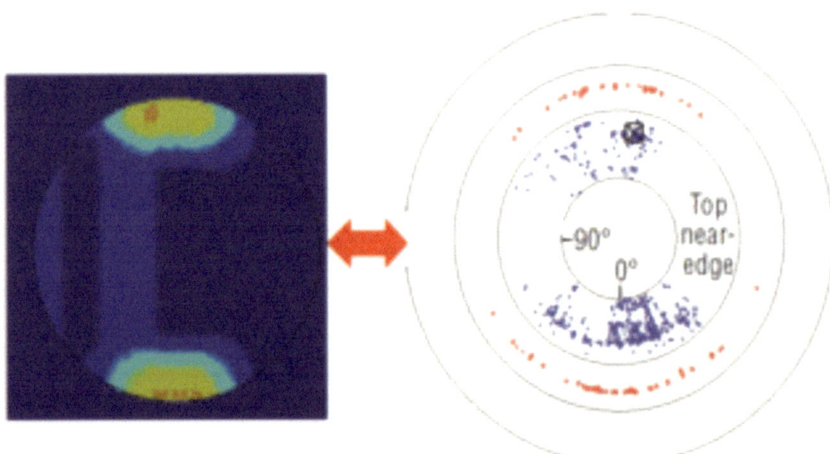

Fig. 6. Immersion characterization (quantification of edge flakes at wafer edge); the flakes on the apex and top near edge can be quantified by composite analysis of SideScan and TopScan signals.

channel of the SideScan image. For re-deposited defects in the top near-edge region, a combination of signals in the specular and scatter channels gave more accurate classification. Once all the measurement parameters for both areas are fixed, they can be combined in a single measurement sequence that provides defect classification and mapping for all the wafer edge areas of interest. (Fig. 6).

8. Immersion process characterization and optimization

Having qualified the inspection to classify and map edge flake defects, we used our results in a design of experiment to improve our understanding of this kind of defect source and its key impact parameters.
As indicated above, Resist A tends to generate flakes when the IH is passing over its edge. The non-optimized coating process left residues for two other resists, resist B and C; however the residue morphology was different.
When the same immersion exposure was used, significantly fewer edge flakes were detected in the near top region for Resist B and C than for Resist A (Fig. 7). Moreover, the residual defects were less confined to the exposure zone, so some of these defects might be caused by coating and wafer handling. In the TopScan images, no clear sign of damage was seen. Clearly the choice of resist chemistry can be important to prevent these kinds of defects.
As indicated earlier, resist residues can be optimized by changing the EBR recipe on the coat track. Resist A showed several hundred defect flakes with the regular (short) EBR sequence. After optimization, this resist achieved defect values similar to the background values obtained with the non-flaking resists B and C.

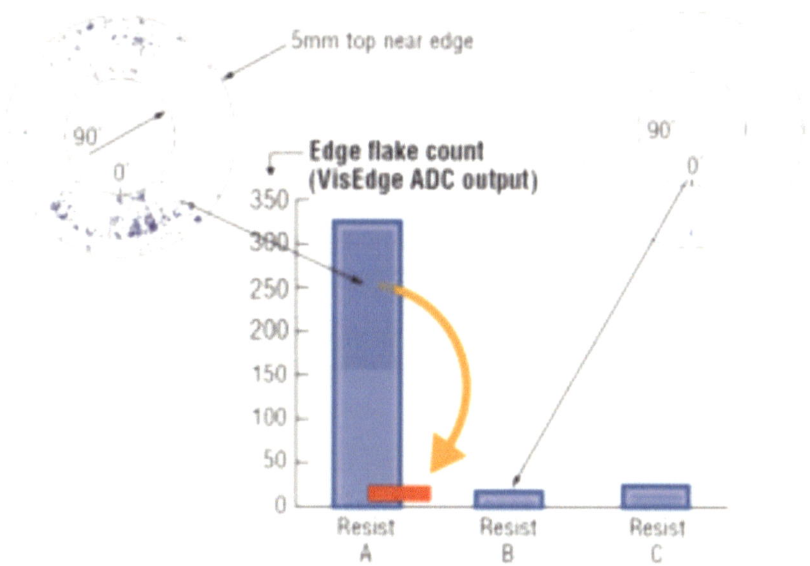

Fig. 7. Edge flake defects as a function of resist chemistry and EBR recipe.

9. Further wafer edge challenges

More kinds of defects besides the edge region flakes can be important in immersion litho. This section discusses other possible defect sources.

10. Wafer handling marks and resist rework process

A variety of artifacts were seen even in fresh Si wafers, primarily on the bevel and apex region. These wafers had very limited processing and handling, but damage was visible in the form of particles in the apex/bevel region. This introduces an additional concern with transport-related artifacts, and illustrates the need for an assessment of wafer edge quality and handling before introduction to the immersion process.

11. Resist rework processes

At IMEC, resist work is typically done by a combination of a dry ashing step, followed by a wet clean. In some cases, rework may be indicated to address an out-of-spec condition. Wafers used for monitoring of focus/dose/CD or overlay processes may be reworked daily. Limited rework typically results in an increased presence of scratches (typically at the lower bottom bevel), and an overall increase in reflectivity variation, indicating degraded surface quality. When wafers are reworked ~10 times or more, the bevel/apex area is much more affected. These defects could pose a risk when the immersion hood is passing over the wafer.

12. Conclusion

In this paper, we investigated the impact of immersion lithography on wafer edge defectivity. In the past, such work has been limited to inspection of the flat top part of the wafer edge due to the inspection challenges at the curved wafer edge and lack of a comprehensive defect inspection solution. Our study used a new automated edge inspection system that provides full wafer edge imaging and automatic defect classification.
The work revealed several key challenges to controlling wafer edge-related defectivity, including choice of resist, optimization of EBR recipes, and wafer handling.

13. Acknowledgments

The authors thank Diziana Vangoidsenhoven, Christie Delvaux, Bart Baudemprez, and Tom Vandeweyer for help in processing and wafer selection; Thomas Hoffmann for help in immersion soak time simulations; and Philippe Foubert, Dieter Van Den Heuvel, Shinichi Hatakeyama (TEL), Kathleen Nafus (TEL), Sean O'Brien (TI), Mireille Maenhoudt, and Richard Bruls (ASML) for helpful discussions on immersion tools and related defectivity. VisEdge is a trademark of KLA-Tencor Corp.

14. References

[1] B. Smith, H. Kang, A. Bourov, F. Cropanese, Y. Fan, "Water Immersion Optical Lithography for the 45nm Node," *Proc. SPIE*, Vol. 5040, p. 679-689, 2003.
[2] M. Kocsis et al., "Immersion-specific Defect Mechanisms: Findings and Recommendations for Their Control," *Proc. SPIE*, 6154, 6154-180, 2006.
[3] M. Maenhoudt et al., *Jour. of Photopolymer Science and Technology*, 19, 585, 2006.
[4] M. Ercken et al., *Jour. of Photopolymer Science and Technology*, 19, 539, 2006.
[5] I. Pollentier et al., *Proc. SPIE*, 5754, 129, 2005.

4

Femtosecond Laser Nonlinear Lithography

Hiroaki Nishiyama and Yoshinori Hirata
Osaka University
Japan

1. Introduction

Micro-optical elements with precisely controlled three-dimensional surfaces are expected to play important roles in the fields of functional photonic devices, micro-opto-electro-mechanical systems, lab-on-a-chip devices, and so forth. For example, blazed gratings exhibit more than two times higher diffraction efficiencies than those of their standard binary counterparts. Aspheric lenses are effective for precise control of light refraction for compensation of optical spherical aberration. Sub-wavelength periodic structures can markedly suppress Fresnel-reflection losses−generated at the interface between lens materials and air−over a wide spectral bandwidth and a large field of view.

Most micro-optical elements are manufactured using semiconductor fabrication technology including lithography and various etching methods. Although such a technology is an effective means for arbitrary nano/microstructures with high quality on a wafer, the cross-sectional profiles of the structures are limited to binary shapes. Consequently, the fabrication of three-dimensional surfaces such as slopes and hemispherical shapes remains a challenging task. For obtaining slope structures, a gray-scale mask method (Suleski & O'Shea, 1995), a multiple mask technique (Herzig, 1997), and a moving mask process (Hirai et al., 2007) are often used. Electron-beam direct writing with dose control can also be utilized (Fujita et al., 1982). However, these techniques entail strong restrictions in spatial resolution and cross-sectional profiles of the resist. Alternatively, highly precise control of exposure conditions and stage alignments are required.

In addition to this limitation, the semiconductor fabrication technology has another problem: this technology typically cannot be applied to the microfabrication on non-flat substrates because of its planar nature. When photoresists are coated onto non-flat substrates, the resist thickness varies from area to area because of the surface tension. The focal position and the light power are modulated by the resist thickness. Consequently, the pattern sizes are strongly affected by the depth profiles of underlying substrates. The spray-coating technique was reported to obtain uniform distribution of the resist thickness even on non-flat substrates (Singh et al., 2005). However, its long-focal-depth optics is unsuitable for achieving high spatial resolution. Some research groups have demonstrated the lithographic fabrication onto convex or concave lenses using specially modified stage systems with ultraviolet laser sources (Radtke & Zeitner, 2007). Although this technique is useful, complex exposure systems are necessary.

To overcome these difficulties of current semiconductor fabrication technology, we have proposed a combined process of nonlinear lithography and plasma etching, and have demonstrated the fabrication of three-dimensional surfaces of various optical materials (Nishiyama et al., 2008). This approach is designated as femtosecond-laser lithography-assisted micromachining (FLAM). When femtosecond laser pulses are tightly focused into transparent materials including photoresists, nonlinear optical phenomena such as two-photon absorption (TPA) occur only near the focal region. Such a unique property of nonlinear processes enables us to induce photochemical reactions directly inside the materials, which can be a powerful tool to elevate the current semiconductor technology. In this chapter, we describe fundamentals of femtosecond laser nonlinear lithography and present the microfabrication of functional three-dimensional surface structures using FLAM.

2. Femtosecond laser nonlinear lithography

Figure 1 schematically shows the optical absorption behaviors when photopolymers are exposed to ultraviolet light and near-infrared femtosecond laser pulses. Here, these photopolymers are transparent within the spectral range from visible to near-infrared wavelength. For the former, photon energy of ultraviolet light is absorbed mainly near polymer surfaces following the Lambert-Beer law. Depending on the concentration of photoinitiators, the penetration depth is within few micrometers. Consequently, ultraviolet lithography is useful only to form planar structures. In contrast, for the latter, photochemical reactions such as photopolymerization can be induced only near the focal volume, where the laser intensity is highest, via nonlinear optical processes including TPA.

The principle of near-infrared femtosecond laser lithography for the formation of three-dimensional polymeric structures is presented in Fig. 2. Once a laser spot is scanned inside the resist three-dimensionally, the photo-modification is initiated along the trace of the laser spot. The resist is developed to reveal the desired three-dimensional structures. For a chemically amplified negative-tone resist such as SU-8, the cross-linking reaction of monomers is induced in exposed regions by post-exposure baking. Subsequently, unexposed regions are washed away by development treatment, leaving only the desired freestanding structures. The photo-modified region in the focal spot can be approximated as an ellipsoid as shown in fig. 2(d). The aspect-ratio of the ellipsoid is 3–5 when laser pulses are focused into the polymers via an objective lens of NA of 1.2–1.4, depending on the exposure conditions and photosensitivity of the materials.

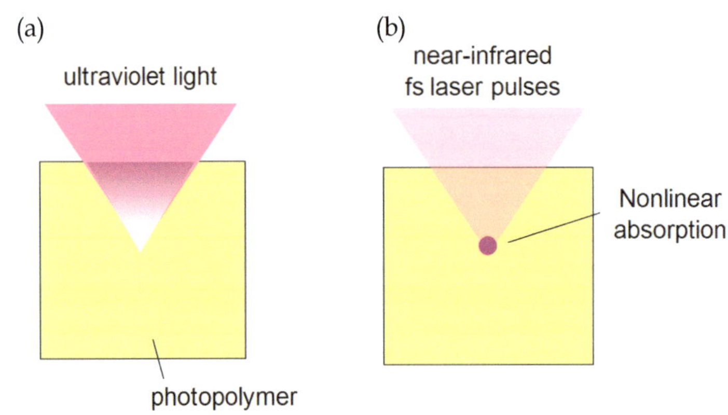

Fig. 1. Optical absorption behaviors of (a) ultraviolet light and (b) near-infrared femtosecond laser pulses.

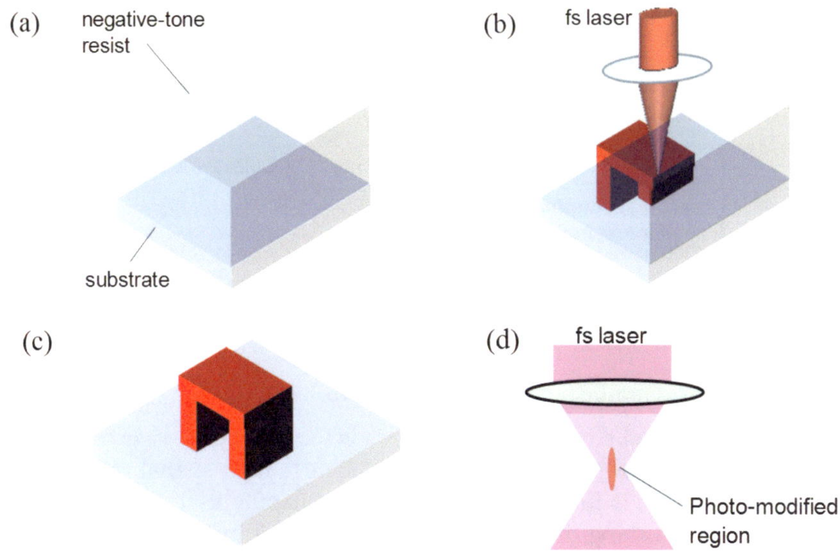

Fig. 2. Femtosecond laser lithography processes in the case of a negative-tone resist. Three-dimensional polymeric structures can be created.

In 1997, Kawata *et al.* first reported the fabrication of full three-dimensional microstructures using femtosecond laser lithography (two-photon lithography) (Maruo et al., 1997). Four years later, the same research group created the world's smallest three-dimensional bulls (Kawata et al., 2001). The bulls were based on polymeric materials and their feature sizes were approximately 10 μm, similar in size to red blood cells. In this report, the lateral spatial resolution was as high as approximately 100 nm, which is below the diffraction limit. After their works, various three-dimensional micro-sculptures such as Venus statues (Serbin et al., 2003), an opera house (Straub, 2004), jars (Park et al., 2005), and gears (Sun et al., 2004) have been demonstrated using this nonlinear lithographic technique. Functional devices such as full three-dimensional photonic crystals (Deubel et al., 2004), plasmonic elements (Reinhardt et al., 2006) and optically-driven microsystems (Maruo et al., 2003) were also reported. Femtosecond laser lithography is a truly three-dimensional process because of its capability of direct exposure of the internal volume, whereas conventional single-photon lithography is based on its planar nature. Depending on laser systems, exposure conditions and material responses, feature sizes of 200–300 nm are typically achievable. Recently, much higher resolution of sub-30 nm was achieved in bridge patterns through careful control of the laser intensity (Juodkazis et al., 2005).

Figure 3 shows a typical exposure setup for micro/nanofabrication using femtosecond laser lithography. Near-infrared femtosecond laser pulses can be obtained from a Ti:sapphire laser system. The laser power and the pulse number are controlled, respectively, by neutral density filters and a high-speed shutter. The repetition rate of the oscillator system is approximately 80 MHz, which is much higher than that of the amplified system (1 kHz). Recently, an Er-doped femtosecond fiber laser is also commercially available as a light source for nonlinear lithography. Compared to Ti:sapphire systems, fiber laser systems have several advantages of beam quality, long-term stability, and cost competitiveness. The fundamental wavelength of Er-doped fiber laser is 1560 nm, and the second harmonic wavelength (780 nm) is obtainable via periodically poled $LiNbO_3$ (PPLN).

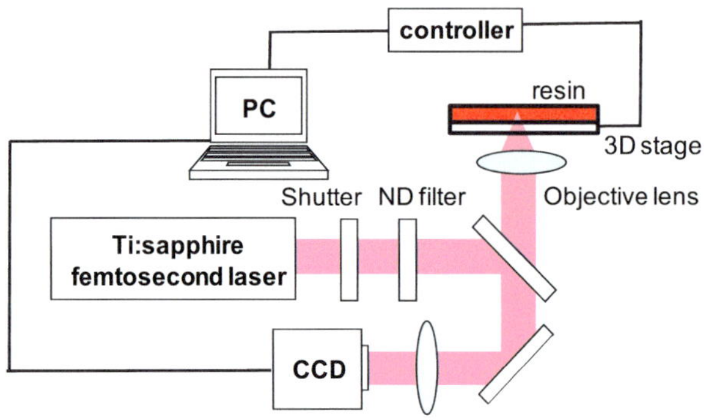

Fig. 3. Femtosecond laser nonlinear lithography system.

In most previously reported studies, the laser beam is focused into resists or photopolymers through an ultra-high-NA objective lens (≥ 1.2) to achieve high spatial resolution. In these cases, a cover glass, which is thin and transparent, is generally used as a substrate because of the extremely short working-distance of the lens. The laser pulses are focused into resins from the back of the substrates. It must be emphasized that, in femtosecond laser lithography, laser irradiation from above the substrate is rarely used because of the short working-distance problem. Direct laser writing of three-dimensional patterns is carried out using a piezo-stage system or a galvano-scanner system using three-dimensional computer-aided design data. Although photopolymers such as SCR-500 were used in early works of femtosecond laser lithography, various negative-tone (Kondo et al., 2005), positive-tone photoresists (Yu et al., 2003) and hybrid materials (Serbin et al., 2004) have been used for the microfabrication during the last several years.

3. Fundamentals of nonlinear absorption

In this section, we describe TPA as a popular approach of femtosecond laser lithography. Two-photon absorption is an interaction between optical waves and materials that occurs in the photo-excitation process of an atom or a molecule from a lower quantum state to an excited state. Two kinds of TPA mechanisms exist, as portrayed in Fig. 4: sequential excitation and simultaneous excitation. The former is an electronic transition to excited energy state via an intermediate state. The electron excited to the intermediate state is pumped further to a higher quantum state by absorbing the second photon with the same energy. For the latter, the electrons are excited to a higher state by absorbing two photons simultaneously in one quantum event with no excited intermediate state. This simultaneous TPA is generally referred to TPA. The life time of the virtual intermediate state for the simultaneous excitation is on the order of several femtoseconds. Two photons can be absorbed simultaneously to transcend the energy band-gap if the second photon arrives before the decay of the virtual state.

The TPA rate is proportional to the square of the laser intensity. In single-photon lithography, photochemical reactions are induced along the entire beam path through the photopolymers, as shown in Figure 1. The number of excited species in any cross-section along the optical axis decreases as the laser beam is absorbed. Compared to this, the integrated intensity squared is markedly increased at the beam waist in two-photon lithography. This quadratic dependence of TPA on laser intensity enables the direct internal

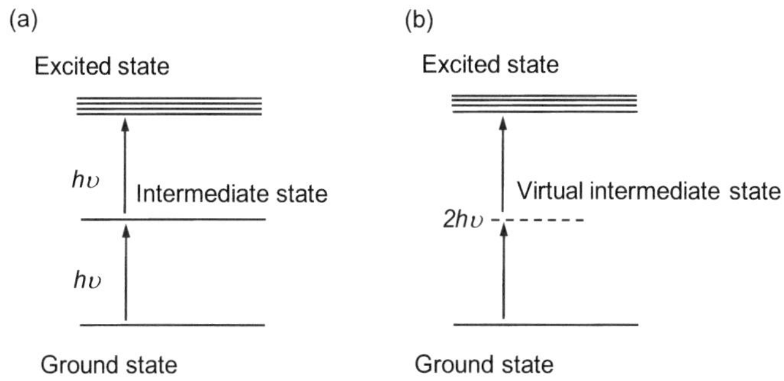

Fig. 4. Two-photon absorption mechanism: (a) sequential two-photon absorption and (b) simultaneous absorption.

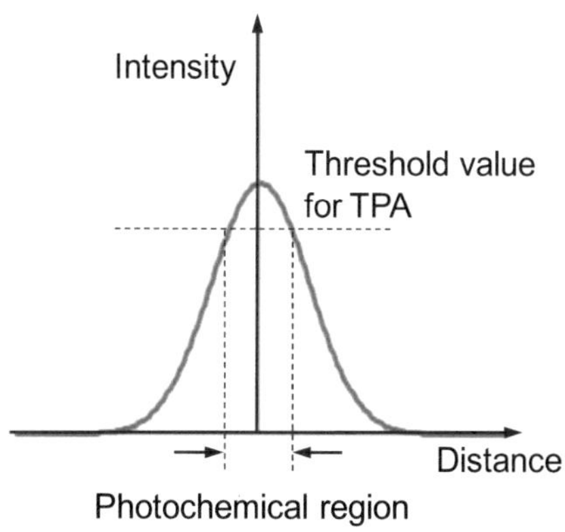

Fig. 5. Laser intensity distribution. Two-photon absorption occurs only near the focal volume where the laser intensity exceeds the threshold value for two-photon absorption.

exposure of materials and the confinement of the photochemical changes into submicron dimensions near the focal point, where the laser intensity exceeds a certain threshold value for TPA. When femtosecond laser pulses are focused into a rhodamine B solution, the localization of photochemical responses can be observed directly as visible fluorescence, which is intuitively understandable for TPA processes (Wu et al., 2006). Figure 5 shows the laser intensity distribution near the focal region schematically. The photochemical reactions can be initiated at the volume below the diffraction-limit by controlling the laser energy and the pulse numbers because of the threshold behaviors of TPA.

The existence of TPA was predicted theoretically in 1931 (Göppert-Mayer). An experimental demonstration was first conducted in 1961 (Kaiser & Garrett). However, widely various applications were not been found for many years until the development of a light source with sufficient intensity. The cross-section for TPA is orders of magnitude lower than that for single-photon absorption. Extremely high laser intensity is required to generate a sufficient concentration of starter species via the TPA process. For example, with a resist of TPA coefficient of 20 cm/TW, intensity on the order of terawatts is necessary to drive TPA.

Two-photon absorption can be induced easily using femtosecond laser with ultra-high peak intensity generated by a novel pulse-compression technique. The typical pulse duration of femtosecond laser is 100–200 fs, and the time between pulses is about 10 ns. Consequently, the optical electric field sufficient for TPA is achievable. Although the average laser power of femtosecond fiber laser is several tens of milliwatts, TPA can be initiated at average power of a few milliwatts when laser pulses are tightly focused using a high-NA objective lens. The half wavelength of the oscillation wavelength of 800 nm is located around absorption bands that are responsible for photochemical reactions of resists. Simultaneously, laser pulses can be focused into the internal region without single-photon absorption because typical resists are transparent in the near-infrared spectral region.

4. Femtosecond-laser lithography-assisted micromachining

4.1 Fabrication of three-dimensional surfaces

Figure 6 presents a process for the formation of three-dimensional surfaces using FLAM (Nishiyama et al., 2008). Here, as an example, we chose microfabrication on non-flat substrates using negative-tone resists. First, resists are spin-coated onto substrates. Subsequently, arbitrary patterns are written directly inside the resists via femtosecond laser-induced TPA. These patterns are transferred to underlying substrates by reactive plasma after post-baking process and development treatment.

Conventional ultraviolet or electron beam lithography is based on a linear (single-photon) absorption process. Therefore, incident energy is absorbed from resist surfaces, which causes difficulty in the dose-control and alignment of the exposure position for deformed resists. In contrast, FLAM enables us to irradiate the internal regions directly through TPA. Previously, three-dimensional lithography was reported using single-photon process with precise control of laser power (Maruo & Ikuta, 2000). Compared to this, high spatial resolution is available in TPA. By combining femtosecond laser nonlinear lithography with plasma etching, we can fabricate complicated three-dimensional surface structures of various optical materials.

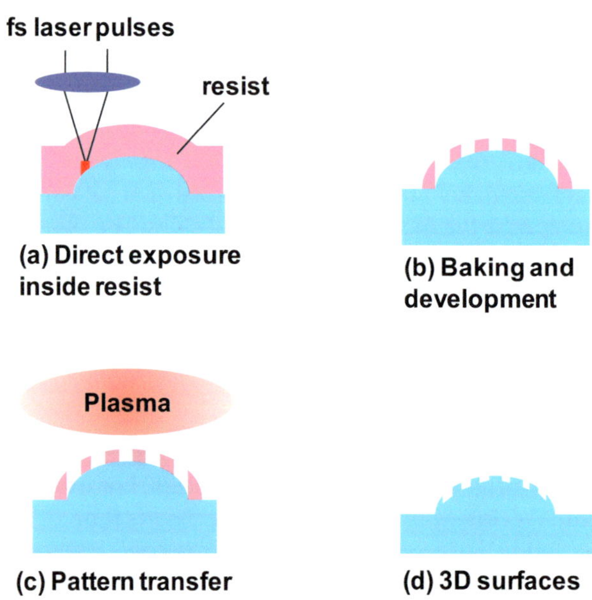

Fig. 6. Femtosecond-laser lithography-assisted micromachining.

We used a femtosecond fiber laser system for lithographic processes in FLAM. The central wavelength, pulse duration and repetition rate are, respectively, 780 nm, 128 fs and 100 MHz. The focusing lens is expected to have sufficient working distance to fabricate structures of various optical materials including SiO_2 and Si. Consequently, the laser pulses were focused into resists through an objective lens with a numerical aperture (NA) of 0.5 and working distance of 15 mm. Resist patterns are finally removed as sacrificed layers after the pattern transfers in FLAM. Therefore, we used a chemically-amplified negative-tone resist KMPR and positive-tone resist PMER because of the lower adhesive strengths than that of SU-8, which was often used in previous works of femtosecond laser lithography. The pattern transfers were conducted using electron cyclotron resonance plasma with CHF_3 and O_2 gases.

4.2 Nonlinear lithographic properties of FLAM

In this section, we describe the formation of high-aspect-ratio patterns, which are observable only when femtosecond laser pulses are focused via a low-NA objective lens into resists (Nishiyama et al., 2008). Figure 7 shows SEM images of KMPR patterns generated using TPA processes. Each filamentary rod was formed by laser irradiation from above without translating the focal spot along the optical axis (z-axis). Here, the resist thickness was 50 μm. The average laser power and exposure time were, respectively, 27 mW and 1 s. The focal position for writing of each pattern is also indicated in Fig. 7, where the z-axis origin is defined as the substrate surface. The dotted line represents the resist surface before development treatment, indicating that all filamentary rods were formed inside the resist. Compared to previous reports of femtosecond laser lithography, much higher rods were observed, as shown in Fig.7. For example, rod heights were reported to be approximately 500 nm for SCR-500 when femtosecond laser pulses were focused using a high-NA (ca. 1.4) lens (Kawata et al., 2001). In our experiments, the maximum rod height was more than 45 μm, which corresponds to three-times larger than the focal-depth of the exposure setup. On the other hand, we could not find a clear threshold of laser power for this filament formation, suggesting that the mechanism was not related to the optical Kerr effect.

Kewitsch *et al.* reported the formation of fiber-like patterns with high aspect ratios, and explained the mechanism in terms of "self-trapping based on photopolymerization" (Kewitsch & Yariv, 1996). Refractive indices of photopolymerized regions increase by cross-linking reactions of monomers through the Lorentz–Lorenz relation. Therefore, the photopolymerized region can works as a waveguide core, leading to channel propagation of subsequent laser pulses. However, unlike photopolymers used in Kewitsch's experiments, the cross-linking reactions are initiated during post-exposure baking using photogenerated acid for chemically amplified resists such as KMPR and SU-8. In short, a marked increase in the refractive index cannot be expected from light exposure alone. On the other hand, Seet *et al.* reported a heat accumulation effect by femtosecond lasers exposure, and also demonstrated fine patternings of SU-8 without post-exposure baking (Seet et al., 2006). The thermal cross-linking temperature of KMPR was measured as approximately 170°C, which is close to that of SU-8. Therefore, the laser-induced cross-linking reaction can be induced in KMPR as it is in SU-8. However, the formation of filamentary rods (channel propagation) cannot be explained sufficiently solely by the cross-linking reactions. When laser pulses are focused using a high-NA objective lens (≥ 1.2), we cannot observe the formation of filamentary rods. The amplitudes of refractive index increases of KMPR are very small (less

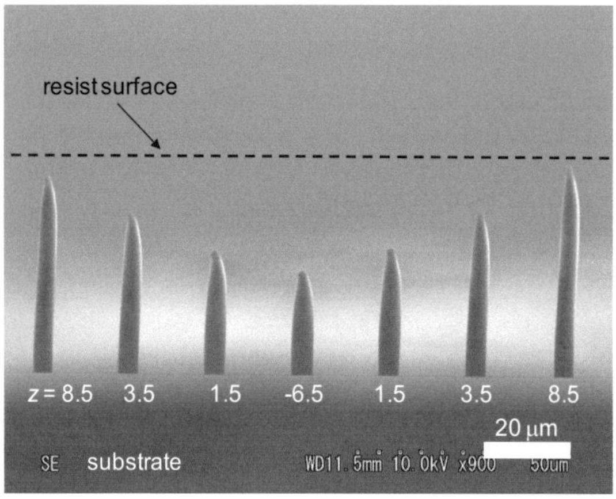

Fig. 7. SEM images of filamentary rod patterns formed using femtosecond laser lithography. Reprinted with permission from Optics Express. Copyright 2008, Optical Society of America.

than 2.0×10^{-3}) before and after post-exposure baking in the case of ultraviolet exposure. Therefore, when laser pulses are focused using a high-NA lens as described in most previous reports, the coupling of incident laser pulses to guided modes of a polymerized waveguide is strongly suppressed because of large mismatch of NA, assuming that laser-induced refractive changes are similar to those by ultraviolet exposure. In contrast, the laser pulses can be coupled to the guided modes effectively in the case of low-NA focusing. The filamentary rods in fig. 7 were most likely to be formed by laser-induced thermal cross-linking reactions of KMPR followed by effective coupling of subsequent laser pulses to a polymerized waveguide. As mentioned in section 2, the photo-modified region, where the laser intensity exceeds a certain threshold value for TPA, is an ellipsoid. The diameters of pointed ends of the ellipsoids were approximately 400 nm, which corresponds to one-fifth of the diffraction-limit 1.9 μm. Using a pointed end of a filamentary rod, high spatial resolution was obtained, even with low-NA focusing.

5. Fabrication of microlenses and micro-Fresnel lenses

In this section, we present the fabrication of three-dimensional surfaces of SiO_2-based glasses on planar wafers using FLAM (Nishiyama et al., 2009). Figure 8 shows a schematic illustration of the laser writing procedure of microlenses. Lens patterns of two types (sample 1 and sample 2) were written directly inside resists. Then, the patterns were transferred to underlying GeO_2-SiO_2 glass films. The GeO_2-SiO_2 thin films were deposited onto SiO_2 substrates using plasma-enhanced chemical vapor deposition. The lenses consist of a series of concentric circles of different radii and heights. The distance *dr* between adjacent circles was 300 nm. When the *dr* was larger than 300 nm, the trace lines of laser-spot were often visible. The average laser power and the laser translation speed were, respectively, 33 mW and 300 μm/s.

Figures 9(a)-9(d) present SEM images of two-dimensional arrays of resist microlenses. The respective diameters and curvature radii of sample 1 and sample 2 were 90 and 200 μm (Figs. 9 (a) and 9(b)) and 38 and 27 μm (Figs. 9 (c) and 9(d)), respectively. It is apparent that uniform patterns were obtained. Although several steps were visible on the surfaces of sample 1 because of the large curvature radius, the lenses of sample 2 have smooth surfaces. Figure 10 portrays optical microscope images of focal spots from GeO_2-SiO_2 microlenses when 633-nm-wavelength He–Ne laser light was normally coupled to the lenses. The curvature radii increased during pattern transfers because of the difference of etching rates between the resist and GeO_2-SiO_2. The beam diameters were 3.0 μm. It is apparent that all lenses exhibited almost identical behaviors.

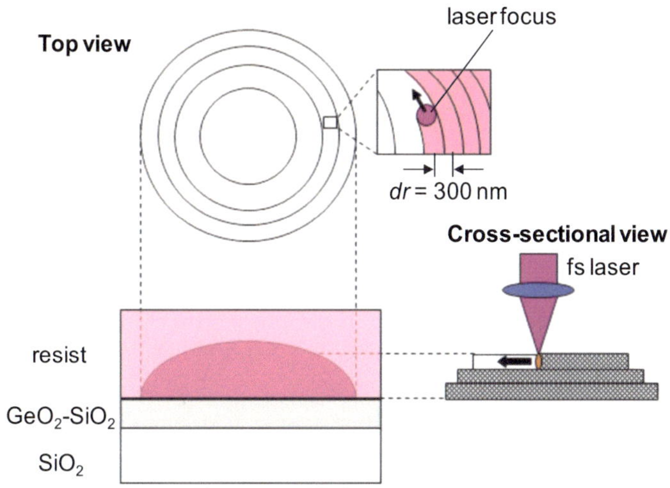

Fig. 8. Direct laser writing procedure for microlenses.

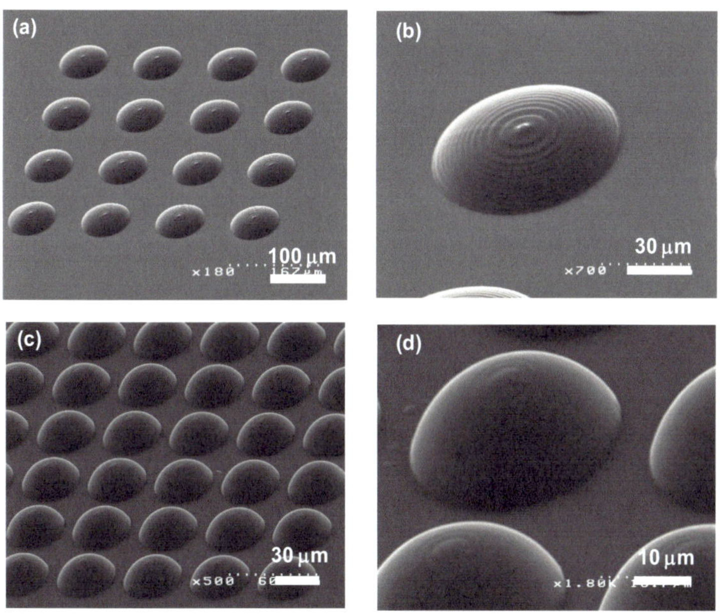

Fig. 9. SEM images of two types of resist microlens. Reprinted with permission from Applied Surface Science. Copyright 2009, Elsevier.

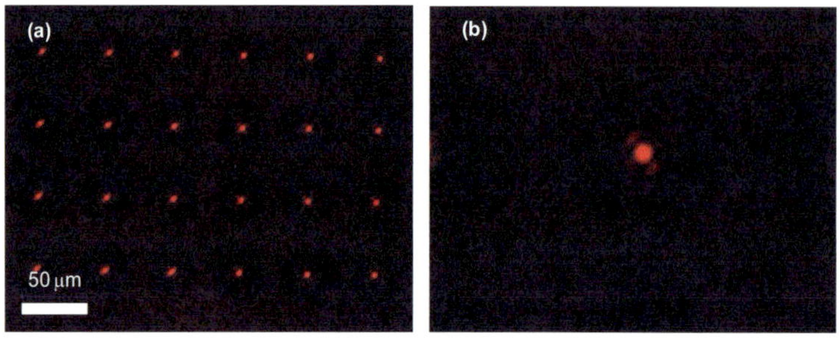

Fig. 10. Optical microscope images of focal spots from a GeO_2-SiO_2 glass microlens array fabricated using femtosecond laser lithography. Reprinted with permission from Applied Surface Science. Copyright 2009, Elsevier.

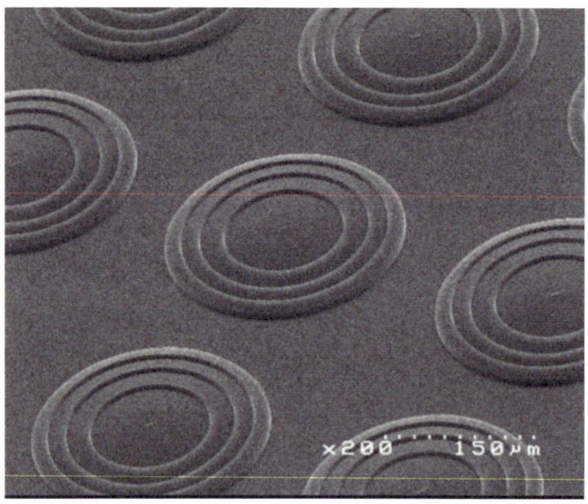

Fig. 11. SEM images of two-dimensional array of resist micro-Fresnel lenses.

Figure 11 presents SEM images of two-dimensional arrays of the resist micro-Fresnel lens with blazed structures. These patterns were transferred to SiO_2 glass substrates using plasma etching. The inner radius r_m of m-th zone and cross-sectional profile are expressed using the following equations.

$$r_m = \sqrt{2m\lambda_0 f + (m\lambda_0)^2} \tag{1}$$

$$S(r) = \frac{m\lambda_0}{n-1} + \frac{cr^2}{1+\sqrt{1-(k+1)c^2r^2}} \tag{2}$$

$$k = -n^2 \tag{3}$$

Therein, r signifies the distance from the center of the Fresnel lens ($r_m \leq r \leq r_{m+1}$); λ_0, n, and f respectively denote the operation wavelength, refractive index, and primary focal length. Using these relations, the maximum radius and the primary focal length of a SiO_2 Fresnel lens were designed, respectively, to be 240 and 2800 μm. The operation wavelength was 633 nm.

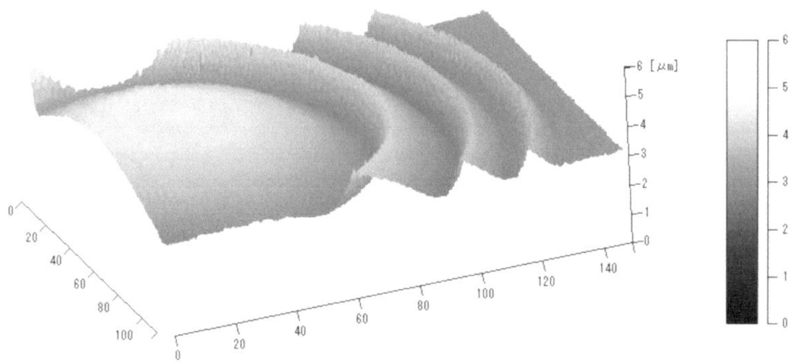

Fig. 12. Laser microscope image of a SiO_2 micro-Fresnel lens. Three-dimensional surfaces such as slopes and convex shapes can be obtained.

Considering that the etch selectivity (SiO_2/KMPR) during pattern transfers is 0.25, the resist heights were determined to be four times higher than those of transferred SiO_2 counterparts. The Fresnel lens patterns were written in KMPR under laser exposure conditions similar to those for GeO_2-SiO_2 microlenses. From laser microscope images of a SiO_2 Fresnel lens after pattern transfers in Fig. 12, well-defined three-dimensional surfaces of slopes and hemispherical shapes can be confirmed. The primary focal length was measured to be 2790 μm for He–Ne laser light, which agreed well with the theoretical value. The diffraction efficiencies were about 80%, which was two times higher than the theoretical maximum efficiencies of binary lenses.

6. Fabrication of diffractive–refractive hybrid lenses

The maximum diameters of the photo-modified region are determined mainly using threshold processes of the laser electric field. Consequently, constant pattern widths are expected to be formed at any position in resists. Simultaneously, high spatial resolution is achievable. These properties are great advantages over conventional single-photon lithography. Figure 13 shows SEM images of Si lines across a 65-μm-deep step (Mizoshiri et al., 2008). Wall-like patterns were written directly inside a positive-tone resist PMER on a step using TPA process. Subsequently, the patterns were transferred to the underlying step using plasma etching. Constant linewidths of approximately 5 μm were observed clearly at every depth position on a step. This linewidth is one order of magnitude less than that formed using the spray-coating method.

The binary Fresnel zone plates were fabricated on convex microlenses for SiO_2 hybrid diffractive–refractive lenses (Nishiyama et al., 2008). The direct laser writing procedure is presented in Fig. 14. The Fresnel zone plates were formed by translating a laser-induced filamentary rod three-dimensionally inside KMPR on convex lenses. The respective diameter, height, curvature radius, and primary focal length of individual microlens were 240, 18.9, 380, and 830 μm. The primary focal length of the hybrid lens can be determined using eq. (4), as

Fig. 13. SEM images of Si lines across a 65-μm-deep step. Reprinted with permission from Applied Physics Express. Copyright 2008, The Japan Society of Applied Physics.

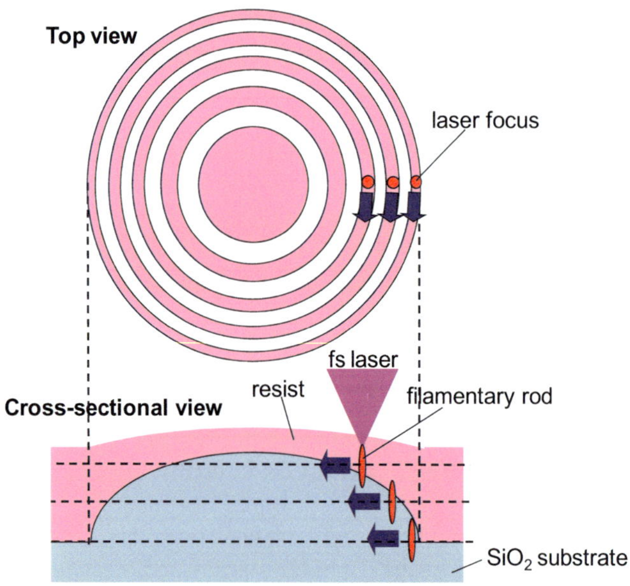

Fig. 14. Direct laser writing procedure for SiO_2 diffractive–refractive hybrid microlenses.

$$\frac{1}{f_h} = \frac{1}{f_m} + \frac{1}{f_F} \tag{4}$$

where f_h, f_m and f_F respectively represent the primary focal lengths of a hybrid lens, a convex microlens, and a Fresnel lens. The inner radius R_m of the m-th zone of the Fresnel zone plate can be expressed as $R_m = \sqrt{m f_F \lambda_0}$. Hybrid lenses with primary focal lengths of 617 μm were obtained by fabricating Fresnel zone plates of the focal length of 2700 μm on convex lenses. Figures 15(a)-15(c) show SEM images of the Fresnel zone plates on the convex lenses before and after the pattern transfers. We could obtain well-defined structures even on non-flat substrates. When He–Ne laser light was normally coupled to a SiO_2 hybrid lens, the primary focal length was shifted by 216 μm because of hybridization. This shift amount was close to the theoretical value of 213 μm, indicating that FLAM is an effective means for fabrication of precisely controlled three-dimensional surfaces. As described in the section 4, the spatial resolution as high as 400 nm is achievable in spite of low-NA focusing. For that reason, FLAM is expected to be useful for the formation of more complicated structures even on non-flat structures.

7. Future prospects

In FLAM, the laser pulses are focused using a low-NA objective lens to fabricate structures of various materials, irrespective of their optical transparencies and thicknesses. Consequently, the aspect ratios of the resist patterns become high, as shown in Fig. 7. Therefore, we cannot obtain high spatial resolution in the vertical direction. On the other hand, the use of high-NA objective lens is not desired because of the strong restrictions of the exposure setup and the choice of target materials by short working-distance. To address such problems, laser beam shaping techniques using diffractive elements and other processes will be of growing importance over the next few years.

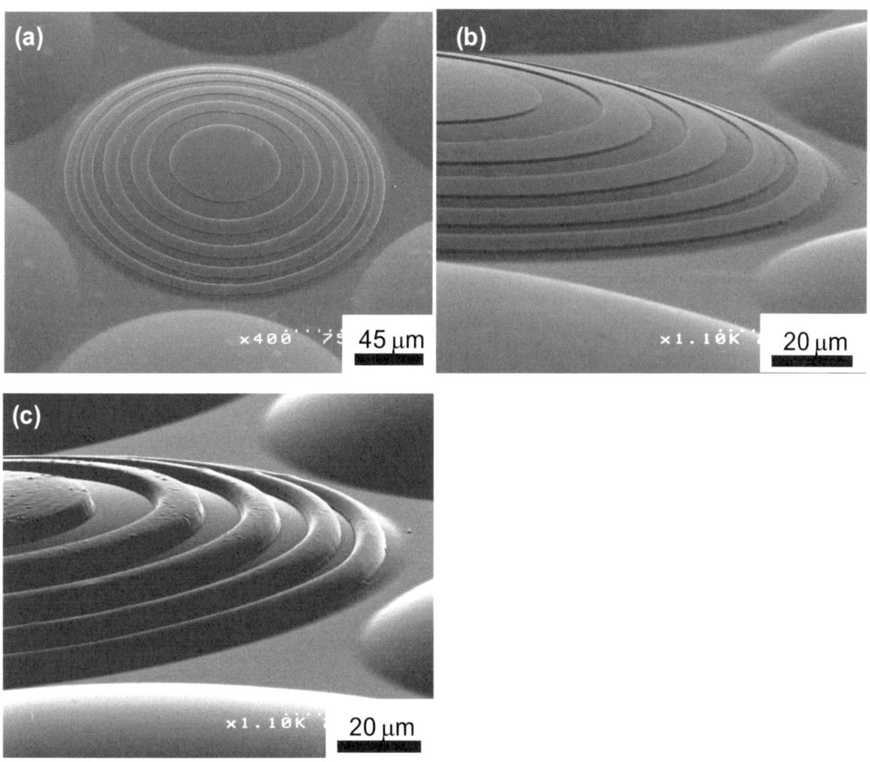

Fig. 15. SEM images: (a) overview, (b) enlarged view of SiO_2 hybrid lenses, and (c) a resist Fresnel zone plate on a convex lens. Reprinted with permission from Optics Express. Copyright 2008, Optical Society of America.

Femtosecond laser lithography is an attractive technique for the formation of three-dimensional polymeric structures. However, this technique is also a serial process by which microstructures are formed one point at a time. Consequently, significant improvement of its throughput is an important issue for scale-up to the wafer-level mass production. Kato *et al.* demonstrated parallel fabrication with femtosecond laser lithography by irradiation through a microlens array (Kato et al., 2005). By scanning multiple focal spots inside a photopolymer, they fabricated individual 227 individual three-dimensional structures simultaneously on a wafer. The fabrication time was as short as 5 min. Fourkas *et al.* reported a replication technique of three-dimensional microstructures generated using femtosecond laser lithography, referred to membrane-assisted microtransfer molding technique (LaFratta et al., 2006). Using this technique, one can reproduce even structures with closed loops such as arches using membranes formed inside original structures and highly elastic PDMS molds. In addition, several approaches using multi-beam interference (Kondo et al., 2003), spatial light modulators (Hayasaki et al., 2005), and phase masks (Jeon et al., 2004) are studied intensively to enhance the throughput of femtosecond laser lithography. Through further studies of these approaches, femtosecond laser nonlinear lithography is expected to play an essential role in the development of novel functional devices in various fields such as photonics, microelectromechanical systems, and microfluidics.

8. Conclusion

In this chapter, we presented the fundamentals of femtosecond laser nonlinear lithography and demonstrated the fabrication of three-dimensional surfaces of inorganic materials using FLAM. We can write complex structures directly inside resists using TPA. Simultaneously, high spatial resolution below the diffraction limit was achievable in spite of low-NA focusing. Using such unique properties, the semiconductor fabrication technology was extended to microfabrication on non-flat substrates. Optimization of lithographic conditions can realize devices with more functionality and with finer and more complicated three-dimensional surfaces.

9. Acknowledgements

The authors thanks Ms. M. Mizoshiri (Osaka Univ.) for the assistance in experiments, and Prof. J. Nishii (Hokkaido Univ.) for helpful discussions.

10. References

Deubel, M.; Freymann, G.; Wegener, M.; Perira, S.; Busch, K. & Soukoulis, C. M. (2004). Direct laser writing of three-dimensional photonic-crystal templates for telecommunications, *Nature* Materials, Vol. 3, 444-447.

Fujita, T.; Nishihara H. & Koyama J. (1982). Blazed gratings and Fresnel lenses fabricated by electron-beam lithography, *Optics Letters*, Vol. 7, No. 12, 578-580.

Göppert-Mayer, M. (1931). Über Elementarakte mit zwei Quantensprüngen, *Annalen der Physik*, Vol. 401, issue 3, 273-294.

Hayasaki, Y.; Sugimoto, T.; Takita, A. & Nishida, N. (2005). Variable holographic femtosecond laser processing by use of a spatial light modulator, *Applied Physics Letters*, Vol. 87, 0311011-0311013.

Herzig, H. P. (1997). *Micro-Optics: Elements, Systems and Applications*, Taylor & Francis, 978-0748404810.

Hirai, Y.; Inamoto, Y.; Sugano, K.; Tsuchiya, T. & Tabata, O. (2007). Moving mask UV lithography for three-dimensional structuring, *Journal of Micromechanics and Microengineering*, Vol. 17, 199-206.

Jeon, S.; Park, J.-U.; Cirelli, R.; Yang, S.; Heitzman, C. E.; Braun, P. V.; Kenis, P. J. A & Rogers, J. A. (2004). Fabricating complex three-dimensional nanostructures with high-resolution conformable phase masks, *Proceedings of the National Academy of Sciences*, Vol. 101, No.34, 12428-12433.

Juodkazis, S.; Mizeikis, V.; Seet, K. K.; Miwa, M. & Misawa, H. (2005). Two-photon lithography of nanorods in SU-8 photoresist, *Nanotechnology*, Vol. 16, 846-849.

Kato, J.; Takeyasu, N.; Adachi, Y.; Sun, H.-B. & Kawata, S. (2005). Multiple-spot parallel processing for laser microfabrication, *Applied Physics Letters*, Vol. 86, 0441021-0441023.

Kaiser, W. & Garrett, C. G. B. (1961). Two-photon excitation in $CaF2:Eu^{2+}$, *Physical Review Letters*, Vol. 7, No. 6, 229-232.

Kawata, S.; Sun, H.-B.; Tanaka T. & Takada K. (2001). Finer features for functional microdevices, *Nature*, Vol. 412, 697-698.

Kewitsch, A. S. & Yariv, A. (1996). Self-focusing and self-trapping of optical beams upon photopolymerization, *Optics Letters*, Vol. 21, 24-26.

Kondo, T.; Matsuo, S.; Juodkazis, S. & Misawa, H. (2001). Femtosecond laser interference technique with diffractive beam splitter for fabrication of three-dimensional photonic crystals, *Applied Physics Letters*, Vol. 79, No. 6, 725-727.

Kondo, T.; Matsuo, S.; Juodkazis, S.; Mizeikis, V. & Misawa, H. (2003). Multiphoton fabrication of periodic structures by multibeam interference of femtosecond pulses, *Applied Physics Letters*, Vol. 82, No. 17, 2758-2760.

LaFratta, C. N.; Li, L. & Fourkas, J. T. (2006). Soft-lithographic replication of 3D microstructures with closed loops, *Proceedings of the National Academy of Sciences*, Vol. 103, No. 23, 8589-8594.

Maruo, S.; Nakamura, O. & Kawata, S. (1997). Three-dimensional microfabrication with two-photon-absorbed photopolymerization, *Optics Letters*, Vol. 22, 132-134.

Maruo, S. & Ikuta, K. (2000). Three-dimensional microfabrication by use of single-photon-absorbed polymerization, *Applied Physics Letters*, Vol. 19, 2656-2658.

Maruo, S.; Ikuta, K. & Korogi, H. (2003). Force-controllable, optically driven micromachines fabricated by single-step two-photon microstereolithography, *Journal of micromechanical systems*, Vol. 12, No. 5, 533-539.

Mizoshiri, M.; Nishiyama, H.; Kawahara, T.; Nishii, J. & Hirata, Y. (2008). SiO_2-based hybrid diffractive-refractive lenses fabricated by femtosecond laser-assisted micromachining, *Applied Physics Express*, Vol. 1, No. 12, 1270011-1270013.

Mizoshiri, M.; Nishiyama, H.; Nishii, J. & Hirata, Y. (2010). Three-dimensional SiO_2 surface structures fabricated using femtosecond laser lithography, *Applied Physics A*, Vol. 98, 171-177.

Nishiyama, H.; Mizoshiri, M.; Kawahara, T., Nishii, J. & Hirata, Y. (2008). SiO_2-based nonplanar structures fabricated using femtosecond laser lithography, *Optics Express*, Vol.16, No. 22, 17288-17294.

Nishiyama, H.; Mizoshiri, M.; Nishii, J. & Hirata, Y. (2009). Microlens arrays of high refractive index glass fabricated by femtosecond laser lithography, *Applied Surface Science*, Vol. 255, 9750-9753.

Park, S. H.; Lee, S. H.; Yang, D.-Y.; Kong, H. J. & Lee, K.-S. (2005). Subregional slicing method to increase three-dimensional nanofabrication efficiency in two-photon polymerization, *Applied Physics Letters*, Vol. 87, 1541081-1541083.

Radtke, D. & Zeitner, U. D. (2007). Laser-lithography on non-planar surfaces, *Optics Express*, Vol. 15, 1167-1174.

Reinhardt, C.; Passinger, S.; Chichkov, B. N.; Marquart, C.; Radko, I. P. & Bozhevolnyu, S. I. (2006). Laser-fabricated dielectric optical components for surface Plasmon polaritons, *Optics Letters*, Vol. 31, No. 9, 1307-1309.

Seet, K. K.; Juodkazis, S.; Jarutis, V. & Misawa, H. (2006) Feature-size reduction of photopolymerized structures by femtosecond optical curing of SU-8, *Applied Physics Letters*, Vol. 89, 0241061-0211063.

Serbin, J.; Egbert, A.; Ostendorf, A.; Chichkov, B. N.; Houbertz, R.; Domann, G.; Schulz, Cronauer, C.; Fröhlich, L. & Popall, M. (2003). Femtosecond laser-induced two-photon polymerization of inorganic-organic hybrid materials for applications in photonics, *Optics Letters*, Vol. 28, No. 5, 301-303.

Serbin, J.; Ovsianikov, A. & Chichkov, B. (2004). Fabrication of woodpile structures by two-photon polymerization and investigation of their optical properties, *Optics Express*, Vol. 12, No. 21, 5221-5228.

Singh, V. K.; Sasaki, M.; Hane, K.; Watanabe, Y.; Takamatsu, H.; Kawakita, M. & Hayashi, H. (2005). Deposition of thin and uniform photoresist on three-dimensional structures using fast flow in spray coating, *Journal of Micromechanics and Microengineering*, Vol. 15, 2339-2345.

Straub, M.; Nguyen, L. H.; Fazlic, A. & Gu, M. (2004). Complex-shaped three-dimensional microstructures and photonic crystals generated in a polysiloxane polymer by two-photon microstereolithography, *Optical Materials*, Vol. 27, 359-364.

Suleski, T. J. & O'Shea, D. C. (1995). Gray-scale masks for diffractive-optics fabrication: I. Commercial slide imagers, *Applied Optics*, Vol. 34, No. 32, 7507-7517.

Sun, H.-B.; Kawakami, T.; Xu, Y.; Ye, J.-Y.; Matsuo, S.; Misawa, H.; Miwa, M. & Kaneko, R. (2000). Real three-dimensional microstructures fabricated by photopolymerization of resins through two-photon absorption, *Optics Letters*, Vol. 25, No. 15, 1110-1112.

Yu, T.; Ober, C. K.; Kuebler, S. M.; Zhou, W.; Marder, S. R. & Perry, J. W. (2003). Chemically amplified positive resists for two-photon three-dimensional microfabrication, *Advanced Materials*, Vol. 15, No. 6, 517-521.

Wu, S, & Serbin, J. & Gu, M. (2006). Two-photon polymerization fro three-dimensional micro-fabrication, *Journal of Photochemistry and Photobiology A: Chemistry*, Vol .181, 1-11.

5

Improving the Efficiency of Pattern Extraction for Character Projection Lithography using OPC Optimization

Hirokazu Nosato, Hidenori Sakanashi,
Masahiro Murakawa and Tetsuya Higuchi
National Institute of Advanced Industrial Science and Technology (AIST)
Japan

1. Introduction

In recent years, with the growth in demand for digital information appliances, system LSI devices for digital information appliances are becoming the mainstream in the semiconductor industry, replacing LSI devices for PCs. For such system LSIs, it is important to be able to manufacture system LSI devices with multi-functionality and high performance with high-variety, low-volume production processes incurring low costs and short turn-around-times (TAT). However, as the sizes of semiconductor devices become increasingly smaller, the accompanying increases in mask costs and TAT are major obstacles preventing rapid advances in the development of digital information appliances with system LSIs. These problems are due to the increasing complexity and data volume of mask patterns. Similarly, while Optical Proximity Correction (OPC) can correct a mask pattern to improve the fidelity of the projected image pattern on the wafer, the method involves huge calculation loads (including runtime, data volume, and processing costs) as the calculations required to correct a mask pattern become ever more complex.

In order to solve mask costs and TAT problems, Character Projection Lithography (CPL) has been devised as a high-throughput electron-beam direct-write technology (Nakayama et al., 1990). CPL has the advantage of reducing the number of electron beam (EB) shots compared to conventional variably-shaped beam lithography (Pfeiffer, 1978), because each character pattern that appears frequently in a layout can be simultaneously written by a single EB shot with a CP aperture mask. In order to achieve such reductions in the number of EB shots, it is important to extract frequently-used character patterns and to prepare CP aperture masks. However, when CPL is applied to OPCed layouts for a system LSI, while CPL can work in reducing the number of EB shots in the memory device, CPL cannot function with the random logic device. Because random logic devices are composed by laying out logic cells randomly, one character pattern can be deformed into many different patterns to correct the Optical Proximity Effect (OPE) from neighboring patterns, and thus require complicated OPC, so they cannot be extracted as a unique CP aperture mask.

In order to overcome the problems of realizing CPL with a random logic device, we propose a method of improving the efficiency of pattern extraction by employing an adaptive OPC method (Matsunawa et al., 2005; Matsunawa et al., 2007). As a new OPC calculation model designed to achieve high-precision correction and reductions in OPC runtimes by reducing the calculation domain, the adaptive OPC method has two important characteristics. One is that OPC is applied to each cell before layout design. Thus, the adaptive OPC method can be regarded as being a cell-based OPC method. The other characteristic is the optimization of variable OPC features in peripheral regions around the OPCed cell due to the OPE from neighboring patterns after layout design.

In this study, we enhance the basic adaptive OPC method by developing two important algorithms: (1) a cell grouping algorithm that categorizes deformed cells and extracts some typical cell groups, and (2) an OPC optimization algorithm that regards the cells in a group as one typical cell and corrects the OPC features of the typical cell to form a CP aperture mask. In conducted verification experiments with two full-chip level OPCed layouts based on actual functional LSI, we successfully achieve an improvement of approximately 60% in extraction efficiency.

This chapter is organized as follows: In Section 2, the problems of realizing CPL with random logic devices are briefly described. Section 3 introduces our proposed improvement method. Section 4 presents the details and results of conducted experiments for a feasibility study and Section 5 shows the results of verification experiments with actual functional LSI data. Finally, Section 6 discusses the results and draws some conclusions.

2. The problems of realizing CPL with random logic devices

CPL is a promising technology as a high-throughput electron-beam direct-write technology. The advantage of CPL is that it has a lower number of electron beam (EB) shots compared to conventional variably-shaped beam lithography, because character patterns that appear frequently in a layout can be simultaneously written by a single EB shot with a CP aperture mask. Accordingly, it is important to extract the sets of frequently-used character patterns and to prepare CP aperture masks in order to reduce the number of EB shots.

However, it has been reported by ASET (ASET, 2006) that although CPL can be beneficial for the memory device in reducing the number of EB shots, CPL cannot function on random logic devices. According to the report for a poly-silicon layer, the single character pattern referred to 1539 times in the memory device is diversified to eight deformed patterns with OPC features. On the other hand, the character pattern referred to 680 times in the logic device is diversified to 660 deformed patterns with OPC features. The reason for the difference in the number of the deformed patterns is that random logic devices are composed by randomly laying out logic cells, and one character pattern can be deformed into many different patterns due to the Optical Proximity Effect (OPE) from neighboring patterns. That is a character pattern in random logic devices will be deformed into many different patterns that require complicated OPC features, and so they cannot be extracted as a unique CP aperture mask. Thus, in order to realize the advantage of CPL to reduce the number of EB shots with random logic devices, the novel OPC method is required to improve the efficiency of pattern extraction with moderate levels of diversification.

3. Improving pattern extraction efficiency for CPL

3.1 Approach

In order to overcome the problems of realizing CPL with a random logic device, we propose a method of improving the efficiency of pattern extraction by employing an adaptive OPC method (Matsunawa et al., 2005; Matsunawa et al., 2007). As a new OPC calculation model designed to achieve high-precision correction and reductions in OPC runtimes by reducing the calculation domain, the adaptive OPC method has two important characteristics. One is that OPC is applied to each cell before layout design. Thus, the adaptive OPC method can be regarded as being a cell-based OPC method. The other characteristic is the optimization of variable OPC features in peripheral regions around the OPCed cell due to the OPE from neighboring patterns after layout design.

In this research, we enhance the basic adaptive OPC method by developing two important algorithms: (1) a cell grouping algorithm that categorizes deformed cells and extracts some typical cell groups, and (2) an OPC optimization algorithm that regards the cells in a group as one typical cell and corrects the OPC features of the typical cell to form a CP aperture mask. Fig. 1 illustrates the basic concept of the proposed method with these two algorithms. *A* character cells are extracted from a cell layout data. Suppose that the *A* character cells would be deformed into five individual cells $A_1 \sim A_5$ after OPC processing. With the cell grouping algorithm, the deformed cells are extracted as two typical cell groups $A_1 \sim A_3$ and $A_4 \sim A_5$. The OPC optimization algorithm optimizes a typical cell, *Aa* or *Ab*, for each cell group based on the cells within each group.

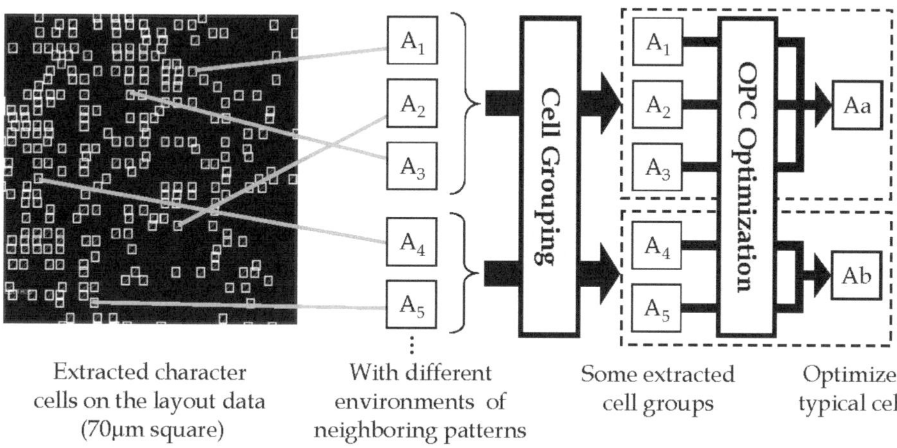

Fig. 1. Framework of the proposed method.

3.2 The cell grouping algorithm

This subsection describes the cell grouping algorithm which can extract a typical cell group. The character cells are inserted into all of the different neighboring patterns. A neighboring pattern refers to a specific layout situation relating to crude density, directionality and design in the peripheral regions of a character cell, as shown in Fig. 2. The OPC features of a character cell are affected by the OPE from the neighboring region. For example, if two character cells are placed in similar neighboring patterns, there will be similarities in their OPC features which can be optimized by the OPC optimization method to produce a typical cell. Thus, it is possible to identify typical cell groups that have similar OPC features by analyzing the neighboring patterns.

In this research, we propose a cell grouping algorithm that involves a two-stage processing of categorizing cells in terms of layout direction and of comparing overlapping patterns by analyzing neighboring patterns. Fig. 3 presents the flowchart for the proposed cell grouping algorithm. In preparation, data for the character cells and their neighboring patterns are loaded from the layout data. In the first stage, the character cells are roughly categorized into several provisional cell groups according to the directionality of neighboring patterns. In the second stage, the multiple cell groups are finely categorized into typical cell groups that can be optimized as one typical cell by comparing the overlaps for neighboring patterns. The two categorization methods are explained in detail in the following subsections.

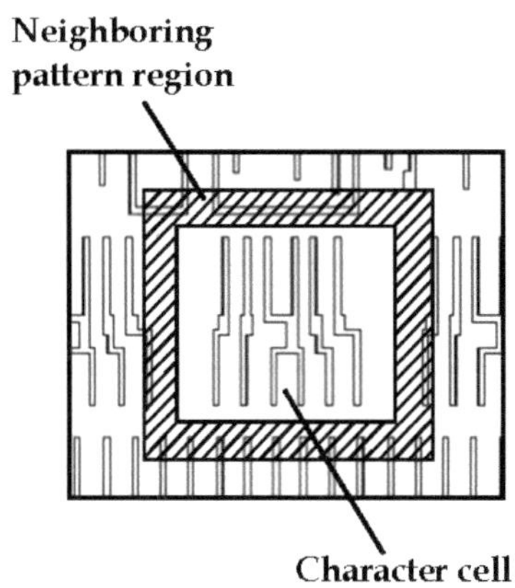

Fig. 2. A character cell with its neighboring pattern. Source: based on Nosato et al., 2007.

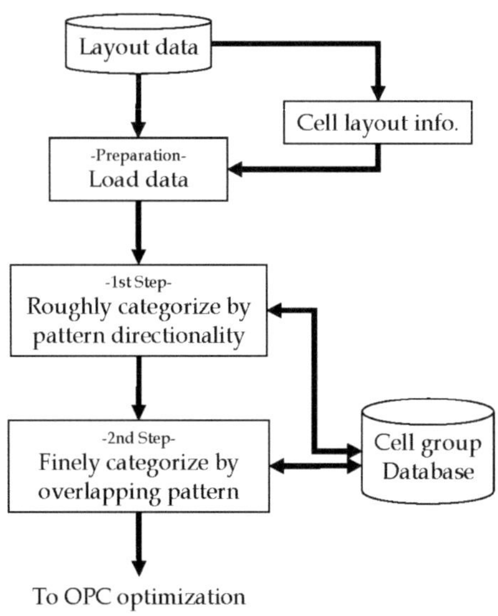

Fig. 3. Flowchart of the cell grouping algorithm.

3.2.1 The rough categorization method according to directionality

The rough categorization method generates provisional cell groups that have similar neighboring patterns. The OPE on a character cell depends on the density of neighboring patterns. Employing this method, we analyzed the neighboring patterns in our test data (which is described in more detail in Section 4.1), and Fig. 4 and Fig. 5 present parts of the analysis results. Fig. 4 presents the crude density of neighboring patterns in the peripheral regions of a character cell. Neighboring patterns are denser in the vertical region (Fig. 4(a)) than in the horizontal region (Fig. 4(b)). Fig. 5 presents results for the directionality of patterns in the vertical region. The horizontal pattern (Fig. 5(a)) is denser than the vertical pattern (Fig. 5(b)) in the vertical region of the character cell. Therefore, in this research, the character cells are roughly categorized into several provisional cell groups according to the criterion of directionality in the vertical regions, as shown in Table 1.

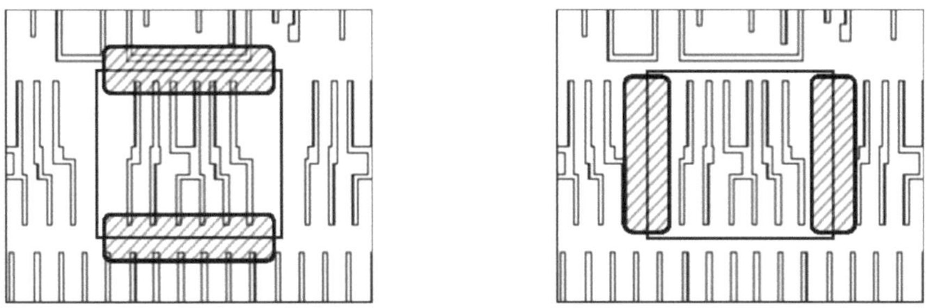

(a) Vertical region of a character cell (b) Horizontal region of a character cell

Fig. 4. Crude densities of neighboring patterns. Source: based on Nosato et al., 2007.

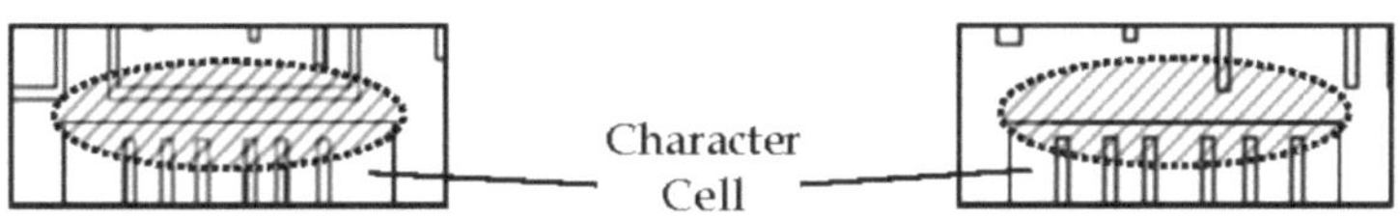

(a) Horizontal patterns (b) Vertical Patterns

Fig. 5. Pattern directionality in the vertical region. Source: based on Nosato et al., 2007.

Provisional cell group	Pattern directionality
1	Horizontal pattern in the upper vertical region
2	Horizontal pattern in the lower vertical region
3	No horizontal pattern in the vertical region

Table 1. The rough grouping criteria for pattern directionality in the vertical region.

3.2.2 The fine categorized method according to overlap comparisons

In order to finely categorize the groups into typical cell groups that can be optimized as a single typical cell, the fine categorization method compares the areas of overlap for neighboring patterns one-by-one for each provisional cell group, as shown in Fig. 6. As Fig. 6 indicates, the overlapping areas show similarities in terms of their neighboring patterns. Therefore, the character cells in the provisional cell groups are finely categorized into typical cell groups based on comparisons for the overlapping areas.

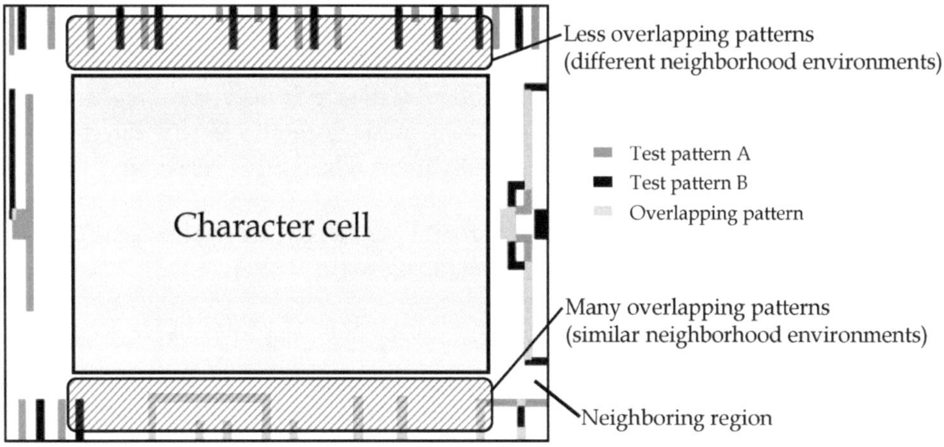

Fig. 6. Comparison of neighboring patterns.

3.3 OPC optimization algorithm

This section describes the OPC optimization algorithm which can adjust the OPC features of a categorized cell group to generate one typical character cell. Fig. 7 shows the flowchart for the proposed OPC optimization algorithm. Initially, categorized cell group data is loaded from the database of cell groups. Next, the OPC features of a character cell are coded as common OPC optimization parameters. These parameters are reflected in all character cells within the same categorized cell group. In the first fitness calculation, initial common OPC parameters are evaluated on each neighboring pattern using a lithography simulation.

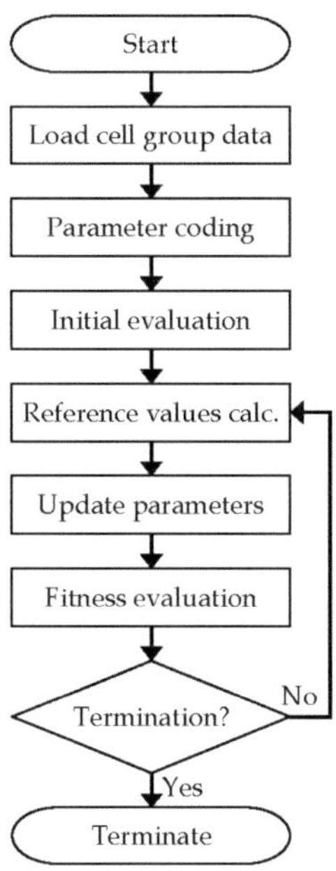

Fig. 7. Flowchart for OPC optimization.

Then, the algorithm calculates which reference values are better to update to the common parameters, based on the fitness values of each neighboring pattern, and updates the common parameters according to these reference values. Finally, the updated common parameters are evaluated again. When fitness evaluations exceed target accuracy for OPC, or evaluation iterations exceed a predetermined number, then the optimization algorithm terminates. We detail the coding of parameters, the fitness evaluation and the reference values calculations below.

3.3.1 Coding of optimization parameters

The common OPC optimization parameters correspond to variables representing the OPC features of an OPCed character cell. The parameters are expressed as a one-dimensional array that consists of the horizontal directions and the vertical directions of the OPC features, as shown in Fig. 8. These parameters represent the polygons for OPC features, and they are expressed as real numbers.

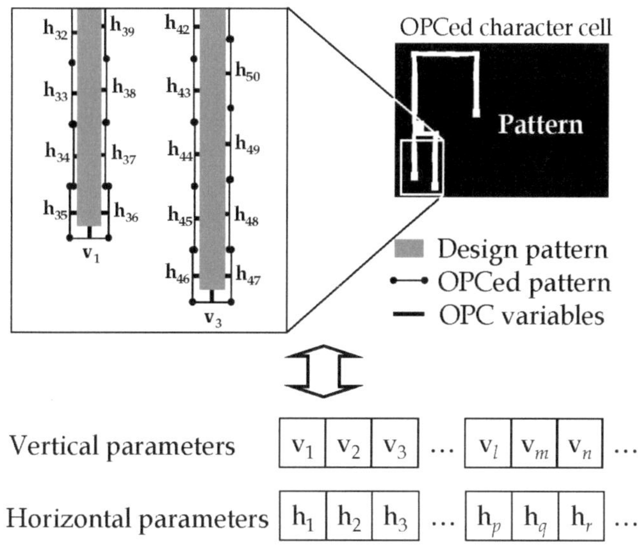

Fig. 8. Coding of optimization parameters.

3.3.2 Evaluation method

In this evaluation method, because it is necessary to simultaneously evaluate the character cells in terms of every neighboring pattern within a categorized cell group, we propose parallel processing based on the master-slave model (Cantú-Paz, 1997) for this evaluation, as shown in Fig. 9. Each slave processor is assigned the task of evaluating each neighboring pattern in a lithography simulation. For example, when there are four cells within a categorized cell group, one master processor and four slave processors would be used for evaluation. Fig. 10 presents the flowchart for processing on the master and the slave processors. Fitness is calculated in terms of differences between the target pattern (design layout pattern) and the lithography image output by the lithography simulator as measured at the evaluation points for each OPC feature.

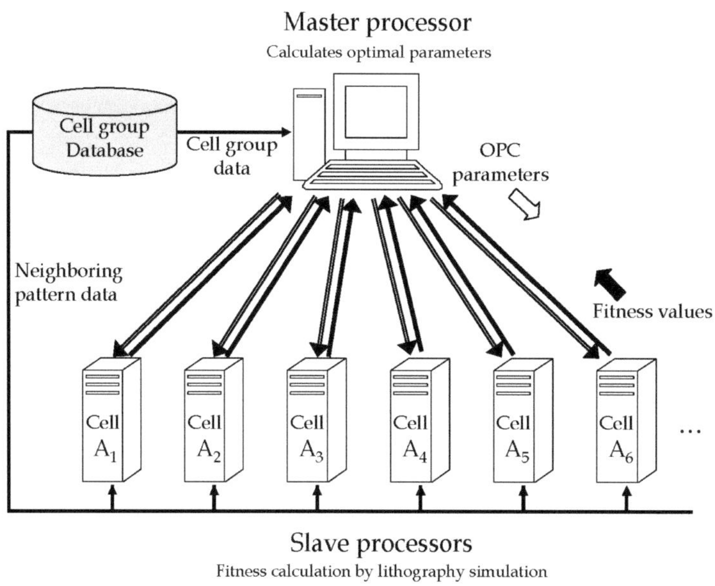

Fig. 9. Parallel processing for OPC optimization.

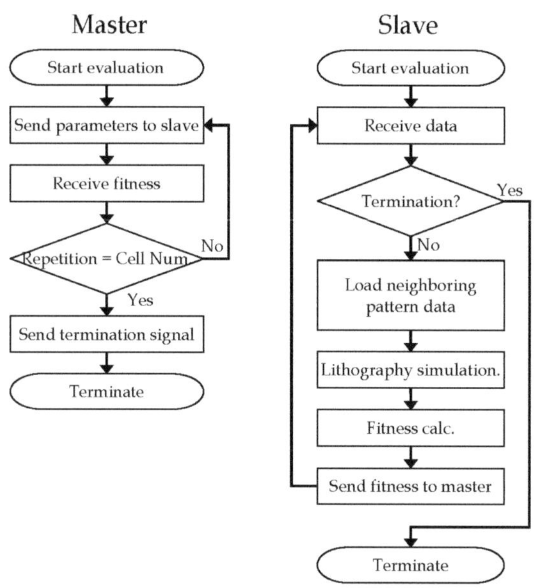

Fig. 10. Flowchart of the evaluation method.

3.3.3 Calculation of reference values

The optimization method utilizes one set of common parameters and multiple sets of fitness value of each neighboring parameter. Thus, it can calculate which reference values are better to update to the common parameters, based on the fitness values of each neighboring pattern, and updates the common parameters accordingly to these references values. The calculation generates the reference value R_j from the fitness F_{ij} according to the following formula,

$$R_j = \sigma \cdot \frac{1}{n}\sum_{i=1}^{n} F_{ij} \quad (1)$$

where i is the index of the cell within a cell group, and j is the position index of the OPC feature for each cell, and σ denotes a control factor for optimization accuracy, which is based on the step function shown in Fig. 11. This factor functions to control the adjustment level, such as coarse adjustment with large fitness values or fine adjustment with small fitness values in terms of the OPC features.

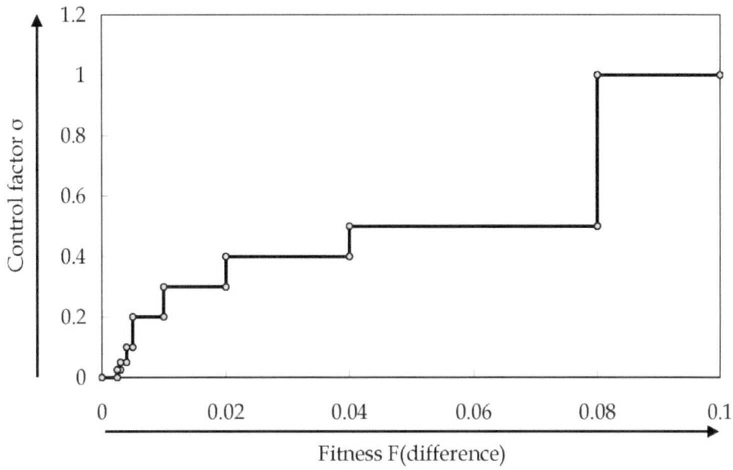

Fig. 11. Control factor for adjustment accuracy.

4. Experimental results and discussion for the feasibility study

4.1 Configuration of the test data

Evaluation experiments were performed using 40 types of test patterns. The test patterns were based on actual functional cell data from the 130nm cell library of STARC (Semiconductor Technology Academic Research Center, Japan). A test pattern consists of one character logic cell in the center and eight logic cells around it, as shown in Fig. 12(a). The character cell was fixed in the center for all 40 test patterns. The eight logic cells were chosen from 26 different logic cells including 13 circuit types, and randomly arrayed around the character cell. That is, arrangements of cell in test patterns imitated different neighboring patterns extracted from the random logic devices, as shown in Fig. 12(b).

During the preparation for the experiments, test patterns according to the 90nm design rule were created by shrinking the 130nm cell library arrayed patterns. In addition, the model-base OPC of Panoramic Technology was applied to the character cell to generate initial OPC parameters as a cell-based OPC before layout.

4.2 Experimental results for the cell grouping algorithm

The experimental conditions are presented in Table 2. The proposed cell grouping algorithm categorized the 40 types of character cells with each test pattern into six typical cell groups. Fig. 13 shows the result tree achieved by the categorization by the cell grouping algorithm. In the first stage of categorization, two provisional typical cell groups were categorized according to the directionality of neighboring patterns, and 19 typical cells were left

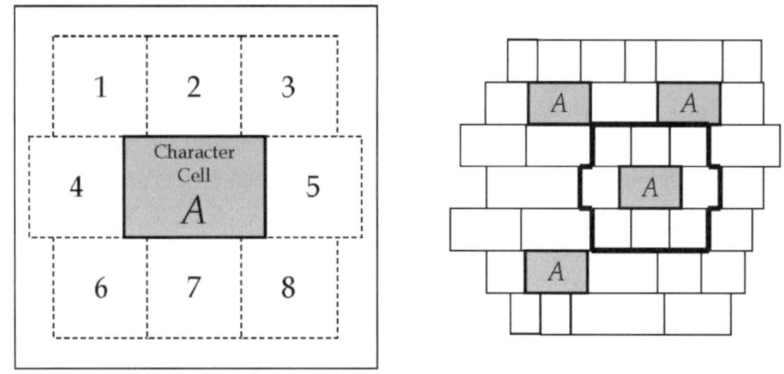

(a) Configuration of test pattern (b) Layout of logic device

Fig. 12. Test pattern for experiments.

OS	WindowsXP Professional x64 Edition Ver. 2003 Service Pack 2
CPU	Inter® Xeon™ 3.00 GHz x2
Memory	2 GB
Simulator	Panoramic Technology EM Suite v5.00
Illumination	λ = 193nm, k_1 = 0.4, NA = 0.85, Annular(0.5/0.75)
Accuracy	0.001 μm

Table 2. Experimental conditions for cell grouping.

uncategorized. In the second stage of categorization, the six typical cell groups were categorized based on comparisons of the overlapping areas in the neighboring patterns. Finally, based on the 6 typical cell groups, sets of multiple types (3, 2, 2, 4, 5, and 2 types) of the character cells for each test pattern were extracted, as shown in Table 3. The extracted typical cell groups were separately corrected as one typical character cell in the subsequent OPC optimization experiment.

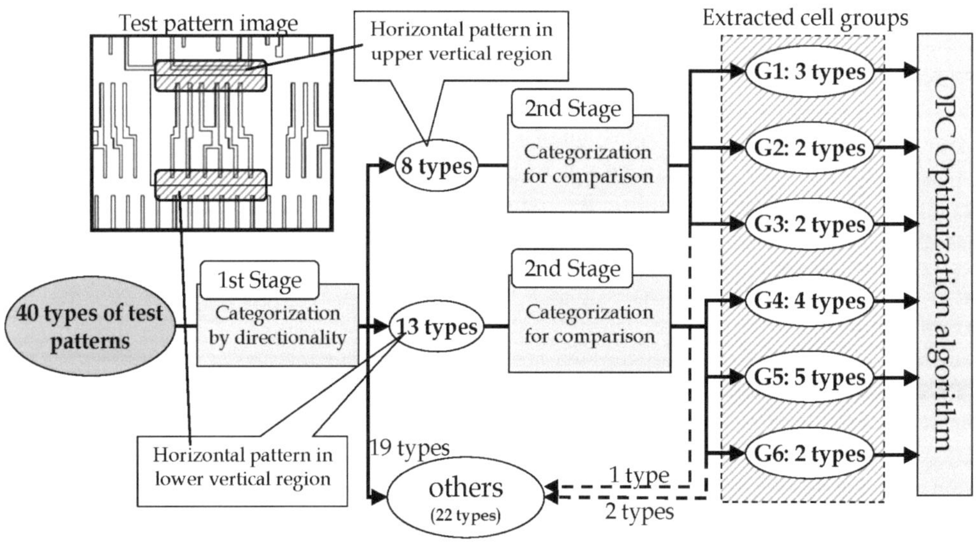

Fig. 13. Categorized tree for cell-group extraction.

Cell group	Type-count	Number of categorized test patterns
G1	3 types	1, 6, 23
G2	2 types	3, 25
G3	2 types	19, 29
G4	4 types	7, 13, 25, 27
G5	5 types	16, 17, 26, 33, 36
G6	2 types	9, 14

Table 3. Categorization results of the experiments.

4.3 Experimental results for OPC optimization

The experimental conditions are presented in Table 4. We applied the proposed OPC optimization algorithm to each of the extracted cell groups. The results of the experiments are presented in Table 5 and the graph shows the largest errors between the sizes of target features and the simulated lithography image. Although the largest errors were approximately 34 nm before optimization, the proposed algorithm decreased the errors to approximately 3 nm. The results before optimization indicate that the application of OPC only for the character cell is insufficient to correct for the character cell within its neighboring patterns. That is, the OPE from neighboring patterns caused differences of approximately 34 nm between the target pattern and the lithography simulated pattern. The differences were corrected by allowing all members in each typical cell group to have common OPC features. Fig. 14 shows how the OPC features were optimized in the experiment. Clearly, the proposed control factor within the optimization algorithm functioned effectively by realizing coarse adjustment with large fitness values and fine adjustment with small fitness values.

OS	Redhut Linux v2.6.9.42.ELsmp
CPU	Intel® Core™ 2 duo 3.00 GHz x2
Memory	4 GB / 8 GB
Compiler	gcc 3.3
Processor	Up to 6
Communication library	Mpich-1.2.6 build by gcc 3.3
Simulator	Lithography simulator from AIST
Illumination	λ = 193nm, k_1 = 0.4, NA = 0.85, Annular(0.5/0.75)
Calculation Area	5.12 x 5.12 μm^2
Accuracy	0.0025 μm
OPC Parameters	279 real values
Fitness points	279 points
Termination	Repetition of evaluation = 200 times

Table 4. Experimental condition for OPC optimization.

Cell group	Before optimization	After optimization
G1	34.3 nm	2.97 nm
G2	34.7 nm	2.68 nm
G3	34.2 nm	2.81 nm
G4	34.3 nm	2.77 nm
G5	34.7 nm	2.84 nm
G6	34.3 nm	3.56 nm

Table 5. Adjustment results for OPC optimization.

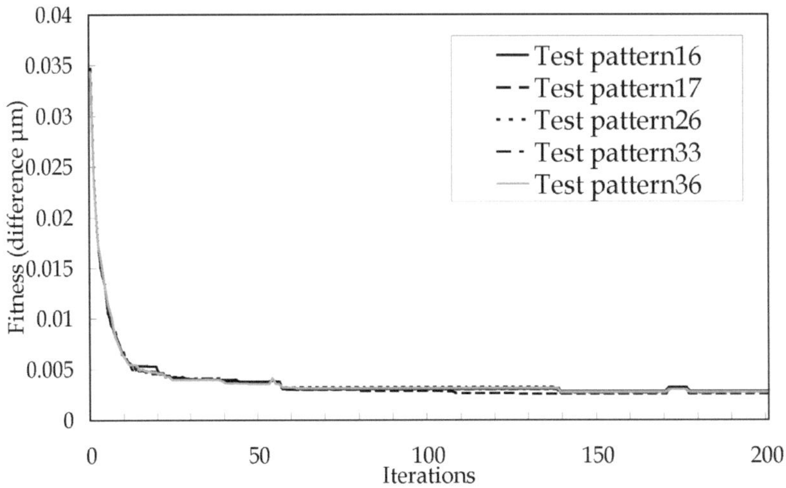

Fig. 14. Fitness as a function of experiment iterations.

4.4 Discussion

As the results of the cell grouping experiment and the OPC optimization experiment, the 40 types of character cells with the each test pattern were reduced to six typical character cells together with a set of 22 unique types, making a total of 28 different types of cells. Accordingly, we successfully achieved a 30% improvement in extraction efficiency. In this research, we proposed a categorization algorithm based on directionality. However, the categorization algorithm remains is a matter for further consideration. If it were possible to discover a more effective approach to categorization, it might be possible to realize further improvements in the pattern extraction efficiency for CPL.

5. Verification experiments with actual functional LSI data

5.1 Configuration of the verification data

In these verification experiments, we verified the proposed method with two full-chips level OPCed layouts based on the actual functional LSI of STARC (Semiconductor Technology Academic Research Center, Japan). There were two kinds of verification data (Data-1 & Data-2), which related to clipped 70 square micrometer areas from different OPCed layouts. Because, Data-1 and Data-2 consist of 1,023 cells and 272 cells respectively, they have different features in terms of cell density.

5.2 Experimental results and discussion

Fig. 15 presents the experimental results for one cell *A* for the cell grouping algorithm applied to Data-1. There were 256 cells for cell *A* within Data-1, as shown in Fig. 15(a). Of these *A* cells, 198 were categorized into 35 cell groups, which are colored and hatched in Fig. 15(b).

Table 6 presents the experimental results for all cells undergoing the cell grouping algorithm for both Data sets. The details for the entries in Table 6 were as follows:

- Total cells: (1)
  Number of all cells in the clipped 70 square micrometer area.

(a) Search result of cell A in the Data-1 (b) Categorization result of cell A in the Data-1

Fig. 15. Categorization results for cell A in the Data-1.

Data	(1) Total cells	(2) Extracted cells	(3) Unextracted cells	(4) Cell groups	(3)+(4) Cell types	{(3)+(4)}/(1) Rate of reduction
Data-1	1023	826	197	206	403	39.4 %
Data-2	272	159	113	27	140	51.5 %

Table 6. Experimental results for all cells undergoing cell grouping algorithm.

- Extracted cells: (2)
  Number of extracted cells for categorizing cell groups.
- Unextracted cells: (3)
  Number of unextracted cells.
- Cell groups: (4)
  Number of categorized cell groups (which become the typical character cell)
- Cell types: (3)+(4)
  Total cell types for both typical character cells and unextracted cells.
- Rate of reduction: {(3)+(4)}/(1)
  Rate of reduction in cell types produced by the proposed method.

In the results for Data A, the reduction rate was 39.4%. In contrast, the results for Data B indicated a 51.5% reduction in cell types. From the results of the verification experiments, clearly the proposed method can reduce the number of cell types in OPCed layouts based on actual functional LSI. Accordingly, we were able to successfully achieve an improvement in extraction efficiency of approximately 60%. Thus, for actual functional LSIs, the proposed method can also greatly reduce the variety in deformed patterns with OPC features.

6. Conclusion

In this research, we have proposed an approach to improving pattern extraction efficiency for character projection lithography on random logic devices by employing OPC optimization. Our proposed method is able to realize reductions in the variety of patterns deformed by the complicated OPC through two developed algorithms: (1) a cell grouping algorithm that categorizes differentiated cells and extracts some typical cell groups, and (2) an OPC optimization algorithm that regards the cells in a group as one typical cell and corrects for the OPC features of a typical cell to form a CP aperture mask. As the results of conducted experiments demonstrate with 40 types test patterns, the proposed method realized a 30% improvement in extraction efficiency. Moreover, in conducted verification experiments with two full-chip level OPCed layouts based on actual functional LSI, we successfully achieved an improvement in extraction efficiency of approximately 60%. In addition, in order to put these research results to practical use for CPL, it is necessary to conduct verifications that compare EB shots.

In the future, as the trend for semiconductor devices to become increasingly smaller in size is likely to continue, it is expected that mask costs and TATs will continue to rise. Technologies that reduce mask costs and TATs are extremely important key factors for the manufacture of system LSIs. We believe that our proposed method may show a new way of overcoming mask problems.

7. Acknowledgments

This work was supported by the Association of Super-Advanced Electronics Technologies (ASET) and the New Energy and Industrial Technology Development Organization (NEDO). The authors would like to thank the Semiconductor Technology Academic Research Center (STARC) for their support with the pattern data.

8. References

Nakayama, Y.; Okazaki, S.; Saitou, N. & Wakabayashi, H. (1990). Electron-beam cell projection lithography: A new high-throughput electron-beam direct-writing technology using a specially tailored Si aperture, *Journal of Vacuum Science & Technology B: Microelectronics and Nanometer Structures*, Vol. 8, No. 6, pp. 1836-1840, ISSN: 10711023.

Pfeiffer, H., C. (1978). Variable spot shaping for electron beam lithography, *Journal of Vacuum Science & Technology B: Microelectronics and Nanometer Structures*, Vol. 15, No. 3, pp. 887-890, ISSN: 10711023.

Matsunawa, T.; Nosato, H.; Sakanashi, H.; Murakawa, M.; Murata, N.; Terasawa, T.; Tanaka, T.; Yoshioka, N.; Suga, O. & Higuchi, T. (2005). The novel approach for optical proximity correction using genetic algorithms, *Proceedings of SPIE*, Vol. 5992, pp. 541-549, SPIE, ISBN: 9780819460141.

Matsunawa, T.; Nosato, H.; Sakanashi, H.; Murakawa, Takahashi, E.; Terasawa, T.; Tanaka, T.; Suga, O. & Higuchi, T. (2007). Adaptive optical proximity correction using an optimization method, *Proceedings of IEEE 7th International Conference on Computer and Information Technology (CIT 2007)*, pp. 853-858, IEEE, ISBN: 0769529836.

ASET. (2006). Frontier Research on Optimization of Design, Drawing and Inspection for 45nmhp System LSI, *FY 2005 Annual Report*, NEDO Japan, (in Japanese).

Nosato, H.; Matsunawa, T; Sakanashi, H; Murakawa, M. & Higuchi, T. (2007). A Fast Method of Adaptive Optical Proximity Correction Using Region Segmentation, *IPSJ Transactions on Mathematical Modeling and Its Applications*, Vol. 48, pp. 75-85, IPSJ, ISSN: 03875806, (in Japanese).

Cantú-Paz, E. (1995). A Summary of Research on Parallel Genetic Algorithms, *IlliGAL Report*, No. 95007, Illinois Genetic Algorithms Lab., University of Illinois.

6

Manufacturing and Investigating Objective Lens for Ultrahigh Resolution Lithography Facilities

N.I. Chkhalo, A.E. Pestov, N.N. Salashchenko and M.N. Toropov
Institute for physics of microstructures RAS, GSP-105, Nizhniy Novgorod
Russia

1. Introduction

Current interest in super-high-resolution optical systems is related to the development of a number of fundamental and applied fields, such as nanophysics and nanotechnology, X-ray microscopy in the «Water window» and the projection nanolithography in the extreme ultraviolet (EUV) spectral range (Gwyn, 1998; Benschop et al., 1999; Naulleau et al., 2002; Ota et al., 2001; Andreev et al., 2000; Cheng, 1987). The great economical importance in applying the EUV lithography which, as expected, should replace the conventional deep ultraviolet lithography in commercial production of integrated circuits with topology at a level of 10-30 nm dictates a level of efforts carried out in the fields related to the technology. In the spectral range of soft X-ray and EUV radiation (λ=1-40 nm) this interest is accompanied by the intensive development of a technology for depositing highly reflecting multilayer interference structures (MIS) (Underwood & Barbee, 1981). In practice, the requirements for the shape of individual optical components and for the spatial resolution of optical systems are imposed on designing projection extreme ultraviolet lithography setups that operate at a wavelength of 13.5 nm (Williamson, 1995). EUV lithography should replace the conventional deep-ultra-violet lithography at 193-nm-wavelength radiation generated by excimer lasers in the commercial production of integrated circuits with a minimum topological-element size of 10-20 nm.

The paper is devoted to the fundamental problems of manufacturing and testing substrates with fine precision for multilayer mirrors which surface shape, as a rule, is an aspherical one and that should be made with a sub-nanometer precision, to characterizing multilayer covers deposited onto these substrates which should not injure the initial surface shape and also to measuring with the sub-nanometer accuracy the wave-front aberrations of high-aperture optical systems, for instance, projective objectives.

The main requirements for the shape and for the micro-roughness values of substrate surfaces attended to depositing MIS on them which are optimized for maximum reflectivity at a 13.5 nm wavelength are considered. The problem of roughness measurement of atomic-level smooth surfaces is discussed. The application of optical interferometry for characterizing the surface shape and wave-front aberrations of individual optical elements and systems is under consideration. A particular attention has been given to interferometers with a diffraction reference wave. The problem of measurement accuracy provided by the interferometers which first of all are connected with aberrations of the diffracted reference wave is under discussion. The reference spherical wave source based on a single mode tipped fiber with a sub-wave exit aperture is fully considered. The results of studying this source and the description of an interferometer with a diffraction reference wave made on the base of the source are given. The application of this interferometer for characterizing spherical and aspherical optical surfaces and wave-front aberrations of optical systems is illustrated.

The achieved abilities of the interferometric measuring the surface shape of optical elements with a sub-nanometer accuracy make possible to develop different methods for correcting the optical element surfaces, initially made with traditional for optical industry precision (root-mean-square deviation from the desired one about $RMS \approx$ 20-30 nm) to the same sub-nanometer accuracy. Two methods of a thin film depositing and an ion-beam etching through the metallic masks produced on evidence derived from the interferometric measurements, are considered for the surface shape correction of optical elements. The dependences of the etching rate and the dynamics of the surface roughness on the ion energy (neutral in the case of fused silica etching) and the angle of the ions incidence to the corrected surface are presented. The final results obtained when correcting substrates for multilayer imaging optics for a 13.5 nm wavelength are reported.

Much attention is paid to the final stage of a mirror manufacturing and depositing a multiplayer interference structure onto the substrate reflecting a short wavelength radiation. Some peculiarities of the deposition technology as applied to the mirrors with ultra-high precision surface shape are discussed. Methods for compensating an intrinsic mechanical stress in the multilayer films developed in IPM RAS are described.

In conclusion, the problems on the way to manufacture atomic smooth ultrahigh precision imaging optics for X-ray and EUV spectral ranges yet to be solved are discussed.

2. The main requirements for the shape and for the micro-roughness of substrate surfaces

According to the Mareshal criterion to achieve the diffraction limited resolution of an optical system it is necessary that a root-mean-square distortion of the system wave front RMS_{obj} must satisfy this ratio (Born & Wolf, 1973)

$$RMS_{obj} \leq \lambda / 14 , \quad (1)$$

where λ is a wavelength of light. Since the errors (distortions) of elements of a complex optical system are statistically independent quantities, the required accuracy RMS_1 of manufacturing an individual optical component is

$$RMS_1 \leq \lambda / (14 \cdot \sqrt{N}) , \quad (2)$$

where N is a number of components in the optical system. For instance in the case of a six-mirror objective typical of EUV lithography at the wavelength of λ=13.5 nm, the reasonable error of individual mirror RMS_1 should not exceed 0.4 nm.

Let us consider the influence of objective aberrations on the image quality by the example of imaging of 150 nm width strips by means of Schwarzschild-type objective made up of two aspherical mirrors and providing linear demagnification coefficient of x5, Fig. 1. The calculation was done with the help of ZEMAX code at a wavelength of 13.5 nm. On the left in Fig. 1 a picture of strips to be imaged and their light emission is given.

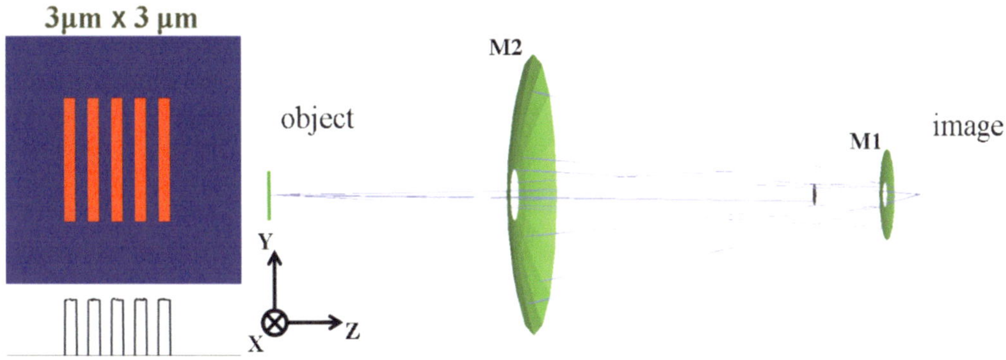

Fig. 1. A projection objective diagram made up of a convex M1 and a concave M2 aspherical mirrors. An exit numerical aperture is NA=0.3 and a linear demagnification coefficient is x5.

The strips images and light intensity distributions in the image plane corresponding to different, from $RMS=\lambda/32$ to $RMS=\lambda/4$, values of the objective wave aberrations are given in Fig. 2. The figure shows that at the aberration of $\lambda/4$ the image has fully disappeared. At $RMS=\lambda/14$ we have the image contrast (ratio of intensities from the minimum to the maximum) at a level of 0.5. At the aberration $RMS<\lambda/24$ the image contrast no longer depends on the aberration and is determined only by a numerical aperture of the objective. In such a manner a reasonable aberration is a value at the level of $RMS\approx\lambda/30$, or 0.45 nm. In view of Eq. (2) the requirements for the quality of individual mirrors are stronger at a level of 0.2 nm.

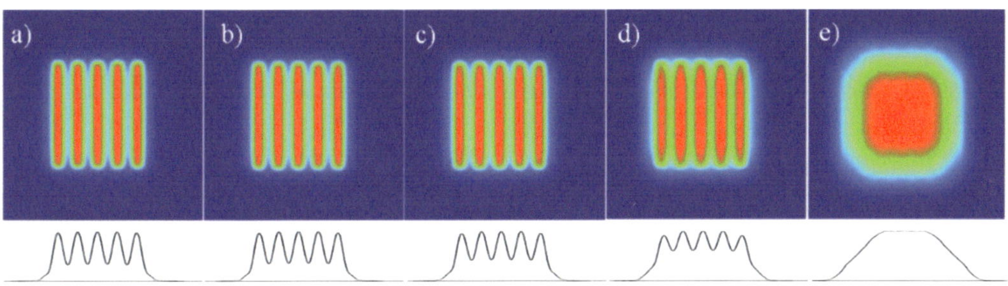

Fig. 2. Images of lines and light intensity distributions in the image plane depending on an objective wave aberration, λ=13.5 nm. a) – root-mean-square aberration $RMS=\lambda/32$; b) - $RMS=\lambda/24$; c) - $RMS=\lambda/14$; d) - $RMS=\lambda/8$; e) - $RMS=\lambda/4$.

The problem of manufacturing mirrors with a sub nanometer surface shape precision for EUV lithography is complicated by a number of factors. First, because of considerable intensity losses due to the reflection of radiation with a wavelength of λ=13.5 nm from MIS a number of mirrors in an optical system must be minimized. For this reason to increase a field of view and to achieve a high space resolution of an objective one has to use aspherical surfaces with high numerical apertures. Second, a small radiation wavelength and a huge number of interfaces in multilayer interference structure participating in the reflection process impose rigid requirements on the interfacial roughness, which in turn is substantially determined by the surface roughness of a substrate (Warburton et al., 1987; Barbee, 1981; Chkhalo et al., 1993).

The interfacial roughness with a root-mean-square height σ effects in decreasing both a reflection coefficient of the multilayer mirror and the image contrast because of the scattering of radiation. The estimation of the total integrated scatter (*TIS*) can be done as follows

$$TIS = 1 - \exp\left[-(4\pi\sigma / \lambda)^2\right], \tag{3}$$

where λ is a radiation wavelength. For instance, if the integral scattering of an individual mirror is to be lower than 10%, the interfacial roughness must be at a level $\sigma \leq 0.3$ nm. The more precise analysis shows, that everywhere over the region of space wavelengths of the Fourier transform of the reflecting surface (from fractions of nanometer up to tens of millimeters) the root-mean-square of the surface distortions should be at a level $RMS \leq 0.2$ nm (Williamson, 1995; Sweeney et al., 1998; Soufli et al., 2001).

3. Surface roughness measurement methods

As is seen from ratio (3) among the factors having effect on reflection coefficients of MIS, a substrate surface roughness plays a significant role. There exist a few methods for surface roughness measurements with heights of nanometer and sub-nanometer level at present. Among them the atomic force microscopy (Griffith & Grigg, 1993) and diffusion scattering of hard X-ray radiation (Sinha et al., 1988; Asadchikov et al., 1998) are mostly developed and widely used. A number of papers report a good agreement of experimental data about the surface roughness obtained by both methods (Kozhevnikov et al., 1999; Stone et al., 1999). It is necessary to mention a recently well-established method of a surface roughness measurement by means of an optical interference microscopy (Blunt, 2006; Chkhalo et al., 2008). In this case the producers of the micro interferometers and opticians who use these instruments, report about roughness measurement precision up to 0.01 nm. In particular it is stated that super-polished quartz substrates produced by General Optics (USA) have the roughness at a level of 0.07 nm (Website GO, 2009). But relatively low lateral resolution characteristics of the methods substantially limit the spectrum of space frequencies of a surface roughness to be registered from the high-frequency side, that impairs their capabilities when measuring super-polished substrates attended for short-wavelength optics. For instance, in Fig. 3, where the angular dependence of a scattered intensity of X-ray radiation with a wavelength of λ=0.154 nm from a fused silica super-polished substrate is presented, one can see that at the angle about 0.6° the scattered intensity is about 10^{-6} of the incident one, down to a detector noise. For the X-rays with λ=0.154 nm the angle 0.6° corresponds to the scattering on surface space frequencies with a wavelength of 2.8 μm. This resolution is comparable with the lateral resolution of the interference microscopes and ranks below the resolution of atomic force microscopes. Therefore, noted in some papers a good agreement of experimental data about the surface roughness obtained by these methods can be explained only as follows. In spite on the fact that the radius of a cantilever of the atomic-force microscope may be at the level of a few tens of nanometers, there exist some other factors, such as the true radius of a probe, the peculiarities of the probe moving, vibrations, mathematic processing of experimental data and others, which are specific for each instrument, for each laboratory which decrease the instrumental lateral resolution.

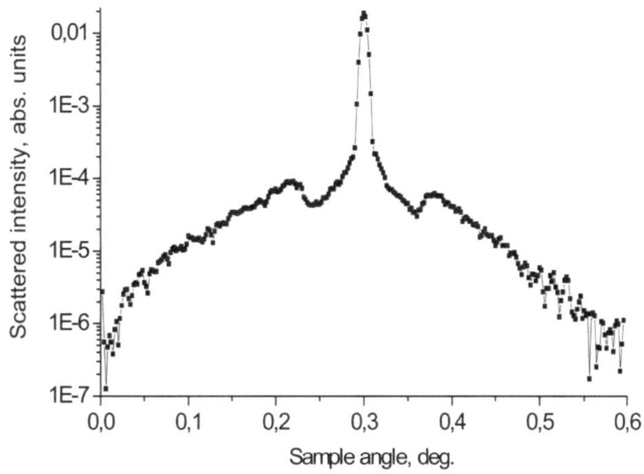

Fig. 3. An angular dependence of scattered intensity of X-ray radiation from a fused silica substrate taken with X-ray diffractometer Philips'Expert Pro at a wavelength of λ=0.154 nm Detector angle is fixed θ_d=0.6°.

The most reliable data about the roughness of atomic-smooth surfaces, as we assume, give a method based on the analysis of angular dependencies of specular reflection coefficients of X-ray radiation. The method was proposed in (Parratt, 1954) and was widely used in Refs. (Chkhalo et al., 1995; Protopopov et al., 1999; Bibishkin et al., 2003). The influence of the roughness σ on the angular dependence of the reflection coefficient $R(\theta)$ was taking into account with the help of attenuation coefficient of Nevot-Croce (Nevot & Croce, 1980) as follows

$$R(\vartheta) = R_{id}(\vartheta)\cdot \exp\left\{-\frac{16\pi^2\sigma^2}{\lambda^2}\cdot \sin\vartheta \cdot \mathrm{Re}\sqrt{\sin^2\vartheta + \chi}\right\}, \tag{4}$$

where $R_{id}(\theta)$ is an angular dependence of the reflection coefficient on the ideal surface and is calculated with the Fresnel's formulae, λ is a radiation wavelength and χ is a dielectric susceptibility of the substrate material. In the calculations the material optical constants were taken from (Palik, 1985). The advantages of this method are of a "small" size of the probe, the radiation wavelength is comparable with the roughness value, and the lack of limitation about the registered spectrum of space frequencies of a surface roughness from the high-frequency side because in "zero" order of diffraction all radiation losses are taking into account. From middle-frequency side the registered spectrum is limited by the width of the detector slit and is about 30 μm. But it is not the principal restriction since in this range the interference microscopes operate well.

An example of applying this method for the characterization of a fused silica substrate produced by General Optics (USA) in 2007 with a factory specified roughness value of 0.08 nm is given in Fig. 4. As one can see the best fit of the experimental curve (dots) is observed for the surface roughness lying in the range of σ = 0.3–0.4 nm. That is 4-5 times more as compared with the specified value. Exactly the same experimental data were obtained in independent measurements made in Institute of crystallography, Russian academy of sciences, in Moscow.

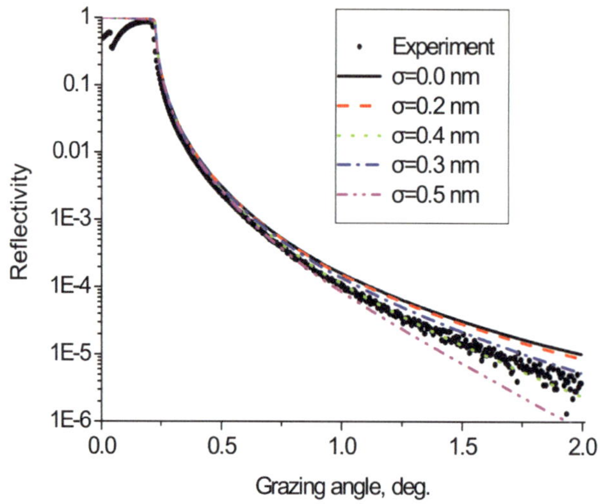

Fig. 4. Angular dependencies of reflection coefficients of radiation with λ=0.154 nm from a fused silica substrate made by General Optics: dots correspond to experiment; solid lines are calculations corresponding to different roughness values.

The investigation of this substrate by means of an atomic-force microscopy carried out in our institute has shown a strong dependence of measured roughness on the probe size. When we use Si-cantilever the measured roughness was about 0.08 nm, but with a wicker the value has increased up to 0.16 nm. So this direct comparison of the application of X-ray reflection and atomic force microscopy for atomic-smooth surface roughness measurement indicates that the latter method gives an underestimated value of the roughness.

A serious disadvantage of the specular X-ray reflection method is that it allows studying only flat samples while components of imaging optics have concave and convex surface shapes. Therefore, in practice for evaluating the surface roughness of nonplanar samples it makes sense to use the atomic force microscopy method taking into account the calibration of X-ray reflection made with flat substrates.

In conclusion it should be noted that a large body of research done with the help of the x-ray specular reflection method to measure a surface roughness showed that a number of substrates fabricated in different countries had a minimal surface roughness 0.2-0.3 nm and included crystal silicon used in electronic industry. The minimal surface roughness of fused quartz substrates were in the range of 0.3-0.4 nm, except for the case (Chkhalo, 1995), where the roughness of 0.25 nm is reported.

4. Investigating the surface shape by means of optical interferometry

Optical interferometry is one of the most powerful and widely used method for measuring a surface shape of optical components and wave front aberrations of complex systems in the industry. The main advantages of the inetrferometry are the simplicity and high accuracy of the measurements. The investigation technique is based on the analysis of a light intensity distribution over interference patterns. In this paragraph the basic principles of a surface shape and wave front distortions of optical component and system reconstruction with the use of data obtained by optical interferometers are described. Both types of interferometers are considered, conventional, utilizing reference surfaces, and diffraction, using as a reference a spherical wave appeared due to the diffraction of light on a wave-sized pin-hole.

4.1 Shape reconstruction and interferometry utilizing reference surfaces

At present a gamma of interferometers is used in an optical industry, including Fizeau and Twyman-Green interferometers which stand out because of their simplicity in operation, high accuracy and universality, is developed. A detailed description of these instruments can be found in many books, for instance (Malacara, 1992; Okatov, 2004). Independently on an optical scheme in this type of instruments the interference patterns are detected which appear as a result of the interaction of two waves, reflected from studied E_S (a surface with defects in Fig. 5) and from reference E_R (top surface in Fig. 5) surfaces. As a result the fringes

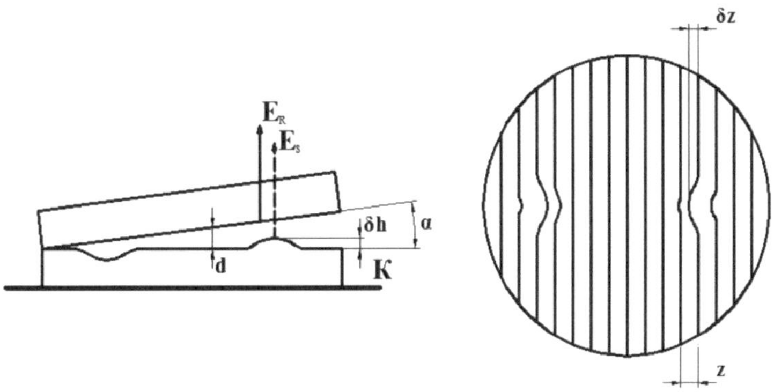

Fig. 5. A fringe pattern formation by interfering of two waves reflected from a surface under study (bottom) and a reference (upper) one.

of so called "equal thickness" are observed. A defect on the surface under study marked in Fig. 5 by symbol z with a height of δh gave rise the phase shift of the reflected light according to

$$\Delta = 2\pi \cdot 2\delta h / \lambda \,, \tag{5}$$

where λ is a wavelength. The corresponding fringe banding δz, induced by the deviation of the studied surface from the reference can be found from the ratio

$$\delta h = (\lambda / 2) \cdot \delta z / z \,. \tag{6}$$

These expressions in particular allow to see that the defect with the height of $\lambda/2$ corresponds to one fringe in the interference pattern. When the defects on the studied surface are absent, the fringes make a system of equidistant straight lines. In such a manner the interference pattern uniquely determines the local shape deviations of the surface under study from the reference. So the task of the surface shape reconstruction needs looking for a mathematical model which will fit the experimental data the best.

$$W(x,y) = \Delta h(x,y) \,, \tag{7}$$

where $\Delta h(x,y)$ is the surface deformation in respect to the reference and being a function of coordinates on the surface, $x,y \in \Omega$, Ω is an operating range of the surface. As a rule, the sought-for function (Braat, 1987) is written in terms of some basis of functions and its description reduces to finding a set of the series coefficients c_k

$$W(\mathbf{r}) = \sum_k c_k \alpha_k(\mathbf{r}), \tag{8}$$

where **r** is a radius-vector of a point on the surface and $\alpha_k(\mathbf{r})$ is a set of basis functions. Mostly common polynomials orthogonal on some area Ω_0 are chosen as the basis. For instance, for the circular area these are Zernike polynomials which are widely used in optics (Rodionov, 1974; Golberg, 2001). The orthogonality of the polynomials results in the fact that each term of the series makes a contribution independent on the remaining terms into the mean-square of the wave front deformation.

Since the representation of the investigated function is global any local dilatation when approximating is smoothed thus distorting global view of the function. Besides when reconstructing the surface shape it is important to evaluate not only the global surface shape, but local errors too. From this it follows that the mathematical model should be oriented on the description of the surface shape not only globally, but locally too.

In the framework of this paper this dilemma was solved by introducing local deformations additionally to the global description (8) of the surface deformation function:

$$W(\mathbf{r}) = \sum_k c_k \alpha_k(\mathbf{r}) + \sum_m \left[\sum_j c_{mj} \cdot \beta_j(\mathbf{r} - \mathbf{r_{0m}}) \right], \tag{9}$$

where c_{mj} and β_j are coefficients and "special" basis functions describe a m-th local deformation with the center in the point with the coordinate of $\mathbf{r_{0m}}$. It is significant that the set of functions $a_k(\mathbf{r})$ and $\beta_j(\mathbf{r}\text{-}\mathbf{r_{0m}})$ are different in general case because the described function $W(\mathbf{r})$ should not be interrupted in the range of definition. It is the reason why the "special" functions $\beta_j(\mathbf{r}\text{-}\mathbf{r_{0m}})$ must be finite, should not have discontinuity and their values at the definition range boundary must be equivalent to zero. An algorithm of searching for the expansion coefficients of the surface model (9) was performed in two stages. At the first stage the global surface shape error is approximated according to the (8) model. At the second stage the residual local surface deformations are approximated by the right part of the expression (9).

In our case the global description of the surface shape errors is performed with the help of Zernike polynomials $W_Z(\rho,\varphi)$ and the residual local dilatations – by using the apparatus of local splines $W_S(\rho,\varphi)$, which advantages when describing the optical surface deformations are analyzed in (Archer, 1997; Sun et al., 1998) in details:

$$\begin{gathered} W(\rho,\varphi) = W_Z(\rho,\varphi) + W_S(\rho,\varphi) = \sum_{n=0}^{Pz}\sum_{m=0}^{n} c_{nm} R_n^m(\rho)\cos(m\varphi) + \sum_{n=0}^{Pz}\sum_{m=0}^{n} s_{nm} R_n^m(\rho)\sin(m\varphi) + \\ + \sum_{i=0}^{k}\sum_{j=0}^{k} B_{i,P_S}(\rho\cdot\sin(\varphi))\cdot B_{j,P_S}(\rho\cdot\cos(\varphi)) P_{ij} \end{gathered} \tag{10}$$

where ρ and φ are polar coordinates of the point on the surface, c_{nm} and s_{nm} are the expansion coefficients, $R_n^m(\rho)$, are radial Zernike polynomials, B_{i,P_S} is the basic function of the B-spline of the order of p_S on i-th interval of y coordinate and B_{j,P_S} – is the basic function of the B-spline of the order of p_S on j-th interval of x coordinate and P_{ij} corresponds to the

points of control of the spline. A number of works (Rodionov, 1995; Swantner & Weng, 1994) are devoted to the methods for determining the coefficients of the mathematical model of the surface shape deformation function (10) by using interferometric data. In present work algorithms and programs developed in (Gavrilin, 2003) are used.
Currently available methods of digital registration and mathematical processing of interferograms allow reconstructing the surface shape (wave front aberrations) in respect to the reference surface, Fig. 5, with the accuracy up to $\lambda/10000$, where λ is an operating wavelength of the interferometer. However, the guaranteed surface shape accuracy of the reference in the root-mean-square (*RMS*) does not exceed $\lambda/30$ – $\lambda/20$ (Website Zygo, 2009) that is two orders of magnitude worse than required, for instance, for projection nanolithography optics.
A substantially better situation is observed when studying surfaces at typical scales lower than 1 mm. At these scales the accuracy of the reference surfaces is better than $\lambda/1000$ that allows measuring the surface shape deformations with the sub-nanometer accuracy. On this basis a number of micro-interferometers with a digital fringe registration has been developed in the last few years (Blunt, 2006; Chkhalo et al., 2008; Website Veeco, 2009; Website Zygo, 2009). Below the spectrum of space wavelengths on the surface under study to be registered and where the surface shape measuring by interferometers with the sub-nanometer accuracy is limited by the lateral resolution of a microscope is given typically lies in the range of 0.3-1 μm.

4.2 Interferometers with a diffraction reference wave

The problem of conventional interferometers using the reference surfaces is solved by the application of interferometers with a diffraction wave as the reference proposed by V.P. Linnik in 1933 (Linnik, 1933). The proposal is based on the theoretical fact, that when a flat electromagnetic wave falls onto a pin-hole with a diameter comparable with a wavelength in an opaque screen, consisting of a zero-thickness superconducting material, diffracted behind the screen wave in a far wave zone is an "ideal" sphere (Born & Wolf, 1973). This feature of the light diffraction was used as the basis for developing interferometers of this type.
The first interferometer attended to the investigation of optics with sub-nanometer accuracy for the EUV nanolithography has been made by G.E. Sommargren in 1996. As a source of the reference spherical wave he used a single-mode optical fiber with a core diameter (exit aperture) of about 4-5 μm. The objective for a projection scheme of a EUV-nanolithographer was tested with the help of the interferometer (Sommargren, 1996). The author evaluated the measurement precision in terms of the *RMS* at the level of 0.5 nm. With a high-coherence laser the interferometer allowed testing a surface shape of individual optical elements. The main disadvantage of the interferometer was a "big" source size. A numerical aperture where the diffracted wave still remains "an ideal sphere" is determined by expression $NA_{ref} < \lambda/d$ and in the case of λ=0.5 μm and d=5 μm corresponds to NA_{ref}<0.1. Partly the low-aperture problem is solved at the expense of stitching measured data over different areas of the studied substrate or objective. Along with the apparent loss of a measurement accuracy when using the stitching procedure, there exists one even more serious disadvantage of using low-aperture interferometers associated with a strong irregularity of interferogram illumination. Fig. 6 demonstrates how the illumination irregularity results in error when determining the position of the fringe extremum. In other words, along with the apparent

loss of measurement accuracy connected with the stitching, low-aperture interferometers have a significant systematical error induced by strong intensity anisotropy of the reference wave. Therefore, the recent trends are toward increasing the use of diffraction on man-made pin-holes in an opaque screen. Modern microelectronic technologies provide manufacturing high-quality, low-edge roughness pin-holes with diameters down to 40-50 nm.

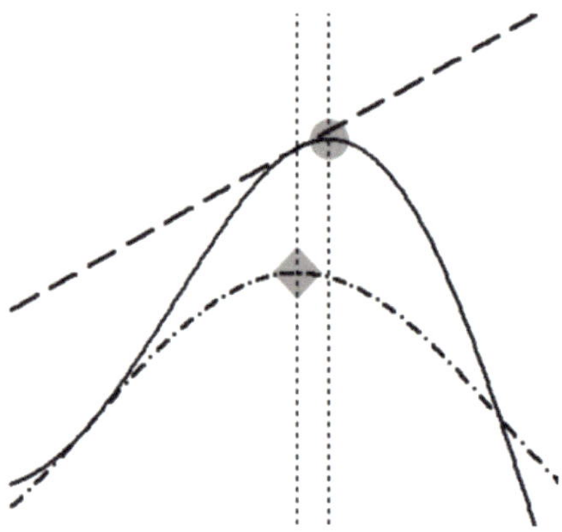

Fig. 6. Shift of determined position of fringe maximum (systematic error) caused by the illumination irregularity.

Such pin-holes are applied in diffraction interferometers operating at the working wavelength of a EUV nano-lithographer, λ=13.5 nm, which are installed on undulators of the last generation synchrotrons in USA and Japan (Naulleau et al., 2000; Murakami et al., 2003). An interferometer circuit installed on ALS (Berkeley, USA) is given in Fig. 7.

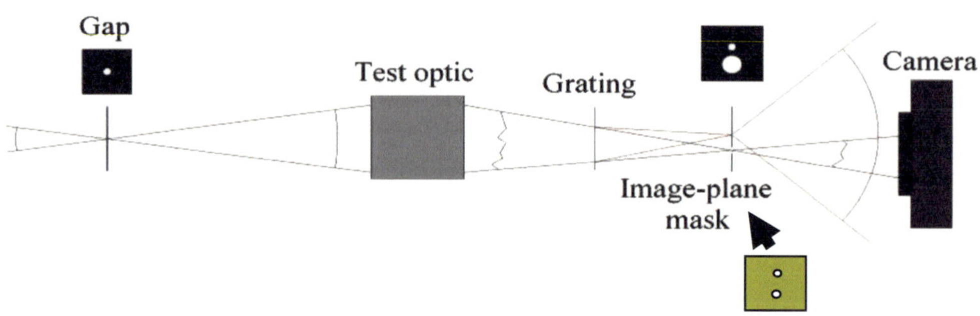

Fig. 7. Optical scheme of interferometer with diffraction reference wave installed on synchrotron ALS in Berkeley, USA. The picture is taken from (Medecki et al., 1996).

The device operates in the following way. The radiation falls onto a small pin-hole Gap, installed on the lens object plane under study. A diffracted "ideal" spherical wave passing through the lens is distorted according to its aberrations and is collected in the image plane. On the radiation way in-between the lens and the image plane a diffraction grating is installed which splits the wave-front into two ones corresponding to zero and to 1-st diffraction orders. The zero-order radiation passes through the hole which diameter is big enough and exceeds a focusing spot thus preventing the distortion of the wave front. The 1-st order beam is focused onto a small-size pin-hole. Due to the diffraction behind the screen "an ideal" spherical wave is generated. The reference wave expands toward a CCD-camera where it interferes with a wave passed through the lens under test. The measurement accuracy provided by the interferometer will be discussed later when comparing with our interferometer.

As we can see such interferometers are efficiently used for the final characterization of wave-front aberrations of objectives for EUV radiation. At manufacturing stages of substrates and mirrors the producers of the optics usually use optical diffraction interferometers. The experimental study of capabilities of point diffraction interferometers operating with a visible light has demonstrated a number of factors reducing the measurement accuracy. Particularly systematic discrepancy of experimental data when studying the same optics with the help of visible and EUV light interferometers is observed (Goldberg et al., 2002). All of this stimulated to study the way how finite conductance and thickness of the screen material have an effect on amplitude-phase characteristics of the diffracted wave behind the screen and, correspondingly, on the interferometer measurement accuracy.

In work (Chkhalo et al., 2008) the calculations of the amplitude-phase characteristics of the diffracted field behind the screen in the far-field zone have been carried out. They used the solution of a task about the field of a point source located above a half-space with arbitrary properties and covered with a film of any thickness and arbitrary optical constants given in (Dorofeyev et al., 2003). Fig. 8 illustrates the statement of the problem. The diffraction of a flat wave on a circular hole with the radius of a in a film with thickness of h_f and the inductivity of ε_f in medias with parameters ε and ε_1 was considered. The field was calculated behind the screen at $Z>0$ on a spherical surface with a fixed radius R_0=10 cm. The radius R_0 was chosen from practical reasons, but its value does not produce any effect on the community of the result since the calculation was done for a case $kr>>1$, where $k=2\pi/\lambda$ is a wave vector of the diffracted wave in a point with the radius-vector r.

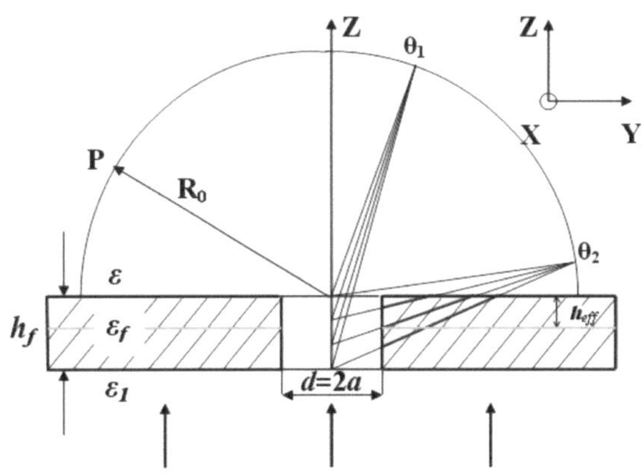

Fig. 8. Sketch of the geometry of the problem, where h_f is the thickness of the film, h_{eff} is an effective film thickness (characterized by the skin-layer), d is the diameter of a hole, ε, ε_1 and ε_f are the dielectric constants of the film and the medias, θ_1 and θ_2 are the angles of diffraction. Incident wave is schematically shown by the arrows.

The results of the calculations as compared with the solution of classical task of diffraction on zero-thickness superconducting screen have shown that most real electrodynamical characteristics of the screen material have an effect on phase characteristics of the diffracted field. The calculation of the phase distortions was carried out for the films of aluminum, molybdenum, tungsten and a number of other materials.

The results of numerical calculations shown in Fig. 9 represent a relative phase deviation on meridional angle of the diffracted scalar field in the case of aluminum different thickness films characterized by dielectric constant $\varepsilon_{Al}\approx$-54.2-i21.8 at the wavelength 633 nm, as compared with the case of a perfect screen $|\varepsilon_{Al}|\rightarrow\infty$. The phase incursion $\Delta\varphi(\theta)$ is expressed in nanometers. The radius of the hole in the films was a=150 nm. A case of free standing ($\varepsilon=\varepsilon_1=1$) films was considered. It is evident that the larger the meridional angle, the larger is the phase deviation of the diffracted field in the case of a realistic screen. We connect the result with an additional phase accumulation due to the field propagation through the real film at larger meridional angles (see Fig. 8 for a qualitative clarification). We did not find a thickness dependence on the phase accumulation in the range of $\delta<h_f<1000$ nm, where δ is a skin-layer, because the phase can be effectively accumulated only inside the skin-layer. The transitions $h_f\rightarrow0$ and $|\varepsilon_{Al}|\rightarrow\infty$, $|\varepsilon_{Al}|\rightarrow\infty$, or $h_f\rightarrow0$ led to the known textbook examples. The corresponding diffraction fields for such idealized problems perfectly coincide with those of a spherical phase front.

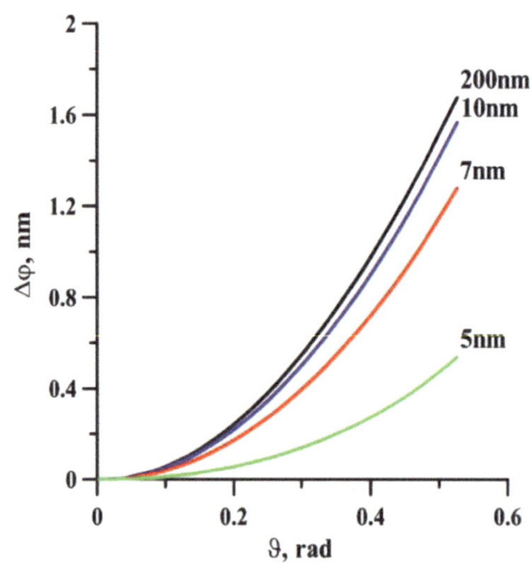

Fig. 9. A relative phase deviation $\Delta\varphi(\theta)=(\varphi(0)-\varphi(\theta))\cdot\lambda/2\pi$ on meridional angle of the diffracted field in a case of Al different thickness film ($\varepsilon_{Al}\approx$-54.2-i21.8) at the wavelength 633 nm on a spherical surface R_0=10 cm, the hole radius is a=150 нм.

Other mechanisms of the phase distortion can be associated with the excitation of waveguide and surface modes in a plane layer and with a diffraction of such modes by a hole followed by their radiative decay. Since the dielectric constant of all materials in soft X-ray and EUV ranges are close to unity, $Re(\varepsilon)\approx1$, corresponding phase distortions are minimal. It is one of the reasons why the quality of the diffracted reference wave in EUV region has a better result in a higher measurement precision.

The carried out calculations have shown (see Fig. 10) that the angular dependencies of the phase distortions on the meridional angle weakly change for different materials. The only exception is osmium for which the phase deformation was extremely low. Optical constants of materials used in these calculations were taken from (Palik, 1985).

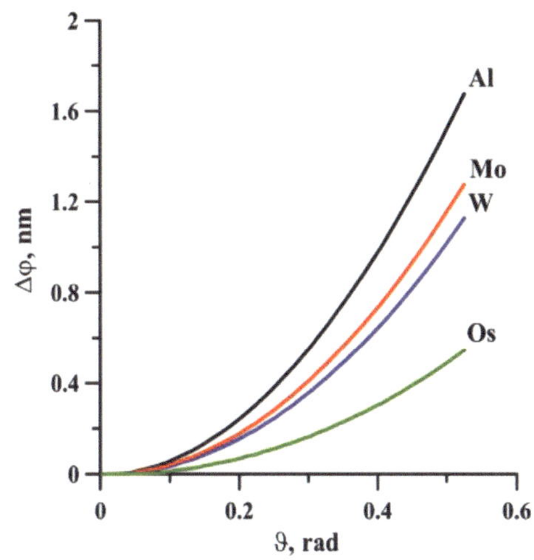

Fig. 10. Angular dependencies of the phase distortions on the meridional angle for different materials.

By this means a fundamental limitation of the point diffraction interferometer measurement accuracy is caused by the aberrations of the reference spherical wave induced by the interaction and propagation of secondary waves through the screen material. These aberrations are liable to reach a few nanometers depending on the angular aperture. If we take into account the polarization characteristics of the light, the aberration of the reference wave increases a few times depending on azimuth angle (Dorofeev, 2009).

The quality of a diffracted wave strongly depends on wave aberrations of preliminary optics focusing the laser beam and the accuracy of adjusting the laser beam axis in respect to the pin-hole center (Otaki et al., 2002). All these limit the working aperture of the interferometer and result in the necessity of testing high-aperture optics zones. The following stitching procedure of the zonal data raises additional errors. All of these disadvantages hinder to widespread use of the point diffraction interferometers in an industry. Taking into account all of these problems and limitations the problem of searching for alternative and more perfect methods for the spherical wave generation becomes very urgent.

5. Spherical wave source on the basis of tipped fiber

The sources of the spherical wave based on tipped down to a sub-wave exit aperture of the single mode fibers are free from the many of mentioned above problems. The application of such fibers as a reference spherical wave source for the point diffraction interferometers has been proposed in (Klimov et al., 2008). In Ref. (Chkhalo et al., 2008) the aberrations of the wave generated by the sources at the wavelength of He-Ne laser λ=633 nm have been studied. These investigations have demonstrated a number of advantages of spherical wave sources as compared with the conventional ones based on a pin-hole in an opaque screen. One of the advantages is connected with the fact, that in the optical fiber the eigenmodes are excited and correspondingly the quality of the diffracted wave does not depend on aberrations and the mechanical adjustment of the preliminary optics.

A high degree of homogeneity of the diffracted wave ($\lambda/d>1$) is combined with a high intensity because the light is input into the fiber using high-efficiency methods through a 5-

μm-diameter core. The convex shape of the source, Fig. 11, decreases the "tip" effect bound to the interaction of the off-axis rays with a metallized part of the fiber along the perimeter of the exit aperture. The lack of a flat screen around the source significantly decreases diffracted wave distortions connected with the light polarization. The well developed methods for handling optical fibers make it possible to easily control the polarization parameters of the diffracted radiation and implement various schemes for interference measurements. For instance, two or even more coherent sources can be organized with the given polarization of each other.

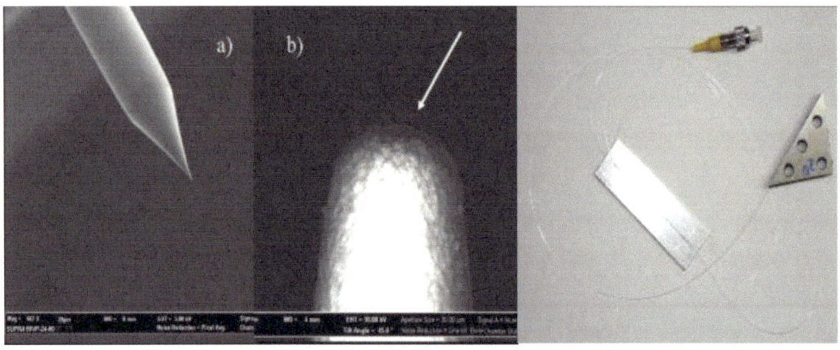

Fig. 11. Photographs of the fiber based source taken with scanning electron microscope (marked as a) and b)) and the source with on optical connector.

The quality of the diffracted wave of the tipped fiber based source was studied with an interferometer (Kluenkov et al., 2008) at two wavelengths: λ=633 nm (He-Ne laser) and λ=530 nm (second harmonic of Nd-YAG laser). As opposed to the former experiments when in the measurement scheme an optical observation system collected the interfering fronts onto the CCD-camera, in present work the data obtained by the direct detection of the fronts on the window-free CCD-camera are given. As the experiments showed both the observation systems and even a thin glass window introduce a significant, at the level of parts of nanometer, error into the measurements.

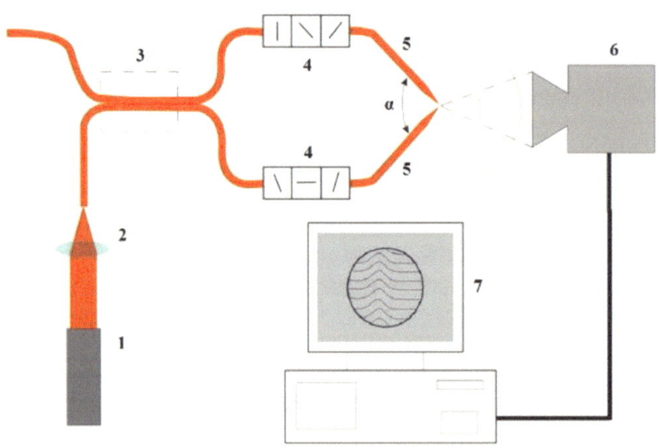

Fig. 12. Experimental setup for studying the wave front deformations of the tipped fiber based sources of spherical wave. 1 is a laser, 2 is a system for input the light into the fiber, 3 is a single mode fiber optics coupler, 4 is a polarization controller, 5 are sources under study, 6 is a CCD-camera and 7 is a computer.

The new experimental setup is given in Fig. 12. Polarization controllers (item 4 in Fig. 12) allow to transform an elliptical polarization of the light at the output of the sources into a linear one as well as to superpose the polarization planes, which provide the best contrast of the interference pattern. The distance between the sources determining the fringe number (a wedge between two interfering wave fronts) was varied in the range of 1-10 µm. The measurements were carried out for different angles *a* between the source axes. The numerical aperture was varied by approaching (moving off) the CCD camera in respect to the sources. A typical interferogram and a wave aberration map obtained in the experiments are shown in Fig. 13. The operating wavelength was λ=530 nm.

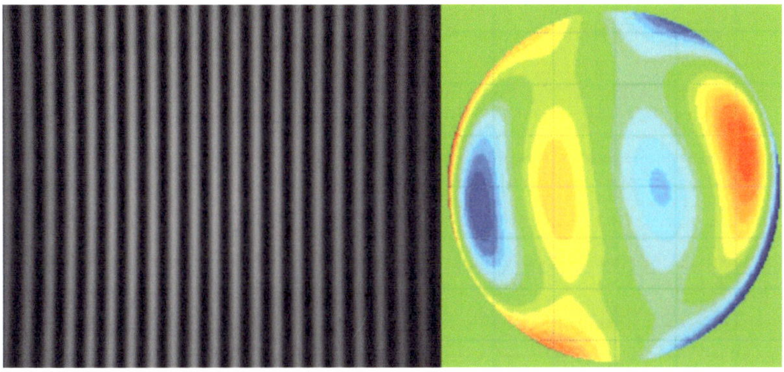

Fig. 13. A typical interferogramm and a wave aberration map observed in the experiments obtained at a wavelength λ=530 nm.

The measured dependencies of the wave deformation *RMS* on numerical aperture *NA* of this couple of sources are given in Fig. 14. The pictures correspond to the measurements in small (left) and in large (right) numerical apertures. Left graphs on both pictures correspond to the experimental data. The data obtained with the interferometer installed on the synchrotron ALS in Berkeley, USA, are marked by stars. The interferometer using a pin-hole in an opaque screen as a reference spherical wave source, operates at the wavelength λ=13.5 nm and is considered as a reference for the measurement accuracy among the point diffraction interferometers (Naulleau et al., 1999). It is clearly seen from the pictures that wave front aberrations of the source, developed in this work, are even less as compared with the ALS-sources despite a longer operating wavelength.

When analyzing the obtained experimental data there is a need to pay attention to two facts:

- Relatively "high", of order $\lambda/1000$, total aberration of the diffracted waves is observed at large numerical apertures (*NA*≈0.3);
- Regular structure (symmetry) of the wave front deformation map is clearly seen.

Such maps and statistical values of the deformations are practically found for all couples of so called "high quality" sources. The wedge direction between interfering fronts (a line passing through the sources) represents itself as the axis of the symmetry. Particularly, the deformation map can be put "horizontally" by turning the wedge through 90 degrees. All of this points to the presence of an aberration, caused by some kind of physical reason rather than by technological fluctuations. This aberration rises with the increase of the observation angle just as it is shown in Figs. 9 and 10. In case of understanding the physical nature of this aberration it can be taken into account when reconstructing the true wave deformation (surface shape) that results in increasing the working aperture of the interferometer by a few times.

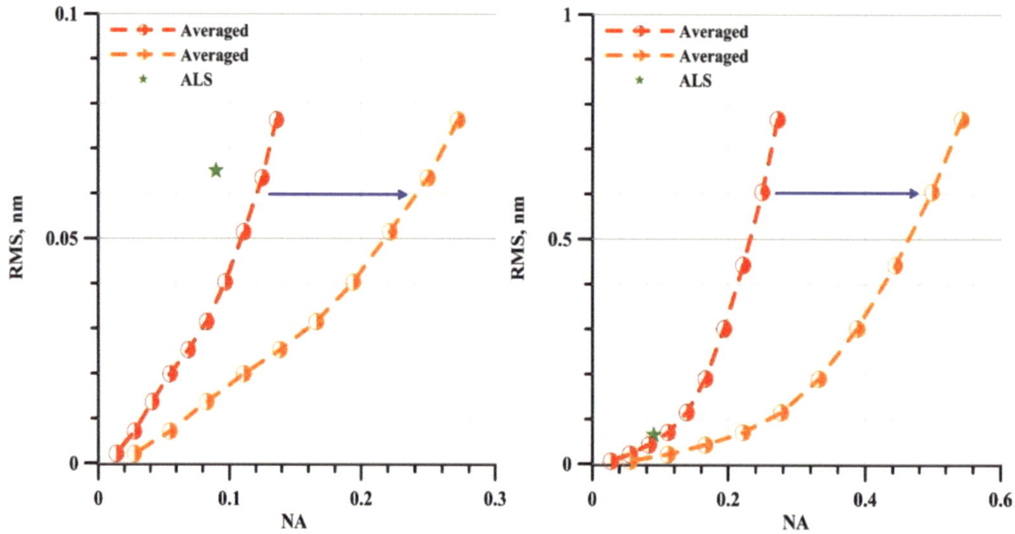

Fig. 14. Averaged over 9 measurements total wave front deformations of couple of the fiber based spherical wave sources depending on numerical aperture (left graphs in both pictures) and ALS-data (Naulleau et al., 1999) (stars). Working wavelength is λ=530 nm. First picture corresponds to lower and right - to the higher numerical apertures. The right graphs in the pictures are corrected according to the geometry of the experiment.

This fact of the uniform angular dependence of the aberration let us interpret in a new manner the measurement data obtained in ALS and in our experiments. In the Young's experiment carried out in ALS the least distorted parts of the diffracted waves round their own axes interfere. Correspondingly *RMS* of the aberration is calculated over the range of the cone, limited by the numerical aperture *NA*. Opposite to that in our case side parts of the wave fronts interfere, Fig. 12, in the range from zero degree up to a double value of the angular aperture of the wave front. In other words, in the search for the wave front aberration *RMS* the averaging is performed over a double numerical aperture, *2NA*. Therefore, the measured dependences presented in Fig. 14 (left graphs) can be corrected in the manner as it is shown by arrows (right graphs). As it is seen from the picture the source of the reference spherical wave based on a tipped single mode fiber has a substantially lower wave front aberration and a higher working aperture as compared with the conventional one based on pin-hole in an opaque screen.

6. Diffraction interferometer based on a single mode fiber with the sub-wave exit aperture

On the basis of a tipped fiber source of the reference spherical wave the laboratory sample of an interferometer has been developed and manufactured, which optical scheme and photo are presented in Figs. 15 and 16. The interferometer is installed in a basement thermostated room on a bearer separated from the house footing. For the additional protection against vibrations, the instrument is installed on 12 bellows under 1.6 bar pressure. Since the measurement accuracy is extremely high, the turbulent air flowing inside the interferometer and effecting optical paths along the rays, can introduce additional errors. To solve these problems, the interferometer is placed into a vacuum chamber which is pumped out down to a pressure of 1 Pa.

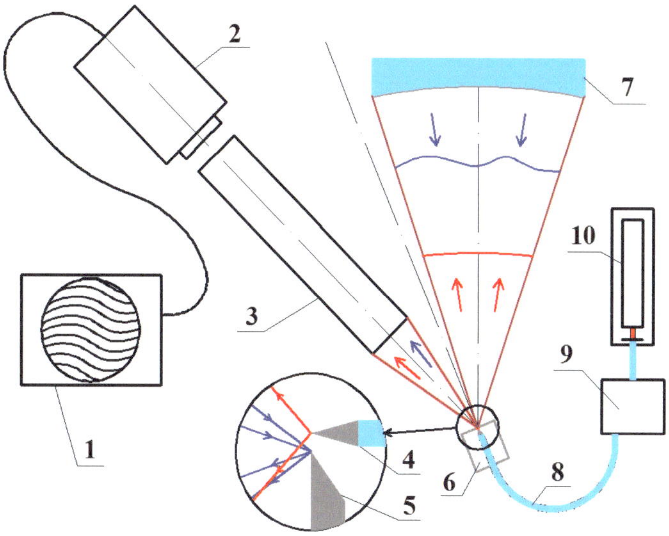

Fig. 15. Optical scheme of the interferometer:
1 - PC, 2 - registering system, 3 - observing system, 4 - source of spherical wave, 5 - flat sharp-edge mirror, 6 - three-axes-controlled precision bench, 7 - tested concave surface, 8 - single mode optical fiber, 9 - polarization controller, 10 - laser.

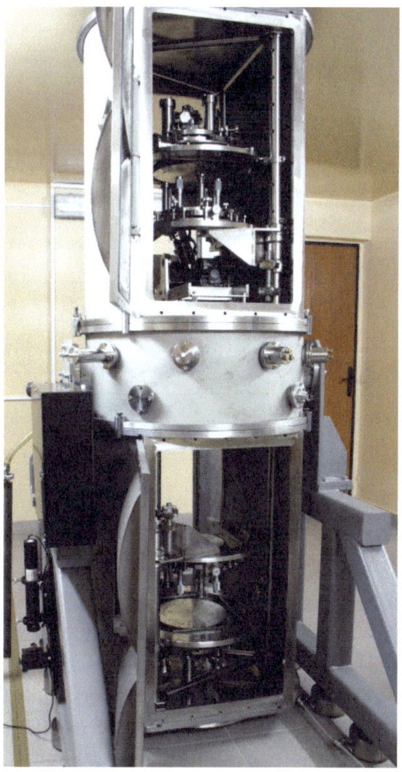

Fig. 16. Photograph of the vertical vacuum interferometer.

The optical scheme shown in Fig. 15 corresponds to the interferometer for studying concave spherical or weakly aspherical surfaces. The spherical wave source (item 4, Fig. 15) is mounted on a high-precision three-axis-controlled bench (item 6, Fig. 15) in the direct, in a distance of several micrometers, proximity from the flat mirror (item 5, Fig. 15). The reference spherical wave, which after reflecting from the studied surface (item 7, Fig. 15) carries the information about its form, is focused on a flat mirror and being reflected from it goes into the CCD registering system, where it interferes with a part of the reference front directly propagating to the detector.

The main disadvantage of this scheme is connected with the fact that the investigated sample is irradiated by the most distorted side part of the wave front that practically halves the interferometer operating numerical aperture. For solution of this problem in (Kluenkov et al., 2008) the scheme of the interferometer with two low-coherent sources of the reference spherical waves has been proposed (Fig. 17). In this scheme one source forms the spherical wave, «Spherical wave №1», irradiating the investigated surface, while the spherical wave of the second source, «Spherical wave №2», is directed into the registering system. The radiation reflected from the investigated surface contains the information on its shape and after reflecting from the flat mirror is directed to the registering system, where interferes with the reference «Spherical wave №2». The main advantage of the scheme is that the axes of the wave fronts coincide with the axes of the investigated surface and the registering system which provides the minimal deformations of reference fronts.

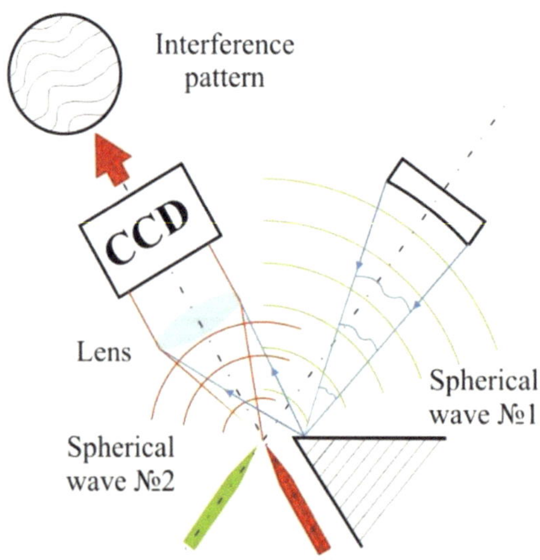

Fig. 17. Scheme of the low-coherence diffraction interferometer with two sources of the reference spherical wave.

As a source of the radiation a super-luminescent diode with a coherence length about 20 microns is applied. It prevents parasitic interferences of wave fronts produced by different elements of the optical scheme. For observing the "correct" interference the optical path of the light of source №2 is increased by a double distance from the investigated surface to the sources. In this case the interference is observed only for the noted above wave fronts. The radiation of the source №1, directly going into the registering system, creates only a background signal on the detector.

As it was already marked, in diffraction interferometers a divergent spherical wave is used as a reference wave. It imposes a number of restrictions on the interferometer capabilities. Strictly speaking, with their help it is possible to investigate only concave spherical and slightly aspherical surfaces and objectives. The measurement of the form of strongly aspherical concave or even spherical, but convex, surfaces demands using additional optical elements, transforming the reference divergent spherical wave to a new one which front is close to the surface shape under study.

Of special interest are interferometric measurements of objective wave front deformations, for example, for final certification of the objective wave aberrations, or for objective adjustment directly in an interferometer, or in the case of the objective application for studying convex spherical surfaces. In Fig. 18 the scheme and photograph of measurements of six-lens objective in the interferometer are shown. In this scheme two coherent sources are used. The measured data of the wave deformation were: PV = 42.3 nm and RMS = 8.2 nm.

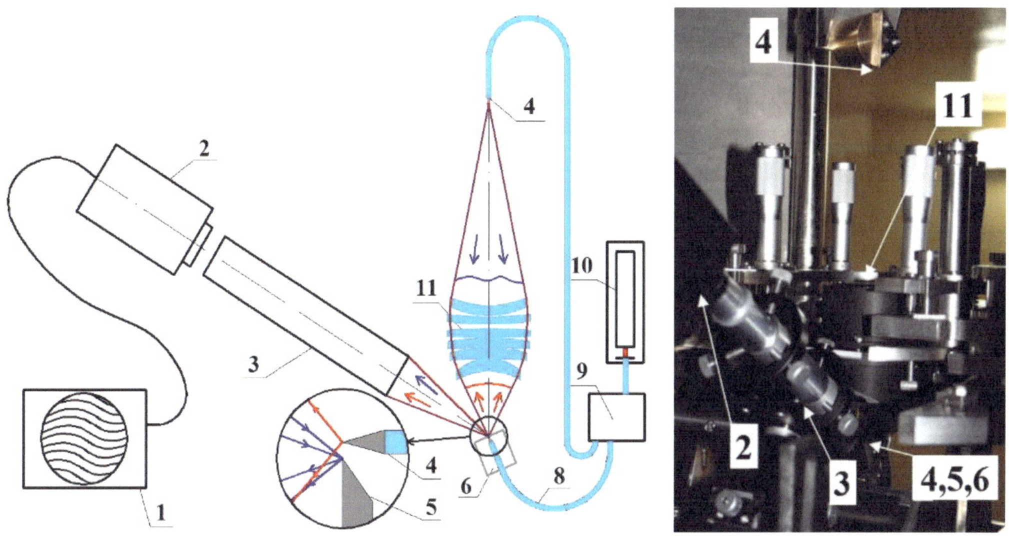

Fig. 18. Scheme of the interferometer with two coherent sources of the reference spherical wave (left) and photograph in the interferometer (right) for testing sixs-lens objective. 1 - PC, 2 - registering system, 3 - observing system, 4 - source of spherical wave, 5 - flat sharp-edge mirror, 6 - three-axes-controlled precision bench, 8 - single mode optical fiber, 9 - polarization controller, 10 - laser, 11 - studied objective.

The certified objective was used for the interferometric measurement (Fig. 19) of a convex spherical surface of a special compensator which in turn later on was applied for studying the concave ashpherical surface for a double-mirror objective. For reconstructing the true shape of the convex surface the expansion coefficients at basis polynomials corresponding to the objective aberrations were deducted from the measured ones.

Aspherical surfaces are used in the majority of high-resolution optical objectives in order to widen the field of view of the optical systems. Direct testing of aspherical surfaces by using an interferometer with the spherical reference wave is often difficult or even impossible. For example, Fig. 20 a) shows the interferogram for a concave aspherical substrate with about 7-μm deviation of the surface from the nearest sphere. The measurements were carried out by using the basic scheme shown in Fig. 15. It is evident, that a large number of fringes makes the position determination of the minima almost impossible. Moreover, the upper part of the interferogram is simply not there, which is explained by a large, about 100 μm, diameter of the focusing spot for which the beam partially propagates above the edge of mirror (item 5, Fig. 15) and misses the measurement system.

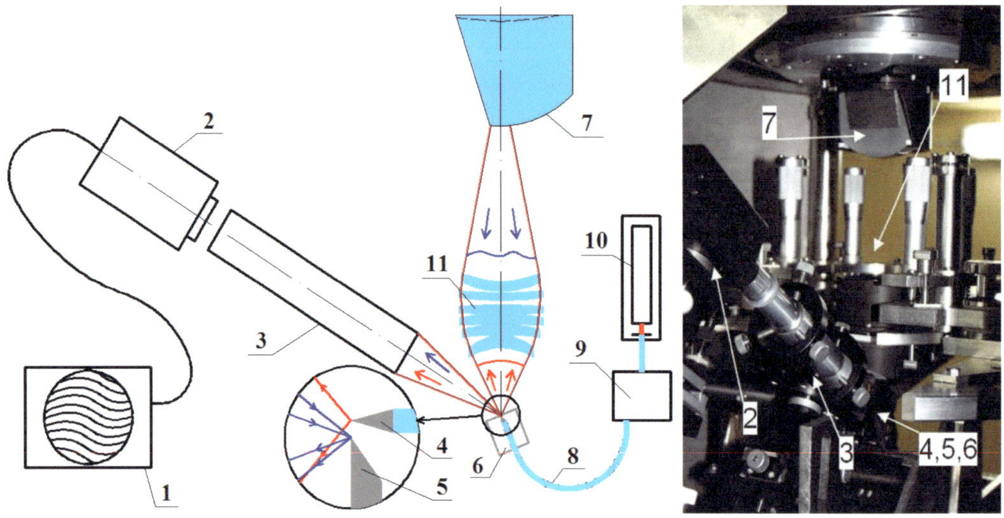

Fig. 19. Optical scheme (left) and photograph of units (right) in the interferometer for certification of convex surfaces. Designations on the scheme are the same, as in Fig. 15.

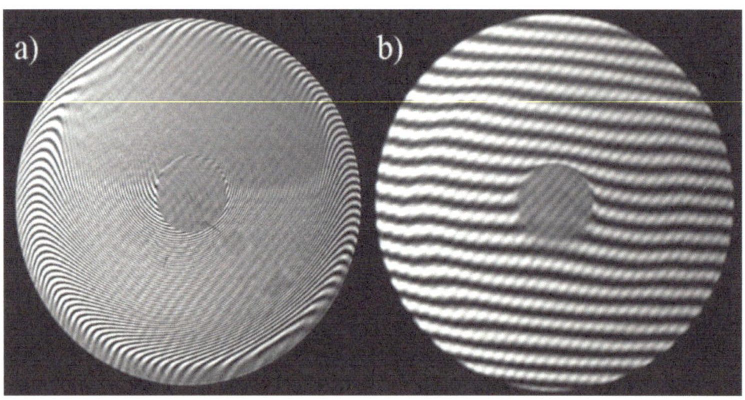

Fig. 20. Interferograms from the aspherical concave mirror M2 obtained without wave-front corrector (a) and using corrector (b).

One of the most effective solution of this problem is the use of wave-front correctors, i.e., special optical elements that transform a reference spherical wave front into an aspherical one which shape is identical to the shape of the studied surface. In order to take into account the wave-front distortions due to errors of the corrector itself, its surfaces should be only spherical. In this case, it is possible to test these surfaces with subnanometer accuracy by using a spherical reference wave interferometer by the techniques stated above.

The scheme for testing the concave aspherical surface by using a corrector is presented in Fig. 21. The parameters of the corrector suitable for testing the concave aspherical surface, i.e. radii of the surfaces, thickness, and the distance from the reference wave source to the top of the first surface, were calculated by minimizing the difference of longitudinal aberrations of the tested aspherical surface and the corrector, described in (Malacara, 1992; Puryaev, 1976) in detail. The calculated root-mean-square deviation of the wave front after the corrector from the investigated aspherical shape was $RMS=\lambda/1810$. This suffices for testing the mirror surface of a nanolithographer with a double-mirror objective. It is clearly seen from Fig. 20 b), that the interferogram from the aspherical surface, obtained by using such a corrector can be reliably processed.

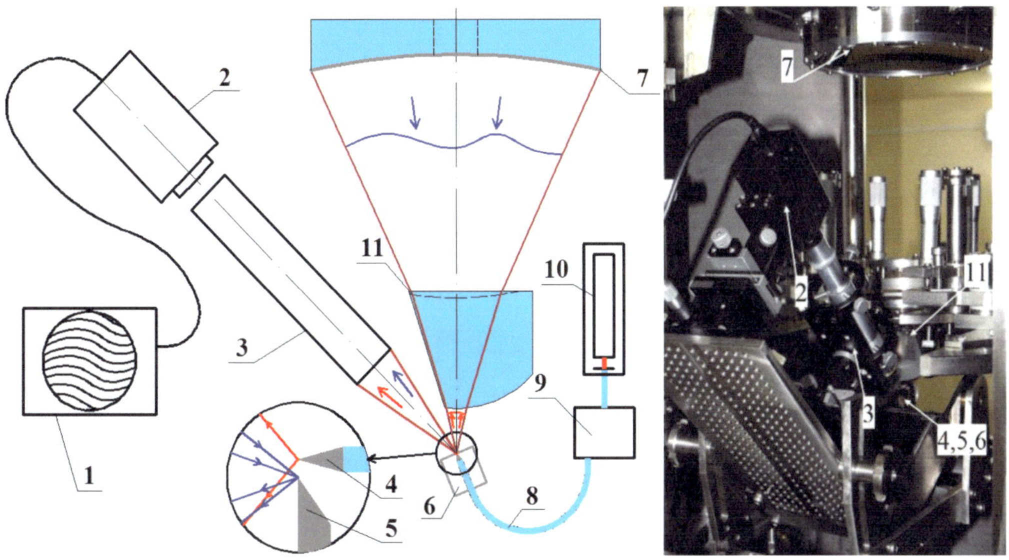

Fig. 21. Optical scheme (left) and photograph of units in interferometer (right) for certification of concave aspherical surfaces using corrector. 1 - computer, 2 - registering system, 3 - observation system, 4 - spherical wave source, 5 - flat mirror, 6 - 3D precision table, 7 - investigated concave aspherical surface, 8 - optical fiber, 9 - polarization controller, 10 - laser, 11 - corrector.

The analysis of the influence of misalignment in the measurement optical scheme and a thickness error of the compensator on "asphericity" of the wave-front passing through the corrector and, accordingly, on the measurement accuracy has shown, that the requirements for these characteristics of the optical scheme are easily performed in practice. In particular, the error of a compensator thickness within the limits of ±0.1 mm leads to a wave front aberration at the level of $\lambda/1000$.

Studying of the convex aspherical surfaces is complicated by the fact that besides the corresponding transformation of a spherical front into an aspherical one, the corrector must provide a converging front of the wave. All of these essentially complicate both the calculation, and the design of the corrector. In particular, in some cases it is necessary to use aspherical surfaces, or more than one lens. In the case of optics for Schwarzschild objectives this problem is solved easily enough. As a corrector the mating concave aspherical mirror, which surface has been preliminary tested by using the corrector system described above and corrected according to the procedure described in the next section, can be used. The measurement scheme is presented in Fig. 22. This scheme can be applied for the final adjustment and certifications of the objective. It is noteworthy to mention, that this method can be applied for a convex spherical surface characterization too besides using additional objectives, as it is shown in Fig. 19.

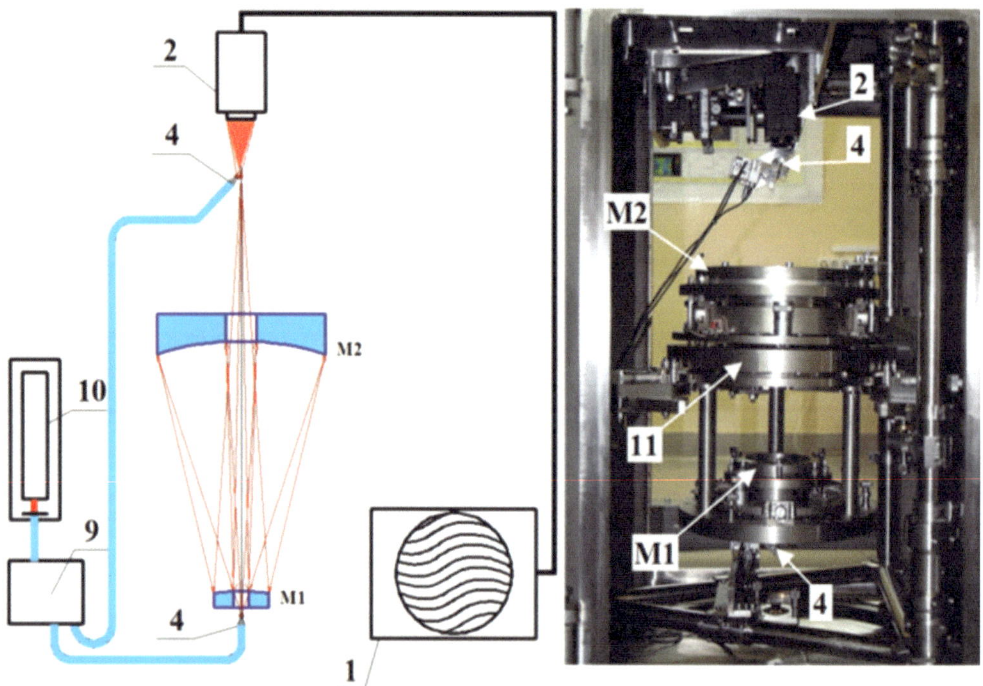

Fig. 22. Measurement scheme of aspherical convex mirror M1 using mirror M2 as a wave-front corrector. 1 - computer, 2 – CCD camera, 4 - spherical wave source, 9-polarization controller, 10 – laser, 11 – Schwarzschild objective, M1 and M2 – convex and concave aspherical mirrors.

The measurement accuracy provided by the interferometer depends on many factors, including the quality of the reference-wave source, the accuracy of determining the positions and a number of interference minima, and the accuracy of mathematical retrieval of the surface (wave-front) shape using a limited number of experimental points, and the number and degree of the polynomials. In turn, the accuracy of determining the minimum positions is strongly dependent on the width of the interference pattern, the degree of pollution of the interference pattern by "parasitic" interference, the method of its rejection, the uniformity and intensity of the interfering fronts, and a number of other unmanageable factors.

In this work, the interferometer measurement precision was evaluated by studying the rotations of the measured surface maps corresponding to the physical turn of the tested detail. The idea of this method is the following. If the interferometer provides zero-error measurements then when the detail is rotated the measured surface map should turn correspondingly and the measured statistical parameters of the surface shape deformations should remain unchanged. The measured surface deformation maps of a spherical substrate with a numerical aperture *NA=0.14* corresponding to different rotations of the detail are presented in Fig. 23. The figures demonstrate that the turn of the maps is well observed, and the statistical parameters of the deformations change within the limits of *RMS*=3.07±0.055 nm. This test confirms the high measurement accuracy provided by the interferometer, which is sufficient for many lithographic applications.

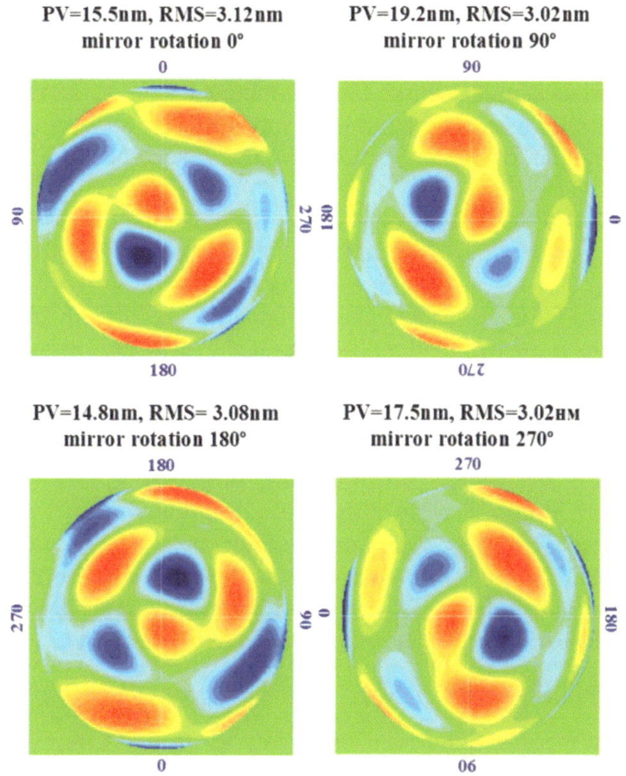

Fig. 23. Maps and statistical parameters of the surface form deformation corresponding to rotations of the spherical mirror by 0, 90, 180 and 270 degrees.

To summarize this section it is possible to ascertain, that the developed interferometer with the diffraction reference wave provides a precision certification of the high-aperture optical surfaces and systems to the subnanometer accuracy.

7. Shape correction of optical surfaces to subnanometer accuracy

The developed precise techniques for measuring the shape of optical surfaces made it possible to pass to the main stage: the correction of the surface shape made with traditional accuracy, $\lambda/2$ - $\lambda/20$, depending on the type of optics, up to $\lambda/1000$ - $\lambda/2000$. In the context of the given work two methods of correction have been developed: a vacuum deposition of thin film coatings and ion etching of previously deposited films through metal masks. An extremely high accuracy of correcting the surface shape imposes specific requirements on the correction method.

1. As a result of the correction process the microroughness of the corrected surface (as a rule, it is initially supersmooth) should remain at the initial level.

2. In the case of a shape correction by means of a local thin film deposition the correcting layers should not distort the substrate surface shape due to intrinsic stress, i.e. they should have a minimally possible stress.
3. The correction method should provide the deposition or the removal of materials in a wide range of thickness (1-300 nm) with the accuracy not worse than several percents. This problem is especially urgent in the last correction stages, when the thickness of the layers to be deposited (removed) is about one nanometer.
4. The film materials should allow their complete removal from the substrate (for example, chemically) without changing the initial surface shape and microroughness. It allows repeating the correction process in the case when the previous correction attempt failed.

Researches on stress compensation in multilayer structures and on a substrate restoration have shown that a material meeting all these requirements is *Cr/Sc* multilayer structure (Salashchenko, 2001). The deposition of coatings was made by the magnetron sputtering method. The scheme of the correction process and the photograph of the installation are shown in Fig. 24. Vacuum in the chamber was (0.7-1)$\cdot 10^{-6}$ Tor, pressure of working gas (argon) was (0.7-0.9)$\cdot 10^{-4}$ Tor. The characteristic power on DC magnetrons was 300-400 W. The film thickness control of the deposited layers is performed by variating the passage speed of the corrected surface above the magnetrons sources. The stabilization of the discharge power and the argon pressure were provided at the level of 0.1 %. The predetermined distribution of the layer thickness over the substrate surface was reached due to the application of figured diaphragms mounted between the magnetron sources and the substrate. In addition the substrate rotates around its axis. Local values of the film thicknesses over the substrates were measured with an x-ray reflectometer at 4.47 nm wavelength. The details of these measurements will be shown below.

Fig. 24. Photograph of a technological facility and scheme of surface shape correction by means of multilayer film deposition through masks.

The operational experience with this technology has shown a number of important advantages. Firstly, the roughness of the film surface practically does not depend on the angle of incidence of the atoms in respect to the substrate that allows to carry out the correction procedure at normal angles. It reduces shadows on the surface, increasing the localness of the correction, correspondingly. Secondly, the high accuracy of the layer thickness control, up to 1 %, provides a high reproducibility of the process. The film-correction method proved to be efficient for initial stages, in which the layers with of tens-hundreds nanometers are deposited in a single correction. The process becomes unstable as the thickness of deposited layers decreases down to the units of nanometers. Correcting films have changed the proper color that is connected with films oxidation at initial stages of deposition. The distribution of oxygen in depth of scandium film obtained with the help of a secondary ion mass-spectroscopy (Drozdov et al., 2009) is given in Fig. 25. The figure shows an increased content of oxygen on the film surface and at the film-substrate interface. Other serious disadvantages of the thin-film-correction method has a discrete and iterative character of the process. After each iteration it is necessary to vent the vacuum chamber, to make a new diaphragm, to pump the chamber again. All this strongly decreases the throughput of the process.

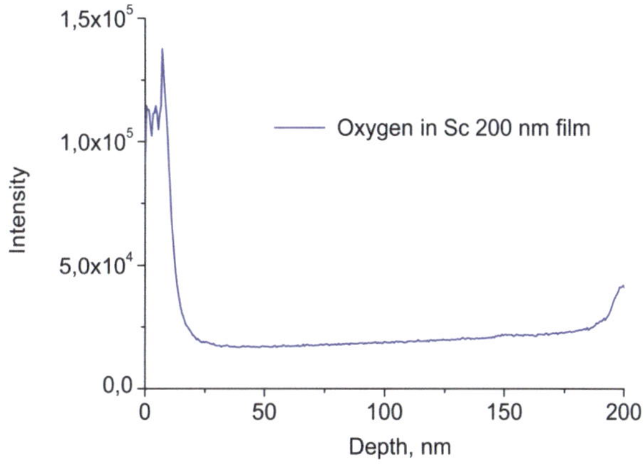

Fig. 25. The distribution of oxygen in depth of scandium film deposited onto Si-substrate.

Therefore the application of ion-beam etching at the final correction stage looks very attractive. This method allows, in particular, when focused ion-beams are used, to realize the "maskless" correction without frequent opening of the installation to atmosphere. Now this method is widely used for substrate form correcting and even for final ion-beam polishing in optical industry (Allen et al., 1991; Chason & Mayer, 1993; Ghigo et al, 2007). Recently an intensive literature has evolved which is devoted to the application of this method for the precision correction of substrates for projection EUV lithography (Dinger et al., 2000; Murakami, 2007).

For the ion beams, applied for the surface shape correction of substrates for EUV and soft X-ray optical elements, a number of requirements are imposed, the basic of which is the preservation of the surface roughness at an initial (supersmooth) level even when the removed layer depth reaches a few hundreds of nanometers. Therefore an optimization of ion energy and incidence angles of the ions on the corrected surface at different depths of material removal is an actual problem.

Other problems inviting a further investigation are connected with some features of the ion-beam etching at an extremely small depth (≈ 1 nm) removal of materials.

The experiments have been performed on the installation (Fig. 26) with the "Cauphman» type ion-beam source KLAN-103M (Website Platar Company, 2009). The basic advantage of this source is the possibility of working with small ion energies, E_i = 50- 300 eV, at a significant current density.

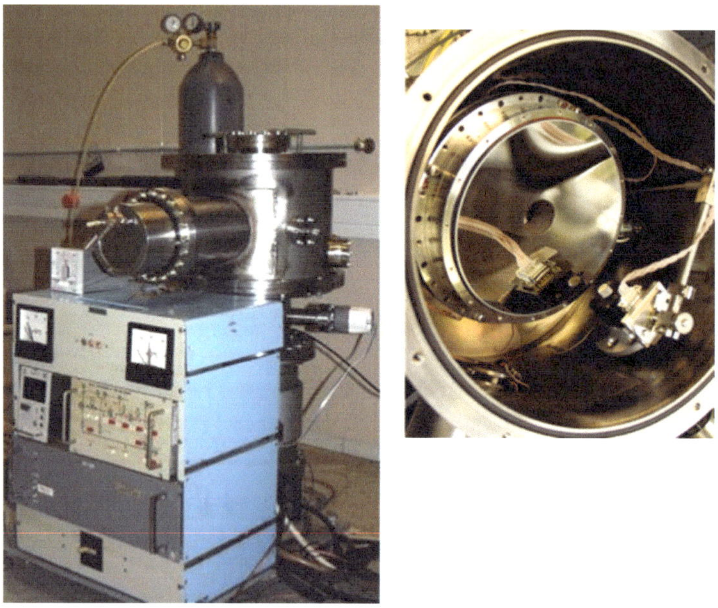

Fig. 26. Photographs of installation for substrates surface shape correction by ion-beam etching.

The main characteristics of the ion source are: the operating gas is argon or other inert gases; the maximal diameter of the ion beam is 100 mm; the maximal ion current is 100 mA; the maximal ion current density is up to 2.3 mA/cm^2; the range of ion energy regulation is 50-1500 eV at the energy spread of ±3 eV; the discharge working voltage is 35-50 V; the thermo-cathode is made of thoriated tungsten (0.3 mm in diameter) filament. Presence of charge-neutralizer (additional thermo-cathode) allows compensating the ion beam charge that provides using the ion source for correction substrates of dielectric materials too.
The measured dependence of the etching rate of *Cr/Sc* coating on the removal depth taken at ion energy E_i=200 eV and grazing incidence angle of θ = 42° is shown in Fig. 27 by dots. It is evident that at the initial stage of the etching, the removal is less than 120 nm, practically a linear increase of the etching rate from the removal depth is observed. At a very small etching depth, h <10 nm, a wide scatter of the etching rate values is observed. Considering the data of the secondary ion mass-spectrometry, Fig. 28, such a behavior of the etching rate can be explained by oxidation of the sample surface. The etching rate of the oxide layer is significantly lower as compared with pure Cr/Sc material. If take this into account the calculated etching rate, a solid line in Fig. 27, matches well the experimental data. The observed instability of the etching rate at h <10 nm might be explained by the variation of the oxidized layer depth in various samples. When samples were etched by a neutralized beam (stars in Fig. 27), no changes of the rate were observed. With the increased ion energy up to 300 eV, Fig. 29, such instability of the etching rate is not observed either. According to atomic-force microscopy measurements the surface roughness remained constant, at the level of 0.2 nm, down to 400 nm removal depths. For the ion energy of E_i=*300* eV when the etching depth exceeds 80 nm an increase of surface roughness is observed (Fig. 30).

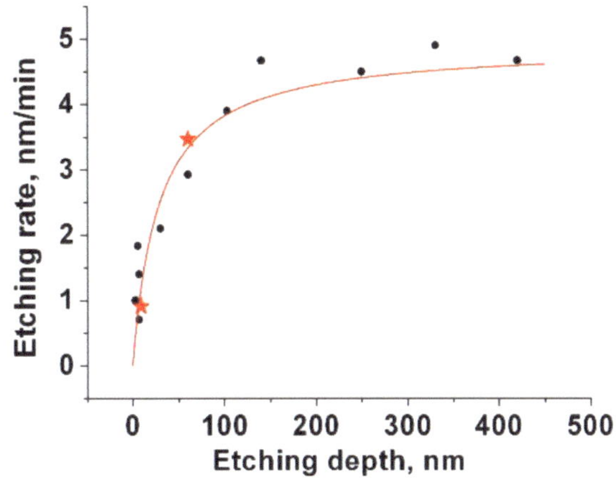

Fig. 27. Dependence of etching rate on removal depth of *Cr/Sc* multilayer structures: dots are experiment with ions, stars are experiment with neutrals and line is calculation accounting etching of the oxidized layer. Ion energy E_i = 200 eV and current I_i=60 mA.

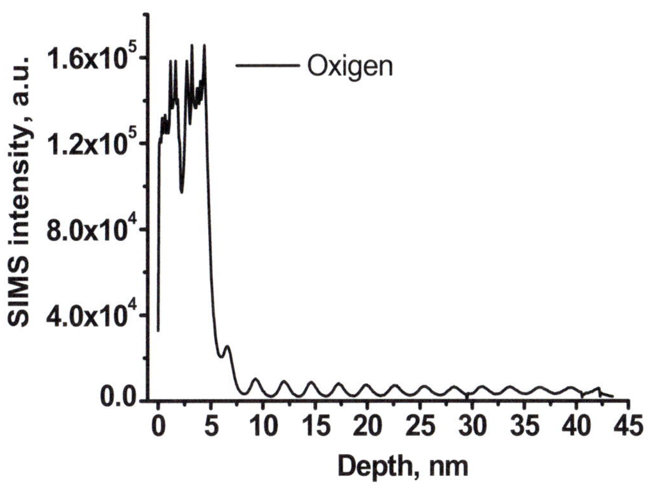

Fig. 28. Typical distribution of oxygen atoms on depth of *Cr/Sc* multilayer structure taken by secondary ion mass-spectrometry.

The dependences of etching depth on an ion dose (the dose is the product of ion current and etching time) both from the ion current at fixed time, and from the time at fixed current, show a good linearity, that is useful from the practical point of view because it allows to easily define the necessary etching time at a given ion current. The example of the dependences of the etching depth on the dose is given in Fig. 31. The dashed line corresponds to the linear approximation of the dependence and line with squares represents the experimental data. It should be noted, that the experiments carried out with greater ion energies, up to E_i=*1.5* keV, have shown a fast growth of the surface roughness, even at small, nearby 20°, grazing angles of incidence of the ions on the surfaces (Chkhalo et al., 2008).

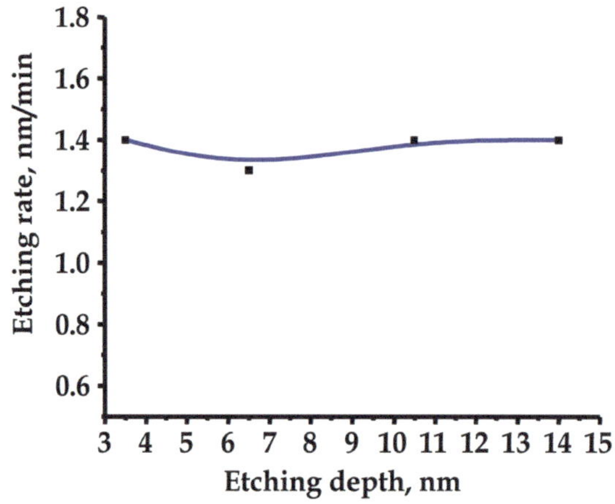

Fig. 29. Dependence of ion etching rate on etching depth of Cr/Sc multilayer structures at ion energy E_i = 300 eV (area of small removal depths).

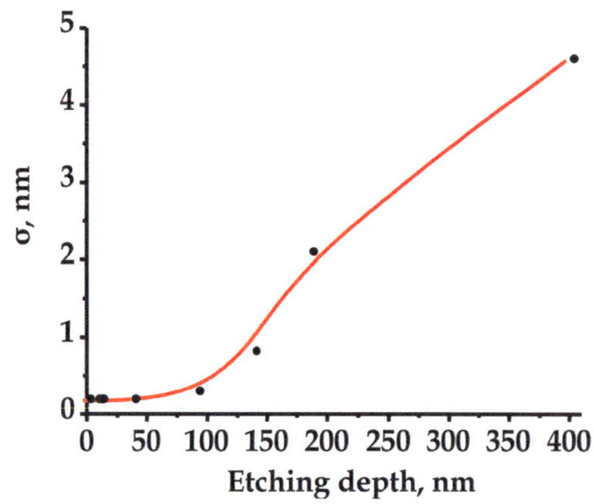

Fig. 30. Dependence of surface roughness of *Cr/Sc* multilayer structures on etching depth at sputtering by argon ions with energy E_i = 300 eV.

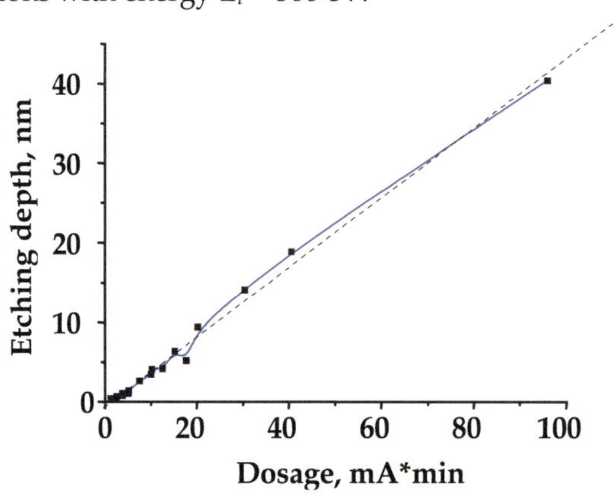

Fig. 31. Dependence of etching depth of the *Cr/Sc* multilayer structures on ions doze, at ion energy E_i = 300 eV. - - - linear approximation of the dependence.

For correcting the surface shape of substrates for imaging X-ray optics when the subnanometer form accuracy is required, the following strategy providing minimal time expenses has been chosen. At the initial stage when peak-to-valley values (*P-V*) of the surface shape are 50-200 nm, the local deposition of the thin films is applied. Then, at the reduction of the *P-V* values down to 10-20 nm the ion beams with the energy of 200 eV are used. At the final correction stage the 300 eV-ions are applied. Figure 32 illustrates the result of using this correction procedure for a spherical substrate with a numerical aperture of *NA*=0.25 (the diameter is 130 mm and the curvature radius is 260 mm), which was initially the standard etalon for a conventional interferometer. The initial surface map is shown in Fig. 32 a). The parameters of the surface were *P-V=42.6* nm and *RMS=7.3* nm. The surface shape improved to *P-V=7.3* nm and *RMS=0.6* nm (Fig. 32 b)). Thus, a root-mean-square deviation of the initial surface shape from an ideal sphere has decreased in 12 times, i.e. in units of the interferometer working wavelength (λ=0.633 μm) it has became better than $\lambda/1000$. The photograph of this mirror mounted on a metal frame is given in Fig. 33.

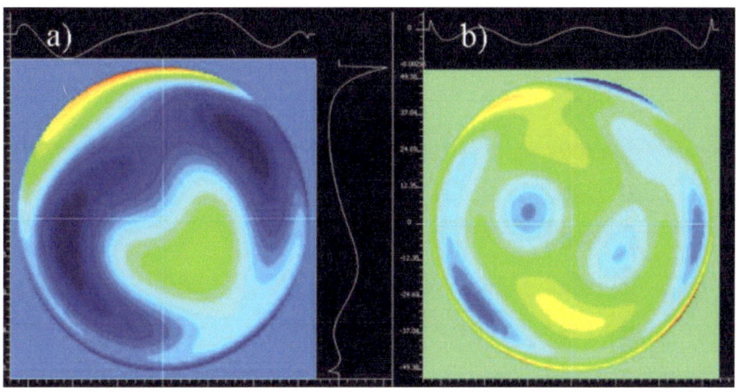

Fig. 32. Maps of deviations of the spherical surface shape from the nearest ideal sphere: a) initial shape (*P-V=42.6* nm, *RMS=7.3* nm); b) after 12-th correction (*P-V=7.3* nm, *RMS=0.6* nm).

Fig. 33. Photograph of the multilayer spherical mirror for 13.5 nm wavelength in the frame. Diameter of the working aperture is 130 mm.

In spite of a greater wavelength of the traditional deep ultraviolet lithography (193 nm), because of a great number of lens and high requirements for the quality of the projective objectives, the methods of correction and certification with a subnanometer accuracy described in this article are urgent for the refracting optics too. In connection with this the researches on etching of fused quartz surfaces by ion beams with a compensated charge have been carried out. As well as in the case of metal surfaces, the main attention was given to the etching rate, to the reproducibility of the etching process and to the surface microroughness depending on the beam characteristics, incidence angles and material removal depths.

In Fig. 34 the dependences of quartz etching depth on the etching dose are presented. The dashed lines in the figure correspond to linear approximations of the dependencies. The ion energy was 200 and 300 eV, the grazing incidence angle was 42°. The comparison with similar dependences for *Cr/Sc*, Fig. 31, shows that the etching rate of fused quartz is almost twice as much. No nonlinearity at the initial stage of the etching is observed, that on the one hand, proves the hypothesis about the effect of the oxide layer in the case of Cr/Sc material and on the other hand, essentially facilitates the correction process.

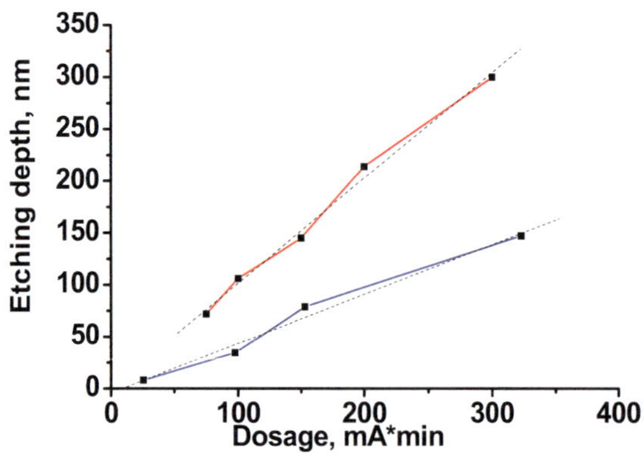

Fig. 34. Dependences of the etching depth on the doze for fused quartz. Etching is performed by argon ions with compensated charge and with energy *Ei* = 200 and 300 eV; - - - - linear approximation of dependences.

The data on the surface roughness of a fused quartz substrate etched by ions (neutrals) with energies of 200 and 300 eV, for various removal depths, measured by means of the atomic-force microscope and by x-ray diffractometry are presented in table 1. The substrates manufactured by General optics (USA) were used in the experiment. The minimal roughness of initial surfaces measured with an atomic-force microscope was 0.13 nm, and with the x-ray reflection it was about 0.4 nm. From the table it is evident that atomic-force microscope measurements have shown a weaker influence of the etching process on the surface roughness, in comparison with the roughness measurements made by the x-ray diffractometry method. A more detailed discussion of this difference is beyond the given article.

Thus, the presented experimental data show that these technologies of vacuum deposition and ion-beam etching in the aggregate allow to correct with a subnanometer accuracy the surface shape of optical elements practically of any materials (metals, dielectrics, semiconductors) with keeping the surface on a supersmooth level. This technology is suitable for manufacturing elements for imaging x-ray optics, and for high resolution systems in visible and deep ultra-violet ranges.

Energy of neutrals in the beam, eV	Etching depth, nm	Etching rate, nm/min	Roughness, nm	
			AFM	X-ray
200	30	7.90	0.13	0.40
200	79	7.90	0.15	0.60
200	157	7.85	0.18	0.65
200	237	7.90	0.20	0.70
300	353	11.70	0.28	1.0
Initial surface roughness, nm			0.13	0.40

Table 1. Roughness of fused quartz substrates depending on neutral (ion) energy and etching depths measured by atomic-force microscopy (AFM) and x-ray diffractometry (X-ray) methods.

8. Reflecting multilayer interference mirrors

The final stage of manufacturing a precise mirror for ultrahigh resolution imaging x-ray optical systems is the procedure of the deposition of multilayer reflecting coating on a substrate. A low polarizability of materials ($\varepsilon \leq 1$) and correspondingly a small jump of the refraction factor Δn=0.0001-0.1 on the vacuum-material interface and a strong absorption of the radiation by all materials completely exclude the application of refractive optics in the extreme ultra-violet and soft x-ray ranges.

Because of a low reflection from one interface the reflection optics of normal incidence can be only interferential and the number of layers demanded for the achievement of a high reflectivity depending on a wavelength can reach a value of 10^2-10^3.

The absorption of the radiation in multilayer structures leads to the fact that the reflectivity is always less than unity. While choosing an optimal pair of materials providing a maximal reflectivity on a given wavelength the circumstance is often used that in the range of wavelengths directly adjoining the absorption edge of the material, $\lambda > \lambda_{AE}$ (λ_{AE} is the wavelength of the material absorption edge) and coinciding with an anomalous dispersion of the optical constants the lowest absorption and the maximal polarizability of materials are observed. For some materials in the field of *K*-and *L*- absorption edges the refraction factor can even exceed unity. In Fig. 35 the calculated spectral dependences of the maximal reflection coefficients for the most attractive pairs of materials in a spectral range of 3 - 16 nm, representing the biggest interest for projection nanolithography and high resolution x-ray microscopy are shown.

The greatest experimental reflectivity, about 70 %, was obtained at a wavelength of 13.5 nm. The spectral range around $\lambda \approx 6.7$ nm (area of boron anomalous dispersion) is of interest for projection x-ray lithography. However the reflectivity obtained here with La/B_4C multilayer mirrors at a normal angle of incidence does not exceed 45-47 % (Andreev et al., 2009) that is not enough for lithographic applications.

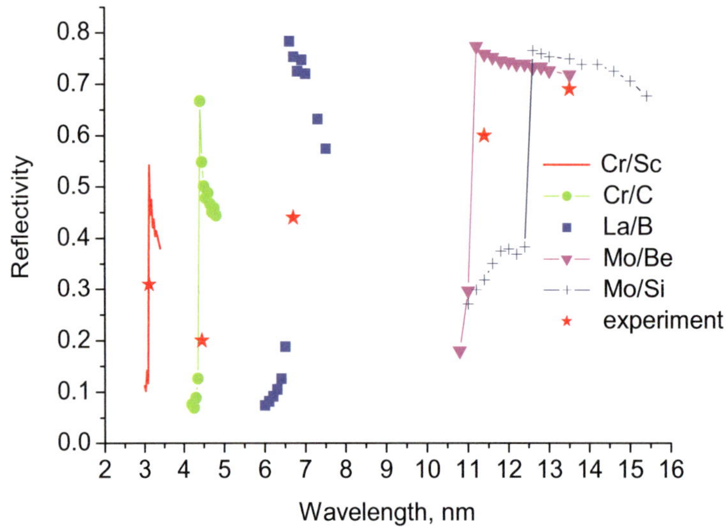

Fig. 35. Calculated (lines with symbols) and experimental (stars, (Website ESRF, 2006; Andreev et al., 2003)) reflection coefficients of mostly attractive pairs of materials in X-ray and EUV ranges.

On the short-wave length side the working wavelength of the normal incidence multilayer optics is limited by technology capabilities for depositing high-quality short period multilayer structures. At present the shortest periods have multilayer structures based on W/B_4C (the period $d \approx 1$ nm) (Vainer et al., 2006; Platonov, 2002) and *Cr/Sc* ($d \approx 1.5$ nm) (Salashchenko & Shamov, 1997; Website ESRF, 2006).

A number of special requirements are imposed on the multilayers which should be deposited onto substrates for imaging optics. In particular, to provide a constructive interference of the waves reflected from different interfaces, the radiation wavelength λ, the local angle of incidence $\theta(\vec{r})$ and the period of the multilayer structure $d(\vec{r})$ should satisfy the Bragg condition

$$d(\vec{r}) = \frac{\lambda}{2\sin\theta(\vec{r})}, \tag{11}$$

where $\vec{r}$ is a radius-vector of a point of incidence of the radiation on the structure. As the incidence angles are a function of the coordinates it results in the necessity to deposit multilayer structures with a variable over the surface period, so-called "graded" structures. In some cases, for example, when the angular divergence of the X-ray beam falling on one point of the mirror surface exceeds the Bragg's peak width, the multilayer structure should have a variable period on depth (Kozhevnikov et al., 2001).

One more important feature of multilayer structures for precise optics can be explained as follows. For multilayer dispersive elements the maximum deviation of the local value of the period δd from the nominal one is determined by the condition of synchronism of the waves reflected from various interfaces which can be expressed through the spectral selectivity of a mirror $S = \delta\lambda / \lambda = \delta d / d \cong 1/N$ (N is a number of periods). In the case of precision multilayer imaging optics when the required surface shape accuracy is at a subnanometer level to the synchronism condition an extra requirement for the preservation of the mirror surface shape is added. For example, for the well-known *Mo/Si* multilayer structure, optimized on the wavelength of λ=13.5 nm, the period is $d \approx 7$ nm at a number of periods $N \approx 40$. Correspondingly the general thickness of the structure is $L=N \cdot d=280$ nm. The local error of the period of 1 % will affect weakly on reflectivity, however the surface shape of the mirror will produce a change of 2.8 nm, which already can essentially produce an influence on imaging properties of the mirror.

From these simple estimations two important conclusions follow. Firstly, when designing and manufacturing the surface shape of a substrate it is necessary to take into account the real thickness of a multilayer coating which would be deposited onto the substrate. Secondly, it is necessary to provide a high, at a level of 0.2-0.3 %, accuracy of local values of the period over the mirror surface. Besides, an intrinsic stress in the multilayer structures, which can also deform the surface shape, must be taken into account. The estimation of the deformation δx of a substrate with a diameter D, a thickness h_{sub}, with the Yung modulus E and the Poisson factor ν on which a h_f thick film with the intrinsic stress σ is deposited, can be performed by means of equations

$$\delta x = \frac{3 \cdot \sigma \cdot h_f}{4 \cdot Y} \cdot \frac{D^2}{h_{sub}^2}. \qquad Y = \frac{E}{1-\nu} \tag{12}$$

For instance, for optimized on maximum of the reflection coefficient *Mo/Si* multilayer structure (portion of *Mo* layer in the period $\gamma = d_{Mo}/(d_{Mo}+d_{Si}) = 0.4$) the stress in our case is about σ =-350 MPa (Fig. 36). At deposition of such a multilayer structure (general thickness is $L=N \cdot d=280$ nm) on a substrate of fused quartz with the diameter $D=220$ mm and thickness $h=50$ mm ($Y=84{,}4$ ГПа) the maximal deformation of the substrate surface would be about 17 nm, that essentially exceeds an admissible level of the surface shape deformation of a mirror for EUV lithography.

Different groups solve this problem in different ways. One the solution is the use of mirrors with a specified thickness ratio of molybdenum layer in the periods $\gamma_{Mo} \approx 0.5 - 0.7$ when the stress is close to zero (Windt et al., 1994; Zoethout et al., 2003). The other solution consists in adding a third material into the periods, for example *Ru* (Oshino et al., 2003). However, all these methods, either because of a stronger absorption of EUV radiation or because of an interface roughness growth have smaller, of 10-20 %, absolute values of the reflectivity.

In IPM RAS a technique of the stress compensation based on the use of special sublayers lying under the basic *Mo/Si* multilayer structure is under development. For *Cr/Sc* multilayer structures the stress is close to zero at $\gamma_{Cr}=d_{Cr}/(d_{Cr}+d_{Sc}) \approx 0.5$ (Fig. 37 (Andreev et al., 2005)). At $\gamma_{Cr}>0.5$ a sign of the stress is positive on the contrary to the *Mo/Si* structures at $\gamma_{Mo}= 0.4$, when it is negative. The total stress σ_{eff} in the structure consisting of *Cr/Sc* and *Mo/Si* multilayer films, can be determined from a ratio

$$\sigma_{eff} = \frac{\left[(\sigma_{Mo/Si} \cdot d_{Mo/Si} \cdot N_{Mo/Si}) + (\sigma_{Cr/Sc} \cdot d_{Cr/Sc} \cdot N_{Cr/Sc})\right]}{(d_{Mo/Si} \cdot N_{Mo/Si}) + (d_{Cr/Sc} \cdot N_{Cr/Sc})}, \tag{13}$$

from which it follows, that optimizing the composition and the thickness of *Cr/Sc* layers it is possible to compensate the stress in the basic *Mo/Si* reflecting multilayer structure.

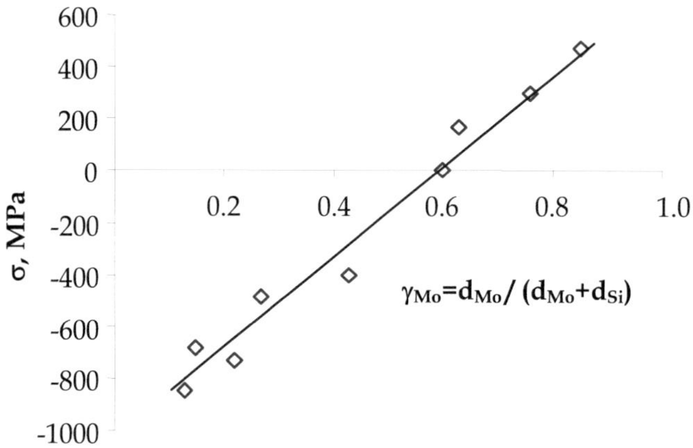

Fig. 36. Stress in *Mo/Si* multilayer structures depending on ratio of *Mo* layer in period $\gamma_{Mo}=d_{Mo}/(d_{Mo}+d_{Si})$ at fixed period d=6.97 nm.

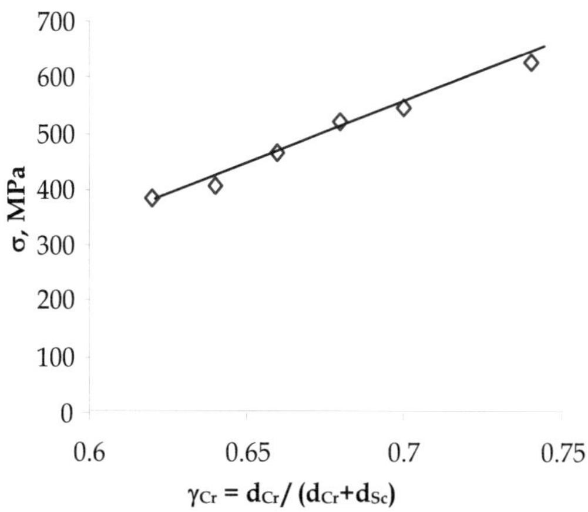

Fig. 37. Stress in *Cr/Sc* multilayer structure depending on ratio of *Cr* layer in period $\gamma_{Cr}=d_{Cr}/(d_{Cr}+d_{Sc})$ at fixed period d=5.4 nm.

In practice it is necessary to consider a process of the stress relaxation which, depending on the structure and materials, is variable over several days. Totally the stress can change by tens percent. An example of such a process is illustrated in Fig. 38 in which the time dependencies of the stress in multilayer structures consisting of anti-stress *Cr/Sc* sublayers and *Mo/Si* reflecting layers are presented. The structures differ by thicknesses of the sublayers.

It is necessary to note, that the stress value is affected not only by ratio of the layer thickness in the period, but also by the period size and conditions in which the films were deposited.

For depositing multilayer structures a method of magnetron spattering is used in IPM RAS. The method allows depositing films onto curved as well as flat substrates. The conditions, under which multilayer structures are deposited, are briefly described in section 6, devoted to the correction of the substrate surface shape by the method of local thin-film deposition. The facilities have from two to six magnetron sources that allows depositing at one technological cycle reflecting, anti-stress and anti-diffusion barrier layers. Now the technologies of depositing more than twenty various pairs of materials are developed. It allows manufacturing optics for a whole x-ray and extreme ultra-violet spectral range. More details about the properties of the multilayer structures can be found in (Andreev et al., 2003).

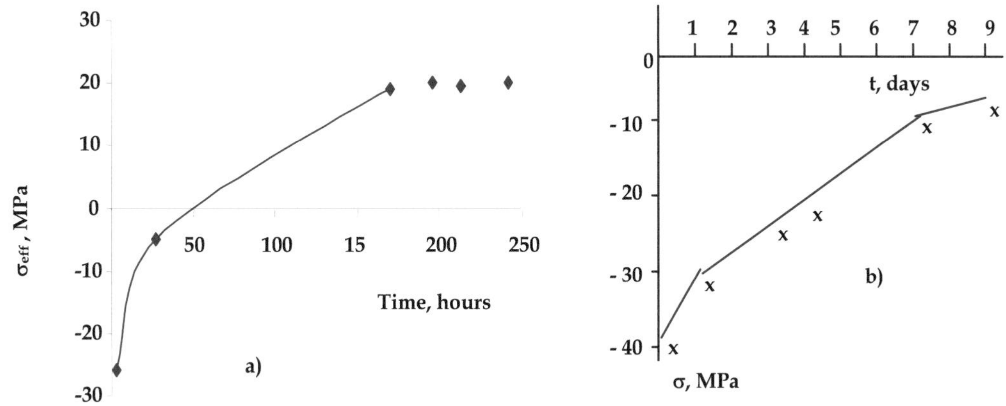

Fig. 38. Stress in *Mo/Si* multilayer structures with anti-stress *Cr/Sc* sublayers versus of time.
a) - [$(Cr/Sc)\times42$; d=5.38 nm; γ =0.67 / $(Mo/Si))\times45$; d=6.98 nm; γ =0.4] and
b) - [$(Cr/Sc)\times40$; d=5.38 nm; γ =0.67/ $(Mo/Si)\times45$; d=6.98 nm; γ =0.4].

A great volume of works, high requirements for the coating quality and for accuracy of the period distribution over a substrate surface dictate the need for an operative information about characteristics of the deposited coatings with the purpose of correcting the technological process. It is most convenient to perform such work with the presence of laboratory reflectometers, allowing the investigation of multilayer structures on working wavelengths and located in the immediate vicinity of technologists. Therefore laboratory reflectometers are constantly being developed and improved recently. In Refs. (Loyen et al., 2003; Miyake et al., 2003; Gulikson et al., 1992) the specialized reflectometers for investigating multilayer optics in the range around 13.5 nm are described. As a source of the soft x-ray and EUV radiation the laser plasma is used, which arises due to the irradiation of solid-state targets by laser light with a power density about 10^9 W/cm^2. For monochromatization the radiation spherical diffraction gratings operating at grazing angles are used. The characteristic fluxes of the photons in the working range of wavelengths of 10-16 nm on the investigated samples with the probe beam size ≈ 2 mm are 10^7 - 10^8 photons / (0.1 nm×s). The results of reflectivity measurements of *Mo/Si* multilayer mirrors in the area of 13.5 nm obtained with the help of these reflectometers as compared with the data obtained with a PTB-reflectometer (synchrotron BESSY-2, Berlin, Germany, which is considered as the reference one), have shown the coincidence of the measured reflection coefficients at a level of 3 % and the resonant wavelength ≈ 0.01 nm.

The basic disadvantages of these reflectometers, in our opinion, are a narrow range of working wavelengths and a low spectral resolution caused by the applied monochromator scheme. The monochromator consists of two stationary slits and spherical (or toroidal) diffraction grating. Scanning across the spectrum is carried out due to the diffraction gratings rotation around its axis. According to this arrangement, the position of the slits and the diffraction grating coincides with the Rowland circle (Rowland, 1893) only for one wavelength that essentially limits the range where the spectral resolution of the monochromator is kept at the acceptable level of $\Delta\lambda_{app} \approx$ 0.01-0.02 nm. The other factor limiting working spectral range of the reflectometer is the type of the used radiation source (laser plasma) which large size does not allow scanning by the entrance slit of the grating spectrometer.

Our group develops multilayer optics for a wide spectral range λ=0.6-50 nm and the monochromator spectral resolution up to 0.003-0.005 nm is frequently required. Therefore in IPM RAS laboratory reflectometers on the basis of high resolution grating monochromators have been developed. These monochromators have a movable entrance slit and diffraction grating (Chkhalo et al., 1998; Zuev & Mitrofanov, 2002; Bibishkin et al., 2004; Bibishkin et al., 2005). In the reflectometrs the monochromators of type RSM-500 (Lukirsky et al., 1967) and McPherson-247 (Website Mcpherson Company, 2009) are used. The photograph and optical scheme of the last generation reflectometer are given in Fig. 39. Owing to such a spectrometer the kinematic scheme slits and diffraction grating are always located in the Rowland focusing circle so that the high spectral resolution of the device in the whole range of wavelengths is provided. Two diffraction gratings with various radiuses of curvature cover this spectral range.

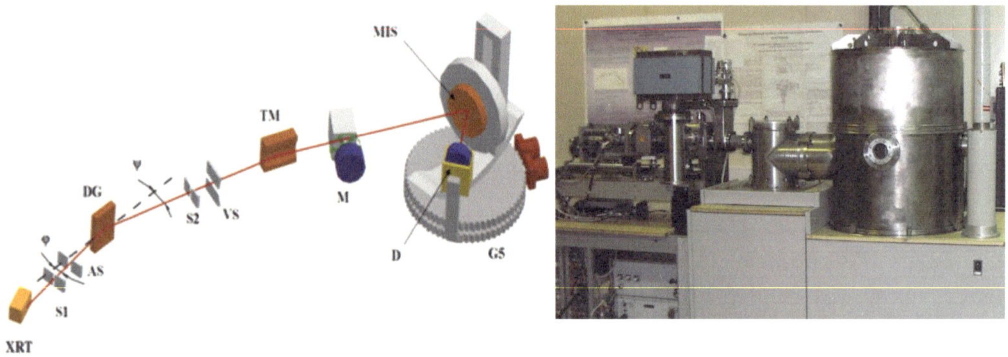

Fig. 39. Optical scheme and photograph of the reflectometer. XRT - x-ray tube; S1, S2 - entrance and exit slits; AS, VS - horizontal and vertical anti-scattering slits; DG - diffraction grating; TM - toroidal mirror; M – probe beam intensity monitor; MIS - investigated sample; D – detector; G5 – five-axes goniometer.

In the reflectometer the radiation from the anode of the x-ray tube (XRT) through the entrance slit of the monochromator (S1) falls on the diffraction grating (DG). The diffracted monochromatic radiation according to the diffraction grating equation $\cos\varphi - \cos\psi = m\lambda/D$ (m - the order of diffraction, λ – a wavelength and D - the grating period, φ and ψ angles of diffraction and incidence, correspondingly) is focused onto the spectrometer exit slit S2. Scanning across the spectrum is carried out by linear moving of the diffraction grating and the exit slit, with a simultaneous turn of the diffraction grating so that the condition $X_l=R_g\cdot \sin\psi$, where R_g - radius and X_l - distance between the diffraction grating center and the exit slit is satisfied. An anti-scattering diaphragm (AS) reduces the intensity of scattered radiation at the output of the monochromator and limits the irradiated area of the diffraction grazing. The monochromatic radiation by means of a toroidal mirror (TM) is focused on the investigated sample. The other important function of the toroidal mirror is connected with a filtration of higher diffraction orders which is especially urgent when working with bremsstrahlung of the tungsten anode. Radiuses of the toroidal mirror are chosen in such a way that in a horizontal plane the exit slit and in a vertical plane the source are imaged on the sample under study.

The studied sample is installed on a five-axis goniometer which photograph and scheme of movements are shown in Fig. 40. The linear translation along X and Z axes, the rotation around them (ω and φ) and the inclination θ, allow to insert any point of the sample under study with a diameter up to 300 mm into the x-ray beam. The detector executes 2φ rotation. The additional inclination of the detector 2θ allows to increase the numerical aperture of the sample which can be tested up to NA=0.5.

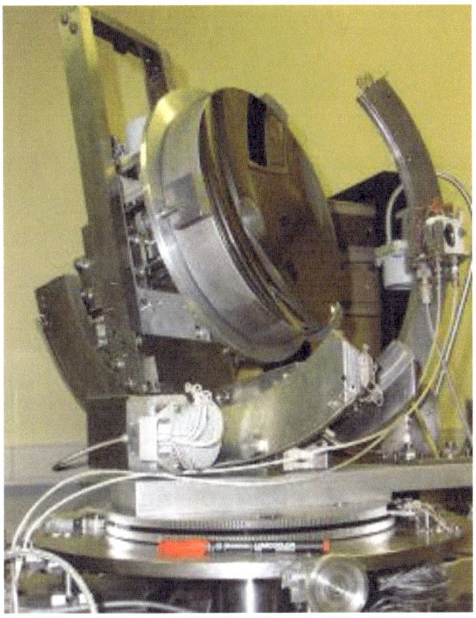

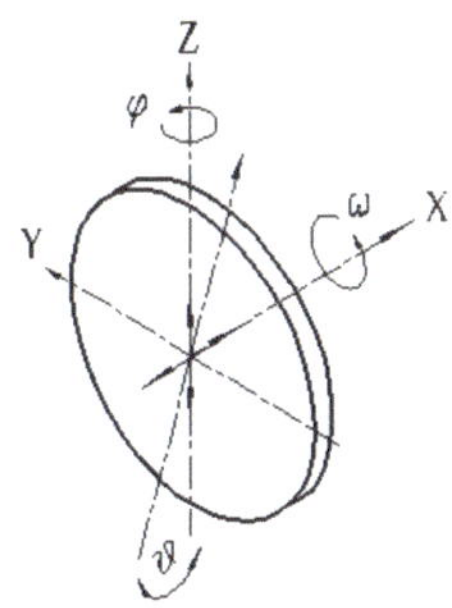

Fig. 40. Photograph of the goniometer and scheme of movements provided by the goniometer.

The concept of a spectrometer-monochromator with a movable slit and diffraction grating became possible only due to the application of the dismountable x-ray tube of an original design as the radiation source. The advantages of classical sealed x-ray tubes are their small dimensions and weight, simplicity in operation, sufficient for reflectometric applications intensity of the x-ray radiation, and long life-time. However the application of these sources for precision reflectometry in soft x-ray and extreme ultra-violet ranges has shown that because of the fast contamination of the anode by products of thermocathode material erosion and hydrocarbons decomposition under a powerful electron-beam, emission characteristics and life-time of the tubes collapsed when comparing with expected values.

For solving this problem in our work the tube was equipped with an ion-beam source for cleaning anodes from contaminations (Bibishkin et al., 2003). It allows to work without opening the spectrometer and the tube to the atmosphere within several months thus approaching its operation mode to the traditional sealed X-ray tubes for hard X-rays. The work is done on K, L and M characteristic lines of target (anode) materials and on continuous bremsstrahlung spectrum of tungsten. On one holder about 8 targets of various materials are installed. It is necessary to note, that frequently it is possible to use a few spectral lines from one target.

As an illustration in Fig. 41 the measured spectral dependence of reflectivity of *Mo/Si* multilayer spherical mirror shown in Fig. 40 is presented. The measurement has been performed with the reflectometer at an incidence angle of the radiation to the sample of 5° (from the local normal).

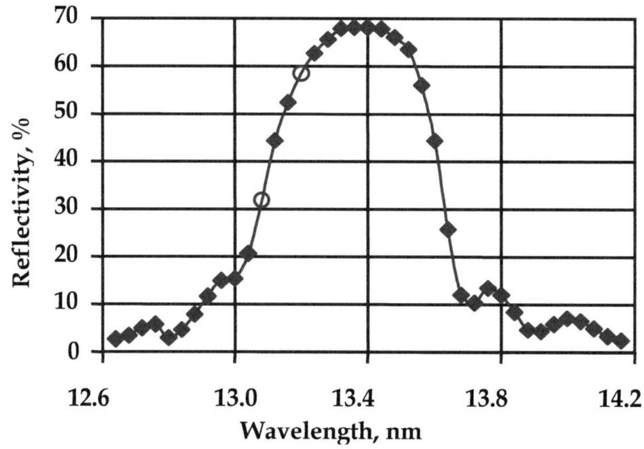

Fig. 41. Spectral dependence of the reflection coefficient of *Mo/Si* multilayer mirror deposited onto spherical substrate. Angle of incidence was 5° from the local normal.

One more important advantage of the x-ray tube as a source of X-ray radiation is that in the field of wavelengths $\lambda<3$ nm because of aberrations of spherical grating the spectral width of the probe beam is $\Delta\lambda_G \geq 0.01$ nm while the spectral resolution of short period multilayers reaches values $\Delta\lambda_{MIS} < 0.01$ nm. Since the natural width of characteristic lines of some materials in this range is essentially less than this value this circumstance provides precision measurements of reflective characteristics of multilayer mirrors, even when monochromator does not provide the demanded resolution.

Thus, the developed in IPM RAS methods of multilayer coating deposition, stress compensation in the film structures, laboratory metrology of the mirrors at working wavelengths, restoration of the substrates without any damage of the substrate surface shape and roughness allow us to make and certify high reflecting multilayer mirrors with a sub-nanometer level surface shape accuracy attended for super-high resolution imaging X-ray optics.

9. Conclusion

In the given article the basic researches and the full cycle of developed in IPM RAS technologies and metrology techniques, from the substrate surface shape correction to the deposition of multilayer coatings on them which are required for manufacturing high reflecting mirrors for diffraction quality imaging in the extreme ultra-violet and in the soft X-ray ranges are described. As a whole, it is evident that the problem of manufacturing such optics for high resolution EUV lithography and X-ray microscopy is already solved as a first approximation. The further activity should be directed toward the increase of both in manufacturing accuracy of optical surfaces up to a level of *RMS*=0.1 nm and in the numerical aperture of objectives up to *NA*=0.4-0.6.

In this connection upgrading the interferometric methods for surface shape measurements of optical elements and wave-front deformations of objectives, first of all, and regarding the increase of their working numerical aperture is a subject of current interest. The diffraction interferometers utilizing the tipped fiber as a source of the reference spherical wave look rather attractive. Owing to the simplicity of the design, high intensity of the diffracted wave, weak sensitivity to the adjustment in an optical circuit and a number of other advantages the interferometers which are simple in operation can be developed for the optical industry.

In the framework of the given article the problem, arising when aspherical surfaces are under investigation, of development and optimization of correctors transforming the reference spherical fronts into the aspherical ones which shapes are identical to the shapes of the studied surfaces was not practically discussed. Meanwhile, traditional lens null-correctors have limited capabilities, rather complicated in manufacturing and expensive and demand complicated calibrating procedures for taking into account the wave-front distortions due to errors of the corrector itself. In this connection diffraction correctors represent a significant interest (Akatov et al., 2004).

A gross potential for the surface shape correction of dielectric materials by means of ion-beams has RF-etching in chemically active environments. The first experiments on etching of quartz substrates by means of the RF-etching have shown that the surface micro-roughness did not change down to the removal depths of a few micrometers (Akhsakhalyan et al., 2008).

The most serious problem on the way of manufacturing the effective optics for the soft X-ray radiation, $\lambda<10$ nm, is that reflection coefficients of the multiplayer mirrors are not sufficiently high. Nevertheless, the results that have been recently obtained in the field of the X-ray optics teams may be considered as a break-through because they lay the foundation of developing such important physico-technological directions as projection EUV nanolithography, X-ray microscopy and extraterrestrial astronomy of the ultrahigh resolution.

10. References

Akhsakhalyan, A.D.; Vainer, Yu.A.; Volgunov, D.G.; Drozdov, M.N.; Kluenkov, E.B.; Kuznetcov, M.I.; Salashchenko, N.N.; Kharitonov, A.I. & Chkhalo, N.I. (2008). Application of reactive ion-beam etching for surface shape correction of X-ray mirrors, Proceedings of Workshop "X-ray optics - 2008", pp. 26-28, Chernogolovka, Moscow region, 6-9 October 2008, p. 26-28, IPTM, Chernogolovka. (in Russian)

Allen, L.N.; Keim, R.E. & Lewis, T.S. (1991). Surface error correction of a Keck 10 m telescope primary mirror segment by ion figuring, Proceedings SPIE, Vol. 1531, 1991, pp. 195–204.

Andreev, S.S.; Bulgakova, S.A.; Gaponov, S.V.; Gusev, S.A.; Zuev, S.Yu.; Kluenkov, E.B.; Luchin, V.I.; Lopatin, A.Ya.; Mazanova, L.M; Prokhorov, K.A.; Sadova, E.N.; Salashchenko, N.N. & Shamov, E.A. (2000). Investigations in field of projection lithography of extreme ultraviolet region in Institute for physics of microstructures RAS, Surface. X-ray, synchrotron and neutron investigations, Vol. 1, 2000, pp. 32-41. (in Russian)

Andreev, S.S.; Akhsakhalyan, A.D.; Bibishkin, M.S.; Chkhalo, N.I.; Gaponov, S.V.; Gusev, S.A.; Kluenkov, E.B.; Prokhorov, K.A.; Salashchenko, N.N.; Schafers, F. & Zuev,

S.Yu. (2003). Multilayer optics for XUV spectral region: technology fabrication and applications, Central European Journal of Physics, Vol. 1, 2003, pp. 191-209.

Adreev, S.S.; Kluenkov, E.B.; Misinov, A.L.; Polkovnikov, V.N.; Salashchenko, N.N.; Suslov, L.A. & Chernov, V.V. (2005). Compensation of elastic deformation of Mo/Si multilayer structures on quartz and zerodur substrates using buffer Cr/Sc technique, Surface investigations, Vol. 2, 2005, pp. 45-48. (In Russian)

Andreev, S.S.; Barysheva, M.M.; Chkhalo, N.I.; Gusev, S.A.; Pestov, A.E.; Polkovnikov, V.N.; Salashchenko, N.N.; Shmaenok, L. A., Vainer, Yu.A. & Zuev, S.Yu. (2009). Multilayered mirrors based on La/B_4C(B_9C) for x-ray range near anomalous dispersion of boron (λ near 6.7 nm), Nuclear Instruments and Methods in Physics Research Sec.A, Vol. 603, 2009, pp. 80-82.

Archer, K.M. (1997). Craniofacial reconstruction using hierarchical b-spline interpolation", A thesis submitted in partial fulfilment of the reguirements for the degree of master of applied science, The university of British Columbia, 1997, p. 113.

Asadchikov, V.E.; Kapabekov, A.Yu.; Kozhevnikov, I.V.; Levashov, V.E. & Sagitov, S.I. (1998). Experimental investigation of roughness correlation of a film and a substrate by means of X-ray scattering, Crystallography, Vol. 43, No. 1, 1998, pp. 119-130. (In Russian)

Barbee, T.W. (Jr.) (1981). Sputtered layered synthetic microstructures (LSM) dispersion elements, Proceedings of Conference "Low energy X-ray diagnostics", pp. 131-145, Monterey, USA, 1981, N. Y.: Amer. Inst. Phys.

Benschop, J.P.H.; Kaiser, W.M. & Ockwell, D.C. (1999). EUCLIDES, the European EUVL program, SPIE Symposium on Micro-lithography, Vol. 3676, 1999, p. 246.

Bibishkin[a], M.S.; Zabrodin, I.G.; Klyuenkov, E.B.; Salashchenko, N.N.; Chekhonadskih, D.P. & Chkhalo, N.I. (2003). A new dismountable tube for soft X-ray radiation, Surface investigations, No. 2, 2003, pp. 41-45. (In Russian)

Bibishkin[b], M.S.; Zuev, S.Yu.; Salashchenko, N.N. & Chkhalo, N.I. (2003). Measuring of microroughness of surfaces using soft X-ray radiation, Surface investigations, No. 1, 2003, pp. 94-96. (In Russian)

Bibishkin, M.S.; Chekhonadskih, D.P.; Chkhalo, N.I.; Klyuenkov, E.B.; Pestov, A.E.; Salashchenko, N.N.; Shmaenok, L.A.; Zabrodin, I.G. & Zuev, S.Yu. (2004). Laboratory methods for investigation of multilayer mirrors in Extreme Ultraviolet and Soft X-Ray region, Proceedings of SPIE, Vol. 5401, 2004, pp. 8-15.

Bibishkin, M.S.; Vainer, Yu.A.; Pestov, A.E.; Prokhorov, K.A.; Salashchenko, N.N.; Fraerman, A.A. & Chkhalo, N.I. (2005). Investigation of characteristics of multilayer X-ray mirrors with ultra-short periods d = 0,7-2,4 нм, Bulletin of the Russian Academy of Sciences: Physics, Vol. 69, No. 2, 2005, pp. 199-206.

Blunt, R. (2006). White light interferometry - a production worthy technique for measuring surface roughness on semiconductor wafers, Proceedings of CEMANTECH Conference, pp. 59-62, Vancouver, Canada, April 24-27, 2006.

Born, M. & Wolf, E. (1973). Principles of optics, second edition, Science, Moscow. (In Russian)

Braat, J. (1987). Polynomial expansion of severely aberrated wave fronts, Optical Society of America, Vol. 4, No. 4, 1987, pp. 643-650.

Chason, E. & Mayer, T. M. (1993). Low energy ion bombardment induced roughening and smoothing of Si02 surfaces, Applied Physics Letters, Vol. 62, No. 4, 1993, pp. 363-365.

Cheng, P. (1987). Instrumentation and Biological Applications, In X-ray Microscopy, Springer Verlag.

Chkhalo, N.I.; Dolbnya, I.G.; Fedorchenko, M.V.; Kruglyakov, E.P.; Zarodyshev, A.V. & Zolotarev, K.V. (1993). Multilayer mirror on copper-nickel substrates for intense X-ray beams. Proceedings of the 4-th International Conference "X-Ray Microscopy 4", pp. 586-592, Russia, Chernogolovka, September 20-24, 1993, IPTM, Chernogolovka.

Chkhalo, N.I.; Fedorchenko, M.V.; Kruglyakov, E.P.; Volokhov, A.I.; Baraboshkin, K.S.; Komarov, V.F.; Kostyakov, S.I. & Petrov, E.A. (1995). Ultradispersed diamond powders of detonation nature for polishing X-ray mirrors, Nuclear Instruments and Methods in Physics Research A, Vol. 359, 1995, pp. 155-156.

Chkhalo, N.I.; Kirpotin, A.N.; Kruglyakov, E.P. & Semenov, E.P. (1998). Reflectometer for precision tests of optical components in the ultrasoft X-ray range, Nuclear Instruments and Methods in Physics Research. A, Vol. 405, 1998, pp. 393-395.

Chkhalo[a], N.I.; Raskin, D.G. & Salashchenko, N.N. (2008). Capabilities of microinterferometer with digital recording of images for studying micro-objects with sub-nanometer resolution, Proceedings of SPIE, Vol. 7025, 2008, p. 702512.

Chkhalo[b], N.I.; Dorofeyev, I.A.; Salashchenko, N.N. & Toropov, M.N. (2008). A plane wave diffraction on a pin-hole in a film with a finite thickness and real electrodynamic properties, Proceedings of SPIE, Vol. 7025, 2008, p. 702506.

Chkhalo[c], N.I.; Klimov, A.Yu.; Rogov, V.V.; Salashchenko, N.N. & Toropov, M.N. (2008). A source of a reference spherical wave based on a single mode optical fiber with a narrowed exit aperture. Rev. Sci. Instrum., Vol. 79, 2008, p. 033107.

Chkhalo[d], N.; Paramonov, L.; Pestov, A. ; Raskin, D. & Salashchenko, N. (2008). Correction of the EUV mirror substrate shape by ion beam, Proceedings of SPIE , Vol. 7025, 2008, p. 702503.

Dinger, U.; Bisert, F.; Lasser, H.; Mayer, M.; Seifert, A. ; Seitz, G.; Stacklies, S.; Stickel, F.-J. & Weise, M. (2000). Mirror substrates for EUV-lithography: progress in metrology and optical fabrication technology, Proceedings of SPIE, Vol. 4146, 2000, pp. 35–46.

Dorofeyev, I.; Jersch, J. & Fuchs, H. (2003). Spectral composition of electromagnetic fluctuations induced by a lossy layered system, Annalen der Physik, Vol. 12, No. 7-8, 2003, pp. 421-437.

Dorofeev, I. (2009). Amplitude-phase characteristics of electromagnetic fields diffracted by a hole in a thin film with realistic optical properties, Physica E, Vol. 41, 2009, pp. 762-770.

Drozdov, M.N.; Drozdov, Yu.N.; Barysheva, M.M.; Polkovnikov, V.N. & Chkhalo, N.I. (2009). Minimization of matrix effects upon layer-by-layer analysis of multilayer metallic nanostructures La/B4C by SIMS using cluster secondary ions, Proceedings of Symposium "Nanophysics and nanoelectronics", Vol. 1, pp. 190-191, Nizhniy Novgorod Russia, 16-20 March 2009, IPM RAS, Nizhniy Novgorod. (in Russian)

Gavrilin, D.A. (2003). Investigations of methods for shape description of complicated optical surfaces at interferometrical control, Ph.D. thesis, St.-Petersburg, Russia, 103 p. (in Russian)

Ghigo, M.; Canestrari, R.; Spiga, D. & Novi, A. (2007). Correction of high spatial frequency errors on optical surfaces by means of Ion Beam Figuring, Proceedings of SPIE, Vol. 6671, 2007, p. 667114.

Goldberg, K.A. & Kevin, G. (2001). Wave-front measurement errors from restricted concentric subdomains, J. Opt. Soc. Am. A, Vol. 18, No. 9, 2001, pp.

Goldberg, K.A.; Naulleau, P.; Bokor, J.; Chapman, H.N. & Barty, A. (2002). Testing extreme ultraviolet optics with visible-light and extreme ultraviolet interferometry, J. Vac. Sci. Technol. B, Vol. 20, No. 6, 2002, pp. 2834-2839.

Griffith, J. E. & Grigg, D.A. (1993). Dimensional metrology with scanning probe microscopes, Journal of Applied Physics, Vol. 74, 1993, pp. R83-R109.

Gullikson, E.M.; Underwood, J.H.; Batson, P.C. and Nikitin, V. (1992). A soft X-ray/EUV reflectometer based on a laser produced plasma source, Journal of x-ray science and technology, Vol. 3, 1992, pp. 283-299.

Gwyn, C. (1998). White paper on Extreme ultraviolet Lithography, Proceedings of EUV LLC, 1998, Livermore.

Klimov, A.Yu.; Rogov, V.V.; Salashchenko, N.N. & Chkhalo, N.I. (2008). A Spherical Wave Source Based on a Probe for a Near-Field Microscope, Bulletin of the Russian Academy of Sciences: Physics, Vol. 72, No. 2, 2008, pp. 204-206.

Kluyenkov[a], E.B.; Pestov, A.E.; Polkovnikov, V.N.; Raskin, D.G.; Toropov, M.N.; Salashchenko, N.N. & Chkhalo, N.I. (2008). Testing and Correction of Optical Elements with Subnanometer Precision, Nanotechnologies in Russia,. Vol. 3, Nos. 9-10, 2008, pp. 602-610.

Klyuenkov[b], E.B.; Polkovnikov, V.N.; Salashchenko, N.N. & Chkhalo, N.I. (2008). Shape Correction of Optical Surfaces with Subnanometer Precision: Problems, Status, and Prospects, Bulletin of the Russian Academy of Sciences: Physics, Vol. 72, No. 2, 2008, pp. 188-191.

Kozhevnikov, I.V.; Asadchikov, V.E.; Duparre, A.; Gilev, O.N.; Havronin, N.A.; Krivonosov, Yu.S.; Ostashev, V.I. & Steinert , J. (1999). Comparative study of the roughness of optical surfaces and thin films using atomic force microscopy, x-ray scattering and light scattering methods, Proceedings of SPIE, Vol. 3739, 1999, pp. 348-354.

Kozhevnikov, I.V.; Bukreeva, I.N. & Ziegler, E. (2001). Nuclear Instruments and Methods in Physics Research Sec. A, Vol. 460, 2001, p. 424.

Linnik, V.P. (1933). A simple interferometer to test optical systems, Bulletin of the Academy of Science of the USSR, Vol. 1, 1933, pp. 210-212.

Loyen, L.; Bottger, T.; Braun, S.; Mai, H.; Leson, A.; Scholze, F.; Tummler, J.; Ulm, G.; Legall, H.; Nickels, P.V.; Sandner, W.; Stiel, H.; Rempel, C.; Schulze, M.; Brutscher, J.; Macco, F. & Mullender, S. (2003). A New Laboratory EUV Reflectometer for Large Optics using a Laser Plasma Source, Proceedings of. SPIE, Vol. 5038, 2003, pp. 12-21.

Lukirsky, A.P.; Brytov, I.A. & Komyak, N.I. (1967). Prototype of an X-ray spectrometer with diffraction grating and ionization detection for spectral region of 10-55 Å, Apparatus and methods of X-ray analysis, No. 2, 1967, pp. 4-19, L.: Engineering. (in Russian).

Malacara, D., ed. (1992). Optical shop testing, 2nd ed., John Wiley & Sons, Inc., ISBN 0-471-52232-5, New York.

Medecki, H.; Tejnil, E.; Goldberg, K.A. & Bokor, J. (1996). Phase-shifting point diffraction interferometer, Optics letters, Vol. 21, No. 19, 1996, pp. 1526-1528.

Miyake, A.; Miyachi, T.; Amemiya, M.; Hasegawa, T.; Ogushi, N.; Yamamoto, T.; Masaki, F. & Watanabe, Y. (2003). LPP-based reflectometer for characterization of EUV lithography systems, Proceedings of SPIE, Vol. 5037, 2003, pp. 647-655.

Murakami, K.; Saito, Y.; Ota, K. et al. (2003). Development of an experimental EUV interferometer for benchmarking several EUV wavefront metrology schemes, Proceedings of SPIE, Vol. 5037, 2003, pp. 75-82.

Murakami, K.; Oshino, T.; Kondo, H.; Chiba, H.; Komatsuda, H.; Nomura, K. & Iwata, H. (2007). Development of optics for EUV lithography tools, Proceedings of SPIE, Vol. 6517, 2007, pp. 65170J.

Naulleau, P.P.; Goldberg, K.A.; Lee, S.H.; Chang, C.; Attwood, D. & Bokor, J. (1999). Extreme-ultraviolet phase-shifting point-diffraction interferometer: a wave-front metrology tool with subangstrom reference-wave accuracy, Appl. Opt., Vol. 38, No. 35, 1999, pp. 7252-7263.

Naulleau, P.; Goldberg, K.A.; Lee, S.H.; Chang, Ch.; Atwood, D. & Bokor, J. (2000). The EUV phase-shifting point diffraction interferometer, Synchrotron Radiation Instrumentation: Eleventh US National Conference, ed. by P. Pianetta, p. 66-72, ISBN 1-56396-941-6, American Institute of Physics.

Naulleau P.; Golberg K.A.; Anderson E.H. et. al. (2002). Sub-70 nm Extreme Ultraviolet Lithography at the Advanced Light Source Static Microfield Exposure Station Using the Engineering Test Stand Set-2 Optic, Jour. Vac. Sci. Technol. B, Vol. 20, 2002, pp. 2829–2833.

Nevot, L. & Croce, P. (1980). Characterisation des surfaces par reflection rasante de rayon Application a l'etude du polissage de quelque verres silicates, Revue de Physique Appliquee, Vol. 15, 1980, pp. 761-779.

Okatov, M.A.; Antonov, E.A.; Baigozhin, A. et al. (2004). Handbook for optical-technologist, St- Petersburg: Politechnica, 2004, 679 p.

Oshino, T.; Shiraishi, M.; Kandaka, N.; Sugisaki, K.; Kondo, H.; Ota, K.; Mashima, K.; Murakami, K.; Oizumi, H.; Nishiyama, I. & Okazaki, S. (2003). Development of illumination optics and projection optics for high-NA EUV exposure tool (HiNA), Proceedings of SPIE, Vol. 5037, 2003, pp. 75-82.

Ota, K.; Murakami, K.; Kondo, H.; Oshino, T.; Sugisaki, K. & Komatsuda, H. (2001). Feasibility study of EUV scanners, Proceedings of SPIE, Vol. 4343, 2001, pp. 60-69.

Otaki, K.; Ota, K.; Nishiyama, I.; Yamamoto, T.; Fukuda, Y. & Okazaki, S. (2002). Development of the point diffraction interferometer for extreme ultraviolet lithography: Design, fabrication, and evaluation, J. Vac. Sci. Technol B, Vol. 20, No. 6, 2002, pp. 2449-2458.

Palik, E.D. (1985). Hanbook of optical constants of solids, Naval Research Laboratory, Washington D.C., 1985, 749 pp.

Parratt, L.G. (1954). Surface studies of solids by total reflection of x-rays, Physical Review, Vol. 35, 1954, pp. 359-369.

Platonov, Yu.; Gomez, L. & Broadway, D. (2002). Proceedings of SPIE, Vol. 4782, 2002, p. 152.

Protopopov, V.V.; Valiyev, K.A. & Imamov, R.M. (1999). Measurement of spatial distribution of superpolished surface roughness and defects in multilayer X-ray mirrors, Surface investigations, No. 1, 1999, pp. 111-119. (In Russian)

Puryaev, D.T. (1976). Methods for control of aspherical surfaces, M: Engineering, 1976, 13 p. (in Russian)

Rodionov, S.A.; Przhevalinsky, L.I. & Shekhonin, A.A. (1974). Application of coefficients of interpolation polynomial for optical system aberrations presentation, Bulletin of the USSR institutions of higher education: Instrument engineering, Vol. 17, No. 10, 1974, p. 104. (in Russian)

Rodionov, S.A. (1995). Mthematical apparatus for ptics control - general approach, Optical journal, No. 8, 1995, pp. 49-51.

Rowland, H. (1893). Gratings in theory and practice, Phil. Mag., Vol. 35, 1893, p. 397.

Salashchenko, N.N. & Shamov, E.A. (1997). Optics communication, Vol. 134, 1997, p. 7.

Salashchenko, N.N. (2001). Optical Element and Method for Recovering the Substrate, Patent DE 101 01 820 A1. 00074P.

Sinha, S.K.; Sirota, E.B.; Garoff, S. & Stanley, H.B. (1988). X-ray and neutron scattering from rough surfaces, Physical Rev. B., Vol .38, No.4, 1988, pp. 2297-2311.

Sommargren, G.E. (1996). Diffraction methods raise interferometer accuracy, Laser Focus World, Vol. 8, 1996, p. 61-71.

Soufli, R.; Spiller, E.; Schmidt, M.A.; Davidson, J.C.; Grabner, R.F.; Gullikson, E.M.; Kaufmann, B.B.; Mrowka, S.; Baker, S.L.; Chapman, H.N.; Hudyma, R.M.; Taylor, J.S.; Walton, Ch.C.; Montcalm, C. & Folta, J.A. (2001). Multilayer optics for an extreme ultraviolet lithography tool with 70 nm resolution, Proceedings of SPIE, Vol. 4343, 2001, pp. 51-59.

Stone, V.W.; Jonas, A.M.; Nysten, B. & Legras, R. (1999). Roughness of free surfaces of bulk amorphous polymers as studied by x-ray surface scattering and atomic force microscopy, Physical Rev. B., Vol. 60, No. 8, 1999, pp. 5883-5894 .

Sun, X.; Chen, S. & Shen, Q. (1998). Geometric calibration of lens using B-spline surface fitting, Wuhan University Journal of Natural Sciences, Vol. 3, No. 4, 1998, pp. 440-442.

Swantner, W. & Weng, W. C. (1994). Gram-Schmidt orthogonalisation of Zernike polynomials for general aperture shapes, Applied optics, Vol. 33, No. 10, 1994, pp. 1832-1837.

Sweeney, D.W.; Hudyma, R.M.; Chapman, H.N. & Shafer, D. (1998). EUV optical design for a 100 nm CD imaging system, Proceedings of SPIE, Vol. 3331, 1998, pp. 2-10.

Underwood, J.H. & Barbee (Jr.), T.W. (1981). Layered synthetic microstructures as Bragg diffractors for X-rays and extreme ultraviolet: theory and predicted performance, Appl. Opt., Vol. 20, 1981, pp. 3027-3034.

Vainer, Yu.A.; Pestov, A.E.; Prokhorov, K.A.; Salashchenko, N.N.; Fraerman, A.A.; Chernov, V.V. & Chkhalo, N.I. (2006). Analysis of Cross-Correlation of Interface Roughness in Multilayer Structures with Ultrashort Periods, Journal of Experimental and Theoretical Physics, Vol. 103, No. 3, 2006, pp. 346-353.

Warburton, W.K.; Rek, Z.U. & Barbee (Jr.), T.W. (1987). Quality Investigation of multilayer covers for X-ray optics, In X-ray optics and microscopy, pp. 222-231, World, Moscow. (In Russian)

Website ESRF. http://www.esrf.eu/events/conferences/past-conferences-and-workshops/pxrms06/Detailed-programm/12_gullikson.pdf

Website GO Corporation. http://www.goochandhousego.com/products/cat/3

Website Mcpherson Company. http://www.mcphersoninc.com/spectrometers/xuhvvuvuv/xuhvvuvuv.htm

Website Platar Company. http://www.orc.ru/~platar/P1R.html

Website Veeco Instruments. http://www.veeco.com

Website Zygo Corporation. http://www.zygo.com

Williamson, D.M. (1995). The elusive diffraction limit, OSA Proceedings on Extreme Ultraviolet Lithography, F. Zernike and D.T. Attwood eds., Vol. 23, 1995, pp. 68-76.

Windt, D.L.; Brown, W.L.; Volkert, C.A. & Waskiewicz, W.K. (1994). J. Appl. Phys., Vol 78, 1994, p. 2423.

Zoethout, E.; Sipos, G.; van de Kruijs, R.W.E.; Yakshin, A.E.; Louis, E.; Mullender, S. & Bijkerk, F. (2003). Stress mitigation in Mo/Si multilayers for EUV lithography, Proceedings of SPIE, Vol. 5037, 2003, pp. 872–878.

Zuev, S.Yu. & Mitrofanov, A.V. (2002). Measurements of parameters of optical elements for X-ray telescopes, Surface investigations, No. 1, 2002, pp. 81-83. (In Russian)

7

Advances in Resist Materials and Processing Technology: Photonic Devices Fabricated by Direct Lithography of Resist/Colloidal Nanocrystals Blend

Antonio Qualtieri[1], Tiziana Stomeo[2], Luigi Martiradonna[1], Roberto Cingolani[1,2] and Massimo De Vittorio[1,2]
Italian Institute of Technology (IIT), Center for Bio-Molecular Nanotechnology National Nanotechnology Lab. of CNR/INFM, Scuola superiore ISUFI Italy

1. Introduction

Lithographic techniques are the major driving force for the development of new photonic nanostructures. A wide range of techniques have been and are being developed for nano fabrication: e.g., deep UV (λ=200–290 nm) and extreme UV (λ <200 nm) photolithography, electron-beam lithography (EBL), focused ion beam (FIB) lithography, X-ray lithography, scanning probe lithography, and others.
The degree of accuracy, reproducibility and resolution guaranteed by lithographic techniques is definitely better than other nanofabrication processes, such as wet and dry etching, sputtering, evaporation and so on. For this reason, "all-lithographic" fabrication protocols are being proposed for the realization of photonic devices based on non-conventional materials such as organic fluorophores or wet-chemical synthesized colloidal nanocrystals (NCs), whose exploitation in nano-optoelectronics is more recent as compared with epitaxially grown semiconductor emitters.
A potential approach for the fabrication of novel nanophotonic and nanoelectronic devices without recurring to traditional semiconductor processing technologies is represented by direct lithography of polymeric resists embedding active materials. For this purpose, colloidal nanocrystals appear as a promising active medium to be embedded into the hosting polymer, since they provide, with a high photochemical stability, emission wavelengths tunable from ultraviolet to infrared spectral region, high fluorescence efficiency, broad excitation spectra and narrow emission bands also at room temperature. Direct lithography of resist/NCs blends has proved to be an efficient strategy for the nanopositioning of NCs while preserving both the optical properties of the active material and the lithographic sensitivity of the hosting resist. The possibility of selectively localize semiconductor NCs on a substrate can lead to the fabrication of high performing photonic devices with novel active materials significantly cheaper than the traditional epitaxially grown active layers. The chapter is devoted to the presentation of several demonstrators of this innovative technique. Moreover, a survey of the fabrication results achieved by

exploiting direct lithography of NCs/resist blends is reported, also discussing present limitations and future development of NC-based photonic devices.

2. Nanocrystals localization and device fabrication

As colloidal nanocrystals are dispersed in liquid solution, in order to be integrated in solid state devices they must be directly deposited on a substrate or embedded in a matrix. In the last years, some interesting results have been shown in literature about the deposition and the realization of nanocrystal-based devices.

Eisler et al. (2002) realized surface-emitting DFB laser structures by spin coating a NC/titania thin film on top of the grating previously realized by standard lithographic and dry etching processes. In this example, the matching between the Bragg condition of the grating and the ASE peak of the nanocrystal/titania film has been reached by modulating the grating periodicities and the volume fraction of nanocrystals by varying spinning speed. By using this strategy single wavelength DFB laser have been easily obtained, while multiple-wavelength lasers are harder to be constructed by simply spin-coating films on top of pre-formed gratings. In this sense, another strategy implemented by Sundar et al. (2004), consists in the typical "bottom-up" approach combining the processing ease afforded by semiconductor nanocrystals with soft lithography. The used scheme is described in Fig. 1. In particular, the NC films have been spin coated on a substrate and then embossed with a poly(dimethylsiloxane) (PDMS) elastomeric stamp. After thermal curing the successfulness of the technique has been verified by assessing the efficient diffraction of the light. Following a progressive assembly of multiple layers with well-matched gain at multiple wavelengths, these layers have been sequentially spin-coated and embossed. As a result, a multiple wavelength distributed feedback (DFB) laser oscillating simultaneously at room temperature has been obtained.

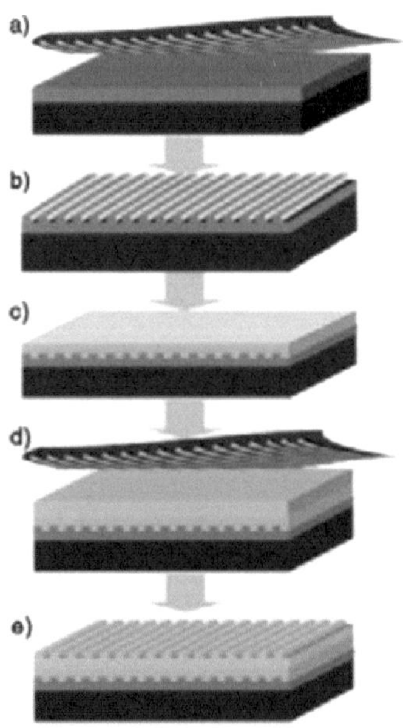

Fig. 1. Schematic illustration of the employed embossing strategy (Sundar et al., 2004).

A more efficient enhancement of the emission from colloidal nanocrystals has been obtained by Wu et al. (2007), by coupling NCs to an advanced photonic structure such as photonic crystal microcavities. In this device the deposition of PbSe quantum dots has been realized by dip-coating the patterned substrate in solutions of as-synthesized nanocrystals (Fig. 2). This strategy allows a significant enhancement of spontaneous emission strictly connected to the microcavity design but it is not able to control the positioning of the nanoemitters on the patterned substrate.

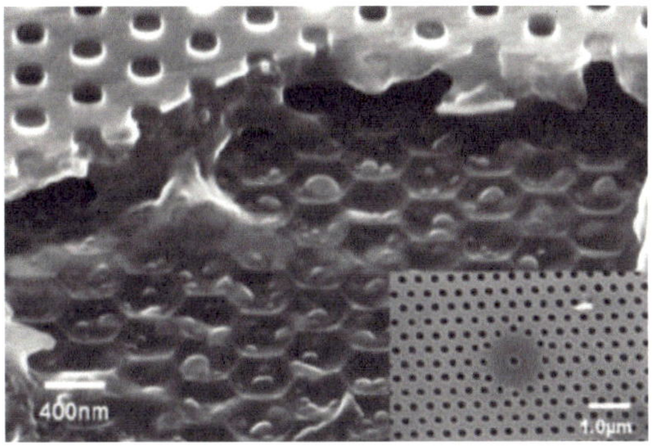

Fig. 2. Scanning electron microscope (SEM) image of a cleaved section of the two-dimensional photonic crystal microcavity showing PbSe quantum dots attached inside. The inset represents a SEM image in plane of the photonic crystal microcavity (Wu et al., 2007).

A better control on the micrometer scale has been achieved by Jun et al. (2006), with the exposure of a nanocrystals film to an UV source for the integration of nanocrystals in light-emitting devices. The direct lithography of nanocrystals films allowed the oleic acid ligands on the surface of the nanocrystals to form an insoluble cross-linked network while the unexposed areas were still soluble to toluene solvent (Fig. 3).

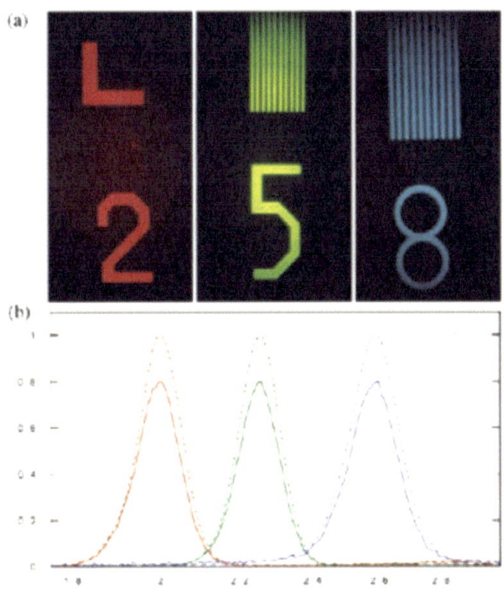

Fig. 3. (a) fluorescence micrograph and (b) PL spectra of the nanocrystal patterns (Jun et al., 2006)

Although, the UV light exposure through a shadow mask, could represent an appealing approach to the positioning of nanocrystals, it is limited due to the presence of a bottom layer and to the non guiding properties of the patterned structures.
These examples, even if not exhaustive, exploit the peculiar properties of the nanocrystals but all suffer from some deficiencies: i.e. they do not provide precise positioning and control on the lateral confinement of the waveguiding structure. Moreover, they require specific adherence properties as a function of the substrates and present low throughput.
The dispersion of these quantum dots into polymeric matrices proved to be a more powerful strategy: as it will be described in the next section, the chemical stability of semiconductor NCs together with their easy functionalization and the optical and electrical characteristics of organic materials lead to a new class of composite materials which better preserve the emission of quantum dots and show a good degree of processability (Fushman et al., 2005; Dick et al., 2004).

3. Nanocrystals-based devices fabricated by direct lithography

3.1 Proof of the technique

To localize colloidal quantum dots by lithographic techniques, the affinity between the nanocrystals and the most common resists exploited in nanolithography has to be proved. Martiradonna et al. (2006a, 2006b) tested for their experiment Polymethylmethacrylate (PMMA) and SU-8. The first is a positive e-beam resist, while the second is a negative resist, both sensitive to e-beam and UV radiation respectively. Moreover, core/shell CdSe/ZnS nanocrystals have been prepared and diluted in chlorobenzene (C_6H_5Cl) using standard methods (Reiss et al., 2002). The core/shell configuration has been preferred due to the higher chemical stability and quantum efficiency. The semiconductor clusters with different size have been dispersed with different molar concentration in PMMA-950K and SU-8 in order to understand the influence of the concentration and size of NCs on the lithographic properties of the blends. Then, a test pattern composed by a two-dimensional triangular lattice of pillars with sub-micron diameter has been defined in the PMMA layer by an e-beam lithography process. The contrast curve of the blend by testing several doses has been measured, in order to derive the sensitivity of the composite material to the beam. By means of an UV exposure, stripes ranging from 2 μm to 100 μm have been patterned on the SU-8 layer. In order to validate the technique, the optical properties of colloidal nanocrystals before and after the lithographic processes have been verified by collecting photoluminescence (PL) spectra at room temperature on the test samples through a confocal laser scanning microscope, with in-plane spatial resolution of 200 nm. A laser diode emitting at λ_{ex} = 405 nm has been used as the excitation source.
As it can be observed by the two PL maps (shown in false colours in Fig. 4a and 4b), no emission has been detected from the regions of the blends which have been removed after development (black areas), thus confirming the removal of nanocrystals together with soluble resist. Moreover, by comparing the emission of the localized quantum dots (red and green lines in Fig. 4c) with the original photoluminescence spectrum (black line), no significant changes have been observed. This means that, thanks to the optical stability of these core/shell emitters, no relevant structural modifications or degradation of the optical properties are detected, also after the exposure to accelerated electrons or ultraviolet radiation.

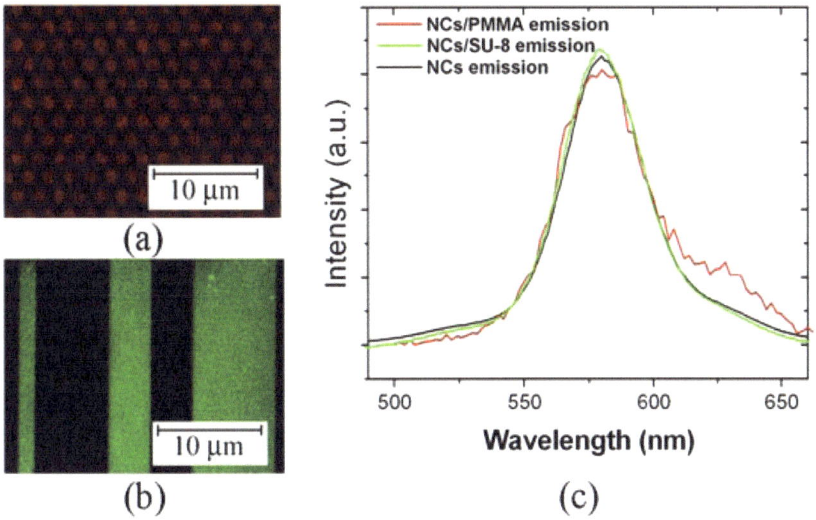

Fig. 4. (a) PL map (in false colours) of PMMA/NCs pillars after the e-beam lithographic process; (b) PL map (in false colours) of SU- 8/NCs stripes after the photolithographic process; (c) Emission spectra detected on the patterned PMMA/NCs pillars (red line) and on the patterned SU-8/NCs stripes (green line), as compared to the emission of pure NCs in liquid solution (black line) (arranged from Martiradonna et al., 2006a, 2006b).

In the following subsections, this technique has been exploited to demonstrate the successful fabrication of the main building blocks of photonic systems such as: multicoloured devices, waveguide structures, suspended stripes and photonic crystal nanocavities.

3.2 Multicoloured devices

The fabrication of multicolored micro- and nano- displays by realigned photolithographic steps or mix-and-match electron beam and photolithographic approaches has been recently demonstrated (Qualtieri et al., 2009a). Fully-customable micro- and nano- patterns can be obtained by a single lithographic procedure. Importantly, the possibility to re-align subsequent lithographic steps on the same substrate (as schematically shown in Fig. 4) enables the localization of different NCs ensembles in confining or superimposed regions. Thus, this technique has been exploited for the realization of RGB pixels on the micrometer scale, by localizing red, green and blue emitting nanocrystals in contiguous regions. The overall photoluminescence of each pixel results in the sum of the intensities of these chromatic components. By controlling the molar concentrations of the NCs and by varying the geometrical parameters of the localized patterns, the authors can finely tune the PL spectrum of these pixels thus obtaining multicoloured and white micro and nanoemitters.
In order to demonstrate the potential of this technique, the authors have reported the multicoloured logo of their affiliation by localizing two different ensembles of NCs. A first logo has been realized by exploiting two photolithographic steps: red-emitting nanocrystals dispersed in SU-8 (M_{NCs}=7.2·10^{-6} mol/l), have been spin-coated on a Silicon substrate and exposed to ultraviolet radiation; after development and thermal curing, a second layer of green emitting NCs dispersed in SU-8 has been deposited on the sample and exposed. The sample has been finally developed and optically characterized by means of a confocal

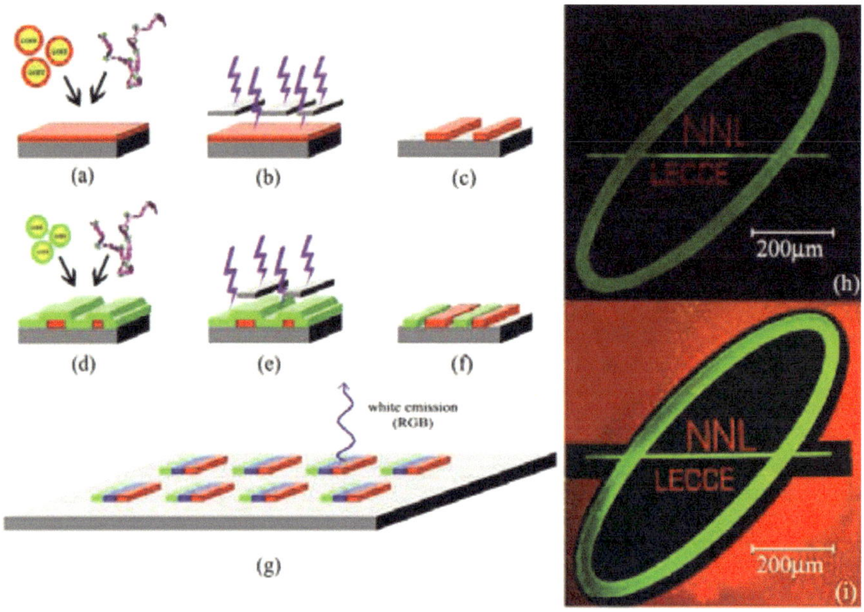

Fig. 4. Scheme of a realignment process for the localization of different ensembles of colloidal nanocrystals on the same substrate with lithographic techniques (Qualtieri et al. 2009a).

microscope. The combined fluorescence map, collected in two separate spectral ranges and shown in Fig. 4(h), demonstrates that no mixing between the two NCs ensembles can be detected, showing that the localization approach is not affected by the two sequential steps of the lithographic process.

Similarly, a second logo has been realized in mix-and-match lithography: a layer of PMMA containing red-emitting nanocrystals has been patterned through an e-beam lithography process, in order to selectively remove the red emitters from the exposed area. After development, a second layer of green/SU-8 blend has been spin-coated on the sample and patterned through photolithography. The photoluminescence map of the processed logo is reported in Fig. 4(i): the solvents used for PMMA and SU-8 polymers (chlorobenzene and cyclopentanone, respectively) has been verified being orthogonal (i.e. the already deposited layers are insoluble in the subsequent solvents), therefore well defined and distinguished colored regions are detected at the end of the process.

Furthermore, the authors also verified the possibility to combine the emission of different NCs ensembles by performing localization on superimposed regions. The CIE (Commission Internationale de l'Eclairage) coordinates of the emitted light have been calculated and has been found that the emission detected from superimposed patterns is a linear combination of red and green NCs emission. By increasing the green-to-red molar concentration ratio per unit area, the corresponding CIE coordinates shift from the red towards the green region, turning to a yellowish colour.

In this way, the possibility to realize multicoloured pattern by exploiting precise localization of colloidal nanocrystals dispersed in polymeric blends sensitive to electron beam or ultraviolet radiation has been demonstrated. This technology has been actually verified with red and green emitting quantum dots, but it can be also extended to blue emitters. In this way, the full RGB colour range can be covered by exploiting just three NCs ensembles, and white emitting pixels can be easily obtained.

Moreover, the possibility to combine this technology with electrical injection already proposed for quantum dots based single-coloured LEDs (Sun et al., 2007), will allows the realization of bright, stable and cheap high resolution multicolour displays, for the next generation of video systems.

3.3 Waveguide photonic devices based on colloidal nanocrystal emitters

The top-down approach based on the dispersion of CdSe/ZnS core/shell NCs into a layer of electro-sensitive resist PMMA, has been exploited for the fabrication of distributed feedback (DFB) structures and distributed Bragg reflectors (DBR) suitable for the realization of in-plane waveguide lasers (Martiradonna et al., 2007).

To this aim, nanocrystals emitting at 640 nm have been dispersed in PMMA-950K positive resist with a molar concentration of $M_{NCs} = 2.51\text{x}\ 10^{-5}$ mol/l. This high loading fraction of the nanoemitters increases the refractive index value of PMMA from 1.49 to 1.62 at λ = 640 nm, thus allowing the index confinement of visible light into PMMA/NCs waveguides fabricated on glass substrates.

Two realigned lithographic steps have been used and a single development of the resist has been carried out after both the lithographic steps with different beam parameters, without recurring to following etching processes. In particular, the optimized electron beam lithography (EBL) parameters have been used on a PMMA/NCs film in order to pattern a 100 µm long, 30 µm wide active ridge cavity and two distributed Bragg reflectors composed by alternating $3\lambda/4$ thick stripes of PMMA/NCs and air. A 20 pairs back DBR and a 10 pairs front DBR have been obtained by exposing 480 nm wide stripes having a period of 780 nm.

The second e-beam lithographic process has been used to define, by a shallow removal of the blend, a DFB on the waveguide by creating a periodical corrugation on the top of the waveguide. A single development step of the resist has been carried out after both the two lithographic steps with different beam parameters.

The active blend has been removed from the sides of the ridge cavity (dark lateral regions) and from the stripes of the DBR (as shown in Fig. 5).

Fig. 5. SEM image of the ridge cavity obtained by localizing the PMMA/NCs by means of EBL processes (arranged from Martiradonna et al 2007).

The main advantage of this all-lithographic localization approach is the extreme versatility, which allows the implementation of mostly any geometry required by the laser designer, with resolutions pushed down to the nanometre scale. Moreover, also the optical properties (optical gain, emission wavelength, refractive index) of the active material can be arbitrarily tuned by simply varying the size and the density of the emitters dispersed in the resist. The possibility to re-align multiple lithographic steps on the same substrate allows also the localization of different active blends on near or even superimposed regions, thus realizing multi-wavelength laser sources on a single device. Importantly, since this technique can be applied to any kind of substrate, also the fabrication of DFB lasers on flexible surfaces can be enabled. As a straightforward application, optical sensors detecting strain-induced variations of the DFB corrugation period, and therefore of the lasing wavelength, could be obtained.

3.4 Colloidal nanocrystals air bridge fabricated by direct lithography

Membrane structures, also called air bridges, are suspended structures completely surrounded by air. Since the bridge is completely surrounded by low index claddings (either air or low-index materials) a strong light confinement and efficient waveguiding can be obtained. Passive suspended waveguides and Photonic Crystals based suspended filters (Carlsson et al., 2002), able to model the transmitted light, have been fabricated. Nanolasers coupling 2D PhC microcavities on semiconductor bridges with active InGaAsP quantum wells (Loncar et al. 2004) are also being developed both as infrared light sources and for optical sensors. In this sense, Qualtieri et al., (2007) demonstrated that the lithography based localization approach proved to be effective in the fabrication of 3D geometries like NCs air bridges by cyclically depositing and exposing different blends of nanoemitters/resists on the same substrate, with no recursion to following etching steps.

To this aim, two different ensembles of core/shell CdSe/ZnS nanocrystals, emitting in the red and green region, have been prepared and dispersed in a SU-8 negative photoresist matrix. Stripes with varying spacing from 6 μm up to 120 μm have been localized by exposing the red active film to the ultraviolet radiation of a KARL-SUSS MJB3 mask aligner (Hg i-line at λ = 365 nm) with an exposure dose of 0.87 J. A post exposure bake (PEB) has been performed on the prototype, but it has not been developed in order to preserve its surface planarity. On this sample a second 1500 nm thick layer of green-emitting NCs has been spin-coated and thermal cured. A second array of perpendicular stripes with varying widths (2 μm - 20 μm) has been localized by means of an UV exposure (exposure dose = 0.87 J). After the PEB, the sample has been developed into SU-8 developer for 1 min. SEM measurements performed on the fabricated prototype tilted at 45° assessed the obtained 3D structures (Fig. 6a).

Moreover, photoluminescence images collected by means of a confocal laser microscope reveal the green and red emitting regions on the sample (Fig. 6(b)) . Red PL collected in the 630–670 nm spectral range shows a bright emission only from the first layer, while green PL (collected in the 530–570 nm spectral range) is detected only from the suspended bridges.

Fig. 6(c) compares the emission spectra coming from a stripe of the first layer (circle line), from a stripe of the second layer (triangle line) and from an unpatterned zone (solid line). As it is evident from the latter spectrum, the localization of both the ensembles of NCs is perfectly controlled with this lithographic approach since in unexposed regions of both layers no PL signal is detected. Thanks to the optical transparency of SU-8 at in the whole

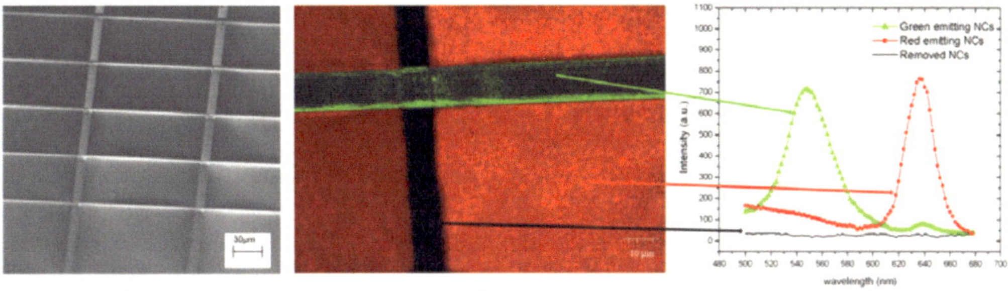

(a) (b) (c)

Fig. 6. a) SEM image of the fabricated air bridge structure; b) Superposition of two (PL) maps of the fabricated sample, collected in the 530 - 570 nm spectral range (green) and in the 630 - 670 nm spectral range (red); c) emission spectra coming from the regions in which only red NCs are localized (circle line), only green NCs are localized (triangle line) and both the blends are removed (square line) (arranged from Qualtieri et al., 2007).

visible spectral range, the red emission coming from the underlying layer is detected also through the green emitting layer (triangle line). The PL increase, shown in the blue range (400 - 500 nm) both in circle and triangle lines, is due to a slight emission coming from the polymeric matrix after its exposure to UV light.
Active 3D photonic crystals and woodpile structures can be implemented in the whole visible spectral range, but also in the ultraviolet and infrared range by properly tuning the optical properties of the NCs dispersed in the blend.

3.5 Active photonic crystal structures fabricated on suspended membranes

The insertion of nanoemitters into photonic crystal (PhC) cavities with localized optical modes having high quality factor (Q) and small modal volume has been exploited to tailor both the spatial and spectral properties of the NCs radiation. In fact, colloidal nanocrystals confined into two-dimensional PhCs are promising candidates for the development of next generation high performing optical devices such as single photon sources, ultralow threshold lasers, and nonlinear devices. At present, several works have been proposed in order to couple colloidal NCs to 2D-PC structures fabricated on Si or AlGaAs membrane layers (Wu et al., 2007; Fushman et al., 2005). In these reported cases, high Q passive nanocavities have been previously fabricated on solid substrates then PbS or PbSe colloidal nanocrystals emitting in the infrared spectral range have been added by dipping or spin-coating techniques on top of the patterned layer. The coupling between the emitters and the quantum-confined optical environment has been demonstrated by the presence of resonant peaks in the photoluminescence spectra collected on the fabricated devices. This kind of approach requires a pattern transfer on the rigid membrane layer by high-resolution lithographic techniques and etching steps, which could introduce fabrication imperfections in the structure i.e. deviation of the hole radii, slightly angled sidewalls, high surface roughness after resist stripping. Moreover, these techniques can be only applied to infrared-emitting colloidal nanocrystals, due to high absorption of Si and AlGaAs in the visible spectral range. Direct lithography of NCs/blends allows the fabrication of semiconductor nanocrystal-based 2D-PhC nanocavities, without requiring pattern transfer from the resist to the underlying membrane layers, since the resist itself acts as the suspended waveguiding layer. The optical quality of the resonating structure is only determined by a single high-resolution lithographic step, being not affected by any etching process. Moreover, the absence of absorbing layers from the visible to the infrared region of the spectrum allows the use of NCs on a large spectral range, thus conferring a broader applicability to this technique. Nevertheless, in this approach, a new type of semiconductor nanoemitters, which are dot-in-a-rod colloidal nanocrystals (DR-NCs) consisting of CdS rods nucleated at high temperatures around CdSe spherical seeds has been exploited (Carbone et al., 2007; Talapin et al., 2007). These elongated core-shell nanocrystals show remarkably high quantum efficiency (up to 70-75%), due to a high quantum yield coupled with a large extinction coefficient in the ultraviolet spectral range, the latter due to the presence of a thick CdS shell. By virtue of their high efficiency and photostability DR-NCs PL emission is easily detectable even at room temperature with no need of highly sensitive detection setup. Due to the low index contrast between the membrane and the surrounding air cladding, in order to increase the mode confinement, an approach similar to the graded square lattice proposed by Srinivasan & Painter (2002) has been followed by Martiradonna et al., (2008) . For the practical realization of the device, a 1 μm thick layer of $Al_{0.7}Ga_{0.3}As$ has been epitaxially grown by metal-organic chemical vapor deposition (MOCVD) on GaAs substrate and a 320 nm thick layer of ZEP/NCs has been spin coated on top of it. Fig 7a shows an in-plane SEM image of the patterned structure after development. In order to remove the underlying sacrificial layer, the sample has been then dipped in a HF:H_2O etching solution at room temperature. It has been then rinsed with isopropyl alcohol with the aim to decrease the surface tension of the solvent during its evaporation in air. A sketch of the final structure after wet-etching is shown in Fig. 7b.

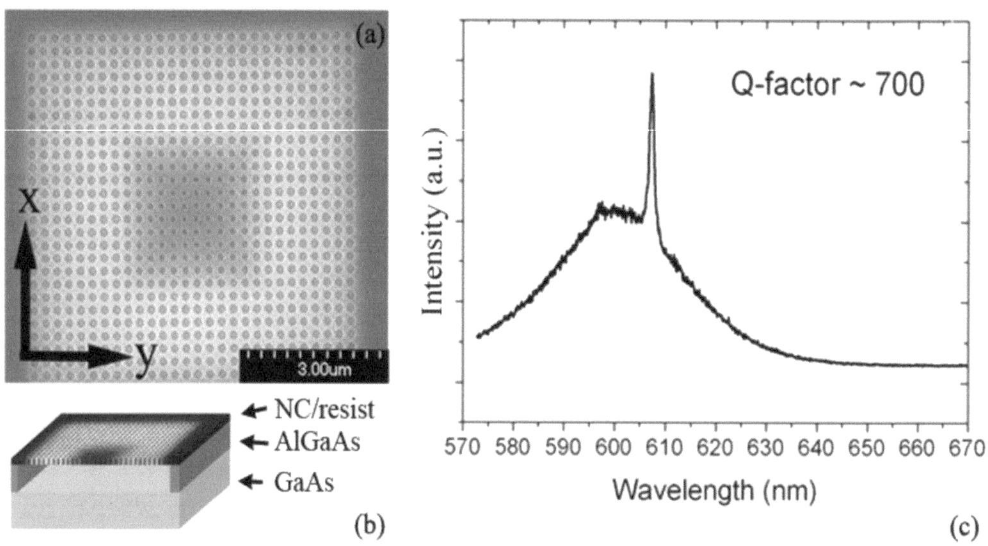

Fig. 7. a) and b) Scanning electron microscope image of the patterned nanocavity and sketch of the final membrane structure, (c) Output spectra of the nanocavity embedding nanoemitters (arranged from Martiradonna et a., 2008)

In order to verify the effectiveness of the fabrication process, micro-photoluminescence measurements at room temperature have been performed on the nanocavities and a Q-factor of 700 has been found (Fig. 7c). This new strategy for the fabrication of high-performing NCs-based devices is very promising for the development of all-optical circuits that can be fully integrated with traditional Silicon technology.

3.6 Single colloidal nanocrystals in microcavity

Secure Quantum Communication Systems (QCS) based on the transmission of crucial information through single photons are among the most appealing frontiers for telecommunications, though their development is still hindered by the lack of cheap and bright Single Photon Sources (SPS) operating at room temperature. In this section we describe a high resolution direct lithographic technique for obtaining photon antibunching behavior at room temperature from single colloidal CdSe/ZnS nanocrystals (NCs) localized in a 1D-PhC microcavity (Qualtieri at al. 2009b).

In order to verify the possibility to localize single nanocrystals by means of electron beam lithography, a dielectric structure, consisting of a SiO_2/TiO_2-based distributed Bragg reflector (DBR) and an additional SiO_2 spacing layer, has been preliminary fabricated. A very low concentration of CdSe/ZnS colloidal nanocrystals emitting at $\lambda = 610$ nm has been dispersed in a high-resolution electron-beam resist, namely Hydrogen Silsesquioxane (HSQ) (Namatsu et al., 1998). A 90 nm thick layer of the blend has been spin-coated on the dielectric structure and subsequently exposed to the electron beam to define an array of active pillars with diameters ranging from 500 nm down to 30 nm. The emission spectrum of the array of nanopillars has been collected by means of a confocal microscope at room temperature. Remarkably, the array of nanoemitters showed different optical properties depending on the different dimensions of the pillars. Structures having a diameter above 100 nm showed a fluorescence typical of an ensemble of nanocrystals with a FWHM of about 30nm (Fig.8a, black line), whilst a narrower emission (FWHM ~ 19 nm) has been collected from the smallest ones (Fig. 8a, blue line). The lorentzian shape of the latter PL spectrum, together with the typical blinking emission in time (Neuhauser et al., 2000), indicate the single photon nature of the source. The clear fingerprint of single photon emission, represented by the reduced probability of two photons being detected within a short time interval and namely the photon antibunching behaviour (Messin et al., 2001) of the luminescence, has been observed by analyzing the auto-correlation function of the emitted signal collected by means of a standard Hanbury-Brown and Twiss setup (Fig.8b).

Having successfully accomplished the control of the localization of single quantum emitters by the high resolution electron beam lithography (EBL) based approach, a complete planar microcavity as sketched in Fig. 9a (the detail of the cavity region is shown in Fig. 9b), cladding the array of pillars between two SiO2/TiO2 DBRs has been realized. The microcavity has been designed to resonate at a wavelength of 610 nm.

Micro-photoluminescence analyses has been performed and the Fig. 10a shows a representative PL spectrum detected from a 90 nm-wide pillar. Its line-width is further reduced down to 12.3 nm with respect to the NCs ensemble luminescence in the cavity, since the presence of a single semiconductor emitter in the pillar causes lower absorption and scattering losses. The evidence of single photon generation is given, also in this case, by the measurement of the autocorrelation function of the emitted light, which reveals an antibunching behaviour at zero delay (see Fig. 10b).

The area of the peak at zero delay is equal to 0.27. After subtracting the signal corresponding to the background, this area is even lower with a value of 0.21. The normalized area of the peaks at nonzero delays goes up to 1.35, reflecting an additional bunching of the

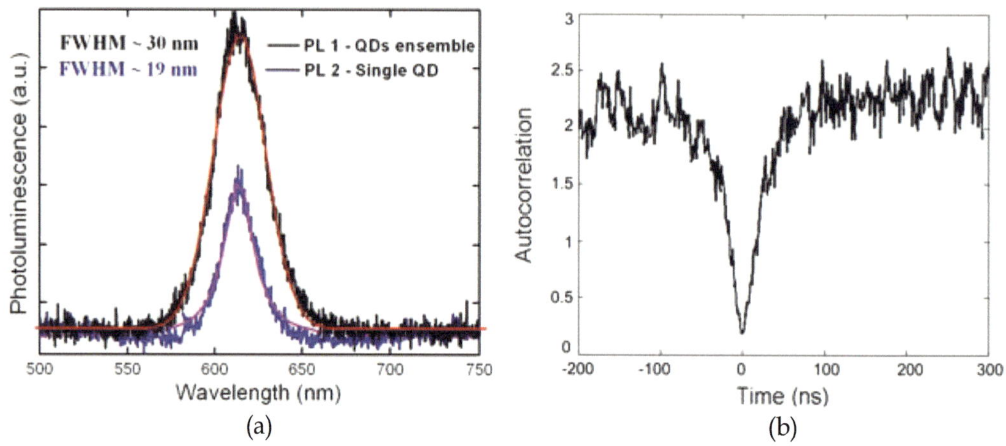

Fig. 8. (a) photoluminescence spectra showing the different line-shape depending on the different dimensions of the pillars (black line collected from a 500 nm wide pillar and blue line collected from a 90 nm wide pillar); (b) auto-correlation trace of the PL arising from a 90 nm wide pillar (normalized with background correction), showing an antibunching behaviour (arranged from Qualtieri et al., 2009b).

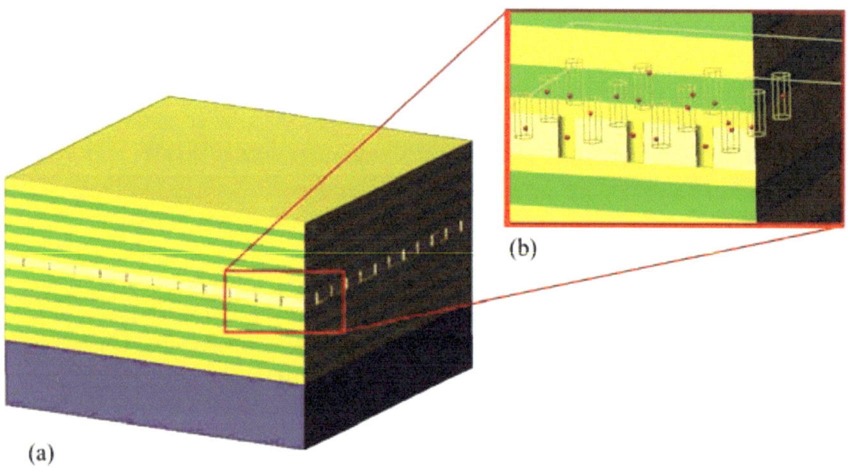

Fig. 9. (a) and (b): 3D- sketches of the whole structure and of the cavity region, respectively (arranged from Qualtieri et al., 2009b).

fluorescence due to the emission intermittency, as explained by Messin et al., (2001). Using the histogram of the coincidences counts, the lifetime of the excited state has been also deduced. The exponential fit of the experimental data collected from several SPSs gives an average radiative lifetime of $\tau_{cav} \sim 9.1$ ns that compared to the decay lifetime of an ensemble of nanocrystals surrounded by HSQ not inserted in a cavity ($\tau_{fs} \sim 23$ ns) showed a shortening of the radiative lifetime due to the wavelength-sized cavity.

The unprecedented photon antibunching behaviour observed from colloidal nanocrystals embedded in a planar microcavity, together with an efficient tailoring of the recombination decay rate at room temperature, confirms the suitability of this approach for the fabrication of SPSs based on colloidal nanocrystals, thus overcoming the limitation in the operating temperature shown by epitaxial quantum dot sources.

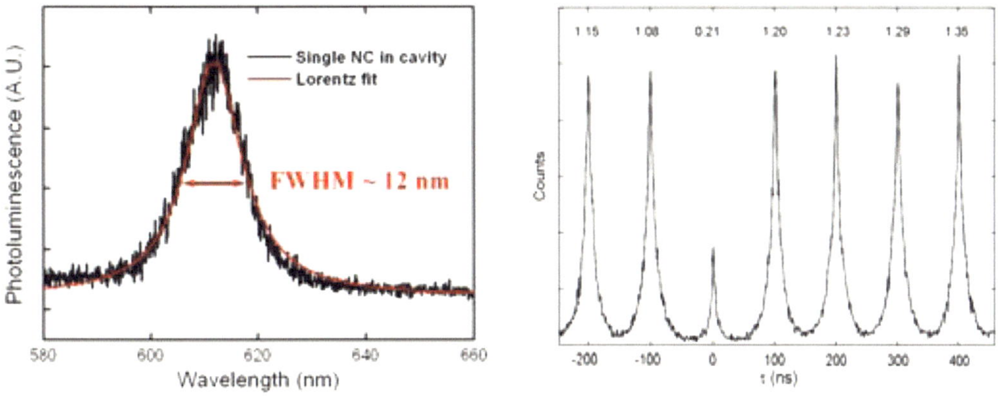

(a) (b)

Fig. 10. (a) Room-temperature photoluminescence of a 90-nm wide pillar. (b) Histogram of coincidence counts of a single QD fluorescence. Above each peak the area normalized to Poissonian photon statistics, with background correction, is given (arranged from Qualtieri et al., 2009b).

4. Conclusion

In this chapter, a novel approach for the direct lithographic fabrication of photonic devices based on colloidal quantum dots has been proposed. Different examples of nanocrystals localization and devices fabrication have been reported in section 2. They exploit the unique optical and spectroscopic properties of colloidal nanocrystals, but all of them suffer from some deficiency: they do not provide precise positioning and control on the lateral confinement of the waveguiding structure. Moreover, they require specific adherence properties in function of the substrates and present low throughput. On the contrary, the precise localization of colloidal NCs through the nanopatterning of active blends, without recurring to any preparation of the substrate, leads to broader applications in the nanophotonic field.

Successful examples of the fabrication approach able to manipulate blends composed of polymeric matrices and colloidal nanocrystals by exploiting peculiar characteristics of the hosting material have been reported:

- multicolour emitting devices obtained by the localization in nearby regions of different PMMA/QDs and SU8/QDs blends, in re-aligned e-beam and photo-lithography steps.;
- PMMA/QDs blend-based in-plane waveguide devices with distributed Bragg reflectors and distributed feed-back patterns, fabricated by the subsequent exposures of the sample to e-beam lithography processes with different acceleration voltages;
- suspended air-bridge structures suitable for the fabrication of highly optical confined active waveguides. The release of the suspended structure has been performed in the development of the exposed sample, with no need of following chemical etching steps.

Moreover, it allowed to fabricate advanced photonic devices based on colloidal nanocrystal for integrated all-optical systems and information processing.

In particular, by exploiting both Photonic Crystals and NCs properties two different devices have been realized:

- 2D-PhC resist nanocavity, with a Q-factor of ~700, aiming to the realization of high-efficiency ultra-small lasers and non-classical light sources in the visible range;
- Single photon source (SPS) device based on 1D-PhC embedding colloidal nanocrystals with an enhancement of the radiative decay of ~2.4 (as expected by theoretical prediction). The precise nanopatterning has been essential to define, by electron beam lithography, the nanometre feature size of the NCs/HSQ blend for SPS. This single photon device demonstrated for the first time antibunching behaviour of a single NC in microcavity at room temperature.

The presence of a controlled concentration of NCs in the resists does not modify their peculiar behaviour thus enabling the fabrication of 1D-2D and 3D active photonic crystals devices without recurring to etching processes.
As a consequence, the dispersion of wet chemically synthesized quantum dot in hosting materials sensitive to photo or electron beam radiation reveals to be a very promising and appealing strategy for the fabrication of innovative photonic devices, fully integrated with traditional Silicon technology.

5. References

Carbone, L.; Nobile, C.; De Giorgi, M.; Della Sala, F.; Morello, G.; Pompa, P. P.; Hytch, M.; Snoeck, E.; Fiore, A.; Franchini, I. R.; Nadasan, M.; Silvestre, A. F.; Chiodo, L.; Kudera, S.; Cingolani, R.; Krahne, R. & Manna, L. (2007). Synthesis and Micrometer-Scale Assembly of Colloidal CdSe/CdS Nanorods Prepared by a Seeded Growth Approach. *Nano Letters*, Vol. 7, No. 10 (September 2007) 2942-2950, ISSN 1530-6992

Carlsson, N.; Ikeda, N.; Sugimoto, Y.; Asakawa, K.; Takemori, T.; Katayama, Y.; Kawai, N. & Inoue, K. (2002). Design, nano-fabrication and analysis of near-infrared 2D photonic crystal air-bridge structures. *Optical and Quantum Electronics*, Vol. 34 (January 2002) 123-131, ISSN 1572-817X

Dick, K. A.; Deppert, K.; Larsson, M. W.; Martensson, T.; Seifert, W.; Wallenberg, L. R. & Samuelson, L. (2004). Synthesis of branched "nanotrees" by controlled seeding of multiple branching events. *Nature Materials*, Vol. 3 (May 2004) 380-384, ISSN 1476-1122

Eisler, H. J.; Sundar, V. C.; Bawendi, M. G.; Walsh, M.; Smith, H. I. & Klimov, V. (2002). Color-selective semiconductor nanocrystal laser. *Applied Physics Letters*, Vol. 80 (June 2002) 4614-4616, ISSN 1077-3118

Fushman, I.; Englund, D. & Vuckovic, J. (2005). Coupling of PbS quantum dots to photonic crystal cavities at room temperature. *Applied Physics Letters*, Vol. 87 (December 2005) 241102-1 - 241102-3, ISSN 1077-3118

Jun, S.; Jang, E.; Park, J. & Kim, J. (2006). Photopatterned Semiconductor Nanocrystals and Their Electroluminescence from Hybrid Light-Emitting Devices. *Langmuir*, Vol. 22, No. 6 (February 2006) 2407-2410, ISSN 1520-5827

Loncar, M.; Yoshie, T.; Okamoto, K.; Qiu, Y.; Vuckovic, J. & Scherer, A. (2004). Planar Photonic Crystal Nanolasers (I): Porous Cavity Lasers. *IEICE Transaction on Electronics*. Vol. E87-C, No. 3 (March 2004) 291-299, ISSN 0916-8516

Martiradonna, L.; Stomeo, T.; De Giorgi, M.; Cingolani, R. & De Vittorio, M. (2006a). Nanopatterning of colloidal nanocrystals emitters dispersed in a PMMA matrix by e-beam lithography. *Microelectronic Engineering*, Vol. 83 (February 2006) 1478–1481, ISSN 0167-9317

Martiradonna, L.; Stomeo, T.; Carbone, L.; Morello, G.; Salhi, A.; De Giorgi, M.; Cingolani, R. & De Vittorio, M. (2006b). Nanopositioning of colloidal nanocrystal emitters by means of photolithography and e-beam lithography. *Physica Status Solidi (b)*, Vol. 243 No. 15 (December 2006) 3972-3975, ISSN 1521-3951

Martiradonna, L.; Qualtieri, A.; Stomeo, T.; Carbone, L.; Cingolani, R. & De Vittorio, M. (2007). Lithographic nano-patterning of colloidal nanocrystal emitters for the fabrication of waveguide photonic devices. *Sensors and Actuators B: Chemical*, Vol. 126, No. 1 (September 2007) 116-119, ISSN 0925-4005

Martiradonna, L.; Carbone, L.; Tandaechanurat, A.; Kitamura, M.; Iwamoto, S.; Manna, L.; De Vittorio, M.; Cingolani, R. & Arakawa, Y. (2008). Two-Dimensional Photonic Crystal Resist Membrane Nanocavity Embedding Colloidal Dot-in-a-Rod Nanocrystals. *Nanoletters*, Vol. 8, No. 1, (December 2007) 260-264, ISSN 1530-6992

Messin, G.; Hermier, J. P.; Giacobino, E.; Desbiolles, P. & Dahan, M. (2001). Bunching and antibunching in the fluorescence of semiconductor nanocrystals. *Optics Letters*, Vol. 26, No. 23 (December 2001) 1891-1893, ISSN 1539-4794

Namatsu, H.; Yamaguchi, T.; Nagase, M.; Yamazaki, K. & Kurihara, K. (1998). Nano-patterning of a hydrogen silsesquioxane resist with reduced linewidth fluctuations. *Microelectronic Engineering*, Vol. 41 (March 1998) 331-334, ISSN 0167-9317

Neuhauser, R. G.; Shimizu, K.T.; Woo, W. K.; Empedocles, S. A. & Bawendi, M. G. (2000) Correlation between Fluorescence Intermittency and Spectral Diffusion in Single Semiconductor Quantum Dots. *Physical Review Letters*, Vol. 85, No. 15 (October 2000) 3301-3304, ISSN 1079-7114

Qualtieri, A.; Martiradonna, L.; Cingolani, R. & De Vittorio, M. (2007). Colloidal nanocrystals air bridge fabricated by direct lithography. *Microelectronic Engineering*, Vol. 84 (February 2007) 1488-1490, ISSN 0167-9317

Qualtieri, A.; Martiradonna, L.; Stomeo, T.; Todaro, M. T.; Cingolani, R. & De Vittorio, M. (2009a). Multicolored devices fabricated by direct lithography of colloidal nanocrystals. *Microelectronic Engineering*, Vol. 86 (April 2009) 1127-1130, ISSN 0167-9317

Qualtieri, A.; Morello, G.; Spinicelli, P.; Todaro, M. T.; Stomeo, T.; Martiradonna, L.; De Giorgi, M.; Quélin, X.; Buil, S.; Bramati, A.; Hermier, J. P.; Cingolani, R. & De Vittorio, M. (2009b). Nonclassical emission from single colloidal nanocrystals in a microcavity: a route towards room temperature single photon sources. *New Journal of Physics*, Vol. 11, No. 3 (March 2009) 033025-033034, ISSN 1367-2630

Reiss, P.; Bleuse, J. & Pron, A. (2002). Highly Luminescent CdSe/ZnSe Core/Shell Nanocrystals of Low Size Dispersion. *Nanoletters*, Vol. 2 No. 7 (June 2002) 781-784, ISSN 1530-6992

Srinivasan, K. & Painter, O. (2002). Momentum space design of high-Q photonic crystal optical cavities. *Optic Express*, Vol. 10, No. 15 (July 2002) 670-684, ISSN 1094-4087

Sun, Q.; Wang, Y. A.; Li, L. S.; Wang, D.; Zhu, T.; Xu, J.; Tang, C. & Li, Y. (2007). Bright, multicoloured light-emitting diodes based on quantum dots. *Nature Photonics*, Vol. 1, No. 12 (November 2007) 717-722, ISSN 1749-4885

Sundar, V. C.; Eisler, H. J.; Deng, T.; Chan, Y.; Thomas, E. L. & Bawendi, M. G. (2004). Soft-Lithographically Embossed, Multilayered Distributed-Feedback Nanocrystal

Lasers. *Advanced Materials*, Vol. 16, No. 80, (December 2004) 2137-2141, ISSN 1521-4095

Talapin, D. V.; Nelson, J. H.; Shevchenko, E. V.; Aloni, S.; Sadtler, B. & Alivisatos, A. P. (2007). Seeded Growth of Highly Luminescent CdSe/CdS Nanoheterostructures with Rod and Tetrapod Morphologies. *Nano Letters*, Vol. 7, No. 10 (September 2007) 2951-2959, ISSN 1530-6992

Wu, Z.; Mi, Z.; Bhattacharya, P.; Zhu T. & Xu, J. (2007). Enhanced spontaneous emission at 1.55 μm from colloidal PbSe quantum dots in a Si photonic crystal microcavity. *Applied Physics Letters*, Vol. 90 (April 2007) 171105-1 - 171105-3, ISSN 1077-3118

8

A Method for Optical Proximity Correction of Thermal Processes: Orthogonal Functional Method

Sang-Kon Kim
Hanyang University
South Korea

1. Introduction

Pattern reduction has created a great deal of interest in finding effective methods to reduce the feature sizes of microelectronic and data-storage devices. These methods are divided between top-down approach such as photolithography and bottom-up approach such as self-assembly. For below 32 nm node technology, top-down approach has obstacles such as diffraction-limited resolution and high cost of ownership and bottom-up approach has obstacles such as the insufficient support of processes and mass production.

Thermal treatment is a new process extension technique using current-day lithography equipment and chemically amplified resists (CARs). In the lithography process, thermal processes are softbake (SB), post-exposure bake (PEB), and thermal reflow. The purpose of SB is to remove excess solvent after spin coating, relieve strain in the solid film, and provide better adhesion to the substrate. The purpose of PEB is to reduce the standing wave effect and, thus, increase linewidth control and resolution. The purpose of thermal reflow is to reduce the pattern size by using thermal heating at temperatures about the glass transition temperature of the resist after development. These three kinds of thermal processes are essentially the same for heat treatment, but they have different effects on CD. Hence, it is required to understand mechanism behaviors that drive photo resist image and to deal with optical proximity effects (OPEs) due to thermal processes. OPE is quite severe as the critical dimensions (CDs) shrink down to the sub-30 nm patterns. Although the simulation parameters are not used to analyze the chemical phenomena of thermal processes, the CD bias after thermal reflow can be predicted in the linear system. Through the optical proximity correction (OPC), the distorted image from OPE can be manipulated back to the original design. However, the thermal affects of CD is the non-linear system, so that the prediction of OPE is not easy. Hence, the understanding of mechanistic behaviors that drive photo resist image can make the physically corrected resist model, and achieve the best prediction of resist images across multiple process conditions.

One of the most critical issues for sub-50-nm patterning is patterning a fine contact hole (C/H). The resolution performance of contact hole patterns is lower than that of line and space patterns because the depth of focus (DOF) for patterning contact holes is insufficient due to the low aerial-image contrast. The resist reflow process is a good method due to its simplicity without the additional complex process steps and due to its efficient technique with the appropriate selection of the resist properties and the thermal loading process conditions. However, it is difficult to predict the results of the thermal flow and the process optimization. Thus, the layout designs are only done by trial and error because there are so many effective parameters, such as baking temperature, baking time, resist original characteristics, resist volume surrounding C/H, initial C/H size and shape, and C/H array. To predict the resist shrinkage of resist reflow, according to the experimental data fitting method, the shrink bias can only be expressed as the functional relationship of the baking temperature, normalized resist volume, and native property of resist. By the quantitative analysis of the shrink bias for the baking temperature, the relationship can be expressed as a linear equation. The simple equation helps in understanding the characteristics of the resist reflow and also shows mechanism for process stabilization.

The objectives of the work presented in this chapter are to make the physically corrected resist model based on the understanding of mechanistic behaviors that drive photo resist image, to achieve the best prediction of resist images across multiple process conditions, and to develop a general simulation approach of OPC to reduce the OPEs during lithography thermal processes. Section 2 describes the modeling of the lithography process, the modeling of thermal processes, and the simulation of those thermal processes. Section 3 describes thermal bias based on the simulation of the models in section 2 and introduces the orthogonal functional method. The simulation results of SB and PEB are used by the comparison of a commercial tool, SOLID-C (Shynopsys Inc.)[1]. Section 4 describes the analysis for the below 45 nm contact hole (C/H) pattern for a KrF 248 nm CAR and 32 nm C/H pattern for a ArF 193 nm CAR. Two stage corrections of model-based OPC are performed by using the orthogonal functional method. Section 5 describes the conclusion of this chapter.

2. Thermal process model

2.1 Simulation benchmarking for the whole resist process

For the semiconductor industry, lithography is the key technology driver and a significant economic factor, currently representing over thirty five percent of the chip manufacturing cost. Significant investments from research and development to device commercialization will improve the infrastructure of this vital technology and maintain industry growth. The cost of experiments is now very large and the complexity of lithographic patterns and processes (e.g. phase shift and chemical amplification) is too big to ignore the simulation. To support this, the availability, accuracy, and easy use of lithography simulation must be improved [2]. As a research tool, lithography simulation performs experiments that would be difficult or impossible to do by any other way.

As a process development and process transfer tool, a simulation tool can quickly evaluate options, optimize processes, or save time and money by reducing the number of experiments in the fab. As a manufacturing tool, it can troubleshoot process problems, determine optimum process settings, and aid in decision making by providing facts to support engineering judgment and intuition. As a learning tool, its proper application allows the user to learn efficiently the lithography process. Hence, performing the physical and molecular simulation can be helpful to analyze, understand, and predict the lithography technologies for gigabit era.

Figure 1 shows the structure of lithography simulation. These simulation models are used for spin coating, soft bake, illumination, exposure, delay, post exposure bake, development,

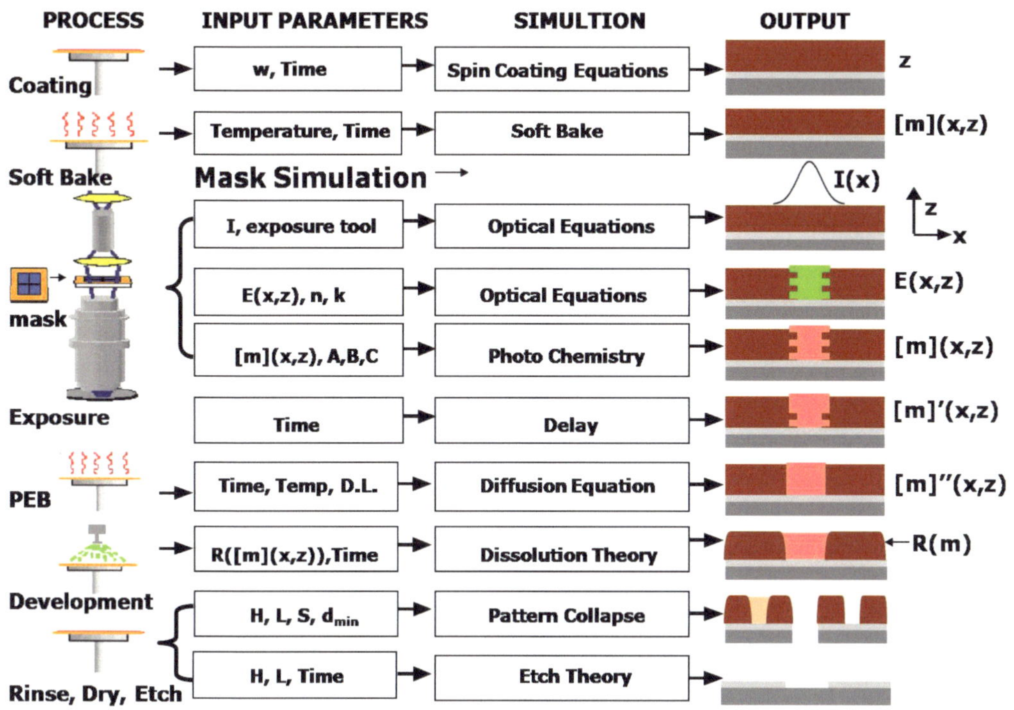

W: spin velocity, I: intensity, E(x,z):exposure, n:refractive index, k:absorbtion index, [m](x,z): photo acid generator (PAG), A, B, C; Dill's parameters, [m]'(x,z): PAG after delay, [m]''(x,z): PAG after PEB, R([m])(x,z): dissolution rate, H: pattern height, L: pattern width, S: line width, d_{min}: minimum line width.

Fig. 1. Simulation modeling

rinse, dry, and etch processes. For the spin coating model, input parameters are spin velocity and time. Resist thickness is calculated by solving the Navier-Stokes equation or the convective diffusion equation. Irregular surfaces caused by spin coating and solvent evaporation are not considered. In the soft bake model, input parameters are temperature and time. The photo acid compound (PAC) is calculated.

The illumination part is modeled as the projection lens system. Aerial image is calculated by solving the optical equations due to illumination parameters as input parameters. In the bulk model, bulk image inside resist is calculated by using the multiple thin layer interference theory and the Berning theory instead of the rigorous solution for the Maxwell's equations. During the exposure simulation, the PAC for the non-CAR (chemically amplified resist) and the photo acid generator (PAG) for the CAR are calculated by Dill's model. In the delay model, the input parameter such as delay time affects the PAC or the PAG. For the post exposure bake (PEB) model, the modified PAC or PAG is calculated by solving diffusion equation due to input parameters such as time, temperature, and diffusion length. In the development model, the dissolution rate is calculated by using the dissolution theory. Resist profile is shown by using the ray tracing algorithm. Etch simulation shows the impact of the final resist profile shape on the quality of the etched pattern in the substrate by using the Hamilton-Jacobi equation of the higher dimensional function. LAVA (the Lithography Analysis using Virtual Access) [3], PROLITH (the Positive Resist Optical LITHography model) [4], and SOLID-C (the Simulation of Optical Lithography in three Dimensions for the personal Computer) [1] are the lithography simulators for the personal computer among the academic and commercial tools.

2.2 Softbake (SB)

SB is a thermal process before exposure. During SB, the occurrence of photo-chemical reactions depends largely on the diffusion ability of a photoactive compound in a photoresist, and the free volume content of a photoresisst significantly affects its lithographic properties. In terms of the process physics, since the solvent of the resist is evaporated, the thickness of the resist shrinks, and the refractive index and the density of the resist are changed. This process can be a combination of solvent evaporation, solvent diffusion, and polymer compaction. The solvent concentration $S(z)$ in the solvent evaporation-diffusion process is

$$\frac{dS(z)}{dt} = -k_{evap} \bullet S(z) \bullet \delta(z) + \nabla D_S \nabla S(z), \tag{1}$$

where k_{evap} is the evaporation rate constant and D_S is the solvent diffusion coefficient. The free volume of the polymer film is

$$v_f = (1 - \chi_S) \bullet v_g + \alpha \bullet (T - T_g) + \chi_S \bullet \beta, \tag{2}$$

where χ_S is the mass fraction of solvent, T_g is the glass transition temperature of the pure polymer, v_g is the free volume of the pure polymer at T_g, α is the differential thermal expansion coefficient above the T_g, and ρ is density-scaled fractional volume parameter. The solvent diffusivity is

$$D = D_o \exp\left(-B\left(\frac{1}{v_f} - \frac{1}{v_g}\right)\right), \quad v_f > v_g, \qquad D = D_o, \quad v_f < v_g, \tag{3}$$

where D_o is the diffusion coefficient at T_g and B is the free volume efficiency parameter. The refractive index of the resist is

$$\eta_{resist} = \left(1 - \sum_i \chi_i\right) \bullet \eta_{polymer} + \sum_i \chi_i \bullet \eta_i, \tag{4}$$

where $\eta_{polymer}$ is the refractive index of the resist polymer, χ_i is the fraction of the i-th resist component remaining after the softbake step, and η_i is the complex refractive index of the i-th component [5].

2.3 Post-exposure Bake (PEB)

A typical positive-type CAR is composed of a polymer resin, a photoacid generator (PAG), and a dissolution inhibitor. When the CAR is exposed to light, acid is generated by conversion of the PAG. The acid concentration is related to the exposure dose through Dill's ABC parameters. During PEB, this initial acid derives the de-protection reaction with a thermal acid-catalyzed reaction that alters the development rate of the resist. Exposing the resist to deep ultraviolet (DUV) light generates acid from the PAG. During a subsequent PEB, the photogenerated acid catalyzes a thermally induced reaction that cleaves the dissolution inhibitor groups (protecting groups), rendering the reacted (deprotected) region soluble in the aqueous developer. Meanwhile, the acid diffuses from a high-dose region to a low-dose region, washing out standing waves and causing the reacted region to be larger than the initially exposed region. In addition, most CARs exhibit volume shrinkage after the PEB step due to desorption of volatile group byproducts that are created during the bake. The PAG concentration changes during exposure and has been described by

$$\frac{d[PAG]}{dt'} = -C[PAG]I \rightarrow [PAG] = [PAG]_o e^{-CIt'}, \tag{5}$$

$$[A]_{dose} = [PAG]_o - [PAG] = [PAG]_o \left(1 - e^{-CE}\right), \tag{6}$$

where A is the acid and exposure dose (E) is the multiplication of intensity (I) to exposure time (t'). The acid concentration changes during PEB and has been described by

$$\frac{d[A]}{dt} = -k_{loss} \bullet [A] - k_{quench} \bullet [A] \bullet [B] + \nabla \bullet (D_{acid} \nabla [A]), \tag{7}$$

$$\frac{d[B]}{dt} = -k_{quench} \bullet [A] \bullet [B], \tag{8}$$

$$\frac{d[M]}{dt} = -k_{amp} \bullet [M] \bullet [A]^n, \tag{9}$$

where M is the protecting group of the polymer resin, B is the base quencher, k_{amp} is the acid-catalyzed deprotection rate, k_{loss} is the acid loss reaction rate, k_{quench} is the acid neutralization rate, and t is the PEB time.

The acid diffusivity is

$$D_{acid} = D_0 \exp\left(\frac{\alpha M}{1 + \beta M}\right), \quad D_0 = A_r \exp\left(-\frac{E_a}{RT}\right), \tag{10}$$

where α and β are constants, E_a is the thermolytic decomposition activation energy, A_r is the Arrhenius coefficient, and R is the Boltzmann constant.

Table I shows the simulated parameters of the 90-nm line and space pattern. 140-nm mask pattern is used for the profile of a 90-nm L/S pattern after the development process.

Figure 2 shows simulation results for CDs for overbake of SB and PEB for isolated 94-nm line and 90-nm L/S patterns. When the SB time is increased in Fig. 2 (a), the CD becomes larger, and then saturate. When the PEB time is increased in Fig. 2 (b), the CD decreases to below 40 nm. In the L/S patterns, the underbake of SB and the overbake of PEB can be used for 45-nm pattern formation. In the active area of Fig. 2, the CD sensitivity of the PEB time (2.86 nm/s) is larger than that of the SB time (0.26 nm/s).

- **Mask Pattern**
  90 nm isolated Line and Space (L/S) pattern
- **Modeling Options**
  Image Calculation Model: Scalar
- **Stepper Parameters**
  Illumination: Convential-partially coherent
  Defocus: 0.0 μm, Wavelength: 193 nm
  Illumination Pupil Shape: Dipole(radius: 0.5, holesize: 0.3, angle: 45)
  Numerical Aperture (NA): 0.8, Aberrations: none
  Flare[%]: 0
- **Film Stack**
  Layer 1: 390-nm resist, refraction : 1.72 - j0.0
  Layer 2: 82-nm $AR^{TM}19$, refraction : 1.79 - j0.4
  Layer 3: Silicon, refraction : 6.522 - j2.705
- **Prebake Parameters**
  Temperature: 120°C
  Time: 60 s
- **Exposure Parameters**
  Exposure Dose: 15 mJ/cm^2
  A: 0.01 (1/μm), B: 0.5 (1/μm) , C: 0.03 cm^2/mJ
- **Post-Exposure Bake Parameters**
  Diffusion Length: 0.035 μm
  Temperature: 130 °C
  Time: 77 s, Q: 0.067
  k_{amp} (1/s): 0.75, k_{loss} (1/s): $2.3x10^{-5}$
  Resist Type: Positive
  Exponent n: 2
  Resist thickness reduction parameter: alpha: 1.77, beta: 0.02
  Activation Energy E_2: 0 eV
- **Develop Parameters**
  Development Model: Enhanced notch model
  Time: 45 s,
  R_{min}: $2x10^{-5}$ nm/s R_{max}: 0.045 μm/s
  R_{Res}: 0.0187 n: 1.61 l: 9.84

Table I. Parameters of the 90 nm line and space pattern.

Although the CD sensitivity of thermal parameters can be different due to other process parameters, the CD sensitivity of the PEB temperature is larger than that of the PEB time in the simulation of the L/S patterns. Both CD sensitivities are larger when the target pattern size is smaller.

Figure 3 shows boundary movements of the acid and the inhibitor concentration due to the extended times of SB and PEB for an isolated 94-nm line pattern. Those resist bulk images are the distributions of acid concentration after exposure and inhibitor concentration after PEB at a SB time of 120 s in Figs. 3 (a) and (c) and at a SB time of 10 s in Figs. 3 (b) and (d).

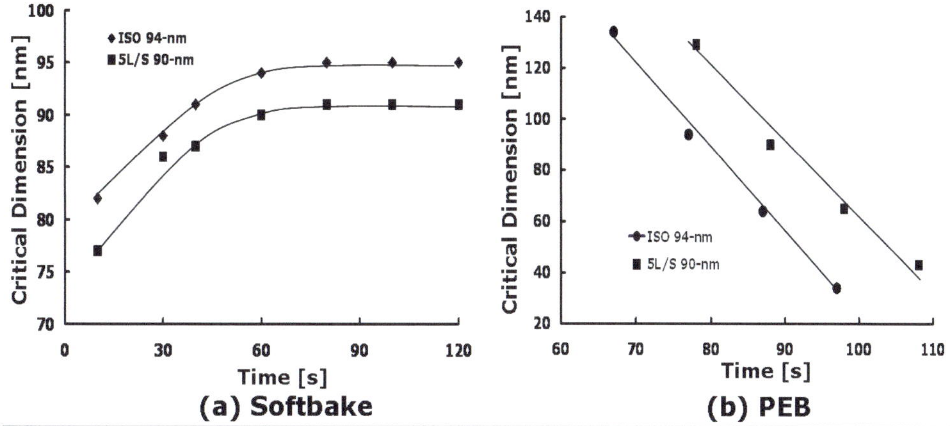

Fig. 2. Simulation results for the critical dimensions (a) SB and (b) PEB for isolated 94-nm line and 90-nm line and space (L/S) patterns.

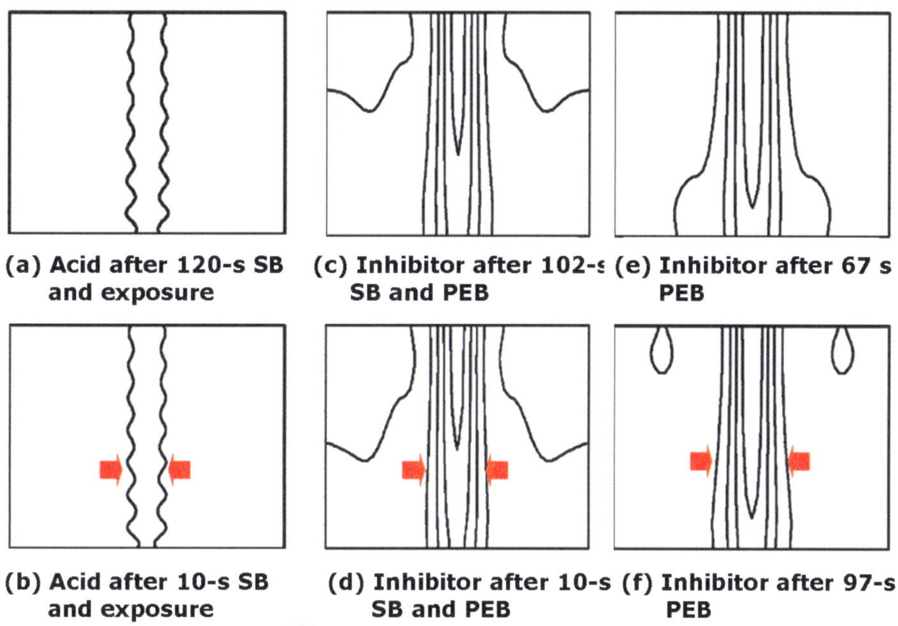

Fig. 3. Resist bulk images of (a) acid concentration after exposure and (c) inhibitor concentration after PEB at a SB time of 120 s, (b) the acid concentration after exposure and (d) inhibitor concentration after PEB at a SB time of 10 s, and inhibitor concentrations at PEB times of (e) 67 s and (f) 97 s.

Figures 3 (e) and (f) show inhibitor concentrations at PEB time of 67 s and 97 s, respectively. Figures (b), (d), and (f) for small patterns can be distinguished from Figs. (a), (c), and (e) for large patterns. Hence, for small pattern formation, the boundaries of acid and inhibitor concentrations move to the center of horizontal axis and become dense. A small SB time and a larger PEB time make the boundaries of the acid and the inhibitor concentrations dense at the center of the horizontal axis.

2.4 Thermal reflow process

Thermal reflow, as a resolution-enhancement technique, is a good method without any additional complex process steps and with appropriate selection of the resist properties and thermal loading process conditions. The thermal reflow process is used to reduce the pattern size of a resist by thermally heating the resist over its glass transition temperature after the development process. When the resist is heated over its glass transition temperature, the bonding of the synthesized resist is reduced and its mobility is improved. The three-dimensional structure of the synthesized resist is changed, and the contact hole (C/H) pattern size is shrunk due to the additional thermal energy.

However, it is difficult to predict the results of thermal reflow and process optimization, one reason being the optical proximity (or process) effects (OPEs), such as the impact of the image distortion. The OPEs of thermal reflow becomes quite severe as the critical dimensions shrink down to sub-50 nm patterns. If the physics governing a particular OPE is known through experiment and simulation, the distortion can be predicted and modeled. Even when the physics that govern the systematic effects are not fully understood, it still could be modeled if sufficient simulated data are available. Although the simulation parameters are not used to analyze the chemical phenomena of thermal reflow, the critical dimension bias after resist reflow can be predicted in a linear system by using the fitting function of experimental data. Through the optical proximity correction (OPC), the distorted image can be manipulated back to the original designed image.

2.4.1 Boundary-movement method

Boundary movement represents the boundary as a geometric model and solves the partial differential equations for boundary motion. The level-set method is more accurate and efficient than the Lagrangian-type method. The level-set function $\psi(x,y,t)$ is defined as the distance of a point (x,y) to the boundary of the geometry under consideration. The point (x,y) in the geometry boundary is

$$\psi(x,y,t) = 0. \tag{11}$$

Assuming that each boundary point moves along its surface normal,

$$\frac{\partial \psi(x,y,t)}{\partial t} + V\left|\nabla \psi(x,y,t)\right| = 0, \tag{12}$$

where V is velocity (μm/s) of movement. The basic equation system of fluid flow consists essentially of the continuity equation and the conservation equations of momentum and energy. Under the condition that fluid is incompressible and Newtonian, the continuity equation in two dimensions is

$$\frac{\partial v}{\partial x} + \frac{\partial u}{\partial y} = 0, \tag{13}$$

where v and u are the velocity components in the x and y directions, respectively. The conservation equations of momentum are

$$\rho\left(\frac{\partial v}{\partial t} + v\frac{\partial v}{\partial x} + u\frac{\partial v}{\partial y}\right) + \frac{\partial p}{\partial x} = \mu\left(\frac{\partial^2 v}{\partial x^2} + \frac{\partial^2 v}{\partial y^2}\right) + \rho g_x, \tag{14}$$

$$\rho\left(\frac{\partial u}{\partial t}+v\frac{\partial u}{\partial x}+u\frac{\partial u}{\partial y}\right)+\frac{\partial p}{\partial y}=\mu\left(\frac{\partial^2 u}{\partial x^2}+\frac{\partial^2 u}{\partial y^2}\right)+\rho g_y, \tag{15}$$

where p is pressure and ρ is density. The components of gravitational acceleration in the x and y directions are denoted by g_x and g_y, respectively. The symbol μ is fluid viscosity [kg/m•s]. When the velocity component in the y-direction is negligibly small compared to that in the x-direction, from Eqs. (13) and (14)

$$\frac{\partial v}{\partial x}=0, \tag{16}$$

$$\rho\left(\frac{\partial v}{\partial t}\right)+\frac{\partial p}{\partial x}=\mu\left(\frac{\partial^2 v}{\partial x^2}\right)+\rho g_x. \tag{17}$$

If the gravitational acceleration g_x is negligible, then Eqs. (16) and (17) above can be approximated as

$$\rho\left(\frac{\partial v}{\partial t}\right)=\mu\left(\frac{\partial^2 v}{\partial x^2}\right). \tag{18}$$

If we assume that Eq. (18) is the standard diffusion equation, the thermal reflow length is

$$L=\sqrt{\frac{\mu}{\rho}t}. \tag{19}$$

The boundary point from Eq. (12) moves such that

$$\psi(x,y,t+\Delta t)=\psi(x,y,t)-L\left|\nabla\psi(x,y,t)\right|, \tag{20}$$

where the thermal reflow length (L) is $\Delta t \bullet v$, because the length can be represented as the product of velocity and time.

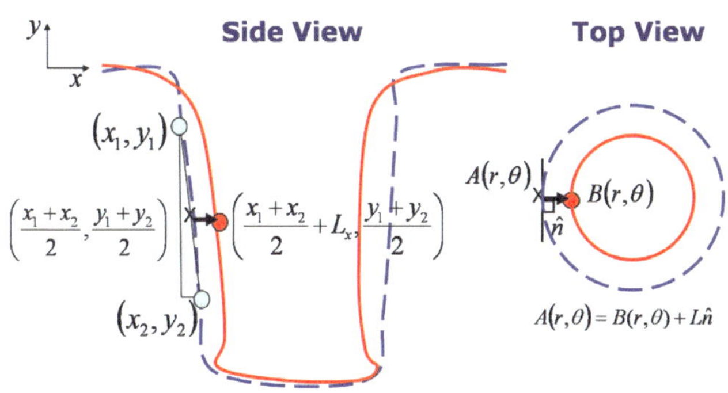

Fig. 4. Thermal flow length.

Figure 4 shows the movement of point (x_1, y_1) and point (x_2, y_2) using the thermal reflow length. After the midpoint of point (y_1, y_1) and point (y_2, y_2) is calculated, this midpoint is moved to thermal reflow length due to time and temperature.

Piecewise-linear	**Polynomial**
$\mu(T)=\mu_n+(\mu_{n+1}-\mu_n)/(T_{n+1}-T_n)*(T-T_n)$ (a)	$\mu(T)=A_1+A_2T+A_3T^2+\dots$ (e)
Piecewise-polynomial	**Sutherland's law**
For $T_{min,1}<T<T_{max,1}$ $\mu(T)=A_1+A_2T+A_3T^2+\dots$ (b)	$\mu(T)=C_1T^{3/2}/(T-C_2)$ (f) or
For $T_{min,2}<T<T_{max,2}$ $\mu(T)=B_1+B_2T+B_3T^2+\dots$ (c)	$\mu(T)=\mu_0(T/T_0)^{3/2}(T_0+S)/(T+S)$ (g)
Power law	**Standard Arrhenius form (supposed)**
$\mu(T)=BT^n$ (g) or $\mu(T)=\mu_0(T/T_0)^n$ (d)	$\mu(T)=\mu_0\exp(E/(RT))$ (h)

Table II. Viscosities as a function of temperature.

Viscosity (μ) is related to the fluid's resistance to motion and is an important aspect of rheology. Viscosity is determined by relating the velocity gradient in fluids to the shear force causing flow to occur. To define the viscosity equation as a function of temperature, viscosity equations in Table II are fitted to the experiment data.

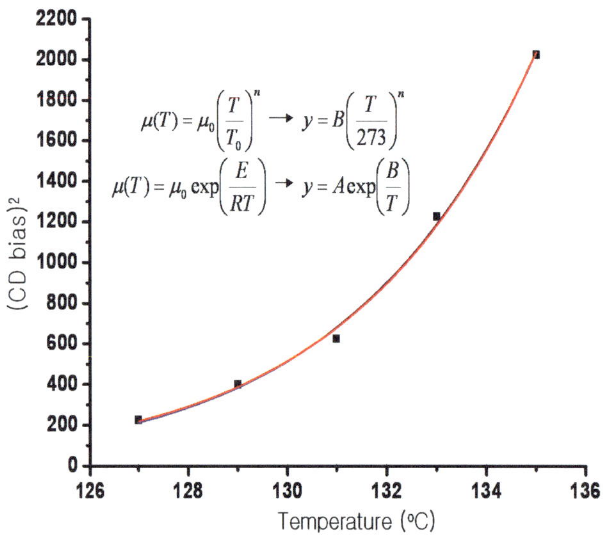

Fig. 5. Fitting experimental data by using viscosity equation of a function of temperature.

Figure 5 shows the experimental relation of temperature and the viscosity, which is from Eq. (19),

$$L^2=\frac{t}{\rho}\mu \rightarrow \mu(T)\propto(CD\,bias)^2. \tag{21}$$

Two equations are fitted when $A=3.0062\times10^{14}$, $n=36.52117$, $A=7.5727\times10^{18}$, and $B=-4840.2$. Hence, two functions can be used as the viscosity function.
The function of viscosity can be

$$\mu(T)=\mu_0\exp\left(\frac{E}{RT}\right), \tag{22}$$

- **Mask Pattern**
  140-nm × 140-nm isolated rectangle pattern
- **Modeling Options**
  Image Calculation Model: Scalar
- **Stepper Parameters**
  Illumination: Convential-partially coherent
  Defocus: 0.0-μm, Wavelength: 193-nm
  Illumination Pupil Shape: Dipole (radius: 0.5, hole size: 0.3, angle: 45°)
  Numerical Aperture (NA): 0.7, Aberrations: none
  Flare[%]: 0
- **Film Stack**
  Layer 1: 250-nm resist, refraction : 1.72 - j0.02
  Layer 2: 82-nm AR™19, refraction : 1.79 - j0.4
  Layer 3: Silicon, refraction : 6.522 - j2.705
- **Prebake Parameters**
  Temperature: 120°C
  Time: 60 s
- **Exposure Parameters**
  Exposure Dose: 19 mJ/cm^2
  A: 0.01 (1/μm), B: 0.08 (1/μm), C: 0.02-cm^2/mJ
- **Post-Exposure Bake Parameters**
  Diffusion Length: 0.035-μm
  Time: 80 s, Q: 0.067
  k_{amp} (1/s): 0.75, k_{loss} (1/s): 2.3×10^{-5}
  Resist Type: Positive
  Exponent n: 1
  Resist thickness reduction parameter: alpha: 1.74, beta: 0.02
  Activation Energy E_2: 0 eV
- **Develop Parameters** Development Model: Enhanced Mack
  Time: 150 s,
  R_{min}: 6.5×10^{-5}-nm/s R_{max}: 0.1-μm/s
  R_{Res}: 0.0187 n: 0.01 l: 0.73
- **Resist Reflow Parameters**
  Temperature: 129°C, Time: 90 s

Table III. Parameters of the 140-nm contact hole pattern after resist reflow.

where E is the activation energy, μ_o is the pre-exponential factor, and R is the universal gas constant. A fitted viscosity function, $\mu = (4.8414 \times 10^{16}) \bullet \exp(-4840.2 / T)$, is used for the simulation when the density is constant due to temperature.

Table III shows the simulated parameters of the 140-nm contact hole after resist flow. A 140-nm mask pattern is used for the profile of a 180-nm C/H pattern after the development process. After the resist flow process of 129 °C for 90 seconds, the 180-nm developed profile is shrunk to the 140-nm resist profile.

Figure 6 shows the simulated and experimental resist profiles after resist flow at 129 °C and 90 seconds. The critical dimension bias is 40 nm. The bulk image of simulation modeled by Eqs

(21) and (22) is shown as similar to the experimental bulk image. Figure 6 shows the comparison of CDs and CD biases after resist flow at 90 seconds in experiment and simulation according to temperature. Our simulated results agree well with experimental results.

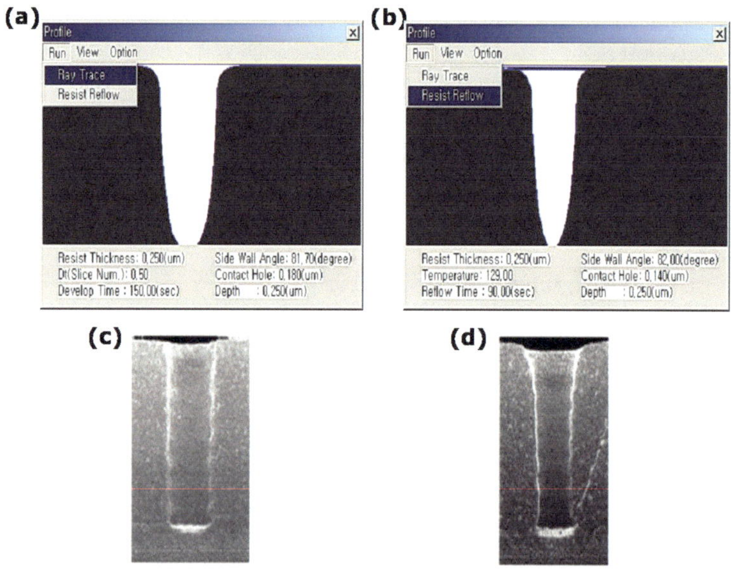

Fig. 6. Simulation results of (a) a 180-nm contact hole pattern before resist reflow, and (b) a 140-nm contact hole after resist reflow. Experimental results of (c) 180-nm contact hole pattern before resist reflow, and (d) 140-nm contact hole pattern after the resist reflow process.

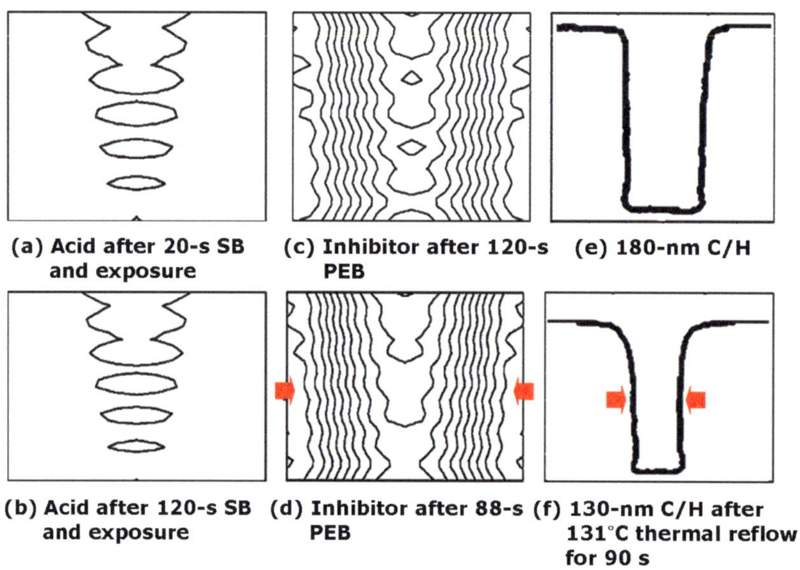

Fig. 7. Resist bulk images of acid concentration after exposure at SB times of (a) 20 s and (b) 120 s, inhibitor concentrations at PEB times of (c) 120 s and (d) 88 s, (e) the 180-nm contact hole before thermal reflow, and (f) the 130-nm contact hole after thermal reflow at 131 °C for 90 s.

Figure 7 shows boundary movements for a C/H in SB, PEB, and thermal reflow. These resist bulk images are the distributions of acid concentrations after exposure at a SB time of 20 s in Fig. 7 (a) and at a SB time of 120 s in Fig. 7 (b). Figures 7 (c) and (d) show inhibitor concentrations at PEB times of 120 s and 88 s, respectively. Figures 7 (e) and (f) show the 180-nm C/H before thermal reflow and the 130-nm C/H after thermal reflow, respectively, at 131 °C for 90 s. For the extended SB time, the boundaries of the acid concentration are same in Figs. 7 (a) and (b), so the SB time is not sensitive to shrinkage of the contact hole. For the extended PEB time, the boundaries of the inhibitor concentration move to the center of the horizontal axis, and become dense. During the thermal reflow in Figs. 7 (e) and (f), the boundary of the C/H shrinks due to temperature and time.

2.4.2 Solution of the Navier-Stokes equation

The resist reflow at temperature above its glass transition temperature can be assumed to be an ideal fluid, which is an incompressible fluid, and to have a constant density (ρ) and force ($\rho\hat{n}\delta S$) exerted across a geometrical surface element $\hat{n}\delta S$ within the fluid. Hence,

$$\frac{D\vec{u}}{Dt} = -\frac{1}{\rho}\nabla P + g\,,\quad \nabla \bullet \vec{u} = 0\,, \tag{23}$$

where $\vec{u}(u,\phi,w)$ is the fluid velocity, g is the gravitational acceleration, ρ is the density, and P is the pressure. At this point, the element of the stress tensor (T_{ij}) in an incompressible fluid is

$$T_{ij} = -p\delta_{ij} + \mu\left(\frac{\partial u_j}{\partial x_i} + \frac{\partial u_i}{\partial x_j}\right),\quad \nabla \bullet \vec{u} = 0\,. \tag{24}$$

where μ is fluid viscosity. By Eqs. (23) and (24),

$$\rho\frac{Du_i}{Dt} = -\frac{\partial P}{\partial x_i} + \mu\frac{\partial}{\partial x_j}\left(\frac{\partial u_j}{\partial x_i} + \frac{\partial u_i}{\partial x_j}\right) + \rho g_i\,' \tag{25}$$

$$\frac{Df}{Dt} = \frac{\partial f}{\partial t} + u\frac{\partial f}{\partial x} + v\frac{\partial f}{\partial y} + w\frac{\partial f}{\partial z} \rightarrow \frac{Df}{Dt} = \frac{\partial f}{\partial t} + (\vec{u} \bullet \nabla)f \cdot \tag{26}$$

Hence, the Navier-Stokes equation is

$$\frac{\partial \vec{u}}{\partial t} + (\vec{u} \bullet \nabla)\vec{u} = -\frac{1}{\rho}\nabla P + v\nabla^2\vec{u} + g\,' \tag{27}$$

$$\nabla \bullet \vec{u} = 0\,, \tag{28}$$

where $v = \mu/\rho$ is kintic viscosity. In the x and the z directions,

$$\rho\left(\frac{\partial u}{\partial t} + u\frac{\partial u}{\partial x} + w\frac{\partial u}{\partial z}\right) = -\frac{\partial P}{\partial x} + \mu\left(\frac{\partial^2}{\partial x^2} + \frac{\partial^2}{\partial z^2}\right)u + \rho g_x\,' \tag{29}$$

$$\rho\left(\frac{\partial w}{\partial t}+u\frac{\partial w}{\partial x}+w\frac{\partial w}{\partial z}\right)=-\frac{\partial P}{\partial z}+\mu\left(\frac{\partial^2}{\partial x^2}+\frac{\partial^2}{\partial z^2}\right)w+\rho g_z, \tag{30}$$

$$\frac{\partial u}{\partial x}+\frac{\partial w}{\partial z}=0\cdot \tag{31}$$

When $\vec{u}\bullet\nabla\vec{u}\langle\langle\,\mu\nabla^2\vec{u}$ for the slow flow equation and when the gravitational acceleration (g_x) and pressure are neglected, the Navier-Stokes equation becomes the diffusion equations

$$\frac{\partial H}{\partial t}=\nu\frac{\partial^2 H}{\partial x^2}, \quad \text{if } u\approx H(x,t), \tag{32}$$

where H is a geometric boundary. In the shallow-water approximation, the Navier-Stokes equation becomes

$$\frac{\partial u}{\partial t}+u\frac{\partial u}{\partial x}+w\frac{\partial u}{\partial z}=-\frac{1}{\rho}\frac{\partial P}{\partial x}, \tag{33}$$

$$\frac{\partial w}{\partial t}+u\frac{\partial w}{\partial x}+w\frac{\partial w}{\partial z}=-\frac{1}{\rho}\frac{\partial P}{\partial z}-g_z, \tag{34}$$

$$\frac{\partial u}{\partial x}+\frac{\partial w}{\partial z}=0\cdot \tag{35}$$

From Eq. (35),

$$w=-\frac{\partial u}{\partial x}z+f(x,t) \tag{36}$$

and $f(x, t) = 0$ when w = 0 at z = 0. If $F(x, z, t) = z - H(x, t)$ in the kinematic condition at the free surface,

$$\frac{\partial F}{\partial t}+(\vec{u}\bullet\nabla)F=0, \text{ at } z=H(x,t). \tag{37}$$

From Eqs. (36) and (37),

$$\frac{\partial H}{\partial t}+u\frac{\partial H}{\partial x}+H\frac{\partial u}{\partial x}=0\cdot \tag{38}$$

By using $c(x,t)=(gH)^{\frac{1}{2}}$ and Eq. (33),

$$\left[\frac{\partial}{\partial t}+(u-c)\frac{\partial}{\partial x}\right](u-2c)=0\cdot \tag{39}$$

When $u+2c=2c_0$, $z=-3c+2c_0$, and $z\approx c\approx H(x,t)$,

$$\frac{\partial H}{\partial t}+H\frac{\partial H}{\partial x}=0\cdot \tag{40}$$

For an analytical solution, the gravitational acceleration (g_x) in the x-direction is zero and $\vec{u} \bullet \nabla \vec{u} \langle\langle \mu \nabla^2 \vec{u}$ for the slow flow equations. From Eq. (29),

$$0 = -\frac{1}{\rho}\frac{\partial P}{\partial x} + \nu \frac{\partial^2 u}{\partial z^2} \quad \text{for x-coordinate,} \tag{41}$$

where kinetic viscosity is $\nu = \mu/\rho$. From Eq. (30),

$$0 = -\frac{1}{\rho}\frac{\partial P}{\partial z} - g \quad \text{for z-coordinate.} \tag{42}$$

The net upward force per unit area of surface is, by Eq. (41),

$$P - P_0 = -\Im \frac{\partial^2 H}{\partial x^2} \rightarrow P = -\rho g H - \Im \frac{\partial^2 H}{\partial x^2}, \tag{43}$$

where P is the pressure in the fluid just below the surface, P_0 is the atmospheric pressure, and $\Im$ is the surface tension force. If the geometric boundary function (H) is a power series, $H = 1 + x + x^2 + \cdots$ and if x is smaller then 1, $H \approx 1 + x$ and

$$P = -\rho g(1 + x) - \Im \frac{\partial^2 H}{\partial x^2} \cdot \tag{44}$$

From Eqs. (41) and (44),

$$u = -\left[\frac{g}{\nu}\frac{\partial H}{\partial x} + \frac{\Im}{\nu\rho}\frac{\partial^3 H}{\partial x^3}\right]\frac{1}{2}z^2 \cdot \tag{45}$$

From Eq. (31),

$$w_{z=H} = \frac{\partial}{\partial x}\left[\frac{g}{\nu}\frac{\partial H}{\partial x} + \frac{\Im}{\nu\rho}\frac{\partial^3 H}{\partial x^3}\right]\frac{1}{6}H^3 \cdot \tag{46}$$

The boundary condition is $F(x,z,t) = z - H(x,t) = 0$, so that

$$\frac{\partial H}{\partial t} = \frac{\partial}{\partial x}\left[\frac{H^3}{6}\left(\frac{g}{\nu} + \frac{\Im}{\nu\rho}\frac{\partial^3 H}{\partial x^3}\right)\right], \tag{47}$$

$$\frac{\partial H}{\partial \tau} = \frac{1}{6}\frac{\partial}{\partial X}\left[\Omega^{-2}\left(\frac{\partial^3 H}{\partial X^3}\right)H^3 + H^3\right], \quad X \equiv \frac{x}{w}, \quad \tau \equiv \frac{gt}{w\nu}, \quad \Omega^2 \equiv \frac{\rho g w^3}{\Im}, \tag{48}$$

where H is the film geometry, w is the feature width, v is the kinematic viscosity of the fluid, and $\Im$ is the surface tension. A quasi-steady state of resist reflow can be obtained by dropping $\partial H/\partial \tau$ from Eq. (48):

$$\left(\frac{\partial^3 H}{\partial X^3}\right)H^3 + \Omega^2 H^3 = \Omega^2 \cdot \tag{49}$$

Analytical solutions with the dimensionless parameter (Ω^2) can be obtained. In the isolated contact hole, the entire domain can be divided into subdomain I of the left ridge side, subdomain II of the inside ridge, and subdomain III of the right ridge side. The film geometries are

$$H_I = 1 + b_1\phi_2 + c_1\phi_3, \quad H_{II} = 1 \pm \frac{d}{1+\Omega^2} + a_2\phi_1 + b_2\phi_2 + c_2\phi_3 + \phi_4, \quad H_{III} = 1 + a_3\phi_1, \tag{50}$$

where *a*, *b*, and *c* are coefficients, which are calculated by using the boundary conditions, *d* is the height of resist, and ϕ is the linearly independent homogenous solutions. The three linear independent homogenous solutions are

$$\phi_1 = \exp(-\lambda x), \quad \phi_2 = \exp(\lambda x/2)\cos(\lambda x\sqrt{3}/2),$$

$$\phi_3 = \exp(\lambda x/2)\sin(\lambda x\sqrt{3}/2), \quad \phi_4 = \frac{1}{3}\left[\left(1 \pm \frac{d}{1+\Omega^2}\right) - \left(1 \pm \frac{d}{1+\Omega^2}\right)^4\right], \tag{51}$$

where in subdomains I and III, $\lambda = (3\Omega^2)^{1/3}$ and , in subdomain II, $\lambda = \{3\Omega^2\{1 \pm d/(1+\Omega^2))^{-4}\}^{1/3}$. In the boundary condition of x = -1/2,

$$H_I \pm d = H_{II}, \quad \frac{\partial H_I}{\partial X} = \frac{\partial H_{II}}{\partial X}, \quad \frac{\partial^2 H_I}{\partial X^2} = \frac{\partial^2 H_{II}}{\partial X^2}. \tag{52}$$

In the boundary condition of x = 1/2,

$$H_{III} \pm d = H_{II}, \quad \frac{\partial H_{III}}{\partial X} = \frac{\partial H_{II}}{\partial X}, \quad \frac{\partial^2 H_{III}}{\partial X^2} = \frac{\partial^2 H_{II}}{\partial X^2}. \tag{53}$$

If wafer has a topology, the thickness of resist after thermal reflow is different due to reflow conditions. The surface tension of resist varies due to the non-uniformity of resist thickness. In the dense contact holes, the film geometries are

$$H_I = 1 + \frac{d}{1+\Omega^2} + a_2\phi_1 + b_2\phi_2 + c_2\phi_3 + \phi_4, \tag{54}$$

$$H_{II} = 1 + a_2\phi_1 + b_2\phi_2 + c_2\phi_3. \tag{55}$$

The boundary conditions at x = 1/2 are

$$H_I = H_{II} + d, \quad \frac{\partial H_I}{\partial X} = \frac{\partial H_{II}}{\partial X}, \quad \frac{\partial^2 H_I}{\partial X^2} = \frac{\partial^2 H_{II}}{\partial X^2}. \tag{56}$$

The boundary conditions between x = - 1/2 and x = a + 1/2 are

$$H_{II(X=z+1/2)} + d = H_{I(X=-1/2)}, \quad \frac{\partial H_I}{\partial X_{(X=-1/2)}} = \frac{\partial H_{II}}{\partial X_{(X=a+1/2)}}, \quad \frac{\partial^2 H_I}{\partial X^2{}_{X=-1/2}} = \frac{\partial^2 H_{II}}{\partial X^2{}_{(X=a+1/2)}}. \tag{57}$$

Figure 8 shows the numerical result of the diffusion equation in Eq. (32). The isolated pattern is wide and symmetrical due to the time of thermal reflow. The numerical calculation methods are the second predictor-corrector scheme, quick scheme, the forward time and centered space (FTCS) scheme, and the Kawamara-Kuwahara scheme.

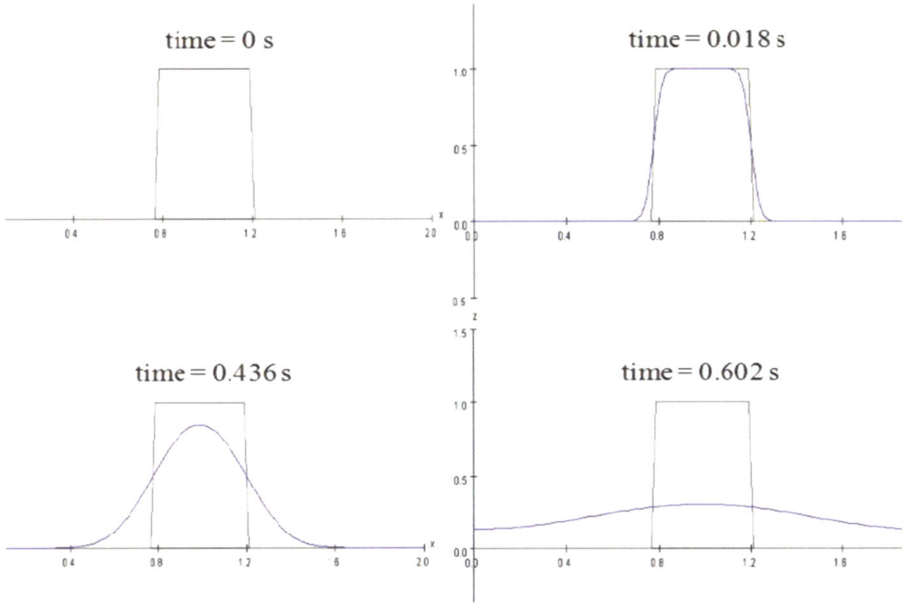

Fig. 8. Simulation results of the diffusion equation in Eq. (32) by using are the second predictor-corrector scheme, quick scheme. The isolated pattern is wider and symmetrical due to the time of thermal reflow.

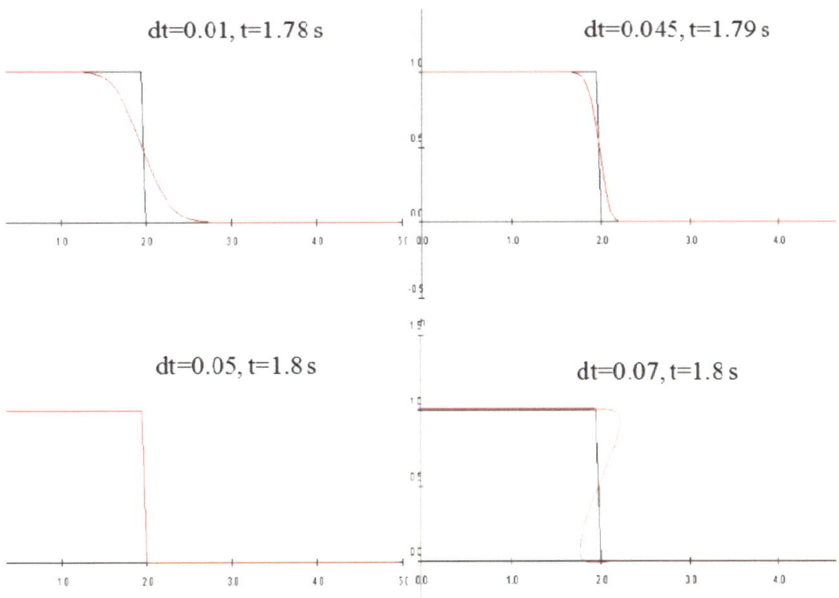

Fig. 9. Simulation flow of Eq. (40) due to Δt of the courant number, which is reflow time interval.

Figure 9 shows pattern reflow due to Δt of the courant number, which is reflow time interval in Eq. (40). As this time interval is larger, pattern side forms the slop of h = - x, a vertical slop, and the slop of h = x. These results are same as experiment results. It can be assumed that the temperature of thermal reflow makes molecular of resist moving, this high temperature corresponds to the large time interval, which is rapid reflow. The numerical calculation methods are the upstream scheme and Lax-Wendroff scheme.

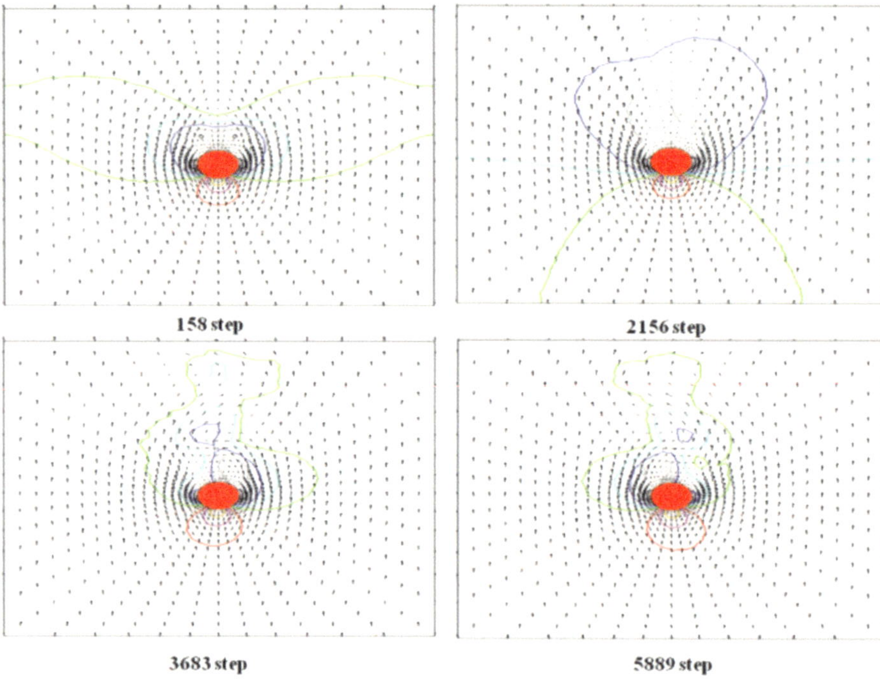

Fig. 10. Reflow velocity and pressure distributions in Eqs (29-31) by using a finite element method. These results are obtained by modifying the simulation of Prof. Tomomi Uchiyama in Nagoya University, Japan.

Figure 10 shows the reflow velocity and pressure distributions of Eqs (29-31) without the gravitational acceleration by using a finite element method. These results are obtained by modifying the simulation of Prof. Tomomi Uchiyama in Nagoya University, Japan. The boundary conditions are $v_i = \bar{v}_i$ on bottom, left, and right sides, and $\left(-p/\rho\delta_{ij} + v\partial v_i/\partial x_j\right)n_j = \bar{t}_i$ on top side where t is surface tension. In Fig. 10, the step is corresponding to the time of thermal reflow, reflow velocity is described as an arrow, and pressure is described as contour in the two dimensions. Due to steps, various velocity directions and pressure contours are shown in Fig. 10.

Figures 11 (a)-(c) show wet profiles of thermal reflow according to the parameter Ω in Eq. (49). The simulation parameters are resist thickness (μm) = 0.1 and trench width (μm) = 0.1. The larger the dimensionless parameter (Ω_1^2 = 7.54 < Ω_2^2 = 426.09 < Ω_3^2 = 7538.46, which are surface tension $\Im_3 < \Im_2 < \Im_1$ due to Eq. (48)) is, the sharper the slope of trench area is. Figure 11 (d) shows the SEM digitization lines due to the temperature of the resist reflow process [6]. The higher the temperature is, $T_3 < T_2$, the shaper the corner edges of resist profile are. These corners become round and collapse at temperature above T_1 (> T_2). Surface tension is the magnitude of the force that is a force of attraction between molecules in liquids and controls the shape of the liquid. When temperature is increased, the surface tension is reduced because the attractive force between molecules in liquids is decreased, and density is reduced. As the change of density is larger than that of surface tension due to temperature, the simulation results of the dimensionless parameters ($\Omega_1 < \Omega_2 < \Omega_3$) in Fig. 11 (a-c) can be described the simulation results of temperatures ($T_1' > T_2' > T_3'$). The simulation results of temperatures can be described the geometry boundary movements of the experimental results in temperatures ($T_1 > T_2$). In comparison with experiment in Fig. 11 (d), the simulation result is for a quasi-steady-state and a wet resist state of resist reflow, so that this result excludes time dependency and the evaporation of resist materials.

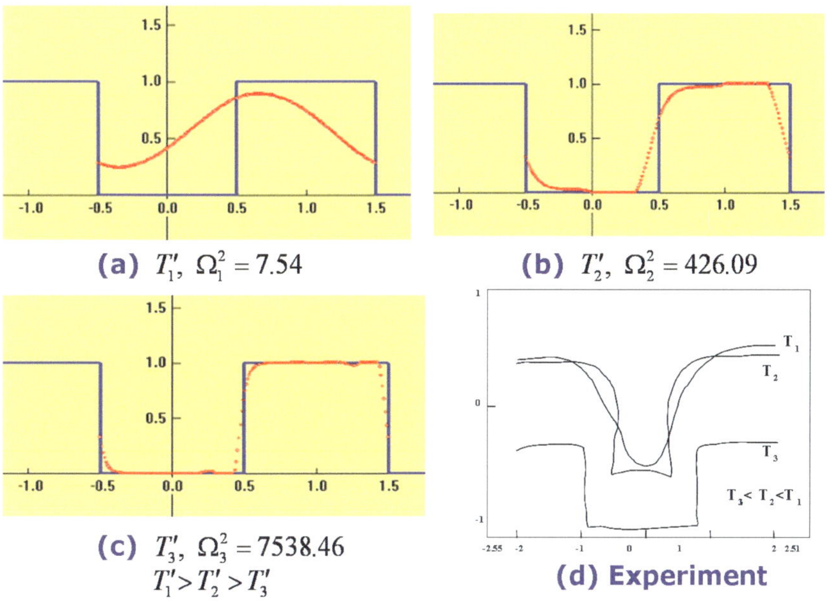

Fig. 11. Simulation results of (a) wet thermal reflow Ω_1, (b) wet thermal reflow Ω_2, (c) wet thermal reflow Ω_3, and (d) SEM digitization [6].

3. Thermal bias

During lithography process, image distortion is produced. Intensity distribution from the exposure system is different due to pattern size, wavelength, and illumination condition. Figure 12 shows the simulated optical process effects of the line and space (L/S) patterns and contact hole (C/H) patterns. The simulated resists of Figs. 12 (a) and (b) are V2 Shipley [7] and XP-99185 Shipley [8] respectively. The dose parameter is only controlled to get a L/S pattern with 80° side wall angle in V2 Shipley and a C/H pattern with 89.5° side wall angle in XP-99185 Shipley. In case of L/S patterns in Fig. 12 (a), a similar increased behavior in both conventional illumination and annular illumination is shown according to pitch size. However, it is not enough to decide this behavior. In case of C/H patterns in Fig. 12 (b), it is hard to find symmetric OPE effects for C/H patterns in annular and conventional illuminations. OPE effects of C/H patterns are more random than OPE effects of L/S patterns. OPE effects according to pitch size are differently depended on illumination conditions, pattern shapes, and mask size.

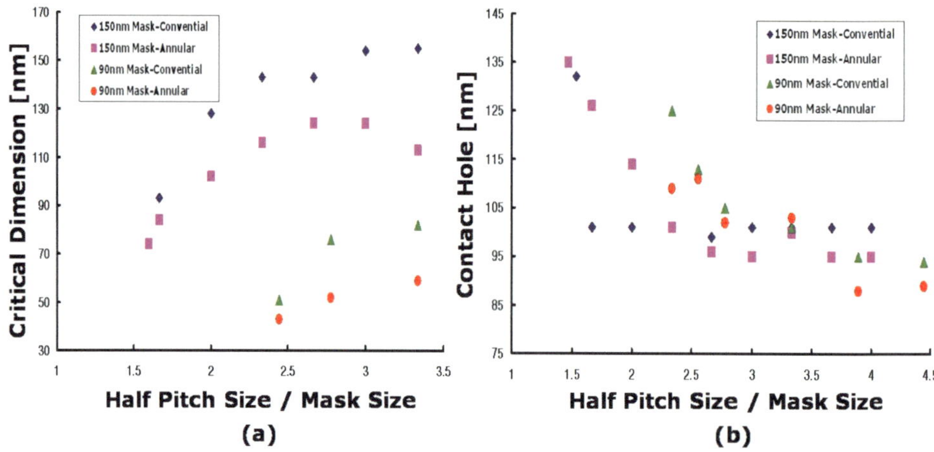

Fig. 12. Simulated optical process effects: (a) the graph between critical dimension and the ratio of haft pitch size and mask size at the L/S patterns and (b) the graph between contact hole and the ratio of half pitch size and mask size at the C/H patterns due to different mask size and illumination conditions.

Figure 13 shows the simulation results for the dependence of the size of the C/H on the pitch size for various PEB times and thermal reflow. Due to pitch size, C/H patterns between 140 nm and 114 nm are shaped in the solid line of Fig. 13 with triangle points by using a 140 nm mask and a PEB time of 100 s. These patterns, which depend on pitch size, can be reduced to C/H patterns between 97 nm and 43 nm in the dotted line of Fig. 13 with circle points by using a PEB time of 94 s and a thermal reflow of 138 °C. However, the optical proximity effects (OPE) are severe in Fig. 13.

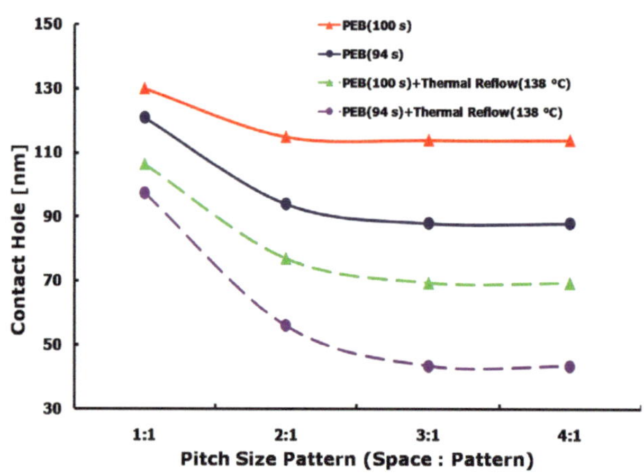

Fig. 13. Simulation results showing the dependence of the size of the contact hole on the pitch size for various PEB times and thermal reflow.

During the lithography process, image distortion is produced. The intensity distribution from the exposure system is different due to pattern size, wavelength, and illumination conditions. In the lithography process including thermal reflow, the MEEF can be divided into the MEEF after development and the MEEF after thermal reflow:

$$MEEF = [MEEF]_{develop} \bullet [MEEF]_{thermal}, \tag{58}$$

$$\frac{\Delta CD_{wafer}}{\Delta CD_{mask}/M} = \left[\frac{\Delta CD_{develop}}{\Delta CD_{mask}/M}\right]_{develop} \bullet \left[\frac{\Delta CD_{wafer}}{\Delta CD_{develop}}\right]_{thermal}, \tag{59}$$

where M is the imaging reduction ratio. The MEEF after development means the MEEF before thermal reflow.

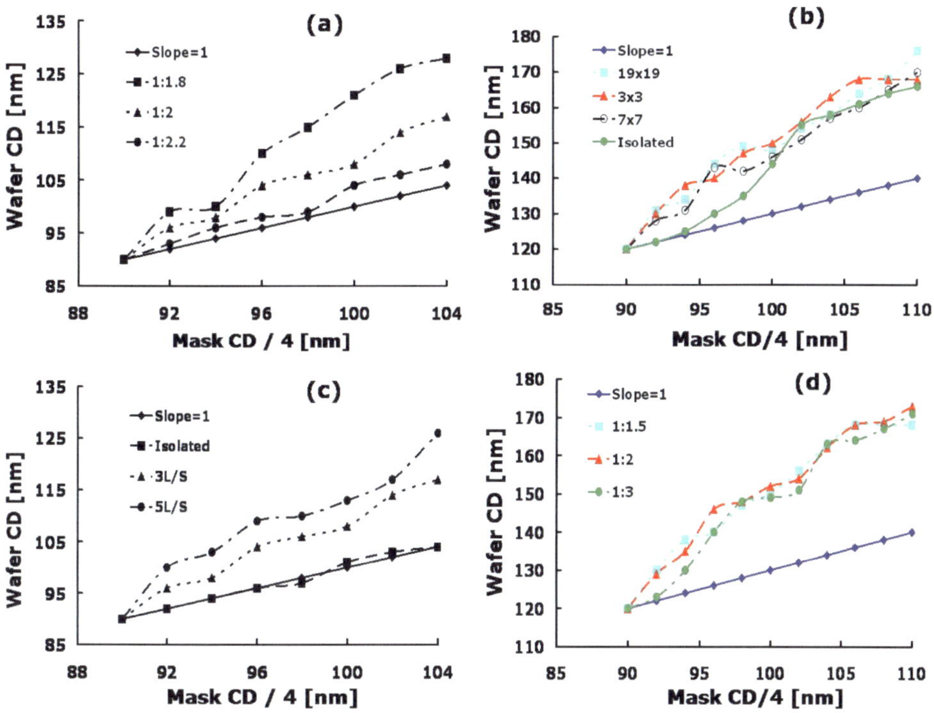

Fig. 14. Simulated MEEF after development for optical process effects: (a) MEEF due to the L/S pattern density, (b) MEEF due to the C/H pattern density, (c) MEEF due to the L/S pitch size, and (d) MEEF of the C/H pitch size.

Figure 14 shows the simulated optical process effects of the line and space (L/S) patterns and the C/H patterns. The simulated resists are the V2 Shipley resist [7] for L/S patterns and the XP-99185 Shipley resist [8] for C/H patterns. The dose parameter is only controlled to get a L/S pattern with an 88° side wall angle for the V2 Shipley resist and a C/H pattern with an 89.5° side wall angle for the XP-99185 Shipley resist. For a target 90-nm critical dimension, the MEEF value is large for the dense 1:1.8 pattern, and the MEEF value is approximately 1 for the isolated line in Fig. 14 (a). When the pitch is smaller in Fig. 14 (c), the MEEF value is larger, and the pattern is difficult to form. For a 90-nm contact hole, the MEEF factors after development are not much different due to the pattern density of contact hole in Fig. 14 (b). However, variations of these OPEs with the pattern density is not symmetric. When the pitch of a 3 x 3 C/H pattern is smaller, the MEEF after development is larger in Fig. 14 (d). The OPEs of C/H patterns are more random than the OPEs of L/S patterns. The higher-order MEEF factors are the MEEF of the C/H, the L/S, and the isolated patterns. The OPEs very with the line number and the pitch size for L/S and contact-hole patterns. Hence, different OPC methods are needed for those patterns.

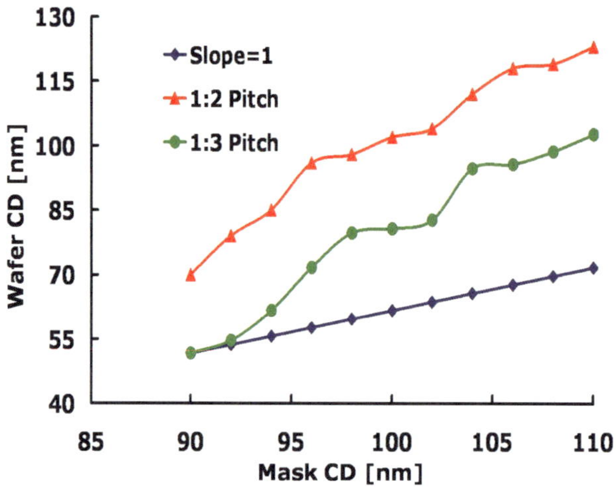

Fig. 15. Simulated MEEF of the contact hole after thermal reflow due to pitch size.

Figure 15 shows the simulated MEEF factors after thermal reflow for contact holes. The MEEF factors after thermal reflow at 140 °C are different due to pitch size in Fig. 15. Thus, different OPC methods must be applied for temperature and pitch size.

Thermal bias means the reduction of CDs before and after thermal processes. SB is a thermal process before exposure. During SB, the reduction of the free volume content in CAR significantly affects its lithographic properties. When CAR is exposed to the light, acid is generated by conversion of the photoactive generator. During a subsequent PEB, the photogenerated acid catalyzes a thermally induced reaction that cleaves the dissolution inhibitor groups (protecting groups), rendering the reacted (deprotected) region soluble in aqueous developer. Meanwhile, the acid diffuses from a high dose region to a low dose region washing out standing waves and causing the reacted region to be larger than the initially exposed region. For SB and PEB, the function of thermal biases ($CD_{SB\ (or\ PEB)}$) can be represented as

$$\Delta CD_{SB(or\ PEB)} = CD_{T=T_{target}} - CD_T \,, \tag{60}$$

where $CD_{T=T_{target}}$ is CD after development with SB (or PEB) at temperature T = T_{target} and CD_T is CD after development with SB (or PEB) at temperature T. Thermal reflow is an extra bake and chill step to reduce the pattern size of resist after the final images are developed. When resist is heated over its glass transition temperature, the bonding of synthesized resist is reduced and its mobility is improved. The three-dimensional structure of the synthesized resist is changed, and C/H is shrunk due to the additional thermal energy. For thermal reflow, the function of thermal bias can be represented as

$$\Delta CD_{\text{Thermal reflow}} = CD_{\text{dev}} - CD_{\text{wafer}}. \tag{61}$$

Those three kinds of thermal processes have similar effective parameters such as baking temperature (T_b), baking time (t_b), resist original characteristics (K_r), resist volume surrounding (V_R), initial pattern size and shape, and pattern array. For the simple thermal assumption, the actual thermal bias of each thermal process can be approximated as an orthogonal function,

$$\Delta CD = f(T_b, t_b, V_R, K_r) = f(T_b, t_b) \bullet f(V_R) \bullet f(K_r), \tag{62}$$

where $f(T_b, t_b)$ is a thermal bias due to temperature and time, $f(V_n)$ is a thermal bias due to resist volume surrounding C/H, and $f(K_r)$ is a thermal bias due to resist characteristics.

4. Comparison to experiment results

By matching with experiment results [9], simulation conditions are optimized. For a 193 nm ArF CAR, 6% transmittance attenuated PSM, quaterpole off-axis illumination (OAI), ArF 193 nm illumination, 0.75 NA, 0.35 μm CAR thickness, and 0.082 μm BARC are used. For a 248 nm KrF CAR, 6% transmittance attenuated PSM, dipole OAI, KrF 248 nm illumination, and 0.8 NA are used. Photoresist (PR) type is SL120H, PR thickness is 0.45 μm, and BARC is 0.06 μm.

For thermal reflow, an antireflective layer of 80-nm-thick resist is coated over the silicon wafer prior to the resist process. The coated thickness is 0.37 μm. An ethylvinylether-based polymer is coated and prebaked at 100°C for 60 s. The exposure system is an ASML-700 with a numerical aperture (NA) of 0.6, a partial coherency (σ) of 0.4 and an attenuated phase-shift mask. Exposed wafers are baked at 110°C for 60 s on a hot plate and developed in 2.38 wt % tetramethyl ammonium hydroxide (TMAH) aqueous solution for 60 s. The time of resist reflow is 90 s.

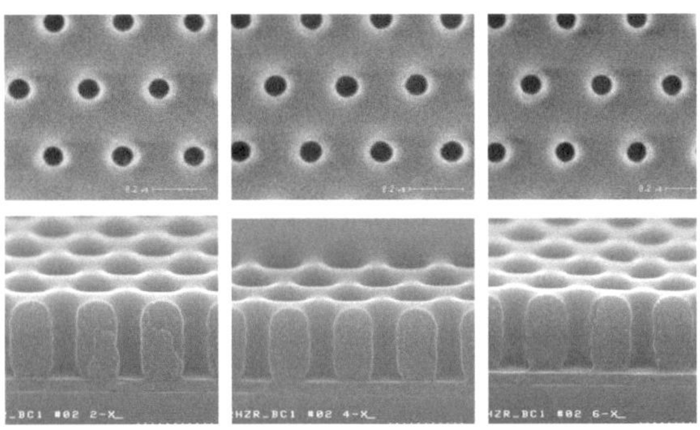

Fig. 16. SEM images of contact holes for various temperatures and duty ratios [9].

Figures 16 and 17 show experimental results of Hynix Semiconductor Inc. due to various temperature and duty ratios for KrF 248 nm [9] and ArF 193 nm illumination [10, 11]. The contact holes become smaller as the duty ratio becomes larger and as the temperature becomes larger.

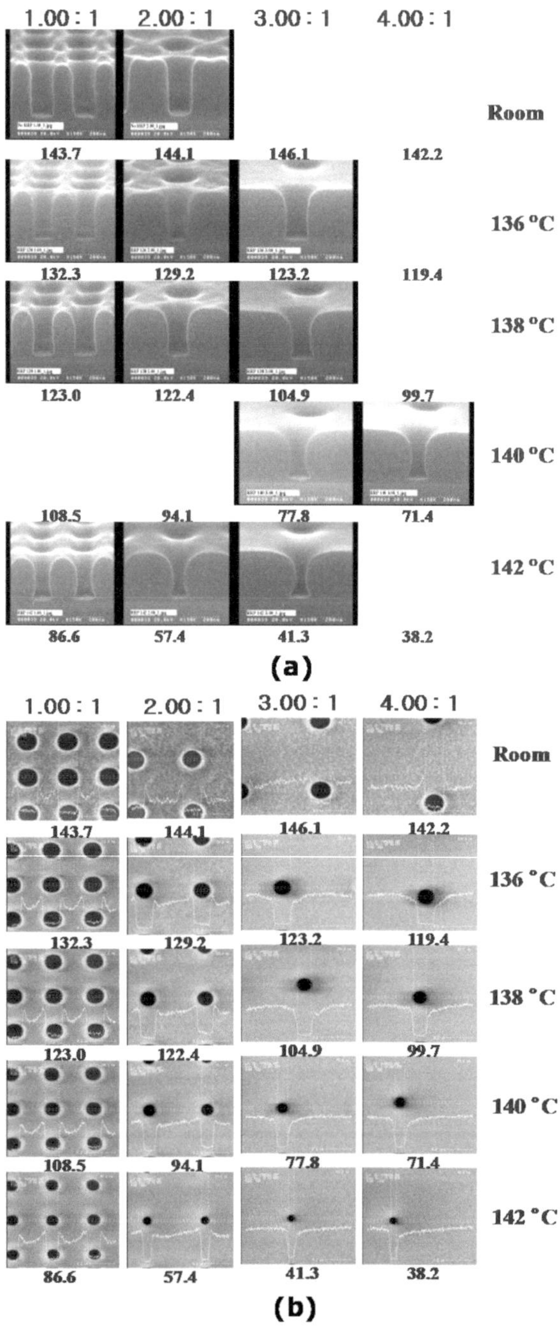

Fig. 17. Experimental results of (a) side view and (b) top view for various temperature and duty ratios in Hynix Semiconductor Inc. [10, 11]

Figure 18 shows simulation results of C/H pattern shrinkages after thermal processes as a function of time for 248 nm KrF and 193 nm ArF CARs. Both of CARs have similar thermal effects. The underbake PEB and the thermal reflow can reduce CD. However, during SB, CD is unchanged. During thermal reflow, CD is saturated quickly due to time. SB is soft bake, PEB is post exposure bake, and TR is thermal reflow.

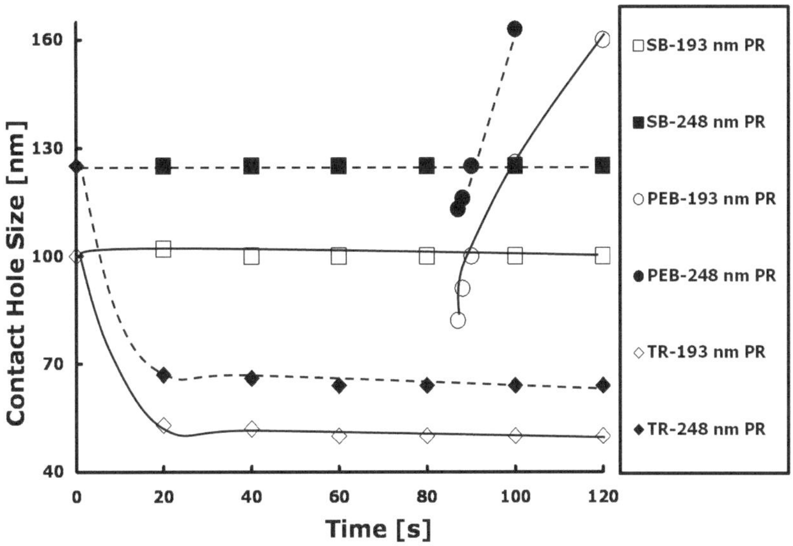

Fig. 18. Simulation results of pattern shrinkages after thermal processes as a function of time for ArF 193 nm and KrF 248 nm CARs.

	Thickness	CD sensitivity on temp.	CD sensitivity on time
Softbake	$d_y \approx \chi_S$	≈ 0.26 nm/°C	≈ 0 nm/s
PEB	$d_y \approx C_{as}$	≈ 3 nm/°C	≈ 2.3 nm/s
Thermal Reflow	$d_x = \sqrt{\frac{\mu}{\rho}t}$	≈ 10 nm/°C	≈ 2.3 nm/s

Table IV. Comparison of thermal processes (d_y: the thickness reduction in y-direction, d_x: the thickness reduction in x-direction, χ_S: the mass fraction of solvent, C_{as}: the concentration of deprotection resin, μ: the fluid viscosity, ρ: the density, t: the time, and temp.: the temperature)

Table IV shows simulation parameters with thickness variation of thermal processes, and the CD sensitivities of these process temperatures and times for a C/H pattern. Among the thermal processes, thermal reflow is the most effective method for 45-nm pattern formation. The CD sensitivity of a C/H pattern can be improved to 4.6 nm/s by using the PEB time and thermal reflow. However, due to the pitch size, these thermal processes produce OPEs, such as the impact of the image distortion. The OPE is quite severe when those thermal process are used for 45-nm patterns, so model-based OPC must be performed.

Figure 19 shows simulation results for thermal SB bias. CD shrinkage is dependent on pitch sizes. Thermal SB biases are depended only on two kinds of pattern types such as isolated pattern and dense pattern:

$$f_{SB}^2(V) \approx f_{SB}^2(R) = \begin{cases} 1 & \text{for isolated pattern }, \\ \alpha_1 & \text{for dense patterns} \end{cases} \quad (63)$$

where R is the ratio of contact hole and half pitch size and α_1 is a constant value. The thermal PEB bias of pitch size includes thermal SB bias because PEB process is a following process after SB process: $f_{SB}(R) \le f_{PEB}(R)$. A 193 nm resist has larger thermal bias than a 248 nm resist in Fig. 19.

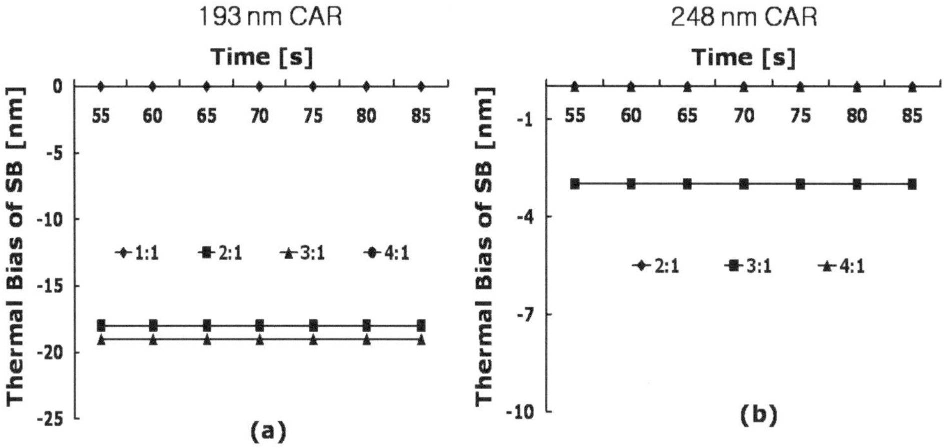

Fig. 19. Simulation result for thermal biases of SB. For both of a ArF 193 nm CAR in Fig. 19 (a) and a KrF 248 nm CAR in Fig. 19 (b), CD shrinkage is independent on SB time and SB temperature, but dependent on pitch sizes.

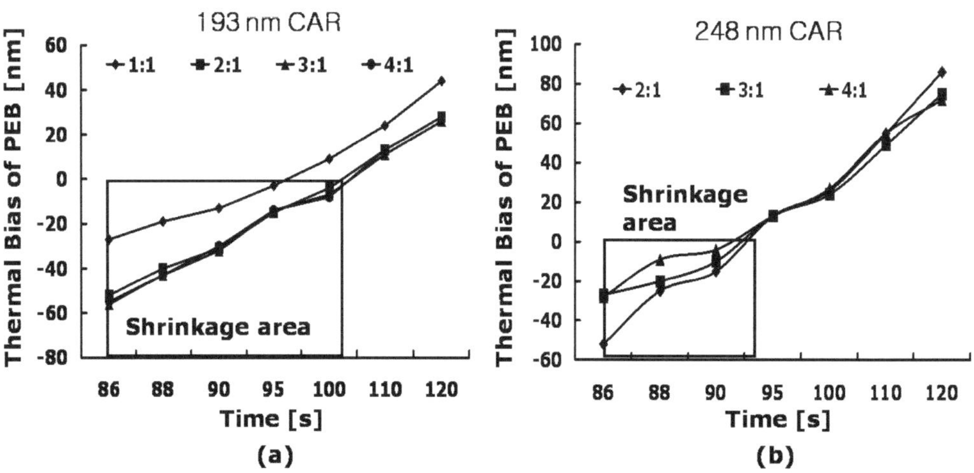

Fig. 20. Simulation results for thermal biases of PEB. In both of a ArF 193 nm CAR in Fig. 20 (a) and a KrF 248 nm CAR in Fig. 20 (b), the C/H pattern of 1:1 line and space has different PEB effects with other C/H patterns

Figure 20 shows simulation results for thermal PEB bias. In a 193 nm CAR in Fig. 20 (a) and a 248 nm CAR in Fig. 20 (b), PEB effects for C/H pattern of 1:1 line and space are different with those of other patterns. In both of CARs, PEB time is more controllable than PEB temperature for CD shrinkage. Thermal PEB biases depended on pattern types is same as thermal SB biases, so that

$$f_{PEB}^2(R)=\begin{cases}1 & \text{for isolated pattern},\\ \alpha_2 & \text{for dense patterns}\end{cases} \tag{64}$$

where α_2 is constant. When PEB temperature is at a target temperature, the function of PEB biases depended on PEB time is an exponential function of PEB time, so the thermal bias of PEB can be represented as:

$$f_{SB\,(or\,PEB\,)}^2(t)=\left[\alpha_3\bullet\exp\left(\alpha_4 t\right)\right], \tag{65}$$

where α_3 and α_4 are constant values. Hence, For SB and PEB, thermal biases ($\Delta CD_{SB\,(or\,PEB)}$) of Eq. (60) based an orthogonal function can be represented as

$$(\Delta CD)_{SB}^2=f_{SB}^2(R), \tag{66}$$

$$(\Delta CD)_{PEB}^2=f_{PEB}^2(R)\bullet f_{PEB}^2(t), \tag{67}$$

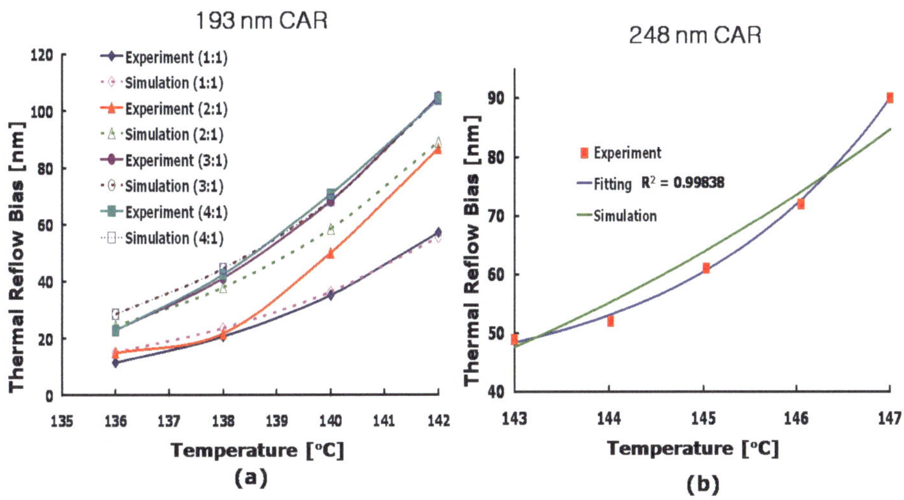

Fig. 21. Comparison between experiment results and simulation results for thermal reflow biases depended on temperature and pitch sizes in a ArF 193 nm CAR and a KrF 248 nm CAR.

Figure 21 shows the comparison between experiment results and simulation results of thermal reflow biases depended on temperature and pitch sizes in a 193 nm CAR in Fig. 21 (a) and a 248 nm CAR in Fig. 21 (b). Simulated results agree well with the experimental results in a small error range. The simulated function of thermal bias for a 193 nm CAR is

$$\begin{aligned}(\Delta CD)_{Thermal\ reflow}^2&=\left[\alpha_1\bullet\exp(\alpha_2/T)\right]\left[\alpha_3\bullet\exp(\alpha_4 R)\right]\\&=\left[3.696\times10^{29}\bullet\exp(-8347.64393/T)\right]\left[1.88152\bullet\exp(-1.90217\bullet R)\right]\end{aligned}. \tag{68}$$

The simulated function of thermal bias depended on temperature for a 248 nm CAR is

$$(\Delta CD)^2_{Thermal\ reflow}(T) = \left[\alpha_1 \bullet \exp(\alpha_2 / T)\right] = \left[6.9796 \times 10^{25} \bullet \exp(-7429.2622 / T)\right]. \quad (69)$$

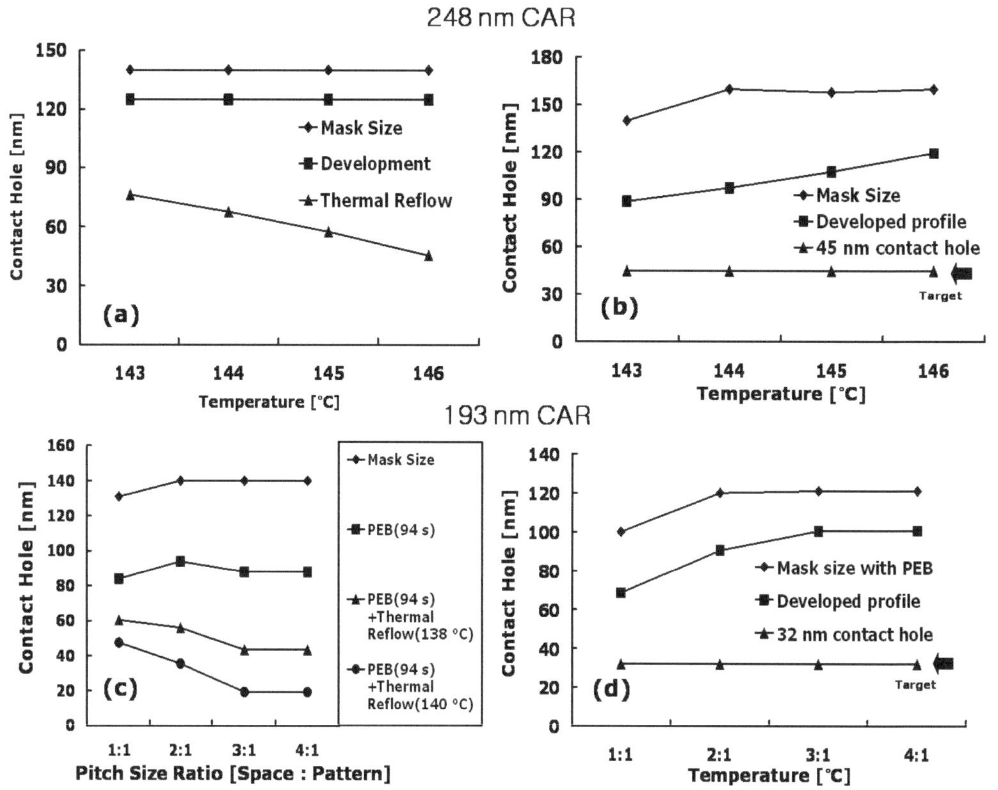

Fig. 22. Simulation results by using the orthogonal function method as an OPC method for a KrF 248 nm CAR in Figs. 22 (a) and (b) and a ArF 193 nm CAR in Figs. 22 (c) and (d).

Figure 22 shows simulation results by using the orthogonal function method as an OPC method. For a 248 nm CAR, 140 nm mask pattern size is shrunk down between 76.2 nm and 45.5 nm depended on the temperature of thermal reflow in Fig. 22 (a). When target pattern size is 45 nm in Fig. 22 (b), mask pattern size can be between 140 nm and 160 nm due to the temperature of thermal reflow. For a 193 nm CAR, mask pattern sizes between 131 nm and 140 nm can be shrunk down between 60.41 mm and 88 nm depended on pitch size ratio by using 138 °C temperature of thermal reflow in Fig. 22 (c). Also, those mask patterns can be reduced down between 47.46 nm and 19.42 nm depended on pitch size ratio by using 140 °C temperature of thermal reflow. When target pattern size is 32 nm in Fig. 22 (d), mask pattern size can be between 100 nm and 121 nm due to pitch size ratio.

5. Conclusion

Thermal treatment is a new process-extension technique using current-day lithography equipment and a chemically amplified resist. Thermal processes, such as softbake, post-

exposure bake, and thermal reflow process, are modelled from the chemical and physical behaviors of the CAR. For each thickness reduction, the simulated results agree well with the experimental results. For small pattern formation, the thermal effects move the boundaries of the resist bulk images to a horizontal point and make the boundaries dense. For line and space patterns, an underbake of SB and an overbake of PEB can be used for 45-nm pattern formation. For contact hole patterns, an underbake of PEB and thermal reflow can reduce the critical dimension. Although the CD sensitivity of the thermal parameters can be different due to other process parameters, thermal reflow is more effective method than the other two thermal processes. However, the optical proximity effect is quite severe when thermal processes are used for 45-nm patterns, so a model-based optical proximity correction is required for each thermal processes.
Hence, the orthogonal functions of thermal biases depended on temperature (or time) and pitch size are assumed and defined by using experimental data. Simulated results of thermal reflow for contact hole agree well with the experimental results in a small error range according to the baking temperature, time, and pitch size. For C/H patterns, underbake PEB and thermal reflow can be formed 32 nm pattern for a ArF 193 nm CAR and 45 nm pattern for a KrF 248 nm CAR. An orthogonal functional method is useful to deal with OPC of thermal effects.

6. References

[1] Sigma-C: http://www.solid-c.de.
[2] C. A. Mack, "Inside PROLITH: A Compreshensive Guide to Optical Lithography Simulation", FINLE Technologies (Austin, TX: 1997).
[3] S. I. Lee, K.C. Ng, T. Orimoto, J. Pittenger, T. Horie, K. Adam, M. Cheng, E. Croffie, Y. Deng, F. Gennari, T. Pistor, G. Robins, M. Williamson, B. Wu, L. Yuan and R. Neureuther, "LAVA web-based remote simulation: enhancements for education and technology innovation," Proc. SPIE 4346, 1500 (2001): http://cuervo.eecs.berkeley.edu/Volcano.
[4] KLA-Tencor, FINLE Division, "Introduction to the PROLITH Toolkit,": http://www.kla-tencor.com.
[5] J. D. Byers, M. D. Smith, C. A. Mack, and J. J. Biafore, "Modeling soft-bake effects in chemically amplified resists", SPIE 5039, 1143 (2003).
[6] G. Thallikar, H. Liao, T. S. Cale, and F. R. Myers, "Experimental and simulation studies of thermal flow of borophosphosilicate and phosphosilicate glasses," J. Vac. Sci. Technol. B 13, 1875 (1995).
[7] D. Kang, S. Robertson and E. Pavelchek, "Simulation of 193 nm photoresists based on different polymer platforms," SPIE 4345, 936 (2001).
[8] D. Kang, S. Robertson, M. Reilly and E. Pavelchek, "Effects of mask bias on the mask error enhancement factor (MEEF) of contact holes," SPIE 4346, 858 (2001).
[9] H.-R Kim, Y.-B. Ahn, J. Kim, S. Kim, D. Park and Y.-S. Kim, "Implementation of contact hole patterning performance with KrF resist flow process for 60nm node DRAM application," Proc. SPIE 6154, 615441 (2006).

[10] C-W. Koh, D-H. Lee, M-S. Kim, S-N. Park, and W-K. Kwon, "Characterization of resist flow process for the improvement of contact hole CD uniformity," SPIE 5039, 1382 (2003).

[11] J-S. Kim, J-C. Jung, K-K. Kong, G. Lee, S-K. Lee, Y-S. Hwang, and K-S. Shin, "Contact hole patterning performance of ArF Resist for 0.10 μm technology node," Proc. SPIE 4690, 577 (2002).

EUV and X-Ray Lithography

9

CO_2 Laser Produced Tin Plasma Light Source as the Solution for EUV Lithography

Akira Endo
Friedrich-Schiller University
Institute of Applied Physics,
Jena, Germany

1. Introduction

High average power EUV light source has been the "most critical" issue in the research and development of the EUV lithography system in one decade. EUV LLC and International Sematech significantly stimulated the global research community to work seriously to advance plasma technology in achieving the goal of the EUV source, required by the semiconductor industry. It is instructive to look into the EUV lithography source workshop held in October 2001 in Matsue, Japan. MEDEA+ project consortium "EUV source" had started already in June 2001 in Europe including many public and private research organizations (Stamm, 2002). EUV source workshop was organized in Japan several times by ASET in 2000-2001 to evaluate the technical possibility to develop the required EUV source (Okazaki, 2001), which was then succeeded by EUVA project in 2002.
ASML and NIKON talked in the Matsue workshop as the required EUV power was more than 80W with faster than 5 kHz repetition rate, assuming the resist sensitivity as $2mJ/cm^2$. The requirement came from 80 wafers/hour throughput. The wavelength was confirmed at 13.5nm to optimize to the peak reflectivity of Mo/Si coated mirrors. This caused a serious concern on the scaling of the once established method based on Nd:YAG laser irradiated xenon gas plasma, in which the peak conversion efficiency was at 11nm. Nd:YAG laser irradiation of a gas puff target was a typical laboratory method in laser applications like higher harmonics generation, short wavelength generation, fast ion generation and so on (e.g., Fiedorowicz, 1999). TRW had been working closely with EUV LLC from 1997 to develop the first generation EUV light source, based on Nd:YAG laser irradiated xenon gas puff plasma. The obtained conversion efficiency was around 0.2%, with laser power of 500W by a single beam (Ballard, 2002). It was suggested to increase the density of xenon at longer distance from xenon nozzle to improve the conversion efficiency, together with higher laser power of better beam quality. It seemed still possible to work following the proposed direction, but experiments showed practical limitations around this approach. The author discusses in the following sections on each limitation in the history of the EUV source development, and describes each obtained solution, overcoming the limitation to realize the required source performance finally. The major work was conducted during the project of "Extreme Ultraviolet Lithography System Development Association (EUVA)" in Japan from 2002 to 2009.

2. Liquid xenon jet and fast ions

Typical conversion efficiency (CE), from input laser pulse energy to in band EUV energy (2π sr, 2% bandwidth), was 0.1% from a Nd:YAG laser irradiated xenon gas puff target. Cluster phase xenon was shown to generate higher EUV flux in the same experimental arrangement. Clusters were formed in the gas puff with higher back pressure of xenon in the nozzle (Mori et.al., 2001). Cryogenic liquid jet target was proposed by H.M.Hertz of Royal Institute of Technology, Sweden, as a useful tool for higher CE EUV target (Berglund, et.al., 1998). Advantages were; longer distance between nozzle tip and plasma, which was essential to avoid nozzle erosion due to plasma heat, initially higher density, smaller plasma size and fast jet speed to sustain more than 10 kHz laser irradiation. One of the disadvantages was excessive xenon gas injection into vacuum chamber due to evaporation of continuous liquid xenon jet, which in turn absorbed generated EUV light. Figure 1 is the concept of the xenon liquid jet. Continuous droplet target was proposed for liquid xenon to reduce the injected mass, so called mass limited target from TRW. Xenon jet and droplet targets were irradiated at 10 kHz repetition rate, with a short pulse Nd:YAG laser.

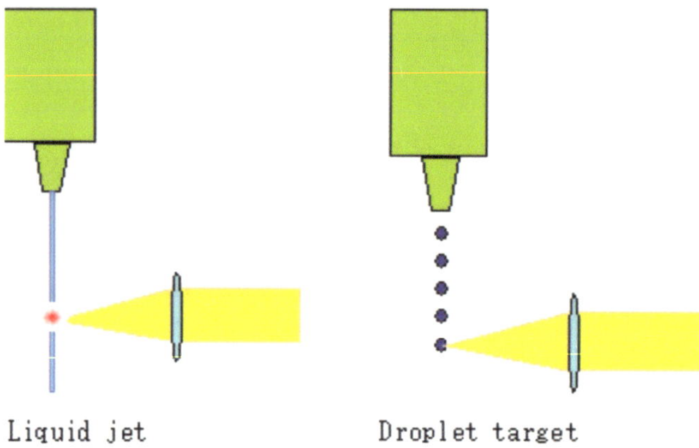

Fig. 1. Concept of liquid xenon jet and droplet targets for a laser produced plasma EUV source

Condenser mirror (C1) lifetime is the most critical technical issue in a commercial EUV light source for the sake of lower cost of ownership (CoO). Multilayer erosion was not significant in the initial model of ETS EUV source, in which gas puff or cluster xenon target was employed. It was reported in February 2003 that one layer of Mo/Si coating was lost by 300 million laser shots of 500W average power, with the distance between plasma and condenser mirror as 15cm (Ballard, 2003). The life time of C1 was estimated as six months in this operational mode. Drastic erosion speed was reported after half year from EUV LLC in the case of liquid xenon spray jet target, in the 2nd international EUV Lithography Symposium in Antwerp, Belgium in September 2003 (Klebanoff, 2003). One layer was lost by 15 million laser shots, and the erosion was not geometrically symmetric. It strongly suggested an influence of fast particle emission during laser irradiation of dense spray target. Fast ions were already measured at 10cm from cryogenic liquid jet target by short pulse laser irradiation at 10^{16}W/cm^2 (Wieland, 2001). The kinetic energy of the emitted ions was more than 100keV from a nitrogen liquid jet target. The plasma temperature is typically

30eV for EUV emission, and the reason of the fast ion generation is not a simple thermal expansion. There were another two talks in the Antwerp Symposium on this subject. J.D.Gillaspy of NIST reported on the significant effect of highly charged xenon ions like Xe^{10+} responsible for the multilayer damage (Gillaspy, 2003). H.Komori of EUVA reported on the measurement of xenon ion distribution with Xe^{2+} as the main ion species and the dominant energy as 3keV at 10^{11}w/cm^2 laser irradiation of 8ns pulse width on a liquid jet xenon target (Komori, 2003). Etching rate was also reported by using a xenon ion gun with data confirming the multi layer loss rate in the ETS light source. Figure 2 is the result of the erosion of a test multilayer sample with an equivalent ion beam flux.

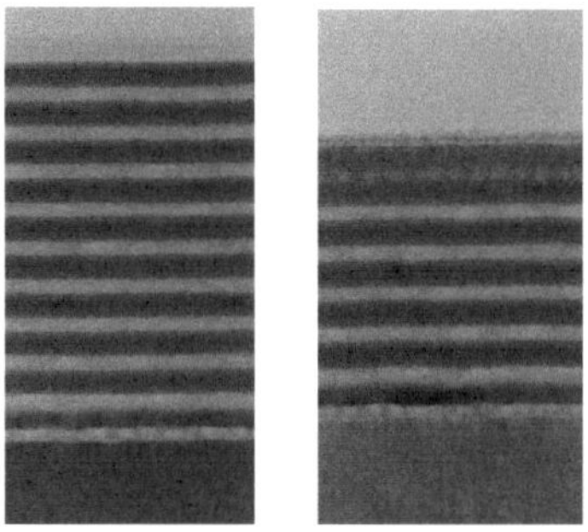

Fig. 2. Cross sections of a sample 10 layer Mo/Si mirror, before (left) and after (right) ion sputtering Several layers are lost and mixing of Mo/Si are observed

CE was improved from 0.2% by a gas puff xenon target to 0.5% by a spray xenon target, but the C1 erosion was increased almost 10 times faster. It was clear for many researchers that any high density target like liquid xenon jet was not a simple solution for a practical EUV light source.

3. Scaling of laser power

It is useful to recall the discussion in Matsue Source Workshop on the scalability of the component technology for the required EUV source. TRW predicted as a 10kW laser power was needed to achieve 70W collectable EUV power with 1.4% CE from xenon liquid jet target. Xtreme assumed 60kW laser power for 200W EUV power with 1% CE. Mentioned laser power was something unimaginable as a short pulse solid state laser at the time of the workshop.

The first industrial CW, kW Nd:YAG laser was introduced into the market in the late 80's. Main component was a flash lump pumped laser rod with water cooling in a series arrangement for high average power operation (Hodgson et.al., 2005). The beam quality was not good enough for fine focusing at high average power due to thermal depolarization and lens effect. Typical beam parameter product was almost 10 times worse than a CW CO_2 laser at the same average power. Adaptive optics technology was studied to improve the beam quality, like phase conjugation or deformable mirror, by compensating the beam distortion inside the rod (Druon, et.al., 1998). After many works on the adaptive optics, fundamental solution was thought as to design an efficient cooling of the laser media, and new types of fiber (Nilsson et.al., 1993), thin disc (Giesen et.al., 1994) and slab (Shine et.al., 1995) lasers were started to be developed from early 90's together with laser diode pumping for higher efficiency. Significant progress was achieved in the development of CW, high beam quality laser by slab, fiber, and thin disc configuration with Yb material together, in the last ten years achieving more than 10kW average power. Short pulse amplification is not as easy as CW operation due to higher peak power inside the optical components. Typical average power of high beam quality solid state laser is less than kW level in short pulse, high repetition rate mode in slab, fiber and thin disc types at the time of Photonics West 2009.

Initial generation of high average power, high repetition rate, short pulse Nd:YAG laser was started to be developed for EUV plasma ignition, based on the same technology of the series arranged rod amplifiers, from late 90's. Beam distortion was also much significant inside the laser rod due to higher peak power in pulsed operation. It was experimentally shown that the practical maximum average power, typically at 10 kHz with 10 ns pulse width, was around 1kW from a single beam, even by laser diode pumping, based on the rod amplifier technology.

Multiplexing of laser modules was thought as a possible approach. Exulite program employed this scheme by using 500 W modules composed of two rods, from Thales SA, (Fay, 2003). 10 laser modules were arranged to focus onto a xenon gas jet target. It was shown that the arrangement was too complicated for alignment in a busy target chamber. The laser module was driven by two acousto-optic modulators with pulse length more than 30ns at 10 kHz.

An alternative approach for the driver laser was definitely required for high average power EUV source. One of the candidates as a high average power laser was DUV excimer laser, which was already used in lithography, flat panel display manufacturing and ink jet nozzle drilling. Typical output power was 300W in MOPA configuration at 300Hz repetition frequency for liquid crystal annealing application. Cymer tried to use injection seeded XeF laser for liquid Li target (Hansson et.al., 2004). EUVA used KrF laser for xenon jet target (Abe et.al., 2005). Both trials were abandoned due to two reasons. The first one was the shorter wavelength of the DUV lasers. Short wavelength optical wave penetrates deep into the plasma, and significant reabsorption of generated EUV light and cluster blow off are the problems. The second reason was unrealistic scaling of the DUV excimer laser towards multi 10kW output power.

The remaining candidate was the CO_2 laser only, for multi 10kW average power in pulsed mode. CW CO_2 lasers are the most frequently used lasers in industry due to their comparably low initial and operational costs, as well as their robustness and reliability with environmentally non toxic nature. CO_2 lasers are also operated in long pulsed mode for certain applications. RF-excitation is the most commonly employed scheme in axial flow or conduction cooled slab/waveguide configurations. The design guideline of a multi kW short pulse CO_2 laser system is described in this section. Its main characteristics are high repetition rate, high pulse energy, high amplification efficiency and high beam quality. The system is based on commercial high average power CO_2 laser modules, used as amplifiers.

Medium average power pulsed CO_2 laser systems are very successful tools for various applications ranging from material processing of metals, glass, ceramics and epoxy, paint removal and medical or spectroscopic applications, to the generation of laser produced plasmas as UV, EUV and soft X-ray sources. One drawback is the limited repetition rate of TEA CO_2 laser based source, another drawback is limited controllability of the pulse width in low pressure microwave excited lasers. Attempts were reported in early 90's to operate microwave excited CO_2 laser modules in a Q-switched oscillator mode of CW 2kW device (Sakai et.al., 1994) and an oscillator-amplifier mode of CW 7kW system (Bielesch et.al., 1992). Typical performances were at the repetition rate of 4 kHz with output average power of 680 W with pulse energy of 170 mJ and pulse width in full width half maximum (FWHM) of 250 ns, and at the repetition rate of 10 kHz with average power of 800W, with pulse energy of 70 mJ, and 35 ns pulse width, respectively. Laser extraction efficiencies, however, were not very high in both cases in the short pulse mode. Commercially available short pulse CO_2 laser oscillator was known typically as EOM-10 from De Maria Electro Optics Systems, Inc (now Coherent Inc). The specification was average power of 10W at 100 kHz repetition rate with 15ns pulse width. Systematic laser development was started in the EUVA laboratory in Hiratsuka, Japan from 2004 aiming at short pulse, 10kW level CO_2 laser technology (Endo et.al., 2006).

A short pulse oscillator was installed as the seeder for the amplifiers. The laser was an EO Q-switched, 15~30 ns, single P(20) line, RF pumped waveguide CO_2 laser with 60 W output at a repetition rate of 100 kHz. The repetition rate was tunable as 10~140 kHz.

Commercial 5 kW and 15 kW CW CO_2 lasers were installed in the laboratory as amplifiers. Every unit is 13.56 MHz RF-excited, fast axial flow lasers from Trumpf Inc. Lasers were modified as amplifiers by replacing both cavity mirror with ZnSe windows. The 5 kW laser used a standard gas composition of CO_2:N_2:He=5:29:66 at 120 Torr gas pressure. The axial gas flow speed was sufficiently high to keep the laser gas temperature low inside the operational condition. The length of a single gain region was 15 cm, and 16 cylindrical gain regions were connected in series in one laser unit; the tube inner diameter was 17mm. The total length of the optical pass inside the laser was 590 cm. The laser operated at 5 kW CW output power with a M^2 =1.8 beam quality. The electrical input power was 36 kW. The 15 kW laser as the main amplifier, used a standard gas composition of CO_2:N_2:He=2:10:48 at 150 Torr gas pressure. The length of a single gain region was 28 cm, and 16 active cylindrical gain regions were connected in series; the tube inner diameter was 30 mm. The total length of the optical pass inside the laser was 890 cm. The maximum electrical input power was 88 kW. The key parameters of the amplifier are the extraction efficiency and beam quality. A series of experiments were performed to clarify these parameters to estimate the final possible values (Hoshino et.al., 2008).

The maximum average output power of 8 kW was obtained at a repetition rate of 100 kHz with 3kW input power to the main amplifier. Efficient short pulse amplification with RF-pumped gain modules, requires that parasitic oscillations and/or optical coupling between amplifier modules do not exist, and which was experimentally verified. It was successful to extract 5kW power in pulsed mode from CW 15kW laser. The extraction efficiency ((output power-input power)/ CW output power) was over 30%. Filling factor is the parameter to occupy the amplifier active volume with injected laser beam. Extraction efficiency was measured with the filling factor as the parameter. Once the average power was restricted about 6kW, extraction efficiency depended on the filling factor. It indicated a possibility to extract more power from preamplifier stage by optimizing the input flux.

The laser beam quality was measured with a ZnSe lens of 508mm focal length and a slit-scan type beam profiler (Photon Inc., NanoScan). The laser beam size at the lens focus was measured for the oscillator and amplifier, resulting in a beam quality factor M^2 as 1.1. Especially, the laser beam size was identical before and after amplification, i.e. the amplification did not cause any phase distortion. Fig.3 shows a typical beam profile. Fig.4 shows the temporal laser pulse profile of the amplified laser output. The pulse duration was 20 ns (FWHM) and the pedestal was below 10% of the total pulse energy. A pedestal and/or tail of the seed laser pulse could be amplified and reduce the laser gain. There was no harmful further amplification of back scattering light from tin (Sn) solid target in the amplifiers after full depletion of laser gain.

After the series of laser experiments, it was concluded that the reasonable output laser power is 20kW from a single amplifier chain (Endo, 2007). Additional amplifier of CW 20kW power device increased the output power to 13kW (Endo, et.al., 2008). The critical issue is the thermal distortion of the optical components (windows and mirrors) inside the laser system. Present effort is now focused on the improvement of the extraction efficiency, by using slab amplifier and multiline amplification (Nowak et.al, 2008).

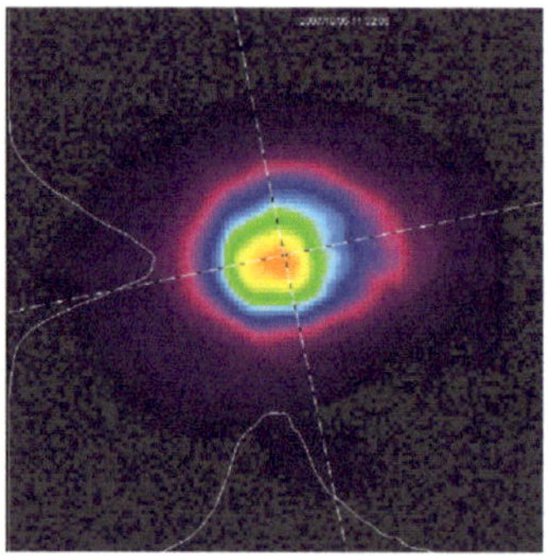

Fig. 3. CO_2 laser beam of M^2=1.1

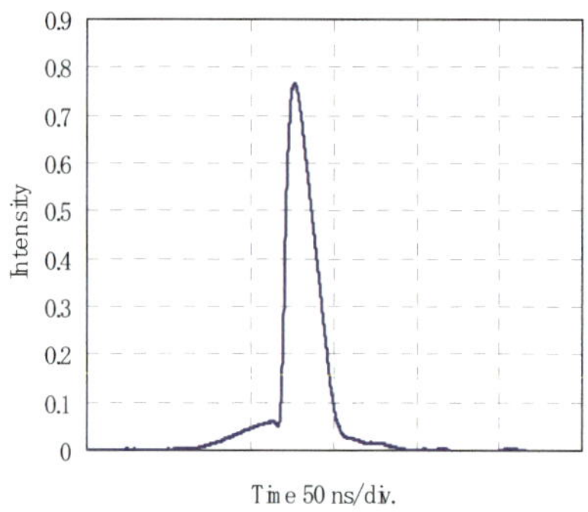

Fig. 4. CO_2 laser pulse shape with low pedestal

4. Laser wavelength and EUV conversion efficiency

The conversion efficiency (CE), from the input laser pulse energy to the generated EUV pulse energy at 13.5nm (2% bandwidth, 2π sr), is the major parameter for improvement in high average power EUV light source for better economy. The next step, which was necessary after the laser power availability by CO_2 laser, was to confirm the practical conversion efficiency by 10 μm wavelength laser pulse. Pulsed CO_2 laser was often employed in laser plasma experiments until 80's, but gradually disappeared from laboratories after improvements of solid state pulsed lasers. It was once employed as a driver of a plasma X-ray laser from a carbon target (Suckewer, et.al., 1983).

The first report of the CE measurement by a pulsed CO_2 laser appeared in 2004 from a research group of Kyushu University (Tanaka et.al., 2004). They observed a Nd:YAG comparable CE from xenon gas and cryogenic target by long (100ns) CO_2 laser pulse. The results were more or less unexpected for many EUV source researchers. It was believed in general that the CO_2 laser wavelength was too long to heat up the plasma efficiently. More systematic experiments were performed in EUVA Hiratsuka Laboratory by using a xenon liquid jet target and Nd:YAG pre-pulse together with a short pulse width CO_2 laser as the main heating source (Komori, et.al., 2006), (Ueno , et.al., 2007). Optimization of the delay time between pre-pulse and main pulse was adjusted for the highest CE from a xenon liquid jet target. It was measured that the density optimization of the initially dense xenon target by a pre-pulse, was essential to raise the CE by a short pulse length CO_2 laser. Maximum CE of 0.6% was measured in the experiment. Another report was given from Kyushu University group that a tin target gave 2% CE by CO_2 laser irradiation. The combination of CO_2 laser wavelength with tin indicated the best matching for higher CE. CO_2 laser gave the maximum CE at the laser irradiation intensity lower than Nd:YAG laser case (Tanaka, et.al.,2007). Further improvement of CE was reported in a case of tin cavity target close to 5% (Ueno,et.al., 2007). A cavity of 200 μm depth on a planar tin was irradiated by a 10 Hz short (10ns) CO_2 laser pulse with a ZnSe lens by 60mm focusing length. The focal spot diameter was 100 μm, and the irradiation intensity was 1-3 x 10^{10} W/cm^2. It was concluded from the experiments on the origin of higher CE compared to Nd:YAG irradiation, that the self absorption of the generated EUV emission by surrounding plasma reduced the CE. CO_2 laser wavelength is ten times longer than the one of Nd:YAG laser, and the plasma density is hundred times lower, which in turn results in a lower EUV self re-absorption. Plasma spectrum is less affected by the self absorption in the case of CO_2 laser irradiation of tin. The results were confirmed experimentally by UCSD group independently (Tao et.al., 2009). It is now established that CO_2 laser power is scalable to 10kW level, and the CE is high enough to achieve more than 100W intermediate focus (IF) EUV power from a reasonable scale device.

Numerical simulation by a radiation hydro code gave a higher CE up to 8% in the best optimization of the pre-plasma condition by a pre-pulsed tin plasma (Nishihara et.al., 2008). The increase of the CE was explained due to increased CO_2 laser absorption in a broader critical density region generated by an optimum pre-pulse irradiation. The numerical model predicted 10ns laser pulse length as the optimum one for the highest CE in 2-dimensional model. Figure 5 shows the measured CE from tin plasma generated from a cavity target by Nd:YAG and CO_2 lasers. Figure 6 shows the spectrum from tin plasma generated by Nd:YAG and CO_2 laser. Both spectra were overlapped with its peaks as the same value.

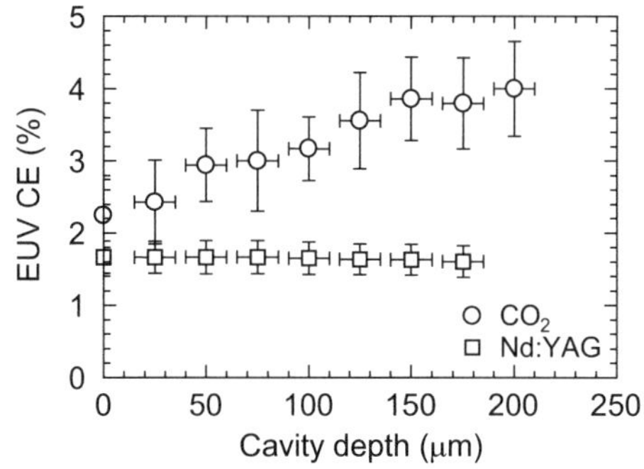

Fig. 5. CE from tin cavity target

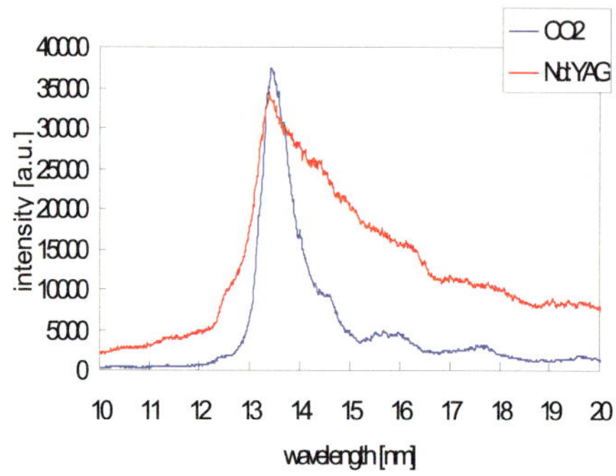

Fig. 6. Spectrum from tin plasma

5. Wavelength dependence of ablation particles

Laser ablation generates fast nano particles, vapors, fragments etc through a complex physical process. Wavelength dependence of the process was studied intensively in 2006 (Ueno, et.al., 2007). The motivation of the work was to evaluate the effect on the life time of the large EUV collector (C1) mirror.

The experiment was performed to evaluate the behavior of a QCM (Quartz Crystal Micro Balance) detector located from a plasma plume generated by a tin plate target by a 6ns Nd:YAG laser and 10ns TEA CO_2 laser. The QCM signal increased continuously during Nd:YAG laser irradiation, which meant a continuous tin particle deposition on the surface. It was observed at all angles, but the amount of deposition increased faster at a position closer to the laser beam axis, i.e. closer to the target normal. Completely different behavior was observed in the case of CO_2 laser irradiation on the QCM located close to the laser irradiation axis, which meant a continuous sputtering. The absolute signal amplitude was smaller than the case of Nd:YAG irradiation. Measurement of a silicon sample surface after irradiation confirmed the phenomena. Physical reason of the behavior was the difference of the wavelength of lasers. The critical density is two magnitudes lower in the case of CO_2 laser irradiation, which in turn benefits the lower number of ablated tin particles.

Ion signals were measured by Faraday cup detectors placed inside the vacuum chamber. The ion signals were almost identical for both Nd:YAG and CO_2 laser irradiation in the same EUV in-band energy. No major difference of the total ion current was observed. This measurement indicated the sputtering effect of a QCM surface was identical for both case. It was then concluded that the generated neutral particles were 1000 times more in the case of Nd:YAG laser irradiation.

The difference of the plasma behavior is summarized based on these measurements as table 1. CO_2 laser generated tin plasma has a higher CE, narrower spectral bandwidth, 1000 times fewer neutral particles.

6. Magnetic field control of plasma particles

A magnetic confinement of ablation plasma was tested in earlier experiments (Nd:YAG, CO_2 laser + xenon jet) with a coil pair. The maximum magnetic field was 0.6 Tesla, where reduction of the QCM surface was observed for both cases (Komori, 2006). A new series of experiments were performed to evaluate tin ablation effect on the surrounding surface in

Items	CO2		Nd:YAG
CE	2.5 - 4.5%		1.5 - 2.5%
In band spectral bandwidth (nm, FWHM)	0.8		2.5
Debris	1	:	1000
Fast ion	1	:	2
Out-of-band		<	
Cost			
Initial	1	:	2 - 3
CoO (exclude electricity)	1	:	5
Mo/Si reflectivity at laser wavelength	> 90%		30%

Table 1. Comparison of CO_2 and Nd:YAG laser generation of EUV tin plasma

higher magnetic field (Ueno et.al., 2008). A vacuum chamber of 50mm active height and 530mm diameter was fabricated and installed inside a normal conductance magnetic field of up to 1.1Tesla. A movable tin plate target was positioned at the center of the chamber, with Mo/Si test mirror or QCM detector 60mm apart from the plasma, and Faraday Cups 100mm apart. The vacuum level was kept as 10^{-3} Pa during the experiment. The plasma behavior was measured as 2 dimensional phenomena in this experiment. CO_2 laser irradiation was at 1-3 x 10^{10}W/cm^2 focused intensity on the normal incidence tin plate target typically with 2.0%-2.5% CE.

Time of flight ion signal showed peak energy of 0.9keV, with maximum energy of 4keV. Deposition of tin vapor was observed on the sample mirror located off axis of the laser irradiation, as 3nm/10^5 laser shots, and evaluated as the results of the balance between sputtering and deposition of fast and slow ions. The ion signal vanished completely with 1 Tesla magnetic field under signal noise level. Tin deposition speed was decreased to 0.6nm/10^5 laser shots, which was 20% of the case of no magnetic field. It was concluded that ablation plume consisted of more than 80% as ions in this experiment. Recent experiment indicates that recombination of tin ions plays additional contribution to the neutral generation during plasma flow.

A large volume vacuum chamber was fabricated to evaluate the behavior of plasma plume in a 3 dimensional space in higher magnetic field (Endo, et.al., 2008). A superconducting magnet of maximum 3 Tesla installed the vacuum chamber inside the magnet bore. A tin plate was fixed on the magnetic axis at the edge of the magnet, and irradiated by a Nd:YAG laser with 30mJ pulse energy, 5ns pulse duration to give the intensity up to 1.6 x 10^{11} W/cm^2. A well collimated plasma flow was observed along the magnetic field. A large diameter (44mm) Faraday Cup was installed on the beam pass of the plasma stream at 300mm from the plasma generation position. The measured plasma current had pulse

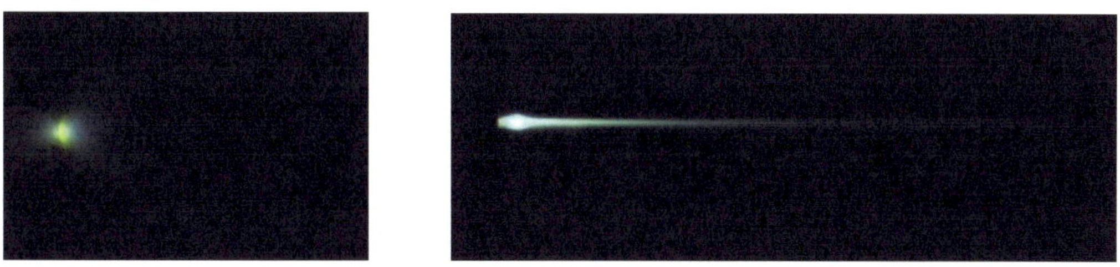

Fig. 7. Plasma images without (left) and with (right) horizontal magnetic field

duration of 20 μs with peak current of 5A. The total charge was 0.1mC. It was concluded from the measurement that ions generated in the ablation process were efficiently transported along the magnetic field line to the exhaust. The remaining contamination of the surroundings is concluded as neutral tin vapor deposition. The origin and behavior of the neutral tin vapor is the key to fully control the chamber contamination.

7. Behavior of tin vapor after ablation of a droplet target

Charged particles are fully controllable with a magnetic field with less than 1 Tesla, and exhausted to a plasma collector installed outside of the active region. More detailed observation of solid state tin behavior relating with its rapid phase change, is necessary to evaluate its deposition inside the EUV chamber and on the collector (C1) mirror.
A mass limited, micro droplet of liquid tin is employed as the plasma target in an actual EUV source. Initial conditioning of the density distribution of tin particles is essential for higher CE and less contamination, and a small pre-pulse is used to disperse tin particles before main CO_2 laser irradiation. Breakup of liquid droplet by a Nd:YAG laser as a pre-pulse, was observed as "fragmentation" of liquid tin droplet (Nakano et.al., 2008). The measurement was by a shadow graph method as the size of the fragments is less than μm diameter. The typical speed of the fragments was quite slow around 10m/s. Fig. 8 shows a typical fragmentation of a 60 μm diameter liquid tin droplet (selected from droplet train) 16μs after irradiation by a Nd:YAG laser with a few mJ pulse energy. Laser irradiation was from right hand side to the droplet.

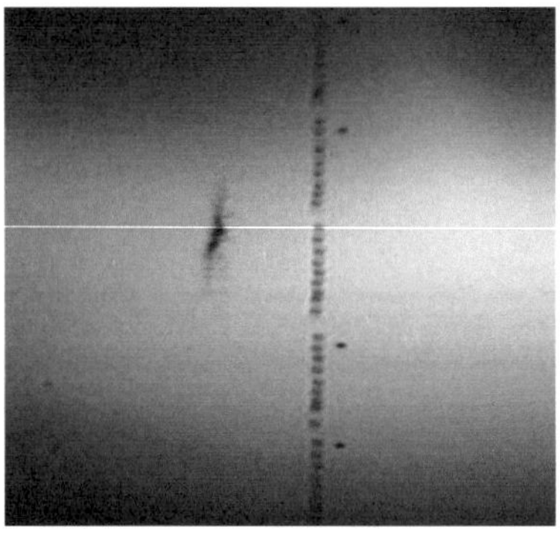

Fig. 8. Droplet fragmentation

Kyushu University group reported on the measurement of the ablation dynamics of a tin micro droplet by double pulse laser irradiation (Nakamura et.al., 2008). Laser induced fluorescence imaging (LIF) was employed to observe the behavior of neutral particles after pre-pulse irradiation by a Nd:YAG laser of a 30μm diameter solid tin droplet. The neutral tin atoms showed isotropic expansion at fast speed up to 40km/s (less than 1keV), and LIF signal continued more than 5μs. Main CO_2 laser irradiation was optimized with time delay of 1μs for higher CE and full ionization of tin particles.
Further experiments were performed by EUVA group by using a liquid tin droplet as the target (Ueno, et.al., 2009). LIF signal continued longer than the solid target case with higher signal level. The observed behavior is similar to the "Spallation" of metal foil target, but the speed of neutral particles is faster than the typical spallation (Coq-Lelandais, et.al., 2009). Laser induced shock wave inside a spherical object, and fast electron motion in an

electrically isolated metal droplet, may play the essential role in this phenomena (Sokollik, et.al., 2009). Spallation neutral tin atoms have kinetic energy lower than sputtering threshold of Mo/Si multilayer, and contribute to tin nano-deposition inside the chamber. Fig. 9 is a LIF image of neutral tin atoms after 10 μs from a pre-pulse laser irradiation by spallation from a 100 μm diameter liquid droplet. Diffusion is spatially uniform.

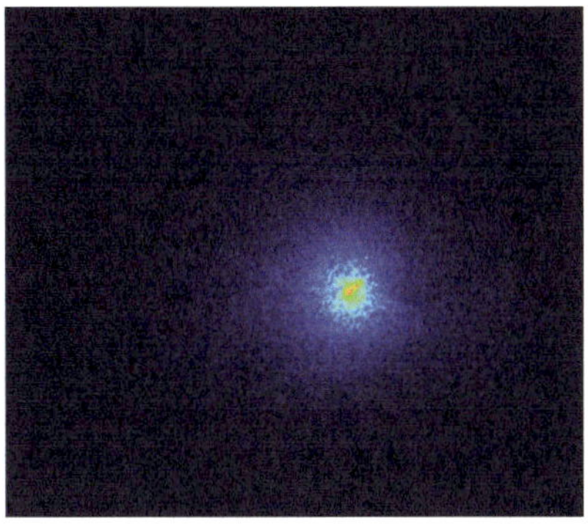

Fig. 9. LIF image of tin neutral atoms

8. Modeling of tin vapor in microscopic and macroscopic approach

Modeling of laser ablation has a long tradition in the field of laser-matter interaction. Analytical modeling of a hydrodynamic system is a conventional method so far. Target is a planar shape metal or non metal material. More precise atomic modeling has been emerging as a molecular dynamics method (MD), based on the interaction between particles (Zhigilei, et.al., 1998). The situation in the spherical liquid tin droplet ablation is complex enough to treat only by a standard numerical method. A microscopic MD modeling has been developed based on a modified Lennard-Jones potential for tin to predict the rapid behavior of tin phase change (Masnavi et.al., 2009). The heating speed was assumed as a modeling parameter around 10^{13} k/s. The calculated P-V diagram showed a strong temperature hysteresis of heating and cooling of tin. The calculation gives the equation of state (EOS) for tin, which is then used in a 1D numerical hydrodynamic calculation for the optimization of pre-pulse irradiation for moderate neutral particle diffusion. The resulting macroscopic

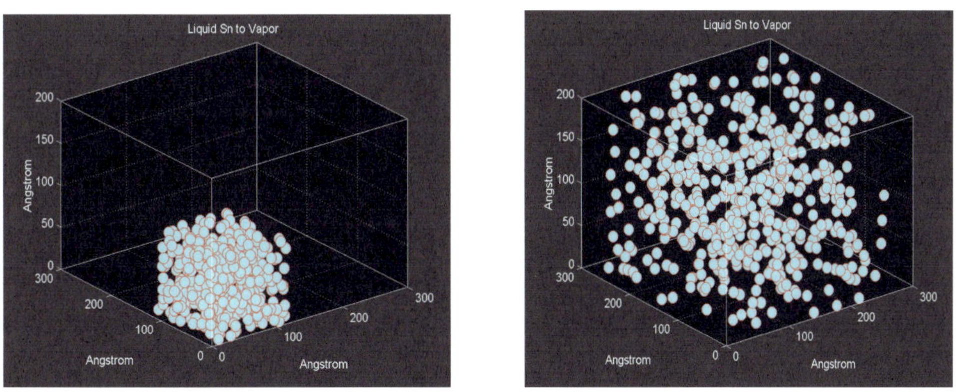

Fig. 10. MD simulation of fast (left) and slow (right) heating of tin to 3500K

fluid motion gives the impulse to the remaining droplet to move opposite to the laser irradiation into dispersion.
A macroscopic behavior of the fragmentation of the tin droplet is modeled by the emerging Moving Particle Semi-implicit (MPS) method (Koshizuka. et.al., 1996) With a given large scale parameter like surface tension, density etc, fragmentation of the droplet is modeled with the impulse as a given parameter. The calculated behavior of the test tin droplet has a good agreement with experiments.
Optimization of the pre-pulse irradiation, time delay to the main CO_2 laser pulse, and main pulse irradiation itself, is possible through the analysis of the laser ablation process of tin droplet, which has unique characteristics as a laser ablation.

9. Remaining tasks

The basic architecture of high average power EUV source is established, and the main effort is now shifting to an engineering demonstration of the system in real work places. Cymer is the leading company in this field, and trying to deliver multiple systems to global semiconductor industry (Fomenkov, et.al., 2009). It is now intuitive to notice a similarity among different categories of high average power, pulsed laser driven applications, namely laser inertial fusion and laser isotope separation. The system size is extremely different compared to these huge energy applications, but the key configuration has interesting analogies. It is discussed in the High Average Power Laser Program (HAPL), which is aiming at engineering design of a laser fusion reactor at 10Hz of 1MJ laser pulse energy, on a magnetic intervention to keep off the ions from the wall. Fusion target is injected as a pellet of 1mm diameter at 10 Hz. Averaged material injection speed is similar to the case of EUV source, in which 30μm diameter droplet is injected at 100 kHz repetition rate. Fuel recovery cycle is also designed (Sethian, 2006).
Ion sputtering (initially the fatal obstacle for the EUV light source design), is now fully avoided by the magnetic plasma guiding. Metal vapor generation was studied experimentally and numerically, and understood well on the behaviour. Slow tin vapor deposition inside the chamber is the last subject to be treated to prevent the loss of reflectivity of the C1 mirror. Laser isotope separation was once studied intensively more than a decade ago, in which metal vapor handling was the main technological subject. Isotope separation is based on the selected laser resonant ionization of different isotopes of the same atomic species (Bokhan, et.al., 2006). Resonant ionization technology was reported on tin vapor for efficient ion gun design (Liu. et.al., 2006). Full ionization of tin vapor, generated during the droplet dispersion, seems the straightforward way to exhaust tin atoms from the active region of the chamber. It seems also meaningful to learn from experiences of metal vapor laser engineering on the protection of windows from contamination (Little. 1999).

10. Conclusion

Brief review on the evolution of the laser driven, tin plasma EUV source, is described in this article. Tremendous efforts of many researchers contributed to this extremely difficult research successful after one decade. The established architecture is scalable to multi 100 watts EUV power (Endo, et.al., 2009), which is sooner or later required from semiconductor industry for lower cost production during the course of the next decade. Further engineering effort is required on the improvement of the EUV source, especially on the stability and robustness. Chemical and mechanical engineering becomes more important in the high temperature condition by continuous 20kW pulsed CO_2 laser irradiation.

The author deeply expresses his thanks to his colleagues in this challenging program. Early study on fast ions and sputtering of multilayer coating was efficiently performed by Dr. H.Komori. Solid state and CO_2 laser technology was successfully developed by Dr. T.Miura. Dr.Y.Ueno has greatly contributed to discover the plasma and EUV emission physics of CO_2 laser produced tin target, to convince semiconductor business to select this architecture as the final solution for the high volume manufacturing EUV lithography.

11. References

Abe, T.; Suganuma, T.; Moriya, M.; Yabu, T.; Asayama, T,; Someya, H.; Ueno, Y.; Soumagne, G.; Sumitani, A. & Mizoguchi, H. (2007). Small Field Exposure Tool (SFET) Light Source, *Proceeding of SPIE Emerging Lithographic Technologies*, vol. 6517, ISBN: 9780819466365, San Jose CA, February 2007, SPIE

Bielesch, U.; Budde, M.; Fischbach, M.; Freisinger, B.; Schaefer, J.H.; Uhlenbusch, J. & Viol, W. (1992). Q-switched multi kilowatt CO2 laser system excited by microwaves, *Proceedings of SPIE 9th International Symposium on Gas Flow and Chemical Lasers,* pp. 57- 60, vol.1810, ISBN: 9780819410108, August 1992, Heraklion, Greece, SPIE

Berglund, M.; Rymell, L. & Hertz, H.M. (1998). Cryogenic liquid jet target for debris-free laser-plasma soft x-ray generation. *Rev.Sci. Instrum.* 69 (6), pp. 2361-2364

Ballard, W.P. (2002). High power laser produced plasma source operation in the ETS, *1st International EUV Source Symposium*, October 2002, Dallas, TX, International Sematech

Ballard, W.P.; Tichenor, D.A.; O'Connell, D.J.; Bernardez Jr, L.J.; Lafon, R.E.; Anderson, J.M.; Leung, A.H.; Williams, K.A.; Haney, S.J.; Perras, Y.E.; Jefferson, K.L.; Porter, L; Knight, D.; Barr, P.K.; Van de Vreugde, J.L.; Campiotti, R.H.; Zimmerman, M. D.; Buchenauer, A.; Replogle, W.C.; Smith, T.G.; Wronosky, J.B.; Darnold, J.R.; Blaedel, K.L.; Chapman, H.N.; Taylor, J.S.; Hale, L.C.; Sommergren, G.E.; Gullikson, E.M.; Naulleau, P.P.; Goldberg, K.A.; Lee, S.H.; Shields, H.; St. Pierre, R.J. & Ponti, S. (2003). System and process learning in a full-field high power EUVL alpha tool, *Proceedings of SPIE Emerging Lithographic Technologies,* vol.5037, ISBN: 9780819448422, February 2003, Santa Clara, CA, SPIE

Bokhan, P.A.; Buchanov, V.V.: V Fateev, N. V.; Kalugin, M. M.; Kazaryan, M. A.; Prokhorov, A. M.; Zakrevskii, D. E. (2006). *Laser isotope separation in atomic vapor*, Wiley-VCH, ISBN-10: 3-527-40621-2, Berlin

Coq-Lelandais, J.P.; Boustie, M.; Berthe, L.; de Resseguier, T. ; Combis, P. ; Colombier, J.P. ; Nivard, M. & Clavenie, A. (2009) Spallation generated by femtosecond laser driven shocks in thin metallic targets. *J.Phys.D. Appl.Phys.* 42 (6), 065402

Druon, F.; Cheriaux, G.; Faure, J.; Nees, J.; Nantel, M.; Maksimchuk, A. & Mourou, G. (1998). Wave-front correction of femtosecond terawatt lasers by deformable mirrors. *Opt.Lett.* 23 (13), pp1043-1045

Endo, A.; Hoshino, H.; Ariga, T. & Miura, T. (2006). High power pulsed CO_2 laser for EUV lithography, *EUV Source Workshop,* May 2006, Vancouver, B.C. Canada, International Sematech

Endo, A. (2007). High power CO_2 laser, *EUV source workshop,* May 2007, Baltimore, MD, International Sematech

Endo, A.; Hoshino, H.; Suganuma T.; Nowak, K.M.; Yanagida, T; Yabu, K.; Asayama, T.; Ueno, Y.; Moriya, M.; Nakano, M.; Someya, H.; Nishisaka, T.; Abe, T.; Soumagne, G.; Komori, H.; Mizoguchi, H.; Sumitani, A. & Toyoda, K. (2008). CO2 laser-produced Sn-plasma source for high-volume manufacturing EUV lithography, *Proceeding of SPIE Emerging Lithographic Technologies*, vol. 6921, ISBN: 9780819471062, February, 2009, San Jose CA, SPIE

Endo, A.; Komori, H.; Ueno, Y.; Nowak, K.; Yabu, T.; Yanagida, T.; Suganuma, T.; Asayama, T.; Someya, H.; Hoshino, H.; Nakano, M.; Moriya, M.; Nishisaka, Abe, T.; Sumitani, A.; Nagano, A.; Sasaki, Y.; Nagai, S.; Watanabe, Y.; Soumagne, G.; Ishihara, T.; Wakabayashi, O.; Kakizaki, K. & Mizoguchi, H (2009). Laser produced plasma source development for EUV lithography, *Proceedings of SPIE Alternative Lithographic Technologies,* vol. 7271, ISBN: 9780819475244, February, 2009, San Jose CA, SPIE

Fay, B. & Exulite project team (2003). Modular laser produced plasma source for EUV lithography, *EUV source workshop,* Antwerp, Belgium, September 2003, International Sematech

Fiedorowicz, H.; Bartnik, A.; Szczurek, M.; Daido, H.; Sakaya, N.; Kmetik, V.; Kato, Y.; Suzuki, M.; Matsumura, M.; Tajima, J.; Nakayama, T. & Wilhein, T. (1999). Investigation of soft X-ray emission from a gas puff target irradiated with a Nd:YAG laser. *Opt. Comm.* 163 (1-3), pp. 103-114

Fomenkov, I.V.; Brandt, D.C.; Bykanov, A.N.; Ershov, A.I.; Partlo, W.N.; Myers, D.W.; Boewering, N.R.; Farrar, N.R.; Vaschenko, G.O.; Khodykin, O.V.; Hoffmann, J.R.; Chrobak, C.P.; Srivastava, S.N.; Golich, D.J.; Vidusek, D.A.; Dea, S.D. & Hou, R.R (2009). Laser-Produced Plasma Light Source for EUVL, *Proceedings of SPIE, Alternative Lithographic Technologies,* vol.7271, pp 727138- 1 -13, ISBN: 9780819475244, San Jose, CA, February, 2009, SPIE

Giesen, A.; Huegel, H.; Voss, A.; Wittig, K.; Brauch, U. & Opower, H. (1994). Scalable concept for diode-pumped high power solid-state lasers. *Appl.Phys.* B58 (5), pp. 365-372

Gillapsy, J.D. (2003). Study of EUV source collector damage mechanism, *2nd International EUV Lithography Symposium,* Antwerp, Belgium, September 2003, International Sematech

Hansson, B.A.M (2004). Progress in development of a high power source for EUV lithography, *3rd International Symposium on EUV Lithography,* October 2004, Miyazaki, Japan, International Sematech,

Hodgson, N & Weber, H (2005). *Laser resonators and beam propagation: Fundamentals, Advanced Concepts and Applications 2nd edition,* ISBN-10: 0387400788, Springer, Berlin

Hoshino, H.; Suganuma, T.; Asayama, T.; Nowak, K.; Moriya, M.; Abe, T.; Endo, A. & Sumitani, A. (2008). LPP EUV light source employing high power CO_2 laser, *Proceedings of SPIE Emerging Lithography,* vol.6921, ISBN: 9780819471062, San Jose, CA, February, 2008, SPIE

Kosizuka, S. & Oka, Y. (1996) Moving particle semi-implicit method for fragmentation of incompressible fluid. *Nucl.Sci.Eng.* 123 pp.421-434

Klebanoff, L. (2003). Condenser erosion observations in the ETS, *2nd International EUV Lithography Symposium,* Antwerp, Belgium, September 2003, International Sematech

Komori, H. (2003). Study of ion damage on EUV light source collector mirrors, *2nd International EUV Lithography Symposium,* Antwerp, Belgium, September 2003, International Sematech

Komori, H.; Ueno, Y.; Hoshino, H.; Ariga, T.; Soumagne, G.; Endo, A. & Mizoguchi, H. (2006) EUV radiation characteristics of a CO_2 laser produced Xe plasma. *Appl.Phys.*B83 (2), pp213- 218

Little, C.E. (1999). *Metal Vapor lasers: Physics, Engineering and Applications,* John Wiley & Sons, Ltd. ISBN 0-471-97387-4, Chichester, West Sussex, England

Liu, Y.; Baktash, C.; Beene, J.R.; Bilheux, H.Z.; Havener, C.C.; Krause, H.F.; Schultz, D.R.; Stracener, D.W.; Vane, C.R.; Brück, K.; Geppert, Ch.; Kessler, T. & Wendt, K. (2006). Laser ion source tests at the HRIBF on stable Sn, Ge and Ni isotopes. *Nucl.Instrum and Meth.* B243 (2), pp. 442-454

Masnavi, M.; Nakajima, M. & Horioka, K. (2009). Molecular dynamics investigation on tin, *International Workshop on EUV Lithography,* July 2009, Oahu, Hawaii, USA, EUV Litho, Inc

Mori, M.; Shiraishi, T.; Takahashi, E.; Suzuki, H.; Sharma, L.B.; Miura, E. & Kondo, K. (2001). Extreme ultraviolet emission from Xe clusters excited by high intensity lasers. *J.App.Phys.* 90 (7), pp 3595-3601

Nakamura, D.; Akiyama, T.; Okazaki, K.; Tamura, K.; Takahashi, A & Okada, T. (2008) Ablation dynamics of tin micro-droplet irradiated by double pulse laser used for extreme ultraviolet lithography source. *J.Phys. D:Appl.Phys* .41 (24), 245210

Nakano, M.; Yabu, T.; Someya, H.; Abe, T.; Soumagne, G.; Endo, A. & Sumitani, A. (2008). Sn droplet target development for laser produced plasma EUV light source, *Proceedings of SPIE, Emerging Lithography,* vol.6921, ISBN 9780819471062, San Jose, CA, February, 2008, SPIE

Nishihara, K.; Sunahara, A.; Sasaki, A.; Fujioka, S.; Shimada, Y.; Numamo, M.; Tamura, H.; Murakami, M.; Aota, T.; Fujima, K.; Furukawa, H.; Nishikawa, T.; Koike, F.; More, F.; Kato, T.; Zhakhovskii, V.; Gamata, K.; Ueda, H.; Nishimura, H.; Yuba, Y.; Nagai, K.; Miyanaga, N.; Izawa, Y. & Mima, K. (2008). Advanced laser-produced EUV light source for HVM with conversion efficiency of 5-7% and B-field mitigation of ions, *Proceedings of SPIE, Emerging Lithography,* vol.6921, ISBN: 9780819471062, San Jose, CA, February, 2008, SPIE

Nilsson, J. & Jaskorzynska, B. (1993). Modeling and optimization of low repetition rate high energy pulse amplification in cw-pumped erbium-doped fiber amplifiers *Opt. Lett.* 18 (24), pp. 2099 -2101

Nowak, K.M.; Suganuma, T.; Endo, A.; Sumitani, A.; Goryachkin, D.A.; Romanov, N.A.; Sherstobitov, V.E.; Kovalchuk, L.V.; Rodionov, A.Y. (2008). Efficient and compact short pulse MOPA system for laser-produced-plasma extreme-UV sources employing RF-discharge slab-waveguide CO2 amplifiers, *Proceedings of SPIE High-Power Laser Ablation,* vol.7005, ISBN: 9780819472069, Taos, NM, April 2008, SPIE

Okazaki, S. (2001). *Report of domestic EUV source workshop*, July 2001, ASET (in Japanese)

Sakai, T. & Hamada, N. (1994). Q-switched CO2 laser using intense pulsed RF discharge and high speed rotating chopper, *Proceedings of SPIE Gas Flow and Chemical Lasers: Tenth International Symposium*, vol.2502, pp. 25-30, ISBN: 9780819418609, Friedrichshafen, Germany, September 1994, SPIE

Sethian, J. (2006). The high average power laser program, *15th HAPL meeting,* September 2006, San Diego, CA

Shine Jr, R.J.; Alfrey, A.J. & Byer, R.L. (1995) 40-W cw, TEM_{00} –mode, diode-laser- pumped, Nd:YAG miniture-slab laser. Opt.Lett. 20 (5), pp. 459-461

Sokollik, T.; Schnuerer, M.; Ter-Avetisyan, S.; Steinke, S.; Nickles, P.V.; Sandner, W.; Amin, M.; Torcian, T.; Willi, O. & Andreev, A.A. (2009). Proton imaging of laser irradiated foils and mass limited target, 2nd *International Symposium on Laser driven relativistic plasmas applied to science, industry and medicine,* Kyoto, Japan, January 2009.

Stamm, U. (2002). Recent advances in extreme ultraviolet sources and technology, Invited Talk CthB1, *Conference on Lasers and Electro Optics 2002,* May 2002, Long Beach, CA, Optical Society of America

Suckewer, S.; Skinner, C.; Voorhees, D.; Milchberg, D.; Keane, C. & Semet, A. (1983). Population inversion and gain measurements for soft X-ray laser development in a magnetically confined plasma column, *IEEE-QE* 19, pp.1855-1860

Tanaka, H.; Akinaga, K.; Takahashi, A. and Okada, T. (2004) Emission characteristics of extreme ultraviolet radiation from CO_2 laser produced plasma. *Jpn.J.Appl.Phys.* 43 L585

Tanaka, H.; Matsumoto, A.; Akinaga, K.; Takahashi, A & Okada, T. (2007) Comparative study on emission characteristics of extreme ultraviolet radiation from CO_2 and Nd:YAG laser produced tin plasma. *Appl.Phys.Lett.* 87 (4), 041503

Tao, Y.; Tillack, M.; Sequoia, K.; Yuspeh, S. & R. Burdt, R. (2009). Investigations on Plasma Physics for an Extreme Ultraviolet Lithography Source Based on Laser-Produced High-Z Plasmas, Invited talk IO3A-1, *36th International Conference on Plasma Science and 23rd Symposium on Fusion Engineering,* May 2009, San Diego, CA, IEEE

Ueno, Y.; Ariga, T.; Soumagne, G.; Higashiguchi, T.; Kubodera, S.; Pogorelsky, I.; Pavlishin, I.; Stolyarov, D.; Babzien, M.; Kusche, K. & Yakimenko, V. (2007). Efficient extreme ultraviolet plasma source generated by a CO_2 laser and liquid xenon microjet target. *Appl.Phys.Lett.* 90 (19), 191503

Ueno, Y.; Soumagne, G.; Sumitani, A.; Endo, A. & Higashiguchi, T. (2007) Enhancement of extreme ultraviolet emission from a CO_2 laser produced Sn plasma using a cavity target. *Appl.Phys.Lett.* 91 (23), 231501

Ueno, Y.; Hoshino, H.; Ariga, T.; Miura, T.; Nakano, M.; Komori, H.; Soumagne, G.; Endo, A.; Mizoguchi, H.; Sumitani, A. & Toyoda, K. (2007). Characterization of various Sn targets with respect to debris and fast ion generation, *Proceeding of SPIE Emerging Lithographic Technologies,* vol. 6517, ISBN: 9780819466365, San Jose CA, February 2007, SPIE

Ueno, Y.; Soumagne, G.; Sumitani, A.; Endo, A.; Higashiguchi, T. & Yugami, N. (2008). Reduction of debris of a CO_2 laser produced Sn plasma extreme ultraviolet source using a magnetic field. *Appl.Phys.Let.* 92 (21), 211503

Ueno, Y.; Yanagida, T.; Suganuma, T.; Komori, H.; Sumitani, A.& Endo, A. (2009). Characterization of Tin vapor from CO_2 laser produced EUV light source, *Proceedings of SPIE Damage to VUV, EUV, and X-Ray Optics II,* vol.7361 ISBN: 9780819476357, April 2009, Prague, SPIE

Wieland, M.; Wilhein, T.; Faubel, M.; Ellert, Ch.; Schmidt, M. & Sublemontir, O. (2001). EUV and fast ion emission from cryogenic liquid jet target laser-generated plasma. *Appl.Phys.* B72 (5), pp.591-597

Zhigilei, L.V.; Kodali, P.B.S. & Garrison, R.J. (1998) A microscopic view of laser ablation. *J.Phys.Chem.* B1998, 102, pp. 2845-2853

10

Grazing Incidence Mirrors for EUV Lithography

Mariana Braic, Mihai Balaceanu and Viorel Braic
National Institute for Optoelectronics
Romania

1. Introduction

The ongoing development of the knowledge-based society has brought about new challenges derived from the increasing volume and complexity of the information resources available. For a faster and more affordable transfer and processing of information, further progress is resting upon the development of increasingly smaller, higher density, integrated microelectronics, and their availability at lower prices.
The actual industrial standard governing the manufacture of microprocessors, employing today's deep-ultraviolet lithography (DUVL), is limited by the minimum wavelength of 193 nm.
As microprocessor feature size continues to decrease to sub-45 nm levels, greater spatial resolution is needed to be obtained. In an optical lithographic system the critical dimension is determined by the Rayleigh criterion, so that in order to obtain a smaller critical dimension one can increase the numerical aperture while maintaining the conventional lithographic wavelength of 193 nm. This approach permitted to obtain feature size less than 45-nm [Rothschild et al., 2005]. However, the increased demands on the miniaturisation have already reached the limits of the DUVL. Considering the Rayleigh criterion, another possible approach to minimize the critical dimension is to decrease the radiation wavelength. This idea is materialised as the Extreme-Ultraviolet Lithography (EUVL), using a wavelength of 13.5 nm [Stulen & Sweeney, 1999].
In Europe, ASML and its main co-developer Carl Zeiss, are heavily investing in the development of key technological issues, aiming for the realization of an EUVL α-tool [Meiling et al., 2003], [Meiling et al., 2005], for subsequent world-wide commercialization [ASML, 2009], while in USA, SEMATECH organization has the leading role [SEMATECH, 2009].
EUVL is a domain undergoing intensive research at the moment, as one of the leading candidates among emerging lithography techniques, enabling 35 nm half-pitch patterning and providing also extendibility to 22 nm, using a solely reflective optics.
The requirement for the replacement of transmitting lenses with reflecting mirrors within the patterning tool is due to the strong absorption of the EUV light in most of materials - gases included, a high reflection coefficient being difficult to obtain at either near grazing or near normal incidence angles. The requirement for an entirely reflective optics, placed in vacuum, introduces numerous technological challenges, as well as the requirement to efficiently generate 13.5-nm light, with high intensity and high reliability.
The EUV source specifications are derived, accordingly to the existing standards in visible and DUV lithography, from the customer requirements: high throughput - more than 100 wafers /hour, imaging quality and cost of ownership. To generate EUV light, hot and dense

plasmas are required. EUV radiation is generated using gas-discharge-produced plasmas (DPPs), laser produced plasmas (LPPs) with ultra-fast lasers, synchrotron or X-ray radiation [Stamm, 2004].
EUVL systems using LPPs and DPPs sources rely on optical configurations including collection and projection modules composed by grazing and near incidence reflective mirrors [Singh & Braat, 2000], [Bakshi, 2006]. In a LPPs-based system, EUV light is produced by bombarding a sliver of Sn with a high-power laser, while in a DPPs system the EUV light emerges from Xe plasma, enriched with metal EUV radiators, as Li and Sn [Banine & Moors, 2004]. The EUV light produced by either source is being collected by specially engineered EUV mirrors, which then focus the EUV beam in the EUV scanner to produce microchip patterns.
The main aspects to be addressed by the source specification are: the operating wavelength, the EUV power, the hot spot size, the collectable angle, the repetition rate, the pulse-to-pulse repeatability and the debris induced lifetime of components. While the first requirements are addressing only the EUV plasma source engineering, the last one deals also with the engineering of the optical system, mainly with the collector system - optical design and materials to be used. The mirrors responsible for collecting the light are directly exposed to the plasma and are therefore vulnerable to the damage done by the high-energy ions [Komori et al., 2004], [Hansson et al., 2002], and other debris [Srivastava et al., 2007]. These damage associated with the high-energy process of generating EUV radiation have precluded the successful implementation of practical EUV light sources for lithography. This chapter addresses the issues related to the attainability of high reflectivity grazing incidence collection mirrors with extended life time, to be used in EUVL systems.

2. Requirements for the collection optics in EUVL

In the lithographic process of patterning fine-scale structures onto a substrate, the radiation used to selectively expose the recording medium (resist) can be optical, e-beam, X-ray or ion beam. If the optical radiation is used, including DUV and EUV, the lithographic system consists of four main elements, integrated in an unique optical system [Jaeger, 2002]:

a. light source;
b. mask containing the patterns corresponding to the structures to be fabricated;
c. exposure system to generate an aerial image of the mask pattern and
d. resist, for recording the image generated by the exposure system.

In the EUV lithographic systems, there are two more modules to be integrated in the optical system: the collector and the projection modules [Bakshi, 2006].
Concerning the light source, up to date, the most used EUV light sources are the DPPs ones. These sources are reported to generate more power, consume less energy, being less expensive. As a result, this type of source was integrated into alpha-level EUVL scanners. Also, the low power version of DPPs source has been used in EUV micro-exposure tools, in industrial EUV metrology and EUV resist development projects [Lebert et al., 2003], [Song et al., 2006], [Bolanti et al., 2003], [Zuppella et al., 2009] and [Fiedorowicz et al., 2005]. This type of source was used also in the development and testing of the grazing incidence mirrors presented in the following [Choi et al., 2004].
The part of the EUV optical system including the mirrors, which are designed to collect as much as possible EUV light produced by the source, is known as the collector module, while the mirrors designed to focus the EUV light on the resist, are part of the projection module.

Apart from the mirrors, the optics may comprise also filters for cutting the wavelengths higher than 40 nm. The mask is part of the exposure system, including the resist exposure stage, with movement control in the Angstrom range [Itani, 2009], [Wallace et al., 2007].
Due to the strong absorption of EUV light in any material, including gases (even at pressures in the Pa range), in an EUVL system all the components must reside in high and clean vacuum. The EUV light source it-self represents the cause of important demands on the EUV optics geometry and materials to be used, as the heat and debris emerging from it can seriously damage the optics facing the source. Usually the EUVL optics is protected from the energetic debris from the source using gas (He) curtains, electrostatic and mechanical shields [Vargas-Lopez et al., 2005], [Bakshi, 2006].
The collection mirrors must have high reflectivity at grazing angles, high resistance to the energetic particle bombardment, high adhesion to the substrate and a high stability in a wide range of temperatures, as the heat load from the entire radiation emitted by the EUV source is important. The overall reflectivity of EUV mirrors, especially of those in the collection module, is under continuous degradation due to erosion and contamination from within the EUV source, as Xe, Li and Sn are the conventional, currently used, EUV light fuel [Allain et al., 2008]. This is a matter of great concern because it directly affects the available power of the EUV source, and thus the final cost of production [Neumann et al., 2007], [Bakshi, 2006].
The collector mirrors are facing a continuous bombardment of debris emerging from EUV light sources, fast ions, neutrals, off-band radiation, droplets, and background impurities (i.e., H, C, N, O), as well as the heat load generated by the sources themselves, all of them inducing serious damage to the nearby collector mirrors. The challenge is to obtain a collection mirror exhibiting high reflectivity at grazing incidence angles, high resistance to the bombardment done by energetic particles, micro-chemical stability at high temperatures and corrosion resistance, combined with small roughness, so as to prevent significant radiation loss via scattering.

3. Materials for EUV mirrors

The materials used for manufacturing the EUVL mirrors must have some valuable properties, as [Shin et al., 2009], [Hecquet et al., 2007]:
- manufacturing design freedom of shape and size;
- sustaining of polishing procedures, up to a tenth of a nanometre for the final roughness, because the roughness became a very sensitive parameter to be considered, as the wavelength decreases;
- low coefficient for thermal expansion (CTE), in order to reduce the optics distortion due to geometrical factors.

There are several types of such materials, as the well known silica and quartz, or the recently developed zerodur (lithium aluminosilicate glass-ceramic) and ULE (a titania-silica binary glass with zero CTE). The bare surfaces of these materials are presenting lower reflectivity values for grazing incidence, as compared to most of the metallic surfaces, considering the same surface roughness.
An all reflective optical system can have either only near normal incidence mirrors, or a combination of grazing incidence and near normal incidence mirrors. In order to choose the best materials, modelling the EUV light interaction with matter is a valuable tool. The wavelength domain to be considered is centred near-by the value where the maximum EUV throughput is obtained from the EUV sources, e.g. 13.5 nm [Bakshi, 2006]. In the following all the materials will be considered with respect to their properties at this specific wavelength.

The interaction with matter of the soft X-ray and EUV radiation (in the range 1 - 40 nm), in terms of transmission or reflectivity, can be "experienced" by using the free on-line connection to the ***Center for X-Ray Optics,*** a facility within the Materials Science Division at Lawrence Berkeley National Laboratory, USA [CXRO, 2009]. The details about the interactions of the soft X-rays with matter (photoabsorption and the coherent scattering) are clearly explained [CXRO, 2009]. The basic assumption made is considering the condensed matter as a collection of non-interacting atoms, a condition fulfilled for energies sufficiently far from the absorption thresholds, while in the threshold regions, the specific chemical state become important, so that direct experimental measurements are to be used. The inelastic Compton scattering cross section has not been included in the reflectivity calculations, as the Compton cross section contribution is significant only for light elements ($Z < 10$) at energies higher than 10 keV [Hubbell et al., 1975]; Hubbell et al., 1977].
Using the CXRO facility, one can easily obtain the ideal values for the EUV radiation reflectivity on a certain surface. For near-normal incidence mirrors, an enhancement of the reflectivity is obtained if we consider a typical Bragg reflector, in which the thickness of each layer is approximately a quarter-wave. Although the reflectivity of a single transition from a layer to another one is very small, the addition of multiple reflections results in a saturation to a maximum reflectivity. In EUV optics there are used multilayers (ML) formed by materials with alternate high and low absorption coefficients [Gloecker & Shah, 1995]. In EUV, the light absorption is directly linked with the Z value of the considered material. The best known near-normal incidence mirrors for EUVL are made from 40 to 60 bi-layers of Mo (on top) and Si (on bottom part), the bi-layer period being 6.9 nm. Another important parameter to be considered is the ratio of the bottom layer thickness to the overall bi-layer thickness, which in this case have the value $\Gamma = 0.4$. The ideal reflectivity of such a mirror is about 72% at normal incidence (with respect to the surface, not to the surface normal, as in VIS optics) [Benoit et al., 2006], [Wang et al., 2006], [Fiegl et al., 2006], [Schroeder et. al., 2007].
However, these mirrors are very sensitive to the oxidation, contamination, and the reflectivity is decaying in time, due to the mixing of the individual layers, the mechanism being diffusion driven. In fig. 1 is presented, comparatively, the ideal reflectivity obtained by simulation [CXRO, 2009] for a Mo/Si multi-layered mirror, with 60 bi-layers, with no inter-diffusion (Fig. 1a - s=0) and with layers inter-diffusion of 1 nm (Fig. 1b - s=1). It can be observed that not only the maximum reflectivity value is decreased, but also the maximum is shifting towards lower wavelengths, generating an overall loss of EUV light in the system. Due to the specific pattern obtained for the reflectivity of this multilayered mirror, it's role in the EUVL system is also to create a highly monochromatic radiation, by the repetitive reflection on several mirrors. As presented above, at this wavelength is also obtained the maximum EUV intensity from the Xe, Li or Sn radiators used in either DPPs or LPPs EUV sources.
Concerning the reflectivity at grazing incidence, from the system's geometry, it results that the angle to be considered for the reflectivity modelling is about 6^0, with respect to the surface. In Fig. 2 are presented the ideal values obtained by using the CXRO facility, for the reflectivity at $\lambda = 13.5$ nm, at 6^0 incidence with respect to the surface, for the most used materials in optics (silica, quartz, zerodur and ULE), and in Fig. 3 are presented the ideal reflectivity of some thick metallic layers. As expected the metallic surfaces present a much higher reflectivity for EUV light, as compared with the materials used as substrates, due to the different electronic configurations of these types of materials.

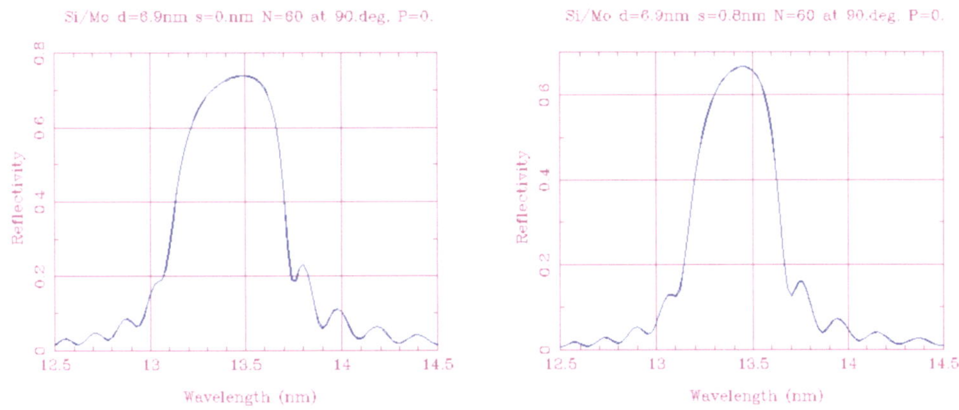

(a) (b)

Fig. 1. Ideal reflectivity at normal incidence of a Mo/Si multi-layered mirror, with 60 bi-layers, with (a) no inter-diffusion (s=0 nm) and (b) an inter-diffusion zone of 1 nm (s=1 nm).

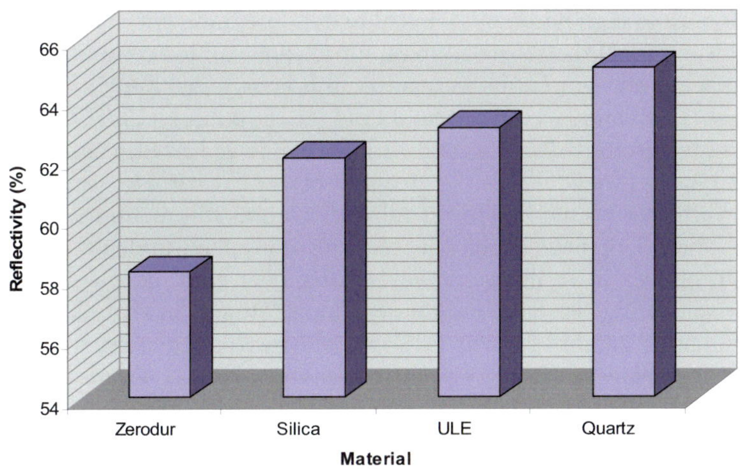

Fig. 2. Ideal reflectivity values, at λ = 13.5 nm, 6^0 incidence with respect to the surface, for silica, quartz, zerodur and ULE

Up to date, the EUV mirrors are usually made from special substrate materials as ULE or zerodur, subsequently coated with metallic thin films selected to be highly reflective in 12-15 nm domain, having a relatively high native oxidation resistance, e.g. palladium, ruthenium, and rhenium [Bakshi, 2006], [Alman et.al., 2007].

The new approach considered by the authors is to substitute the metallic coatings with covalent type materials, such as transitional metal carbides and nitrides thin films, with low surface roughness values (< 0.5 nm), which exhibit high reflectivity at 13.5 nm [Braic et al., 2005], [Braic et al., 2004], [Braic et al., 2008].

The nitrides and carbides of the transitional metals are well known for their stable micro-chemical properties at high temperatures, high oxidation resistance, high melting point, high hardness, high toughness and Young's modulus, high electric conductivity, excellent chemical stability, together with good wear resistance and high adhesion onto different

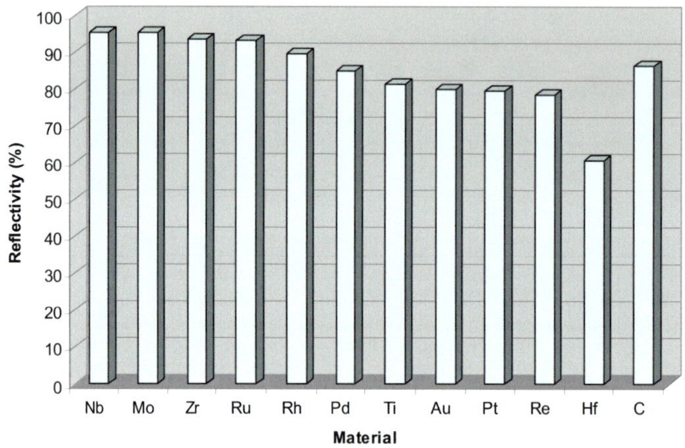

Fig. 3. Ideal reflectivity values, at $\lambda = 13.5$ nm and 6^0 incidence with respect to the surface, for some thick metallic layers

substrates [Barshilia & Rajam, 2006], [Ducros & Sanchette, 2006], [Braic et al., 2006]. Their chemical inertness and high hardness is linked to their predominant covalent type bonding. Group IV - VI transitional metals are forming nitride and carbide compounds characterized by a large number of nitrogen/carbon vacancies. It is known that the nitride/carbides of the group IV metals (Ti, Zr and Hf) can crystallize only in the cubic NaCl (FCC) structure, while the metals form the group V (Nb, Ta) and VI (Mo) can form carbides either in cubic (MeC) or in hexagonal forms (Me_2C). While the group IV of metallic carbides can accommodate up to about 50% vacancies on the non-metal sub-lattices, and still retain their cubic structure, the groups V and VI of metallic nitride/carbides crystallize under different structures, switching from hexagonal to the more stable cubic one, with increased nitrogen/carbon content, the process being also temperature dependent [Hugosson et al., 2001].
Due to their superior stability in their FCC crystallographic structure, there were selected the nitride and carbide of the group IV transitional metals. From the group V, Nb compounds are good candidate, but they also present high Young modulus values, so that the adhesion to a non-metallic substrate may develop undesired problems. From the group VI, Mo may also represent a good candidate, but at lower temperatures (<1000^{0}C) the FCC structure of its compounds tend to transform into a hexagonal or orthorhombic one. Also, the mechanical properties and oxidation resistance are less performing if compared to those exhibited by the compounds of the group V transitional metals, probably due to its large ionic component, compared to the covalent one [Kanoun et al., 2007].
In fig. 4 are presented the ideal values obtained by using the CXRO modelling facility, for the reflectivity at $\lambda = 13.5$ nm, at 6^0 incidence angle with respect to the surface, for the nitride and carbide compounds of Ti, Zr and Hf.
From the results presented above, it results that the best coatings to be used as grazing mirrors in the EUV collector system, are the Zr based carbide and nitride films. However, one major problem to be tackled in this application is the requirements to obtain coatings with high adhesion to the substrate, and with a reduced mechanical intrinsic stress, which is known to build-up during the growth process. There are several strategies used for reducing the stress: ion bombardment during growth [Vladescu et al., 2007], moderate substrate temperature during film deposition, use of a substrate with a CTE similar to

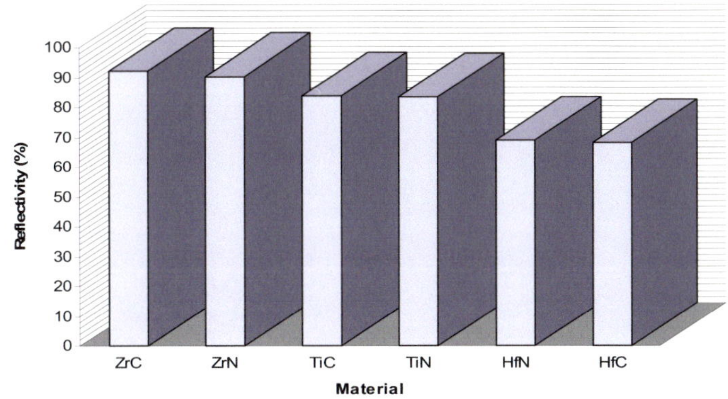

Fig. 4. Ideal reflectivity values at λ = 13.5 nm and 6^0 incidence with respect to the surface, for Ti, Zr and Hf nitrides and carbides

that of the film to be deposited. All these are currently used during films' deposition, however the films are still not stress-free.

Another approach is to use multi-layered (ML) films, as this kind of structure is known to inhibit cracks propagation and to have a low level of residual stress, a high adhesion strength and very high microhardness values, superior to those of the component layers [Helmersson et al., 1987], [Yashar & Sproul, 1999], [Musil, 2000], [Vladescu et al., 2008]. This direction is worth to be explored, so that the ideal reflectivity values, at λ = 13.5 nm and 6^0 incidence with respect to the surface, for some ML based on ZrN/TiN and ZrC/TiC were calculated, the results being presented in Fig. 5.

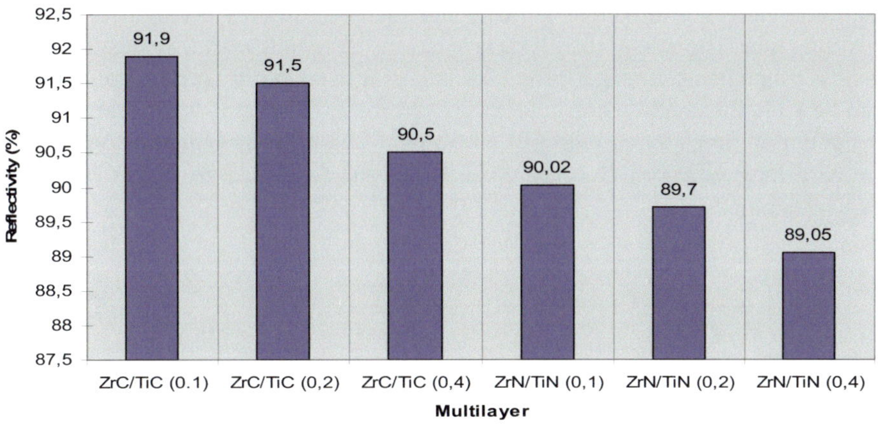

Fig. 5. Ideal reflectivity values at λ = 13.5 nm and 6^0 incidence with respect to the surface, for multilayered structures based on ZrN/TiN and ZrC/TiC, for Γ = 0.1; Γ = 0.2 and Γ = 0.4.

To illustrate the reflectivity dependence on the surface and interface roughness in Fig. 6 are presented the ideal reflectivity for a ZrN/TiN multilayer, n=40, Λ= 7 nm, Γ=0.1, either perfectly smooth (a - rms=0 nm) or with a higher roughness at the interfaces between the individual layers (b - rms=0.1 nm).

From the presented results, it clearly results the superiority of Zr based coatings to be used for surface finishing of the collector mirrors used in EUV lithographic systems.

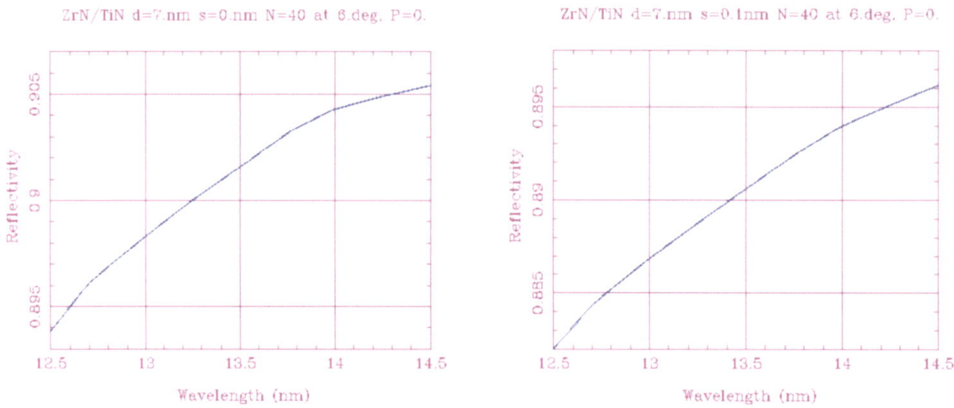

(a) (b)

Fig. 6. Ideal reflectivity of a ZrN/TiN multilayered mirror, 300 nm thickness, with surface roughness rms=0 nm (a) and rms=0.1 nm (b).

4. Collector module design with grazing incidence mirrors

A straightforward approach for the design of the collection optics, based on a wide emission angle of the EUV source, consists of an ellipsoidal configuration, the EUV source being in one focus and in the other one the image for the projection mirror is formed. The collected EUV radiation is focused as a narrow circular shape of illumination at a near-normal incidence angle on to a reflecting Mo/Si spherical mirror. The focused beam resulted from the projection mirror is directed towards the resist through a Si_3N_4 or a Zr filter to a transmission mask. The role of the filters is to cut the radiation with wavelengths > 20 nm.
The design of the first approach for the EUV collector used in our work is presented in Fig.7. As it can be seen, the central cone emerging from the EUV source, which contains the most part of the debris (fast atoms, ions and electrons), is lost. Minimizing the central cone and using a very long ellipsoid (1 m between focal points at 0.12 m ID), a high collecting efficiency is obtained.

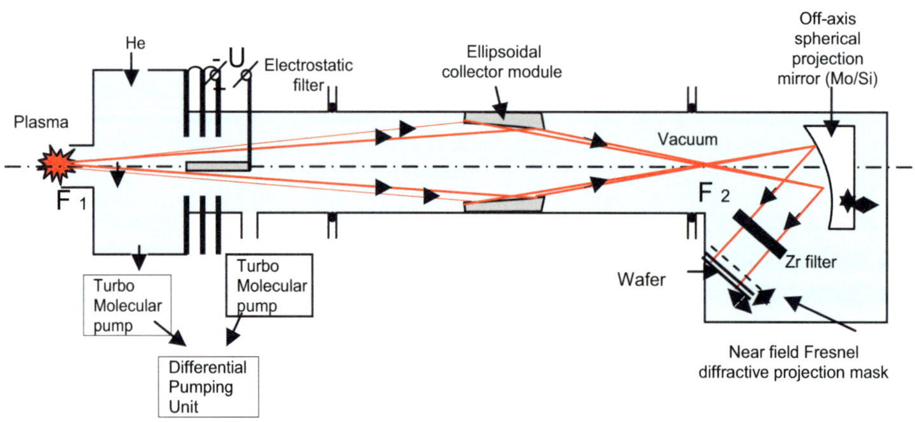

Fig. 7. The design of the first approach for the EUV collector, with ellipsoidal mirrors

However, in most of the used PDDs EUV sources, the emission of light and debris is directed mainly forward, in a narrow angle. So, an off-axis collection optics allowing the collection of the whole cone is required. In the same ellipsoidal geometry the source is oriented at approx. 10^0 off-axis and points towards a 210 mm long and 60 mm wide area, near by the central section of the ellipsoid. The image of the source is formed in the same position (focal point) as in the on-axis configuration. The total length of the optical path, from EUV source to the projection spherical Mo/Si mirror, is approx. 1.4 m, imposing the same length for the vacuum chamber, which has to be pumped according with its volume, with turbo-molecular pumps [Braic et al., 2008].

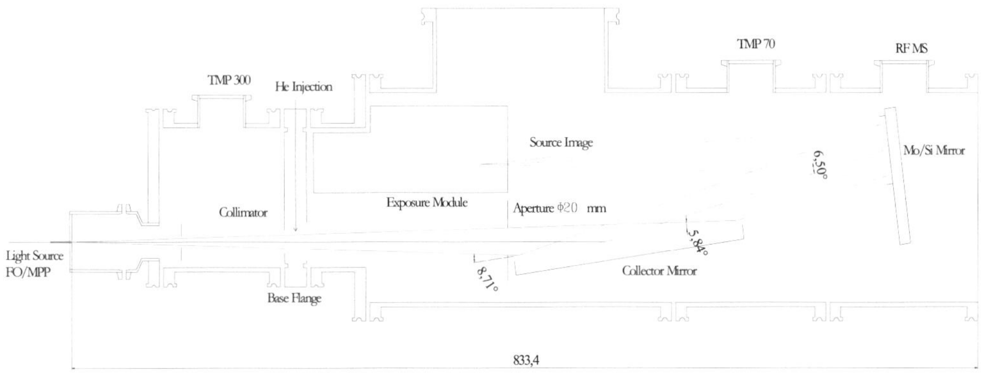

Fig. 8. Schematic view of the off-axis optics and vacuum chamber

The off-axis configuration opened a new opportunity to reduce the optical path, placing the Mo/Si mirror between the grazing incidence ellipsoidal surface and the focal point, significantly shortening the length of the optical path to approx. 0.8 m, also offering a significant reduction of the vacuum chamber volume and pumping unit (Fig. 8). The projector is a Mo/Si coated mirror with 13.5 nm centred wavelength. In the presented configuration, the diameter of the mirror is 2" and the curvature radius is 500 mm.

Vacuum chamber is designed to be assembled from standard high vacuum stainless steel components ISO 100 and ISO 160. Sealing of the parts is done with fluoro-elastomer gaskets, with very low out-gassing rate. The high and also clean vacuum environment is considered appropriate for the exposure tests. Checking of the residual gas composition is done by an RF mass spectrometer connected to the exposure chamber. The UHV technology was considered not appropriate, because of the exposure module components, which are not suitable for high temperature baking (up to 250^0C). The cleanliness of a vacuum environment is determined by the components and their out-gassing rate under vacuum. The chamber is differentially pumped by turbo pump located near the EUV source exit (mainly for heavy gases - Xe, Ar removing) and by another turbo pump located nearby the optics and exposure module (for He, resist and other components effluents). The turbo pumps are backed by mechanical dry rotary pumps.

One requirement from the EUVL optics is to minimize the contamination of the Mo/Si sensible mirrors due to the debris or effluents resulted from the used components, e.g. resist, fixture materials, etc [Shin et al., 2009]. As a result, the use of any coating including carbon was dismissed for further work related to EUV mirrors [Braic et. al., 2009] despite their valuable properties [Balaceanu & Braic, 2005].

5. Coatings' deposition for the grazing incidence mirrors

The coatings were deposited on optical glass, Si, plain carbon steel and high-speed steel substrates by bi-bipolar pulsed reactive magnetron sputtering method, using Ti and Zr cathodes and a reactive atmosphere consisting of a mixture of N_2 and Ar gases. The deposition set-up is presented schematically in Fig. 9.

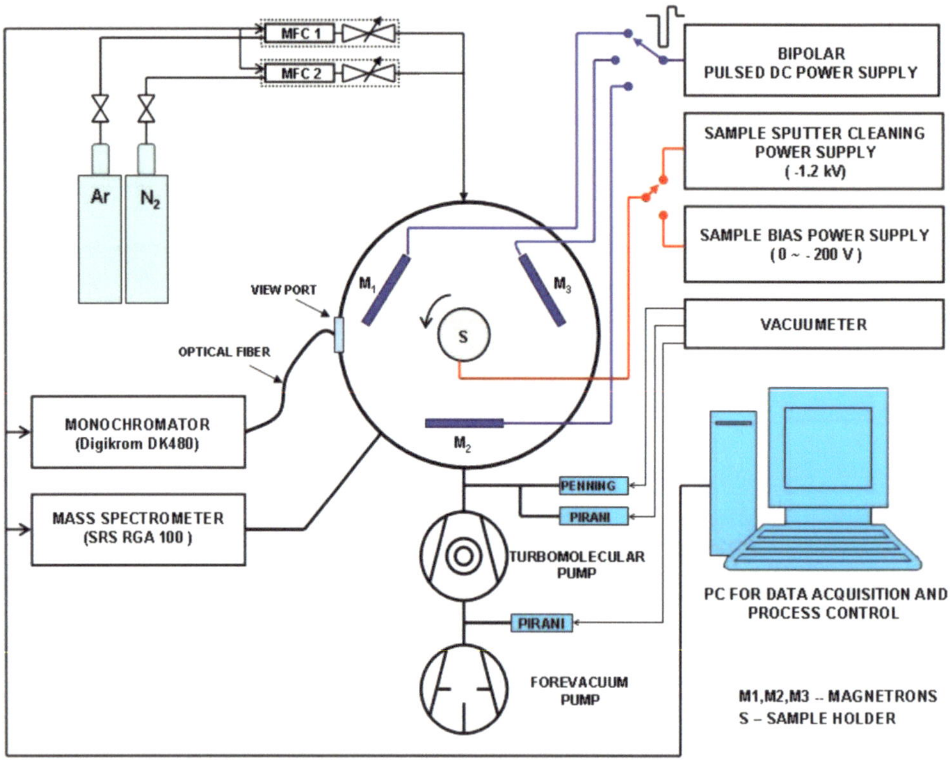

Fig. 9. Deposition chamber set-up for the multilayered coatings

The deposition cylindrical chamber (300 mm internal diameter and 500 mm height) has three rectangular (140*250 mm) magnetron targets.

The Ti and Zr cathodes were fed using a pulsed bipolar generator – type ENI RPG5, to avoid the problems related to the target poisoning during reactive deposition. The magnetron discharges are fed by a two channel Brooks 502 mass low controller with argon and nitrogen. The argon is introduced by pipes in front of each magnetron target and the nitrogen is directed towards the substrates.

The substrate holder assembly includes a thermocouple for temperature measurement during the deposition, is electrically insulated and provides the rotation of the individual sample holders. The rotation is provided by a stepper motor in order to have a computer controlled rotation speed and also to move the substrates in front of the two different targets.

The main deposition parameters were: deposition pressure - $4 \cdot 10^{-1}$ Pa, substrate bias – 50 V, ratio of N_2 and Ar partial pressures – 0.37.

The distance from any of the cathodes to the substrate holder is 10 cm. The UHV vacuum system ensures a base pressure in the deposition chamber was of about $1x10^{-5}$ Pa. The absolute pressure was measured with a MKS 626 Barocel capacitance manometer.

To obtain alternated ZrN/TiN films, two shutters placed in front of each magnetron were used. By periodically opening and closing the shutters and rotating the substrates in front of the active magnetron, Ti and Zr ions/atoms sputtered from the magnetrons were alternatively introduced in the deposition atmosphere for a predetermined time. Depending on the deposition rates, previously measured for each type of monolayer (TiN and ZrN), and on the rotating speeds of the shutters, various multilayer configurations (with different bilayer thicknesses Λ and Γ parameter) were prepared.

Following the information obtained by simulation, multilayer with a constant Γ value, equals to 0.1, and different thicknesses of the bilayer, were deposited. In order to differentiate between the different types of ML, the notation ZrN/TiN-n/Λ will be used, where "n" is the number of bi-layers and " Λ " is the bi-layer thickness.

Different overall thicknesses of the coatings were deposited, depending on the tests taken into account: for EUV reflectivity measurements and surface roughness evaluation, the thickness was of about 280 nm (n = 40; Λ =7 nm), while for elemental, structural and mechanical characterisation the total thickness was much greater, of about 3500 nm (n = 500; Λ = 7 nm).

Prior to deposition, specimens to be coated were chemically cleaned in an ultrasonic bath with isopropyl alcohol. Both the substrates and the magnetron targets were sputter cleaned in vacuum by Ar ion bombardment (1200 eV) for 10 minutes, in order to remove any residual impurities.

To ensure the deposition of stoichiometric layers, the optical emission spectroscopy (OES) was done, by on-line monitoring of the Zr, Ti and N_2^+ lines emitted by the discharge, by using a monochromator (Digikrom DK480), equipped with an R446 Hamamatsu photomultiplier. The acquisition, storage and processing of the spectra were performed by an Advantech 818 data acquisition system. The light signal emitted by the plasma was transmitted to the monochromater, through a quartz window and a collimator positioned in front of the target, by using an optical fibre. Concomitantly, the process gases were monitored by RF mass spectrometry (SRS RGA 100), in order to maintain a fine balance between the argon and nitrogen in the deposition chamber.

In magnetron sputtering deposition of nitrides (TiN, ZrN), the ratio of Ar/N_2 partial pressures in the working gas determines two working regimes with different sputtering yield of the cathode's material under ion bombardment, affecting the film's stoichiometry. During the deposition, the compound deposited onto the substrates is also deposited on the cathode, so the decrease of cathode sputtering yield, of about 5 times, is observed.

The transition between the “metallic” and the “ceramic” modes takes place abruptly at a small change of nitrogen flow (small increase of the nitrogen content in the working atmosphere). The continuous monitoring of the relative intensity of a metallic line (Ti or/ Zr) shows a sharp decrease in the transition region. By continuously monitoring the metallic lines, any small tendency of decrease, is followed by a command to either decrease the nitrogen flow or increase the current fed on magnetron. The optimized conditions were established for TiN and ZrN single layers. In Fig. 10 is presented the variation of the Ti line intensity (Ti - λ = 468.19 nm) at the nitrogen flow increase in the deposition system.

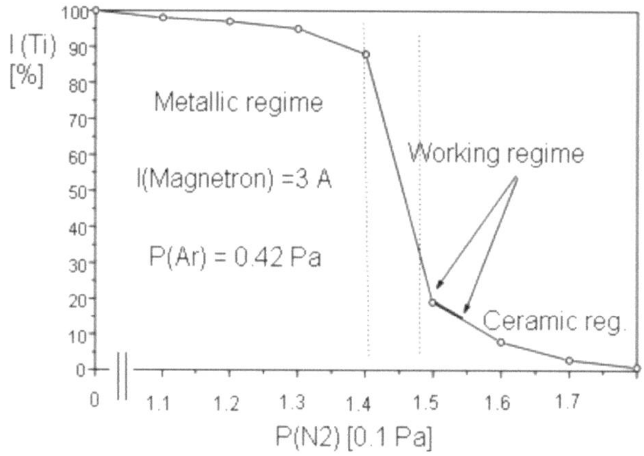

Fig. 10. Relative intensity of the Ti (λ=468.19nm) line vs. nitrogen partial pressure in the magnetron discharge

6. Coatings' characterisation

The chemical composition of TiN, ZrN and ZrN/TiN coatings was determined by energy - dispersive X-ray (EDX) spectroscopy, by means of a XL-30-ESEM TMP scanning electron microscope.

The coatings' texture, phase composition and bilayer period Λ were determined by high angle X-ray diffraction (XRD), with Cu K_α radiation, using a Rigaku MiniFlex II device.

Auger electron spectroscopy (AES) technique was used to determine the elemental composition of the films by using a PHI Model 3017 AES PC-Based System, equipped with an ion gun (for sputter cleaning and etching) in the range 3 ÷ 5 keV. The N/Ti ratio was determined from the positive slope of the nitrogen line located at 377 eV and the negative slope of the Ti peak at 418 eV.

The coatings' resistance to ion beam bombardment (5 keV, Ar^+) was done in the AES system, using a collimated ion gun for in-depth analyses. The assessment was done by surface roughness evaluation by AFM.

Films thicknesses were determined by a surface profilometer - Dektak 150. The surface morphology was observed by an atomic force probe microscope Veeco - Innova AFM/SPM, operating in tapping mode.

RBS spectra were obtained using a 2.7 MeV He^+ ion beam, revealing the elemental composition and the modulation periodicity of the multilayer, with Λ bilayer values ranging from 80 to 160 nm. The backscattered particles were detected by surface barrier detectors placed at 165^0 to the beam direction.

The deposited samples were measured for their reflectivity in the range $\lambda \in [11, 17]$ nm at the synchrotron facility from the National Institute of Standards and Technology [NIST, 2009].

Microhardness (Vickers) measurements were performed with a microhardness tester at 0.15 N load. Scratch tests under standard conditions (10 N/min*mm), using an indentor tip with electronic control of x, y, z position, were done to estimate the coatings adhesion. The critical load (L_c) values were determined by optical microscopy, L_c being defined as the load where film flaking starts.

Typical EDX spectra are shown in figures 11 - 13. The elemental composition of the films, as resulted from the EDX analyses, is presented in Table 1. It can be seen that the single layer coatings are almost stoichiometric: N/Zr = 0.9; N/Ti= 1.1. The presence of a small amount of oxygen is probably due to the contamination during sample handling in open atmosphere before the composition analysis. For the multilayer, the EDX analysis revealed a much higher Zr content as compared with that of the Ti. This result is due to the differences in the ZrN and TiN individual layer thicknesses [Braic et al., 2006].

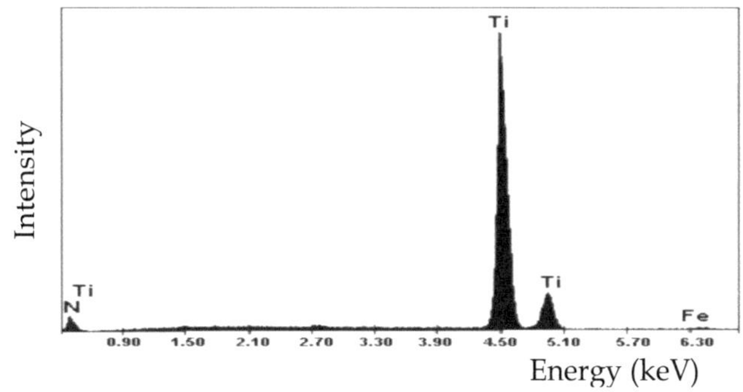

Fig. 11. EDX spectrum for a TiN coating

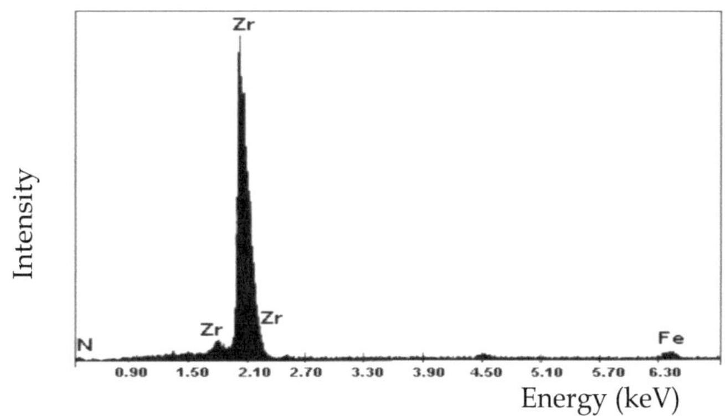

Fig. 12. EDX spectrum for a ZrN coating

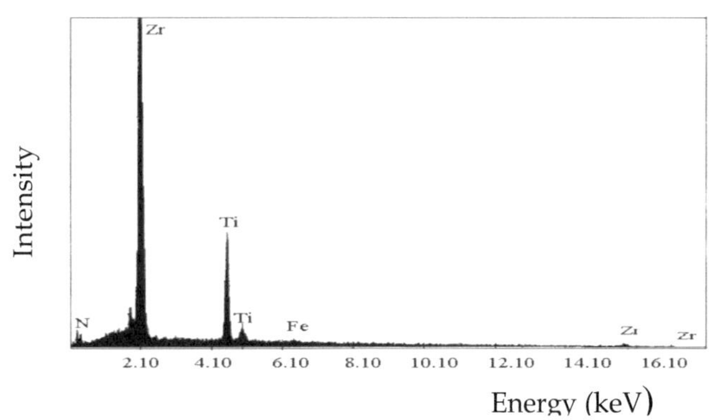

Fig. 13. EDX spectrum for a ZrN/TiN multilayer coating

Film type	Elemental concentration (at.%)				
	Ti	Zr	Fe	N	O
TiN	44.9	-	0.7	51	3.4
ZrN	-	49.6	0.9	44.8	4.7
TiN/ZrN-900/7	15.8	31.1	0.7	48.1	4.1

Table 1. Elemental composition of TiN and ZrN coatings

Three typical diffraction patterns of ZrN/TiN multilayer deposited on optical glass with different bilayer periods Λ are shown in Fig. 14a, 14b and 14c. As in the case of the single layer coatings (TiN - Fig. 15 and ZrN - Fig. 16), the diffraction patterns for the multilayer exhibit a strong (111) preferred orientation.

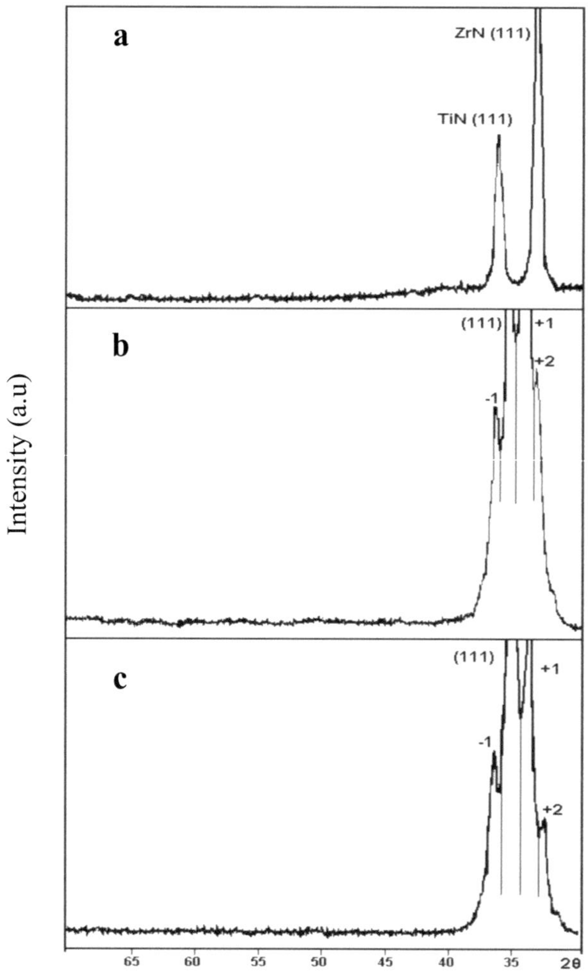

Fig. 14. X-ray diffraction patterns for ZrN/TiN multilayer

For the ZrN/TiN ML coatings with a large bilayer period (Λ = 550 nm, Fig. 14a), the diffraction lines belong both to the ZrN and to the TiN films, from which the coating is composed. It is worth to note that the diffraction pattern is not similar with that exhibited by a TiZrN layer (Fig. 17).

For the nanometre scale ZrN/TiN multilayer ($\Lambda < 10$ nm), the pattern generally consists of a Bragg peak located at the average lattice spacing of the multilayer surrounded by equally spaced satellite peaks, as can be seen in Fig. 14b and Fig. 14c. The bilayer period Λ was calculated, as in ref. [Yashar & Sproul, 1999], from:

$$\sin\theta_{\pm} = \sin\theta_B \pm \frac{m\lambda}{2\Lambda} \quad (1)$$

where $\theta_{\pm}$ are the positions of the m-th order positive (+) and negative (-) satellite peaks, θ_B is the position of the main Bragg reflection and the λ is the X-ray wavelength.

In the cases illustrated in figures 14b and 14c, the calculations lead to the Λ values of 9.2 nm and 7.2 nm, in excellent agreement with the values determined from the measurement of coating overall thickness and deposition time (9.1 nm and 7.0 nm).

The AES analyses were done on ZrN (Fig. 18a), TiN (Fig.18b) and TiZrN (Fig.19) mono layers, in order to obtain the specific pattern of the individual layers in multi-layered structure and also of the possible mixture of them. For this purpose, in order to have a clear answer, there were investigated multi-layers with two large Λ values (40 and 180 nm) and $\Gamma = 0.5$. The multilayers with high Λ values were studied in order to determine the etching rate of TiN and ZrN layers. The elemental analyses were done after cleaning the surface by ion bombardment etching (Ar^+, 3 keV) for 10 min.

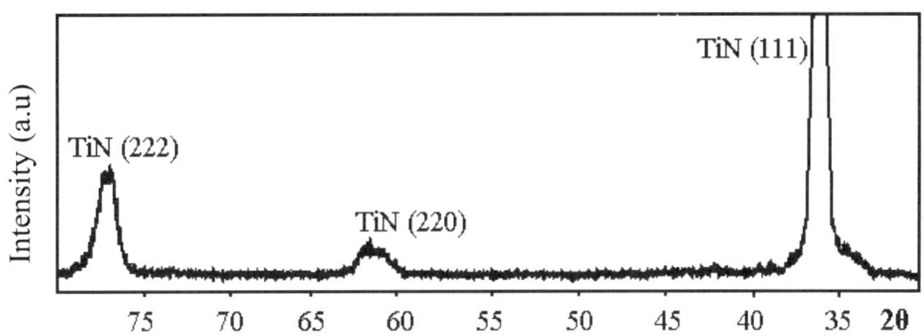

Fig. 15. X-ray diffraction patterns for TiN

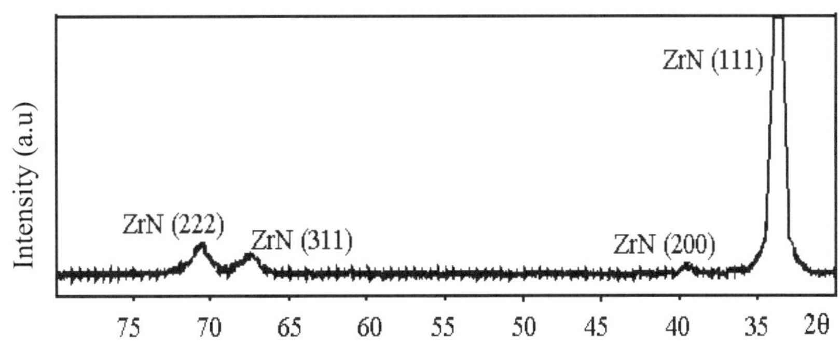

Fig. 16. X-ray diffraction patterns for ZrN

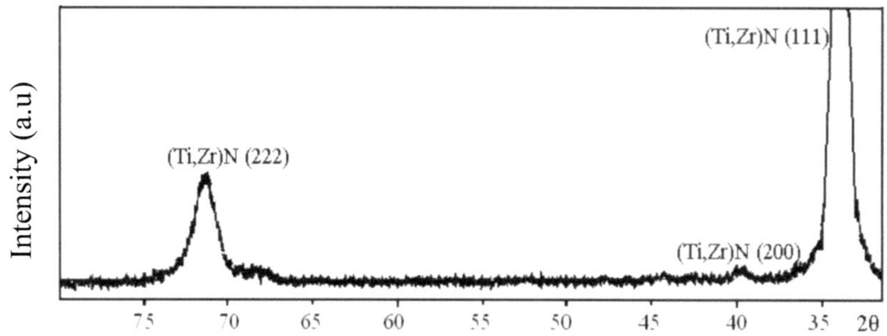

Fig. 17. X-ray diffraction patterns for (Ti,Zr)N

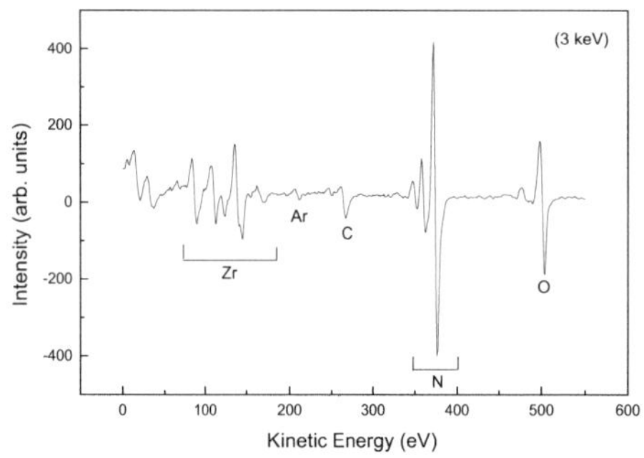

(a)

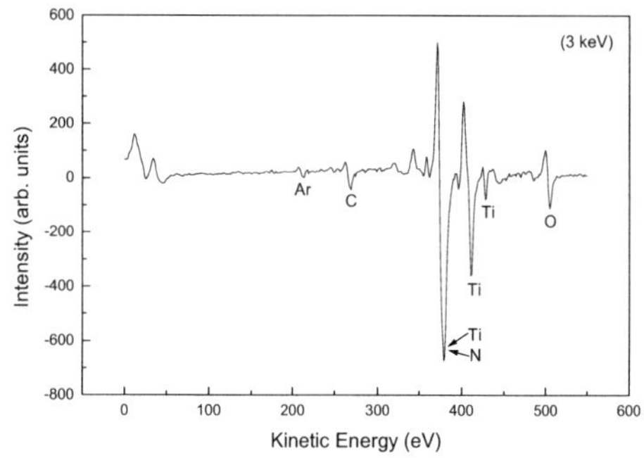

(b)

Fig. 18. Elemental composition done by AES of ZrN (a) and TiN (b) stoichiometric single layers

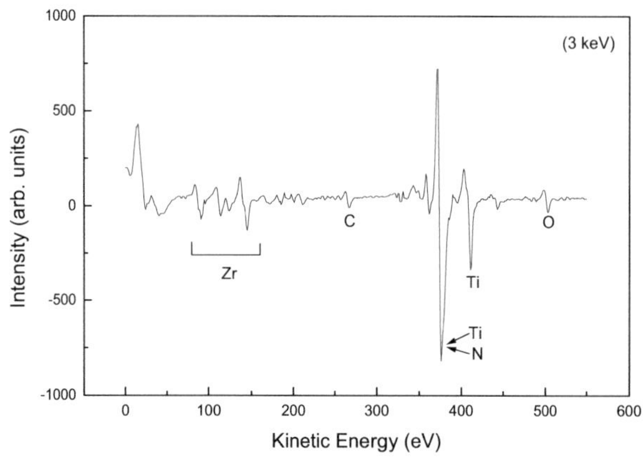

Fig. 19. Elemental composition done by AES of ZrTiN single layers (Zr:Ti=1.0)

For the ML structure with Λ= 40 nm and Γ = 0.5, AES signature of the top ZrN layer is presented in Fig. 20a, while in Fig. 20b is presented the bottom TiN layer. From the analysis it results that the individual TiN and ZrN layers are well defined in the ML, the obtained AES patterns being practically identical with the AES patterns obtained for single layers (Figs. 18a and 18b), with no "signature" for layer intermixing.

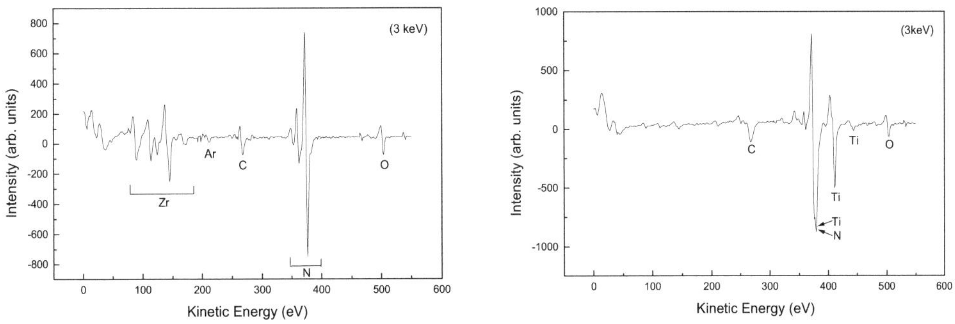

(a) (b)

Fig. 20. AES analyses of the ZrN top layer (a) and TiN bottom layer (b) in ZrN/TiN

Selected images of some MLs deposited on Si and optical glass substrates are presented in Figs. 21. The roughness values of the uncoated and ML coated (n = 40 nm, d = 7 nm) specimens are summarized in Table 2, where R_a is the roughness average, R_q is the rms roughness and R_t is maximum height of the profile in the investigated area:

$$R_a = (\text{nm}) = \left|\frac{\sum(x - \bar{x})}{N}\right| \tag{2}$$

$$R_q = (\text{nm}) = \sqrt{\frac{\sum(x - \bar{x})^2}{N}} \tag{3}$$

$$R_t = (\text{nm}) = x_{Max.} - x_{min.} \tag{4}$$

These results do not show any changes in the roughness of coated surfaces as compared with uncoated ones. The experiments showed that the increase of the number of bilayers n, accompanied by the overall thickness increase, lead to roughness increase. A comparison between the roughness of ML coatings with different thickness - as resulting from variations in the number of bilayers, is presented in Table 3.

Roughness parameters	Si	Glass	ML/Si	ML/glass
R_a (nm)	0.5	0.5	0.5	0.5
R_q (nm)	0.5	0.5	0.5	0.6
R_t (nm)	10.3	11.8	10.9	9.0

Table 3. AFM determined roughness for uncoated and ZrN/TiN-40/7 coated specimens

Roughness parameters	ML/Si		ML/glass	
	n=100	N=40	n=100	n=40
R_a (± 0,5 nm)	1,0	0,5	0,8	0,5
R_q (± 0,5 nm)	1,4	0,5	1,2	0,6
R_t (± 0,5 nm)	27,9	10,9	26,8	9,0

Table 4. AFM determined roughness for uncoated and ZrN/TiN-40/7 coated specimens

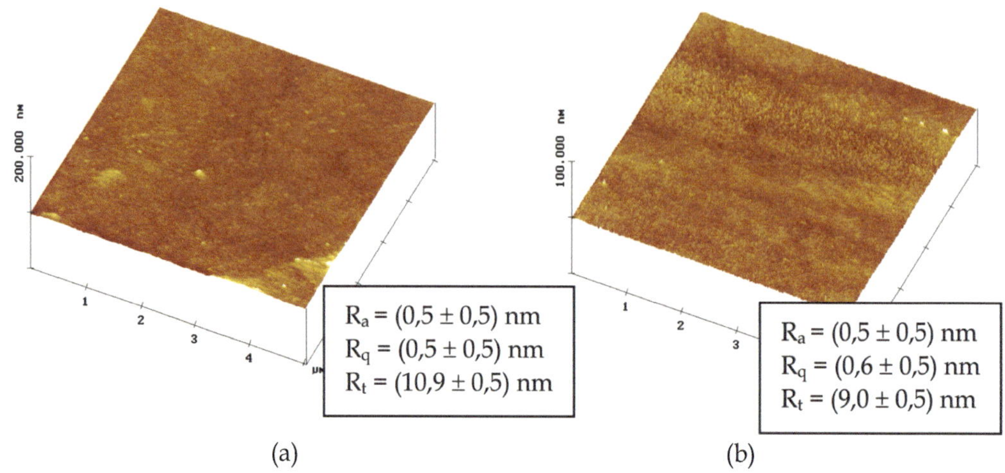

Fig. 21. ZrN/TiN-40/7 films deposited on Si (a) and on optical glass (b)

For the assessment of ion beam damage induced on coatings, relatively thick ZrN monolayered films were used (800 nm), as it is important that the top layer with the highest EUV reflectivity to be preserved. The averaged roughness values R_a decreased from the initial value of 1.1 nm to 0.8 nm, after ion bombardment with 5 keV Ar^+ ions, for 30 minutes. The RBS spectra of the ZrN/TiN multilayer n = 5, Λ = 80 nm, Γ = 0.5, presented in Fig. 22, are experimental data and simulated curves. For the coatings with large bilayer periods (80 - 160 nm), a good agreement between the experimental and theoretical curves was obtained. The individual layers may be clearly distinguished and their thicknesses were accurately determined. On the other hand, the in-depth resolution of the method does not allow observing the multilayer structure for bilayer periods less than 40 nm. The aim of the presented BRS analyses was to evidence the well structured multilayered coatings.

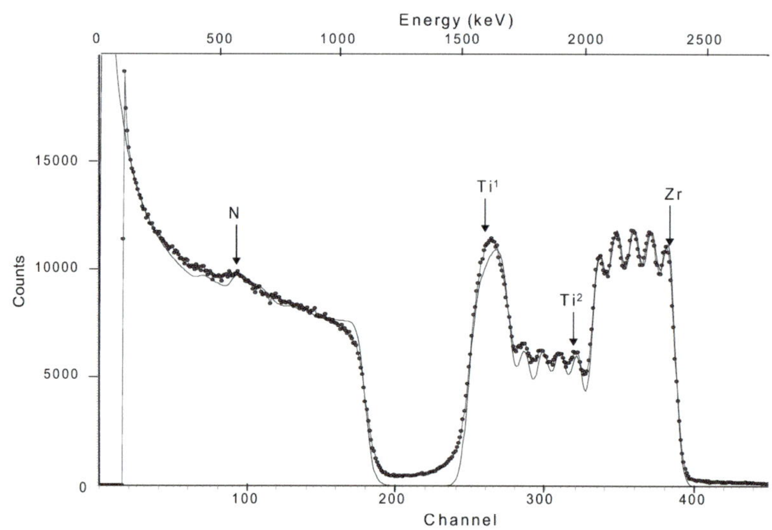

Fig. 22. RBS spectrum from a ZrN/TiN multilayer; n = 5; Λ = 80 nm, Γ = 0.5 (experimental-thick line; simulated data-thin line)

For the ZrN/TiN-40/7 ML type, the measured reflectivity at λ = 13.5 nm, at grazing incidence angles ranging from 6° to 15° (Fig. 23) proved to be in good agreement with the values obtained by modelling (Fig. 5).
The observed decreasing of the EUV reflectivity, as compared to the ideal values, is due to the surface contamination in open air during sample manipulation.

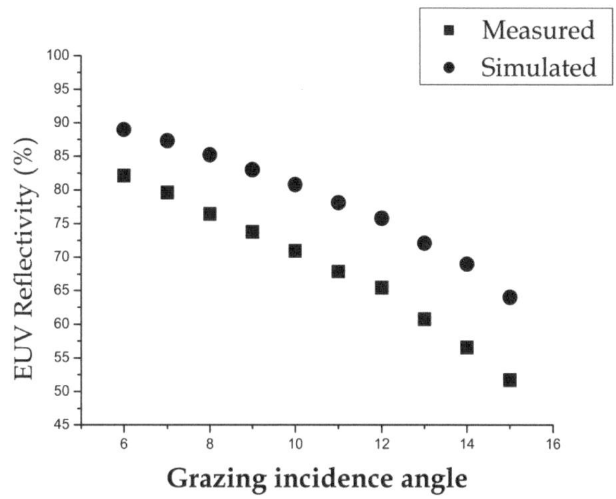

Fig. 23. Simulated (modelled) and measured reflectivity of ZrN/TiN/40/7 samples at 6° - 15° grazing incidence angles at λ = 13.5 nm

The microhardness and critical load (L_c) values of different coatings, when deposited on high-speed steel substrates are given in Table 5.
The existence of a narrow range of Λ values for which the hardness reaches its maximum value was already found for other ML with low Λ values, also known as superlattices [Shih & Dove, 1992], [Yang et al., 2002], [Barshilia et al., 2006], [Balaceanu et al., 2005],[Braic at al., 2006], [Abadias et al., 2007], being explained as the result of the combined effects of different strengthening mechanisms as Hall-Petch and Koehler models, coherency strain effects [Helmersson et al., 1987], [Yashar & Sproul, 1999], [Musil, 2000].

Coating	$HV_{0.015}$ (GPa)	L_c (N)
TiN	24	48
ZrN	21	56
ZrTiN	28	47
ZrN/TiN - 1100/6	23	52
ZrN/TiN - 700/9	34	58
ZrN/TiN - 900/7	35	57

Table 5. Mechanical characteristics of the films

As for the adhesion, the critical failure loads L_c for the monolayers in the scratch test were in the range of 47 - 56 N, the highest value being measured for ZrN layer. A higher adhesion strength was found for the ZrN/TiN - 700/9 coatings (L_c = 58 N). The result might be accounted for by the reduction in residual stress by the multilayered structure, as commonly reported for various multilayers, e.g. [Oh & Je, 1993]. It can also be seen that the adhesion decreased with decreasing bilayer period. This finding could be attributed to the role played by the interfacial bonding, coherency strain and interfacial delamination [Abadias et al., 2007], which is more pronounced as the number of layers, and hence interfaces, increases.

7. Conclusions

ZrN/TiN reflective hard coatings with bilayer periods Λ ranging in the nanometre range were successfully deposited on Si, optical glass and other test substrates using the pulsed bipolar magnetron sputtering method. The monolayer films of ZrN and TiN were almost stoichiometric (N/Zr=0.9 and N/Ti=1.1). The XRD patterns of the ML with small Λ values exhibited a pattern typical for superlattice coatings, consisting of a main Bragg peak surrounded by satellite peaks. Ion bombardment (5 keV Ar^+) of the MLs, intended to mimic the bombardment with EUV source debris, showed for the ZrN layer a slight decrease of the rms roughness with 0.3 nm after 30 minutes of ion bombardment. The multilayers with bilayer period Λ in the 7 - 9 nm range were the hardest (~ 35 GPa) and exhibited the best substrate adhesion, but good adhesion values were obtained for all the other coatings as well. The deposited films with a multilayered architecture led to an enhanced ion bombardment resistance, presenting promising reflectivity values for 13.5 nm EUV radiation. The obtaining of dense, adherent and highly reflective coatings for grazing incidence is a valuable research direction, using the bipolar pulsed magnetron sputtering deposition method.

As future challenges, a diversity of problems such as the ability to create a reliable high power EUV source, the maintenance of EUV mirrors through the use of debris mitigation schemes and the cleaning of contaminants, the development of a resist with low line edge roughness, and of a defect-free EUV mask, need to be solved. Each challenge needs to be overcome for EUVL to be a viable candidate for high volume manufacturing. The combination of efficient debris mitigation schemes, innovative methods for mirrors quick cleaning in adequate gases, as well as new coatings with high toughness, high adhesion and chemical inertness will provide the optimal design for EUVL systems, as reliable tools for a cost effective process for the mass production of nano-electronics components.

8. Acknowledgement

The work was supported under the European project SOARING, National Kern Projects TINOSOL and OPTOSOL.

The authors want to kindly acknowledge the help of Dr. Fred Bijkerk (FOM-Netherlands) for EUV reflectance measurements.

9. References

Abadias, G., Michel, A.S., Tromas, C., Jaouen, C. and Dub, S.N. (2007). Stress, interfacial effects and mechanical properties of nanoscale multilayered coatings, *Surface and Coatings Technology*, 202, 24 May 2007, pp: 844-853, ISSN 0257-8972

Allain, P.P., Nieto, M., Hassanein, A. (2008) Specular reflectivity of 13.5-nm light from Sn islands deposited on grazing incidence mirror surfaces, *Applied Physics,* Vol. A 91, April 2008, pp: 13–16 ISSN 0947-8396

Alman, D.A., Qiu, H., Spila, T., Thompson, K.C., Antonsen, E.L., Jurczyk, B.E. and Ruzic, N.D. (2007). Characterization of collector optic material samples exposed to a discharge-produced plasma extreme ultraviolet light source, *J. Micro-Nanolith. MEMS MOEMS*, Vol. 6, pp:013006.1-013006.5, ISSN 1817-6461

ASML (2009) http://www.asml.com/euv/asml_160_technical.html.

Bakshi, V. (2006), *EUV Sources for Lithography*, Vol: PM149, 23 February 2006, SPIE Press Book, ISBN: 9780819458452

Balaceanu, M., Braic, M, Braic, V. and Pavelescu, G. (2005). Properties of arc plasma deposited TiCN/ZrCN superlattices coatings, *Surface and Coatings Technology*, Vol. 200, Issues 1-4 , 1 Oct. 2005, pp:1084-1087; ISSN 0257-897

Balaceanu, M., Braic, M., Braic, V., Negrila, C.C. (2005). Surface chemistry of plasma deposited ZrC hard coatings, J.Optoel. Adv.Mat., Vol. 7/5, pp: 2557-2560, ISSN 1454-4164

Banine, V., Moors, R. (2004). Plasma sources for EUV lithography exposure tools, *J. Phys. D Appl. Phys.* Vol. 37 Dec.2004, pp: 3207-3212, ISSN 0022-3727

Barshilia, H. C. and Rajam, K.S. (2006). Reactive sputtering of hard nitride coatings using asymmetric-bipolar pulsed DC generator, *Surface and Coatings Technology*, Volume 201, Issues 3-4, 5 oct. 2006, pp: 1827-1835, ISSN 0257-8972

Barshilia,H.C., Rajam, K.S., Jain, A., Gopinadhan, K. and Chaudhary, S. (2006), A comparative study of the structure and properties of the nanolayered TiN/NbN and TiAlN/TiN multilayered coatings prepared by reactive direct current magnetron deposition, *Thin Solid Films*, Vol.503 May 2006, pp: 158 - 166 ; ISSN 0040-6090

Benoit, N., Yulin, S., Feigl, T., Kaisers, N. (2006). EUV multilayer mirrors with enhanced stability, *Proceedings of the SPIE*, Vol.6317, Aug.2006, no. 63170K, ISSN 0277-786X.

Bollanti, S., Bonfigli, F., Burattini, E., Di Lazzaro, P., Flora, F., Grilli, A., Letardi, T., Lisi, N., Marinai, A., Mezi, L., Murra, D. and Zheng, C. (2003). High-efficiency clean EUV plasma source at 10–30 nm, driven by a long-pulse-width excimer laser, *Applied Physics B: Lasers and Optics*, Volume 76, Number 3 / March, 2003, pp: 277-284, ISSN 0946-2171

Braic, M., Choi, P., Dumitrescu, C., Balaceanu, M., Braic, V. (2004). ZrN/TiN multilayers as debris shields for EUV mirrors grown by magnetron sputtering, *IX Int. Conf. on Plasma Surface Engineering*, Garmisch-Partenkirchen, Germany, 13 - 17 Sept., 2004

Braic, M., Balaceanu, M., Kiss, A., Vladescu, A., Braic, V. (2005). Zr based hard coatings as EUV grazing collection mirrors in EUV lithography Fourteenth International Summer School on Vacuum, Electron and Ion Technologies, *VEIT 2005, Abstracts, pag. 77-78,* 12-16 September 2005, Sunny Beach, Bulgaria

Braic, M., Balaceanu, M., Vladescu, A. , Kiss, A., Braic, V., Purice, A., Dinescu, G., Scarisoreanu, N., Stokker-Cheregi, F., Moldovan, A., Birjega, R. and Dinescu, M. (2006). TiN/ZrN heterostructures deposition and characterisation , *Surface and Coatings Technology*, Volume 200, Issues 22-23, 20 June 2006, pp: 6505-6510, ISSN 0257-8972

Braic, V., Balaceanu, M., Braic, M. (2008) Grazing incidence mirrors for EUV lithography, *Intl. Semiconductor Conf. Proceedings,* pp: 267-270, IEEE Catalog no. CFP08CAS-PRT, ISBN: 9781424420049

Braic, M., Braic, V., Balaceanu, M., Thin film reflectors for EUV (10-20 nm) radiation, Romanian Patent no.6/004/2009

Choi, P., Aliaga-Rossel, R., Braic, V., Braic, M., Dumitrescu, C., Leblanc, C., Mengtong, Q., Novikov, V.G., Ware, K.D., Yuan, A., Zakharov, S., Zakharov, V., Zhen, L. (2004). The SOARING. Micro Plasma Pulse Discharge EUV Source., Sematech-Lithography Meetings, *EUV Source Workshop*, Santa Clara, California, USA, 22 February 2004, http://internationalsematech.org/meetings/archives/litho/euvl/20040222b/posters/18Choi.pdf

CXRO (2009) http://henke.lbl.gov/optical_constants/

Ducros C. and Sanchette, F. (2006). Multilayered and nanolayered hard nitride thin films deposited by cathodic arc evaporation - Mechanical properties and cutting performances, *Surface and Coatings Technology*, Volume 201, Issues 3-4, 5 October 2006, pp:1045-1052, ISSN 0257-8972

Feigl, T., Yulin, S., Benoit, N. and Kaiser, N. (2006). EUV multilayer optics, *Microelectron. Eng*. Vol. 83, pp: 703-706, ISSN 0167-9317

Fiedorowicz, H., Bartnik, A., Jarocki, R., Kostecki, J., Krzywiński, J., Mikołajczyk, J., Rakowski, R., Szczurek A. and Szczurek, M. (2005). Compact laser plasma EUV source based on a gas puff target for metrology applications, *Journal of Alloys and Compounds*, Volume 401, Issues 1-2, 29 September 2005, pp: 99-103, ISSN 0925-8388

Hequet, C., Ravet-Krill, M.-F., Delmotte, F., Jerome , A., Hardouin, A., Bridou, F., Arniere, F., Roulliay, M., Bourcier, F., Desmarres, J.-M., Costes, V., Berthon, J., Rinchet, A. and Geyl, R. (2007). Design, conception and metrology of EUV mirrors for aggressive environments, *Proceedings of SPIE*, vol. 6586, pp. 65860X.1-65860X.10, Damage to VUV, EUV, and X-ray optics Meeting, 18-19 April 2007, Prague, Czech Republic, ISSN 0277-786X

Helmersson, U., Todorova, S., Barnett, S.A., Sundgren, J.E., Markert, L.C. and Greene, J.E. (1987). Growth of single-crystal TiN/VN strained-layer superlattices with extremely high mechanical hardness , *J. Appl. Phys*., Vol. 62 Jul1987, pp: 481 - 484; ISSN 0021-8979

Hubbell, J.H., Veigele, W.J., Briggs, E.A., Brown, R.T., Cromer, D.T. and Howerton, R.J. (1975), Atomic Form Factors, Incoherent Scattering Functions, and Photon Scattering Cross Sections, *J. Phys. Chem. Ref. Data*, Vol.4, pp: 471-538 ; ISSN 0047-2689

Hubbell, J.H., Veigele, W.J., Briggs, E.A., Brown, R.T., Cromer, D.T. and Howerton, R.J. (1977), Erratum at Atomic Form Factors, Incoherent Scattering Functions, and Photon Scattering Cross Sections, *J. Phys. Chem. Ref. Data*, Vol.6, pp: 615-616, ISSN 0047-2689

Hugosson, H.W., Eriksson, O., Jansson, U., and Johansson, B. (2001), Phase stabilities and homogeneity ranges in 4*d*-transition-metal carbides: A theoretical study, *Phys.Rev.B*. Vol. 63 March 2001,pp: 134108.1 - 134108.11, ISSN 0163-1829.

Glocker, D., Ismat Shah, S., Editors, Spiller,E.,-author (1995). *Handbook of Thin film Process Technology, F5 – Multilayered structures for X Ray mirrors,* IOP Publishing, ISBN 0 7503 0311 5, Bristol, UK

Itani, T. (2009) Recent status and future direction of EUV resist technology, *Microelectronic Engineering,* Volume 86, Issue 3, March 2009, pp: 207-212, ISSN 0167-9317

Jaeger, R. C., (2002), *Lithography-Introduction to Microelectronic Fabrication,* ISBN 0-201-44494-7, Upper Saddle River: Prentice Hall

Kanoun, M.B., Goumri-Said, S. and Jaouen, M. (2007). Structure and mechanical stability of molybdenum nitrides: A first-principles study, *Phys. Rev. B,* Vol. 76, pp: 134109.1-134109.4 , ISSN 0163-1829

Lebert, R., Juschkin, L., Jägle, B., Meisen, M., Berman, V., Bieberle U. (2003). Inband EUV Open Frame Resist Exposer TEUVL, *International EUVL Symposium,* 30.9.–2.10.2003, Antwerp, Belgium

Meiling, H., Banine, V., Kurz, P., Blum, B., Heerens, G. and Harned, N (2003).The EUV program at ASML: an update, *Proceedings of the SPIE* ,Vol. 5037 June 2003, pp: 24-35, ISSN 0277-786X.

Meiling,H., Banine, V., Harned, N., Blum, B., Kürz, P. and Meijer, H. (2005). Development of the ASML EUV alpha demo tool, *Emerging Lithographic Technologies IX.* Ed. by Mackay, R. Scott. *Proceedings of the SPIE,* Vol. 5751 May 2005, pp. 90-101, ISSN 0277-786X.

Musil, J. (2000), Superhard nanocomposite coatings, *Surface and Coatings Technology,* Vol. 125 March 2000, pp: 322 - 330. ISSN 0257-8972

Neumann, M.J., DeFrees, R.A., Qiu, H., Ruzic, D.N, Khodykin, O., Ershov, A. and Bristol, R.L. (2007). Plasma cleaning of lithium off of collector optics material for use in extreme ultraviolet lithography applications, *J. Micro-Nanolith. MEMS MOEMS,* Volume 6, Issue 2, 2007, pp: 023005.1- 023005.8 2007, ISSN 1817-6461

NIST. (2009). http://physics.nist.gov/MajResFac/euv.html

Oh, U.C. and Je, J.H. (1993). Effects of strain energy on the preferred orientation of tin thin films, *Journal of Applied Physics,* 74, 1 August 1993, pp. 1692-1696, ISSN 0021-8979

Rothschild, M., Bloomstein, T.M., Efremow, N.Jr., Fedynyshyn, T.H., Fritze, M., Pottebaum, I. and Switkes, M. (2005). Nanopatterning with UV Optical Lithograpy. *MRS Bull.* Vol.30 Dec. 2005, pp: 942-946, ISSN 0883 - 7694

Schroeder, S., Feigl, T., Duparre, A. and Tuennermann, A. (2007). EUV reflectance and scattering of Mo/Si multilayers on differently polished substrates, Optics Express, Vol. 15, Issue 21, pp. 13997-14012, 11 October 2007, ISSN: 1094-4087

SEMATECH.(2009). http://www.sematech.org/corporate/news/releases/20090211.htm

Shih, K.K. and Dove, D.B. (1992), Ti/Ti-N Hf/Hf-N and W/W-N multilayer films with high mechanical hardness, *Applied Physics Letters,* Vol. 61 Aug. 1992, pp: 654 - 656, ISSN 0003-6951

Shin, H., Sporre, J.R., Raju, R. and Ruzic, D.N. (2009), Reflectivity degradation of grazing-incident EUV mirrors by EUV exposure and carbon contamination, *Microelectronic Engineering,* Volume 86, Issue 1, January 2009, pp: 99-105, ISSN 0167-9317

Song, I., Kobayashi, Y., Sakamoto, T., Mohanty, S.R., Watanabe, M., Okino, A., Kawamura, T., Yasuoka, K., Horioka K. and Hotta, E. (2006). Performance of gas jet type Zpinch plasma light source for EUV lithography, *Microelectronic Engineering*, Volume 83, Issues 4-9, April-September 2006, pp: 710-713, ISSN 0167-9317

Stamm, U. (2004) Extreme ultraviolet light sources for use in semiconductor lithography-state of the art and future development, *J. Phys. D Appl. Phys.*Vol 37 Dec. 2004, pp: 3244-3253, ISSN 0022-3727

Stulen, R. and Sweeney, D. (1999). Extreme ultraviolet lithography, *IEEE J. Quantum Electron.* Vol. 35, pp: 694-699, ISSN 1063-7818

Zuppella, P., Reale, A., Ritucci, A., Tucceri, P., Prezioso, S., Flora, F., Mezi , L. and Dunne, L. (2009)Spectral enhancement of a Xe-based EUV discharge plasma source, *Plasma Sources Sci. Technol.*, 2009, pp: 025014.1-025014.18, ISSN 0963-0252

Vargas Lopez, E., Jurczyk, B.E., Jaworski, M.A., Neumann M.J. and Ruzic, D.N. (2005), Origins of debris and mitigation through a secondary RF plasma system for discharge-produced EUV sources, *Microelectronic Engineering*, Volume 77, Issue 2, February 2005, pp: 95-102, ISSN 0167-9317

Vladescu, A., Balaceanu, M., Braic, V., Cotrut, C., Kiss, A. and Braic, M.(2007) Effect of energy ion bombardment on characteristics of Ti based biocompatible coatings, *Plasma Process. Polym.*, 4 (S1), S406–S410, April 2007, ISSN 1612-8850

Vladescu, A., Kiss, A., Popescu, A., Braic, M., Balaceanu, M., Braic, V., Tudor, I., Logofatu, C., Negrila, C.C. and Rapeanu, R., (2008), Influence of Bilayer Period on the Characteristics of Nanometre-Scale ZrN/TiAlN Multilayers , *Journal of Nanoscience and Nanotechnology* , Vol. 8, No.2, February 2008, pp: 717–721, ISSN, 1533-4880

Wallace, J., Cheng, Y.-C., Isoyan, A., Leonard, Q., Fisher, M., Green, M., Bisognano, J., Nealey, P. and Cerrina, F. (2007), A novel EUV exposure station for nanotechnology studies, *Nuclear Instruments and Methods in Physics Research Section A: Accelerators, Spectrometers, Detectors and Associated Equipment*, Volume 582, Issue 1, 11 November 2007, pp: 254-257, ISSN: 0168-9002

Wang, S., Zhu, J.T., Wang, F.L., Zhang, Z., Wang, H.C., Qin, S.J. and Chen, L.I. (2006) Development of Multilayer Optics for EUV, Soft X-Ray and X-Ray Regions in IPOE , Springer Proceedings in Physics - *X-ray Lasers*, Volume 115, pp: 401-408, Springer Netherlands, January 18, 2008, ISSN 0930-8989; ISBN 978-1-4020-6017-5

Yang, O., He, C., Zhao, L.R. and Immarigeo, J.P. (2002), Preffered orientation and hardness enhancement of TiN/CrN superlattices coatings deposited by reactive magnetron sputtering, *Scripta Materialia*, Vol.46 Feb 2002, pp: 293 - 297; ISSN 1359-6462

Yashar, P.C. and Sproul, W.D. (1999). Nanometer scale multilayered hard coatings, *Vacuum*, Vol.55 Dec.1999, pp: 179 - 190; ISSN 0042-207X

11

Steady-state and Time-dependent LPP Modeling

White, Dunne, and O'Sullivan
University College Dublin
Ireland

1. Introduction

A primary goal in developing extreme ultraviolet lithography (EUVL) is the modeling of plasma-based light sources, created either by intense lasers or high-current pulsed discharges, which have applications in semiconductor lithography, nanotechnology, and plasma diagnostics (Attwood, 2004, Derra et al., 2005). Such modeling can be the key factor to important scientific and technological solutions in EUVL source optimization. Radiation hydrodynamic modeling is also important in astrophysics and inertial confinement fusion.

As stated in the *International Technology Roadmap for Semiconductors* (2008), "Extreme ultraviolet lithography is expected to provide a single exposure solution for use in manufacturing starting at 22 nm half pitch and possibly for 32 nm half pitch." To match the proposed multilayer Mo/Si mirror imaging system (~70% refelctivity at 13.5 nm), the EUVL wavelength of choice for high-density, low-cost integrated circuits is 13.5 nm, created either by a tin containing laser-produced plasma (LPP) or discharge produced plasma (DPP). The source power must be greater than 180 W at intermediate focus.

In-band EUV emission (13.365–13.635 nm) in a range of overlapping tin ions (Sn^{4+} to Sn^{13+}) has previously been identified as originating from 4d-4f, 4p-4d, and 4d-5p transitions (O'Sullivan & Carroll, 1981). The source conditions for optimum brightness are in the computationally challenging non-local thermodynamic equilibrium (non-LTE) (*a.k.a.* collisional-radiative or CR) regime, and the emission is highly anisotropic in spectral shape and intensity (Hayden et al., 2006) for all DPPs and most LPPs (i.e., slab or liquid jet targets). Xenon and lithium sources have been proposed as possible targets, but are no longer considered viable (Al-Rabban et al., 2005).

There are numerous challenges to develop and integrate efficient and cost-effective high-flux plasma EUV sources; in particular the improved conversion efficiency (CE) (ratio of 13.5 nm in-band emission to input laser energy) of the proposed source, and the characterization and mitigation of debris (fast ions, neutrals, and nanoparticles). Of primary importance is to identify optimum source parameters (laser wavelength, λ, pulse duration, τ, power density, Φ, and pulse shape) and material composition (e.g., target shape and ion concentration) for maximum CE, particularly with the use of reliable modeling tools.

Laser-plasma interaction is complex, but the fundamentals are as follows. A laser is incident on a target, producing an ablation front, an over-dense region up to 10^3 times solid, which creates a shockwave in the target as well as heating and ionizing the target to produce an

expanding plasma, a process that continues throughout the pulse duration. Photons are absorbed by inverse Bremsstrahlung, and re-emitted via recombination, where up to 80% of the incident energy can be converted to EUV radiation. The plasma expands (typical ion velocities are up to 10^6 cm/s) but as electron density decreases, less laser energy is absorbed. The plasma is a self-regulating regime of generation, heating, and expansion over the pulse duration, where electrons equilibriate on a very small time scale compared to pulse duration. Opacity effects are important since emitted radiation can be significantly reabsorbed within the plasma. A wide range of electron densities and temperatures exist, which require challenging atomic and plasma hydrodynamic models (Greim, 1964, Key & Hutcheon, 1980, Carroll & Kennedy, 1981, Attwood, 1999, Al-Rabban et al., 2005).
Section 2 presents a background survey of some typical LPP models, citing conditions for optimum emission where possible. Major issues in modeling steady-state and time-dependent plasmas are discussed in Section 3. Firstly, a straightforward steady-state model in the optically thin regime (emitted radiation not reabsorbed) is used, which provides a simple estimation of optimum electron temperature at maximum in-band emission, from which important information about the complex atomic physics in LPPs can be determined. Secondly, a more sophisticated 1D model that includes radiation transport within an optically thick plasma (emitted radiation reabsorbed) is used, from which the conversion efficiency using different laser parameters can be calculated. Some 2D results are also presented which take into account lateral expansion. Section 4 presents current trends and future challenges in the field of LPP modeling and EUV source optimisation.

2. Modeling background

To give a background to LPP modeling, a number of cases are cited that highlight a variety of codes, laser parameters (wavelength, pulse duration, power density), target material and geometry, and dimension, as well as recent results related specifically to EUVL. The choice is by no means exhaustive, but is intended to give an idea of the different approaches to optimising LPP parameters. For brevity, nomenclature is used without introduction, but can be found in Section 3. Note that the following survey is intended as background only and the reader is directed to the literature for more detail.
Atomic structure codes such as the HFCI (Hartree Fock with Configuration Interaction) code of Cowan (Cowan, 1981), GRASP (Grant et al., 1980), HULLAC (Bar-Shalom et al., 2001), and FAC (Gu, 2003), among others, have been used to determine cross sections and transitions involved in spectral emission. A number of plasma codes exist to model hydrodynamics, such as the steady-state, 0D, CR code of Colombant & Tonon (1973), the 1D Lagrangian code MEDUSA (Christiansen et al., 1974) and the 2D codes CASTOR (Christiansen & Winsor, 1979), LASNEX (Zimmerman & Kruer, 1975) and Z* (Zakharov et al., 2005), which all use a simplified single electron model that excludes detailed atomic term structure. These codes are discussed more fully as applicable below.
The postprocessor FLY code (Lee, 1995) is a time-dependent, single-cell, hydro-dynamics, CR-based model that solves the differential rate equation using a 1st-order escape probability approximation, valid for photons that are absorbed locally or escape without interactions. SWARM (Sondhauss et al., 2001) is a multi-cell extension to FLY using the Average Atom (AA) model which accounts for non-local re-absorption in the plasma. Angle-resolved spectra can be calculated for planar, cylindrical, or spherical target geometries that account for Doppler-shifted anisotropic radiation. FLYCHK is an online

extension to FLY which incorporates HULLAC atomic data for non-H-, He-, and Li-like plasmas (Chung et al., 2005). However, accuracy is limited as the AA model is essentially a Bohr atom approach and ion level energies are assumed to be *l*-degenerate, a major drawback when dealing with medium to high Z species such as Sn.

SCROLL (Bar-Shalom, 1997) is a super configuration, non-LTE CR model, which modifies a simpler LTE model by splitting supershells to populate new superconfigurations proportional to their partition function (> 10^{-5} of total population). Levels up to n = 8 were considered in optically thin selenium (Z = 34) and lutetium (Z = 71) plasmas.

To account for *nl*-splitting, the atomic model of Mirone et al. (1997) assumes thermodynamic equilibrium for population levels using the AA, screened-charge model (a reasonable assumption where collisions are dominant) to reconstruct the one-electron atomic potentials. The authors also used the LTE hydrodynamic code MULTI (λ = 0.53-μm, τ = 0.5-ns, 3-μm diameter germanium target) to highlight the effect of radiation loss in non-LTE conditions. Their simulations showed that a non-LTE plasma model produces a hotter corona, reduced radiative heat wave, and more penetration than in an LTE plasma, though less so with their reconstructed versus hydrogenic model.

The kinematics of supersonic ionization fronts and radiation transport was studied in Gumbrell et al. (1998) for ps-pulses, where measured plasma velocities are up to 40 times greater than for ns-pulses. A 1D laser-plasma hydrodynamics code based on MEDUSA was used to describe heating within the target.

Dürsterer et al. (2001) conducted experiments on oxygen-containing targets (Nd:YAG, 700-mJ, τ = 8-ns, ~10-μm diameter mass-limited water droplets and 20-μm diameter solid glass SiO_2 spheres). The authors noted that laser energy absorption is always less than 100% but is insufficient for pulses of either too short or too long duration: for shorter pulses, absorption is reduced; for longer pulses, greater expansion reduces electron density below critical (where the trailing edge of a long pulse is not absorbed at all). They noted a logarithmic increase in CE over 5 orders of magnitude of pulse duration (200 fs–6 ns), and that energy was independent of pulse duration for a finite drop for optimum EUV emission because of the fixed number of atoms. They also noted that a mass-limited droplet expands much faster and isotropically in 3D compared to an essentially 1D expansion in a bulk target, and suggested that energy and pulse duration be independently optimized in mass-limited targets rather than intensity (sufficient for bulk targets). Using a steady-state CR model they found the optimum electron temperature was 30 eV (reasonably agreeing with a blackbody model of 95 eV). From MEDUSA simulations, they noted that spherical targets cool faster and that the position of critical density moves slower than in bulk targets.

Sasaki et al. (2004) use the parametric potential atomic code HULLAC and CR Whiam code to calculate Xe and Sn emission spectra. The authors assume LTE and a spherical plasma to simplify the rate equations. For xenon, a calculated spectrum (n_e = 10^{21} cm^{-3} at T_e = 25 eV) is compared to experiment (Nd:YAG, τ = 8-ns, Φ = 10^{12} W/cm^2, gas jet target), and for tin, a calculated spectrum (n_e = 10^{20} cm^{-3} at T_e = 22 eV, uniform sphere of radius 30 μm) is compared to experiment (Φ = 1–9.6 10^{11} W/cm^2). The authors note the effect of satellite lines in sufficiently dense plasmas (on the long wavelength side of the UTA) and the considerable opacity effects in a higher density plasma. They also included configuration interaction (CI) effects to describe the atomic physics in both HULLAC and GRASP. They comment that a plasma is in quasi-steady-state if the temporal evolution of the EUV spectral intensity is identical to the input laser pulse shape.

Fujioka et al. (2005) used a 6-beam, Gekko II laser (λ = 1.053-μm, τ = 5-ns, Φ = 1 x 10^{13} W/cm^2), and observed that the expansion velocity of 3.8 x 10^6 cm/s compared well with that predicted by ILESTA-1D. The authors noted that the calculated UTA width was broader than the experimental UTA width (see also Mandelbaum et al. (1997)). Simple modeling using just *gA* or *f*-value distributions for reproducing spectra based on the assumption that levels are populated uniformly within a particular configuration does not take into account the energy dependence of the excitation rate coefficients which strongly influence the UTA shape, however, and accurate term specific rate coefficients are essential for very accurate modeling. The atomic physics was modelled using HULLAC with CI included, although the authors noted that disagreements between experimental and calculated spectra result from the number of configurations used. They noted that opacity effects are a function of plasma size, and that there are two well-known ways to change plasma size: 1) by incident wavelength—a shorter wavelength heats higher density regions producing a larger plasma, 2) by pulse duration—a longer duration produces a larger plasma. They also noted that satellite emission originated from deeper (higher density) layers than the UTA emission. The experimental CE versus power density was shown to be qualitatively consistent with their calculated CE (using the 1D code STAR and HULLAC), and that a shorter pulse duration leads to a higher CE.

Tao et al. (2005) conducted a comparison between experiment (λ = 1.064-μm, τ =10-ns, 220-μm focal spot diameter, Φ = 0.1-3 x 10^{11} W/cm^2 at normal incidence) and a 1D simulation (Φ = 1 x 10^{11} W/cm^2), which showed reasonable agreement, noting that the difference was because of lateral expansion not accounted for in the 1D code. The authors noted that because of a finite focal spot size, which was comparable to the plasma size, lateral expansion occurs, which removes plasma energy, reduces ion velocity, and reduces ion density. They commented that due to opacity effects, most of the EUV radiation comes from the under-dense, coronal region, which they observed using two interferometers and two probe beams (at 266 nm and 532 nm) to profile electron density along the centre of the plasma. They observed that at Φ = 1 x 10^{11} W/cm^2 the electron temperature in the coronal region is close to optimum value, whereas at higher intensities it is too hot.

Yamaura et al. (2005) used the 12-beam Gekko XII facility (Nd:YAG, λ = 1064-nm and 4ω 266-nm, τ= 6.5-ns and 10-ns, spatial flattop, Φ = 10^{10}-10^{11} W/cm^2 at normal incidence, tin target) and large spot sizes (~500 μm) to exclude the effect of energy loss from lateral expansion. The authors reported an absorption dip at 13.5 nm for the 266-nm wavelength due to greater opacity in the lower-wavelength, higher-density plasma, which was reproduced in a 1D code. They also noted an angular dependence of $\cos^{1.3}\theta$ at 266 nm and $\cos^{0.5}\theta$ at 1064 nm, and reported that the shifted-peak 4ω spectrum (to 13 nm) could be used with a modified Mo/Si mirror system.

Shimada et al. (2005) used the 12-beam Gekko XII laser (τ =1.2-ns, target diameter varied from 300-700-μm, Φ = 0.2–1 x 10^{11} W/cm^2, dodecahedral configuration), primarily used in fusion research, to uniformly irradiate a spherical tin target and thus remove 2D effects to compare with a 1D code. Using a EUV pinhole camera, they observed that the diameter of emission expanded to several tens of microns. They noted that at Φ = 1.2 x 10^{11} W/cm^2, the temporal EUV profile precisely matched the laser profile, whereas at Φ = 9.6 x 10^{11} W/cm^2, the spectral maximum occurred much later during the delayed recombination phase. A maximum CE of 3% occurred at Φ = 0.5–1 x 10^{11} W/cm^2.

MacFarlane et al. (2005) used the 1D Lagrangian code HELIOS-CR and postprocessor SPECT3D to study LPPs and z-pinches. The material equations of state are based on SESAME or PROPACEOS tables and frequency-dependent opacities on non-LTE level populations. Atomic cross sections were calculated using ATBASE and oscillator strengths from a HFCI model, where in Sn^{10+}, for example, ~500 higher n-shell configurations (to n = 9) and 57 doubly excited configurations were included. More than 5 x 10^6 oscillator strengths and energy levels from Sn I to Sn XX were computed. In planar geometry, radiation is transmitted along a single ray at an angle θ with incidence, and in spherical geometry along a multi-ray conical path. The authors compared HELIOS-CR output to data from the 12-beam Gekko XII (λ = 1.06-μm, τ = 1.2-ns, Φ = 0.9–9 x 10^{11} W/cm^2, 700-μm diameter CH spheres coated with a 1-μm tin layer), which showed good agreement with tin UTA evolution. Planar tin foil experiments were also conducted (λ = 1.06-μm & τ = 1-ns, λ = 0.35-μm & τ = 10-ns) giving a maximum CE of 4.5% for the 1.06-μm & 1-ns case. They also noted that CEs were higher for targets in front of the laser focus because the plasma couples to a larger effective laser spot size as it expands outward.

Zakarov et al. (2005) used the 2D RMHD Z* to model EUV spectra in a number of scenarios, the preprocessing code THERMOS to calculate the spectral and plasma transport coefficients and material and mixtures EOS database, and the postprocessor code RAY which includes the effect of complex level kinetics. Z* results (λ = 1064-nm, τ = 15-ns, flat pulse, 30–300-mJ, 40-μm diameter focussed spot on a 30-μm diameter solid Sn or cryogenic Xe droplet) showed a CE of about 3% for tin and 0.65% for xenon. The authors noted that the total emission solid angle is less than 4π because of the target shadow, and that the plasma is in the shape of a conical shell, which consists of hotspots. Further pre-pulse-pulse simulations (Zakarov et al., 2007) (Nd:YAG: 2.5-mJ & τ = 10-ns pre-pulse; CO_2: 50-mJ & τ = 15-ns pulse; on 20-μm Sn droplet) at varying delay times (25–125 ns), showed CE as a function of delay time, CO_2 pulse duration and energy (τ = 20–100 ns, 20–200 mJ), and CO_2 intensity (Φ = 0.03–2 x 10^{11} W/cm^2).

The Laser Plasma Laboratory at the University of Central Florida (Al-Rabban, 2005) modelled oxygen (liquid water droplets) in spherical geometry using MED103 (λ = 1064-nm, τ = 10-ns, 80-μm spot size, Φ = 4.5 x 10^{11} W/cm^2), which predicted a maximum electron temperature of 56 eV. Line emission was modelled using the LTE code Spectra and atomic data from the astrophysical Opacity Project database (Z = 1 to 26), which produces synthetic spectra from 1 to 1000 nm. The authors also used the CHIVAS hydrodynamic code and a CR ionization model to model xenon droplets, where results showed that "important droplet expansion occurs at the beginning of the laser pulse, which results in a rather inefficient overall coupling between the laser and the spray."

Ando et al. (2006) derived a scaling law for absorption, showing that optical depth is a function of laser wavelength, pulse duration, and power density. The authors calculated that a 3–7-ns duration pulse produces an optical depth of 1 cm for a λ = 1064-nm, Φ = 1 x 10^{11} W/cm^2 laser. Their experiments (Nd:YAG, λ = 1064-nm, τ = 2–9-ns (Pockels cell), τ = 1-2-ns (SBS pulse compressor), 300–900-μm spot size with 100-nm Sn layer, Φ =0.1–10 x 10^{11} W/cm^2) showed that the EUV emitting region is thinner for shorter pulses, and that emitted EUV intensity increases as pulse duration decreases (8.5 to 2.3 ns) but decreases beyond that (to 1.2 ns). A maximum CE of 2.2% at 2.3 ns and Φ = 5.0 x 10^{11} W/cm^2 was observed. They also noted that an absorption dip at 13.5-nm decreases as the pulse duration decreases, and that the electron temperature was lower for shorter pulses due to lower absorption.

Rollinger et al. (2008) used the 2D axisymmetric hydrodynamic code POLLUX (which incorporates the steady-state CR code of Colombant & Tonon) and the atomic code HULLAC to determine ion level populations and spectra (as well as assess the limits of LTE). The authors noted that 1D codes such as MED103 misrepresent electron temperature and that non-LTE calculations lower the plasma temperature for optimum CE by 3 eV. 2D target simulations (Nd:YAG, τ =1–15-ns, Φ = 0.1–30 x 10^{11} W/cm^2, 40-μm focal spot size, 50-μm thick planar Sn target) showed that optimum CE occurs for 3–10 x 10^{11} W/cm^2, noting that power density primarily determines the electron temperature and thus effects CE more than pulse duration. They noted that a longer pulse duration produces a larger plume, which reduces CE because of heat transfer from the hotter core. They also noted that spherical targets produce lower velocities and a reduced though more 2D plume.

A number of recent experimental results have been reported that have not yet been fully modelled, but are important with regards to current trends and future challenges in modeling (see also Section 4). Ueno et al. (2008) noted a 4% CE for a CO_2 laser incident on a tin cavity target (200 μm). Fujioka et al. (2008a) noted that the target size should equal the laser spot size to suppress OOB radiation and that debris is reduced by using mass-limited targets. Sequoia et al. (2008) noted dips in the angular distribution of the in-band EUV emission at 0° and 30°, which they attributed to 2D plasma expansion. They also noted that lateral and longitudinal expansion are of similar scale for smaller spot sizes but that expansion is entirely longitudinal for larger spot sizes (more than a few hundred microns) and suggested that for higher CE, small focal spot sizes are required to match the target size. Takahashi et al. (2008) compared the kinetic energy and particle emission for CO_2 and Nd:YAG lasers at the same energy (50 mJ), observing that the CO_2 laser produces higher ion kinetic energy (~ 4x) but lower particle emission (1/4) than the Nd:YAG. They commented that the CO_2 laser was not considered in earlier EUVL research because the deposited energy was considered too small, but noted that the dominant absorption process for longer-wavelength lasers is Joule heating ($\propto \lambda^2$) and that once the plasma has formed energy is absorbed in the plasma surface for a CO_2 laser rather than the target surface, in contrast to the more penetrating Nd:YAG.

Fujioka et al. (2008b) observed the effect of laser spot size and microdroplet diameter in pre-pulse-pulse experiments (Nd:YAG pre-pulse, τ = 8-ns, Φ = 3 x 10^{11} W/cm^2, 50-μm focal spot size, CO_2 main pulse, τ = 40-ns, 1 x 10^{10} W/cm^2, 250-μm focal spot diameter, with optimum delay of 1 μs) and observed a 4% CE (greater than the 2.5% observed from a single CO_2 pulse on a planar Sn target). They noted that the 20-μm minimum-mass droplet was too small for optimum laser coupling with the Nd:YAG pre-pulse but was sufficiently expanded prior to the CO_2 main pulse. They noted that in a 1D simulation of a single CO_2 pulse that about one half of the incident energy is reflected by the Sn plasma surface and that a pre-pulse forms a low-density, expanded target which enhances absorption. They also noted that about 1/3 of the emitted EUV radiation reaches IF, and thus 3 times the IF power is required at source (545 W into 2π).

In summary, therefore, maximum CEs in the range of 3-4% are predicted for Nd:YAG lasers operating at power densities of ~ 10^{11} W/cm^2 with pulse durations of a few ns, while for CO_2 systems, because of reduced opacities, higher CEs should be obtainable, this time in the ~ 10^{10} W/cm^2 range.

3. Major Issues

In this chapter, the theoretical analysis and computational methods used to design and create a LPP light source at 13.5 nm are discussed, with emphasis on tin as the source target. The Hartree-Fock configuration interaction (HFCI) Cowan code (Cowan, 1981), which calculates transitions between atomic configurations, and the laser parameters and resultant plasma conditions needed to produce a tuned LPP light source (O'Sullivan & Faulkner, 1995), are investigated. Unresolved transition array (UTA) statistics are calculated to characterise spectra, which result from hundreds of thousands of unresolved in-band (13.5 nm ±1%) transitions from a range of tin ions (White et al., 2005).
A steady-state model (Colombant & Tonon, 1973) is presented, which characterises the plasma by electron temperature, electron density, and average charge state (or ion stage distribution) from input parameters of laser wavelength, pulse duration, and power density. To quantify in-band emission, a source metric is presented, which convolves the in-band emission with ion densities and mirror reflectivity.
Time-dependent, spatially resolved plasmas are analysed with a 1D, hydrodynamic, laser-plasma interaction code MED103 (Christiansen et al., 1974), where level populations are calculated using an energy functional method (White et al., 2007). A radiation transport model for optically thick plasmas is presented to determine the conversion efficiency of laser energy into useable EUV radiation (Eq. 1).

$$CE = \pi \int \int_{13.5-1\%}^{13.5+1\%} \int_{0}^{\tau} I_{out}(\lambda, t) dt d\lambda dA / E_{tot} \qquad (1)$$

where I_{out} is the emitted spectral output and E_{tot} is the input laser energy.
The effect of laser pulse wavelength (Nd:YAG λ = 1.064 μm and CO_2 λ = 10.6 μm) especially with respect to the opacity of the plasma is calculated using the 1D model and the anisotropic nature of the plasma using 2D simulations.

3.1 Atomic physics

Bauche & Bauche-Arnoult (1988a) reported that a transition array is "the totality of lines between the levels of two electronic configurations." Peyrusse (1999) reported that broadband line emission features are present in plasma emission for mid- to high-Z elements and Jin & Richardson (1995) that "high-Z solid targets, for example, those elements in the vicinity of tin, characteristically emit broadband spectra that come from many excited levels. These energy levels are so close that the radiation they generate in the EUV range can be considered a continuum." Salzmann (1998) noted that as the number of bound electrons increases, the density of lines increases such that "their spacing is smaller than their width."
In a tin plasma, hundreds of thousands of 4d-4f, 4p-4d, and 4p-5d transitions, from near-degenerate $4p^64d^{N-1}4f^1$ + $4p^54d^{N+1}$, and $4p^64d^{N-1}5p^1$ (N = 9-1) excited energy levels in Sn^{5+}–Sn^{13+} ions, overlap to form a UTA in the in-band EUVL region (O'Sullivan & Carroll, 1981). To aid the computationally prohibitive calculations, UTAs can be treated statistically, using moments of the weighted oscillator strength (gf) versus wavelength (λ) distribution. The statistical approach is relevant because of the extreme complexity of the UTA features (Bauche & Bauche-Arnoult, 1992).
The HFCI Cowan code (Cowan, 1981) solves the multi-electron Schrödinger equation using the Slater-Condon theory of atomic structure. Configuration interaction (CI) effects are

known to be considerable between $4p^64d^{N-1}4f$ and $4d^54d^{N+1}$ configurations and must be included. The *gf* versus λ distribution is shown in Figure 1, from the HFCI Cowan code for tin ions in the 12.5–14.5-nm (7.4%) range (the full Mo/Si mirror response region).

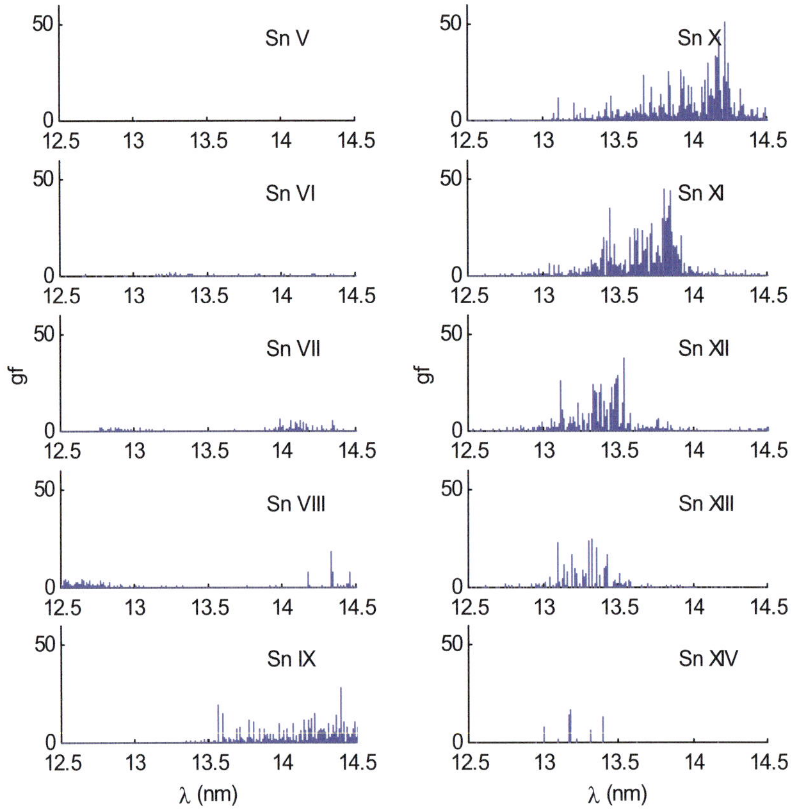

Fig. 1. Sn V–Sn XIV weighted oscillator strength versus wavelength (12.5–14.5 nm)

The statistical parameters used to describe a UTA are the weighted mean, μ_1, variance, v, skewness, α_3, and kurtosis, α_4, determined from the *n*th-order centred moments of the distribution (moments about the mean), μ_n^c (Bauche & Bauche-Arnoult, 1992). Eq. 2 gives the *n*th-order, non-centred moments, expressed by Q_i, the energy difference between two levels, and w_i, the strength of line *i* (Eq. 3).

$$\mu_n = \frac{\sum_{i=1}^{N} (Q_i)^n w_i}{\sum_{i=1}^{N} w_i} \tag{2}$$

$$w_i = \left|\left\langle \alpha J \,|\, \mathrm{D}^{(1)} \,|\, \alpha' J' \right\rangle\right|^2 \tag{3}$$

where $\mathbf{D}^{(1)}$ is the electric dipole operator and αJ and $\alpha' J'$ are the upper and lower levels.

The non-centred moments can be expressed in terms of the energy E_i (or wavelength, λ_i) and weighted oscillator strength, gf_i, for each line i (Svendsen & O'Sullivan, 1994) (Eq. 4). The gf values are used instead of intensities, assuming that the populations in the upper level are proportional to the statistical weight 2J + 1 (Bauche & Bauche-Arnoult, 1988b). The non-centred moments are centred, from which the UTA statistics are calculated.

$$\mu_n = \frac{\sum_{i=1}^{N} (\lambda_i)^n gf_i}{\sum_{i=1}^{N} gf_i} \tag{4}$$

Skewness is positive ($\alpha_3 > 0$) when the distribution is biased towards the right tail (lower energy or longer wavelength, in this case) and negative towards the left tail (higher energy or shorter wavelength). Kurtosis compares the distribution to a Gaussian (or normal) distribution, where $\alpha_4 = 3$ for a perfect Gaussian. The full width at half maximum (FWHM) instead of the variance can also be used as a measure of width (Eq. 5) as can the standard deviation, σ, the square root of the variance.

$$FWHM = \Gamma = 2\sqrt{2\log(2)}\sigma = 2.35\sigma \tag{5}$$

A UTA can thus be represented as a Gaussian by spectral position and width, in terms of the mean (μ_1) and standard deviation (σ) (Eq. 6), and convolved with an area Gaussian equal to the Σgf to produce a plot of relative intensity versus wavelength for a given ion stage. The UTA statistics from a full HFCI calculation can be calculated for each transition type based on the leading eigenvalue percentages (White et al., 2005).

$$f(x) = \frac{1}{\sigma\sqrt{2\pi}} e^{-\frac{(x-\mu_1)^2}{2\sigma^2}} \tag{6}$$

Statistical data from the Cowan code for the 4d-4f, 4p-4d, and 4d-5p transitions are given in Table 1 and the corresponding Gaussian representation in Figure 2, showing how the Gaussian representation can be used to greatly simplify complex UTA data.

Figure 3 shows the mean wavelength versus ion stage (error bars as standard deviation) for the 4d-4f, 4p-4d and 4d-5p transitions, where it is seen that the mean wavelength decreases with ionisation because of increased Coulombic force, as is to be expected. Note that $\Delta n = 0$ transitions are less sensitive than $\Delta n = 1$ transitions to the change in average nuclear charge and that the 4d-5p transition array overlaps with the 4d-4f and 4p-4d UTA in the 13.5-nm region at Sn XIII.

3.2 Steady-state ion distribution

Transition array statistics simplify numerical calculations involving hundreds of thousands of spectral lines to help interpret unresolved LPP EUV spectra. The results can be coupled with a steady-state, CR plasma model (1 electron temperature, 1 electron density) (Colombant & Tonon, 1973) to quantify in-band emission in an optically thin plasma or with a time-dependent hydrodynamic model to calculate CE for an optically thick plasma.

Ion	# lines	Σgf	mean λ (nm)	std (nm)	skew	kurt
Sn V	3	2.63	22.61	0.21	5.54	32.55
Sn VI	81	30.43	19.49	0.47	1.92	7.11
Sn VII	721	185.54	17.40	0.61	2.22	8.22
Sn VIII	2825	614.63	15.77	0.62	3.01	15.22
Sn IX	5470	1032.42	14.84	0.64	4.32	26.44
Sn X	5346	1227.41	14.34	0.70	4.17	20.50
Sn XI	2825	940.50	13.87	0.46	4.81	29.68
Sn XII	721	483.43	13.50	0.32	6.21	62.43
Sn XIII	81	129.50	13.36	0.29	8.54	93.94
Sn XIV	3	13.03	13.41	0.22	10.68	115.16

Table 1a. 4d-4f UTA statistics

Ion	# lines	Σgf	mean λ (nm)	std (nm)	skew	kurt
Sn V	0					
Sn VI	3	2.88	15.37	2.28	0.15	1.03
Sn VII	60	75.20	16.36	0.47	-0.06	2.76
Sn VIII	466	362.98	15.05	0.65	2.22	12.26
Sn IX	1718	904.94	14.30	0.59	3.50	27.04
Sn X	3170	1061.48	13.91	0.64	4.62	34.12
Sn XI	3245	888.11	13.64	0.61	4.67	33.33
Sn XII	1718	471.46	13.39	0.59	5.41	38.80
Sn XIII	466	158.26	13.34	0.69	4.61	26.41
Sn XIV	60	53.95	13.29	0.61	5.15	30.24

Table 1b. 4p-4d UTA statistics

Ion	# lines	Σgf	mean λ (nm)	std (nm)	skew	kurt
Sn V	1	0				
Sn VI	60	10.46	30.19	0.72	-0.97	7.72
Sn VII	466	44.24	25.80	0.70	-0.46	5.68
Sn VIII	1718	107.17	22.48	0.63	-0.14	5.14
Sn IX	3245	158.89	19.90	0.53	0.06	5.46
Sn X	3170	129.21	17.80	0.46	0.13	5.90
Sn XI	1718	97.46	16.07	0.41	-0.42	9.02
Sn XII	466	59.19	14.64	0.33	-0.24	6.49
Sn XIII	60	54.73	13.41	0.17	1.36	10.56
Sn XIV	3	1.05	12.43	0.10	-0.27	3.01

Table 1c. 4d-5p UTA statistics

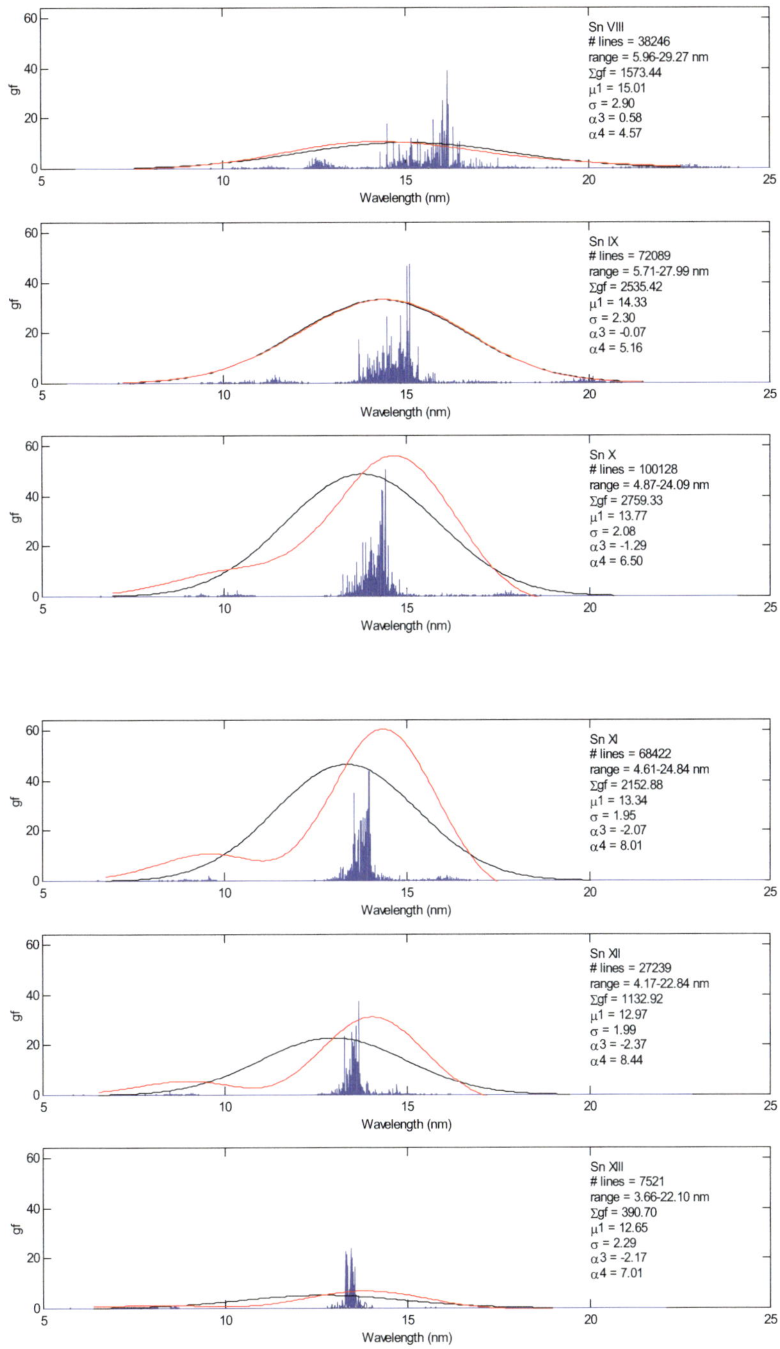

Fig. 2. Calculated discrete Cowan (blue), Gaussian (black), and skewed Gaussian (red) (Sn VIII–Sn XIII)

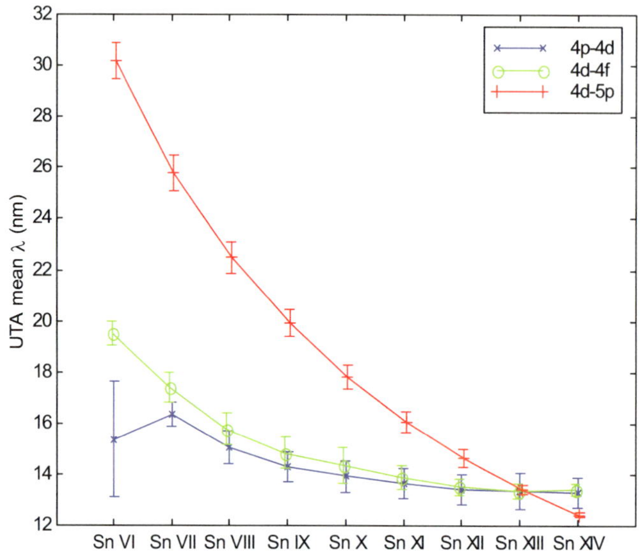

Fig. 3. 4d-4f, 4p-4d, and 4d-5p transitions arrays: mean wavelength versus ion stage

In the steady-state, CR model, collisional and radiative process for each ion state are balanced according to the rate equations for optically thin plasmas (Eq. 7).

$$\frac{dn_{z+1}}{dt} = n_e n_z S(z,T_e) - n_e n_{z+1}\left[S(z+1,T_e) + \alpha_r(z+1,T_e) + n_e\alpha_{3b}(z+1,T_e)\right] + n_e n_{z+2}\left[\alpha_r(z+2,T_e) + n_e\alpha_{3b}(z+2,T_e)\right] \quad (7)$$

where n_z is the density of ion z, n_e the electron density, S, α_r, and $n_e\alpha_{3b}$ the collisional ionisation, radiative recombination, and three-body recombination rate coefficients, and T_e the electron temperature. The temperature is a function of the laser power density (or intensity Φ) and scales as $\Phi^{3/5}$.

The rate equations for the z coupled equations for each ion are determined from cross sections averaged over a free electron distribution function (assumed to be Maxwellian). The semi-empirical rate equations are a function of electron temperature T_e, electron density n_e, charge z, ionisation potential χ, and number of open shell electrons ξ. Dielectronic recombination is not included in this model, but can be included as a percentage of radiative recombination. The electron density is assumed to be equal to the critical electron density (or cut-off density), n_{ec}, where $n_{ec} \approx 10^{21}/\lambda^2$ cm^{-3} (λ in μm).

For a steady state model, $dn/dt = 0$, and thus the ratio of ionisation (S) and recombination processes (α_r and α_{3b}) is equal to the ion density from one ion stage to the next. n_{z+1}/n_z is a recursive relation, which when summed to 1 gives the fractional ion density (f_z) at each ion stage (z), and hence the ion distribution or average charge <z>.

$$f_z = \frac{n_{z+1}}{n_z} = \frac{S(z)}{\alpha_r(z+1) + n_e\alpha_{3b}(z+1)} \quad (7)$$

Figure 4 shows the ion density fractions for tin as a function of electron temperature to 100 eV for the Nd:YAG laser (λ = 1.064 μm) cut-off density of 9.843 x 10^{20} cm^{-3}. The open 4dsubshell ions are indicated and are present from about 5–100 eV in varying fractions. Especially long plateaus correspond to noble gas-like species, for example Kr-like Sn XIV (Sn^{13+}), seen as the dominant species from about 50–70 eV. A range of ion stages are present in a plasma at any given electron temperature, where higher temperatures produce higher ion stages. Note that the dominant species at any temperature comprises about 40 to 50% of the total ion population.

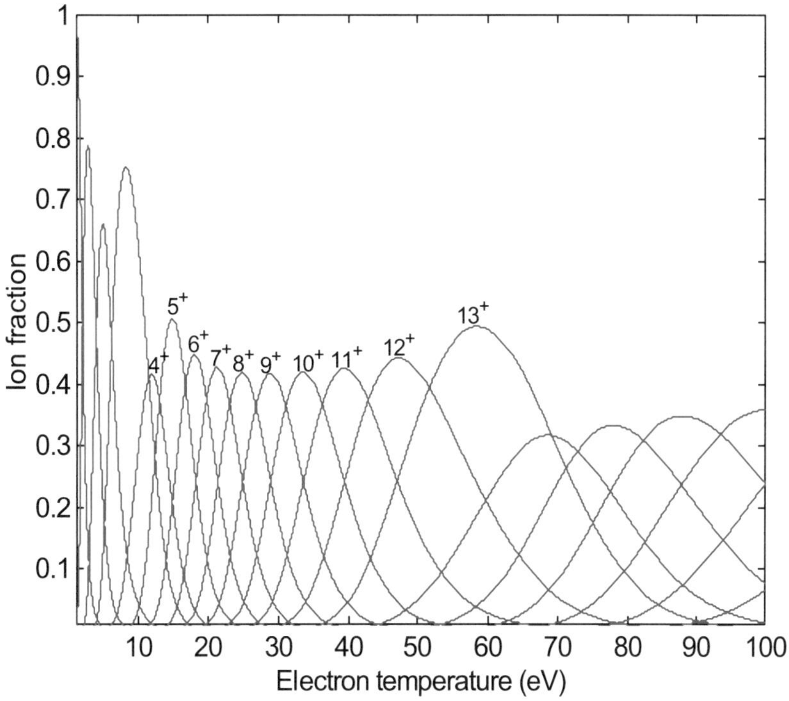

Fig. 4. Steady-state tin ion distribution, f_z, versus temperature (4d-subshell indicated)

To determine the ion distribution, the rate equations can also be solved using the simultaneous equation method (Cummings et al., 2004). This method is more general than the recursive relation method and can be used to include energy level populations within an ion stage, mixed ion species, and a time-dependent solution of the rate equations.

3.3 Steady-state source metric

A theoretical spectrum versus temperature and wavelength can be obtained using the statisitcal UTA as a function of wavelength and the ion fraction as a function of electron temperature, from which an intensity source metric (or figure of merit, F) can be calculated to quantify the in-band oscillator strength within the 2%, 13.5-nm bandwidth.

Figure 5 shows a continuous surface plot versus temperature (White et al., 2005), where the interacting 4d-4f and 4p-4d transitions overlap to form a broad UTA, which narrows with increasing temperature. The overall emission moves to shorter wavelength with temperature because of the higher contributing ion stages. The 4d-5p lines are seen at lower energies and lower temperatures until merging with the UTA. In an optically thick plasma, they will be relatively strong compared to the 4d-4f and 4p-4d lines which self absorb.

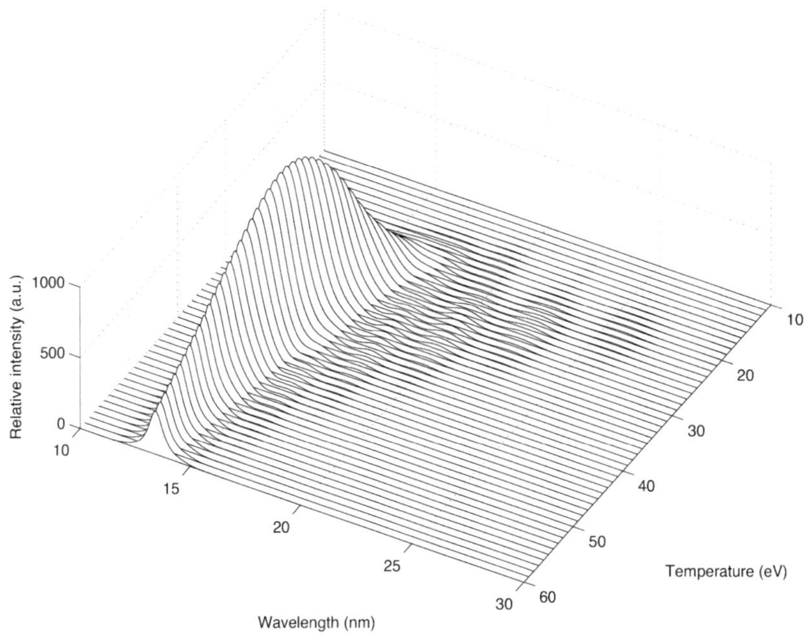

Fig. 5. Relative intensity versus temperature and wavelength

The source metric is taken as the product of the summed *gf* values for all transitions within each ion stage, weighted according to their distribution as a function of electron temperature with the multilayer mirror reflectivity (Eq. 9).

$$F = \sum_{\lambda} \sum_{z} R_{\lambda}^{n} f_{z} g f_{\lambda,z} \tag{9}$$

where R_λ is the wavelength-dependent reflectivity of the Mo/Si multilayer mirror, n is the number of mirrors, f_z is the fractional weighting of an ion of charge z, and $gf_{\lambda,z}$ is the weighted oscillator strength for that ion stage at a wavelength λ. F can be considered without the mirror reflectivity (*i.e.*, $n = 0$).

The calculated metric is shown over three wavelength ranges as a function of electron temperature (Figure 6) to highlight maximum emission as a function of wavelength (or atomic species). The relative in-band contribution increases and then decreases with temperature, showing the UTA diminishing as the 4d-subshell is fully stripped. Maximium 2% in-band emission is at approximately 40 eV. Note that the 7.4% range gives a simple estimate of out-of-band emission (OOB), known to reduce mirror lifetime because of thermal loading (Morris et al., 2007).

The results have been compared to experiment (White et al., 2005), where it is seen (Figure 7) that the long wavelength edge is matched at lower electron temperatures (and thus lower ion stages within the plasma) and the short wavelength edge matched at higher electron temperatures (and thus higher ion stages within the plasma).

Although a simplified UTA approach using a steady-state CR model gives important information about optimum plasma conditions in optically thin plasmas, the plasma is not wholly characterised by one temperature, and a more sophisticated spatially and temporally resolved model is required, that also includes radiation transport.

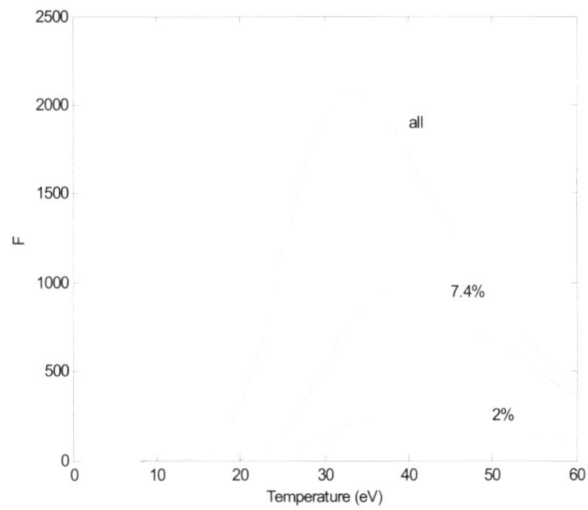

Fig. 6. Tin plasma figure of merit (2%, 7.4%, full spectral range) as a function of temperature

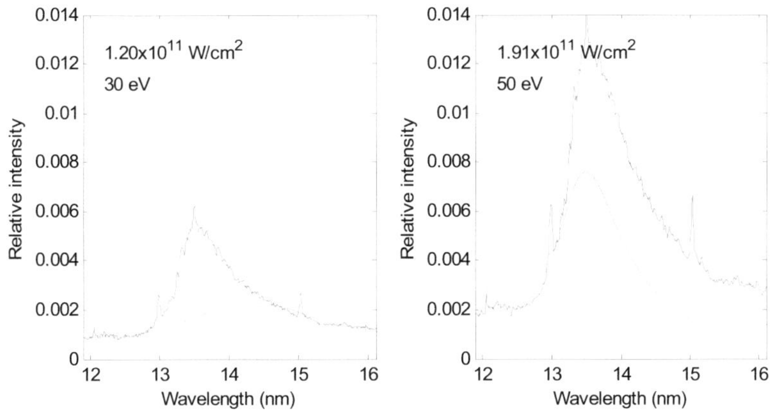

Fig. 7. Statistical UTA weighted by fractional ion density as a function of electron temperature (blue) and compared to experiment 6% tin-doped glass target (red)

3.4 Time-Dependent ion distribution and level populations

Hydrodynamic calculations are required to determine spatial and temporal electron temperature, electron density, ion distribution, and ultimately CE. The 1D Lagrangian, laser-plasma code MED103 (Christiansen et al., 1974) calculates density, velocity, ion temperature, electron temperature as functions of space and time, using the Navier-Stokes equations and can be readily coupled to an atomic model.

To illustrate the complex dynamics within the plasma, simulation results for a Nd:YAG, λ = 1064-nm, 15-ns (FWHM), Φ = 1.3 x 10^{11} W/cm^2 pulse on a solid tin cylindrical target (400 cells) are presented (Figure 8). The electron density (left) and electron temperature (right) versus distance at 3 times (top: before, at, and after the peak of the pulse), and versus time for 3 distances (bottom: at the plasma edge) are shown.

Here the overdense region (left), where the electron density varies from 10^{23} to 10^{19} cm^{-3} at the ablation front, and the effect of plasma heating (right) can be seen.

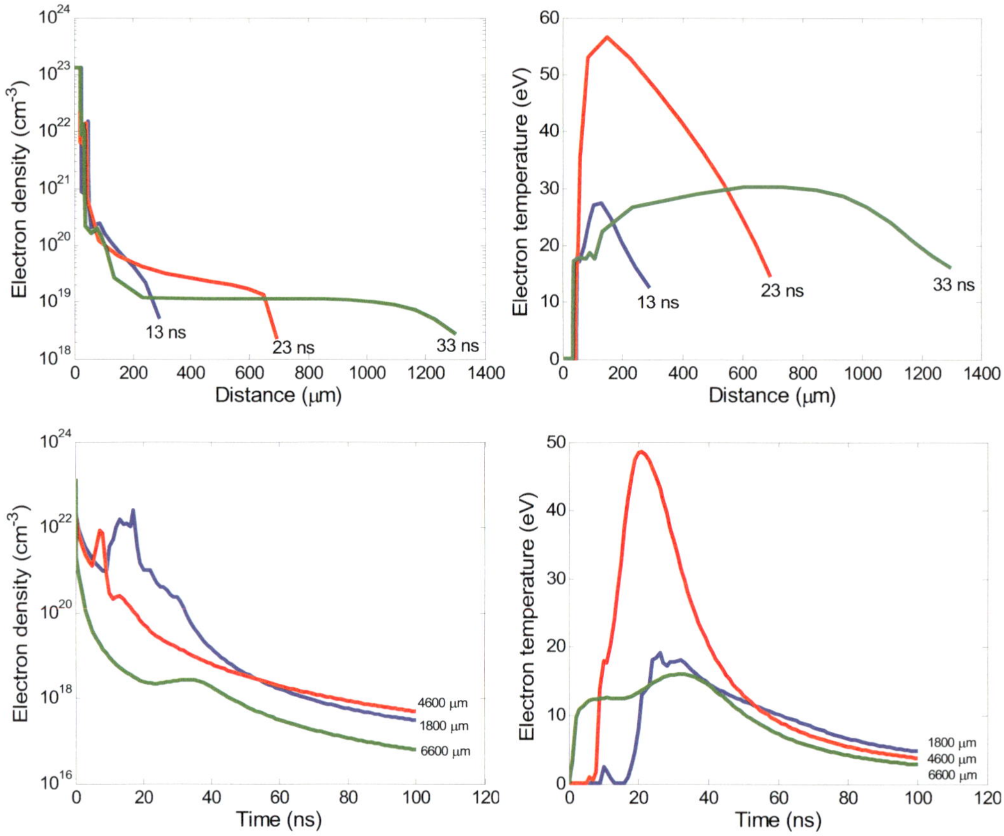

Fig. 8. Hydrodynamics versus space and time (Nd:YAG, 15 ns, .3 x 10^{11} W/cm^2 on solid tin)

Using the average atom (AA) model, the ion distribution and excited (*l*-degenerate) level populations are solved together with the plasma hydrodynamics, and the emission and absorption coefficients calculated independently in opacity tables. The AA model uses a statistical average over all ions to save computation (Djaoui & Rose, 1992), resulting in a fictitious ion with non-integer occupation numbers.

The statistical UTA representation (described above) and an energy functional methodology (Itoh et al., 1987) can be used to remove the *nl* degeneracy for Δn = 0 (4-4) transitions to calculate level populations required for radiation transport modeling (White et al., 2007). The screened hydrogenic model (More, 1982, Perrot, 1989) and UTA statistical data are used to interpolate between energy levels to calculate *nl*-splitting for 4-4 transitions.

Figure 9 shows the ion distribution (left) and a typical plot of population number density versus distance (right, ground state $4p^64d^N$ (black), and excited states, $4p^64d^{N-1}4f^1$ (green), $4p^54d^{N+1}$ (blue), and $4p^64d^{N-1}5p^1$ (red)) at the peak of the pulse (23 ns) for Sn X.

Note that all of the 4d-subshell ion fractions rise, fall, and rise with distance, showing the coupling to the laser pulse on the ablation front (left). Furthermore, the rise, fall and rise of the level populations (primarily a function of ion fraction) is more pronounced at the peak of the pulse as expected with increased coupling to the laser (right). Here, the increased $4p^54d^{N+1}$ (blue) and decreased $4p^64d^{N-1}5p^1$ (red) populations are seen with increased ionisation, indicating the increased 4p-4d and decreased 4d-5p contributions. Note the significant population decrease at the edge of the plasma (last cells).

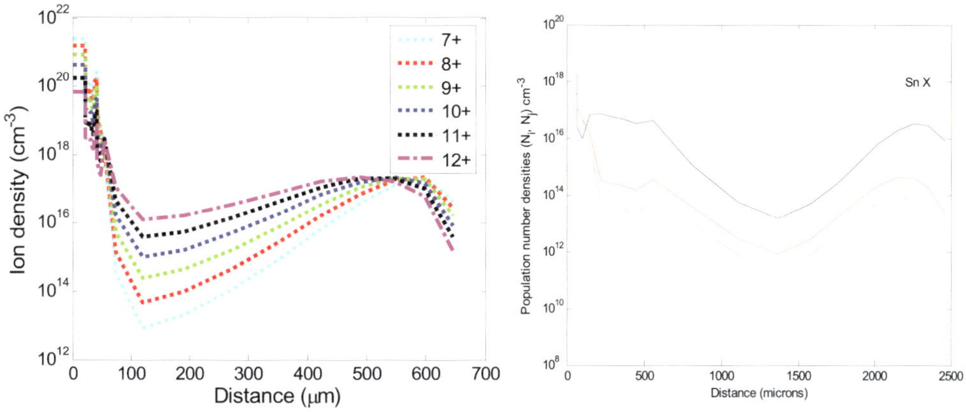

Fig. 9. Ion distribution (Sn^{6+} to Sn^{12+}) (left) and level populations at the peak of the pulse (Sn^{9+}) (right) versus distance

3.5 Radiation transport

The methodology for calculating radiation transport in an optically thick plasma is described elsewhere (Djaoui et al., 1994, Wark et al., 1995, Patel et al., 1997, White et al. 2009) and is summarised here. Radiation transfer involves absorption and emisison of spectral radiation along a path with differing hydrodynamic conditions, which can significantly alter the resultant spectra (line intensities and shape). The radiative transfer equation is given by

$$\frac{\partial I_\nu}{\partial z} = K_\nu I_\nu + E_\nu \tag{10}$$

where K_ν is the absorption coefficient, I_ν is the intensity, E_ν the total emissivity (all functions of frequency ν), and z the distance.

Eq. 10 is computationally intensive and thus an optical thickness, τ_ν, from the point of emission to the plasma boundary is defined to simplify (Djaoui et al., 1994). Thus

$$\frac{\partial I_\nu}{\partial \tau_\nu} = I_\nu - S_\nu \tag{11}$$

where I_ν is the radiation intensity, τ_ν the optical depth, and S_ν the source function, where $S_\nu = E_\nu / K_\nu$. (In the optically thick limit $\tau_\nu >> 1$, and in the optically thin limit $\tau_\nu << 1$.) The solution to the radiative transfer equation is given as

$$I_\nu(\tau_\nu) = I_\nu^0 e^{-\tau_\nu} + \int_0^{\tau_\nu} S_\nu(t) e^{-t} dt \tag{12}$$

where the integral is evaluated from τ_ν to the edge of the material and I_ν^0 is the incident radiation intensity at τ_ν.

For a homogeneous plasma with no incident radiation, the intensity is thus

$$I_m(\nu) = \sum_l S_m^l(\nu)[1 - \exp(\chi_m^l(\nu)\Delta z_m)] \tag{13}$$

where S^l_m (ν) is the source function and χ^l_m (ν) is the opacity of line l and cell m and Δz_m the length of cell m. The sum is taken over each line.
The source function at a given frequency, ν, for line l and cell m is

$$S_m^l(\nu) = \frac{2h\nu^3}{c^2} \frac{1}{\left[\frac{N_i g_j}{N_j g_i} - 1\right]} \tag{14}$$

where S is in W/m^2/sr[/Hz/s], N_i and N_j are the population number densities in cm^{-3}, g_i and g_j are the degeneracies of the ith and jth levels (i lower level and j higher level).
The opacity for line l and cell m is

$$\chi_m^l(\nu) = \frac{\pi e^2}{4\pi\varepsilon_0 m_e c} f_{ij} N_i \left[1 - \frac{N_j g_i}{N_i g_j}\right] \phi_m^l(\nu) \tag{15}$$

where f_{ij} is the absorption oscillator strength and ϕ^l_m (ν) is the area normalised line profile assumed to be the same in emission and absorption (Djaoui et al., 1994).
The emission profile is then recursively attenuated through successive cells, where the observed emission from cell m is

$$I_m^{obs}(\nu) = I_m(\nu)\exp\left(\sum -\chi_n(\nu)\Delta z_n\right). \tag{16}$$

3.6 Laser-plasma simulations

Using the 1D model described above, maximum CE can be determined based on optimum laser parameters of power density and pulse duration. A cylindrical target of 90-μm radius is assumed in all simulations for an incident laser wavelength of 1064 nm (Nd:YAG). Results are compared to experiment (Hayden et al., 2006), where a 15-ns simulated pulse duration was chosen to compare to experiment on a pure tin slab target.
Two simulation surveys are discussed, that vary power density and pulse duration. Power density is varied from 0.5 to 1.5 x 10^{11} W/cm^2 with a 15-ns duration (FWHM), and, then, using the power density with the maximum CE in the first survey (0.8 x 10^{11} W/cm^2), the pulse duration is varied from 7 to 40 ns. The conversion efficiency was calculated for each case and the optimum laser parameters determined (Table 2 and Table 3).
The results of the two surveys showed a maximum CE at 0.8 x 10^{11} W/cm^2 for a 10-ns pulse. Figure 10 shows a simulated and experimental spectra (Hayden et al., 2006) (left) for the maximum CE case and CE versus power density (right) for the 15-ns cases (A1-A9). The calculated results underestimate the experimental power density needed because of the absence of lateral expansion along the target in a 1D model, where the calculated plasma reaches optimum electron temperature (30–40 eV) at a lower power density because no energy is converted into lateral expansion (White et al. 2009).

Run	Duration (ns) FWHM	Power density (x 10^{11} W/cm²)	CE (% per 2π sr)	max <z>
A1	15	0.5	1.03	8.6
A2	15	0.6	1.58	9.9
A3	15	0.7	2.01	11.1
A4	15	0.8	2.21	12.2
A5	15	0.9	2.16	13.2
A6	15	1.0	1.84	14.2
A7	15	1.1	1.48	15.1
A8	15	1.3	1.06	16.7
A9	15	1.5	0.92	18.7

Table 2. Power density survey: 0.5–1.5 x 10^{11} W/cm² at 15 ns

Run	Duration (ns) FWHM	Power density (x 10^{11} W/cm²)	CE (% per 2π sr)	max <z>
B1	7	0.8	1.93	9.4
B2	10	0.8	2.29	11.2
B3	15	0.8	2.21	12.2
B4	20	0.8	1.71	13.6
B5	23	0.8	1.16	14.0
B6	30	0.8	0.51	14.5
B7	40	0.8	0.29	14.7

Table 3. Pulse duration survey: 7–40 ns (FWHM) at 0.8 x 10^{11} W/cm²

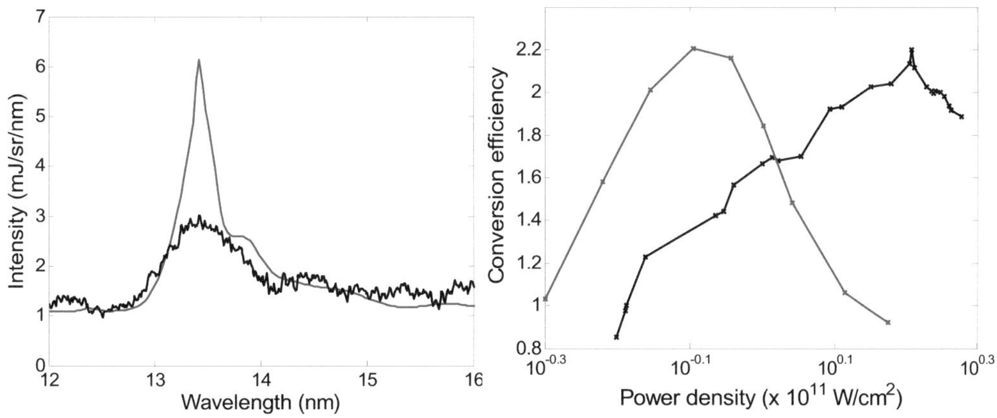

Fig. 10. In-band emission 1064 nm, 15 ns FWHM, 0.8 x 10^{11} W/cm² case (left) and conversion efficiency (right) versus laser power density (calculated (blue) and experiment (black))

3.7 The effect of laser pulse wavelength and pulse duration

The steady-state model can be used to compare the effect of wavelength of a Nd:YAG (1064 nm) which produces a critical electron density, n_{ec}, of 10^{21} cm^{-3}, and the 10,600-nm CO_2 (10 times the Nd:YAG wavelength) which produces a n_{ec} of 10^{19} cm^{-3} (1/100 that of Nd:YAG). The reduced electron density results in a greater average charge at the same electron temperature, because of the ion density dependence on three-body recombination, which decreases with decreased electron density (White et al., 2007). Using the ID model, a 2.2 times maximum CE was calculated for the CO_2 LPP at 7.2 x 10^8 W/cm^2 compared to that for the maximum Nd:YAG LPP CE at 3.3 x 10^{10} W/cm^2, in part because of reduced opacity in the more optically thin CO_2 plasma.

2D results using the RMHD code Z* highlight the effect of spatial pulse profile on opacity (White et al., 2008). Optimal in-band EUV emission occurs primarily from the plasma core (30-40 eV); however, core emission is reduced by self-absorption and absorption in the colder wings because of the high absorption cross section of lower stage ions (Lysaght et al., 2004). Figure 11 shows the larger hot core region for a flat-top spatial profile pulse (left) compared to a Gaussian spatial pulse (right), because of greater coupling.

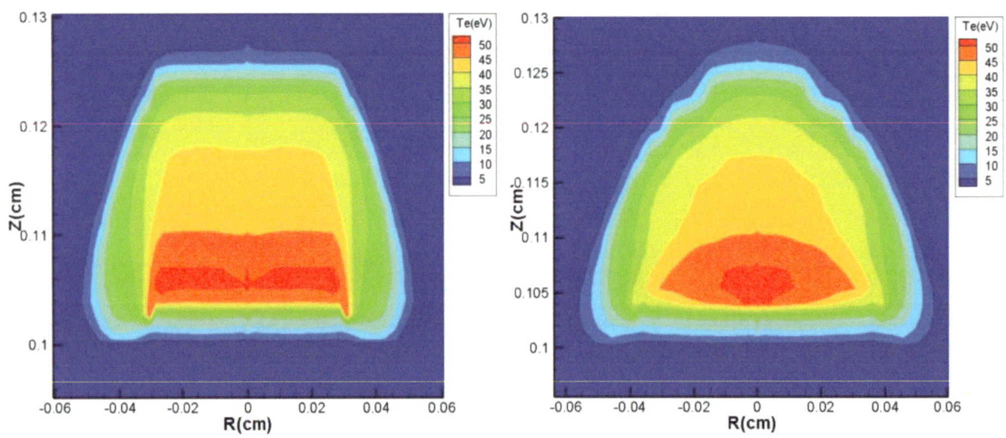

Fig. 11. Electron temperature flattop (left) and Gaussian (right) for a 2.2-ns Nd:YAG

4. Current trends and future challenges

Among the key challenges facing LPP EUV source optimisation are coupling the maximum amount of laser energy into the plasma, and maintaining the plasma in an emitting state for the maximum time possible for a given laser pulse duration. By using a pre-pulse (Dunne et al., 2000), it is possible to prepare a cool plasma on the surface of a solid target, with a density closer to the critical density, which is set by the wavelength of the main driving laser. Thus the bulk of the energy in the main laser pulse is coupled to the cool plasma, increasing the overall CE of the system. Parameters that can be varied include the wavelength of the two lasers, their pulse energies and durations, and the time delay between the two pulses. This latter parameter allows the pre-plasma to expand to the optimum size/density balance and a peak CE obtained at a well-defined delay (Higashiguchi et al., 2006, Ando et al., 2006). The pre-pulse-pulse approach is also valuable with droplet targets where the initial droplet density may be orders of magnitude in excess of the critical density corresponding to the main laser pulse.

A more complex approach involves the tailoring of the main laser pulse shape, away from a simple Gaussian to increase CE. Studies have shown (White et al., 2005) that a sharp fall in the laser pulse soon after the peak intensity can increase the CE, as the tail of the Gaussian pulse does not maintain the plasma at a sufficient temperature to emit EUV radiation, and hence is wasted. Experimental approaches include temporally multiplexing a number of Gaussians of different FWHM, to create a range of pulse shapes, perhaps using a learning algorithm and feedback to assist in the optimisation of the pulse shape. Using CO_2 lasers it is possible to control the pulse duration, and to an extent the pulse shape using a plasma shutter (Hurst & Harilal, 2009). It is simpler to multiplex Nd:YAG lasers, for example, by combining the beams in a single lens setup (Ellwi et al., 2005).

Intelligent algorithms may also be implemented to drive a plasma model, incorporating MHD, radiation transport and plasma expansion, to locate the ideal pulse shape in a complex multi-dimensional parameter space. Potential approaches from control theory include Swarm Intelligence, Multi Object Optimisation and Genetic Algorithms.

The use of large, multilayer, plasma-facing optics as collectors in systems with multi-kW plasma sources raises the issue of debris emission, which both sputters the surface of the mirror and deposits fuel debris on it, in each case degrading reflectivity (Srivastava et al., 2007). A range of mitigation measures has been proposed including magnetic fields, plasma shutters, buffer gases, sacrificial layers, and combinations of these. In addition, the reduction of debris emission by target design has resulted in the proposed use of mass-limited targets, targets containing nanoparticles (Kaku et al., 2008), cavity targets (Ueno et al., 2007), etc. However, it is most likely that a combination of these measures will be required in addition to a cleaning routine based on scavenger gases flushed through the system.

5. Conclusions

In this review, we have introduced computational methods for LPP modeling, and used both a simple, steady-state model in an optically thin plasma and a more sophisticated, 1D model with radiation transport in an optically thick plasma to illustrate a particular calculation. Optimum plasma conditions were estimated, e.g., electron temperature for maximum in-band emission (40 eV) as well as optimum laser parameters for maximum CE for a Nd:YAG LPP (0.8×10^{11} W/cm^2, 10-ns pulse). The effect of laser wavelength on CE was calculated using the steady-state model and the effect of spatial pulse shape using a 2D RMHD code.

The use of different codes is primarily a trade off between computation time and sophistication although complicated atomic theory involving hundreds of thousands of transitions can be investigated with simplified models to benchmark more sophisticated multi-dimensional codes. Much modeling work continues into the search for optimum laser parameters and plasma conditions necessary in HVM EUVL.

6. Acknowledgement

The authors wish to acknowledge support from Science Foundation Ireland under principal investigator grant 07/IN.1/B1771.

7. References

Al-Rabban, M., Richardson, M., Scott, H., Gilleron, F., Poirier, M., & Blenski, T., Modeling LPP sources, *EUV Sources for Lithography*, Ch. 10, Ed. V. Bakshi, SPIE Press (2005).

Ando, T., Fujioka, S., Nishimura, H., Ueda, N., Yasuda, Y., Nagai, K., Norimatsu, T., Murakami, M., Nishihara, K., Miyanaga, N., Izawa, Y., & Mima., K., Optimum laser pulse duration for efficient extreme ultraviolet light generation from laser-produced tin plasmas, *Appl. Phys. Lett.* 89, 151501 (2006).

Attwood, D., *Soft X-Rays and Extreme Ultraviolet Radiation, Principles and Applications*, University Press, Cambridge (1999).

Attwood, D., Extreme ultraviolet light sources for semiconductor manufacturing, *J. Phys. D* 37, 23 (2004).

Bar-Shalom, A., Oreg. J., & Klapisch, M., Collisional radiative model for heacy atoms in hot non-local-thermodynamical-equilibrium plasmas, *Phys. Rev. E* 56, 71–73 (1997).

Bar-Shalom, A., Klapisch, & M., Oreg. J., HULLAC, an integrated computer package for atomic processes in plasmas, *J. Quant. Spectrosc. Radiat. Transfer* 71, 169-188 (2001).

Bauche, J. & Bauche-Arnoult, C., Transition arrays in the spectra of ionised atoms, *Advances in atomic and molecular physics* 23, 131–195 (1988).

Bauche, J. & Bauche-Arnoult, C., Unresolved transition arrays, *Physica Scripta* 37, 659–663, (1988).

Bauche, J. & Bauche-Arnoult, C., Statistical approach to the spectra of plasmas, *Physica Scripta* T40, 58–64 (1992).

Carroll, P.K. & Kennedy, E.T., Laser-produced plasmas, *Contemporary Physics*, 22, 1, 61–96 (1981).

Christiansen, J.P., Ashby, D.E.T.F., & Roberts, K.V., Medusa: a one-dimensional laser fusion code, *Comput. Phys. Commun.* 7, 271–287 (1974).

Christiansen & Winsor, CASTOR 2: a two-dimensional laser target code, *Comput. Phys. Commun.* 17, 397 (1979).

Chung, H.-K., Chen, M.H., Morgan, W.L., Ralchenko, Y., & Lee, R.W., FLYCHK: Generalized population kinetics and spectral model for rapid spectroscopic analysis for all elements, *High Energy Density Physics* 1, 3–12 (2005).

Colombant & Tonon, X-ray emission in laser produced plasmas, *J. Appl. Phys.* 44, 8, 3524 (1973).

Cowan, D., *The theory of atomic structure and spectra*, University of California Press, Berkeley (1981).

Cummings, A., O'Sullivan, G., Dunne, P., Sokell, E., Murphy, N., White, J., Fahy, K., Fitzpatrick, A., Gaynor, L., Hayden, P., Kedzierski, D., Kilbane, D., Lysaght, M., McKinney, L., & Sheridan, P., Variable composition laser-produced Sn plasmas—a study of their time-independent ion distributions, *J. Phys. D* 37, 2376–2384 (2004).

Derra, G., Zink, P., Krücken, T., Weber, A. & Pankert, J., Tin delivery systems for gas discharge sources, *EUV Source Workshop*, San Jose, CA, USA (2005).

Djaoui, A., Rose, S.J., & Wark, J.S., Calculation of the effects of velocity gradients and opacity on line transfer in laser-produced plasmas, *J. Quant. Spect. Radiat. Trans.* 52, 5, 531 (1994).

Dunne, P., O'Sullivan, G., & O'Reilly, D., Prepulse-enhanced narrow bandwidth soft x-ray emission from a low debris, subnanosecond, laser plasma source, *Appl. Phys. Lett.* 76, 1, 34-36 (2000).

Dusterer, S., Schoerer, H., Ziegler, W., Ziener, C., & Sauebrey, R., Optimization of EUV radiation yield from laser-produced plasma, *Applied Physics B* 73, 693–698 (2001).

Ellwi, S., Comley, A., Hay, N., Henderson I., & Brownell, M., Proc. SPIE, 5751, 272 (2005).

Fujioka, S., Nishimura, H., Nishihara, K., Sasaki, A., Sunahara, A., Okuno, T., Ueda, N., Ando, T., Tao, Y., Shimada, Y., Hashimoto, K., Yamaura, M., Shigemori, K., Nakai, M., Nagai, K., Norimatsu, T., Nishikawa, T., Miyanaga, N., Izawa, Y., & Mima, K., Opacity effect on extreme ultraviolet radiation from laser-produced tin plasmas, *Phys. Rev. Lett.* 95, 235004 (2005).

Fujioka, S., Shimomura, M., Shimada, Y., Maeda, S., Sakaguchi, H., Nakai, Y., Aota, T., Nishimura, H., Ozaki, N., Sunahara, A., Nishihara, K., Miyanaga, N., Izawa, Y., & Mima, K., Pure-tin microdroplets irradiated with double laser pulses for efficient and minimum-mass extreme-ultraviolet light source production, *Appl. Phys. Lett.* 92, 241502 (2008a).

Fujioka, S., Nishimura, H., Shimomura, M., Sakaguchi, H., Nakai, Y., Aota, T., Shimada, Y., Sunahara, A., Nishihara, K., Optimum Laser-Produced Plasma for Extreme Ultraviolet Light Source, Journal of Physics: Conference Series 112 042049 (2008b).

Grant, I.P., McKenzie, B.J., Norrington, P.H., Mayers, D.F. & Pyper, N.C., An atomic multiconfigurational Dirac-Fock package, *Comput. Phys. Commun.* 21, 207–31 (1980).

Greim, H. R., *Plasma Spectroscopy*, McGraw-Hill, New York (1964).

Gumbrell, E. T., Smith, R. A., Ditmire, T., Djaoui, A., Rose, S. J., & Hutchinson, M. H. R., Picosecond optical probing of ultrafast energy transport in short pulse laser solid target interaction experiments, Physics of Plasmas 5, 3714 (1998).

Gu, M. F., *Astrophysical Journal*, 582, 1241, 2003.

Hayden, P., Cummings, A., Murphy, N., O'Sullivan, G., Sheridan, P., White, J. & Dunne, P., 13.5 nm extreme ultraviolet emission from tin based laser plasma sources, *J Appl Phys.* 99, 093302 (2006).

Higashiguchi, T., Kawasaki, K., Sasaki, W., Kubodera, S., Enhancement of extreme ultraviolet emission from a lithium plasma by use of dual laser pulses, *Appl. Phys. Lett.* 88, 161502 (2006).

Hurst, N. & Harilal, S. S., Pulse shaping of transversely excited atmospheric CO_2 laser using a simple plasma shutter, *Rev. Sci. Instrum.* 80, 035101 (2009).

Itoh, M., Yabe, T., & Kiyokawa, S., Collisonal-radiative and average-ion hybrid models for atomic processes in high-Z plasmas, *Phys. Rev. A* 35, 233 (1987).

International technology roadmap for semiconductors: http://public.itrs.net.

Jin, F. & Richardson, M., New laser plasma source for extreme-ultraviolet lithography, *Appl. Opt.* 34, 5750–5760 (1995).

Kaku, M., Suetake, S., Senba, Y., Kubodera, S., Katto, M., & Higashiguchi, T., Deposited debris characteristics and its reduction of a laser-produced plasma extreme ultraviolet source using a colloidal tin dioxide jet target, *Appl. Phys. Lett.* 92, 181503 (2008).

Key, M. H. & Hutcheon, R. J., Spectroscopy of laser-produced plasmas, *Advances in Atomic and Molecular Physics* 16, Ed. Bates, Sir David, Academic Press (1980).

Lysaght, M., Kilbane, D., Murphy, N., Cummings, A., Dunne, P. & O'Sullivan, G., Opacity of neutral and low ion stages of Sn at the wavelength 13.5 nm used in extreme-ultraviolet lithography, *Phys. Rev. A* 72, 014502 (2005).

MacFarlane, J.J., Rettig, C.L., Wang, P., Golovkin, I.E., and Woodruff, P.R., Radiation-hydrodynamics, spectral, and atomic physics modeling of laser-produced plasma euv lithography light sources, in *Emerging Lithographic Technologies IX*, Proc. SPIE, Ed. R.S. Mackay, 5751, 588–600 (2005).

Mandelbaum, P., Finkenthal, M., Schwob, J.L., & Klapisch, M., Interpretation of the quasicontinuum band emitted by highly ionized rare-earth elements in the 70–100 Å range, *Phys. Rev. A* 35, 12, 5051–5059 (1987).

Mirone, A., Gauthier, J. C., Gilleron, F., & Chenais-Popovics, C., Non-LTE opacity calculations with *n-l* splitting for radiative hydrodynamic codes, *J. Quant. Spectrosc. Radiat. Transfer* 27, 3, 345–357 (1982).

More, R. M., Electronic energy-levels in dense plasmas, *J. Quant. Spectrosc. Radiat. Transfer* 58, 791–802 (1997).

Morris, O., Hayden, P., O'Reilly, F., Murphy, N., & Dunne, P., Angle-resolved absolute out-of-band radiation studies of a tin-based laser-produced plasma source, *Appl. Phys. Lett.*, 91, 081506 (2007).

Nishikawa, T., Takabe, H., & Mima, K., Atomic modeling and radiation transport in laser produced plasmas, in *Technology Reports of the Osaka University* 41, 2059, 253–266 (1991).

O'Sullivan, G. & Carroll, P. K., 4d–4f emission resonances in laser-produced plasmas, *J. Opt. Soc. Am* 71, 3, 227–230 (1981).

O'Sullivan, G. & Faulkner, R., Tunable narrowband soft x-ray source for projection lithography, *Opt. Eng.* 33, 3978 (1994).

Patel, P.K., Wark, J.S., Heading, D.J., Djaoui, A., Rose, S.J., Renner, O., & Hauer, A., Simulation of x-ray line transfer in a cylindrically expanding plasma *J. Quant. Spect. Radiat. Trans.* 57, 683 (1997).

Perrot, F., Fast calculation of electronic structure in plasmas: the screened hydrogenic model with *l*-splitting, *Physica Scripta* 39, 332–337 (1989).

Peyrusse, O., Atomic configuration averages and nonlocal thermodynamical equilibrium plasma spectroscopy calculations, *J. Phys. B: At. Mol. Opt. Phys.* 32, 683-700 (1999).

Rollinger, B., Bleiner, D., Chokani, N. & Abhari, R.S., Characteristics of a minimum-debris optimum conversion efficiency tin-based LPP source, in *Emerging Lithographic Technologies XII*, Proc. SPIE, Ed. F.M. Schellenberg, 6921, 1–12 (2008).

Sasaki, A., Nishihara, K., Koike, F., Kagawa, T., Nishikawa, T., Fujima, K., Kawamura, T. & Furukawa, H., Simulation of the EUV spectrum of Xe and Sn plasmas, *J. Sel. Top. Quant. Elec.* 10, 1307 (2004).

Salzmann, D., *Atomic physics in hot plasmas*, International series of monographs on physics, Oxford University Press (1998).

Sequoia, K.L., Tao, Y., Yuspeh, S., Burdt, R., & Tillack, M.S., Two dimensional expansion effects on angular distribution of 13.5 nm in-band extreme ultraviolet emission from laser-produced Sn plasma, *Appl. Phys. Lett.* 92, 221505 (2008).

Shimada, Y., Nishimura, H., Nakai, M., Hashimoto, K., Yamaura, M., Tao, Y., Shigemori, K., Okuno, T., Nishihara, K., Kawamura, T., Sunahara, A., Nishikawa, T., Sasaki, A., Nagai, K., Norimatsu, T., Fujioka, S., Uchida, S., Miyanaga, N., Izawa, Y., & Yamanaka, C., Characterization of extreme ultraviolet emission from laser-produced spherical tin plasma generated with multiple laser beams, *Appl. Phys. Lett.* 86, 051501 (2005).

Sondhauss, P., Rose, S. J., Lee, R. W., Al'miev, I., & Wark, J. S., Extension of the code suite FLY to a multi-cell postprocessor for hydrodynamics plasma simulation codes, *J. Quant. Spectrosc. Radiat. Transfer* 71, 721-728 (2001).

Srivastava, S.N., Thompson, K. C., Antonsen, E. L. Qiu, H., Spencer, J. B., Papke, D., & Ruzic, D. N., Lifetime measurements on collector optics from Xe and Sn extreme ultraviolet sources, *J. Appl. Phys.* 102, 023301 (2007).

Svendsen, W. & O'Sullivan, G., Statistics and characteristics of XUV transition arrays from laser-produced plasmas of the elements tin through iodine, *Physical Review A: General Physics* 50, 5, 3710–3718 (1994).

Tao, Y., Nishimura, H., Okuno, T., Fujioka, S., Ueda, N., Nakai, M., Nagai, K., Norimatsu, T., Miyanaga, N., Nishihara, K., Izawa, Y., & Izawa, Y., Characterization of denisty profile of laser-produced Sn plasma for 13.5 nm extreme ultraviolet source, *Appl. Phys. Lett.* 87, 241502 (2005).

Takahashi, A., Nakamura, D., Tamaru, K., Akiyama, T., & Okada, T., Emission characteristics of debris from CO2 and Nd:YAG laser-produced tin plasmas for extreme ultraviolet lithography light source, *Appl. Phys. B* 92, 73–77 (2008).

Ueno, Y., Soumagne, G., Sumitani, A., Endo, A., & Higashiguchi, T., Enhancement of extreme ultraviolet emission from a CO2 laser-produced Sn plasma using a cavity target, *Appl. Phys. Lett.* 91, 231501 (2007).

Wark, J.S., Renner, O, Djaoui, A., Rose, S. J., Missalla, T., Neely, D., Foerster, E., & Hauer, A., Measurements and simulation of the intensity and shape of the hydrogenic Al resonance line ($1s^2S$-$2p^2P$) in laser-produced plasmas *J. Quant. Spectrosc. Radiat. Transfer* 54, 1/2, 419 (1995).

White, J., Hayden, P., Dunne, P., Cummings, A., Murphy, N., Sheridan, P., & O'Sullivan, G., Simplified modeling of 13.5 nm unresolved transition array emission of a Sn plasma and comparison with experiment, *J. Appl. Phys.* 98, 113301 (2005).

White, J., Dunne, P., Hayden, P., O'Reilly, F., & O'Sullivan, G. Optimizing 13.5 nm laser-produced tin plasma emission as a function of laser wavelength, *Appl. Phys. Lett.* 90, 181502 (2007).

White, J., Dunne, P., Hayden, P., & O'Sullivan, G. Simplified one-dimensional calculation of 13.5 nm emission in a tin plasma including radiation transport, *Appl. Phys. Lett.* **106,** 113303 (2009).

White, J., O'Sullivan, G., Zakharov, S., Choi, P., Zakharov, V., Nishimura, H., Fujioka, S., & Nishihara, K., Tin laser-produced plasma source modeling at 13.5 nm for extreme ultraviolet lithography, *Appl. Phys. Lett.* 92, 151501 (2008).

Yamuara, M, Uchida, S., Sunahara, A., Shimada Y., Nishimura, H., Fujioka, S., Okuno, T., Hashimoto, K., Nagai, K., Norimatsu, T., Nishihara, K., Miyanga, N., Izawa, Y., & Yamanaka, C., Characterization of extreme ultraviolet emission using the fourth harmonic of a Nd:YAG laser, *Appl. Phys. Lett.* 86, 181107 (2005).

Zakharov, S.V., Novikov, V.G., Choi, P., Z* Code for DPP & LPP source modeling, *EUV Sources for Lithography,* Ch. 8, Ed. V. Bakshi, SPIE Press (2005).

Zimmerman, G.B. & Kruer, W.L., Numerical simulation of laser-initiated fusion, *Comments Plasma Phys. Control. Fusion* 2, 51–61 (1975).

12

Nano-crystalline Diamond Films for X-ray Lithography Mask

Linjun Wang, Jian Huang, Ke Tang and Yiben Xia
School of Materials Science and Engineering, Shanghai University, Shanghai, China

1. Introduction

At present, the integrated circuits have developed from only dozens of devices on a chip in 20th century 60's to today's one billion devices on each chip. And lithography technology has played a crucial role in its rapid development. As the X-ray wavelength is very short, it meets the needs of the development of ultra-large scale integrated circuit and has drawn a lot of attention in recent years. X-ray lithography is the first alternative to optical lithography in the future

In the X-ray band, the material reduction of atomic absorption coefficient decrease with the atomic number. Although the diamond is consisted of carbon element and the atomic number is greater than the conventional X-ray window materials beryllium, it can be prepared very thin (less than 1 micron), and is also able to withstand 1 atm pressure because of its high strength. Thus, compared to the traditional 8-micron beryllium window, the thickness of prepared diamond films can be nearly 1/20, and a higher X-ray transmission rate can be obtained. In particular, in the longer wavelength soft X-ray band, beryllium window, basically, is not translucent. By this work, it shows that the prepared free-standing diamond film X-ray window has a transmission rate as high as 52.7% in the vicinity of E = 258eV. Thus, compared to a highly toxic beryllium window, the thin film has many application advantages, such as high transmission rate, high strength, high radiation resistance and nontoxicity.

In this paper, hot filament assisted chemical vapor deposition (HFCVD) method was applied to prepare diamond film. While maintaining the concentration of carbon source and changing the deposition parameters, such as reaction pressure, substrate temperature, nano-crystalline diamond (NCD) film with low surface roughness was deposited on silicon substrates. On this basis, hydrogen etching process was applied to further improve the quality and structure of the films. X-ray lithography mask-based material, whose optical transmittance and surface roughness meet the requirements of large integrated circuits (VLSI) was obtained.

2. Preparation of nano-crystalline diamond films

The preparation of nano-crystalline diamond films lies in: 1, the increase of the growth of nucleation density; 2, a very high secondary nucleation rate to suppress the growth of diamond nuclei and obtain films with nanometer-sized grain. In this study, HFCVD apparatus was used to prepare NCD films, with tantalum wire as a heat source, acetone and hydrogen as a reactant, 2 cm × 2 cm p-type polished silicon as substrate. The detailed deposition parameters were shown in Table 1. By changing the deposition temperature, pressure, hydrogen plasma treatment during the process of the hot-filament chemical vapor deposition (HFCVD), we studied the new technology for the preparation of NCD films. The microstructure and quality of the NCD film were characterized and analyzed.

Pressure (kpa)	Gas flow rate (acetone: hydrogen)	substrate temperature (°C)	Substrate bias/current (V/A)
1-3	50:180	600-700	+18V/4A

Table 1. HFCVD deposition parameters

It is noteworthy that, tantalum wire was polished with sand papers to clean the surface before the experiment. Pretreatment was carried out by importing hydrogen and acetone about 30 minutes to remove the surface oxide layer impurities of tantalum wire and form a layer of carbide coating in the surface to prevent pollution of prepared diamond film surface because of the evaporation of the tantalum.

In addition, before the preparation of diamond films, a series of pretreatments of Si substrate was conducted: first, Si substrate was placed in HF solution for 5 minutes to remove the oxide layer, then handly grinded for 15 minutes in the mixture of ultra-fine diamond powder with 100 nm particle size and glycerol. After these, the substrate was ultrasonically cleaned in baths of acetone and DI water, for 15 and 10 minutes, respectively. The above cleaning steps were repeated until the substrate surface was very clean.

2.1 The effects of deposition temperature on the quality of nano-crystalline diamond films

(a) Use Custom Size Format, Width = 17cm, Height = 24cm, (b) top margin is set to 2,5 cm and bottom margin is set to 3,0 cm; left and right margins are set to 2,0 cm on the manuscript format 17x24 cm, (c) use the whole space of all pages, don't leave free space, (d) the text must finish exact at the bottom of the last page. The manuscript has to be submitted in MS Word (*.doc) and PDF format. If you use other word editors and can not transfer it in Word and PDF please contact us.

Generally, the deposition temperature and pressure of NCD films are lower than the microcrystalline diamond films. In this experiment, the substrate temperature is 600-700°C. If the temperature is too low, the catalysis of hydrogen and hydrocarbons doesn't work and it's difficult to form diamond film. If the temperature is too high, the carbon tantalum alloys formed on the surface are volatile, resulting in the pollution of the matrix. In this study, while other conditions (acetone and hydrogen flow ratio of 50/180, pressure of 1 kPa) were maintained, the deposition temperatures were changed (detailed parameters shown in Table 2).

Samples	T1	T2	T3	T4
Temperature/°C	600	640	680	700

Table 2. Deposition temperature of diamond films

Figure 1 shows the surface morphology of the samples observed under scanning electron microscopy (SEM). It can be found that as the temperature increases, the grain size increased slightly, but not significantly, and all the grain sizes are basically below 50nm. In this study, a low deposition pressure (1 kPa) was applied and in the temperature range it was easy to achieve conditions beneficial for secondary nucleation. However, with the increase of temperature, the surface roughness trend to become larger.

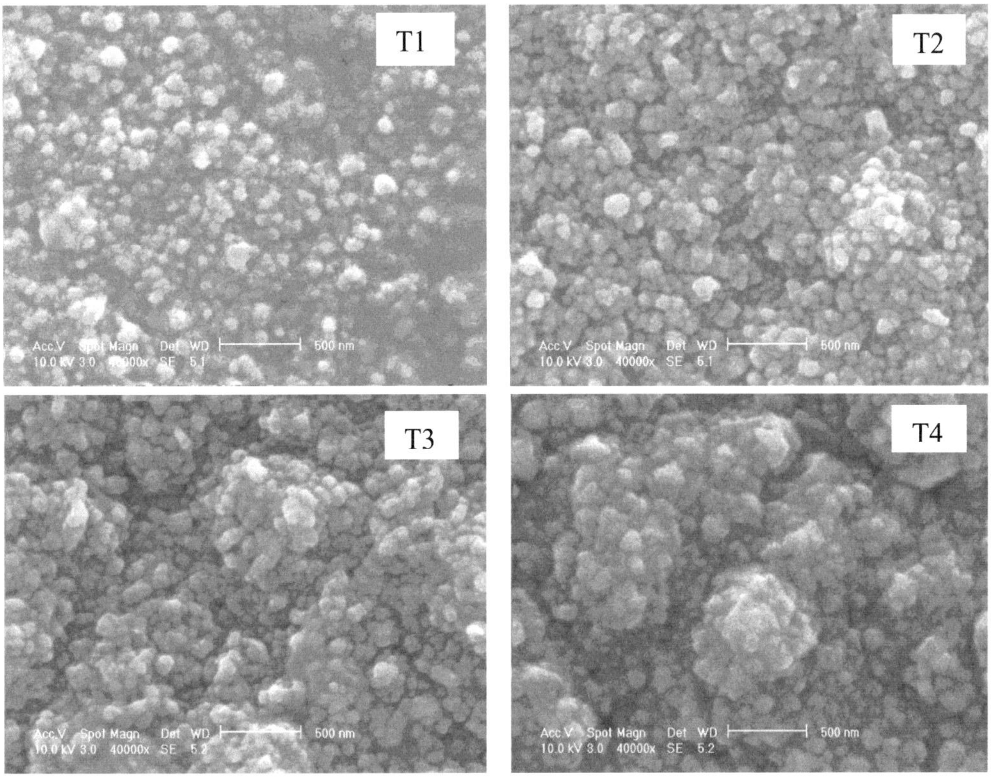

Fig. 1. SEM images of NCD films deposited under different temperature (T1. 600°C,T2. 640°C ,T3. 680°C ,T4. 700°C)

The atomic force microscope (AFM) images shows the surface root mean square roughness (RMS), which are 35 nm, 36.1 nm, 45 mm and 50.2 mm for the sample T1, T2, T3 and T4, respectively. Compared with the surface of micron diamond films, NCD films' surface is very smooth with RMS only in tens of nanometers, which not only solves the issue of diamond films polishing, but also greatly reduced light scattering of the film surface. And with the decreasing of temperature, the surface roughness of the samples decreased rapidly. Raman spectroscopy (HORIBA JOBIN YVON HR800 UV) was applied to characterize Raman spectra of the films. The excitation wavelength is 514.5 nm(Ar^+) with spectral measurement range of 1000~1800cm^{-1}.

The Raman spectrum of the samples given in figure 2 revealed several features that may be linked to NCD characters. First, the diamond peak at 1332cm^{-1} was very weak and significantly broadened, which was well known to be caused by the decrease of the grain size to the nanometer scale. Second, the bands at approximately 1140cm^{-1} and 1480cm^{-1} were presented, which had been assigned to trans-polyacetylene lying at grain boundaries in recent works and always been observed in chemical vapor deposition(CVD) NCD films as well as in hydrogenated amorphous carbon(a-C:H) containing nanocrystalline diamond. They were often regarded as a marker of NCD phase. Third, the scattering intensity at approximately 1350cm^{-1} and 1560cm^{-1},corresponding to sp^2- carbon(graphite D and G bands, amorphous carbon),was obviously enhanced.

As the temperature increased, the diamond peaks reduced and broadened. When the temperature gets close to 700°C, the intensity of characteristic peaks of diamond and graphite are very close, indicating that the increase of temperature leads to the increase of disordered graphite in the thin films. When the temperature is too high, the growth rate is great, and the first nucleation of grains grow soon which restrains the formation of other diamond nuclei, making the nucleation density on the matrix not high and more greater holes between grains. Thus the quality of the film declines. At 640°C, characteristic peaks of diamond of the samples is more obvious and diamond/graphite peak intensity reaches the maximum. It's considered that at this temperature there are a relatively small number of impurities and defects in the films. Above 600°C, the film quality declines because low deposition temperature makes the activity of atoms H decrease, and etching effect not strong. Meanwhile the gasification of graphite and other ingredients weakened, leading to an increase of non-diamond phase.

As the temperature increases, the proliferation and resettlement of various types of activated free radicals, quickens, the effect of active hydrogen enhances, which increase the diamond growth rate and the electric performance. On the contrary, the high deposition temperature makes the distortion from sp^3 carbon bond to sp^2 carbon bond combined group (i.e., graphitization). Moreover, because of the high deposition temperature, the activity of hydrogen atoms is too strong, resulting in the etching of sp^3 hybrid carbon bonds which form the structure of diamond while etching sp^2 carbon bonds. In short, too high or too low substrate temperature will change the effects of the atoms of hydrogen and carbon combined groups on the substrate, affecting the diamond film deposition processes and making amorphous carbon in the thin films increase.

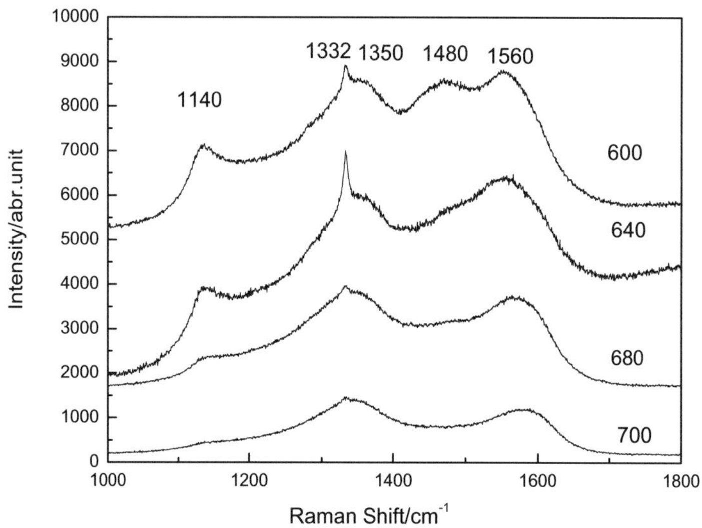

Fig. 2. Raman spectra of nano-crystalline diamond films under different deposition temperature

Optical transmittance of diamond films mainly lies on optical absorption of the film and light scattering on the film surface, grain boundary and film-substrate interface. The optical absorption of diamond films is caused by the imperfection of crystal lattice and impurities in the films, such as non-diamond-like carbon, hydrogen, nitrogen, etc, which can improve by adjusting the growth parameters and processes of CVD. Light-scattering loss mainly depends on the surface roughness of the films.

2.2 The effects of deposition pressure on the quality of NCD films

Selecting the appropriate pressure range is also one of the keys to prepare high-quality diamond films. To prepare NCD films, the reactant gas pressure is usually reduced. This is because, under the setted growth conditions, the nano-diamond particles are difficult to grow when becoming nuclei and the secondary nucleation rate in the film is very high. When preparing nano-crystalline diamond films, high concentrations of carbon source will make more carbon groups. At the same time, high H/H_2 is easy for nucleation. In addition, due to the low gas pressure, free path of the atoms is longer and the atoms will have higher energy. Their impacts with the growing diamond grains surface cause defects, leading to secondary nucleation while impeding normal growth of the original grains.

In this work, the deposition temperature is 640 °C.The NCD films were prepared by maintaining the constant gas flow rate (50/180 sccm) and changing the deposition pressure. SEM photographs of the diamond films surface synthesized at different pressure were depicted in Figure 3. It can be seen from Figure 3, with the gas pressure changes, the surface morphology of the films change significantly. When the pressure is 3kPa, the film surface is rough and the diamond crystals are well developed, indicating a low rate of secondary nucleation of diamond under this deposition condition. At the pressure of 2.5 kPa, the crystal grains of diamond film are no longer complete and there are many large grains surrounded by small particles. And compared with 3 kPa, the secondary nucleation rate increased significantly. When the pressure is down to 2 kPa, the film has a marked decline in grain size and film is nano-scale grain based, while its surface structure reaches nanometer scale. When the gas pressure is further reduced below 1.5kPa, the diamond grain size decreases even more evidently.

The above experimental results show that the reactant gas pressure have a very strong impact on the morphology and structure of the films. The concentrations of hydrogen atoms and carbon combined groups such as CH_2, CH_3, C_2H_2 in the reaction atmosphere as well as the energy state of these groups, play very crucial roles on the structure of the films. When the reactant gas pressure is high, the active hydrogen atoms and carbon-combined groups have short free path and low energy. They collide with the substrate and the generation of secondary nucleation is not easy, leading to the easy growth of diamond particles. When the reactant gas pressure is low, it will increase substrate temperature and H_2 dissociation rate. Meanwhile, the free path of various particles in the reaction chamber increase as well. These two factors will increase the velocity and energy that the particles reach the substrate. When the particles collide with the substrate surface, the energy are transferred to the surface. Thus the activity of the adsorbed particles is enhanced, making the carbon clusters form quickly on the substrate surface and improve the secondary nucleation rate. Besides, since the impacts of the particles on substrate reinforce, it's difficult for diamond grains to grow, resulting in increase of secondary nucleation rate.

Figure 4 shows RMS changes of NCD films under different deposition pressure. When deposition pressure is below 2.5 kPa, the surface roughness changes little. But when the deposition pressure increase to 3 kPa,the surface roughness of the film increases. As the pressure increases, the growth conditions come close to micron diamond growth conditions. The grain size of the film gradually become larger and the surface become rough.

Fig. 3. SEM images of nano-crystalline diamond films under different deposition pressure (P1. 1 kPa, P2. 1.5 kPa, P3. 2 kPa, P4. 2.5 kPa, P5. 3 kPa)

Figure 5 is Raman spectra of diamond films deposited under different pressure. In the spectra of low-pressure synthesized nano-crystalline diamond film, there are two weak broadband scattering peaks around 1350 cm^{-1} and 1580 cm^{-1}, which are D peak and G peak.

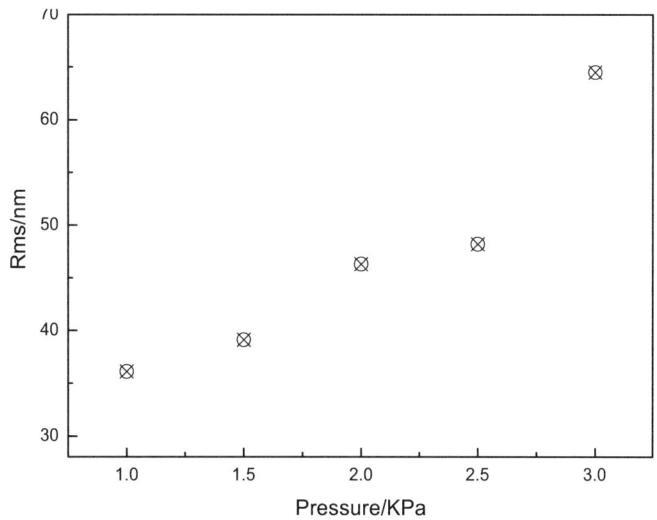

Fig. 4. RMS changes of nano-crystalline diamond films under different deposition pressure

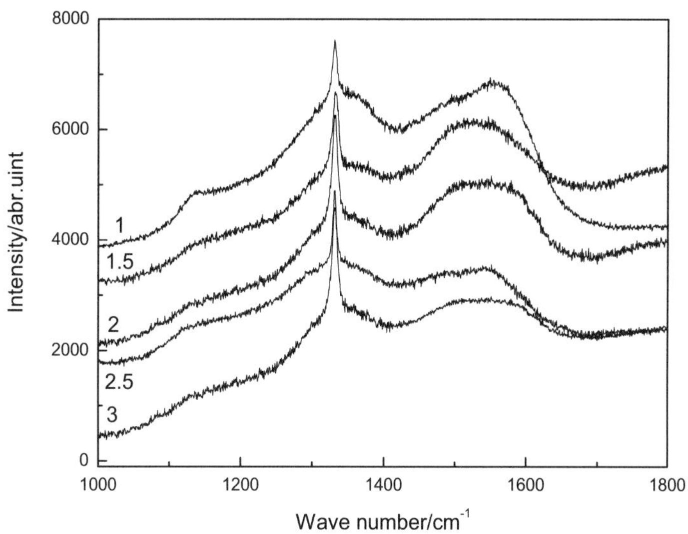

Fig. 5. Raman spectra of nano-crystalline diamond films under different deposition pressure

The diamond peak intensity of 1332cm^{-1} is small and wide. With the increase of gas pressure in the chamber, the scattering intensity of D peak and G peak gradually weakens while the the diamond peak at 1332 cm^{-1} becomes sharp. In addition, when the pressure reaches 2.5kPa, the band scattering background diminishes greatly in the range of 1100~1600cm^{-1}. The peak intensity of graphite decreases while the diamond peak is more clear, which shows good quality of the film.

In the 2 cm × 2 cm sample surface, paraffin was coated as a protective layer. In the middle, a circular window about 1 cm in diameter was left. Afterwards, it was put into the solution with HF: H_2NO_3 = 1:1. The silicon substrate was the corroded and a free-standing structure of diamond film was formed. During the corrosion process, the uniformity and thickness of the paraffin coat and the ratio of etching solution must be noted.

Optical transmittance of NCD films under different deposition pressure are shown in figure 6. In Figure 6, when the deposition pressure is 2.5kPa, the transmittance of the thin film on has reached 42.37% at 632.8 nm. By the above analysis of the surface roughness, we can see that, when the pressure is below 2.5 kPa, the surface roughness of the samples changes little. Here the quality of the thin film is the main factor that affects the transmittance. At 3kPa, although the samples shows good quality in the Raman analysis and the absorption of impurities weakens, the roughness increases while the transmittance decreases slightly. Therefore, transmittance of the film lies on the quality and surface roughness.

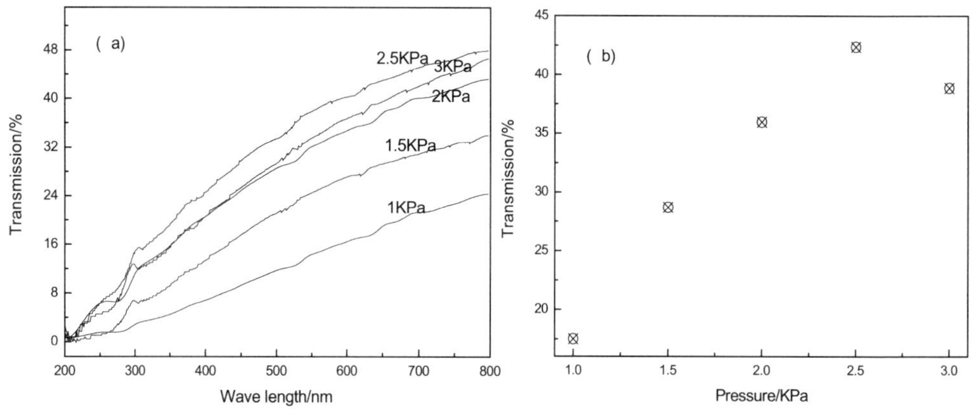

Fig. 6. Optical transmittance of NCD films under different deposition pressure
(a)The UV and visible band (b) Optical transmittance of the film at 632.8nm

2.3 The effects of etching on the electrical and optical properties of thin films

In CVD, the atomic hydrogen plays an important role in the diamond film growth process has been confirmed: atomic hydrogen stabilizes carbon with diamond structure while etching the carbon with graphite structure. The existence of hydrogen contributes to the growth of high-quality diamond films. It's commonly believed that hydrogen has a very important role which is the selective etching of graphite phase. Etching of H atoms for graphite and amorphous carbon are much stronger than for diamond. And this selective etching results in the non-equilibrium gas phase growth of diamond.

In this experiment, intermittent import of acetone gas and strengthening of hydrogen etching of graphite phase were applied to study improvement of the film quality during the process of the chemical vapor deposition (CVD) diamond film deposition. According to the parameters of the samples in previous experiments, the closing time and frequency of acetone were changed to increase the etching time.

In this work, the sample was deposited at pressure of 2.5Kpa, temperature of 640°C and unchanged total growth time. Sample A wasn't hydrogen etched throughout the deposition process. Sample B was etched for 15 minutes every half an hour and the total etching time was 1.5 hours. Sample C was etched for 15 minutes after 15 minutes and the total etching time was 3 hours.

Figure 7 shows surface morphology of the thin films under different eching time. Observed by SEM, without hydrogen etching, the film formed into a flake with large grain of about 120 nm. And with the increase of etching time, the grain of the film decrease. After etching for 3 hours, the grain of the film was reduced to 55 nm. As the grain size decreased, the surface roughness of the film increased. The SEM cross-section morphology of the films shows that the thickness of the samples is about 1μm, so it's unnecessary to consider the impact of film thickness on the optical properties.

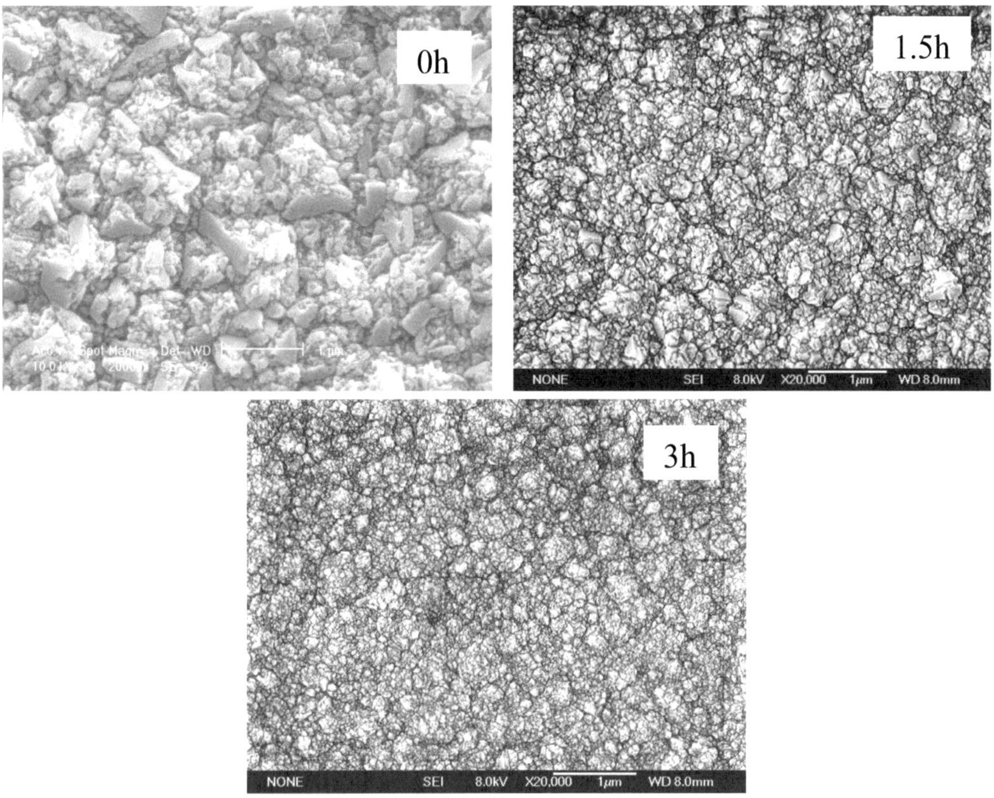

Fig. 7. Surface morphology of the samples under different eching time
(the etching time are 0 h, 1.5 h, 3 h, respectively)

The AFM images also shows that as the etching time increases, the diamond particles decreases, while the surface roughness of the film decreased gradually as well, from 47.2 nm to 14.8 nm. The phenomenon can be explained from the deposition mechanism: in the growth process of thin film, the positive ions and neutral groups in the chamber moves towards the substrate surface where the surface adsorption and reaction happens. Meanwhile, the hydrogen atoms in the chamber will etch the film. In the growth process, the energy of hydrogen atom is relatively low and the etching effect on the film is relatively small. Therefore the film obtained has more organic phase with relatively loose and rough surface. After strengthening the etching, the bombardment time of hydrogen atom is enhanced, leading to the proliferation of surface atoms and the surface etching of the loose structure in the film-forming process. Thus the surface of the film is smooth and dense.

As can be seen from figure 8, after the strengthening of etching, the diamond characteristic peak begins to widen and the FWHM increases from 5.31 cm^{-1} to 27.88 cm^{-1}. By etching, the grain size of the film becomes smaller and smaller. In addition, as the etching time increases, the atoms bombardment increases the deposition temperature, prompting the transformation from sp^2 bonds to the sp^3 bonds. The intensity of the diamond peak increases with graphite peak decreasing. Id/Ig increases and content of sp^2 decreases. As the etching time increases, a scattering peak emerges at ~1130 cm^{-1}, which is usually referred to as the signature of NCD films. And this is widely cited in the literature. However, some recent studies show that this may be wrong as the location of this peak changes with the excitation photon energy used and the scattering intensity decrease with the increase of incident photon energy. This indicates the typical sp^2 hybrid characteristics and was identified as the C-C bond stretching (stretching) and twist (wagging) mode of polyacetylene (trans-polyacytylene) by Ferrari and Robertson.

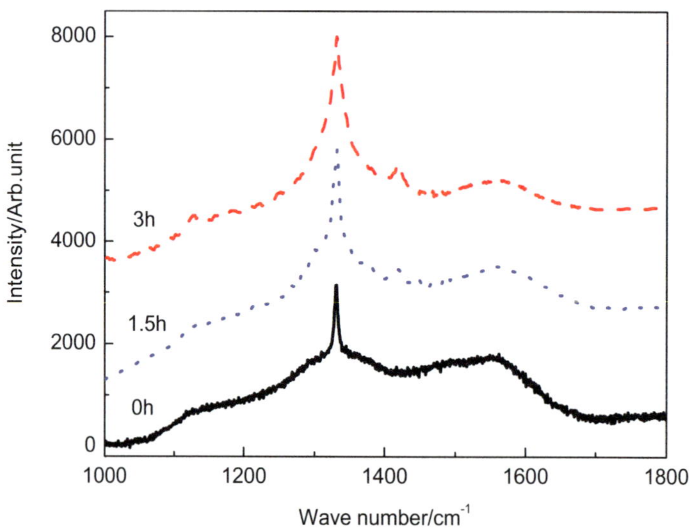

Fig. 8. Raman spectra of diamond thin films under different etching time

The prepared diamond thin films were processed into free-standing diamond films. Infrared spectrometer (LAMDA900) was used to measure the infrared spectra of the samples, which were shown in Figure 9. The measuring wavelength range is 500 ~ 2500nm.

Sample C with 3 h etching time, has the maximum transmittance and its transmittance is greater than the other two samples' in the whole band (500~2500nm). The transmittance curve is an oscillation curve in the low frequency due to the interference effect. This effect is caused by the multiple reflection beams interference, leading to a series of mini/max change on the transmittance spectra. Three samples have relatively large oscillation amplitudes in the whole wavelength region, indicating relatively smooth surface of the samples. In the short wavelength band, infrared light transmittance of the films decreases with the decrease of the wavelength because of the scattering effects caused by surface roughness. However, in the long wavelength band, the scattering losses are not large and the losses of transmittance are mainly due to the absorption within the film, such as absorption of graphite.

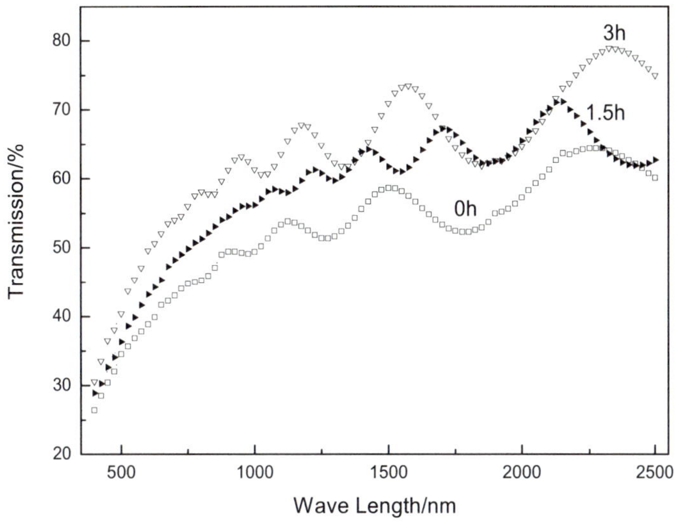

Fig. 9. Infrared transmittance of diamond thin films under different etching time

Sample A has the minimum infrared light transmittance and the diamond-related peak is also the weakest in the Raman spectra, indicating the absorption of non-diamond-like carbon in the film has a large impact on the infrared light transmittance of the sample. After 3 hours of hydrogen etching, the optical transmittance of the thin film at 632.8 nm reaches 51%. By etching, the quality of the thin film has been greatly improved.

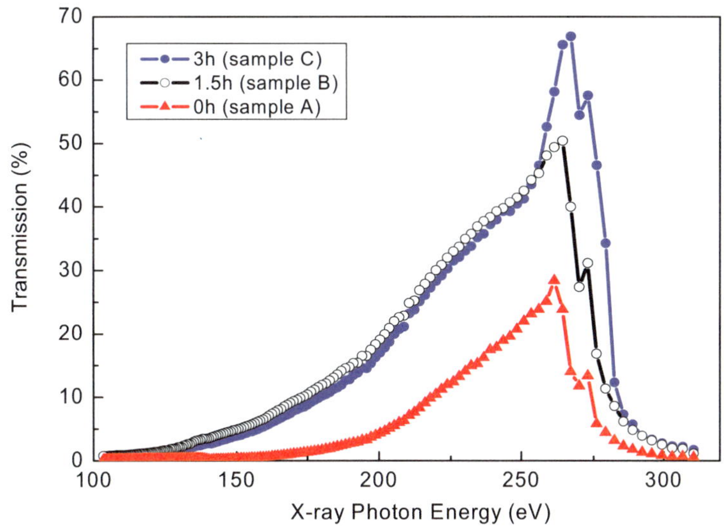

Fig. 10. Transmittance of diamond thin films in the soft x-ray under different etching time

Figure 10 shows the x-ray transmittance of diamond thin films under different etching time. The x-ray photon energy range is 100 ~ 310eV (a wavelength of 4 ~ 12nm). Taking the impact of the limited absorption on the film transmittance into account, we have chosen the transmittance at 258eV to measure the X-ray transmission characteristics. The transmittances of the three samples at 258eV are 52.7%, 47.8% and 24.8%, respectively while the traditional 8μm thick beryllium window in this energy is completely opaque. X-ray transmittance is not only related to the surface roughness, but also to the quality, structure and grain size of the film. By hydrogen etching process, the compactability of the film structure can be improved. Moreover, the internal defects (especially graphite content) and surface roughness decreases while the transmittance increases. However, etching can't make unlimited changes in quality of the film. The improvement of the film's performance is limited and the growth conditions are key factors in the quality of the film.

3. Conclusions

In this work, NCD films are prepared by HFCVD method. While maintaining the concentration of carbon source and changing the deposition parameters, such as reaction pressure, substrate temperature, NCD film with low surface roughness and high quality is deposited on silicon substrates. On this basis, hydrogen etching process was applied to further improve the quality and structure of the films. X-ray lithography mask-based material, whose optical transmittance and surface roughness meet the requirements of large integrated circuits (VLSI) was obtained. The main conclusions of the work are as below:

1. With the decreasement of deposition temperature (600°C~700°C) in the process of diamond thin films deposition, the secondary nucleation increased and the thin films grain size decreased. At the deposition temperature of 640°C, NCD films with the higher diamond quality, the lower surface roughness and the higher optical transmittance in wavelength about 632.8nm was achieved.
2. The influence of deposition pressure variation on the NCD quality, structure and optical properties under the deposition temperature of 640°C is investigated. With the increasement of deposition pressure, grain size of diamond grown up. At relative low deposition pressure, variation of surface roughness of thin films are not so obviously, but increased intensely at 3kPa. The transmittance reaches 42.37% at the wavelength of 632.8nm when deposition pressure is about 2.5kPa.
3. By using the fixed the deposition parameter, such as reaction pressure, substrate temperature, etc, and modifying the duration of hydrogen etch, the NCD films with low roughness and high optical transmittance are successfully deposited on silicon substrate. With the increasement of the duration of hydrogen etch (0~3h), the grain size decreased, and the surface roughness (RMS) reduced from 47.2nm to 14.8nm. The Raman scattering spectra indicated that as the etch duration increased, the contents of impure phase in the diamond thin films such as graphite decreased, the existence of hydrogen in thin films is propitious to the stability of sp3 bond, and could improve the quality of thin films in some extent.
4. The NCD films' transmittance for soft X-ray band at wavelength 4~12 nm has been measured using a synchrotron radiation device. It reveals that when grain sizes are

about nanometer scale, the transmittance of thin films is increased much intensely with the increasement of the particle size, and is slightly lowered with the variation of surface roughness. By modified the frequency and duration of hydrogen etch process, the influence of hydrogen etch on thin film transmittance of X-ray has been studied. With the increasement of etch duration, the transmittance of thin films are increased. After processed by hydrogen etch for 3 h, the transmittance of thin film at 258eV reaches to 52.7%.

5. In this work, the NCD film with low surface roughness and high transmittance of X-ray, which can meet the requirements of large integrated circuits (VLSI), is developed. However, for the optical transmittance of diamond films are depend on both surface roughness and film quality, there still need to further optimize the deposition process to simultaneously obtain smoothed surface and lower content of carbon sp2 bond, and then achieve the preparation of high quality optical window.

4. Acknowledgments

The work described in this paper has been Supported by National Natural Science Foundation of China (60877017), Program for Changjiang Scholars, Innovative Research Team in University (No:IRT0739), Innovation Program of Shanghai Municipal Education Commission (08YZ04) and Shanghai Leading Academic Disciplines (S30107)

5. References

M. F. Ravet, F. Rousseaux, (1992). Status of diamond as membrane material for X-ray lithography masks. *Diamond and Related Materials,* 5: 812-818, ISSN: 0925-9635

B. R. Huang, J. T. Sheu, C. H. Wu, et al. (1999). Bilayer SiNx/diamond films for X-ray lithography mask. *Jpn. J. Appl. Phys.,* 38(4): 6530-6533, ISSN: 0021-4922

P. P. Burggraaf, (1999). Optical lithography to 2000 and beyond. *Solid State Technology,* 42 (2): 31-41, ISSN: 0038-111X

P. W. May, (2000). Diamond thin films: a 21st-century material. *Phil. Trans. R. Soc. Lond. A,* 358: 473-495

G. Drazic, E. Sarantopoulou, et al. (2002). X-ray microanalysis of optical materials for 157nm photolithography, *Crystal Engineering,* 5: 327–334, ISSN: 1463-0184

Y. Ma, N. Wassdahl, P. Skytt, et al., (1992). Soft-x-ray resonant inelastic scattering at the C K edge of diamond. *Phys. Rev. Lett.* , 69 (17): 2598-2610, ISSN 0031-9007

B. R. Huang, C. H. Wu, K. Y. Yang, (2000). Polycrystalline diamond films for X-ray lithography mask. *Materials Science and Engineering B,* 75: 61-67, ISSN: 0921-5107

Mao Minyao, Kang Changjun, Wang Tianping et al. (1995). Optical transmission studies of diamond thin films prepared by microwave plasma chemical vapor deposition. *Actaphysica sinica,* 44(9): 1509-1515, ISSN:1000-3290

A .C. Ferrari, J. Robertson, (2001). Raman and infrared modes of hydrogenated amorphous carbon nitride. *Journal of Applied Physics,* 189 (10): 5425-5430, ISSN 0021-8979

N. Jiang, S. Kujime, et al, (2002). Growth and structural analysis of nano-diamond films deposited on Si substrates pretreated by various methods. *Journal of Crystal Growth*, 281: 265-271, ISSN: 0022-0248

E-Beam and Ion Beam Lithography

13

High-energy Electron Beam Lithography for Nanoscale Fabrication

Cen Shawn Wu[1], Yoshiyuki Makiuchi[2] and ChiiDong Chen[3]
[1]*Department of physics, National Changhua University of Education, Changhua, 500,*
[2]*Elionix Inc. 3-7-6 Motoyokoyamacho Hachioji, Tokyo 192-0063,*
[3]*Institute of Physics, Academia Sinica, Nankang, 115, Taipei,*
[1,3]*Taiwan*
[2]*Japan*

1. Introduction

Recently, there has been much research in the field of nanostructure technology. The objective of this article is to explore the basic physics, technology, and applications of ultra-small structures and devices with dimensions in the sub-100-nm range. Nanostructure devices are now being fabricated in many laboratories to explore various effects, such as those created by downscaling existing devices, quantum effects in mesoscopic devices, tunneling effects in single electron transistors, and so on. In addition, new phenomena are being explored in an attempt to build switching devices with dimensions down to the molecular level. Today, the minimum size of semiconductor production devices is down to 45 nm or less. Miniaturization and performance improvements are allowing the electronics industry to shrink the future size of semiconductor devices. For the last three decades, technological advances in optical lithography have led the semiconductor industry. In the present day mass-production line, deep ultraviolet (DUV 193 nm) optics is used for critical dimensions approaching 90 nm. A further decrease in immersion lithography has pushed the production line into the 32-nm era. However, as wavelengths decrease, optical lithography begins to suffer from difficulties associated with shallow focus length and materials (including lenses, masks and resists) to do with dispersion relationships. Therefore, non-optical lithographic techniques such as electron-beam direct write (L. Pain et al., 2006), electron-beam projection (J. Yamamoto et al., 2000), ion-beam projection (Y. Lee et al., 1998), and soft X-ray extreme ultraviolet (EUV 13 nm) (S. Hector, 1998; C. W. Gwyn et al., 1998) are being increasingly focused in order to replace or "mix & match" with optical lithography. Electron beam direct write is the only technology that does not require masking. Electron beam lithography (EBL) is one of the versatile lithographic tools widely used for nanostructure fabrication.

In EBL, the critical dimension of the exposure patterns is limited by electron scattering in both electron-beam resists and substrates. The demand for fine patterns calls for EBL systems with high acceleration voltages. These have the advantage of having smaller forward scattering angles in the resist and a wide secondary-electron spread in the substrates. The electron scattering process depends greatly on incident electron energy and resist/substrate properties and is a very complex issue; thus the resultant energy-intensity distribution in the resist has to be calculated using Monte Carlo simulation. The calculated distribution is of Gaussian shape, and the contribution of secondary-electron exposure is exponentially suppressed with increased incident beam energy. Due to the long penetration depth in the resist, high-energy EBL allows for the exposure of very thick resists, which are useful for forming nanostructures with large height-to-width ratios. In multi-layer resists of different exposure sensitivities, the linewidth of each layer can be controlled separately by adjusting the development time and by using different developers after a single e-beam exposure. It will be shown that even after many development steps; the linewidth in the top layer remains unchanged. Precise control of the lower layers' linewidths makes the fabrication of sophisticated three-dimensional (3D) structures possible. In this chapter, we will first give an introduction to the state-of-the-art high-energy EBL technique. This will be followed by discussions on electron-optics, Monte-Carlo calculations of energy-intensity distribution, resist profile engineering, and mix-and-match techniques. Finally, we will give some examples to illustrate fabrications of nanoscale electronics and 3D structures, and discuss issues that have to be taken into account when using a 100keV EBL system.

2. Introduction to high-energy EBL system

There are two types of electron beam lithography systems: the point-beam type and the variable-shaped beam type. This chapter introduces some of the basics about the former. The point-beam EBL system, because of its ability to draw patterns in extremely small dimensions, is widely used for cutting-edge research and development, and because of its flexibility in pattern change, is also used in the high-mix, low-volume production of various devices. The point-beam EBL system can generate a high-resolution electron beam with a large electric current on a nano-Ampere order that facilitates the high-speed drawing of fine patterns on the substrate. The high-energy EBL system consists of two major components: the main body and the control system. The former includes the electron-optical system (electron-optical column), high-precision stage, and vacuum pumps; the latter includes data files, the controller computer, and the CAD system. Figure 1 provides an overview of the high-energy EBL system. Each unit has the following respective functions:

1. **Electron-optics control system:**
   It controls the focusing position and the intensity of the electron beam. It consists of the electron gun for generating the electron beam, the blanker for turning on/off the electron beam at high-speed, the electromagnetic lens for sharpening the beam, the beam alignment system for adjusting the beam with the optical axis, the astigmatism correction system, and the electrostatic deflection system.
2. **High-precision stage control system:**
   It controls the high-precision movement of the sample/substrate to be processed. It consists of the X-Y-Z stage for sustaining and moving the sample/substrate, laser interferometer system, specimen chamber which houses the stage and the specimen changing chamber.
3. **Lithography control system:**
   It sends the lithography control data to the main body. It consists of the hardware including the lithography pattern generator and the software for system operations.

4. **Evacuation control system:**
   It controls the vacuum degree at the electron optical system, the lithography chamber housing the stage, and the specimen changing chamber according to the required steps of the lithography operation.

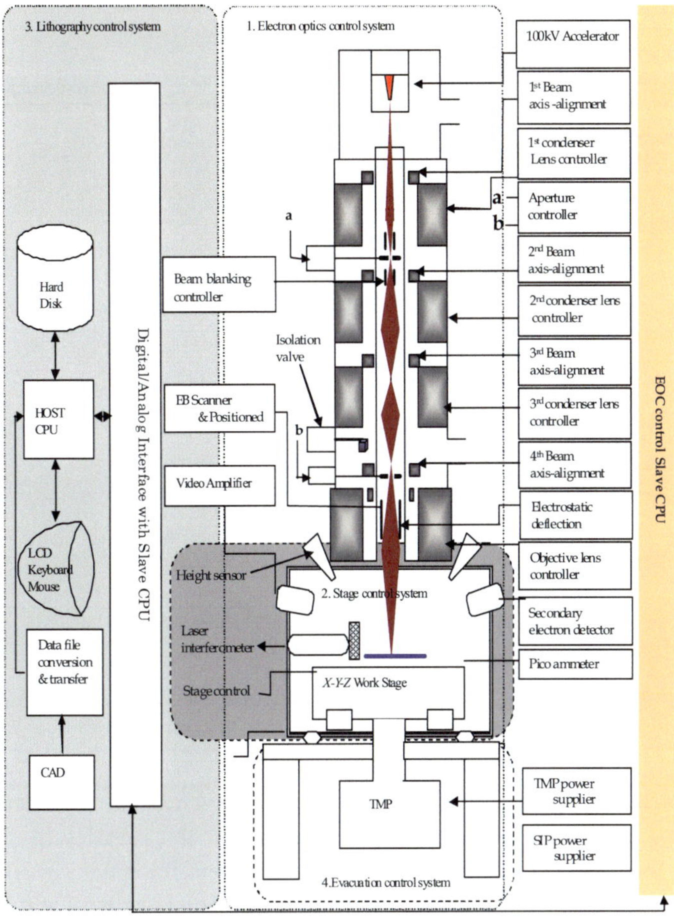

Fig. 1. Schematic drawing showing the major components of a high-energy electron beam lithography system

3. Electron-Optical control System (1)

The electron-optical control (EOC) system, whose structure is shown in Figure 2, is the heart of the high-energy EBL system. It generates a tightly-focused electron beam with a large current density, allowing for the drawing of ultra-fine patterns. While the focusing system of an optical microscope consists of optical lenses, the focusing system in an electron beam optical column comprises a set of electromagnetic lenses which focuses the electron beam and is used in electron microscopes. The EOC system consists of four-tier electromagnetic lenses as shown in Fig. 2. After going through these four electromagnetic lenses, the beam spot diameter will be reduced to almost one tenth: from 15 ~ 20 nm at the starting point to less than 2 nm when hitting the specimen. There are seven main components of the EOC system as indicated in the diagram. This section will describe items 1 - 3, which are relevant to beam generation and focusing.

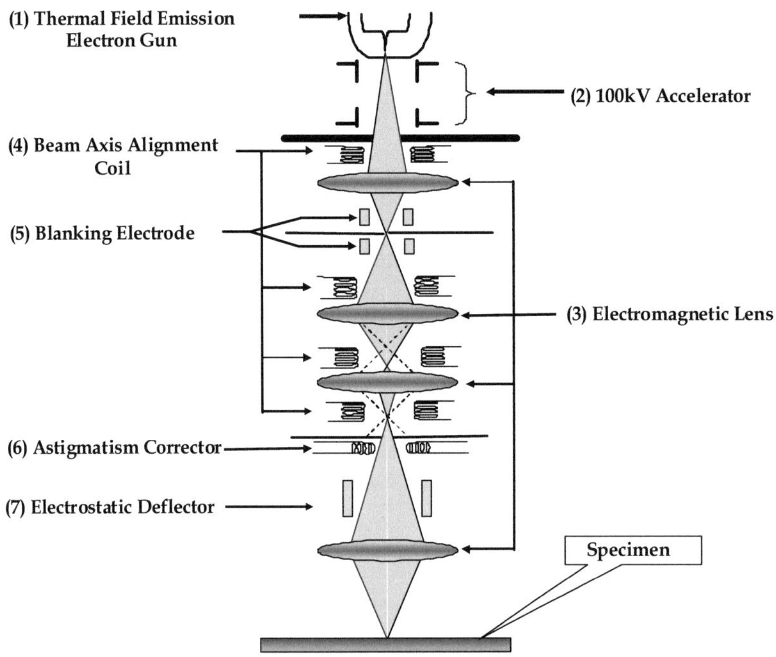

Fig. 2. Cross-section drawing of an electron-optical control system along with a ray-trace of the electrons as they pass through a series of electron-optical components

(1) Thermal Field Emission Electron Gun

When an extremely strong electric field (about 5×10^9 V/m) is applied to a metal surface in the direction of the metal's interior, the free electrons in the metal will penetrate the potential barrier at the metal's surface and will be emitted into the outer vacuum space due to the quantum mechanical tunneling effect. This phenomenon is called Field Emission. To initiate the field emission, the tip of field emission electron gun cathode is manufactured so that its curvature radius becomes as small as approximately 1μm, and a voltage of about 3kV is applied at the tip against the anode. There are two types of field emission electron guns: the cold type, whose cathode is used at room temperature, and the thermal type, whose cathode is heated to about 1800K when in use. The former has the advantage that the energy of the electrons is kept relatively constant, whereas the latter has the following advantages: 1, the degree of tolerance of the environmental vacuum during operation is large; 2, the angular current density is about ten times as large as that of the cold type; 3, the necessary field intensity is relatively low; and 4, the electron beam can be kept highly stable over a long period.

Because of these advantages, the thermal field emission cathode is generally used in the EBL system. The needle-shaped cathode of the thermal field emission (TFE) gun (Figure 3) is made of a single tungsten crystal, to the bottom of which a heater line made of polycrystalline tungsten is welded. The center of the needle-shaped cathode is coated with zirconium dioxide (ZrO), which spreads over the surface of the single tungsten crystal up to the cathode tip, drastically reducing the potential barrier at the tip's surface. Figure 4 shows the structure of a TFE gun, whose cathode is equipped with a suppresser and an extractor electrode for extracting the electron beam.

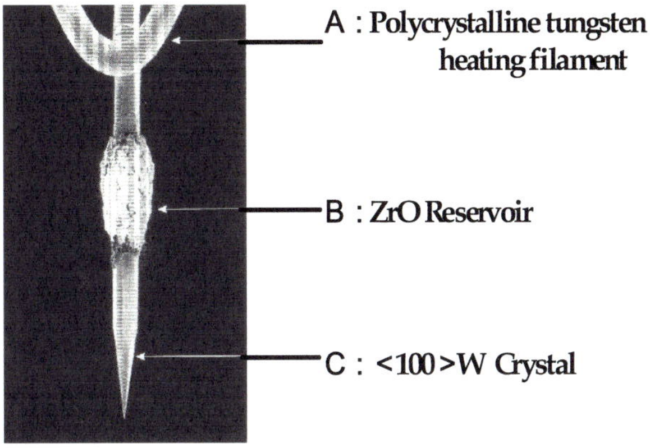

Fig. 3. A ZrO/W Thermal field emission electron source. The singe crystal tungsten is coated with a layer of zirconium oxide to reduce the work function barrier.

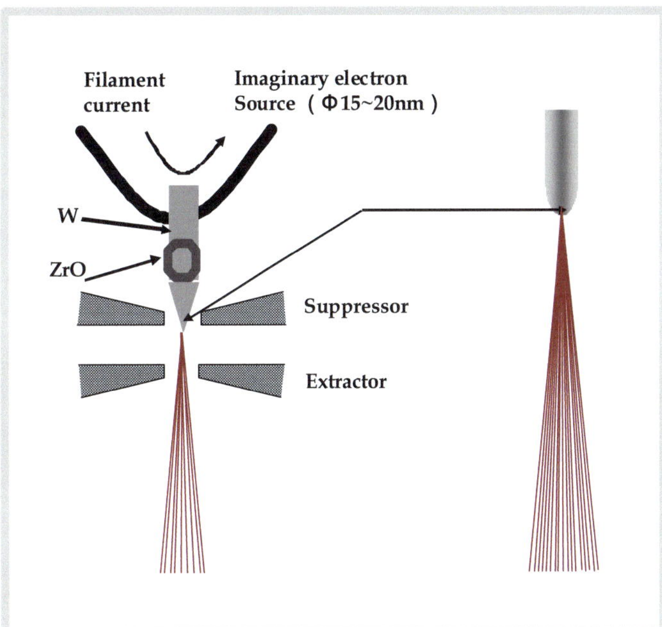

Fig. 4. A thermal field emission gun with a suppresser and an extractor electrode. The electrodes are circularly symmetric about the optical axis. The suppressor is negatively biased with respect to the cathode, while the extractor is positively biased.

(2) 100kV Accelerator

Of the electron beam emitted from the electron gun, the high-quality beam at the flux core (φ15 - 20nm) will enter into the acceleration system, where it will be accelerated up to a specified energies (Figure 5). The higher the acceleration voltage, the finer the patterns the system can draw uniformly. The lithography examples with different acceleration voltages are shown in Figure 6. By accelerating the beam up to 100kV, patterns finer than 5nm can be drawn even using commercially available resists. The stability of the high-voltage system is

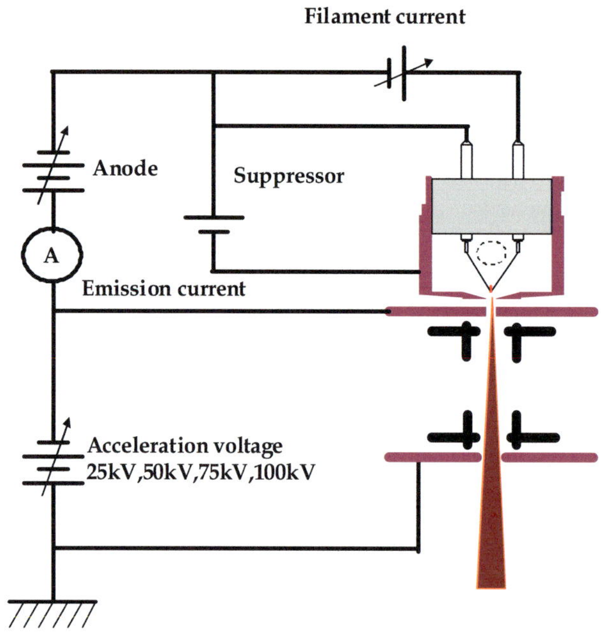

Fig. 5. Biasing circuit for a thermal field-emission electron gun

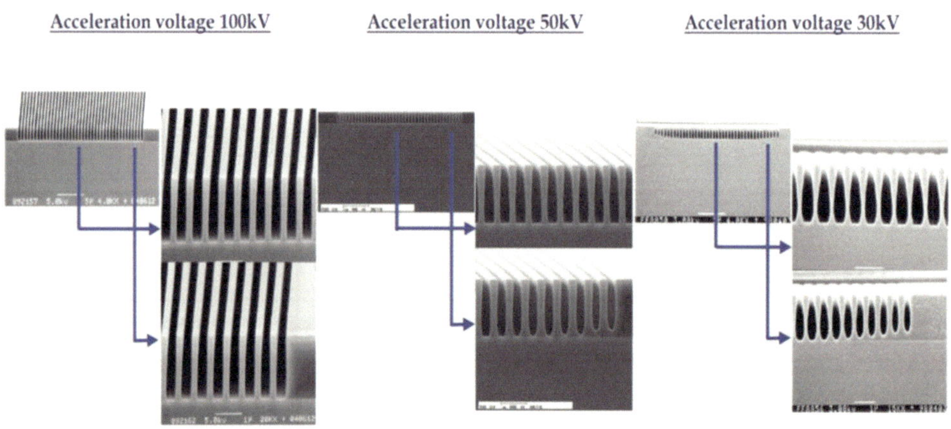

Fig. 6. Example of EBL resists patterns with different acceleration voltages. From left to right, the acceleration voltages are 100kV, 50kV and 30kV.

most important, as any fluctuation in the acceleration voltage directly affects the electron beam energy, changing the focus distance and the deflection angle of the electron beam. This in turn causes defocusing or a skewness in the drawn patterns. The micro-discharge during the drawing session will also occasionally cut the electron beam, leaving some parts of the pattern undrawn. To prevent these problems, the high-voltage system must be carefully designed and manufactured. Such prerequisites make realizing high acceleration voltage technically challenging.

(3) Electromagnetic Lens

Electromagnetic lenses are used to focus the electron beam. The beam focusing mechanism of the lithography system is basically the same as that of the electron microscope. The only major difference from the microscopic system is that the second and third electromagnetic lenses are placed after the focusing lens in order to fix the focus point on the fourth electromagnetic lens, which works as the objective lens. This configuration is designated as a zoom lens, and allows for the easy adjustment of the reduction ratio of the beam diameter and the irradiating current without changing the excitation of the objective lens. By fixing the excitation of the objective lens, stability and repeatability, critical elements of the high-energy EBL system, are achieved. This allows the stable drawing of patterns from a few nm to several hundreds nm (Figure 7).

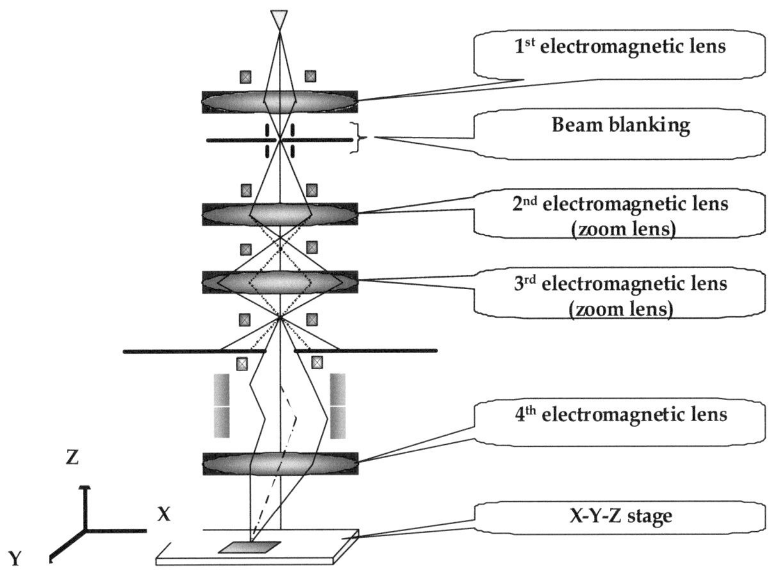

Fig. 7. A schematic of electromagnetic lens system. By applying current to the wire coiled around an iron cylindrical core, a magnetic field is created which acts as a lens. The EBL system takes electrons from a source through the condenser and objective electromagnetic lenses and focuses the beam on a spot on the specimen.

4. Electron optical system – 2

This section describes items 4 - 7 in Fig. 2, which are relevant to controlling the focusing position of the electron beam.

(4) Beam Axis Alignment Coil

Alignment of the electron beam center axis to the optical axis of the electromagnetic lens is essential. Beam axis alignment coil is employed to correct the micron-order deviation between the lenses. There are 4 sets of alignment coils; they are placed immediately above each of the four electromagnetic lenses. Each alignment coil consists of two coils which generate a horizontal magnetic field in the X-Y plane, deflecting the electron beam slightly by the Lorentz force, as shown in Fig. 8.

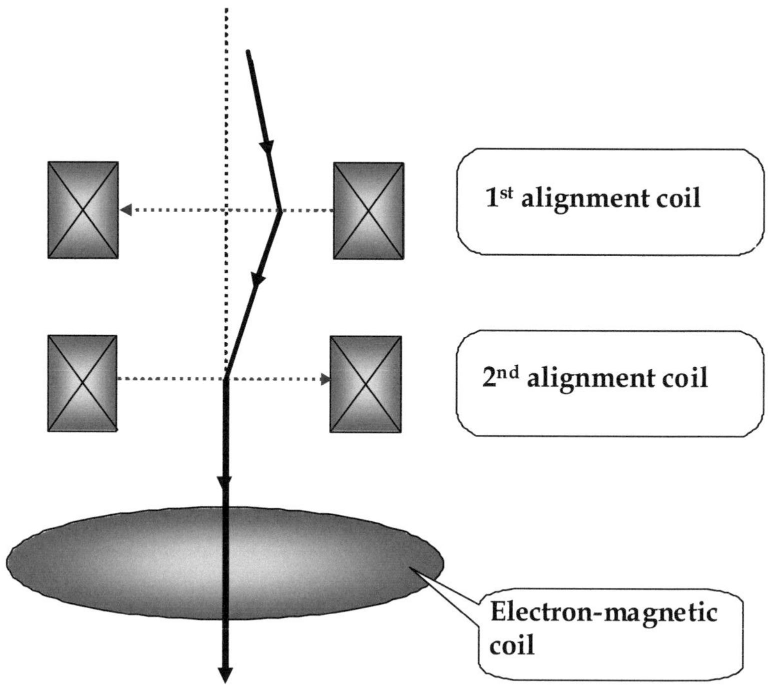

Fig. 8. Beam axis-alignment coils. Each alignment coil consists of two coils which generate a horizontal magnetic field in the X-Y plane, deflecting the electron beam slightly by the Lorentz force.

(5) Blanking Electrodes

In both vector and raster scan model, the electron beam is "turned off" between the designed entities. This is done by a beam blanker which deflects the electron beam far away from the optical axis so that the beam will not reach the specimen. This is accomplished by applying an electric field perpendicular to the optical axis (Figure 9). However, if the beam is on the surface of the specimen when it is deflected from the axis, an unwanted whisker-like line is drawn on the specimen. To overcome this issue, the beam blanker consists of two deflectors placed above and below the blanking aperture. The two deflectors generate electric field of the same strength at the same time and bend the beam by the same amount. Using this method, a virtual electron optical source is formed and the beam is shifted with no lag.

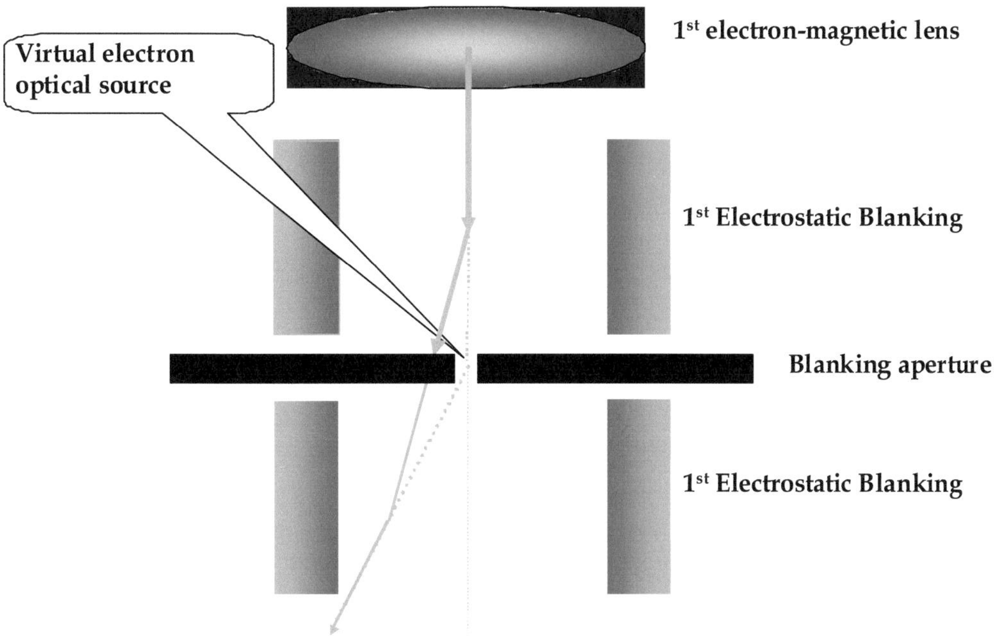

Fig. 9. A beam blanker which consists of two sets of electrostatic deflector. The blanking systems involving multiple sets of plates prevented beam motion lag between the blanking and unblanking processes

(6) Astigmatism Corrector

Although the apparent beam-form on the specimen should ideally be a perfect circle, the beam always has astigmatism (Figure 10). Possible causes of astigmatism are: (a) Asymmetry of the electromagnetic field on the beam path due to manufacturing imperfection, which deviates the lenses and the pole piece from the rotational symmetry. (b) contamination in the optical column. Such astigmatism is corrected by a special astigmatism corrector, which consists of a pair of coils, each comprising four poles, and placed against each other at a 45-degree rotation. This means that the corrector has eight poles in total, as shown in Figure 11. When the electron currents I_1 and I_2 are applied to the respective coils, they generate magnetic fields, which will be unified as a synthetic field. As the electron beam progresses downward perpendicularly to the beam cross-section, the electric current flows upward. Therefore, according to Fleming's Law, the beam receives a force toward the direction indicated by the arrow, producing astigmatism. By applying this electrically-created astigmatism vertically in the direction of the intrinsic astigmatism of the electron-optical system, the effects of the two astigmatisms cancel each other out, thus eliminating any astigmatism from the electron-optical system.

(7) Electrostatic Deflector

The electron-optical system described above creates an excellent electron beam spot (focusing spot) on the specimen. Lithography is performed by irradiating the beam spot accurately on the intended position of the specimen. The electrostatic deflectors shown in figure 12 do the job. Similar to the beam blanker, the electrostatic deflectors control precisely

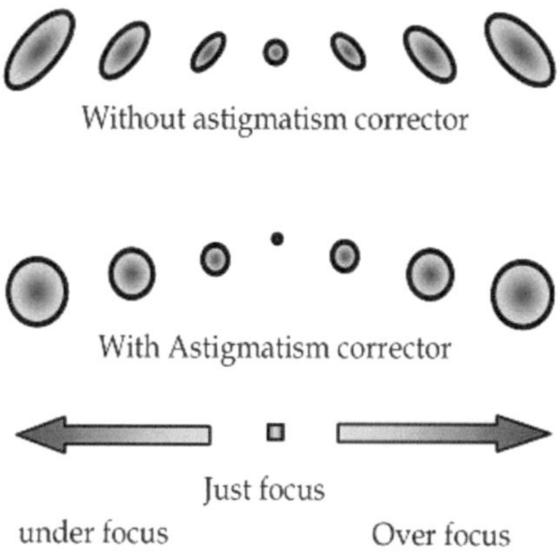

Fig. 10. A comparison of apparent beam-form on the specimen

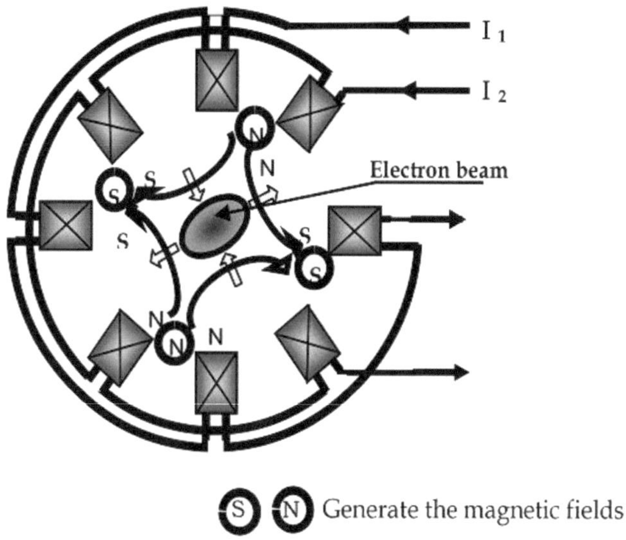

Fig. 11. Schematic drawing of the magnetic field distribution in astigmatism corrector. The corrector eliminates the astigmatism by generating a magnetic field which shifts the beam back into its optimum shape.

the deflection angle of the beam, which determines the positional accuracy of the exposure pattern. In addition, a tight focus is important for ultra-fine patterns. To this end, the distance between the specimen and the objective lens, which is the last tier of the magnetic lens system, must be as short as possible. Because of this restriction, the electrostatic deflector is positioned above the objective lens. At the same time, to minimize any aberration the axis of the beam has to be aligned precisely to the center of the objective lens. To do this, high-energy EBL systems specialized for ultra-fine patterns are equipped with a two-tier deflector by which the beam is first deflected in one direction and then is deflected back. To move the beam to the designed location on the specimen, the two-tier deflection is independently applied in both X and Y directions. The preciseness of the beam positioning depends on the precision of the voltage applied to the deflector plates. The voltage is usually controlled at a precision of 1/240,000 using an 18-bit DAC. There are two types of beam-scan methods that use the electrostatic deflector: the vector-scan and the raster-scan. The vector-scan is a method in which the beam is deflected only over the entities to be exposed. The raster-scan is a method in which the beam scans at constant speed, while turning on/off the beam according to the presence or absence of a pattern (Figure 13). There are special scan methods to draw arcs and circular patterns, too. In these methods, the ordinary X-Y orthogonal coordinate system is only used for beam positioning, and a special scan generator is used to scan the beam on the circumferential coordinate system.

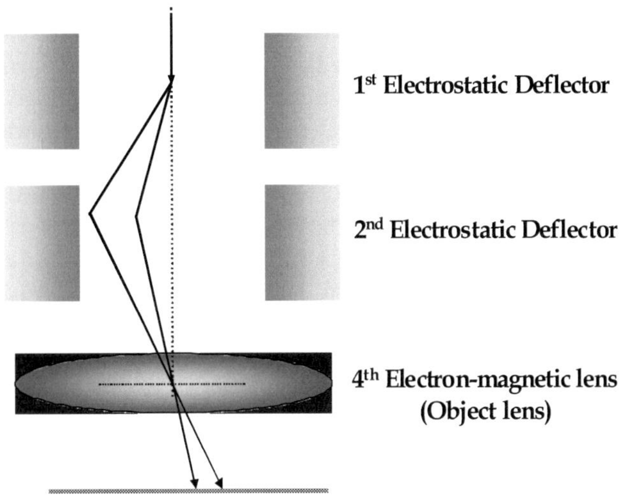

Fig. 12. Schematic drawing of an electrostatic deflectors which controls precisely the deflection angle of the beam. The system consisted of two-tier deflector to minimize any aberration the axis of the beam.

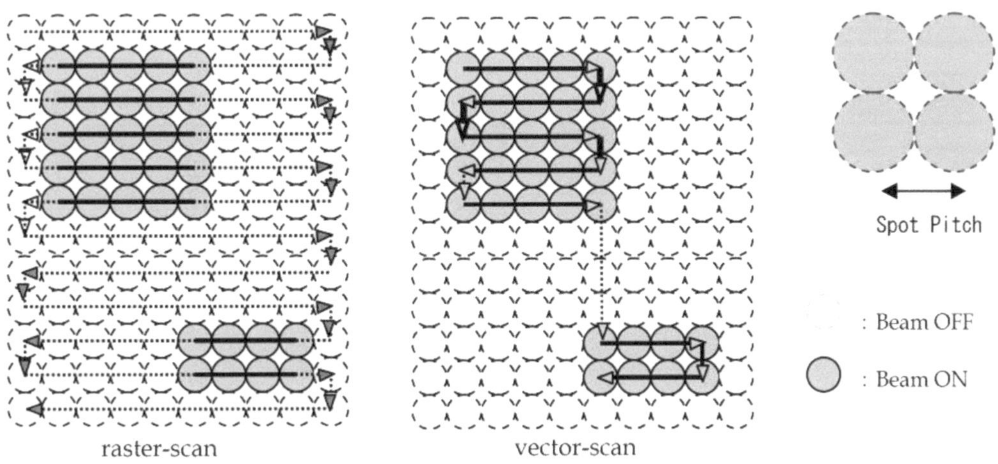

Fig. 13. A comparison of beam scanning methodologies: raster scan (left) and vector scan (right)

5. Proximity effects and energy intensity distribution

Resolution in optical lithography is limited by the wavelength of light. Since the wavelength of an electron beam is much shorter, this is not an issue in the case of electron-beam lithography. When the electron beam bombards the resist, many of the electrons experience small-angle forward scattering, which tends to broaden the primary beam size. As the electrons penetrate the resist and enter the substrate some of them undergo large-angle scattering events, leading to backscattering, in which these electrons return back to the resist in places far from the spot where the primary beam entered. The forward scattering arises from electron-electron interactions, which deflect the primary electrons by a typically small angle, thus statistically broadening the e-beam in the resist (and further in the substrate). The majority of the electrons are not terminated in the resist but penetrate the substrate. These electrons can still contribute to the exposure of the resist by scattering back into the resist, causing subsequent inelastic exposure. This backscattering process originates, for example, from a collision with a heavy particle such as a substrate nucleus, leading to wide angle scattering of the electrons. The collisions take place in an area several micrometers from the substrate surface. This causes additional, usually undesirable, exposure in the resist and is known as the electron beam lithography proximity effect (McCord & Rooks, 1997). The proximity effect turns out to be the most significant factor limiting the linewidth of electron-beam lithography -- it broadens the effective area of exposure in the resist, thus increasing the minimum attainable feature size. In terms of this, high-energy electron beams helps to minimize the effect. The electron scattering trajectories are calculated using Monte Carlo simulation and the results are presented in Figure 14. It shows trajectories in the resist

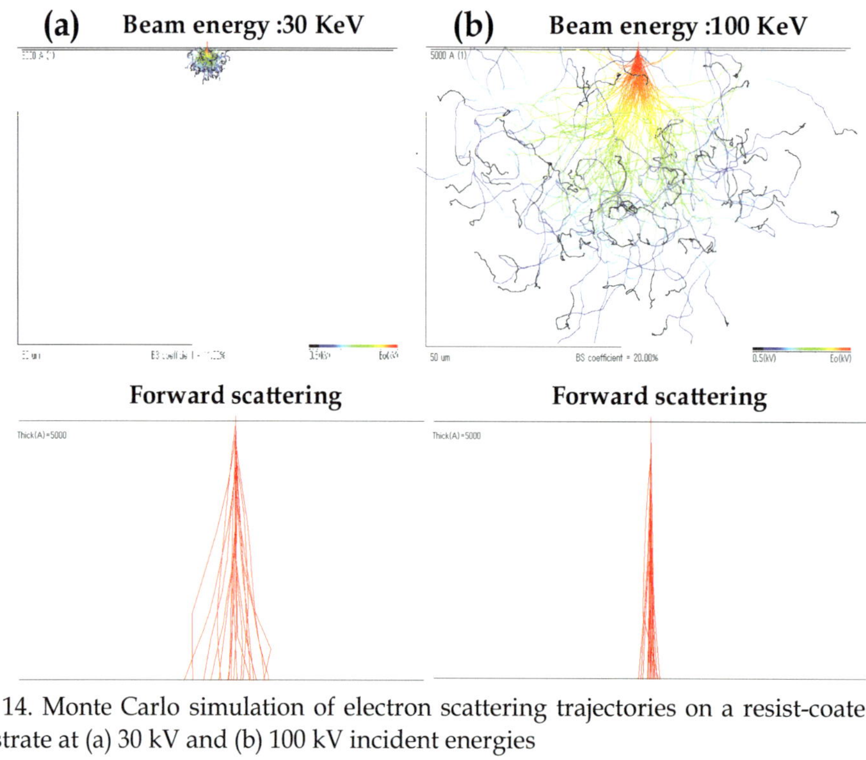

Fig. 14. Monte Carlo simulation of electron scattering trajectories on a resist-coated silicon substrate at (a) 30 kV and (b) 100 kV incident energies

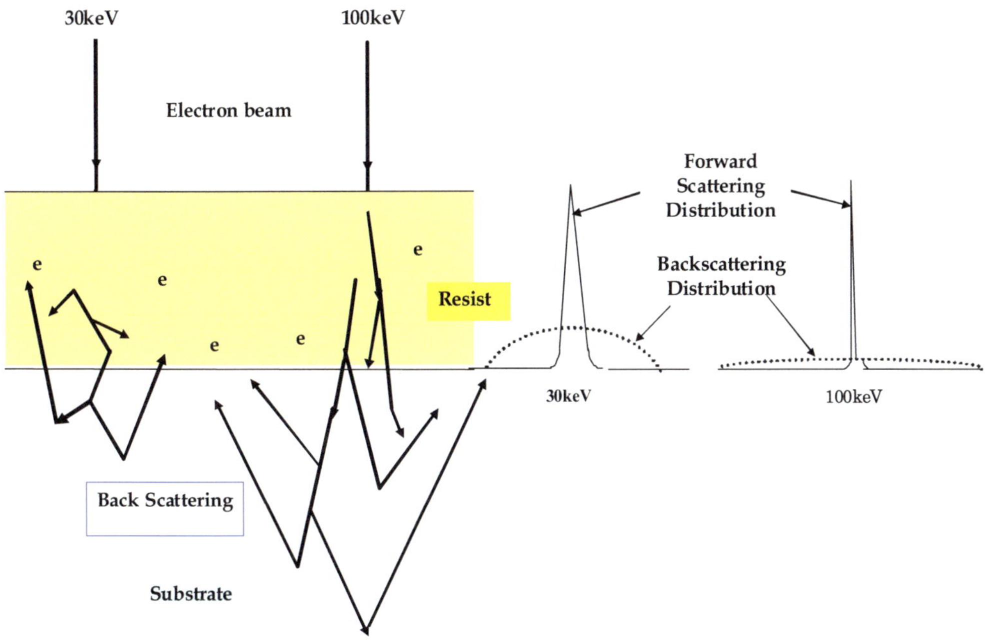

Fig. 15. A schematic drawing showing the electron scattering behavior for electron beam energies of 30 keV and 100 keV on a resist-coated silicon substrate. The exposure is made by the electrons that are scattered from the substrate back to the resist.

and in the silicon substrate for electron beam energies of 30 keV and 100 keV. As the beam energy increases, the forward scattering is reduced and the back scattering area gets deeper and wider, as shown in Figure 15. We note that both forward scattering and backword scattering distributions are of a Gaussian shape. The energy intensity distribution (EID) is often approximated by the sum of two Gaussian distributions:

$$EID(r) = \frac{1}{1+\eta}\left(\frac{1}{\pi\sigma_f^2}\exp(-\frac{r^2}{\sigma_f^2}) + \frac{\eta}{\pi\sigma_b^2}\exp(-\frac{r^2}{\sigma_b^2})\right) \tag{1}$$

where η is the ratio of the backscattered energy to the forward scattered energy, σ_f is the forward scattering range parameter which is associated with the exposure resolution, and σ_b is the backscattering range parameter which affects the proximity effect. Although the backscattering is widely spread for a 100keV e-beam, the electrons (called secondary electrons) that return to the resist from the substrate possess energies that are lower than the threshold for breaking the polymers in the resist, and do not make effective exposures. The exposure is mainly contributed by the primary electron, which produces a much narrower exposure width.

The proximity correction algorithm is another approach for minimizing the unwanted stray exposure, which fine tunes the exposure dose at the edges so as to reveal the designed pattern. For realization of this approach and successful application of the technique to EBL a thorough understanding of electron scattering in solids as well as precise data for spatial distribution of absorbed electron energy intensity in the resist are needed. A proximity effect correction algorithm (T.H.P. Chang, 1975) requires an accurate knowledge of the energy intensity distribution profile in the resist layer caused by a point spread. For a planar and homogeneous substrate, an important property of these profiles is that the shape is independent of dose and position. As shown in Figure 16, thin resist films with a low electron density will reduce forward scattering and thus the achievable feature size. Similarly, thin membrane substrates lower the amount of backscattering and help to reduce the exposure area. In addition, substrates made of light nuclei will reduce backscattering. On substrates with 'heavy' films, such as gold coatings, electron backscattering increases significantly, but the details also depend on the substrate's thickness.

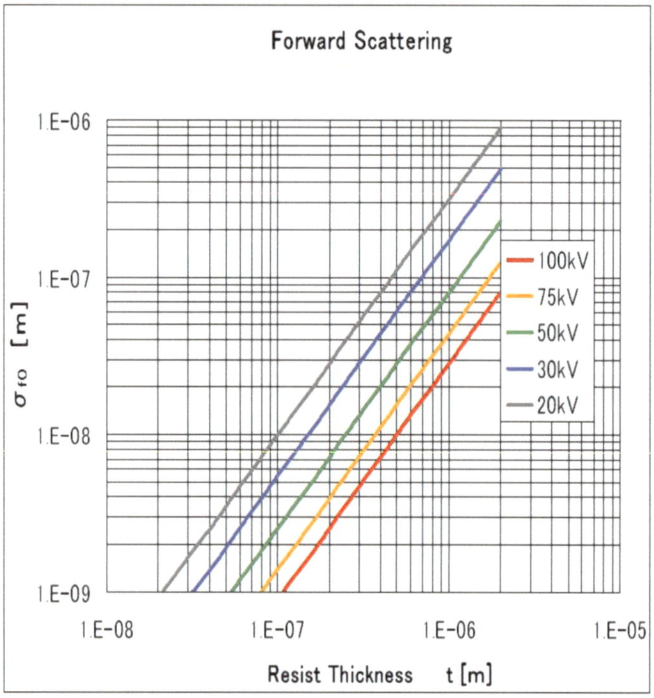

Fig. 16. Forward scattering range parameter as a function of resist thickness at different acceleration voltages.

6. Resist profile reengineering

The lift-off process is one of the most widely used methods for transferring patterns from the resist layers to the material to be structured. A sufficiently large undercut resist profile is a prerequisite for clear separation between the evaporated films on the resist and on the substrate. In view of this, high-energy electron beam exposure is not advantageous as it produces a straight resist profile due to dominating forward electron beam exposure. Here we show that this problem can be eliminated in the PMMA/LOR resist system. When the lift-off resist (LOR), which is based on a Polymethylglutarimide (PMGI) platform, became available, it immediately attracted increasing attention because it is dissolvable in alkali chemicals such as CD26 (from Shipley) but inert in most organic solvents and acids. Further, it is insensitive to electron beam exposure. Since PMMA is inert to CD26, the combined PMMA/LOR by-layer resist is an ideal system for creating a clear undercut for the lift-off process (Y. chen et al., 2002). In addition, the size of the undercut in the LOR layer can be independently controlled by the development time. Shown in Figures 17 (a-c) are the profiles of the PMMA/LOR bi-layer resist with LOR resist development times of 25 sec, 35 sec and 45 sec, respectively. Note that while the undercut size changes from 100 nm to 325 nm, the linewidth on the PMMA layer remains at 100 nm. The dependence of the undercut size on the development time is depicted in Figure 18. Because of the high selectivity in the developer, the linewidth in the PMMA layer can be very small even for a long LOR development time. This is best seen in Figure 19 where the undercut is as large as 430 nm, but the linewidth in the PMMA trench is only 13.5 nm. The inset shows 18 nm-wide

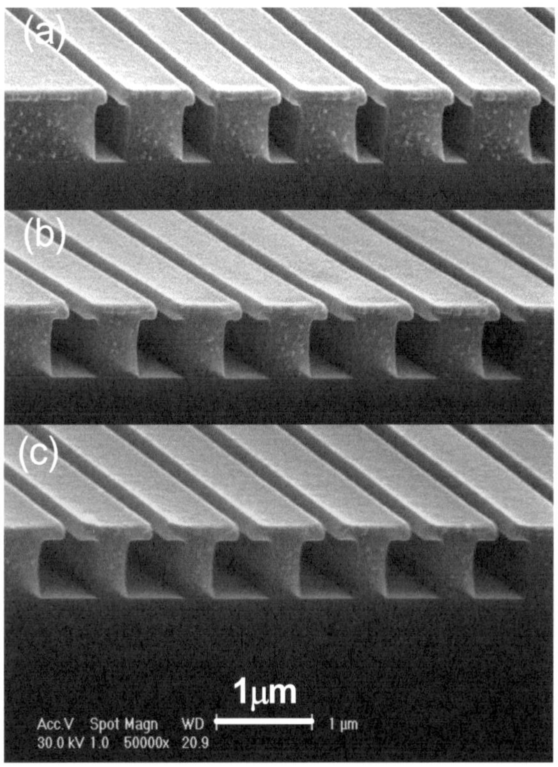

Fig. 17. (a-c) SEM image of the PMMA/LOR bi-layer resist profiles with LOR resist development times of 25 sec, 35sec and 45sec, respectively.

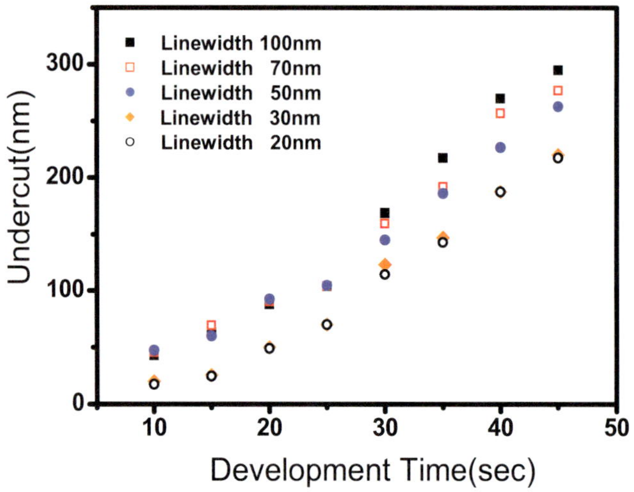

Fig. 18. The dependence of undercut size on the development time for different linewidths of the PMMA layer.

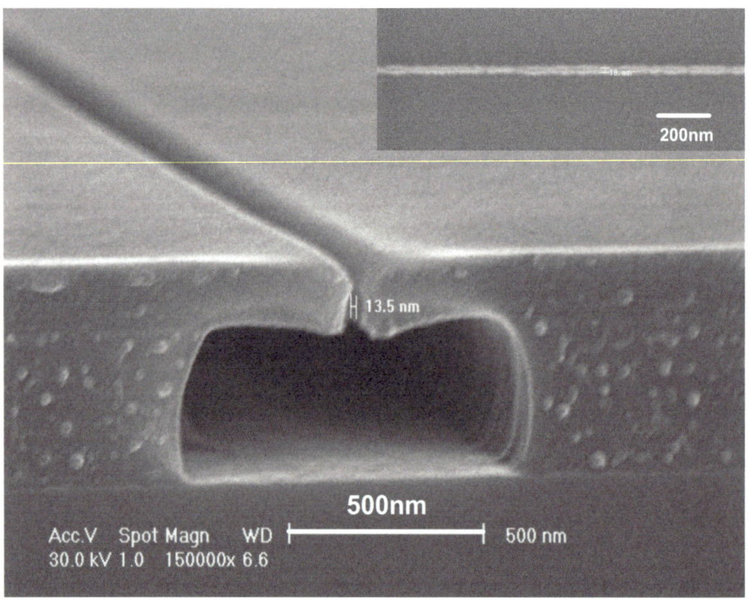

Fig. 19. SEM image of the resist profile with narrow linewidth in the top PMMA layer and a large undercut in the LOR bottom layer. The undercut is 430nm while the linewidth in the PMMA trench is 13.5nm .The inset shows 18nm-wide aluminum wire made by using this bi-layer as the lift-off mask.

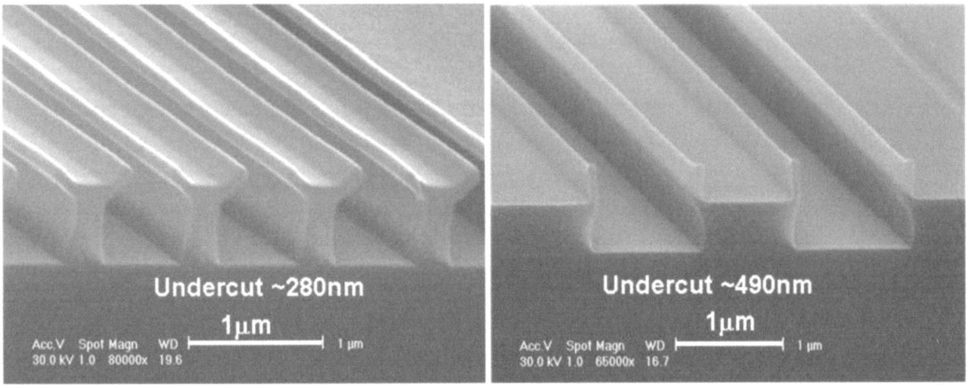

Fig. 20. SEM images of PMMA/LOR bi-layer resist. The larger the undercut in the LOR bottom layer, the more the bend in the PMMA top layer.

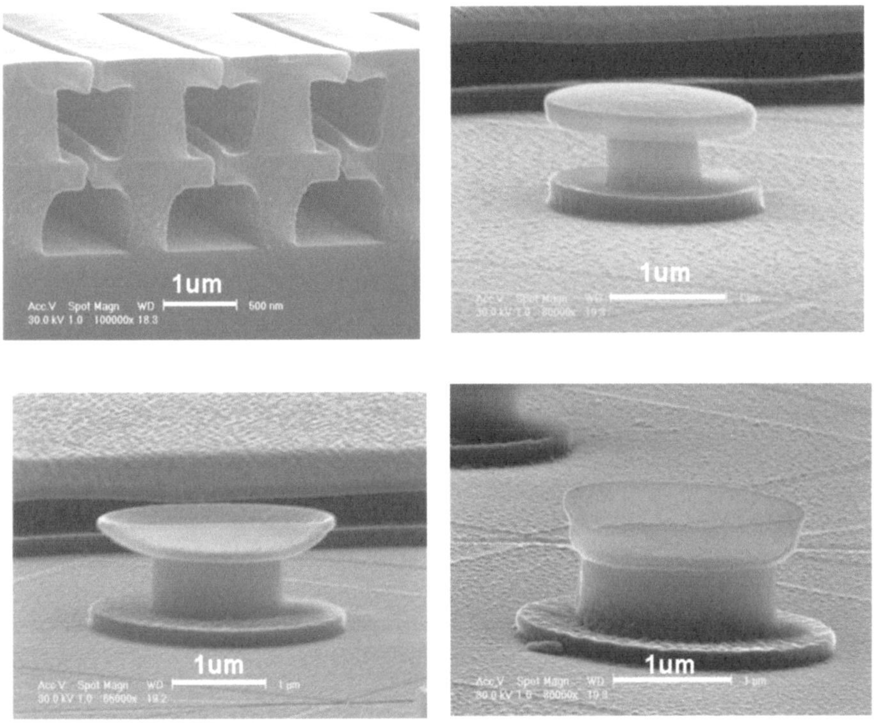

Fig. 21. SEM images of various types of 3D nanostructures made using the layer–by-layer spin-coating technique.

aluminum wire made by using this bi-layer as the lift-off mask. This resist system can also be used as a hanging-bridge mask in the two-angle evaporation process. In addition, due to tension in the PMMA, suspended PMMA usually bends. As the LOR undercut increases, so the PMMA becomes more pliable, as shown in Figure 20. Furthermore, with long penetration depth in the resists provided by a high-energy electron beam and the large contrast in exposure sensitivity between PMMA and LOR, we were able to make multilayer structures. This approach fully utilizes the self-alignment feature and requires only a single lithography step. Figure 21 shows different types of three-dimensional structures made using the layer-by-layer spin-coating technique.

7. Mix-and-match technique

Mix-and-match lithography continues to gain acceptance as a valuable strategy for reducing capital costs and increasing throughput productivity in semiconductor manufacturing (see Figure 22). This is a technique for drawing an EBL pattern overlapping the photo-lithographically-made pattern already drawn on the specimen. The successful implementation of mix-and-match lithography requires consideration of the unique characteristics of both systems. The main issues include design of the alignment marks and alignment strategy. In the case shown in Figure 22, crisscross Au alignment marks were made with optical lithography at the beginning of the process flow. Since the alignment marks are to be covered by an e-beam resist for the following EBL, they should be thick enough to give off a clear secondary electron image. By the same token, high-atomic weight metals such as gold (Au) are preferred as an alignment mark metal. For high energy EBL,

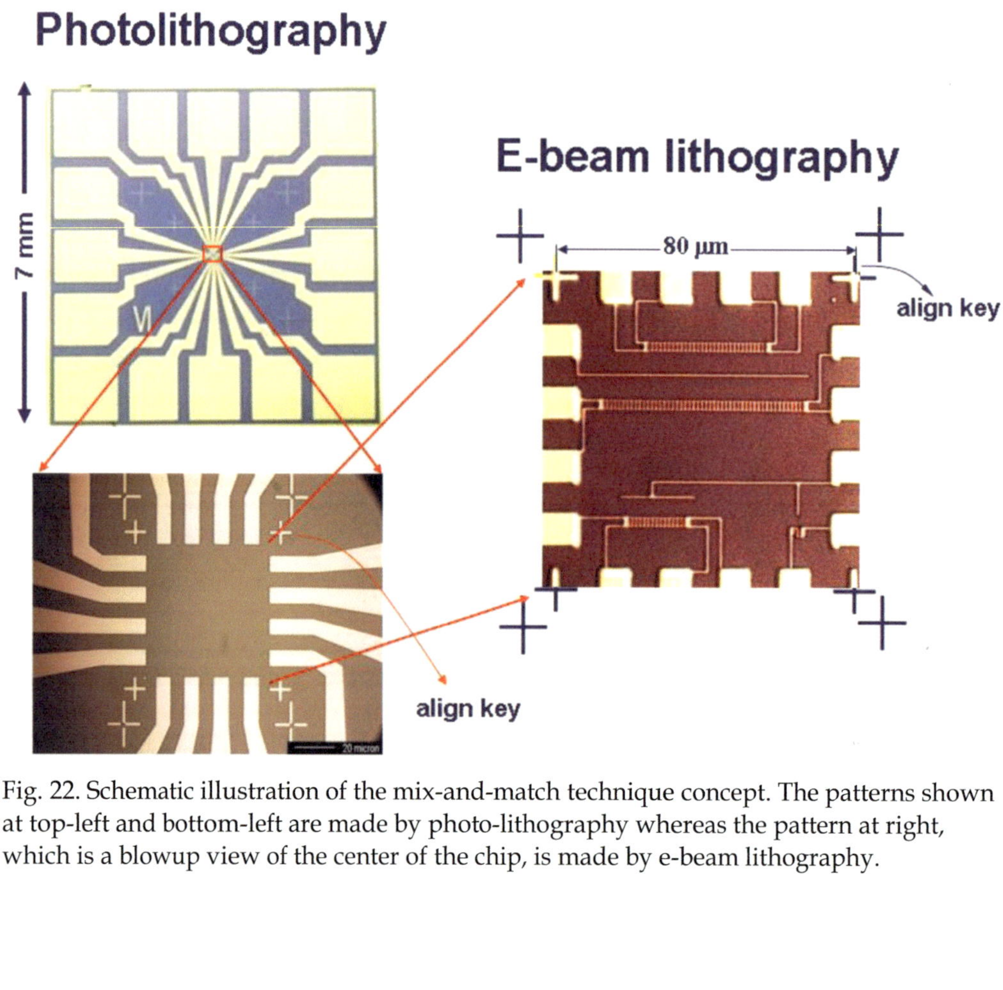

Fig. 22. Schematic illustration of the mix-and-match technique concept. The patterns shown at top-left and bottom-left are made by photo-lithography whereas the pattern at right, which is a blowup view of the center of the chip, is made by e-beam lithography.

the secondary electron signal is relatively weak so thicker Au align marks are usually required. In cases where metal alignment marks are not allowed, such as in the fabrication of semiconducting devices in which metal contamination is a concern, alignment marks in the form of deep trenches on the semiconductor substrate can be considered. To correct the rotational displacement, a set of alignment marks (called chip marks) on the specimen and at least an alignment mark on the stage (called a global mark, see Figure 23) are required.

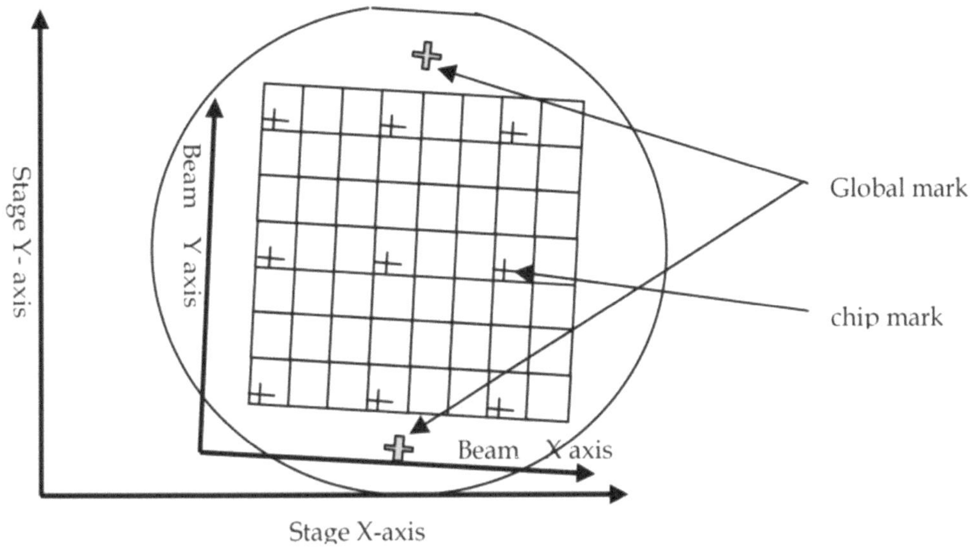

Fig. 23. Global marks and chip marks used in the mix-and-match technique. The global marks are used for acquiring the stage coordinates. To correct the rotational displacement, a set of chip marks on the specimen and at least a global mark on the stage are required.

The global marks are used for acquiring information about the stage coordinates. In this process, the system is operated in the SEM (scanning electron microscope) mode so that the secondary electron images of the marks can be captured and registered. If more than two alignment marks on the specimen are registered, both positions and direction of the specimen within the X-Y field will be calculated. With the information of all alignment masks on both specimen and stage, the coordinates of the specimen can then be re-calculated and aligned in respect to the electro-optics of the EBL system. Figure 24 shows an example of the overlay lithography. In this example, the first set of the Au line-array was made by EBL exposure followed by a lift-off process and the second set was made by the same method but over laying the first one with a high degree of accuracy.

8. Examples of nanodevices

The fabrication processes for making several nanodevices are presented in this section to illustrate the different techniques and applications involved.

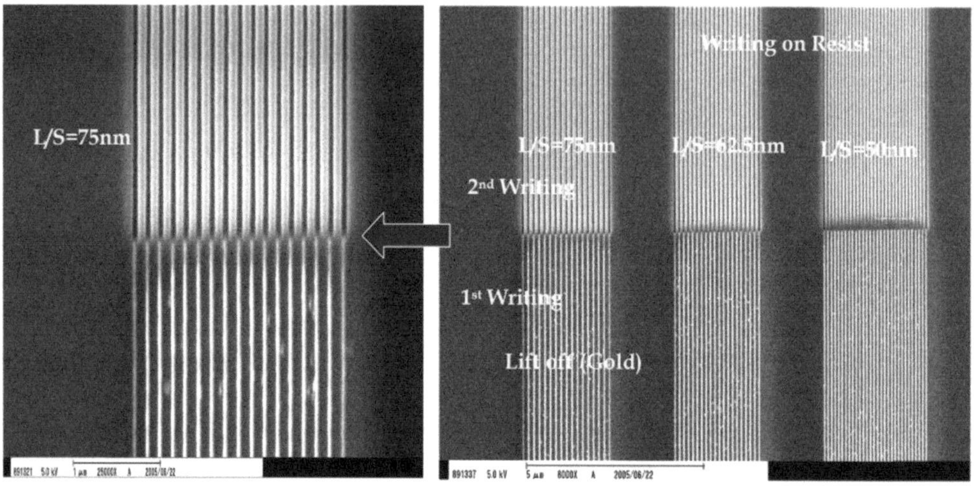

Fig. 24. An example illustrating overlay accuracy. The first set of the Au line-array (bottom array) was made by EBL and the second set of line array patterns was exposed and aligned to the first one with high accuracy.

1. Quasi three-dimensional suspended polymer photonic crystals

Two dimensional (2D) photonic crystals, which can manipulate the propagation of in-plane light at telecommunication wavelengths, is nowadays an area that receives much attention (S. Noda et al., 2000). This device consists of a 2D array of air holes embedded in a high dielectric slab constructed on top of a low dielectric substrate and can be made using present-day semiconductor fabrication technologies. A polymer based photonic crystal with a photonic band gap is a good candidate for optical integrated circuits because of its low propagation loss at telecommunication wavelengths and its compatibility with optical fibers. Polymethylmethacrylate (PMMA) is a widely used polymer material because it is a standard resist for e-beam lithography and can also be used in nanoimprint lithography (Clivia M Sotomayor Torres , 2004) owing to its low glass transition temperature (T_g=105°C). For photonic crystal application, the PMMA layer should have good verticality, a large aspect ratio, and high dimensional accuracy. For fabrication of polymer suspended photonic crystal structures with sub-100 nm resolution, the undercut of the sacrificial layer should be accurately controlled. To this end, a PMMA/LOR bi-layer polymer system is employed because the lift-off resist (LOR) is a polydimethylglutarimide-based polymer and has a much higher charge-sensitivity than that of the top PMMA layer (C. S. Wu et al., 2007). With a prolonged development time, the LOR layer can be removed completely and the top PMMA becomes suspended. Figure 25 shows an SEM image of a suspended polymer photonic crystal slab with a hexagon array of air holes in which the LOR layer serves as a sacrificial layer. Another important feature of the bi-layer polymers process is that the suspended structures can be made on virtually any kind of substrate, which is not possible in the case of semiconductor-based quasi-3D photonic crystals.

Fig. 25. SEM image of a suspended PMMA quasi-3D PC slab with a hexagonal array of air holes. The hole radius is 260nm, and the lattice constant is 800nm. The inset shows top view of a PMMA photonic crystal slab.

2. Three-dimensional polymer photonic crystals

Although quasi-3D photonic crystal slabs are useful for specific applications, they have limited potential in applications requiring confinement in all directions, which can only be achieved with 3D photonic crystals. Fabrication of 3D photonic crystals for short wavelength applications presents a great challenge, even for present-day semiconductor fabrication technology, as it requires definition and alignment of submicron features in the vertical. Multilayer structures can be made by using long penetration depth in resists provided by the high energy electron beam and the large contrast in exposure sensitivity between PMMA and LOR. This approach fully utilizes the self-alignment feature and requires only a single lithography step. Figures 26 (a) and (b) show respectively the cross-section view of a ten-layer line-array structure and the top view of a ten-layer pillar structure made using the layer–by-layer spin-coating technique. It is worth noting that due to electron beam forward scattering in the thick polymer the diameter of the LOR posts at the bottom is somewhat narrower than for those at the top. Since the top few layers have been developed nine times more than those of the lower layers, the small variation in feature sizes also suggests that the process window is wide regarding the developing time. The multilayer pillar structure shown in Figure 26 (b) has tetragonal lattice symmetry with a basis consisting of a square. The multilayer pillar structure shown in Figure 26 (b) has tetragonal lattice symmetry with a basis consisting of a square "roof" supported by a square "pole", and a photonic crystal band gap is possible. Furthermore, inverted 3D pillars made of high-reflective-index dielectric materials would help in enlarging the local band gaps, making an overlap between gaps more possible. The 3D pillars can be constructed by filling (e.g. electroplating [Y. W. Su et al., 2003]) the air gap using the multilayer polymer structure as a mold.

3. Magnetic Single-electron transistors

Electronic transport in nonmagnetic (NM) tunnel junctions has been extensively studied in the past two decades. However, it was only a few years ago that attention was drawn to electron tunneling in magnetic systems (J. S. Moodera et al., 1995; H. Imamura et al., 2000), which became possible owing to recent progress in nanotechnology. Magnetic multi-

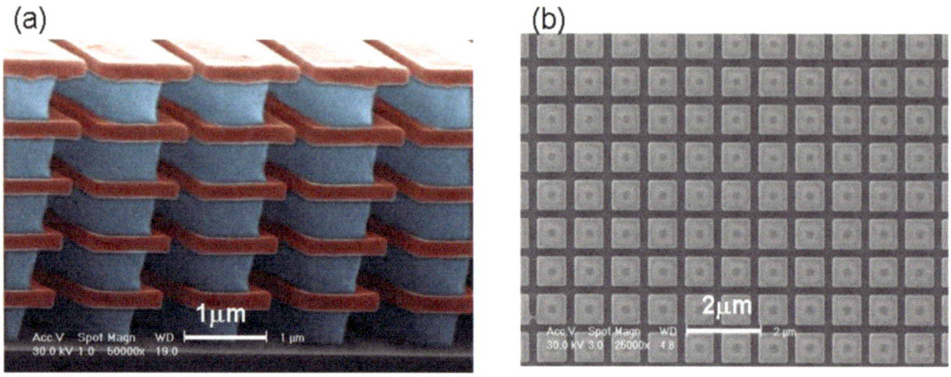

Fig. 26. (a) Cross-section view of a 10-layer line-array structure (b) Top view of a 3D PMMA/LOR pillar structure. The square PMMA roofs are 770nm on each side and 200nm in thickness. The LOR pillars are 550nm on each side and 500nm in height. The inter-roof distance is 230nm and the lattice constant is 1μm.

junctions seem to be of great importance for future electronics based on spin-dependent electronic transport (spintronics) and for information storage technology. Suspended shadow-mask evaporation is a simple, robust technique for fabricating tunnel junction device structures using electron-beam lithography. The basic process entails the fabrication of an undercut structure in a resist bilayer to form a suspended "bridge" followed by two angle evaporations of aluminum material with a brief oxidation step in between. The result is two overlapping wires separated by a thin layer of oxide. In the shadow evaporation, to prevent shorts between the source and the drain electrodes, the minimum possible length of the island electrode is limited by the available minimum line separation in the hanging-bridge mask, which is most likely not less than 50 nm due to the proximity effect. However, high-energy EBL can minimize the proximity effect due to a wide secondary electron spread in the substrate, which helps to reduce pattern broadening. We have demonstrated in Figs. 18 and 19 the high controllability of the undercut profile in the lift-off process using the PMMA/LOR bi-layer resist. By tuning these processing parameters, an undercut length of between 20 nm and 300 nm can be precisely controlled, which paves the way for the reliable lift-off processes for sub-50 nm patterning. For instance, metallic magnetic single electron transistors can be fabricated using a PMMA/LOR resist together with the four angle shadow evaporation technique (as shown in Figure 27). With four evaporations at different tilt angles, one can obtain four patterned structures shifted in respect to each other. As the tilt angles can be controlled to a high degree of precision, the four structures can be shifted to desired locations to form small overlapping regions, which act as source and drain tunnel junctions.

4. Ultra-small Silicon quantum dots

Quantum-dot single electron transistors (QD-SETs) provide an interesting playground for unraveling the rich physics that exists on a mesoscopic scale (D. Heiss, 2005). Semiconductor quantum dots nicknamed 'artificial atoms' provide a discrete energy level spectrum (M. Bayer et al., 2000). When incorporated with charging effect in the form of QD-SETs, they are capable of probing electron dynamics through quantized levels, and are all the more exciting and fascinating. The device was fabricated on a 220 nm thick top silicon layer of a

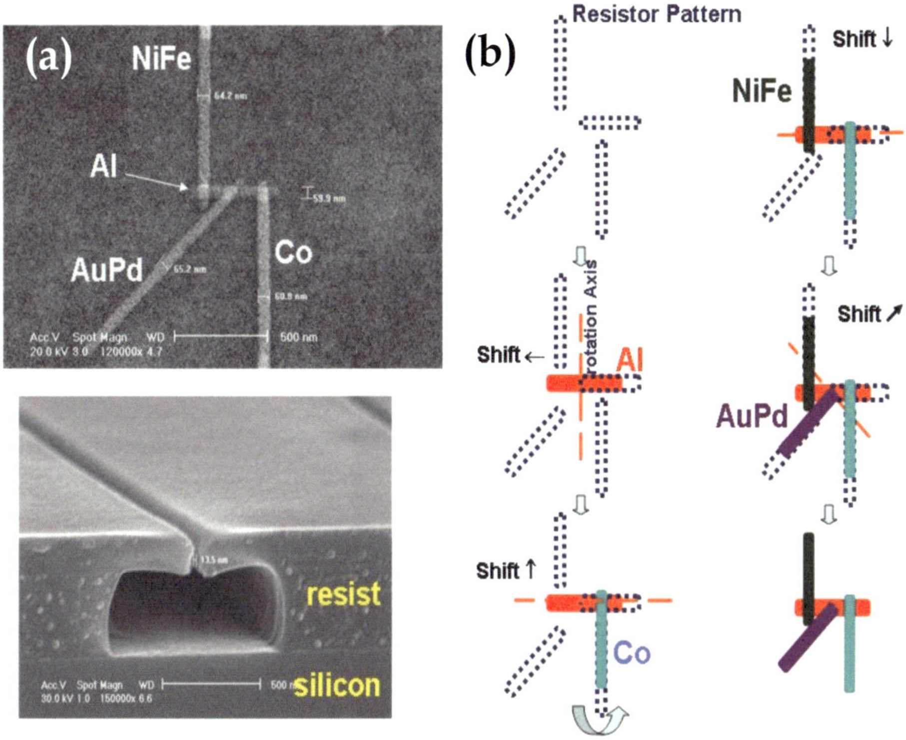

Fig. 27. (a) SEM image of a ferromagnetic single-electron transistor. (b) Schematic of the four-angle shadow evaporation steps; Al (red), Co (cyan), NiFe (black) and AuPd (purple) are subsequently deposited.

Silicon-On-Insulator (SOI) wafer with a 400 nm thick buried silicon-dioxide layer. In order to fabricate narrow constrictions, the top silicon layer was thinned down to 60nm using thermal oxidation, followed by etching in buffered HF solution. Photolithography was used to define a micrometer scale electrode pattern together with the positioning markers needed for the top silicon layer. Inductivity-coupled-plasma etching (ICP) with SF_6+C_4F_8 and 3:5 mixture gases, was then employed to etch the unprotected silicon and subsequently transfer the electrode patterns. This was followed by a second photolithography procedure to define the Ohmic contact pads. The island and constrictions (see Figure 28) were then defined and patterned by high-energy EBL and ICP etching, respectively. After the formation of the device structure, dry thermal pattern-dependent-oxidation at 900°C for 1hr was used to further narrow down the constrictions (K. Yano et al., 1994) between the island and the source/drain electrodes to ~5nm (M. C. Lin et al., 2007). The oxidized silicon formed in the process on the surface of the device was then chemically etched away using buffered HF solution and, as a result, nano-constrictions and the large island were floated. The elimination of the underlying SiO_2 layer helped to increase charging energy and to deduce possible random offset charges embedded in the substrates. Finally, Ohmic contacts (Ni/Au, 150nm/400nm) were patterned by the lift-off technique and served as the source and drain leads for electron transport measurements.

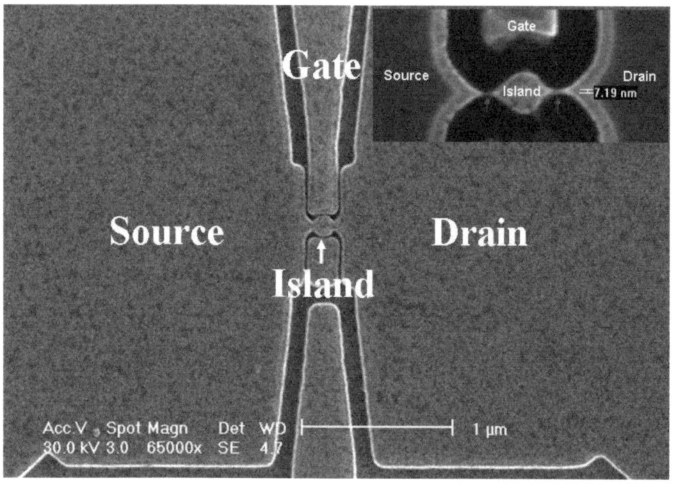

Fig. 28. SEM image of a silicon quantum-dot single-electron-transistor before removal of SiO_2. Inset shows a blowup of the center region after removal of SiO_2. The device contains a big island connected to leads via small dots present in the nano-constrictions (indicated by red arrows).

5. Hard x-ray Fresnel zone plates

Synchrotron radiation sources are the preferred light sources for X-ray microscopes. The intensity of the synchrotron X-ray is a million times higher than the conventional X-ray. High intensity X-ray gives great transmission efficiency and can shorten the exposure time. The transmission X-ray microscope (TXM) is a valuable tool in research fields such as physics, materials science and biology. The Fresnel zone plate (FZP) is used to focus the X-ray and is an essential component for determining the resolution of TXM. In hard-X-ray imaging, the high X-ray photons could penetrate thicker metals and FZP with thick high-Z metals is needed to impede these photons. Fabrication of large-area, fine metal structures with high aspect ratios presents a great challenge to EBL. For example, for a 30nm image resolution with multi-keV X-rays, the Au FZP must be at least 450nm thick and the aspect ratio should be higher than 12. The high-energy electron beam direct writer can define patterns with high resolution (< 10 nm) and small stitching error (< 20 nm). Moreover, development of such a pattern is a crucial step. The FZP patterns require a thick resist structure where the feature size is smaller than 50nm. However, this cannot be achieved by using commonly used developers such as a 1:3 ratio mixture of MIBK (methylisobutylketone) and IPA (2-propanol). It is shown (Y. T. Chen et al., 2008) that a specific mixing IPA and water can produce sufficient contrast and precision for high-aspect-ratio FZP patterns. The final metal nanostructure was obtained by gold electro-deposition in the open trenches of the resist pattern. Details of the electrodeposition procedure—similar to an LECD (localized electrochemical deposition) approach—has been reported elsewhere (G. C. Yin et al., 2006). The period of zones is 60 nm, including a 30 nm outermost zone width, and the thickness of FZP is 450 nm (see Figure 29). FZP devices were employed to image the Siemens Star pattern, and 20 nm-resolution was demonstrated by TXM at 8 keV X-ray (Y.T Chen et al., 2008).

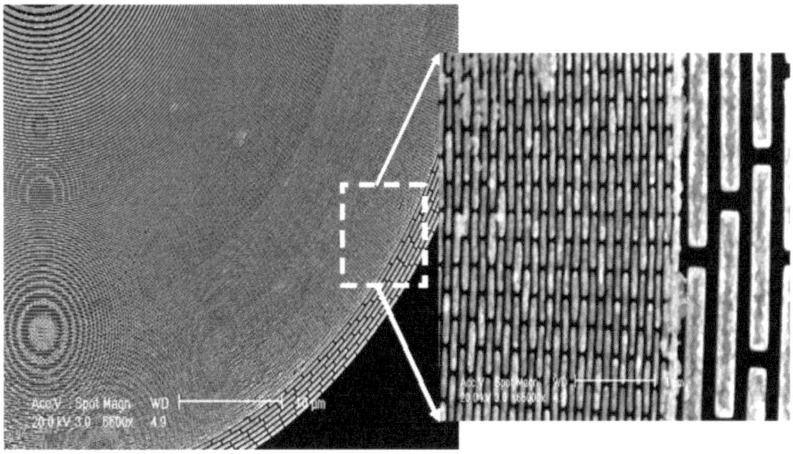

Fig. 29. SEM image of a 450nm-thick Au FZP with 30 nm outermost zone. The pattern is generated by using an Elionix ELS-7000 system operating at 100 keV and 10 pA.

9. Conclusion

High-energy electron beam exposure tools possess several key features required for high fidelity lithography including high flexibility, high resolution, accurate linewidth control and pattern overlay. Using a vector-scanning electron beam, it requires no mask and has the ability to write a variety of pattern geometries giving it significant advantage over other lithographic techniques. However, the electron beam exposure is an ingredient of EBL in which the resist treatment and development process are equally important. EBL is inherently slow compared to photolithography due to the serial pattern writing method used. However, thanks to advances in electron beam lithography technology, such as variable shaped electron beams and the use of microcolumn arrays (E. Yin et al., 2000), throughput can be significantly improved. Nevertheless, the direct writing EBL has been the most flexible system for making a variety of nanodevices and prototypes with critical dimensions below 10 nm. The direct writing approach will continue to play a major role in nanofabrication and remain the de facto technique for making masks for other advanced lithographic processes.

Special thanks are due to Ming-Chou Lin, Jia-Hong Shyu, Fou-E Tang, and Teik-Hui Lee for providing SEM images of their samples. Technical support from NanoCore, the Core Facilities for Nanoscience and Nanotechnology at Academia Sinica, is acknowledged.

10. References

C. S. Wu, C.F. Lin, H.Y. Lin, C. L. Lee & C. D. Chen. (2007). Polymer-based photonic crystals fabricated with single-step electron beam lithography, Advanced Materials, Vol. 19, pp.3052-3056

Clivia M Sotomayor Torres edited. (2004). Alternate Lithography: Unleashing the Potentials of Nanotechnology, Kluwer Academic, Plenum publishers, New York

C. W. Gwyn, R. Stulen, D. Sweeney, & D. Attwood. (1998) Extreme ultraviolet lithography, J. Vac. Sci. Technol. B, Vol. 16, pp. 3142-3149

D. Heiss. (2005). *Quantum Dots: Doorway to Nanoscale physics, Lecture Notes in Physics*, Vol. 667, pp. 97–130, Springer, Berlin

E. Yin, A. D. Brodie, F. C. Tsai, G. X. Guo, &N. W. Parker.(2000). Electron optical column for a multicolumn, multibeam direct-write electron beam lithography system, J. Vac. Sci. Technol. B, Vol. 18, pp.3126-3131

GC Yin, YF Song, MTTang, F R Chen, K S Liang, F W Duewer, M Feser, W Yun & H P D Shieh .(2006). 30 nm resolution x-ray imaging at 8 keV using third order diffraction of a zone plate lens objective in a transmission microscope, Appl. Phys. Lett. Vol.89, pp. 221122

H. Imamura, J. Chiba, S. Mitani, K. Takanashi, S. Takahashi, S. Maekawa, &H. Fujimori.(2000). Coulomb staircase in STM current through granular films, Phys. Rev. B, Vol.61, pp.46-49

J. Yamamoto, F. Murai, Y. Someda, & S. Uchino. (2000). Fine Pattern Fabrication below 100 nm with 70 kV Cell Projection Electron Beam Lithography, Jpn. J. Appl. Phys., Vol. 39, pp. 6854-6860

J. S. Moodera, L. R. Kinder, T. M. Wong, & R. Meservey.(1995). Large Magnetoresistance at Room Temperature in Ferromagnetic Thin Film Tunnel Junctions, Phys. Rev. Lett. , Vol. 74, pp. 3273-3276

K. Yano, T. Ishii, T. Hashimoto, T. Kobayashi, F. Murai,& K. Seki.(1994). Room-temperature single-electron memory ,IEEE Trans. Electron Devices, Vol. 41, 1628-1638

L. Pain, B. Icard , S. Manakli , J. Todeschini , B. Minghetti , V. Wang ,& D. Henry. (2006). Transitioning of direct e-beam write technology from research and development into production flow, Microelectronic Engineering, Vol. 83 , pp. 749–753

M. C. Lin ,K. Aravind , C. S. Wu , Y. P. Wu, C. H. Kuan , Watson Kuo & C. D. Chen.(2007). Cyclotron Localization in a sub-10nm Silicon Quantum Dot Single Electron Transistor, Applied Physics Letters, Vol. 90, 032106

M. Bayer, O. Stern, P. Hawrylak, S. Fafard, &A. Forchel. (2000). Hidden symmetries in the energy levels of excitonic artificial atoms, Nature, Vol. 405, pp.923-926

McCord M. A. & Rooks, M. J.(1997) "Electron Beam Lithography," in Handbook of Microlithography, Micromachining, and Microfabrication, Vol. 1, Ch. 2, ed. by P. Rai-Choudhury, pp. 139-249, SPIE Optical Engineering, Bellingham, WA

S.Noda, A.Chutinan & M. Imada. (2000). Trapping and emission of photons by a single defect in a photonic bandgap structure ,Nature,Vol.407,pp.608-610

S. Hector. (1998). Status and future of X-ray lithography, Microelectronic Engineering, Vol. 41/42, pp. 25-30

T.H.P. Chang. (1975), Proximity effect in electron-beam lithography, Journal of Vacuum Science Technology, Vol. 12, pp.1271-1275

Yu-Tung Chen et al. (2008). Full-field hard x-ray microscopy below 30 nm: a challenging nanofabrication achievement, Nanotechnology, Vol. 19, pp.395302 (5pp)

Y. W. Su, C. S. Wu, C. C. Chen, &C. D. Chen. (2003). Fabrication of Two-Dimensional Arrays of CdSe Pillars Using E-Beam Lithography and Electrochemical Deposition, Advanced Materials, Vol.15, 49-52

Y. Chen, D. Macintyre, E. Boyd, D. Moran, I. Thayne, & S. Thoms (2002).Fabrication of high electron mobility transistors with T-gates by nanoimprint lithography, J. Vac. Sci. Technol. B, Vol. 20, pp.2887.

Y. Lee, R. A. Gough, W. B. Kunkel, K. N. Leung, J. Vujic, M. D. Williams, D. Wutte, & N. Zahir. (1998). Multicusp sources for ion beam projection lithography, Rev. Sci. Instr. Vol. 69, pp.877-879

14

Optimal Design and Fabrication of Fine Diffractive Optical Elements Using Proximity Correction with Electron-beam Lithography

Masato Okano
NALUX Co., Ltd.
Japan

1. Introduction

In recent years, diffractive optical elements (DOEs) are widely used for applications to optical disc pickup, array generators, spectrum filtering, wavelength division multiplexing (WDM), and many others. The DOEs with small feature sizes can be fabricated by the direct-writing electron-beam lithography.(Kley, 1997) To predict the electron dosage to obtain a desired resist profile, the effects of electron scattering in a resist layer and resist characteristics of development process must be considered.(Neureuther et al., 1979) Determining the optimum electron dose is called the proximity correction.

Some methods of the proximity correction were proposed,(Maker & Muller, 1992; Daschner et al., 1995; Nikolajeff et al., 1995) which are classified according to estimation methods of resist profile. One popular method is based on a point-spread function of the electron scatter and a resist contrast curve.(Daschner et al., 1995) The point-spread function is the density distribution of absorbed energy for a focused electron-beam. The function can be assumed to be the sum of two Gaussian distributions due to forward scattering and back scattering. The absorbed energy density is expressed by a convolution of the electron dose distribution and the point-spread function. The resist profile after developing is estimated from the absorbed energy density and a resist contrast curve. The resist contrast curve shows the remaining resist thickness as a function of the absorbed energy density. The point-spread function and the resist contrast curve are determined experimentally. This estimation method is widely used because of the simple process of calculation. However the contrast curve method is not valid for short-period gratings with deep grooves, because the absorbed energy density is not uniform vertically.(Okano et al., 2000) We had reported a precise proximity correction method based on the electron-beam lithography simulator,(Hirai et al., 2000) which consists of an electron scatter simulation by the Monte Carlo method(Kyser & Murata, 1974) and a resist development simulation with the cell removal model.(Dill et al., 1975) This correction method is suitable for short-period gratings and chirped period diffraction gratings such as a diffractive lens.(Okano et al., 2003)

Diffraction efficiencies of the periodic diffraction gratings can be calculated by numerical analysis such as the rigorous coupled wave analysis (RCWA) method,(Moharam & Gaylord, 1981) the differential method,(Vincent, 1980) and the finite difference time domain (FDTD) method.(Yee, 1966) The diffraction efficiency is defined as a ratio of an incident and the diffracted beam powers. It is known that when the grating period is several times as long as the light wavelength (<5λ), the saw teeth profile is not the solution to obtain the highest efficiency of the first order diffracted light.(Shiono et al., 2002) There are some reports to calculate the grating profile to obtain the highest diffraction efficiency using the rigorous grating analysis.(Noponen et al., 1992; Sheng et al., 1997; Kallioniemi et al., 2000; Testorf & Fiddy, 2001) The diffraction efficiencies of the optimized gratings were taken very high values. Noponen et al. found the grating profile with the efficiency of 85.7% in case the diffraction grating has a period of 2.5λ.(Noponen et al., 1992) If the grating profile is the simple saw teeth, the efficiency will be ~45%. However, such the short period gratings are not easy to be fabricated with the direct-writing electron-beam lithography, because the optimized grating profiles have sharp peaks and narrow valleys. The proximity effect of electron scattering restricts the grating profiles. A small difference of the grating profile from the designed one causes a decrease of diffraction efficiency.

For chirped-period diffraction gratings, the boundary element(Hirayama et al., 1996) and FDTD(Prather et al., 2001) methods are used for calculating the diffraction efficiency. Prather et al. showed that a multilevel diffractive lens with a small *F*-number (<1.5) should be analyzed by rigorous electromagnetic wave analysis.(Prather et al., 2001) In some reports, an optimizing design for highly efficient diffractive lenses with a small *F*-number by simulated quenching(Prather et al., 1998) and Gerchberg-Saxton algorithm(Di et al., 2003) is described. Moreover, a diffractive lens consisting of a subwavelength grating structure is suggested.(Prather et al., 1998) These optical design methods are based on only an electromagnetic analysis without restrictions on fabrication limits; thus, it may be difficult to fabricate the designed grating profile, particularly for short-period gratings.

In this chapter, we demonstrate a unified design method optimizes the electron dose and grating profiles simultaneously to obtain the desired diffraction efficiency under the restriction of the proximity effect.(Okano et al., 2004; 2007) The optimization is made with the rigorous electromagnetic wave analysis and the resist development simulator. We design two cases of short periodic diffraction grating and the diffractive lens using the unified design method.

2. Optimal design method within proximity correction and grating analysis

Figure 1 shows the optimal design flow of the DOEs by the electromagnetic grating analysis and the electron-beam lithography simulation. After assuming the electron dose profile D on the resist layer, the resist profile P is calculated by electron-beam lithography simulation. Then the diffraction efficiency η_i of the resist profile P is calculated by the RCWA or the FDTD method, where the suffix i is the diffraction order. The merit function ϕ is defined as a sum of mean square of the differences (η_i-η^d_i). Here, η^d_i is a desired diffraction efficiency. If the value of merit function ϕ is larger than a certain small value $\Delta\phi$, the exposure dose profile D is modified. These calculations are iterated until the value of the merit function ϕ is less than $\Delta\phi$ or it is converged well. Consequently, we will obtain the optimized grating profile P, the diffraction efficiencies η_i and the electron dose profile D for the fabrication, simultaneously.

This optimal design method searches an optimum grating profile under the restriction of the proximity effect, that is, we find the fabricatable best solution. In the following simulation, we take unity for all of weight W_i. The modified electron-dose profile D is expressed as a polynomial function of 9th-order in each grating period. Therefore the

number of parameters to be optimized is 10. We use the simulated annealing method(Kirkpatrick et al., 1983) as an algorithm of the optimization method.
Several parameters of the electron-beam resist are required for the lithography simulation. The resist parameters must be determined from a primary experiment.(Okano et al., 2003) We used the resist parameters of OEBR-1000 (Tokyo Ohka Kogyo) for the simulation. An electron-beam resist of 1.7 μm thickness is coated on a 1.0-mm-thick silica glass plate. The acceleration voltage of the electron beam is 50 kV, and the development time is 120 s. The residual resist thickness after development is measured for different electron doses. The resist solubility rate R can be expressed as follows using Neureuther's formula(Neureuther et al., 1979):

$$R = R_0 \left(R_1 + \frac{E}{E_0} \right)^{\gamma}. \tag{1}$$

Here, E is the absorbed energy density. The parameters R_0, R_1, E_0, and γ are determined so as to satisfy the above-mentioned experimental results. The absorbed energy density is calculated by electron scatter simulation. The determined parameters R_0, R_1, E_0, and γ are 0.27 nm/s, 1.0, 1.8×10^9 J/m^3, and 4.5, respectively.

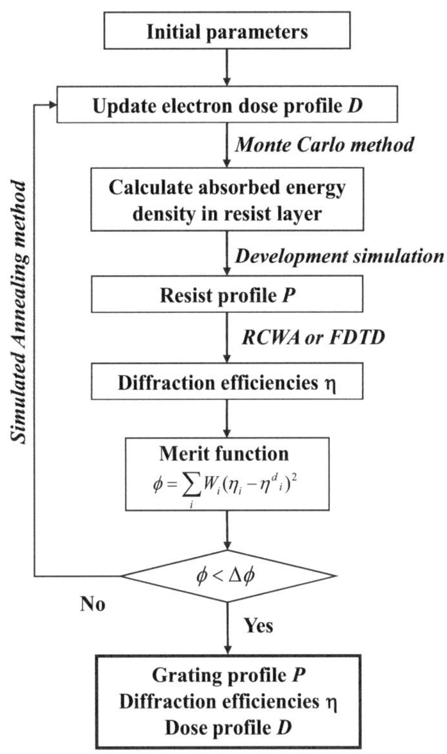

Fig. 1. Optimization flow of DOEs by the electromagnetic grating analysis and the lithography simulation. The optimal design method is called the unified method. The diffraction efficiency is determined from the electric field calculated by the RCWA or the FDTD method. The grating surface profile P and electron dose profile D are optimized simultaneously.

3. Optical design and fabrication results

3.1 Short period diffraction grating

We designed a short period diffraction grating of the electron-beam resist. The grating period is 1.4 μm. The refractive indices of the resist and substrate were 1.49 and 1.46, respectively. The light wave of 633 nm wavelength was normally incident onto the grating, and the light wave is polarized parallel to grating grooves. The resist profile was optimized to obtain the maximum diffraction efficiency for the +1st order diffraction wave. For such a short grating period, the saw-teeth blazed profile does not yield the highest diffraction efficiency.

Figure 2 (a) shows the grating profile designed with the unified method. The calculated diffraction efficiency is 67%. It took about 3 hours to optimize the dose profile by a computer with Compaq Alpha21264 (667 MHz) CPU. The number of iteration was 1000.

We have also designed the grating profile with other designed methods. Figure 2 (b) shows the grating profile optimized only with the RCWA, i.e. the optical optimization process. And the calculated diffraction efficiency is 82%. This diffraction efficiency is higher than that of the suggested optimizing method. However, the grating profile is not easy to be fabricated accurately because of sharp peaks and narrow valleys. Figure 2 (c) is the grating profile designed with the series process, that is, the profile has been optimized under the restriction of the proximity effect as the object profile is Fig. 2 (b). The diffraction efficiency is reduced to 48%, which is about 30% lower than that of Fig. 2 (b). A conventional saw-tooth blazed grating shown as Fig. 2 (d) has the diffraction efficiency of 42%. It is obvious that the grating profile designed with the optimal design method (Fig. 2 (a)) is easy to be fabricated compared with that of Fig. 2 (b). And the diffraction efficiency is superior to those of the saw-tooth blazed grating and the series method.

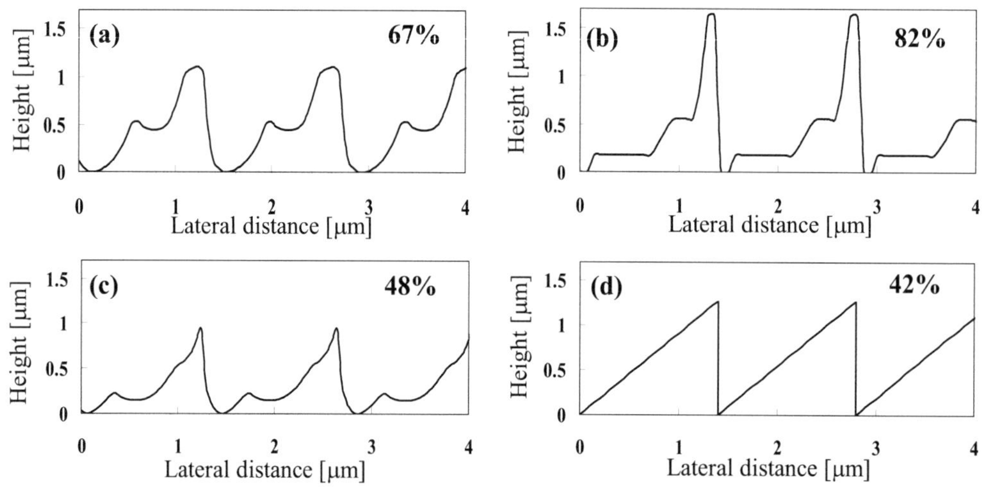

Fig. 2. Grating profiles designed with the different methods; (a) the unified method, (b) the optical optimization process, (c) the series method and (d) a saw-tooth blazed grating, respectively. All of the grating periods are 1.4 μm.

Figure 3 shows the grating profiles designed with the unified method for different grating periods, (a) 1.0 μm, (b) 1.4 μm, (c) 2.6 μm and (d) 6.5 μm, respectively. For the long period grating shown as Fig. 3 (d), the profile becomes the saw-tooth-like blazed profile. On the other hand, for short period gratings, although the optimized grating profile has the round valleys, the diffraction efficiencies are higher than those of the saw-tooth blazed gratings.

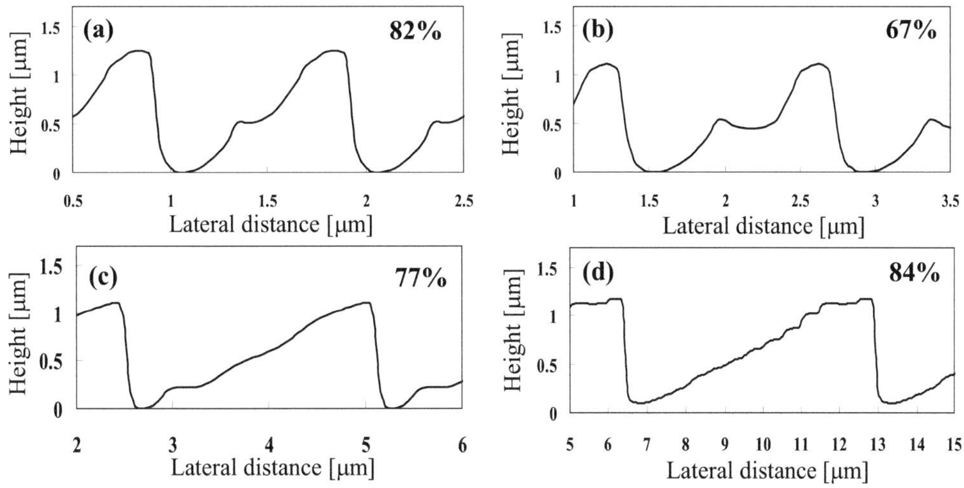

Fig. 3. Grating profiles calculated with the unified method for various grating periods. The grating periods are (a) 1.0 μm, (b) 1.4 μm, (c) 2.6 μm and (d) 6.5 μm, respectively.

The diffraction gratings designed in the previous descriptons have been fabricated with the direct-writing electron-beam lithography. The grating periods were 1.0 μm and 1.4 μm, that is, the grating profiles were Fig. 3 (a) and (b), respectively. We used the electron-beam writer JEOL JBX-5000SI with the acceleration voltage of 50 kV and the beam diameter of 20 nm. The resist thickness was 1.7 μm and the development time was 120 s. The silica glass substrate was 1.0 mm in thickness.

Figure 4 (a) and (d) show the electron-dose profiles calculated with the unified method for the grating periods of 1.0 μm and 1.4 μm, respectively. Though the difference of periods is 0.4 μm, the estimated dose profiles are so different from each other. In the actual fabrication, the electron dosages were according to the calculated dose profiles. The scanning electron microscope (SEM) pictures of the fabricated resist gratings are shown in Fig. 4 (c) and (f). Their surface profiles measured with the atomic force microscopy (AFM) are shown in Fig. 4 (b) and (e), respectively. The grating profiles calculated with the unified method are shown in the figures, too. The grating heights of the AFM profiles are agreed well with those of the calculated profiles. The maximum differences of the profiles are ~50 nm. The measured diffraction efficiency was 70% for the 1.0 μm-period grating, and 58% for the 1.4 μm-period grating. These efficiencies are lower by ~10% than the calculated efficiency.

The decrease in diffraction efficiencies is caused by small difference of the grating profiles. In a case of the 1.4 μm-period grating, the diffraction efficiency calculated from the AFM profiles is lower by ~4% than the designed efficiency. The other reason is the stitching error of subfields in the electron-beam writer. The stitching error decreases by a few percents of the diffraction efficiency. To fabricate the grating profile accurately, the resist parameters must be determined more precisely. And the development process also has to be controlled carefully. The stitching error will be reduced by tuning the electron-beam writer.

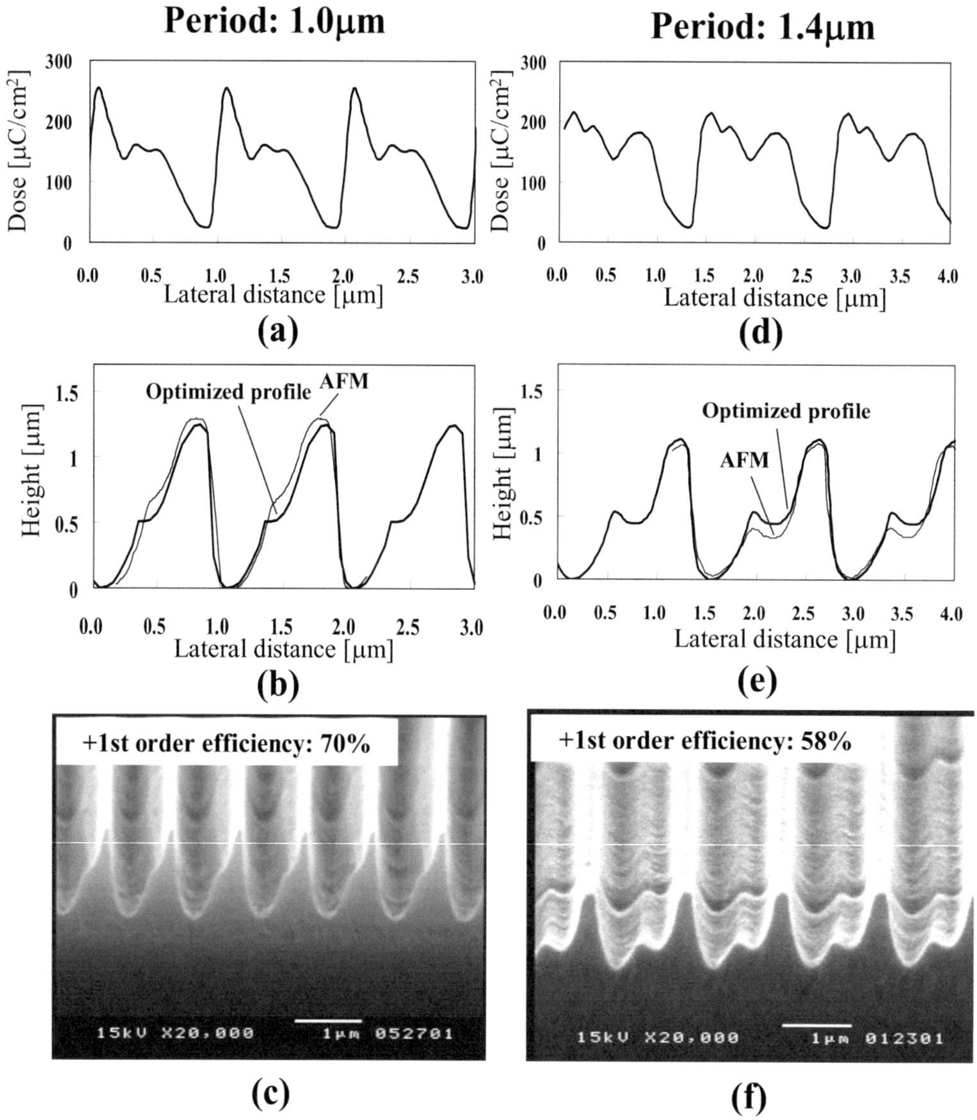

Fig. 4. Fabricated grating profiles. (a) the dose profile calculated with the unified method for the 1.0 μm-period grating. (b) the grating profile measured with AFM and the profile designed with the unified method. (c) the SEM picture of the fabricated grating. (d), (e) and (f) are results for the 1.4 μm-period grating, respectively.

3.2 Diffractive lens

In this section, we designed the diffractive cylindrical lens of an electron-beam resist using the unified method. The diffraction pattern on the focal plane is calculated from the resist profile by the FDTD method. Figure 5 shows the calculation model of the diffractive lens. (x, y) are the coordinates of the field area. To reduce the calculation time and computer memory, the lens focal plane ($y=y_1$) is not included in the calculation area of the FDTD method. The FDTD method yields the complex amplitude $u(x, y_0)$ immediately above the diffractive lens.

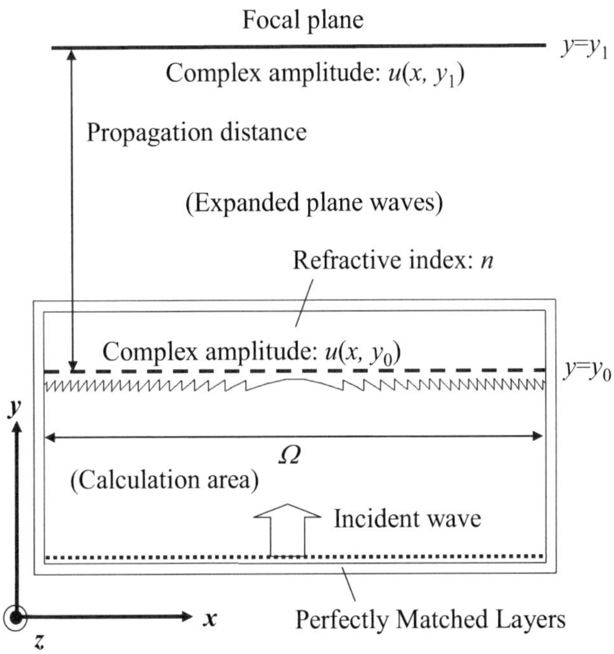

Fig. 5. Estimation of light intensity distribution on focal plane. The incident direction is taken from the substrate to air. The complex amplitude $u(x, y_0)$ of the electromagnetic wave is calculated by the FDTD method. The complex amplitude $u(x, y_1)$ on the focal plane ($y=y_1$) is determined as the integral of Fourier-expanded plane waves.

The electromagnetic field on the focal plane $u(x, y_1)$ is calculated from the amplitude $u(x, y_0)$. $u(x, y_0)$ can be expanded into plane waves by Fourier transform. The complex amplitude of each plane wave $U(k_x)$ is a function of the x component of wave number, k_x, which is expressed as

$$U(k_x) = \int_{-\Omega/2}^{\Omega/2} u(x, y_0) \exp(ik_x x) dx . \tag{2}$$

Ω is the width of the diffractive cylindrical lens. The complex amplitude on the focal plane $u(x, y_1)$ becomes

$$u(x, y_1) = \frac{1}{2\pi} \int U(k_x) \exp[ik_y(y_1 - y_0)] \exp(-ik_x x) dk_x . \tag{3}$$

k_y is the y component of wave number expressed as

$$k_y = \sqrt{(2\pi n / \lambda)^2 - k_x^{\ 2}} , \tag{4}$$

where λ is the wavelength of light in a vacuum and n is the refractive index of light-propagating media. The intensity distribution $I(x)$ on the focal plane is the absolute square of $u(x, y_1)$.

We defined the merit function as an integral of the mean square of the difference:

$$\phi = \int [I(x) - I_d(x)]^2 \, dx , \tag{5}$$

where $I_d(x)$ is the desired intensity distribution. We set the diffraction limit of the cylindrical lens as the desired intensity profile $I_d(x)$, which is expressed as

$$I_d(x) = \mathrm{sinc}^2\left(\frac{\Omega n}{f\lambda}\pi x\right). \tag{6}$$

Here, sinc(x)=sin(x)/x. The diffraction-limit width w is given by

$$w = 2\frac{f\lambda}{\Omega}, \tag{7}$$

where f is the focal length. The focus efficiency is defined as the ratio of the energy contained in the diffraction-limit width w to the energy incident to the cylindrical lens.

If the merit function ϕ is greater than a certain tolerance value, the electron dose profile D is modified. It is described that the simulated annealing algorithm can be used for the modification of electron dose. The above-mentioned calculations are iterated until the merit function ϕ is less than the admissible value or until it converges well. The outputs of the optimal design are the electron-dose profile D, resist surface profile P and intensity distribution $I(x)$ on the focal plane.

The designed diffractive cylindrical lens is 50 μm in width and the focal length is 25 μm for the 650-nm-wavelength light. The diffraction-limit width w becomes 0.65 μm. Since the designed lens has an F-number of 0.5, the middle zone has a parabolic surface profile of 11.4 μm width. The second zones next to the middle are 2.43 μm wide, and both end zones are 0.93 μm wide. The total number of grating periods and the parabolic surface profile is 31. A polarized light wave is normally incident to the diffractive lens from the substrate. The electric field of the incident light is normal to the x-y plane. The refractive indices of the substrate and grating material are 1.46 and 1.49, respectively.

The grid size of FDTD calculation is 50 nm square. The calculation area consists of 1199×109 cells and is surrounded by perfectly matched layers. The electron dose is modulated by 9-58 levels, which depends on the grating period. The calculation time for one process from giving the dose profile to evaluating the diffraction efficiency is 11 min. After 120 iterations, the diffraction efficiency practically converges. The total calculation time for the optimization is about 22 h using a computer with an AMD Athlon 1.2 GHz CPU.

Figure 6 shows the results of the optimal design. The dose and calculated resist profiles are shown in (a) and (b), respectively. The dose profile is complex and the calculated profile is very different from the conventional wrapped parabolic surface profile. Figure 6 (c) shows the calculated intensity distribution on the focal plane. The focus efficiency is 49%. Figure 7 shows the conventional wrapped parabolic surface grating and its intensity distribution. Its focus efficiency is 37%. The efficiency of the optimized diffractive lens is significantly higher than that of the conventional diffractive lens. The focus efficiency of the optical optimization without the lithography simulation is 53%.

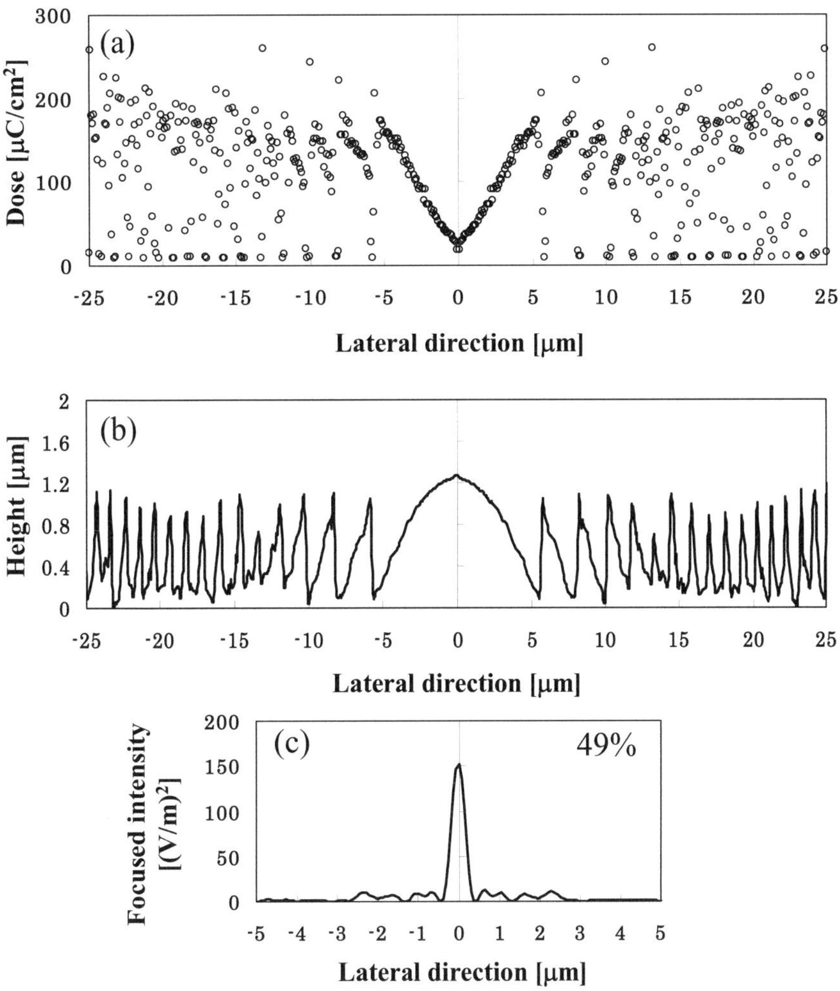

Fig. 6. Diffractive cylindrical lens designed by the unified method: (a) estimated dose profile, (b) calculated resist profile, and (c) light intensity profile on focal plane.

4. Conclusions

We suggested the unified design method of grating profiles with the electromagnetic grating analysis and the electron-beam lithography simulator. The unified method gives us the optimum grating profile and the electron dose profile under the restriction of the proximity effect of the electron scattering. This method is useful especially for short-period gratings. The conventional design method based on only the electromagnetic grating analysis yields the theoretical highest diffraction efficiency. However, the calculated grating profiles may be too complex to be fabricated, so that such a high efficiency will not realized.

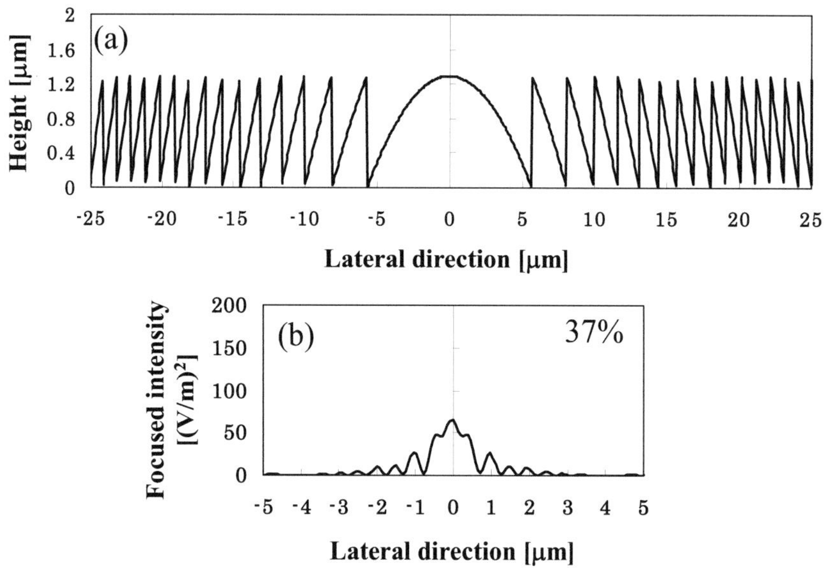

Fig. 7. Conventional wrapped parabolic diffractive lens: (a) cross-sectional profile, and (b) calculated light-intensity profile on focal plane.

On the other hand, the unified method gives us the fabricatable grating profile and the highest diffraction efficiency under the restriction of the proximity effect.

Moreover, the unified method enables to calculate nonperiodic DOEs as diffractive lenses using the FDTD method. The focus efficiency of diffractive lens with small *F*-number is significantly higher than that of the conventional diffractive lens with the saw-tooth blazed grating profiles. However, a long calculation time is required. Parallel computation is suitable for wide diffractive lenses, because both the FDTD method and the lithography simulation are appropriate for parallel computation.

The author thanks Prof. H. Kikuta and Prof. Y. Hirai with Osaka Prefecture University for many discussions.

5. References

Daschner, W.; Larsson, M. & Lee, S. H. (1995). Fabrication of monolithic diffractive optical elements by the use of e-beam direct write on an analog resist and a single chemically assisted ion-beam-etching step, *Appl. Opt.* 34, 2534-2539.

Di, F.; Yingbai, Y.; Guofan, J.; Qiaofeng, T. & Liu, H. (2003). Rigorous electromagnetic design of finite-aperture diffractive optical elements by use of an iterative optimization algorithm, *J. Opt. Soc. Am. A* 20, 1739-1746.

Dill, F. H.; Neureuther, A. R.; Tuttle, J. A. & Walker, E. J. (1975). Modeling projection printing of positive photoresists, *IEEE Trans. Elec. Dev.* 22, 456-464.

Hirai, Y.; Kikuta, H.; Okano, M.; Yotsuya, T. & Yamamoto, K. (2000). Automatic dose optimization system for resist cross-sectional profile in an electron beam lithography, *Jpn. J. Appl. Phys.* 39, 6831-6835.

Hirayama, K.; Glytsis, E. N.; Gaylord, T. K. & Wilson, D. W. (1996). Rigorous electromagnetic analysis of diffractive cylindrical lenses, *J. Opt. Soc. Am. A* 13, 2219-2231.

Kallioniemi, I.; Ammer, T. & Rossi, M. (2000). Optimization of continuous-profile blazed gratings using rigorous diffraction theory, *Opt. Comm.* 177, 15-24.

Kirkpatrick, S.; Gelatt, C. D., Jr. & Vecchi, M. P. (1983). Optimization by simulated annealing, *Science* 220, 671-680.

Kley, E. -B. (1997). Continuous profile writing by electron and optical lithography, *Microel. Eng.* 34, 261-298.

Kyser, D. F. & Murata, K. (1974). Quantitive electron microprobe analysis of thin films on substrate, *IBM J. Res. Develop.* 18, 352-363.

Maker, P. D. & Muller, R. E. (1992). Phase holograms in polymethyl methacrylate, *J. Vac. Sci. Technol. B* 10, 2516-2519.

Moharam, M. G. & Gaylord, T. K. (1981). Rigorous coupled-wave analysis of planar-grating diffraction, *J. Opt. Soc. Am.* 71, 811–818.

Neureuther, A. R.; Kyser, D. F. & Ting, C. H. (1979). Electron-beam resist edge profile simulation, *IEEE Trans. Elec. Dev.* 26, 686-693.

Nikolajeff, F.; Bengtsson, J.; Larsson, M.; Ekberg, M. & Hard, S. (1995). Measuring and modeling the proximity effect in direct-write electron-beam lithography kinoforms, *Appl. Opt.* 34, 897-903.

Noponen, E.; Turunen, J. & Vasara, A. (1992). Parametric optimization of multilevel diffractive optical elements by electromagnetic theory, *Appl. Opt.* 31, 5910-5912.

Okano, M.; Yamamoto, K.; Yotsuya, T.; Hirai, Y. & Kikuta, H. (2000). Validity and limitation of a proximity-compensation method for fabricating diffractive optical elements using the direct-writing electron-beam lithography, *Jpn. J. Opt.* 29, 566-572 [in Japanese].

Okano, M.; Kikuta, H.; Hirai, Y.; Yamamoto, K. & Yotsuya, T. (2003). Proximity correction for fabricating a chirped diffraction grating by direct-writing electron-beam lithography, *Jpn. J. Appl. Phys.* 42, 5602-5606.

Okano, M.; Kikuta, H.; Hirai, Y.; Yamamoto, K. & Yotsuya, T. (2004). Optimization of diffraction grating profiles in fabrication by electron-beam lithography, *Appl. Opt.* 43, 5137-5142.

Okano, M.; Kikuta, H. & Hirai, Y. (2007). Optimal design for nonperiodic fine grating structure controlled by proximity correction with electron-beam lithography, *Jpn. J. Appl. Phys.* 46, 627-632.

Prather, D. W.; Mait, J. N.; Mirotznik, M. S. & Collins, J. P. (1998). Vector-based synthesis of finite aperiodic subwavelength diffractive optical elements, *J. Opt. Soc. Am. A* 15, 1599-1607.

Prather, D. W.; Pustai, D. & Shi, S. (2001). Performance of multilevel diffractive lenses as a function of f-number, *Appl. Opt.* 40, 207-210.

Sheng, Y.; Feng, D. & Larochelle, S. (1997). Analysis and synthesis of circular diffractive lens with local linear grating model and rigorous coupled-wave theory, *J. Opt. Soc. Am. A* 14, 1562-1568.

Shiono, T.; Hamamoto, T. & Takahara, K. (2002). High-efficiency blazed diffractive optical elements for the violet wavelength fabricated by electron-beam lithography, *Appl. Opt.* 41, 2390-2393.

Testorf, M. E. & Fiddy, M. A. (2001). Efficient optimization of diffractive optical elements based on rigorous diffraction models, *J. Opt. Soc. Am. A* 18, 2908-2914.
Vincent, P. (1980). Differential Methods, In: *Electromagnetic Theory of Gratings*, Petit, R. (Ed.), pp. 101-121, Springer-Verlag, Berlin.
Yee, K. S. (1966). Numerical solution of initial boundary value problems involving Maxwell's equations in isotropic media, *IEEE Trans. Antennas Propag.* 14, 302-307.

15

Independent-exposure Method in Electron-beam Lithography

Do-Kyun Woo and Sun-Kyu Lee
Gwangju Institute of Science and Technology
Republic of Korea

1. Introduction

Lithography technologies can be distinguished by mask lithography and scanning lithography so-called direct-writing process. The representative for mask lithography is photo-lithography, while the representative for scanning lithography is electron-beam lithography. Electron-beam lithography as scanning lithography has three major processing in the lithography procedure: (1) resist deposition, (2) alignment & exposure and (3) development.

Electron-beam lithography is one of the most important fabrication technologies for the top-down miniaturization and high-accuracy 2D and 3D surface profile because the wavelength of electron-beam calculated by De Broglie's wavelength is much shorter than that of the source of other lithography (Sinzinger & Johns, 2003; Kley, E.B & Schnabel, B. 1995; Hirai et al., 2000).

In the photo-lithography, there is the limitation of pattern resolution by optical diffraction, but electron beam lithography does not have the limitation of pattern resolution by diffraction due to the much shorter wavelength. Compared to photo-lithography, electron-beam lithography has some attractive advantages including: (1) small spot size under 10nm, (2) no mask, (3) the precise control of electron beam by electrostatic or magnetic lens, (4) lower defect densities (Madou, 2002).

However, electron beam lithography also has some disadvantages including: (1) strong scattering of electrons in solid affecting the increase of pattern resolution, (2) high cost of system and maintenance, (3) more complex machine system than photo-lithography system due to the requirement of vacuum and electron optics (Madou, 2002).

For the fabrication of micro optical elements such as a multi-level lens and a binary structure in electron beam lithography, it was suggested that the alignment method should be used. It consists of a sequence of binary scanning pattern with L step which results in $N=2^L$ number of level (Dammann, 1970; Goodman & Silvestri, 1970; Kong, et al., 2004; Woo et al., 2009).

Fig. 1 shows the alignment method for the fabrication of a 4-level lens in which L step is 2. Thus, there are 2 times of repetitive processes between electron beam lithography and etching process. In addition, alignment marks and technology in this method have to be required in order to reduce the fabrication error. Thus, it is also important to design and fabricate the alignment structure. Due to the sequence of binary fabrication step and repetitive process, it requires more time and cost for the fabrication of such a multi-level lens.

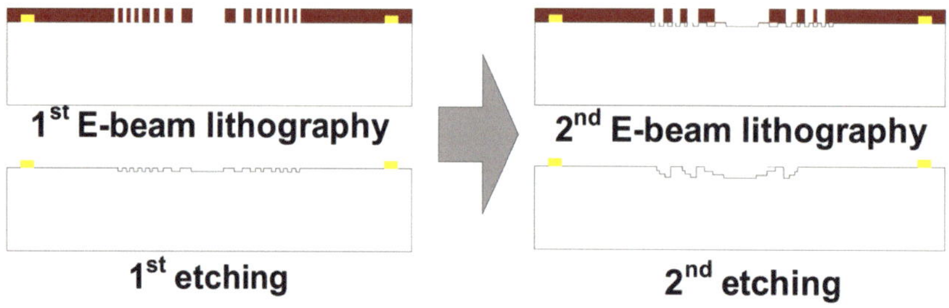

Fig. 1.alignment method for the fabrication of a 4-level lens

However, the independent-exposure method suggested in this chapter can reduce such a fabrication processes, which results in reduction of fabrication time and cost. We will discuss the fundamentals of the new fabrication technique in electron-beam lithography called Independent-exposure method which can be available for the structure of planar substrates with the reduction of the fabrication time and cost.

2. Independent-exposure method

2.1 Principle

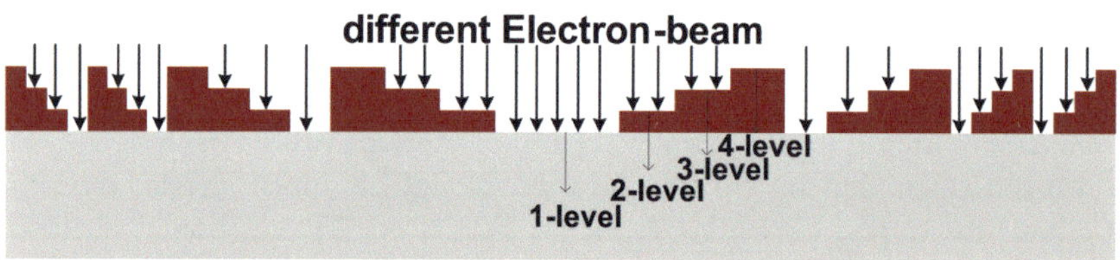

Fig. 2. Schematic of Independent-exposure method (Woo et al., 2008)

It was known that the electron beam resist thickness can be controlled by the modulation of exposure dose in electron beam lithography (Hirai et al., 2000). Based on this, the principle of independent-exposure method for the fabrication of a multi-level lens is to make the resist profile identical to that of a multi-level lens by different and appropriate electron beam dose (Woo, et al. 2008). Fig. 2 shows the schematic of independent-exposure method for a 4-level lens in electron beam lithography. After the independent-exposure lithography, we can obtain the mould of the 4-level lens through a Fast Atom Beam (FAB) plasma etching process with a 1:1 etching ratio of resist and substrate. Thus, the independent-exposure method for the fabrication of a multi-level lens or binary structure consisting of precise patterns has the major advantage of non-repetitive fabrication process between electron beam lithography and etching. The ultimate goal of this approach is to reduce the fabrication time and expense.

In this method, however, the developed resist thickness is not linearly proportional to the amount of electron beam dose due to the proximity effect which results from scattered electrons. When the electrons interact with the resist and substrate, the electrons can be scattered; the former is called "forward scattering", and the latter is called "backward scattering". The proximity effect is created by both forward and backward scattering. Due to the proximity effect, the area of the resist which is not intended can be partially exposed by these scattered electrons. Thus, the proximity effect in the electron beam lithography affects the increase of pattern resolution as well as the developed resist thickness (Chang, 1975).

In order to fabricate such a multi-level lens or binary structure using the independent-exposure method, we have to keep two crucial considerations in mind: (1) The resist thickness should be greater than that of a multi-level lens we designed because of 1:1 etching ratio in etching process, (2) the relation between the electron beam dose and the thickness of developed resist should be definitely clarified (Woo et al. 2008)

2.1 Spin coating

In order to obtain the exactly designed pattern through electron beam lithography, it is also important for the resist to be uniformly deposited on substrate. The spin coating for the deposition of resist with uniform thickness has been considered as the general way. In the spin coating, there are three important parameters to uniformly deposit resist: (1) the enough highly spin speed, (2) the viscosity of the coated resist, and (3) the clean substrate.
If the spin speed is low, the thickness of electron beam resist is not uniform as that shown in Fig. 3 (a) and if the substrate and circumstance is not clean, resist can not be uniformly deposited as that shown in Fig. 3 (b).

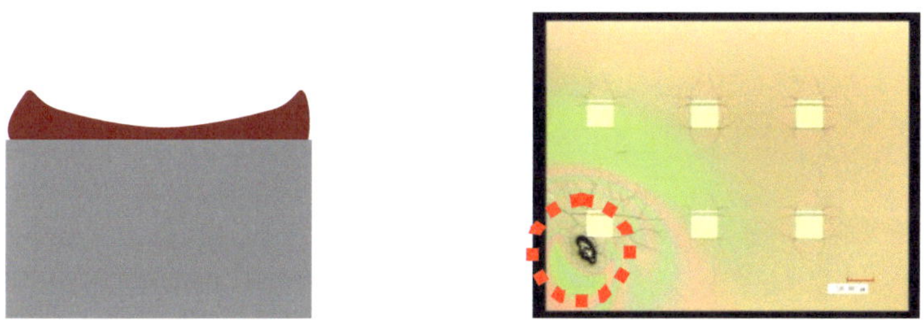

(a) (b)

Fig. 3. factors on non-uniform thickness of resist: (a) low spin speed, (b) Impurity of substrate and circumstance

Because the relation of spin speed and the deposited thickness for each resist is not same, the company producing electron beam resist will provide the data sheet about the relation between the resist thickness and spin coating speed. In this work, we used the ZEP 520A as the positive electron beam resist by ZEON Corporation. It has high resolution, high sensitivity and dry etching resistance. Fig. 4 shows the relation between spin speed and the thickness of deposited resist in the technical report provided by ZEON Corporation (ZEON, 2003). Thus we can uniformly control the thickness of deposited resist with spin coating based on the spin speed curve.
As early mentioned, the independent-exposure method requires that the thickness of electron beam resist be more than that of a designed multi-level lens. In general, the thickness of a multi-level lens is in range of 600nm and 1200nm. However, the thickest thickness of ZEP520A resist with 2000rpm as the lowest spin speed shown in Fig. 4 is approximately 500nm, which is less than that of a multi-level lens.
In this chapter, the way to deposit ZEP520A as electron beam resist with the thickness over 1000nm will be described. It was not simply possible to deposit ZEP520A with the thickness over 1000nm by twice consecutive spin coatings of 2000rpm because the resist in second spin coating cannot easily adhere to the resist deposited in the first spin coating.
However, we could obtain the resist layer satisfying the uniform thickness of approximately 1100nm by inserting an accelerant which is able to improve the adhesion of electron beam resist and substrate. In this work, we used the OAP (Tokyo Ohka Kogyo Co., LTD) as an

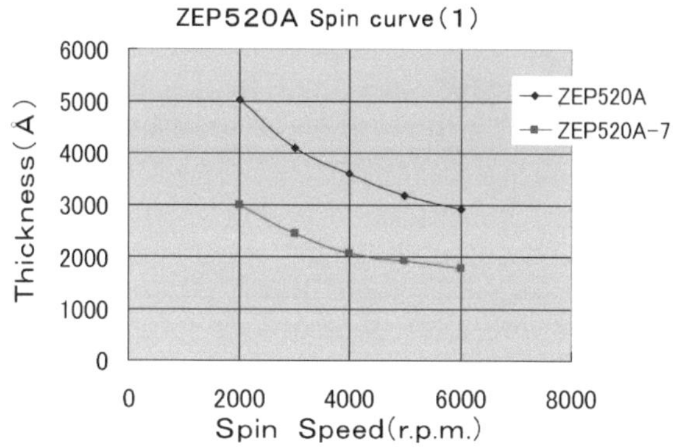

Fig. 4. ZEP520A Spin curve in technical report by ZEON corporation (ZEON, 2003)

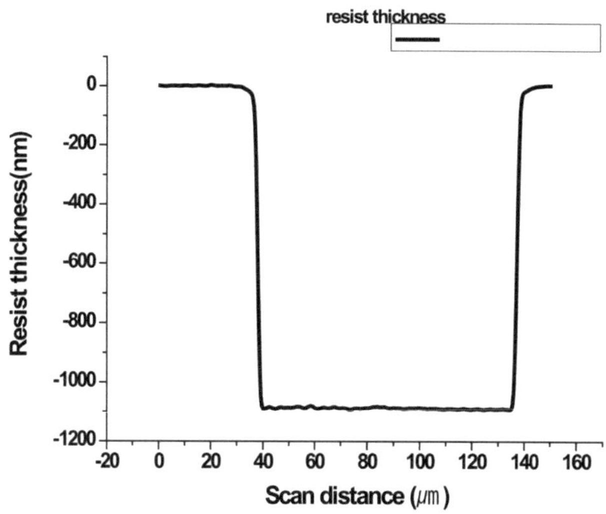

Fig. 5. Resist thickness with twice spin coatings of 2000rpm

accelerant. It is verified through repetitious experiments with the same condition that the range of the thickness of deposited resist is from 1070nm to 1140nm as shown in Fig. 5. Compared to the thickness of a multi-level lens, it is adequate to make an experiment with such a deposited resist for the fabrication of a multi-level lens by the independent-exposure method. Table 1 represents the conditions of the spin coating for the deposition of resist with a thickness of approximately 1100nm.

1. pre-baking	180°C(3min)	4. OAP spin coating	1200rpm(5s) + 3000rpm (30s)
2. ZEP 520A spin coating	500rpm(3s) + 2000rpm (60s)	5. ZEP-520A spin coating	500rpm(3s) + 2000rpm (60s)
3. Soft baking	180°C(3min)	6. Soft baking	180°C(3min)

Table 1. The conditions of the spin coating for independent-exposure method

2.3 Relation between electron beam dose and resist

Before fabricating a multi-level lens by independent-exposure method, we should have another important preliminary experiment about relation between electron beam dose and the developed resist because the thickness of developed resist can be controlled by the modulation of electrons. In addition, it is better for us to certify whether there is another relation between resist and electron beam dose or not. To do this, we used two patterns with a different width of 10μm and 80 μm as shown in Fig. 6 under the experiment conditions in table 2. In this experiment, we controlled the electron beam dose from 30μC/cm^2 to 100μC/cm^2 with a step of 10μC/cm^2 (Woo et al., 2008)

Pattern width (μm)	10 μm, 80 μm
Electron beam dose (μC/cm^2)	30, 40, 50, 60, 70, 80, 90, 100

Table 2. Experiment conditions for relation between the exposure dose and the developed resist thickness (Woo et al., 2008).

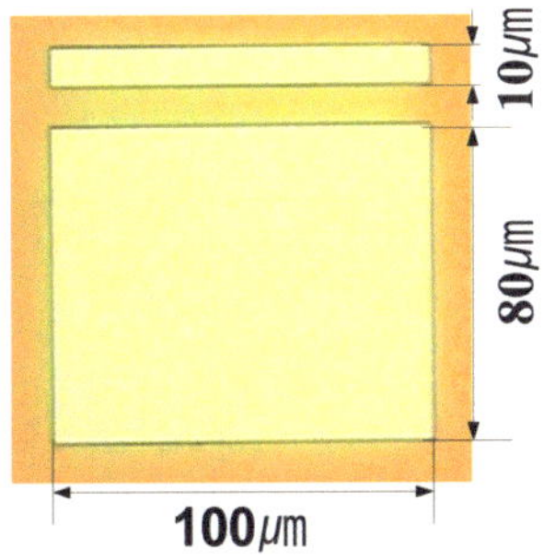

Fig. 6. Patterns for the experiment about the relation between EB dose and the developed resist thickness (Woo et al., 2008).

Fig. 7 shows the experiment results about the relation between electron beam dose and the developed thickness of resist, which represents that if the pattern area is different, the developed thickness of resist was also different despite the same electron beam dose. It means that the developed thickness of resist is related to the area of pattern as well as electron beam dose. Here, we could have a question. Why is the developed thickness of resist different according the pattern size despite the same electron beam dose?

Fig. 8 gives us the answer for the question. The reason for the above phenomenon can be explained by proximity effect. Because the electron beam dose is defined by the amount of exposure dose per unit area, the amount of electron beam dose is proportional to pattern area, i.e, if the pattern area is larger despite of the same electron beam dose, the pattern can be actually exposed to more amount of electron beam dose. Thus, the proximity effect such as forward and backward scattering could be strongly generated due to the increase of electron beam dose (Woo et al., 2008).

Thus, this experiment led us to two conclusions; (1) the thickness of the developed resist is proportional to electron beam dose in the identically given pattern size and (2) the thickness of the developed resist is proportional to the pattern size in the identically given electron dose.

From the Fig. 7, it was seen that the appropriate range of electron beam dose for fabricating a multi-level lens is from 55μC/cm^2 to 80μC/cm^2. To successfully fabricate a multi-level lens

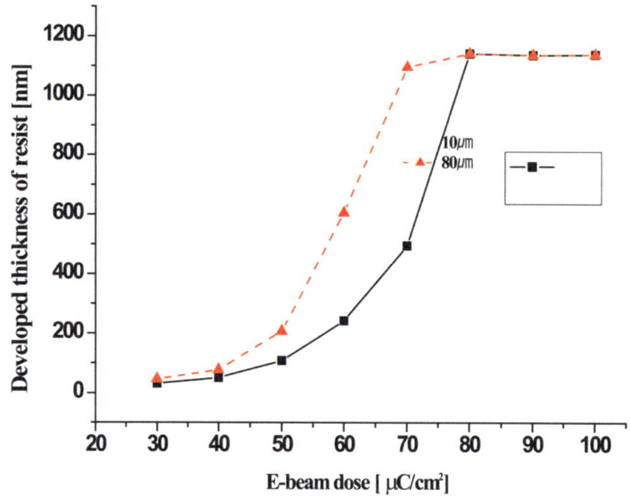

Fig. 7. The developed resist thickness with respect to exposure dose (Woo et al., 2008).

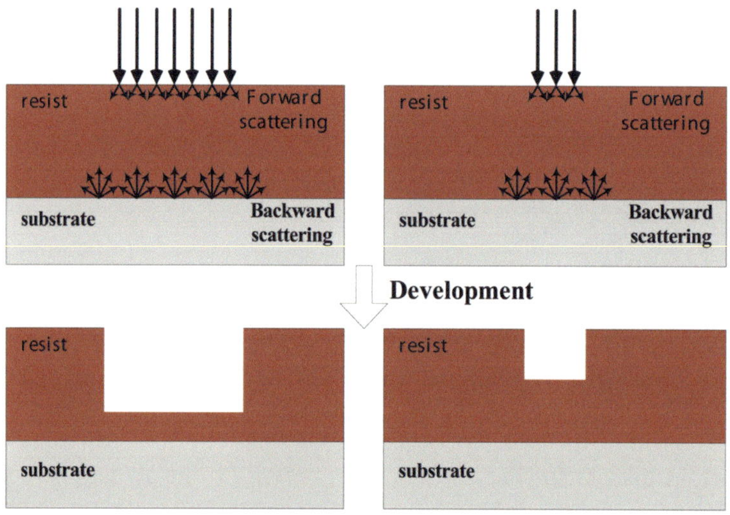

Fig. 8. Proximity effect on different pattern sizes (Woo et al., 2008).

by independent exposure method, additional experiment with an electron beam dose range from 55μC/cm² to 80μC/cm² was required with respect to a variety of pattern size. Based on a lot of experiments with this electron beam range according to the pattern width from 10 μm to 1 μm, it has been verified that the resist on the patterns under a width of 5 μm was developed with similar thickness. Fig. 9 shows the experiment result for the relation of electron dose from 55μC/cm² to 80μC/cm² and the developed resist thickness on the pattern width of 10 μm. Fig. 10 shows the cross-section view of the developed resist on a pattern width of 1 μm. Both Fig. 9 and Fig. 10 support the independent-exposure method by which a multi-level lens can be fabricated in range of electron beam dose from 55μC/cm² to 80μC/cm².

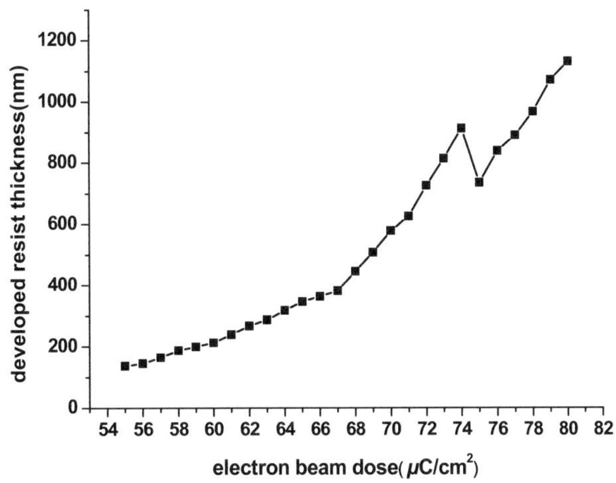

Fig. 9. The developed resist thickness with respect to electron beam dose from 55μC/cm² to 80μC/cm² on a pattern width of 10 μm.

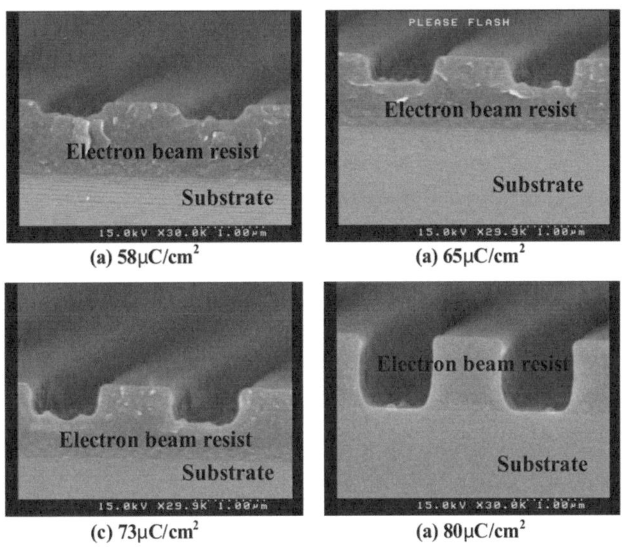

Fig. 10. The developed resist with respect to electron beam dose on a pattern width of 1 μm.

3. Fabrication of a multi-level lens using Independent-exposure method

3.1 Design of a multi-level lens

Before applying the independent-exposure method to the fabrication of a multi-level lens, we should design a multi-level lens. Fig. 11 explains the multi-level lens in Fresnels zone and the relation between the radii of pattern and the focal length of the mult-level lens. A mult-level lens can be designed as stair shape in a Fresnel zone, and radii $R_{j,i}$ and thickness d of a multi-level lens can be calculated geometrically using Eq. (1),

$$R_{j,i} = \sqrt{2\left(j - \frac{N-i}{N}\right)\lambda f + \left(\left(j - \frac{N-i}{N}\right)\lambda\right)^2}$$
$$d = \frac{(N-1)\lambda}{N(n_{PMMA} - 1)} \tag{1}$$

where, $R_{i,j}$ is the radius of i level in the jth pattern and N-level lens, d is the thickness of N-level lens, λ is the wavelength and N is the number of levels. (Turunen & Wyrowski, 1997; Woo, et al., 2008).

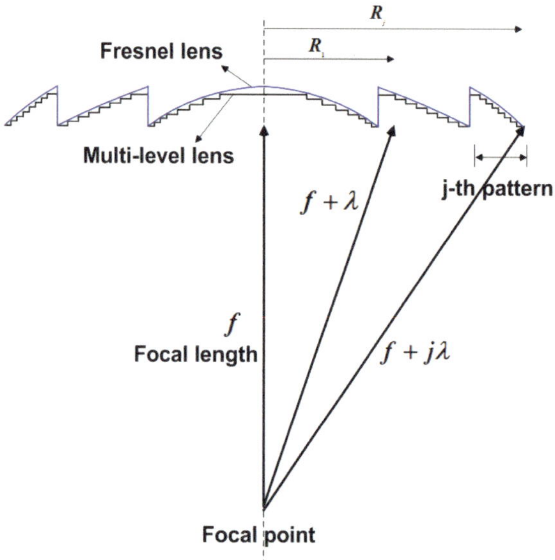

Fig. 11. A multi-level lens in Fresnel zone and relation between focal length and radii of it (Woo et al., 2009)

In this work, it is determined to fabricate a 4-level lens including a focal length of 714.5 μm and a diameter of 127.4 μm at the wavelength of 480nm. By substituting these value into Eq. (1), the the 4-level lens can be designed with a maximu pattern width of 26.204 μm, a minimum width of 0.966 μm, and a thickness of 0.723 μm. Depth for each level is approximately 0.241 μm. Fig. 12 shows the design results of the 4-level lens which has the 11 patterns.

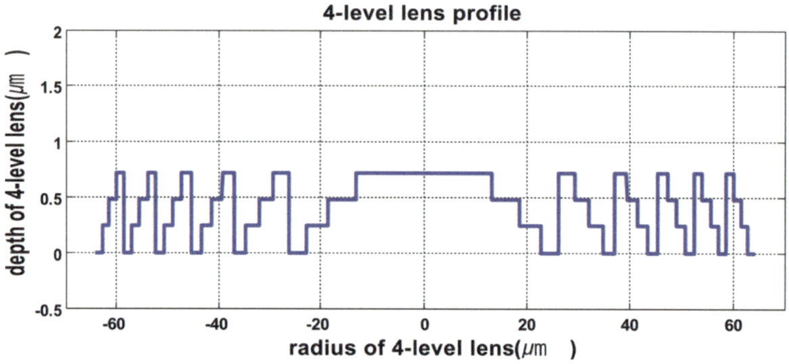

Fig. 12. The designed 4-level lens profile

Because a mutl-level lens has a surface of stairstep, such a lens can be called a mutli-level lens or a multi-phase lens. Based on the Eq. (1) and Fig. 11, moreover, it was known that the width of lens suface at the egde of lens is smaller than that at the centre of lens.

3.2 Fabrication process

For the fabrication of a multi-level lens, the alignment method repeating lithography and etching process was widely used, but the independent-exposure method eliminating this repetitive process will be tried in this chapter. As early mentioned, there are two requirements in the fabrication of a multi-level lens by the independent-exposure method and FAB etching with etching ration of 1:1. One is the thickness of resist should be over the thickness of the designed multi-level lens, and the other is to define the relation of electron beam and the developed resist thickness.

Fig. 13 represents the fabrication procedure of the designed 4-level lens using independent-exposure lithography consisting of four main processes: (1) spin-coating, (2) electron beam lithography, (3) FAB plasma etching, and (4) hot-embossing. As shown in Fig. 13, such an independent-exposure method can contribute to a reduction of fabrication time and cost (Woo, et al, 2008).

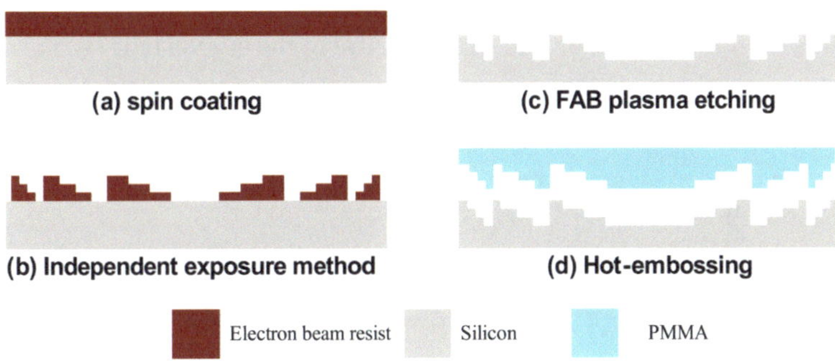

Fig. 13. The fabrication procedure of designed 4-level lens using independent-exposure method in electron beam lithography (Woo et al., 2008)

The purpose of the spin coating is to deposit electron beam resist with uniform thickness more than the designed 4-level lens.

In the electron beam lithography, the surface of resist should be made identical to that of the 4-level lens with an appropriate electron beam dose. As shown in Fig. 12, it can be easily found that the patterns in the centre of 4-level lens are much larger than those in the other side of 4-level lens. As noted earlier, the developed resist thickness can be affected by pattern size as well as by the electron beam dose, and thus appropriate electron beam dose at the centre of the 4-level lens should be considered. In order to find the appropriate electron dose for the pattern in the centre of 4-level lens, another experiment was carried out with the same pattern to that of the centre of 4-level lens shown in Fig. 14. The constant electron beam dose of 65μC/cm^2 was given to the $R_{0,2}$ pattern, the another constant electron beam dose of 58μC/cm^2 was applied on the $R_{0,3}$ pattern, and then the electron beam dose applied on the $R_{0,1}$ pattern whose width is largest in the 4-level lens was changed from 30μC/cm^2 to 65μC/cm^2. As a result of this experiment, the resist of centre pattern was completely developed with electron beam dose of 65μC/cm^2, and the appropriate electron dose to the $R_{0,1}$ pattern ranges from 60μC/cm² to 65μC/cm². In addition, it was found that the developed resist thickness of $R_{0,2}$ pattern given the constant electron beam dose of 65μC/cm² increased as the electron beam dose on the $R_{0,1}$ pattern increases. In the other words, the amount of electron beam dose can affect the adjacent pattern due to the proximity effect. Fig. 15 shows the relation between the electron beam dose and the developed resist thickness of adjacent pattern. Thus, it is also found that there is the interference of the electron beam dose between near patterns in fabrication of a multi-level lens by the independent-exposure lithography (Woo, et al., 2008).

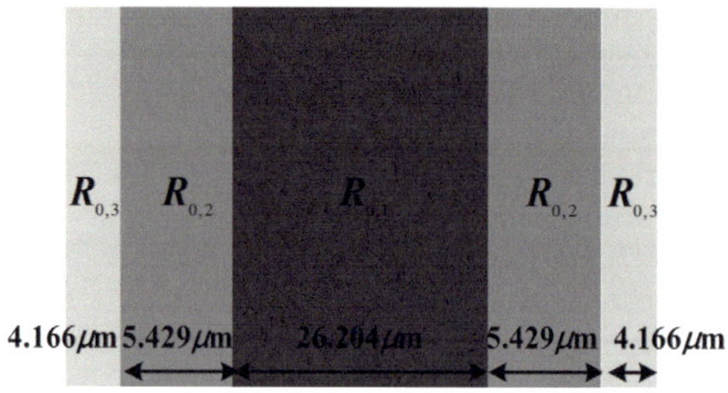

Fig. 14. Patterns with the same width to that of centre in the 4-level lens (Woo, et al., 2008)

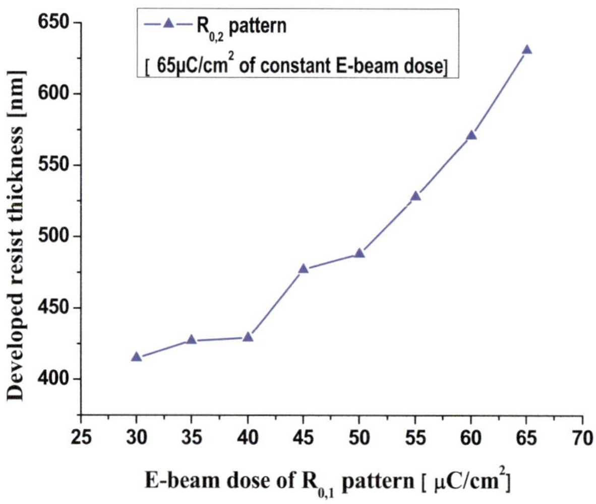

Fig. 15. Developed resist thickness of $R_{0,2}$ pattern with respect to the electron beam dose of $R_{0,1}$ pattern (Woo, et al., 2008)

Thus, there are three considerations in order to efficiently use the independent-exposure method in electron beam lithography: (1) the relation between the electron beam dose and the developed resist thickness on the same pattern, (2) the relation between the pattern size and the developed resist thickness on the same electron beam dose, (3) the interference of the electron beam dose between neighbouring patterns.

Based on these results, the designed 4-level lens was successfully fabricated by the independent exposure lithography under the conditions: 62μC/cm² for the 1-level step $R_{0,1}$

of centre pattern, 74μC/cm² for the 1-level step $R_{j,1}$ of the other pattern, 68μC/cm² for the 2-level step $R_{j,2}$, and 58μC/cm² for the 3-level step $R_{j,3}$. The $R_{0,1}$ pattern has to actually have the same thickness of resist to $R_{j,1}$, but the reason the electron beam dose was differently given, compared to $R_{j,1}$ pattern, is that the $R_{o,1}$ pattern has significantly the wide width which is the largest in the 4-level lens.

After obtaining the resist whose shape is identical to the shape of the 4-level lens through the independent-exposure method, a mould of the 4-level lens can be fabricated through FAB plasma etching with 1:1 etching ratio and with the etching rate of 22~23nm/min.

Fig. 16 shows the fabrication results for the 4-level lens by independent exposure method.

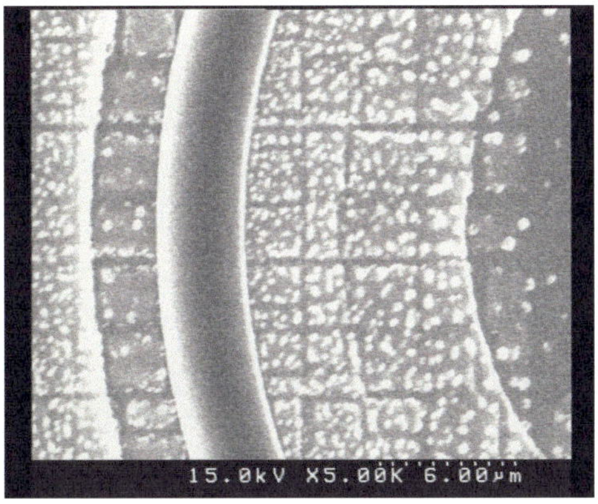

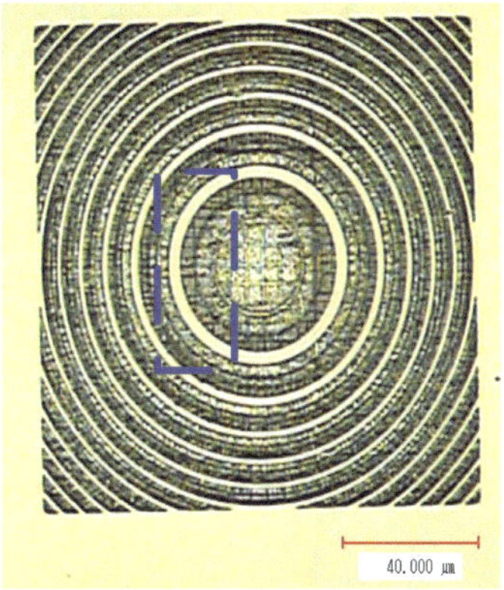

Fig. 16. The mould of 4-level lens fabricated by the independent-exposure method and FAB plasma etching (Woo, et al, 2008)

In order to obtain the 4-level lens by the fabricated mould, Hot-embossing technology with PMMA as stamp process was selected for replication. Hot embossing technology can translate the micro & nano structure of mould into polymers with satisfaction of high accuracy and low-cost (Anke, et al., 1996; Jahns, et al., 1992).

3.3 Fabrication results

This chapter introduced the new fabrication method called independent-exposure method in electron beam lithography which can help significantly to reduce fabrication process. The 4-level lens designed with a maximum width of 26.204μm, a minimum width of 0.966 μm, and a thickness of 0.723 μm was successfully fabricated by the independent-exposure method which is fundamentally based on the relation among electron beam dose, pattern size and electron beam resist.

As mentioned earlier, the new method was strongly affected by the proximity effect. As shown in Fig. 16, it can be found that the surface of fabricated 4-level lens has some swellings as scattering marks owing to proximity effect.

This proximity effect has been considered as negative effect for the others process as well as independent-exposure method in the electron beam lithography. Owen introduced some interesting methods to reduce the proximity effect: (1) the precise control of the electron beam energy, (2) the calculation and pre-compensation of the energy deposition and (3) the use of multilayer resist coatings (Owen, 1990). In addition, if the resist with low sensitivity to electron beam is used, it will be also good solutions.

3.4 Optical evaluation

In order to verify whether the independent-exposure method is suitable to fabricate micro optics components such as a multi-level lens and a binary structure, the optical evaluation of the fabricated 4-level lens has been performed by measuring focal length and diffraction efficiency of it.

Fig. 17 shows an overall schematic of experiment setup for the measurement of focal length and diffraction efficiency of the fabricated 4-level lens. In this evaluation, the incident light travelling to the 4-level lens has to be collimated in order to correctly measure focal length and diffraction efficiency.

For the measurement of focal length, microscope was focused on the 4-level lens, and then the stage moves vertically until the focal point of the 4-level lens clearly appears in the monitor through the microscope and CCD. This moving distance of stage is identical to focal length of the 4-level lens. As a result, the measured focal length of the 4-level lens is 712.4 μm with a depth of focus of approximately 60 μm, while the designed focal length of it is 714.5 μm. Fig. 18(a) is the measured focal point of the 4-level lens (Woo, et al., 2008).

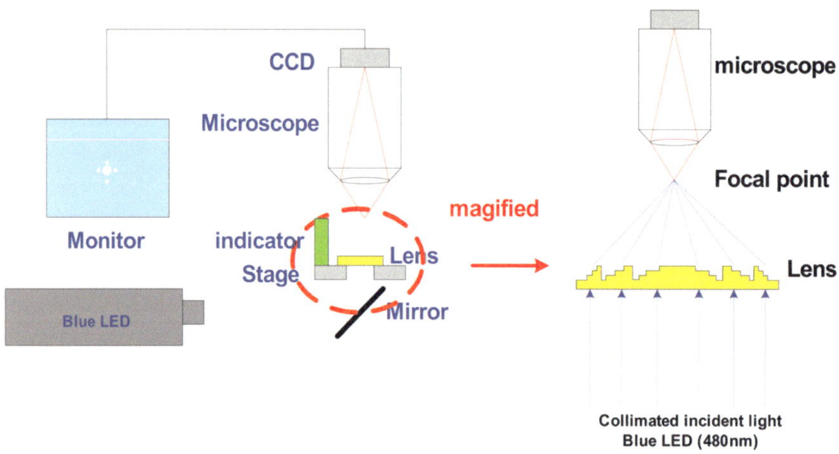

Fig. 17. Overall schematic of experiment setup for the measurement of focal length and diffraction efficiency (Woo, et al., 2008)

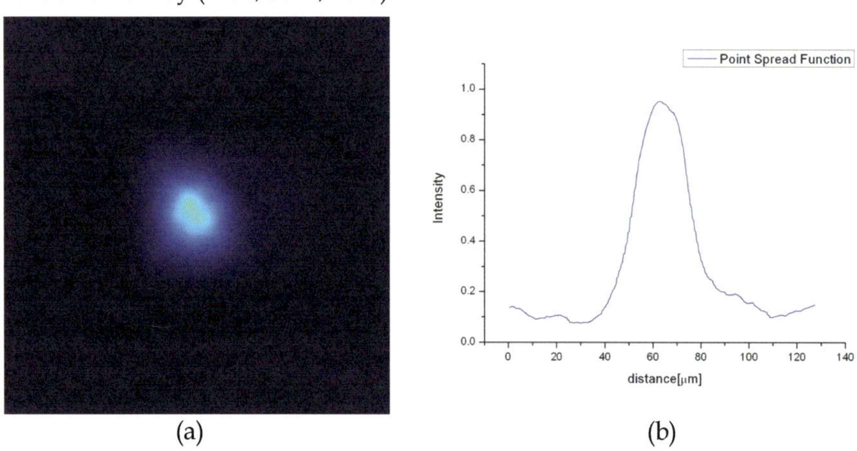

Fig. 18. Optical evaluation results: (a) the focal point of the 4-level lens, (b) point spread function of the 4-level lens (Woo et al., 2008)

The diffraction efficiency is important factor for the evaluating diffractive optical elements such as a multi-level lens. As shown in Fig. 11, the principle of the focusing of a multi-level lens is to diffract an incident light to the focal plane of it. As the diffracted light travels to the focal plane, the diffraction efficiency can be defined as the power of diffracted light at focal point through a multi-level lens divided by the power of incident light.
Such a diffractive efficiency of a multi-level lens is dependent on the number of phase level and can be theoretically calculated by Eq. (2) (Damman, 1979; Jahns & Walker, 1990; Woo, et al., 2008). By Eq. (2), the theoretical diffraction efficiency for a 4-level lens is 81%. From the measurement of diffraction efficiency, the real efficiency of 74.7% can be obtained. Compared to the theoretical efficiency of 81%, error of diffraction efficiency results from the Fresnel reflection loss on the surface of the 4-level lens and fabrication error which includes the period and thickness error and the swelling of the 4-level lens.

$$\eta = \left|A_n^2\right| = \mathrm{sinc}^2(1/N) = \left\{\frac{\sin(\pi/N)}{\pi/N}\right\}^2 \tag{2}$$

4. Conclusion

In this research, the new approach called independent-exposure method in electron beam lithography for the fabrication of micro optical elements such as a multi-level lens was suggested. To sum up, this method can significantly reduce the fabrication process, i.e., a multi-level lens can be fabricated by only one process of both lithography and etching. However, before applying this method to the fabrication of a multi-level lens, proximity effect should be considered and several preliminary experiments or consideration should be required.

1. The electron beam resist should be deposited with the thickness over that of a multi-level lens.
2. The developed resist thickness is proportional to electron beam dose.
3. The developed resist increases as the pattern size increases on identically given electron beam dose.
4. There is the interference of electron beam given to adjacent patterns.

Based on the fabrication and optical results, it was evident that the independent-exposure method is suitable to fabricate a multi-level lens with non-repetitive process.

5. Acknowledgement

This work was conducted with the kind collaboration of the members of the Hane Laboratory in Tohoku University, and was also financially supported by the Korea Science and Engineering Foundation (KOSEF) through National Research Laboratory Program grant funded by the Korea government (MEST) (No. ROA-2008-000-20098-0)

6. References

Anke, I.; Kley, E.B.; Hubner, H.; Schnabel, B. & Pohlmann, R. (1996) Replication of microoptical profiles in ORMOCER and other polymers. *Proc. SPIE*, Vol. 2783, pp. 325-332

Chang, T. H. P. (1975). Proximity effect in electron-beam lithography. *J. Vac. Sci, Technol.B* Vol. 8, No. 6, pp. 1271-1275

Dannan, H. (1970). Blazed synthetic phase-only holograms. *Optik,* Vol. 31, pp.95-104

Goodman, J. and Silverstri, A. (1970). Some effects of Fourier-domain phase quantization. *IBM J. Res. Dev.*, Vol. 14, pp.478-484

Hirai, Y.; Kikuta, H.; Okano, M.; Yotsuya, T. & Yamamoto, K. (2000). Automatic Dose Optimization System for resist Cross-Sectional Profile in a Electron Beam Lithography, *Jpn. J. Appl. Phys.* Vol. 39, pp. 6831-6835

Jahns, H.; Brenner, K.H.; Daschner, W.; Doubrava, C. & Merklein, T. (1992). Replication of dffractive micoroptical elements using PMMA moulding technique. *Optik,* Vol. 89, pp. 98-100

Jahns, J & Walker, S. J. (1990). Two-dimensional array of diffractie microlenses fabricated by thin film deposition. *Appl. Opt.* Vol. 29, pp. 931-936

Kley, E.B & Schnabel, B. (1995). E-beam lithography: a suitable technology for fabrication of high-accuracy 2D and 3D surface profiles. *Proc. SPIE¸* Vol. 2640, pp. 71-80

Kong, L.; Yi, X.; Lian, K.; & Chen, S. (2004). Design and optical performance research of multi-phase diffractive micro lens array. *J. Micromech. Micoreng,* Vol. 14, pp. 1135-1139

Madou, M. J. (2002). *Fundamentals of MICROFABRICATION,* CRC PRESS, ISBN0-8493-0826-7

Owen, Geraint. (1990). Methods for proximity effect correction in electron lithography. *J. Vac. Sci. Technol. B,* Vol. 8, No. 6, pp. 1889-1892

Sinzinger, S. & Jahns, J. (2003). *Microoptics,* WILEY-VCH, ISBN3-527-40355-8

Turunen, J. & Wyrowski, F. (1997). *Diffractive optics for industrial applications.* Akademic Verlag, Berlin, ISBN3-05-501733-1

Woo, D.K.; Kane, K. & Lee, S.K. (2008). Fabrication of a multi-level lens using independent-exposure lithography and FAB plasma etching, *J. Opt. A: Pure Appl. Opt,* Vol. 10. No. 044001, pp. 1-6

Woo, D. K.; Hane, K.; Cho, S.C. & Lee, S.K. (2009). Development of an integral optics system for a slim optical mouse in a slim portable electric device, *J.Vac. Sci. Technol. B,* Vol. 27, No. 3, pp.1422-1427

Zeon, Corporation. (2003). *Technical Report ZEP520A, ZEONREX Electronic Chemicals.* Ver.1.01. http://www.zeon.co.jp/

16

The Interdependence of Exposure and Development Conditions when Optimizing Low-Energy EBL for Nano-Scale Resolution

Mohammad A. Mohammad[1], Taras Fito[1,2], Jiang Chen[2], Steven Buswell[3], Mirwais Aktary[3], Steven K. Dew[1] and Maria Stepanova[2]
[1]Department of Electrical and Computer Engineering, University of Alberta,
[2]National Institute for Nanotechnology NRC,
[3]Applied Nanotools Inc.
Canada

1. Introduction

Electron beam lithography (EBL) is the major direct-write technique to controllably fabricate nanoscale features. A focused beam of electrons induces a chemical change in a layer of radiation sensitive material (resist), such as chain scissioning in positive tone polymethylmethacrylate (PMMA) polymer photoresist. The localized fragmented region is rendered more soluble in a suitable developer solution and removed. In negative tone resists, such as hydrogen silsesquioxane (HSQ) or calixarene, the radiation damage eventually results in bond cross-linking, generating structures locally more resistant to dissolution. Limitations of the technology are related largely with unwanted exposure of the resist away from the impact of the focused electron beam due to scattering of the primary electrons in the resist (often described as the forward scattering), generation of secondary electrons, and backscattering from the substrate (the proximity effect). The exposure and development processes have been optimized and routinely used for fabrication of submicron features. However, as requirements for lithography have progressed toward the sub-20 nm regimes, major challenges have emerged of introducing controllable radiation-induced changes at molecular-size scales, within a reasonable tradeoff with the applicability of the standard materials, as well as cost and simplicity of the processes. Due to the proximity effect, this becomes particularly demanding when dense patterns with closely positioned features must be fabricated. Achieving deep nanoscale resolutions in high density patterns at industrially-relevant throughputs requires new approaches to EBL.

Novel EBL processes that would extend capabilities of the technology significantly into the deep nanoscale regime entail new approaches to resist design, exposure strategies, and development techniques (Häffner et al., 2007; Liddle et al., 2003; Ocola & Stein, 2007; Word et al., 2003). To achieve this will require a much more detailed understanding of the molecular mechanisms involved in both the electron-resist interaction and in the polymer dissolution (development) stages of the nanolithography process. Despite a significant research effort and vast literature on electron beam lithography, the detailed molecular mechanisms are still inadequately understood. Published modeling studies address extensively the processes of electron penetration, scattering, and energy deposition in resist and substrate materials (Kyser & Viswanathan, 1975; Murata et al., 1981; Lee et al., 1992; Raptis et al, 1993; Raptis et al, 2001; Zhou & Yang, 2006), however, the analyses of exposure of resists are mostly limited to conversions of the average amounts of energy deposited by electrons into the average number of relevant molecular events, such as the bond scissions, through the empirically determined radiation chemical yield (Chapiro, 1962; Greeneich, 1974; Han et al., 2003; Kyser & Viswanathan, 1975; Raptis et al., 2001). Furthermore, the detailed molecular processes occurring during dissolution of the most useful resists have been under-addressed if not overlooked so far. As a consequence, understanding of the trends of resist development currently resides mostly on published experimental results (Cord et al., 2007; Dial et al., 1998; Ocola & Stein, 2006; Yan et al., 2008; Yasin et al., 2002; Yasin et al., 2004).

Within the last decade, numerous research groups have invested a significant effort to explore experimentally the resolution limits of deep nanoscale EBL. In particular, optimizing the development conditions such as the developer formula (Dial et al., 1998; Yan et al., 2008; Yasin et al. 2002) and development temperature (Cord et al., 2007; Ocola & Stein, 2006; Yan et al., 2008) have been found effective to achieve improvements in resolution below the 20 nm mark. However, these works were conducted by employing exposures with relatively high beam energies of at least 30 keV or higher. Increasing the acceleration voltages allows for easier fabrication of high-resolution structures due to reduced forward scattering and reduced aspect ratio requirements; however, this approach has its disadvantages. High voltage processes are generally more expensive and have a lower throughput due to the increased exposure dose requirement. High energy electrons also tend to penetrate deep in the substrate causing unwanted substrate damage, as well as give rise to a significant proximity effect. In contrast, ultra low voltage electrons in the 1-3 keV regimes deposit most of their energy within the resist, resulting in less substrate damage and decreasing dramatically the proximity effect (Lee et al., 1992; Schock et al., 1997). Furthermore, exposures employing voltages below 10 keV require lower doses roughly in proportion to the acceleration voltage (Schock et al., 1997). Strong forward scattering of low energy electrons, which is routinely believed to be the major resolution-limiting factor, may alternatively be employed to create nanoscale three-dimensional profiles in the resist (Brünger et al., 1995). Realizing this potential, however, requires a thorough, systematic understanding of the intricate interplay of the numerous process control parameters including the accelerating voltage, optimum exposure dose, and the favorable conditions of development. Clearly, this systematic understanding should rely on a solid knowledge of the physico-chemical molecular-level processes behind the resist exposure to electrons as well as the subsequent post-exposure development stages.

Over the last several years we have been investigating thoroughly, both experimentally and by numeric modeling, the impact of the major EBL process factors on the quality of dense nanoscale patterns in the popular positive-tone resist, PMMA (Aktary et al., 2006; Mohammad et al., 2007; Mohammad et al., 2009, Fito et al., 2009). In this chapter, we outline the methodologies that we have developed for fabrication and visualization of nanopatterns in PMMA, as well as our model for the exposure, fragmentation, and dissolution of positive resists. We discuss the results of our systematic study of the impact of the exposure dose, duration of development, temperature of developer, and other process conditions on the quality and process sensitivity of dense nanoscale gratings fabricated using low to medium (3-30keV) exposure energies. The focus of the discussion comprises dense patterns of 15-35 nm features in 45-70 nm thick PMMA resist. We demonstrate that, in addition to the exposure conditions that are routinely optimized in standard EBL techniques, post-exposure resist processing is also a crucial factor and should be co-optimized when fabricating the dense nanopatterns. We analyze the tradeoff of the dose requirements and pattern quality when employing low-voltage exposures combined with cold development, and discuss possible optimized solutions to ultimately push the limit of EBL toward the nanometer-scale resolution. We also discuss the research challenges related with the molecular mechanisms of development, which should be understood thoroughly in order to efficiently optimize the EBL process.

2. Fabrication, visualization, and quality control of nanostructures in PMMA

In our work, we have studied systematically the morphologies that we fabricated in the most popular positive resist, PMMA, using low to medium energy EBL. We consider dense arrays of periodic lines (gratings) as a convenient benchmark nanostructure. Employing these grating patterns, we analyze typical morphology regimes, which we visualize by SEM, as a function of the various process conditions. We interpret the conventional characteristics of the EBL process, such as the resolution and sensitivity, in terms of the quality of the fabricated nanoscale morphologies. We believe that the morphology-bound interpretations of the sensitivity and resolution adopted in our work are more relevant to characterize EBL nanofablication than the often-employed studies of the resist contrast curves, since the first approach is directly representative of the quality of fabricated nanostructures.

2.1 Nanoscale morphologies of dense gratings in PMMA

The experimental methodology that we have employed has been described in detail elsewhere (Mohammad et al., 2007). In brief, cleaned silicon substrates were spin coated with 950k mol. wt. PMMA providing a nominally 65 nm thick layer. As the next stage, sets of periodic grating patterns were generated by electron beam lithography (Raith 150). The incident electron energy used was typically 10 keV with a beam diameter of 2 nm. The layout of our grating design represented an array of grating units, each of which consisted of a set of parallel, 2 nm wide exposed lines of a given pitch. We have generated multiple sets of parallel lines with the spacing varying from 20 nm to 100 nm, and various exposure doses. Each set of parallel lines covered a lateral distance of 2 μm in width, in order to ensure a uniform cumulative exposure due to backscattered electrons in the central portion of the set. The exposed patterns were developed using a standard procedure of dipping the substrates in a 1:3 MIBK: IPA developer followed by an IPA stopper rinse and drying the substrates with a nitrogen gun. As a final step before imaging our samples (Hitachi S-4800 FE-SEM), we coated our substrates by sputtering (Kurt J. Lesker Co.) an approximately 6 nm thick Chromium as an anti-charging layer.

In the discussion below, we employ two descriptors for the exposure dose: the *average area dose* and the *line dose*. For gratings, these dose descriptors are related by $d_{(area)} = d_{(line)}/\lambda$, where λ is the interline distance (grating pitch). Employing the area dose is more convenient for comparisons of dense nanopatterns of various shapes, whereas the line dose is more straightforward for use when analyzing process conditions for gratings of a given width. Below we employ the different approaches according to the subject discussed.

Fig. 1 shows a representative selection of morphologies that we have obtained experimentally after 5 sec development of high-density gratings with various interline spacings generated with various exposure doses using 10 keV electrons. From the figure, it is evident that the potential to fabricate a quality grating is strongly dependent on both the exposure dose and the interline distance. Thus, for 70 nm pitch, all the images present well developed gratings, with the exception of Fig. 1(d) that corresponds to the high area dose of

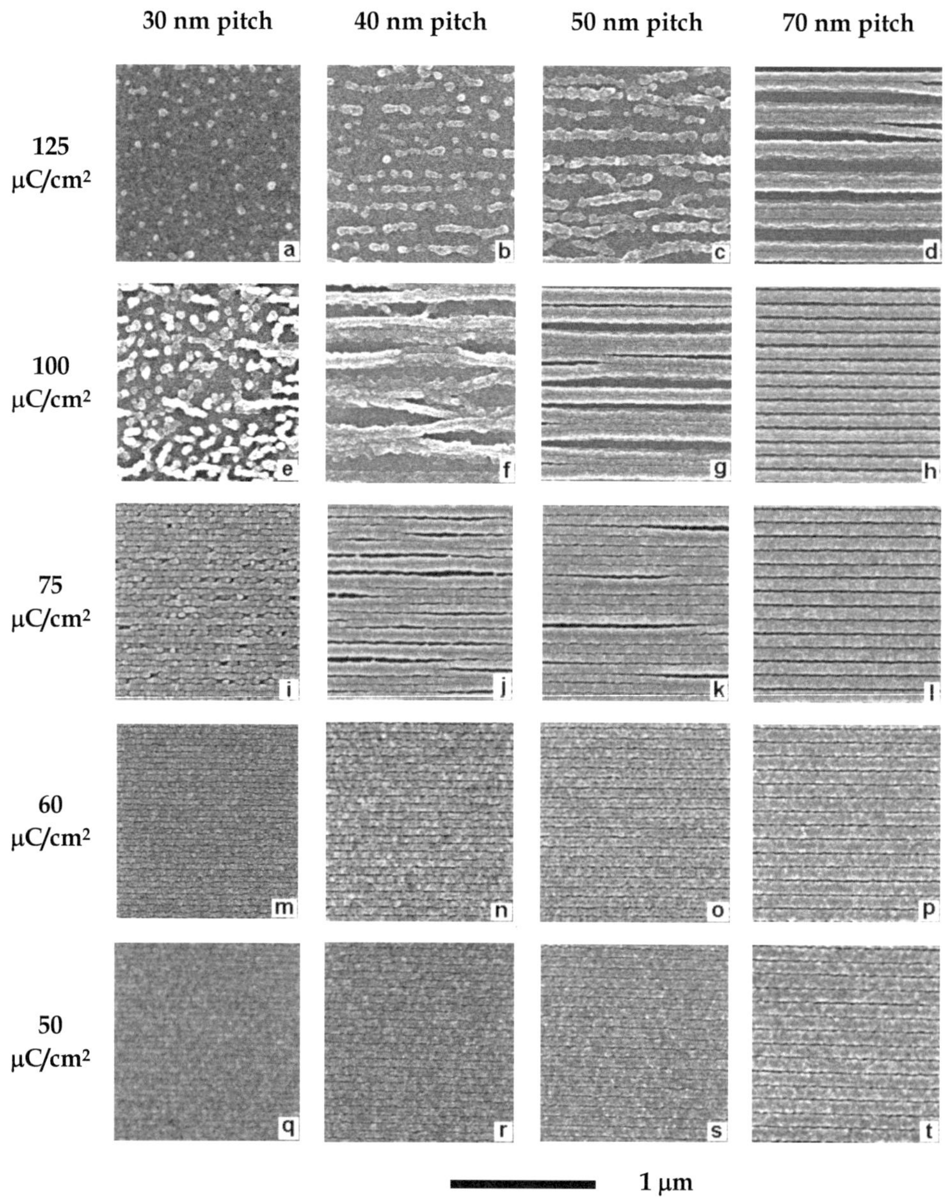

Fig. 1. SEM images of 30 nm, 40 nm, 50 nm, and 70 nm gratings fabricated in a 65-nm thick PMMA layer on a silicon substrate, with 10 keV electrons, at the various area doses. The gratings were developed for 5 sec. in a 1:3 MIBK:IPA solution at room temperature. The lateral size of all images is 1 μm×1 μm (Mohammad et al., 2007).

125 μC/cm². For 40 nm and 50 nm gratings, promising morphologies are seen in Figs. 1(n), 1(o), and 3(k) for the doses from 50 to 75 μC/cm². For 30 nm interline distance, only Fig. 3(m) for 60 μC/cm² exhibits a promising morphology. The other images in Fig.1 show various damaging influences. The first and well-known one is underexposure, when the clearance depth of exposed lines is insufficient to generate a grating pattern. In our SEM images, this is manifested by a low contrast as shown in Figs. 1(q), 1(m), and 1(r). The other extreme regime is overexposure, when the pattern is damaged by excessive clearance of PMMA. Strongly overexposed are the gratings in Figs. 1(a) and 1(b), which were obtained with the area dose of 125 μC/cm². Mild signs of overexposure are also visible in Figs. 1 (c) and (e) for 125 and 100 mC/cm², respectively. The next typical kind of morphology is the collapse of interline walls. Examples of collapsed gratings are given in Figs. 1(d), 1(f), and 1(g). Some collapse is also apparent in Figs. 1(j) and, to a lesser extent, in Fig. 1(k). It is noteworthy that collapse only occurs in gratings with 40 nm and larger pitch. In 30 nm gratings the limiting mechanism is different as illustrated in Fig. 1(e) and 1(i). In these cases, the grating is partly or entirely destroyed by redistribution of PMMA that tends to form islands at irregular locations. Similar morphologies have been reported in the literature for PMMA uniformly exposed to electrons and then developed (Hasko et al., 2000; Yasin et al., 2004). It is well known that formation of globular islands or percolation networks may occur in immiscible liquids that undergo a phase separation generating such morphologies as seen in Fig. 1(e). Since mixtures of PMMA fragments with most common EBL developers have regimes of limited miscibility, a phase separation can occur. The underlying mechanism would comprise relocation of individual molecules of PMMA driven by thermodynamic forces associated with the surface tension between the insoluble PMMA fraction and the developer. In our works we denote this mechanism as micellization, to emphasize the expected involvement of solventophobic interactions resulting in phase separation (Mohammad et al., 2007).

To summarize, we observe four factors that may damage the resist morphology. Two of those factors, underexposure and overexposure, are directly related to the electron dose applied, whereas the other two, micellization and collapse, are post-exposure processes of resist degradation. In the first case, degradation seems to occur through redistribution of individual PMMA molecules, whereas in the second case, entire interline walls are bent or displaced from their initial locations.

In addition to plan-view imaging of the grating morphologies in PMMA, we also visualized the cross-sectional profiles of such structures. Our methodology comprises exposure of long (up to 2 mm) grating patterns, which are then developed and blown dry using nitrogen gas. To generate crisp cross sections, the samples were manually cleaved when dipped in liquid nitrogen to achieve a clean fracture and prevent the resist deformation. Before SEM imaging, we coated our substrates with 4-6 nm of chromium.

A SEM image of a cleaved sample is shown in Fig. 2, and Fig. 3 demonstrates crisp cross sections representing a well-done grating pattern, an underexposed pattern, and a collapsed pattern With this information on three-dimensional structure of the typical surface morphologies, the mechanisms of the resist's development as well as degradation, such as collapse, can be understood significantly better than from plan-view images alone.

2.2 Applicable exposure dose windows, resolution, and sensitivity

Fig. 4 summarizes the observed morphologic regimes for various grating periods and exposure doses. In the low dose regime, the limiting factor is underexposure, whereas at

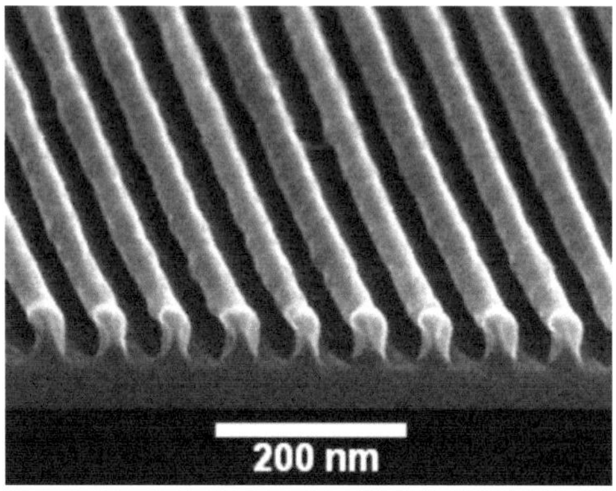

Fig. 2. SEM micrograph of a cleaved grating in PMMA on a Si substrate.

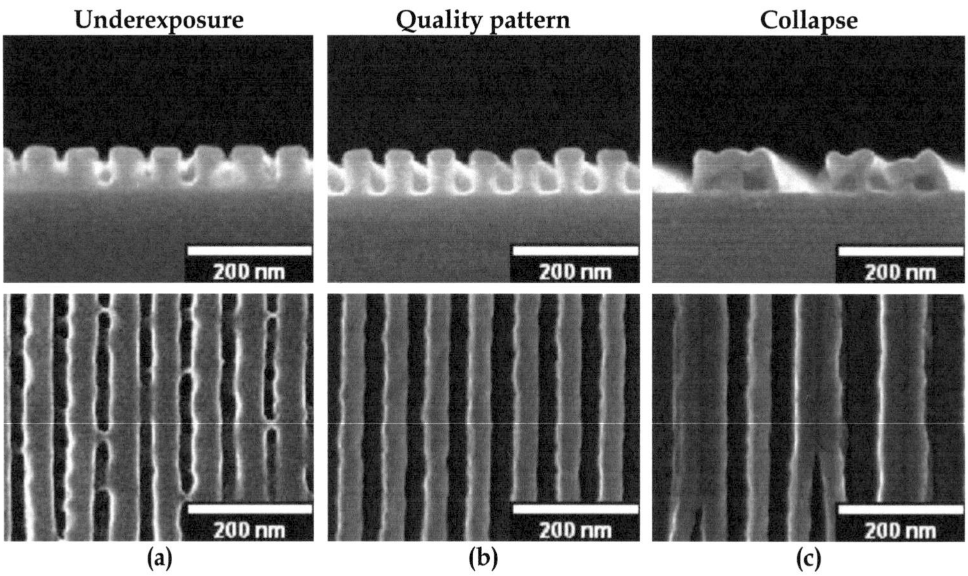

Fig. 3. Examples for (a) under-exposed, (b) well-done, and (c) collapsed gratings in PMMA, imaged by cross-sectional profiles (top) and plan-views (bottom).

increased doses, the patterns degrade through micellization or collapse. The densest gratings, with periods of 20 and 30 nm, degrade through micellization, and the gratings with periods of 40 nm and larger rather tend to collapse. At even higher doses, we observe overexposed patterns that are also micellized for all gratings pitches.

It can be seen that the window of favourable exposure doses at which quality gratings can be expected decreases rapidly when the grating pitch decreases. The explanation is that at the deep nanoscale, the electron beam broadening becomes comparable with the interline distance, and the unwanted exposure of the resist between the lines becomes stronger. Generally, the width of the applicable dose window represents the robustness of the process. A large dose window means that the fabrication process has improved reproducibility, and also has greater controllability over the line width and aspect ratio. The minimum size of nanostructures for which a reasonable dose window is attained

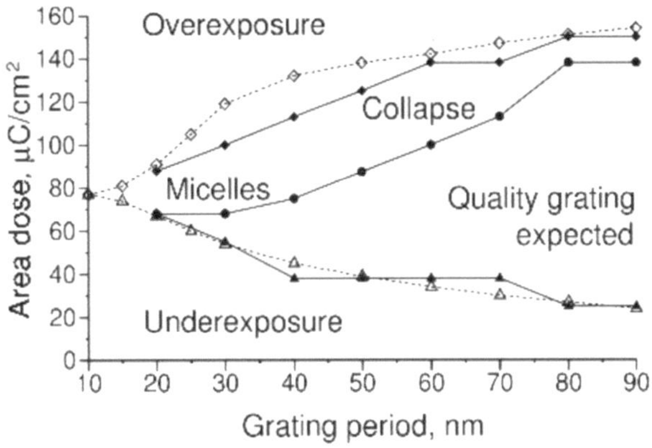

Fig. 4. Diagram of the characteristic morphologies in PMMA for various grating periods and area exposure doses using 10 keV voltage. Filled symbols show experimental results for the conditions as in Fig.1: triangles denote the boundary for underexposure (insufficient clearance); diamonds denote the boundary for overexposure (excessive clearance), and circles indicate the boundary for collapse or micellization. Open symbols show the results of numerical modeling of clearance as discussed in Sect. 3.2. (Mohammad et al., 2007).

characterizes the resolution. From Fig. 4 it follows that the highest resolution attainable at the experimental conditions considered corresponds to approximately 30 nm pitch (or approximately 15 nm half-pitch representing the average line width).

The minimum applicable doses can be associated with the sensitivity of the EBL process. In Fig. 4, the boundary dose for underexposure decreases when the grating pitch increases. The explanation is that in periodic gratings, the exposure dose per line at which the depth clearance is achieved can be viewed as independent of the interline distance. This leads to the corresponding dose per unit area decreasing roughly in proportion to the inverse of the period (Mohammad et al., 2007).

2.3 Cold development of PMMA broadens the applicable dose windows

Next, we have studied the influence of the development temperature on the nanoscale resolution and sensitivity of EBL using PMMA as the resist (Mohammad et al., 2009). We employed 47-55 nm thick layers of 950k PMMA on a Si substrate, a 10 keV exposure voltage, an 1:3 MIBK:IPA mixture for developer, and IPA for stopper. We varied the developer and stopper temperature between -15 °C and 22 °C using a cold plate, and experimented with development times from 5 sec to 20 sec. The stopper temperature was the same as that of the developer. Our approach was to determine the applicable dose windows for fabrication of quality grating patterns by varying the exposure doses and investigating the corresponding plan-view SEM images for various conditions of development.

Figs. 5 (a) and (b) present favorable windows for *line* exposure doses as functions of the development time, at various developer temperatures, for gratings with 70 nm and 50 nm pitches, respectively. In the figures, the solid lines show dose boundaries for underexposure (see Fig. 3(a)), and the dashed lines show the boundaries for collapse (Fig. 3(c)). The regions between the solid and dashed lines represent the favorable dose windows where quality gratings may be fabricated. It can be seen that in a grating with 70 nm pitch, decreasing the development temperature from room temperature (RT) to -15 °C results in an increase in the

applicable line dose window by an order of magnitude. In gratings with 50 nm pitch, the dose window increases approximately twice when the development temperature changes from -5 °C to -15 °C. This indicates a strong improvement in the EBL process resolution when decreasing the development temperature from RT to –5-15°C. Thus, the temperatures of -5°C and less were the only regimes at which we could fabricate 15-20 nm wide trenches in gratings with 40-50 nm pitch. Examples of the highest resolution structures that we could obtain at various temperatures are given in Sect. 4. Another trend revealed by Fig.5 is that the cold development requires higher exposure doses, i.e. decreases the sensitivity of the process. Thus, the decrease in the development temperature from RT to -15°C in Fig. 5(a) has required an approximately 4 times higher line dose to obtain clearance of the resist.
To conclude, cold development results in a dramatic increase in the applicable dose window. However, this advantage is accompanied by a drop in the process sensitivity. Decreased temperatures therefore seem to be a preferred solution when maximizing the resolution is the priority.

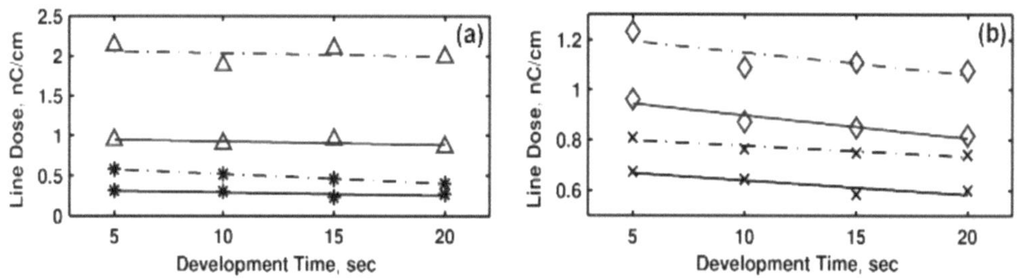

Fig. 5. The applicable dose windows for 70nm pitch (a) and 50nm pitch (b) gratings showing minimum (solid lines) and maximum (dashed lines) line doses for quality patterning. The symbols indicate the temperature of development: room temperature (stars), -5°C (crosses), and -15°C (triangles and diamonds) (adapted from Mohammad et al., 2009).

2.4 Low energy exposures require co-optimized development conditions

Our next target is to explore the EBL process when varying the energies of electrons from ultra low to moderate voltage regimes. Fig. 6 shows the examples of cross-sectional profiles for 70 nm pitch gratings fabricated with 3 keV, 10 keV, and 30 keV exposure energies. The grating exposed with 3 keV voltage shows pronounced undercuts because of strong forward scattering of the low-energy electrons (Lee et al., 1992). With increasing energy, the undercut decreases. 30 keV exposures produce almost straight interline walls.
Fig. 7 presents results of our analysis of the applicable dose windows when the exposure voltages of 3 keV, 10 keV, and 30 keV are employed to fabricate a 70 nm pitch grating. It can be seen that increasing the electron energy from 3 keV to 30 keV results in a dramatic increase of the dose windows. The reason of this is the decrease of the undercut when electrons' energy increases. The large applicable process window allows for more control over properties of the fabricated pattern, such as the aspect ratio. Thus, Figs. 8 demonstrates that in 70 nm gratings fabricated with 30 keV exposures, the variation of the line dose within the applicable window allows one to vary the width of the interline trenches from less than ~15 nm to 45 nm, and Fig. 9 shows the corresponding aspect ratio changes by a factor of two. The aspect ratio of PMMA walls turns out to be less sensitive to the change of the exposure dose because of a decrease of the wall height (see Fig. 8). However, according to

Fig. 7, the minimum applicable line doses increase roughly in proportion to the electron energy, resulting in a significantly lower sensitivity for 30 keV voltages. In contrast, 3 keV exposures provide a better sensitivity. Furthermore, strong undercuts that occur in the ultra-low energy processes may facilitate 3D shaping of the resist required for metallization and lift-off of high-resolution nanostructures, as will be demonstrated in Sect. 4. The complementary strengths of the low- and moderate-voltage exposures motivate further in-depth research of the corresponding process conditions.

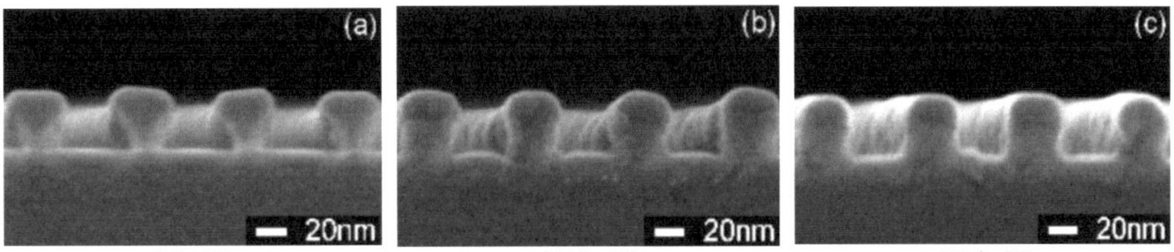

3 keV **10 keV** **30 keV**

Fig. 6. Cross sectional profiles in PMMA gratings exposed using various voltages.

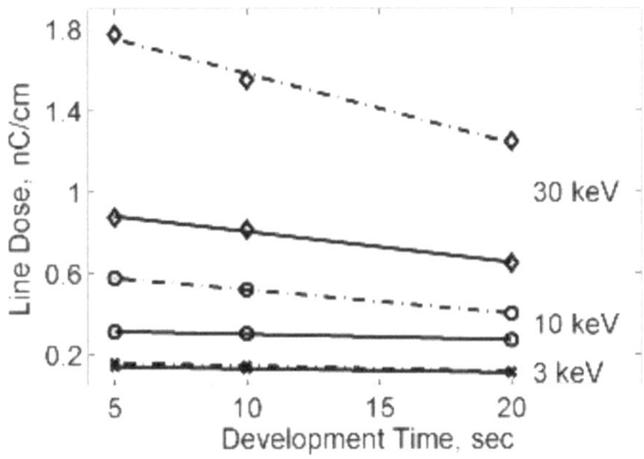

Fig. 7. Applicable dose windows for 70nm pitch gratings using the exposure voltages of 3 keV (crosses), 10 keV (circles), and 30 keV (diamonds). The samples were developed at RT, and the initial PMMA thickness was 55 nm. The meaning of the solid and dashed lines is as in Fig. 5.

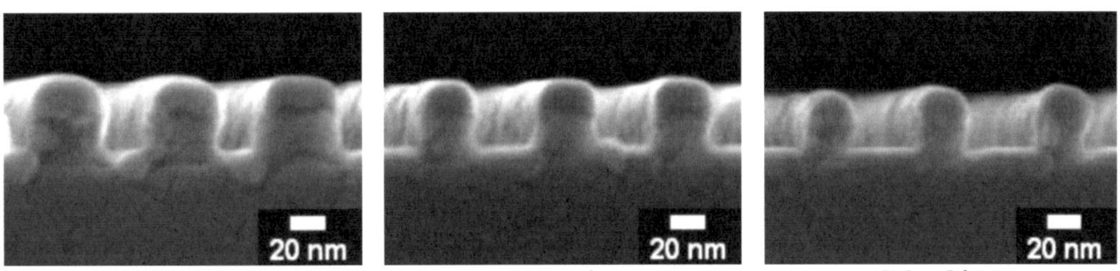

2.0 nC/cm **4.5 nC/cm** **7.0 nC/cm**

Fig. 8. Cross sectional profiles in 70 nm pitch PMMA gratings fabricated using 30 keV voltage and various line doses. The samples were developed at -15°C for 15 sec., and the initial PMMA thickness was 55 nm.

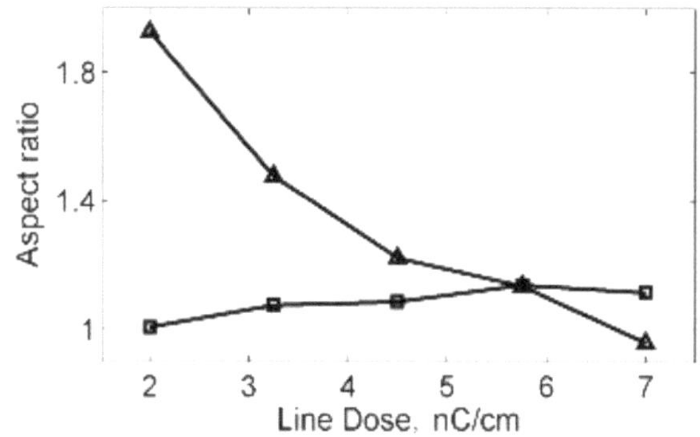

Fig. 9. Aspect ratios for the interline gaps (triangles) and PMMA lines (squares) as a function of the exposure dose, determined from cross-sectional SEM images for 70 nm pitch gratings fabricated using 30 keV electrons. Other process conditions were as in Fig. 8.

Tables 1 and 2 compare the process resolutions and sensitivities, expressed in terms of the applicable line dose boundaries d_{min} and d_{max}, for various exposure energies, development temperatures, and durations, when fabricating gratings with 70 nm and 50 nm pitches, respectively. From the table it is evident that obtaining reasonably broad dose windows with low energy exposures may require decreasing the development temperature, which

Voltage, develop. temperature, & initial resist thickness	Develop. time, sec.	Line doses, nC/cm			d_{max}/d_{min}
		d_{min}	d_{max}	$d_{max}-d_{min}$	
3 keV	5	0.135	0.155	0.020	1.15
RT	10	0.125	0.145	0.020	1.16
55 nm	20	0.110	0.115	0.005	1.05
3 keV	5	0.490	0.665	0.175	1.36
-15°C	10	0.460	0.630	0.170	1.37
55 nm	20	0.400	0.560	0.160	1.40
10 keV	5	0.313	0.578	0.265	1.85
RT	10	0.303	0.520	0.218	1.72
55 nm	20	0.270	0.403	0.133	1.49
10 keV	5	0.875	2.050	1.175	2.34
-15°C	10	0.870	2.040	1.170	2.34
55 nm	20	0.835	2.020	1.185	2.42
30 keV	5	0.875	1.775	0.900	2.03
RT	10	0.815	1.550	0.735	1.90
55 nm	20	0.650	1.250	0.600	1.92
30 keV	5	2.400	6.465	4.065	2.69
-15°C	10	2.400	6.250	3.850	2.60
55 nm	20	2.400	5.715	3.315	2.38

Table 1. The applicable line dose windows for 70 nm pitch gratings.

Voltage, develop. temperature, & initial resist thickness	Develop. time, sec.	Line doses, nC/cm			d_{max}/d_{min}
		d_{min}	d_{max}	$d_{max}-d_{min}$	
3 keV	5	0.370	0.430	0.060	1.16
-15°C	10	0.345	0.405	0.060	1.17
47 nm	20	0.295	0.350	0.055	1.19
10 keV	5	0.960	1.280	0.320	1.33
-15°C	10	0.940	1.260	0.320	1.34
55 nm	20	0.840	1.120	0.280	1.33
30 keV	5	0.750	1.040	0.290	1.39
RT	10	0.725	0.900	0.175	1.24
55 nm	20	0.600	0.695	0.095	1.16
30 keV	5	2.160	3.330	1.170	1.54
-15°C	10	2.200	3.210	1.010	1.46
55 nm	20	2.295	2.920	0.625	1.27

Table 2. The applicable line dose windows for 50 nm pitch gratings.

would however increase both the difference $d_{max}-d_{min}$ and the minimum applicable dose d_{min}. The implication is that using low-energy exposure regimes requires tradeoffs between the process resolution and sensitivity, which may be achieved by a careful co-optimising of the exposure voltage and development temperature. It is elucidative to compare the process employing 30 keV exposure and RT development with that using 3 keV exposures and -15°C. According to Tables 1 and 2, the ultra-low voltage exposure combined with cold development still leads to a strong improvement in the sensitivity. For both 70 nm and 50 nm pitches, the minimum applicable line dose d_{min} is approximately half for the 3 keV, cold development processes than for the 30 keV, RT ones. However, the applicable dose window $d_{max}-d_{min}$ is 2-5 times narrower for 3 keV than for 30 keV in the cases considered. One can conclude that, to profit from the improved sensitivity and 3D capacity with low voltage, cold development exposures, the applied dose should be chosen carefully in order to satisfy the relatively narrow requirements for quality fabrication.

One more control factor to be considered is the duration of development. In Tables 1 and 2 it can be seen that the minimum and maximum boundary doses d_{min} and d_{max} show a minor to moderate decrease with the development time in most cases, and so does the width of the applicable dose window $d_{max}-d_{min}$. In contrast, the ratio d_{max}/d_{min} is largely insensitive to the duration of development, although it depends on the voltage and grating pitch. Further discussion of these trends is given in Sect. 3.4. Noteworthy at this point is the fact that the optimum conditions for exposure (energy, dose) and development (temperature, time), are intimately interrelated and should be selected simultaneously for optimal performance of EBL at the nanoscale.

3. Theory and modelling insights into deep nanoscale-resolution EBL

As demonstrated in Sect. 2, both the exposure and development conditions should be selected carefully to reach the best quality of dense nanopatterns. In practice this may mean that 4 or even more process parameters must be co-optimized for a given resist-substrate-developer combination. To understand and systematize the relevant trends, predict the outcomes, and potentially optimize EBL in-silico, appropriate numeric models should be employed and/or developed. Providing efficient theories and numeric models of the various stages of EBL can substantially facilitate the optimization at the nanoscale. Below we discuss some of the emerging challenges as well as our solutions for numeric modelling of the exposure and development stages of the EBL process.

3.1 Towards new modeling paradigm for exposures of positive resists

Electrons' travel and energy dissipation in resists is believed to be a well explored area of research. Hundreds of publications describing analytic and Monte-Carlo models for handling the interaction of electrons with resists might seem to address every aspect of this. The Reference list to this Chapter features just a few works (Kyser & Viswanathan, 1975; Murata et al., 1981; Lee et al., 1992; Raptis et al, 1993; Raptis et al, 2001; Zhou & Yang, 2006) out of a tremendous amount of literature on theory, modeling, and simulation of EBL exposures. However, the existing models of exposure are mostly limited to computations of the average amounts of energy deposited by primary, secondary, and backscattered electrons in the resist. The resulting distributions of deposited energy are then converted into the probability of bond scissions employing the so-called radiation chemical yield (G-factor), which is the number of main-chain scissions per 100 eV of deposited energy (Greeneich, 1974; Han et al., 2003; Kyser & Viswanathan, 1975). Usually, the G-factor is not

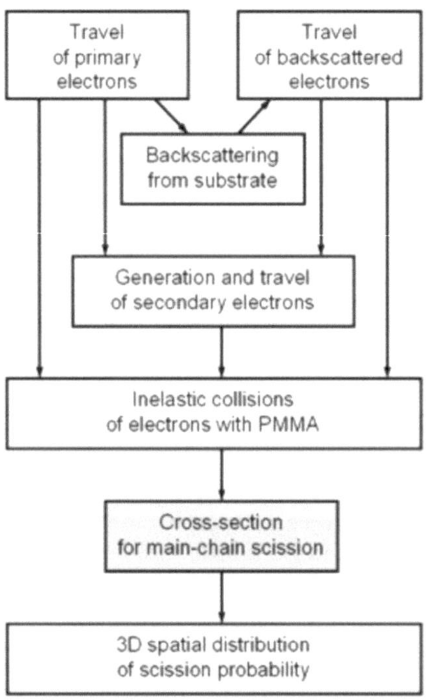

Fig. 10. Outline of our model for resist exposure in EBL. In the model, the probability of main-chain scissions is computed directly through the differential cross-section for inelastic collisions resulting in the scissions (Aktary et al., 2006). The model avoids uncertainties related with the conversion of the distributions of deposited energy into the number of main-chain scissions through the empirical radiation chemical yield.

known precisely and determined from experiments where the resist is exposed to extremely high-energy (1 Mev) electrons or gamma-rays (Chapiro, 1962). Use of such empirical G-factors involves a significant level of uncertainty, since the experimental conditions differ dramatically from those employed in EBL. We have also shown that the G-factor is not necessarily a constant, but may depend on the energies of electrons involved in collisions (Aktary et al., 2006), which should be accounted for. Furthermore, using the distributions of deposited energy as a starting point for the computations of the number of scissions is not an unequivocal choice, since a part of deposited energy is thermalized without scissions involved, whereas the actual yield of scissions is related intimately with the details of individual collisions of electrons with atoms of PMMA.

The uncertainties described above are avoided if one chooses an alternative approach employing a direct computation of the probability of main-chain scissions for each inelastic collision individually (Wu & Neureuther, 2001). By this, the conversion of the cumulative deposited energy into the number of scissions is avoided. We have recently introduced a model that explores this vision (Aktary et al., 2006), see also the outline in Fig. 10. In brief, after focused electrons hit a layer of PMMA on a substrate, they undergo both elastic and inelastic scattering in PMMA, as well as backscattering from the substrate. Travel of primary, secondary, and backscattered electrons in PMMA is described by transport theory, which provides a high robustness of the numerical algorithm. The model employs improved cross-sections for inelastic collisions, which were validated against the dielectric-response modeling (Tan et al., 2005). To account for scissions by the inelastic collisions, we compute the probability of dissociation of the main-chain C−C bonds employing a cross-section defined with an accounting for the molecular properties of PMMA's monomers such as the number of valence electrons and the dissociation energies of bonds. More details on the model are given elsewhere (Aktary et al., 2006). Fig. 11(a) shows an example of a computed 3D distribution of the probability of main-chain scissions in a periodic grating pattern.

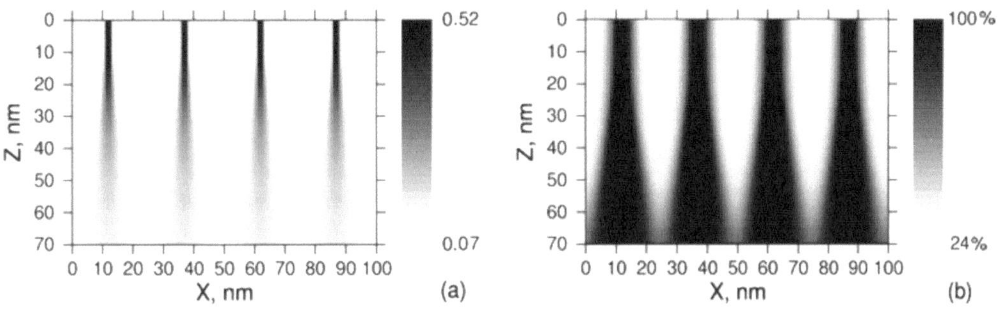

Fig. 11. Simulated probability of the main-chain scission per monomer of PMMA (a) and a corresponding volume fraction of small fragments, n≤12 (b), for a periodic grating with 25 nm pitch exposed with 10 keV electrons. The images present cross-sections of the 3D patterns, where Z is depth and X is width. Periodic boundary conditions were used in the horizontal directions, and Si substrate was located underneath the 70 nm layer of PMMA.

3.2 Fragmentation of PMMA and basic model of resist clearance

After the spatial distribution of the probability of chain scissions is computed, it is possible to identify the corresponding distribution of fragments of various weights. The distributions of the various PMMA fragments are generally believed to be the major factor determining the outcome of the resist development, eventually defining the morphology of the clearance profiles. However, in spite of the understanding of this importance, molecular models of development and clearance were insufficiently addressed in the literature. Most available models of clearance of positive resists relate the distributions of deposited energy, or the corresponding probability of scission, with the local rate of development through various empirical coefficients (Greeneich, 1974; Han et al., 2003; Kyser & Viswanathan, 1975). Obviously, addressing ultra-high resolution processes at the deep nanoscale requires a more explicit accounting for the resist fragmentation to be included in the model. In our works, we have converted the computed 3D distribution of the probability of chain scissions into the spatial distribution of fragments of various weights employing the Poisson distribution (Aktary et al., 2006; Mohammad et al., 2007) or the geometrical distribution (Fito et al., 2009). As the next step, we assume as a tentative approximation that fragments with molecular weight less than a maximum number of monomers, n_{max}, are preferentially soluble at the development stage and thus responsible for the resulting morphology of the resist. This allows dichotomizing the spatial distributions of PMMA fragments into a light and preferentially soluble fraction ($n \leq n_{max}$) and a heavier and less soluble fraction ($n > n_{max}$). Fig. 11(b) shows an example of a spatial distribution of the volume fraction of small fragments for an example of $n_{max}=12$.

It is clear that the overall volume fraction of small fragments in the resist increases with the exposure dose, and the fraction of the heavier fragments decreases accordingly. This is demonstrated by Fig. 12, which shows the maximum percentage of the heavy fraction ($n > n_{max}$), encountered in a dense grating pattern as a function of the area dose, for the various values of n_{max}. It can be seen that the dependencies have a sharp step, whose position depends on n_{max}. It worth to note a qualitative resemblance of the dependencies in Fig. 12 with experimentally measured contrast curves representing the dependence of the remaining resist on the dose (Hasko et al, 2000; Yan et al., 2008). This resemblance implies that the conditions of clearance are intimately related with the size distribution of fragments in the exposed resist. In particular, it is reasonable to hypothesize that during development, PMMA is removed at locations where the content of small, preferentially soluble fragments exceeds a given threshold value C_{min} (Aktary et al., 2006). With this assumption, it is possible to convert spatial distributions of the light fraction of PMMA into 3D profiles of

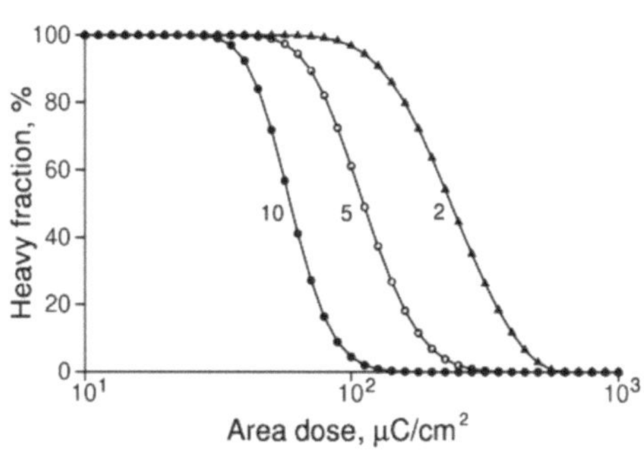

Fig. 12. The maximum percentage of volume fraction of the heavier, less soluble fraction ($n > n_{max}$), in a dense grating pattern as a function of the exposure dose, for the examples of n_{max}= 2, 5, and 10 monomers (adapted from Aktary et al., 2006).

clearance, provided that the parameters n_{max} and C_{min} are available. Fig. 13 presents examples of the clearance profiles for a periodic grating structure exposed with various doses, for the assumed values n_{max}=12 and C_{min}=90% (Mohammad et al., 2007).
With the computed clearance profilers, it is possible to analyze the regimes of clearance as a function of the process conditions. Thus for gratings, the regimes of underexposure (insufficient clearance) and overexposure (excessive clearance) are immediately available from the simulated clearance profiles as demonstrated in Fig. 13.

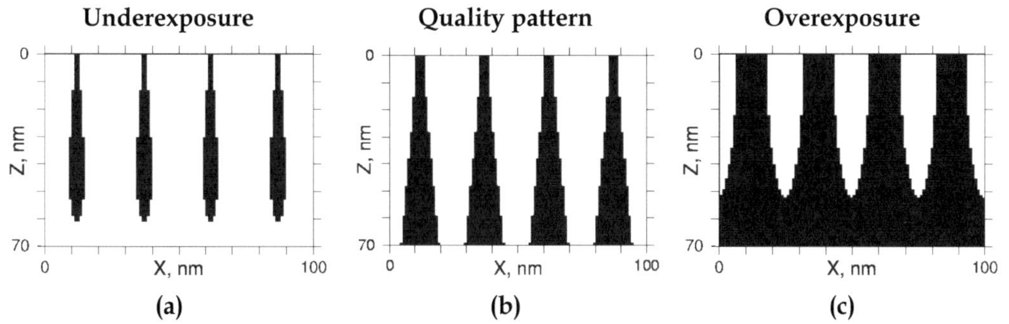

Fig. 13. Simulated clearance regimes for gratings in 70nm thick PMMA exposed by 10 keV electrons with various doses. Black - no resist; white - resist (Mohammad et al., 2007).

In Fig. 4 (Sect. 2.2.) the numerically determined boundary doses for underexposure and overexposure in various gratings are compared with the experiment. The clearance model parameters were as in Figs. 11 and 13. The lower curve in Fig. 4 (open triangles) represents the minimum area dose required for clearance of lines over their entire depth. This dose decreases roughly in proportion to the inverse of the grating period as discussed in Sect. 2.2. The upper curve (open diamonds) shows the boundary dose for overexposure, at which neighbor lines start merging. Since the neighbor lines overlap is less for larger interline distances, higher doses are required to clear the walls between the lines and therefore the boundary dose for overexposure increases with the grating pitch. The overall agreement between the numerical results and experimentally determined boundary doses for underexposure and overexposure is reasonably good, indicating that these limiting factors are well captured by the fragmentation-based model of clearance.
From Fig. 4 it is evident that the theoretical limit of resolution for gratings fabricated in PMMA with a 65 nm initial thickness is close to a 20 nm pitch. It is also possible to identify the area dose at which the best resolution can be reached (which is close to 80 $\mu C/cm^2$ in this case). However, it is also evident that the clearance conditions alone are insufficient to describe the quality of dense nanostructures. The pattern degradation, such as micellization or collapse, may occur at lower doses than over-clearance. As a result, the dose windows determined experimentally are narrower and the actual process resolution is lesser than predicted by the basic model of clearance. Thus, the predicted resolution can be considered as an ultimate theoretical limit, which would only be attainable if resist degradation during development is avoided.
Since the fragmentation-based model of clearance is representative of trends determined by the exposure, it may be useful to analyse the dependence of the EBL process on the applied voltage. Fig. 14 presents the computed boundaries for underexposure and overexposure in

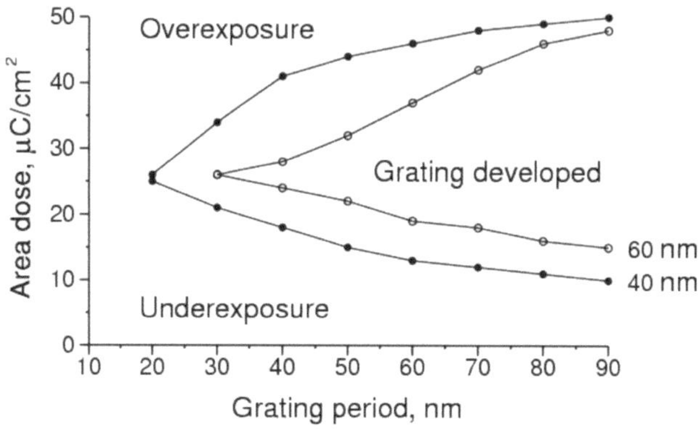

Fig. 14. Simulated clearance regimes for 3 keV exposures of gratings in PMMA with the initial thickness of 60nm (open circles) and 40nm (filled circles), (Stepanova et al., 2007).

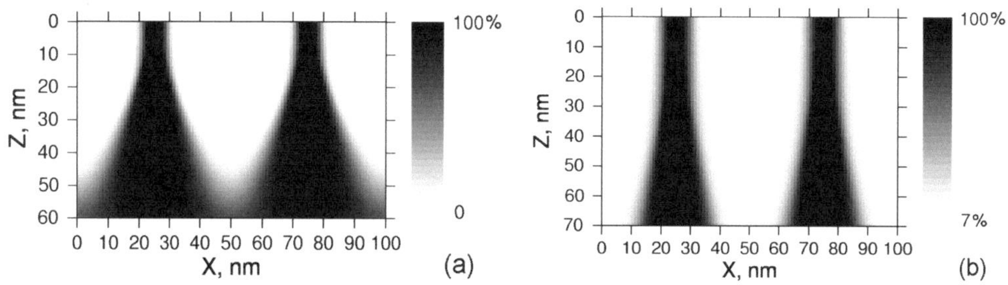

Fig. 15. Simulated distributions of light PMMA fraction (1 to 12 monomers) for a grating with 50 nm pitch, for the exposure voltages of 3 keV and (a) and 10 keV (b). The area doses correspond to the best theoretical resolution as defined from Figs. 14 and 4, i.e. 27 μC/cm^2 and 75 μC/cm^2, respectively (adapted from Stepanova et al., 2007).

gratings for 3 keV electrons energy. It can be seen that the predicted dose windows for 60 nm thick PMMA are dramatically narrower in comparison with modeling results for 10 keV exposures shown in Fig. 4, and the resolution limit is shifted toward sparser gratings. Thus, the theoretical dose window for 50 nm gratings in 60 nm thick PMMA is 10 times narrower for 3 keV exposures than for 10 keV exposures. The origin of this is clarified by Figs. 15 (a) and (b) which compare the spatial distributions of the light PMMA fraction in a periodic grating structure exposed by 3 keV and 10 keV electrons, respectively. It can be seen that the distribution for 3 keV broadens from the top down significantly more than that for 10 keV. The evident explanation is the strong forward scattering of the low-energy electrons. The corresponding broadening of the intensely exposed area leads to the undercuts in the clearance profiles, which is also confirmed by the experiment in Fig. 6(a). As a result the neighbour lines' overlap increases, which limits the resolution. However, a thinner (40 nm) resist layer allows fabrication of smaller features, as well as offers a wider window of favourable doses for high-resolution writing. The reason is that the broadening by forward scattering strongly increases with the depth (see Fig. 15(a)), so that the overlap is less pronounced at lower thickness of the resist. Another trend evident from the comparison of the numerical results in Figs. 4 and 14 is that the predicted area dose for optimum resolution is approximately three times less for 3keV exposures than for 10 keV exposures.

One can conclude that the basic model of clearance represents reasonably the dependencies of the EBL process on the conditions of exposure, such as the applied dose and voltage. Qualitatively, the model predictions are in agreement with the experimentally observed bahaviour of the overall resist clearance on the dose (contrast curves), as well as with the dependencies of the resolution and sensitivity on the conditions of exposure for dense high-resolution nanopatterns. Some of these trends can be predicted quantitatively. For example, the regimes of clearance in dense nanopatterns or the voltage dependence of the optimum area dose for the best resolution (see Figs. 4, 14, and 6). However, the model includes parameters n_{max} and C_{min}, which are not known precisely. Various resists, developers, and development conditions (such as temperature) would generally require a different parameterization. Also this framework does not address explicitly the kinetics of the development, as well as the limiting factors originating from the resist degradation, such as collapse. To extend the capacity of modeling toward a more detailed interpretation of the experimentally observed trends, and efficiently optimize the various aspects of post-exposure stages of EBL, a complementary kinetic analysis of the resist dissolution is required as outlined in the next section.

3.3 On the kinetics of resist dissolution

To analyse kinetics of resist dissolution, we have adopted as a background the mean-field model of diffusion in polymeric systems introduced recently (Scheinhardt-Engels et al., 2003). The model accounts for the molecular mobility, intermolecular interactions, and pressure relaxation in a system containing a solvent and polymer chains of a given length. In contrast to the original work (Scheinhardt-Engels et al., 2003), which addresses only the stationary-state boundary conditions, we have considered a more general case by adopting non-stationary behaviours. This generalization, however, does not affect the basic equations which are similar to those given in (Scheinhardt-Engels et al., 2003), and read:

$$\frac{\partial \varphi_A(\mathbf{r},t)}{\partial t} = -\nabla \mathbf{J}_A(\mathbf{r},t)\,; \tag{1}$$

$$\mathbf{J}_A(\mathbf{r},t) = -\frac{D_A \varphi_A(\mathbf{r},t)}{\sum_C D_C \varphi_C(\mathbf{r},t)} \sum_B D_B \varphi_B \nabla\big(\mu_A(\mathbf{r},t) - \mu_B(\mathbf{r},t)\big)\,; \tag{2}$$

$$\nabla \mu_A(\mathbf{r},t) = \sum_B \left(\frac{\delta_{AB}}{\varphi_A(\mathbf{r},t) N_A} - \frac{1}{N_B} + \chi_{AB} - \sum_C \varphi_C(\mathbf{r},t)\chi_{BC} \right) \nabla \varphi_B(\mathbf{r},t)\,; \tag{3}$$

$$\sum_A \varphi_A(\mathbf{r},t) = 1\,. \tag{4}$$

Here A, B, and C denote the molecules type, i.e. the polymer (PMMA) fragments of various size and the developer; N_A is the number of segments (e.g. monomers) in molecule A; $\varphi_A(\mathbf{r},t)$ is the volume fraction of the component A as a function of the location $\mathbf{r}$ and time t; $\mathbf{J}_A$ is the flux of the component A; $\nabla\mu_A$ is the local gradient of the chemical potential driving the kinetic process expressed by the flux $\mathbf{J}_A$; χ_{AB} is the Flory-Huggins (FH) interaction parameter, D_A is the diffusivity, and δ_{AB} is the Kronecker delta function.

With appropriate initial and boundary conditions, the numeric solution of Eqs. (1-4) would predict spatial re-distribution of the components as functions of time *t*. This, however, would require defining the FH interaction parameters χ_{AB} and the diffusivities D_A for all components in the system, which is a very complex task. Also the numeric solution of the full system of equations in the multicomponent mixture is a relatively time-consuming process, whereas a high numeric efficiency is required for a model that could potentially be employed for in-silico aided optimization of EBL process. For these reasons, we have employed the system of equations (1-4) as a starting point to derive a simpler, but more efficient *ansatz* that describes the kinetic process of development by the movement of the resist-developer interface, and contains less unknown parameters than Eqs. (1-4).
To derive our approximate model of resist dissolution (Fito et al., 2009), we consider PMMA embedded in a developer (solvent). PMMA may contain fragments of various sizes (numbers of monomers) n=1, 2, 3, … . However, at every location $\mathbf{r}$ the polymer melt is represented by one characteristic size of fragments (this may be a monodisperse polymer or a polydisperse system represented by local average fragment size). We denote the developer and PMMA fragments by indexes 0 and i, respectively, and assume that the diffusivity of the developer is higher than the diffusivity of PMMA, $D_0 >> D_i(\mathbf{r})$. Further, we define the position of the resist-developer interface, $\mathbf{r}_{int}$, as the level surface of equal developer concentration, $\varphi_0(\mathbf{r}_{int},t) = p =$ constant, such that $1 - \varphi_0 << 1$ at the interface. Next, we introduce the velocity at which the interface changes its position, $\mathbf{v}=d\mathbf{r}_{int}/dt$, and which is determined by the flux of the developer at the interface,

$$\mathbf{v} \cdot \nabla\varphi_0\big|_{\mathbf{r}=\mathbf{r}_{int}(t)} = \nabla\mathbf{J}_0\big|_{\mathbf{r}=\mathbf{r}_{int}(t)} . \tag{5}$$

Depending on the sign of $\nabla\mathbf{J}_0$ one can distinguish two regimes: prevailing swelling ($\nabla\mathbf{J}_0>0$), and prevailing shrinking ($\nabla\mathbf{J}_0<0$). Considering that swelling of the resist occurs only at the beginning of dissolution, whereas shrinking is the major mechanism responsible for development, we have focused our analysis on the shrinking regime. With these definitions and assumptions, we have obtained the following expression for the velocity of motion of the resist-developer interface (Fito et al., 2009):

$$v = \eta D_i(\mathbf{r}_{int})L^{-1} . \tag{6}$$

In Eq. 6, $D_i(\mathbf{r}_{int})$ is the diffusivity of PMMA at the location $\mathbf{r}_{int}$ and η is a constant coefficient that depends on the function $\varphi_0(\mathbf{r})$ near the interface. For example if the function φ_0 decays asymptotically as $\varphi_0 \sim x^{-1}$, where x is the penetration depth of the developer into the resist, the coefficient is $\eta = 2(1-p)(1-2\chi)$, where χ is the FH parameter describing the interaction of developer with PMMA such that $1-2\chi \geq 0$. L in Eq. 6 is the depth or shrinking of the resist. From Eq. 6, the following expression follows that describes the kinetics of resist shrinking:

$$\frac{dL}{dt} = \tilde{D}_i(\mathbf{r}_{int})L^{-1} , \tag{7}$$

where $\tilde{D}_i$ is the effective diffusivity, $\tilde{D}_i = \eta D_i$. In the case of constant diffusivity D_i, the integration of Eq. 7 leads to the well-known expression for the diffusion length, $L \sim (D_i t)^{1/2}$. For the corresponding rate of resist dissolution, one obtain the relation $v \sim (D_i/t)^{1/2}$.

The important implication of Eqs. 6 and 7 is that the rate of resist dissolution is a function of the entire history of the process of development, and thus depends on development time explicitly. This is different from the framework adopted in most if not all available models of EBL resist development, which assume the existence of a stationary regime that can be described by a constant rate of dissolution (Greeneich, 1974; Han et al., 2003; Hasko et al., 2000). Our analysis demonstrates that this assumption is not applicable in general, and that a more complex kinetics of dissolution, e.g. such as expressed by Eq. 7, should be accounted for to describe resist development at the nanoscale.

Employing Eq. 7, we have implemented an efficient finite-element numeric algorithm to model the process of development of exposed PMMA. As the initial condition, we use our computed 3D distributions of the probability of scission in the resist (see e.g. the example in Fig. 11(a)). We convert the local probability of scission into the local volume fractions of PMMA fragments of various size $\varphi_n(x,y,z)$, where n = 1, 2, 3, ... is the number of monomers in a fragment. For the normalized diffusivity of fragments of size n, we employ the following expression (Masaro & Zhu, 1999; Miller-Chou & Koenig, 2003),

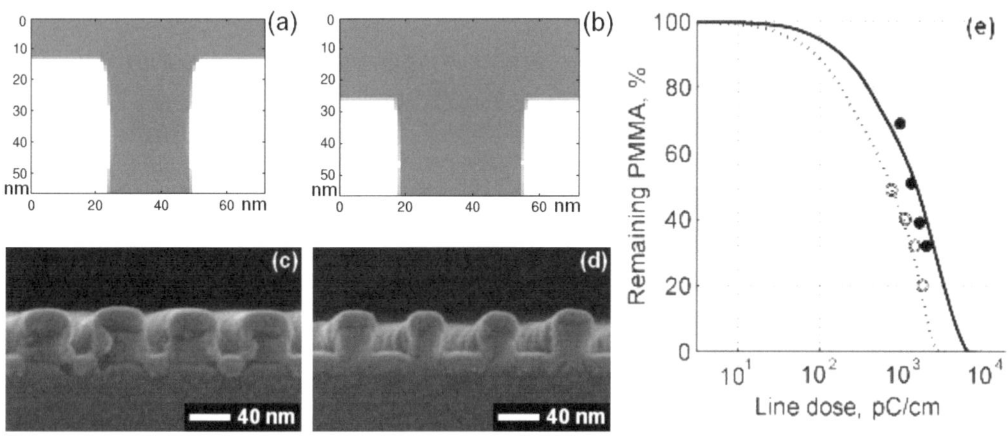

Fig. 16. (a,b) – computed cross-sectional profiles (white - PMMA, grey – no PMMA) in a periodic 70 nm pitch grating exposed by 10 keV electrons with the doses of 700 pC/cm and 1300 pC/cm, respectively, and developed at -10°C during 20 sec; (c,d) – SEM cross-sectional profiles for similar conditions; (e) – computed (lines) and experimental (symbols) percentages of remaining PMMA as functions of the exposure dose, in a 70 nm pitch grating exposed by 10 keV electrons and developed at -15°C during 5 sec. (solid line and filled symbols) and 20 sec. (dotted line and open symbols), (adapted from Fito et al., 2009).

$$\tilde{D}_n = a n^{-\alpha} \exp(-U/kT), \tag{8}$$

where a and α are model parameters, U is the activation energy, and T is temperature. In polydisperse polymers melts, the proportionality $\tilde{D}_n \sim n^{-\alpha}$ can be interpreted as describing the mobility of molecules of size n in a media which properties are represented by power α. Obviously, the mobility decreases with the size of molecules, i.e. smaller molecules are more mobile than larger ones. Since in exposed resist the fragment sizes are represented by the distribution φ_n, a location-dependent average effective diffusivity D can be introduced,

$$D(x,y,z) = \langle \tilde{D}_n \rangle = \left\langle \frac{\beta}{n^{\alpha(x,y,z)}} \right\rangle, \quad (9)$$

where the averaging is performed over the local distribution of fragments, and β=a·exp(-U/kT). In most polymers, the power α varies from 1 in dilute solutions of small molecules to 2 or even higher for denser melts of longer polymer chains, where the mobility is dampened by entanglement (Harmandaris et al., 2003; Masaro & Zhu, 1999; Miller-Chou & Koenig, 2003). We have employed the following dependence for α,

$$\alpha(x,y,z) = \begin{cases} 1 + \langle n \rangle / \gamma, & \langle n \rangle < \gamma; \\ 2, & \langle n \rangle \geq \gamma. \end{cases} \quad (10)$$

We simulate the process of development by a sequence of discrete dissolution steps, where time δt required to dissolve a resist layer of thickness δL is determined by δt=2LδL/D(x,y,z). The simulation provides the location of the 3D resist-developer interface as a function of time. The parameters β and γ can be evaluated by fitting the computed percentages of PMMA left on the substrate to the corresponding experimental results. We obtain these from the SEM cross-sectional profiles for gratings, and parameterize our model by minimizing the mean square difference using the simulated annealing algorithm (Fito et al., 2009). It has been assumed that the parameters β and γ may depend on temperature of development, but not on other process conditions.

Fig. 16 presents our computed cross-sectional clearance profiles for a 70 nm pitch grating (a,b) and the amount of remaining PMMA (e) in comparison with the corresponding experiments. The fit model parameters that we employed in these examples were: Fig. 16(a,b) – $\beta=2.34\cdot10^3$ nm^2/s, γ=23.1; Fig. 16(e) – $\beta=2.38\cdot10^3$ nm^2/s, γ=16.5 (Fito et al., 2009). From Fig. 16 it can be seen that the kinetic model describes quite well the broadening of the clearance trenches in gratings, the thinning of the walls, as well as the shrinkage of the thickness of the resist during development. The dependencies of the total amount of PMMA remaining at the substrate on the exposure dose and development time are also represented with a promising accuracy. However, this simple model does not capture all the details of the shape the cross-sectional profiles, which appear to be smoother at the top, and show stronger undercuts at the base, than predicted numerically. We expect that some of these differences may originate from surface tension forces and/or plastic deformation; the potential impact of these factors still needs to be investigated.

3.4 Kinetic model for applicable dose windows

In this section we extend our kinetic modeling framework to describe the boundary exposure doses for the conditions of resist underexposure and collapse, d_{min} and d_{max}, which determine the applicable dose windows and by this represent the nanoscale resolution and sensitivity of EBL (see Sect. 2).

Our numerical results presented in Figs. 16 (a,b) demonstrate that at a given development time, the locations at which the resist clearance occurs depend on the exposure dose. At lesser doses PMMA is removed preferentially from the intensely exposed trenches where fragments are the most mobile (see Fig. 16(a)), whereas more intense exposures allow for partial dissolution of interline walls containing heavier, less mobile fragments (Fig. 16(b)). According to the kinetic model described in Sect. 3.3, dissolution of the exposed PMMA can be described as a kinetic diffusion-like process, with the effective diffusivities of fragments represented by Eq. 8. For simplicity, we now describe the average local diffusivity in terms of the mean number of monomers in fragments $\langle n \rangle$, $D \sim \langle n \rangle^{-\alpha}$ (Masaro & Zhu, 1999). For moderate exposure doses, $\langle n \rangle$ is inversely proportional to the local probability of scissions (Aktary et al., 2006), which in turn is proportional to the applied dose d. The dose dependence can therefore be approximately accounted for by $\langle n \rangle \sim 1/d$. As a result, the average diffusivity of fragments in exposed PMMA becomes

$$D = c d^{\alpha} \exp(-U/kT), \tag{11}$$

where c and α are, in general, location-dependent parameters. In our recent work (Mohammad et al., 2009) we have assumed that for a given development time and pattern geometry, the minimum and maximum applicable doses for quality nanofabrication, d_{min} and d_{max}, correspond to the same value of the average diffusivity D in the exposed trenches and in the walls, respectively. This assumption results in the following equation for the temperature dependence of these boundary exposure doses d_{min} and d_{max} for a given development time and pattern geometry:

$$d_{\substack{min,\\ max}} = d^{ref}_{\substack{min,\\ max}} \exp\left(\frac{U}{\alpha_{\substack{min,\\ max}} k} (1/T - 1/T^{ref}) \right). \tag{12}$$

Here 'ref' denotes a reference value of the minimum or maximum applicable dose and of the development temperature, and the notation $\alpha_{min,max}$ indicates that the dependencies for d_{min} and d_{max} may correspond to different α values. Here, however, we use a single value U/α=0.22eV, which we evaluated empirically (Mohammad et al., 2009). In Fig 17, we compare the corresponding theoretical results with the experimental temperature dependencies for the boundary line doses in a 70 nm pitch grating exposed with 10 keV electrons. We used the experiments for -15°C for reference.

As it can be seen in Fig. 17, Eq. 12 describes very well the broadening of the applicable dose window with the decrease of the development temperature that we have observed experimentally. According to the model assumptions, the lower boundary of the applicable dose window d_{min} can be interpreted as the dose at which PMMA fragments in the intensely exposed trenches are mobile enough to be removed over the time of development. The maximum applicable dose d_{max} is, in turn, determined by molecular diffusion processes occurring in the walls. The applicable dose window d_{max}–d_{min} is therefore defined by the difference in these diffusivities. It is noteworthy that, if the parameter α is a constant (α_{min}=α_{max}), Eq. 12 predicts that the decrease in temperature would increase the difference d_{max}–d_{min}, but not the ratio d_{max}/d_{min}. This prediction is very well confirmed by our experiments for 50 nm pitch gratings exposed with 30 keV electrons (Table 2). For 70 nm pitches (Table 1) we observe a moderate temperature dependence of d_{max}/d_{min}; however the reasonable agreement of the theory with experiment in Fig. 17 seems to confirm the viability of the approximate model. Further development of the kinetic framework, which is currently in progress, will provide more detailed predictions for the sensitivity and resolution of EBL.

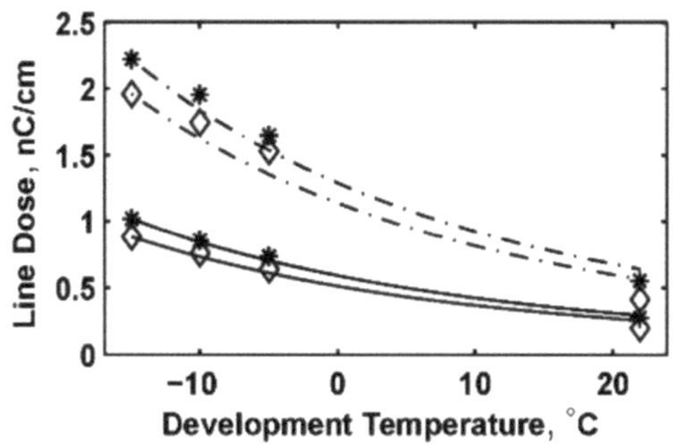

Fig. 17. Comparison of theoretical temperature dependencies for minimum (solid lines) and maximum (dashed lines) applicable exposure doses for a 70 nm pitch grating exposed with 10 keV electrons. The symbols show the experimental data for development times of 5 sec. (stars) and 20 sec. (diamonds) (adapted from Mohammad et al., 2009).

4. The perspective of low-voltage EBL process optimization

From the previous discussion it is evident that fabricating dense nanopatterns with specified characteristics is a complex optimisation task. At least 4 process parameters (the exposure voltage, dose, development temperature, and time) must be co-optimized for a given resist-substrate-developer combination. Our approach comprises a thorough experimental study as well as numerical analysis that allow interpreting and systematising the observed trends. In Sect. 2 we have described the major limiting factors for nanoscale resolution in EBL (underexposure and collapse) and employed the corresponding window of applicable exposure doses as a descriptor for the dependence of resolution and sensitivity on the various process conditions. Based on this framework, we obtained extensive experimental benchmarks representing the dependencies of the boundary doses d_{min} and d_{max} as functions of the size of nanostructures, development time, and temperature. Employing the suite of numeric models described in Sect. 3 has facilitated understanding of the observed trends and allowed us to develop an efficient methodology for optimizing the EBL process at the nanoscale.

The examples in Fig. 18 show the highest resolution grating structures that we obtained by exposing 47-55 nm thick layers of PMMA to 10 keV electrons, for various development temperatures. Thus RT development yields 33 ± 2 nm wide trench lines in a grating with a 70 nm pitch (Fig. 18(a)). With -10 °C development, we could fabricate 20 ± 2 nm lines in a 50 nm pitch grating (Fig. 18(b)). This line width was further improved to 15 ± 2 nm in a 40 nm pitch grating at -15°C temperature (Fig. 18(c)). One can conclude that the minimal resolved feature size significantly decreases with decrease of the development temperature from RT down to -15°C. This is consistent with the observed trend of broadening of the applicable dose windows, and explained by the kinetics of fragment mobility during resist dissolution.

3 keV exposures provide a significantly higher throughput than 10 keV ones, which is accompanied by narrower process windows because of large undercuts of the resist (see Sect. 2.4 and 3.2). However, despite the limitations in resolution of low-voltage EBL, the undercuts facilitate metallization and lift-off for ultrahigh-resolution processes. By combining 3 keV voltage exposures with cold development, we were able to fabricate highly uniform arrays of 20 nm wide metal (Cr) lines in a grating configuration with a 60 nm pitch, such as shown in Fig. 19. The cold development enabled a sufficient robustness of the

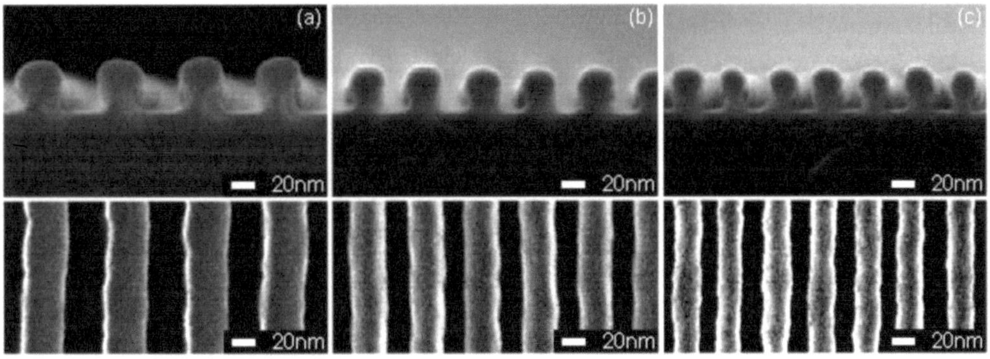

Fig. 18. Cross-section and top-view micrographs showing examples of optimized dense nanoscale gratings fabricated in PMMA using a Raith 150 system with 10keV voltage, and employing various development temperatures: (a) RT, 70nm pitch; (b) - 10 °C, 50nm pitch; and (c) - 15°C, 40nm pitch (Mohammad et al., 2009).

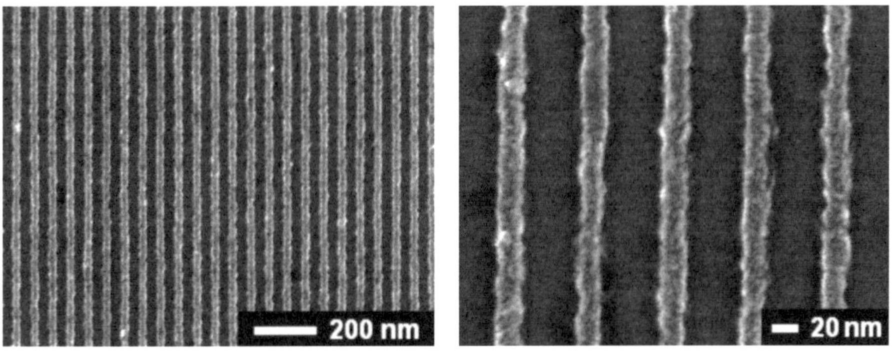

Fig. 19. 20 nm-wide Cr lines in a 60 nm pitch grating, fabricated by 3 keV exposure of a single layer of PMMA-950K, and developed at -15°C. Evaporation of a ~15 nm thick Cr layer was employed followed by ultrasonic lift-off. The substrate is a layer of SiCN on a Si wafer.

process, whereas the undercuts created by the ultra-low energy exposures allowed us to use a single layer of PMMA-950. Overall, the low-voltage based process turned out to be significantly more efficient than the conventional bi-layer schemes that are often used for lift-off at the deep nanoscale.

5. Conclusion

We have studied systematically the resolution and sensitivity of EBL processes for moderate to ultra low voltage exposures. We have employed the conventional resist (PMMA) and developer (MIBK:IPA mixture), however, we have varied the development temperature from RT down to -15°C. We consistently observe that the lowest development temperatures yield the largest applicable dose windows, but also require the highest clearance doses.

Our results indicate that both moderate and low voltage exposures offer great potential for efficient fabrication at the nanoscale dimensions. However, for reaching the ultimate resolution with ultra-low energy exposures, both the exposure and development conditions should be considered as interdependent factors and optimized simultaneously. The major

research challenge is related to the molecular mechanisms of development, which should be understood thoroughly in order to efficiently optimize low-voltage EBL.
We have suggested an improved framework for modeling fragmentation of positive resists, and also described a suite of models for clearance. These models have proven to be extremely useful for understanding and systematising the observed trends. Furthermore, the described models of resist development offer a strong potential for in-silico optimization of EBL. Further effort should address the diverse physico-chemical processes accompanying the resist dissolution, such as gelation, swelling, surface tension, and plastic deformation, whose potential impact on the resolution is not yet understood sufficiently. For strengthening the predictive capacity of the kinetic models, their parameterization employing more detailed molecular simulations should also be implemented.

6. Acknowledgements

The authors thank Marko van Dalfsen for his assistance with the EBL process, and Zsolt Szabò for designing dose test samples and his programming contributions. The work was supported by NINT-NRC, NSERC, Alberta Ingenuity, iCORE, and Raith GmbH.

7. References

Aktary, M., Stepanova, M., & Dew, S.K. (2006) Simulation of the spatial distribution and molecular weight of PMMA fragments in electron beam lithography exposures. *J. Vac. Sci. & Technol. B*, Vol. 24, pp. 768-779.

Brünger, W., Kley, E.B., Schnabel, B., Stolberg, I., Zierbock, M., & Plontke, M. (1995) Low energy lithography; Energy control and variable energy exposure. *Microelectronic Engineering*. Vol. 27, pp. 135-138.

Chapiro, A. (1962) *Radiation Chemistry of Polymeric Systems,* Wiley, ISBN 0470392851, New York, p. 515.

Cord, B., Lutkenhaus, J. & Berggren, K.K. (2007) Optimal temperature for development of polymethylmethacrylate. *J. Vac. Sci. & Technol. B.,* Vol. 25, pp. 2013-2016.

Dial, O., Cheng, C.C. & Sherer, A. (1998) Fabrication of high-density nanostructures by electron beam lithography. *J. Vac. Sci. & Technol. B.,* Vol. 16, pp. 3887-3890.

Fito, T., Mohammad, M.A., Szabó, Zs., Chen, J., Aktary, M., & Dew, S.K., & Stepanova, M. (2009) *manuscript in preparation.*

Greeneich, J.S. (1974) Time evolution of development contours in poly-methylmethacrylate electron resist. *J. Vac. Sci. & Technol.* Vol. 12, pp. 5264-5268.

Häffner, M., Heeren, A., Fleischer, M., Kern, D.P., Schmidt, G., & Molenkamp, L.W. (2007) *Microelectronic Engineering*. Vol. 84, pp. 937-939.

Harmandaris, V.A., Mavrantzas, V.G., Theodorou, D.N., Kröger, M., Ramírez, J., Öttinger, H.C., & Vlassopoulos, D. (2003) Dynamic crossover from Rouse to the entangled polymer melt regime: Signals from long, detailed atomistic molecular dynamics simulations, supported by rheological experiments. *Macromolecules,* Vol. 36, pp. 1376-1387.

Hasko, D.G., Yasin, S. & Mumatz, A. (2000) Influence of developer and development conditions on the behaviour of high molecular weight electron beam resists. *J. Vac. Sci. & Technol. B.* Vol. 18, pp. 3441-3444.

Han, G., Khan, M., & Cerrina, F. (2003) Stochastic modeling of high energy lithographies. *J.Vac. Sci. & Technol. B.*, Vol. 21, pp. 3166-3171.

Jamieson, A., Wlson, C.G., Hsu, Y. & Brodie, A.D. (2004) Low-voltage electron beam lithography resist processes: top surface imaging and hydrogen silisesquioxane bilayer. *J. Microlith., Microfab., Microsyst.* Vol. 3, pp. 442-449.

Kyser, D.K. & Viswanathan, N.S. (1975) Monte Carlo simulation of spatially distributed beams in electron-beam lithography. *J.Vac. Sci. & Technol.* Vol. 12, pp. 1305-1308.

Lee, Y. H., Browing, R., Maluf, N. , Owen, G. & Pease, R.F.W. (1992) Low voltage alternative for electron beam lithography. *J.Vac. Sci. & Technol. B.* Vol. 10, pp. 3094-3098.

Liddle J.A., Gallatin, G.M. & Ocola, L.E., (2003) Resist requirements and limitations for nanoscale electron-beam patterning. *Mat. Res. Soc. Symp.Proc.*Vol.739, pp. H1.5.1-12.

Masaro, L. & Zhu, X.X. Physical models of diffusion for polymer solutions, gels and solids. (1999) *Prog. Polym. Sci.* Vol. 24, pp. 731-775.

Miller-Chou, B.A., Koenig, J.L. (2003) A review of polymer dissolution. *Prog. Polym. Sci.* Vol. 28, pp. 1223-1270.

Mohammad, M.A., Dew, S.K., Westra, K., Li, P. Aktary, M., Lauw, Y., Kovalenko, A., & Stepanova, M. (2007) Nanoscale resist morphologies of dense gratings using electron-beam lithography. *J.Vac. Sci. & Technol. B,* Vol. 25, pp. 745–753.

Mohammad, M.A., Fito, T., Chen, J., Aktary, M., Stepanova, M., & Dew, S.K. (2009) Interdependence of optimum exposure dose regimes and kinetics of resist dissolution for electron-beam nanolithography of PMMA. *J. Vac. Sci. Technol. B. In press.*

Murata, K., Kyser, D.F. & Ting, C.H. (1981) Monte Carlo simulation of fast secondary electron production in electron beam resists. *J.Vac. Sci. & Technol.,* Vol. 52, pp. 4396-4405.

Ocola L.E. & Stein A. (2006) Effect of cold development on improvement in electron-beam nanopatterning resolution and line roughness. *J.Vac. Sci. & Technol. B.,* Vol. 24, pp. 3061-3065.

Raptis, I., Glezos, N., & Hatzakis, M. (1993) Analytical evaluation of the energy deposition function in electron-beam lithography in the case of a composite substrate. *J.Vac. Sci. & Technol. B.,* Vol. 11, pp. 2754-2757.

Raptis, I., Glezos, N., Valamontes, E., Zervas, E. & Argitis, P. (2001) Electron beam lithography simulation for high resolution and high-density patterns. *Vacuum,* Vol. 62, pp. 263-271.

Scheinhardt-Engels, S. M. , Leermakers, F. A. M. & Fleer, G. J. (2003) Lattice mean-field method for stationary polymer diffusion. *Phys. Rev. E.* Vol. 68, Art. 011802, pp. 1-15.

Schock, K.-D., Prins, F.E., Strähle, S., & Kern, D.P., (1997) Resist processes for low-energy electron-beam lithography. *J.Vac. Sci. & Technol. B.,* Vol. 15, pp. 2323-2326.

Stepanova, M., Dew, S.K., Kovalenko, A. & Aktary, M. (2007). Collaborative Project: 3D simulator of exposure and development in EBL. *Interim report.* NINT & University of Alberta, May 2007, Edmonton, Alberta, Canada.

Tan, Z., Xia, Y., Liu, X, & Zhao, M. (2005) Monte-Carlo simulation of low-energy electron scattering in PMMA using stopping powers from dielectric formalism. *Microelectronic Engineering,* Vol. 77, 285-291.

Word, M.J., Adesida, I. (2003) Nanometer-period gratings in hydrogen silsesquioxane fabricated by electron beam lithography. *J.Vac.Sci.& Technol. B.* Vol. 21, pp.L12-L15.

Wu, B. & Neureuther, A.R. (2001) Energy deposition and transfer in electron-beam lithography. *J.Vac. Sci. & Technol. B,* Vol. 19, pp. 2508–511.

Yan, M., Choi, S. Subramanian, K.R.V., & Adesida, I. (2008) The effects of molecular weight on the exposure characteristics of polymethylmethacrylate developed at low temperatures. *J.Vac. Sci. & Technol. B.* Vol. 26, pp. 2306-2310.

Yasin, S. Hasko, D.G., & Ahmed, H. (2002) Comparison of MIBK/IPA and water/IPA as PMMA developers for electron beam nanolithography. *Microelectronic Engineering.* Vol. 61-62, pp. 745-753.

Yasin, S. Hasko, D.G., Khalid M.N., Weaver, D.J., & Ahmed, H. (2004) Influence of polymer phase separation on roughness of resist features. *J.Vac. Sci. & Technol. B.* Vol. 22 pp. 574-578.

Zhou J. & Yang X. (2006) Monte Carlo simulation of process parameters in electron beam lithography for thick resist patterning. *J.Vac. Sci. & Technol.B.,* Vol.24, pp.1202-1209.

17

Computer Simulation of Processes at Electron and Ion Beam Lithography, Part 1: Exposure Modeling at Electron and Ion Beam Lithography

Katia Vutova and Georgi Mladenov
Institute of electronics, Bulgarian Academy of Sciences
Bulgaria

1. Introduction

Increasing the electronic circuit density is a major trend in microelectronics. Lithography is the key technology for the fabrication of very large integrated circuits with smaller device sizes. Today conventional optical lithography is approaching its fundamental physical limits. Electron and ion lithographies are among various candidates capable of sub-150 nm resolution for the new generation of lithographic techniques. Nanometer scale device fabrication rules require tight control of the developed polymer resist profile. Process simulation is a key tool for optimization of the obtained lithography results.

The goal of the computer simulation of the processes at electron and ion beam lithography (EBL and IBL) is the resist profile prediction of developed patterns after exposure of samples, covered by a sensitive polymer resist layer, which is sensitive to irradiation by accelerated particles. The accuracy of the simulated resist profiles strongly depend on the physical knowledge of the processes as well as on the accuracy of the process parameters.

The main step of such modeling: (i) exposure process modeling (latent image) and (ii) simulation of resist developed images at electron and ion beam lithography as well as the peculiarities of the developed Monte Carlo models are described and discussed in these two chapters (part 1 and part 2).

2. Deposited energy in the case of electron beam lithography simulation

The first main step of a complete mathematical model for electron beam lithography (EBL) simulation is to simulate the exposure of polymer resist films. During the exposure process the resist material modifies the local solubility rate. The main goal during the exposure process modeling is the calculation of the absorbed energy space distribution. There exist two types of method to calculate the deposited energy in the resist (latent image): analytical methods (Hatzakis et al., 1974; Hawryluk et al., 1974; Raptis et al., 1998) and numerical methods (Kyser & Murata, 1974; Adesida et al., 1979; Vutova & Mladenov, 1994). The analytical method is based on some particular approximations that simplify the nature of the real process (small-angle and diffusion particle scattering, single-component targets, point source or source of homogeneous cross section, etc). These assumptions do not correspond to the real process of the beam scattering within the target. Nevertheless, the simulation of energy deposition in the resist film coated on patterned substrates using analytical methods is very difficult if not impossible.

The numerical methods, based on the Monte Carlo (MC) technique for the statistical electron trajectory modeling and energy loss calculation, have been extensively developed and have become the most accepted methods in this field. The MC method mirrors the real process and in the case of large-number trajectory modeling, assures high statistical accuracy and satisfactory consistency. On the other hand, the MC method is ideal for parallel processing computers.

In the proposed simulation model the absorbed energy space distributions are calculated using MC algorithm for electron penetration and energy-loss calculation, which has four sub-steps: (i) forming an electron scattering model and calculating the discrete absorbed energy distribution in the resist film due to a point beam, (ii) approximating the absorbed energy using analytical expressions (energy deposition function EDF(r,z)), (iii) convoluting the function EDF(r,z) with the actual current distribution in the electron beam used for exposure, (iv) calculating the spatial distribution of the absorbed energy density in the resist which determines the obtained latent image during the electron beam exposure process of the desired layout. The process resolution is limited by the phenomena of forward deflection and backscattering of the electrons during their passage through the resist layer and the substrate. When a real micro-image is exposed, the absorbed energy in every resist volume point can be calculated by summing up the energy losses, obtained in regions far from the beam incident point (several μm). In this way the scattering of the beam electrons limits the lateral resolution of the exposed lithographic patterns and patterning of dense high-resolution layouts. This phenomenon is known as a proximity effect and the function EDF(r,z) characterizes the so-called proximity effect (undesired exposure dose due to backscattered electrons) (Chang, 1975).

Calculation of energy deposition in the resist film due to a point beam (with a zero-width beam diameter δ-function) requires to investigate thousands electron trajectories, i.e. millions collisions between accelerated electrons and scattering target atoms (an elementary collision sequence with target atoms). We calculate the characteristic changes in the particle motion for each collision, assuming a straight-line trajectory between two collisions (Fig.1).

The scattering atom is presented by a shielded Coulomb potential. The scattering angle θ of the penetrating electron is calculated using the differential scattering cross-section for the penetrating electron (for elastic collisions with target atoms) and assuming a Rutherford shielding potential presenting the scattering atom:

$$\frac{d\sigma}{d\Omega}=\frac{e^4}{(2mV^2 4\pi\varepsilon_o)^2}\left[\frac{Z}{\sin^2(\frac{\theta}{2})+\beta^2}\right]^2=\frac{1.3.10^{-19}}{E^2}\left[\frac{Z}{\sin^2\left(\frac{\theta}{2}\right)+\beta^2}\right]^2, \tag{1}$$

where σ is the differential scattering cross section [mm²], Ω is the solid angle, V is the velocity and E [eV] is the energy of the penetrating electron, Z is the atomic number of the scattering atom. The shielding parameter β characterizes the minimal scattering angle at which the value of $d\sigma/d\Omega$ does not increase more when the value of θ decreases:

$$\beta = 2.33\,(Z^{1/3}/E^{1/2}) \tag{2}$$

and then the total cross section of the elastic scattering is:

$$\sigma = \frac{0.126.10^{-17} Z^2}{E^2 \beta^2 (\beta^2 + 1)} \quad (3)$$

where σ [m^2] if E [eV].

The non-elastic scattering weakly changes the electron trajectory due to its domination only at very small values of the scattering angles. It is taken into account as Z^2 is replaced by Z(Z+1) in Eq.(1) and Eq.(3). The mean energy losses of the penetrating electrons, evaluated per one unit of the trajectory length are:

$$-\left\langle \frac{dE}{d\xi} \right\rangle = \frac{e^4 N_A \rho}{8\pi\varepsilon_0^2 E} \sum_i \frac{c_i Z_i}{M_i} \ln(1.166 \frac{E}{I_m}), \quad (4)$$

where $dE/d\xi$ [eV/Å], the target density ρ [g/cm^3], E [eV], N_A is the Avogadro's number, c_i is the weight portion of the i-kind of the target atoms, M_i is their atomic weight, I_m is the average ionization potential.

The program involves a MC technique to calculate the kind of the atom taking part in the collision, the azimuthal angle value, ect. The probability for a collision with the target atom of k-kind is:

$$P_k = (n_k . \sigma_k) / (\sum_{i=1}^{m} n_i . \sigma_i), \quad (5)$$

where m is the number of the different kinds of the target atoms. The scattering cross-section σ_i is calculated using Eq.(3) for the electron energy E before the collision. The concentration n_k of this kind of atoms is calculated using:

$$n_k = c_k (\rho / \overline{M}) N_A, \quad (6)$$

where $\overline{M}$ is the average target molecular weight. The length of the mean free path between two collisions of the penetrating electron is distributed in $[0, \infty]$ with a probability density:

$$p(\lambda) = \frac{1}{\lambda_0} \exp(\frac{-\lambda}{\lambda_0}), \quad (7)$$

where $\lambda_0 = (\sum_{i=1}^{m} n_i \sigma_i)^{-1}$. The distance that the electron travels between these collisions is:

$$\Delta\xi = - \lambda_0 \ln(1-R_1), \quad (8)$$

where R_1 is a random number evenly distributed in [0,1]. The energy losses of the penetrating electron at the interaction with the target electrons along this path are calculated using the Bethe energy-loss Eq.(4). The scattering angle of the collision is calculated:

$$\cos\theta = 1 - \frac{2\beta_k^2 R_2}{1 + \beta_k^2 - R_2}, \quad (9)$$

The azimuthal angle φ is given by:

$$R_3 = \frac{\varphi}{2\pi}, \tag{10}$$

where R_2 and R_3 are also random numbers. Due to the big difference concerning the weights of the particles, the energy losses at the collision between the penetrating electron and scattering atom are neglected. The values of $\Delta\xi$, θ and φ for each collision are calculated. The calculation is repeated to yield new position in the target for a new set of random numbers until the energy of the electron falls below a predefined value (500 eV) or the electron leaves the target. Then the electron trajectory is calculated (Fig.2).

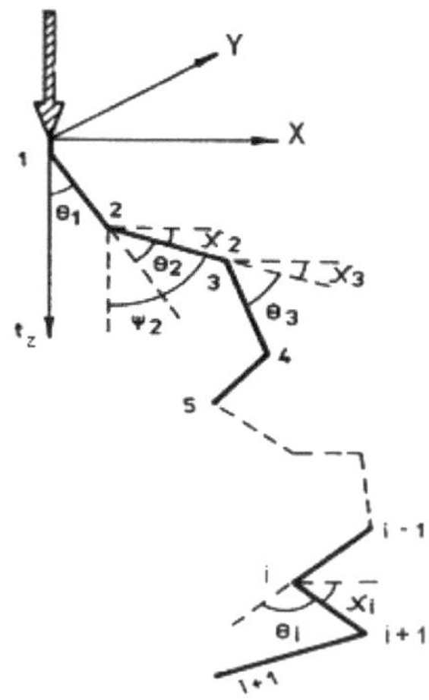

Fig. 1. Geometrical parameters for the electron scattering in the sample.

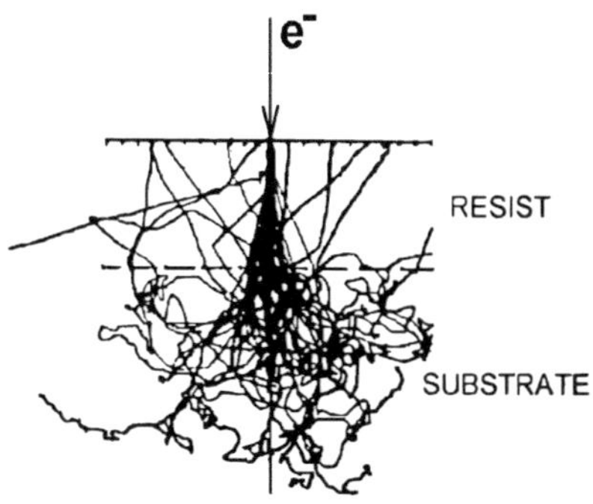

Fig. 2. Simulated trajectories of 100 electrons in PMMA, the electron energy is 30 keV.

The trajectory can be presented in the coordinate system Oxyt_z (easily connected with a cylindrical system, and $r=(x^2+y^2)^{1/2}$), where the axis t_z is parallel to the initial direction of the penetrating electron motion (usually this direction is perpendicular to the target surface) and using the relation:

$$\cos\psi_i = \cos\psi_{i-1}.\cos\theta_i + \sin\psi_{i-1}.\sin\theta_i . \cos\varphi_i. \tag{11}$$

The angles ψ_{i-1}, ψ_i and θ_i are shown in Fig.1. The angle φ_i is the azimuthal angle of scattering for the i-th collision. The point depth t_z for the (i+1)-th collision is calculated using the formula:

$$(t_z)_{i+1} = \sum_{k=1}^{i} \Delta\xi_k \cos\psi_k = (t_z)_i + \Delta\xi_i.\cos\psi_i. \tag{12}$$

The distance to the coordinate axis $0t_z$ is calculated by analogy assuming that the angle χ_i determines the trajectory path projection on the axis $0r$, as well as that the angle ϕ_i corresponds to the azimuthal angle φ_i. The following equations present the relations between these angles:

$$\sin\psi_o = \cos\chi_o \tag{13}$$

$$\cos\psi_o = \sin\chi_o \tag{14}$$

$$\phi_i = \varphi_i - \arccos\frac{-\cos\psi_{i-1}\cdot\cos\chi_{i-1}}{\sin\psi_{i-1}\cdot\sin\chi_{i-1}} \tag{15}$$

$$\cos\chi_i = \cos\chi_{i-1}.\cos\theta_i + \sin\chi_{i-1}.\sin\theta_i . \cos\phi_i \tag{16}$$

$$\sin\chi_i = \sqrt{1-\cos^2\chi_i}. \tag{17}$$

Then:

$$r_{i+1} = \sum_{k=1}^{i} \Delta\xi_k \cos\chi_k = r_i + \Delta\xi_i \cos\chi_i . \tag{18}$$

The energy losses for each segment (a straight line trajectory between two collisions) of the free path $\Delta\xi_i$, is represented as $\left\langle \frac{dE}{d\xi} \right\rangle .\Delta\xi_i$ and can be assumed as the energy losses in the point of the i-th collision for the corresponding elemental volume in the target. We check the energy of the electron and its position in the target. The calculation for this particle stops if the energy falls below the minimal predefined value (500 eV) or the electron leaves the sample. Then the calculation for new electron starts. After summing up the losses corresponding to the all N penetrating electrons, we obtain discrete data (a two-dimensional data array) for a radial energy deposition function (EDF) at various resist depths from the point of beam incidence. It is possible to use various cell dimensions at different radial distances: lower values near the point of beam incidence and higher values far away from this point. When using a Cartesian coordinate system instead of a cylindrical one it is possible to simulate beam incidence inclined to the resist surface (Gueorguiev, 1996). There are procedures for re-calculation of the free path and electron stopping power when a penetrating electron crosses the interfaces in multilayer structures (Gueorguiev et al., 1994). To achieve a satisfactory statistical accuracy, using MC calculations, a large number of particle trajectories are simulated (10-20 x 10^3). A detailed description and obtained results are presented in (Vutova & Mladenov, 1994; Mladenov & Vutova, 2002; Vutova, 2007).

2.1 Results and discussion

We have developed a program which realizes the MC algorithm to model the electron scattering in multilevel multicomponent amorphous targets, named TREM-MV. Using the simulation tool, calculations over a wide range of primary electron beam energies and resist thickness values are performed for different substrates (Vutova & Mladenov, 1994; Gueorguiev et al., 1995; Mladenov & Vutova, 2002; Vutova et al., 2007). In Fig.2 calculated trajectories in the case of a 1 μm thick poly-methyl methacrylate (PMMA) (a widely used polymer resist) on Si substrate are shown. The chemical composition of this resist is $C_5H_8O_2$, the efficient atomic number is Z=3.6, the atomic mass is M = 6.7 and the polymer density is ρ= 1.22 g/cm^3.

Fig.3 presents the radial distribution of the electron energy deposition in the resist film for 0.12 μm thick PMMA on Si due to a point beam by tracing 10 000 electron trajectories (for each simulation). Two characteristic regions can be seen: a narrow one, near the beam axis, with an abrupt drop, which represents the forward scattering electron contribution to the function EDF(r,z) and an wide part with a slightly varying drop, which corresponds to the backscattering electron contribution. As the initial electron energy increases, the area of the forward scattering electrons broadens insignificantly, while the region of the backward scattering electrons undergoes considerable broadening.

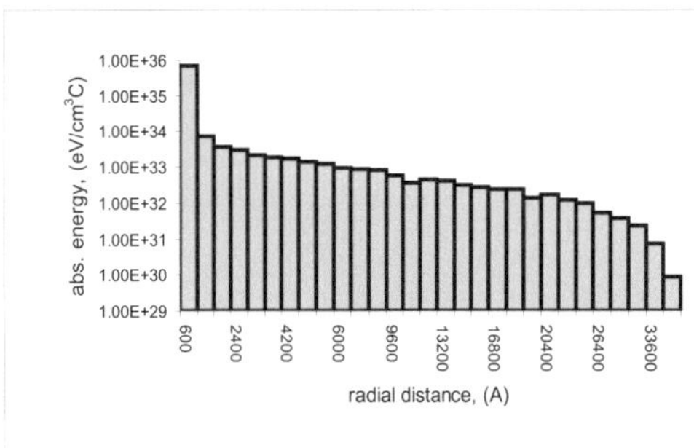

(a)

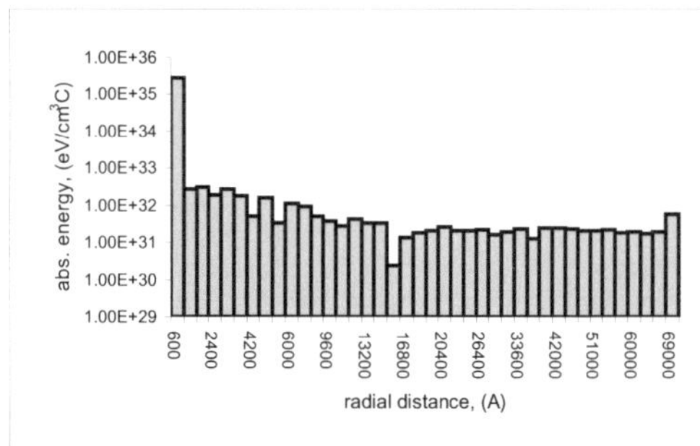

(b)

Fig. 3. (a) - 0.12 μm PMMA/Si , E_0 = 20 keV, (b) - 0.12 μm PMMA/Si, E_0 = 50 keV

Another presentation of the calculation results is by means of the equi-energy density contours as shown in Fig.4. When increasing the polymer thickness (Fig. 4(a)-(c)) a deviation of the most distant equi-energy density contours from the beam axis is observed, which is due to broadening of the forward scattering electron area. Further increase of the resist thickness (Fig.4(d)) results in a deviation of the equi-energy density contours back to the beam axis which is due to the beam intensity decrease. The doses corresponding to the presented equi-energy density contours decrease starting from the beam axis as follows: 1; *1/2; 1/4; 1/10; 1/20; 1/75.* These results can be used both for the latent image profile prediction under particular exposure conditions and for the exposure condition optimization, allowing the desired profile to be obtained. For example, one can see from Fig.4, that to obtain a latent image with vertical walls (primary e-beam energy, 20 keV) it is preferable to use thin polymer film whose thickness is less than 0.4 μm.

The high resolution of EBL however may be degraded by the lateral scattering of incident electrons which causes undesired exposure of unintended regions of the resist. This phenomenon is commonly referred to as proximity effect. In the case of patterning the thin films of the most widely used high temperature superconducting (HTS) material, namely $YBa_2Cu_3O_{7-\delta}$ (YBCO) deposited on $SrTiO_3$ (STO), MgO, ZrO_2: Y_2O_3, $LaAlO_3$, $NdGaO_3$, etc. substrates an enhanced proximity effect has to be taken into account because of their

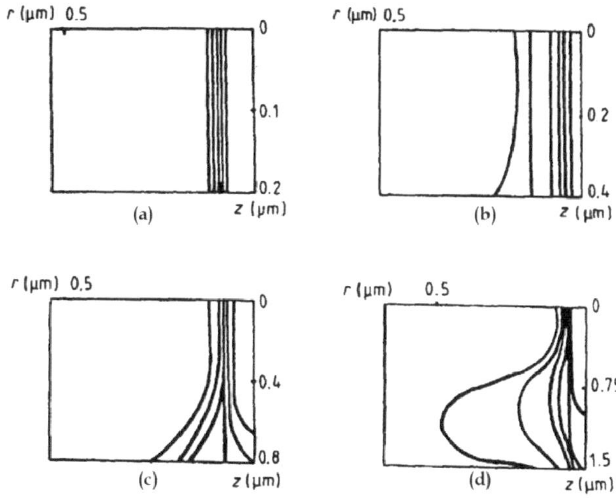

Fig. 4. The calculated equi-energy density contours in a 0.2, 0.4, 0.8 and 1.5 μm PMMA film on Si generated by an electron beam of 20 keV.

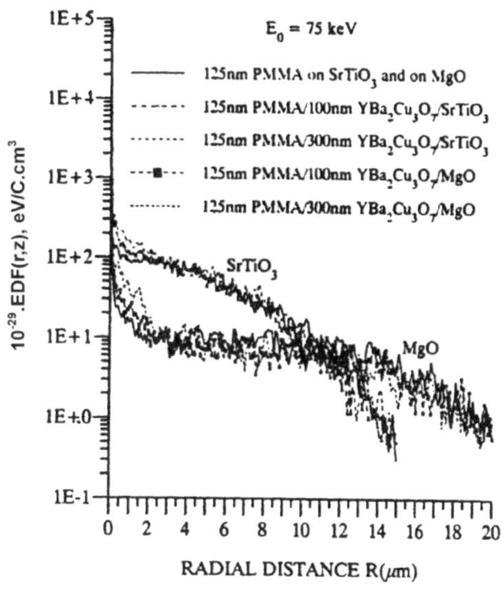

Fig. 5. Radial distributions of the absorbed energy density in the resist at E_0 = 75 keV.

retively high effective atomic numbers (Gueorguiev, 1994). Using our program for MC simulation of the processes of penetration and scattering of accelerated electrons in solids, the radial distributions of the EDF in 125 nm PMMA resist layer coated on structures YBCO thin film/substrate are obtained for an e-beam in the form of a zero-widt δ function, 30 000 electron trajectories, and the following parameters: (i) the substrate material (STO and MgO), (ii) the e-beam energy E_0 (25, 50 and 75 keV), and (iii) the YBCO film thickness d (0, 100, and 300 nm). In Fig.5, the calculated radial distributions of the absorbed electron energy density at E_0 = 75 keV are shown as an example. The results show that the EBL on the above mentioned targets is associated with an enhanced proximity effect in comparison with that on the conventional in microelectronics targets PMMA/Si substrate or PMMA/SiO/Si substrate. Moreover, the HTS thin film causes an additional backscattering of penetrating electrons and, hence, an additional proximity effect (in comparison with the targets PMMA/STO and PMMA/MgO) in the regions close to the incident point of the electron beam. This effect is as greater, as thicker is the film, as lighter is the substrate, as lower is the beam energy and is not completely eliminated even at energies as high as 75 keV, especially for the film thickness 300 nm, as well as for the lighter substrate (MgO).

3. Approximation of the discrete data for the deposited energy in the resist using analytical functions

Due to large lateral scattering of the penetrating electrons, the exposure of many image segments effects the total deposited energy in a specific resist point. This effect, known as a proximity effect, requires high accuracy evaluation of the EDF at large radial distances i.e. far away from the point of incidence. It should be pointed out that the increase of the number of trajectories being modeled, with the purpose of achieving statistical consistency for large lateral distances (characterizing the backward scattered electrons and giving the greatest contribution to the proximity effect) is not quite effective, as only a few trajectories travel through these regions. A MC methodology and a corresponding computer program BET-MK are developed for transformation of the numerical data array, representing the absorbed energy space distribution when exposing one point from the resist surface, into the form of analytical functions (Vutova & Mladenov, 1994). In this way the problem concerning the insufficient statistics of the discrete data for the absorbed energy in the case of large lateral distances is overcome. The main features of this methodology are as follows:

i. Transformation of numerical data for the absorbed energy discrete space distribution in the form of analytical functions. The data arrays containing the energy distribution when exposing one point from the resist surface are obtained using the TREM-MV program.

ii. The absorbed energy at some resist depth in the case of Si substrate is approximated as a sum of two Gaussians:

$$f(r) = k[\exp(-\frac{r^2}{\delta_f^2}) + \eta_E \cdot \frac{\delta_f^2}{\delta_b^2} . \exp(-\frac{r^2}{\delta_b^2})] \tag{19}$$

where k is a normalization constant, δ_f and δ_b are the characteristic widths of the forward and the backward scattering particles, and η_E is the ratio of the energy depth dissipation of the backward scattering particles to that of the forward scattering particles. The input data for the program BET-MK is the 2-D arrays EDF(r,z) containing the absorbed energy distribution values obtained as a result of the trajectory modeling. The first Gaussian (with standard deviation δ_f) dominates for the short lateral distances and describes the energy deposition from the forward scattered electrons. The second Gaussian (with standard deviation δ_b) dominates for the long lateral distances and describes the contribution from the backscattered electrons. The parameters δ_f, δ_b and η_E are called proximity effect parameters.

iii. The parameter values (δ_f, δ_b, η_E) are calculated using an original MC technique, instead of the non-linear least-square method and an arbitrary kind of distribution. The technique comprises the mean square deviation minimization by the interval length decrement for each of the parameters chosen. The minimization is made in an iteration loop.

The main advantages of the MC technique described above are: (i) in contrast to some of the least-squares methods, it does not allow the possibility of an infinite loop in the case of a local minimum; (ii) it enables to approximate an arbitrary kind of distribution of numerical data with a corresponding analytical function.

Fig.6 presents the EDF in the resist film for 0.4 μm thick PMMA on Si for 50 keV beam energy. The standard deviation for forward (δ_f) and backscattering (δ_b) contributions (Eq.19) as calculated from the simulation data are: δ_f = 0.165 μm, δ_b = 7.769 μm, η_E = 1.2.

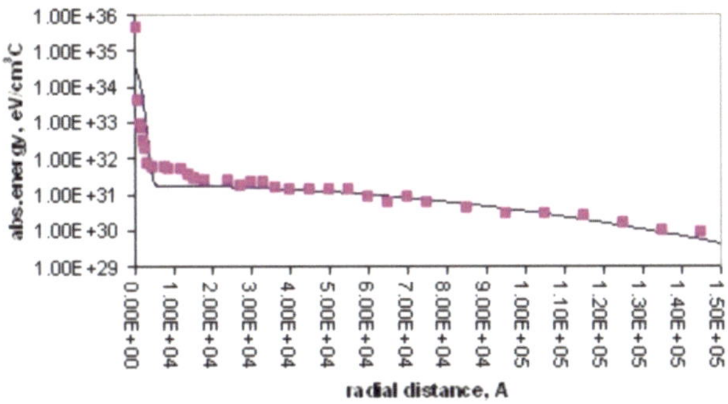

Fig. 6. Comparison between the deposited energy distribution (EDF) in the case of PMMA on Si at the interface resist/silicon and the corresponding analytical fit (Eq.19). The electron beam is focused on a surface point. The symbols (▪) present the results for MC calculation (TREM-MV), the line (-) for f(r).

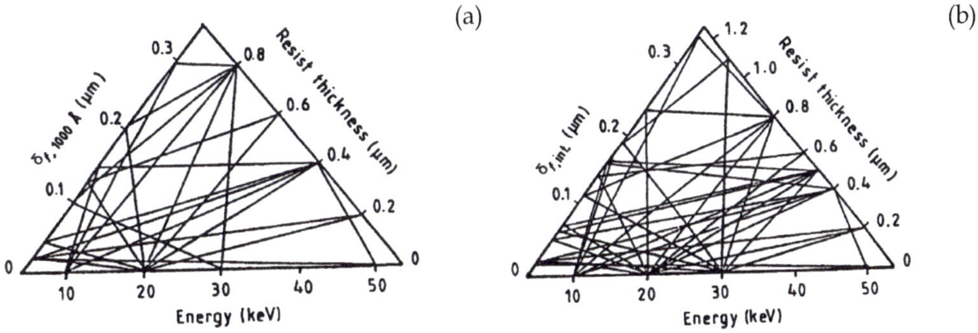

Fig. 7. Dependence of δ_f on incident electron energy and on resist (PMMA) thickness at two characteristic depths: (a) resist surface (1000 Å in depth); (b) film-substrate interface.

The values of the parameters δ_f and δ_b for various electron beam energies and different resist (PMMA) thicknesses are shown in Figs.7-8. In the triangular diagram the points corresponding to the values of δ_f or δ_b, the resist depth, and the beam energy represent vertices of inscribed triangles. Using these results one can approxima determine the δ_f and δ_b values for different beam energies and resist thicknesses. For instance, if the beam energy is 15 keV and the PMMA thickness is 0.8 μm, the δ_f value at the depth of 0.1 μm is in the range of [0.21, 0.29] μm, and that at the resist-substrate interface is between 0.24 and 0.33 μm. If the beam energy is 20 keV, the PMMA thickness is 0.7 μm, the δ_b value at the depth of 0.1 μm is in the range of [0.13, 0.21] μm, and at the resist-substrate interface its value is between 0.14 and 0.23 μm. Similarly, if δ_f (or δ_b) values and the beam energy are known, one can determine the resist thickness and vice versa from known δ_f (or δ_b) and the resist thickness, the beam energy can be determined.

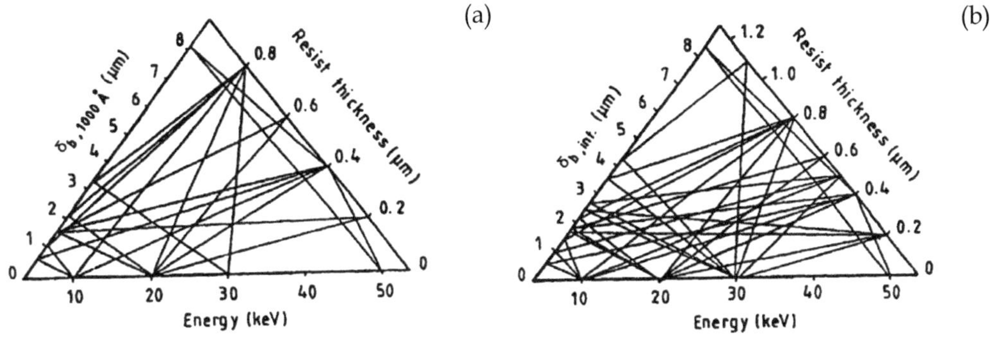

Fig. 8. Dependence of δ_b on incident electron energy and on resist (PMMA) thickness at two characteristic depths: (a) resist surface (1000 Å in depth); (b) film-substrate interface.

According to the simulation strategy, the next step in the modeling is the convolution of EDF(r,z) with the actual beam current distribution j(r,z), giving the single-energy deposition function (SEDF(r,z)). Let the electron beam is Gaussian distributed with a characteristic width δ^*, i.e.

$$j = j_o.\exp(-\frac{r^2}{\delta^{*2}}) = \frac{I_b}{\pi.\delta^{*2}}\exp(-\frac{r^2}{\delta^{*2}}), \tag{20}$$

where I_b is the incident beam current, j and j_o are the current densities at a distance r and at the beam centre, respectively. Then the function for a real e-beam can be calculated by the convolution of (19) and (20). The result from this convolution is the function $f^*(r)$, that has an analytical representation in the form of (19), but its characteristic widths are now modified:

$$\delta_f^* = (\delta_f^2 + \delta^{*2})^{1/2} \tag{21}$$

$$\delta_b^* = (\delta_b^2 + \delta^{*2})^{1/2}. \tag{22}$$

4. A three-dimensional model for absorbed energy calculation

To obtain the absorbed energy space distribution when exposing an arbitrary pattern, using an arbitrary exposure dose distribution, one must take into account the influence of a large number of exposed points. This is due to the fact that the lithography micro-patterning includes many irradiated points. Integrating the data obtained by the computer simulation of the exposure of each point from the resist surface, the absorbed energy space distribution in the case of an arbitrary pattern can be calculated. Due to the large number of calculations, a simplified procedure should be used to calculate the integral space distribution of the absorbed energy. The main features of the procedure proposed in (Vutova & Mladenov, 1991) are as follows.

i. The two-dimensional data array containing the absorbed energy values at some resist depth is presented as (19). If the electron (or ion) exposure is uniformly distributed over an area *A*, then the energy density can be expressed as:

$$F(r) = \oint_A f^*(r).dA \tag{23}$$

If the area *A* is a simple pattern (i.e. a line or a rectangle) the integral (23) can be calculated using the tabulated error function:

$$\mathrm{erf}(t,\sigma) = \int_0^t \exp(-\frac{x^2}{\sigma^2}).dx \cdot \tag{24}$$

ii. In the case of a more complex pattern, it should be divided into simple parts and then the corresponding values of the absorbed energy should be subsequently summed up. The formulas obtained for the absorbed energy density, when exposing either a line, a line segment, or a rectangle (Vutova & Mladenov, 1991), are given below. These simple patterns are sufficient to compose an arbitrary figure.
iii. The procedure takes into account the radial variation of the absorbed energy as well as its modification versus the depth of the resist. To calculate the δ_f, δ_b, and η_E parameter values, the linear approximation along the resist depth is used. In the electron exposure case, two resist depths are used, namely the resist surface and the resist-substrate interface (Table 1). The proposed methodology is realized in a computer program.

Beam energy [keV]	Resist thickness [μm]	Resist depth [μm]	δ_f [μm]	δ_b [μm]	η_E
50	0.4	0.1	0.145	6.67	1.19
50	0.4	0.2	0.152	7.036	1.19
50	0.4	0.3	0.1583	7.403	1.194
50	0.4	0.4	0.165	7.77	1.2
20	0.4	0.1	0.0412	1.51	0.75
20	0.4	0.2	0.0434	1.53	0.759
20	0.4	0.3	0.0458	1.575	0.763
20	0.4	0.4	0.049	1.652	0.768
20	0.8	0.1	0.203	2.08	0.485
20	0.8	0.3	0.206	2.11	0.485
20	0.8	0.5	0.221	2.13	0.497
20	0.8	0.7	0.228	2.13	0.503
20	0.8	0.8	0.235	2.17	0.503

Table 1. Proximity effect parameters (Eq.19) obtained by means of program TREM-MV and BET-MK for a point source.

Let we have Cartesian coordinate system *Oxyz,* such that the axis *Ox* and *Oy* are in the plane of the resist surface, the axis Oz is in its depth, and $x^2 + y^2 = r^2$. Let we expose a line segment and the axis *Oy* coincides with the line segment, whose end points are labeled *a* and *b* (Fig.9).

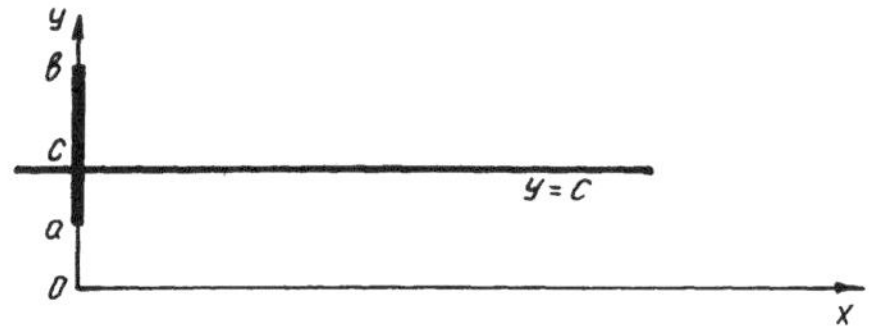

Fig. 9. A case of the exposed segment (the single isolated line of finite lenght).

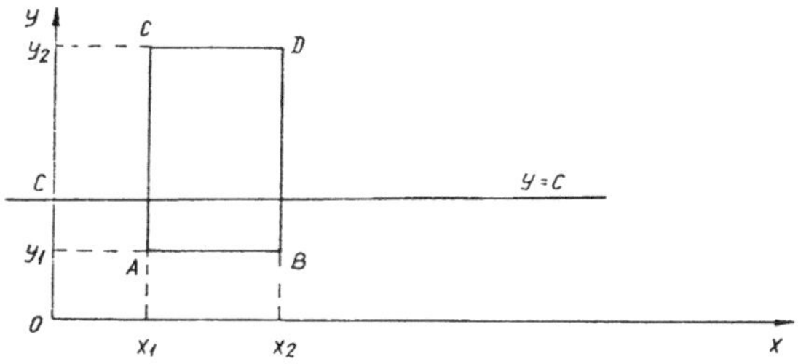

Fig. 10. A case of the exposed rectangle.

We are interested in the cross section that is perpendicular to the segment line and whose analytical expression is $y = c$. From (19) and (23) we obtain:

$$\mathrm{F(r)} = \int_a^b f^*(r)dy = (b-a)f(r) = (b-a)k[\exp(-\frac{x^2+c^2}{\delta_f^{*2}}) + \eta_E \frac{\delta_f^{*2}}{\delta_b^{*2}} \exp(-\frac{x^2+c^2}{\delta_b^{*2}})]. \tag{25}$$

Similarly, we can obtain the expressions for the density of the absorbed energy in a point lying at the $y = \mathrm{c}$ line, when exposing a rectangle ABCD (Fig.10):

a. for $x<x_1<x_2$

$$\begin{aligned} F(r) &= (y_2 - y_1)k\{\exp(-\frac{c^2}{\delta_f^{*2}})[erf(x_2 - x, \delta_f) - erf(x_1 - x, \delta_f)] \\ &+ \eta_E \frac{\delta_f^{*2}}{\delta_b^{*2}} \exp(-\frac{c^2}{\delta_b^{*2}})[erf(x_2 - x, \delta_b) - erf(x_1 - x, \delta_b)]\} \end{aligned} \tag{26}$$

b. for $x_1 \le x \le x_2$

$$\begin{aligned} F(r) &= (y_2 - y_1)k\{\exp(-\frac{c^2}{\delta_f^{*2}})[erf(x_2 - x, \delta_f) + erf(x - x_1, \delta_f)] \\ &+ \eta_E \frac{\delta_f^{*2}}{\delta_b^{*2}} \exp(-\frac{c^2}{\delta_b^{*2}})[erf(x_2 - x, \delta_b) + erf(x - x_1, \delta_b)]\} \end{aligned} \tag{27}$$

c. for $x_2<x$

$$\begin{aligned} F(r) &= (y_2 - y_1)k\{\exp(-\frac{c^2}{\delta_f^{*2}})[erf(x - x_1, \delta_f) - erf(x - x_2, \delta_f)] \\ &+ \eta_E \frac{\delta_f^{*2}}{\delta_b^{*2}} \exp(-\frac{c^2}{\delta_b^{*2}})[erf(x - x_1, \delta_b) - erf(x - x_2, \delta_b)]\} \end{aligned} \tag{28}$$

where $erf(t,\sigma)$ is calculated using (24).

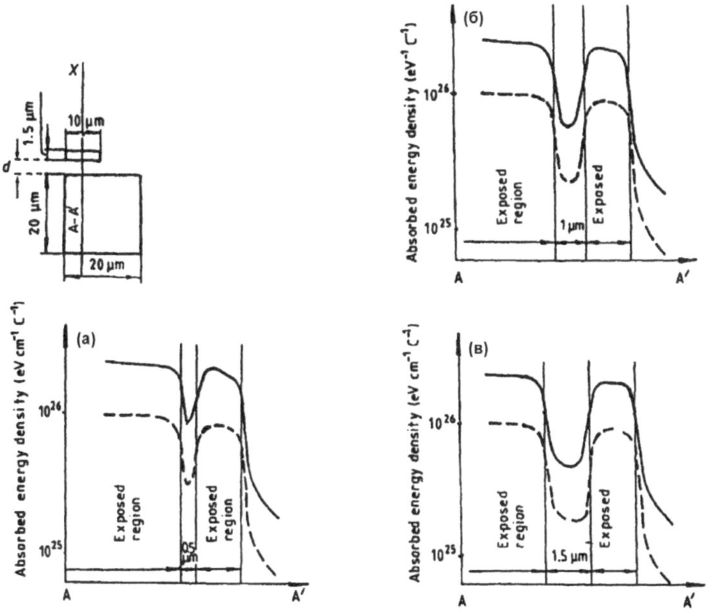

Fig. 11. Variation of the EDF along the line A-A′ at the top layer (1000 Å in depth, broken curves), and at the bottom of the resist (8000 Å, full curves), for a 0.8 µm PMMA on Si for a 20 keV Gaussian e-beam. Inter-element spacings are (*a*) 0.5 µm, (*b*) 1.0 µm, and (c) 1.5 µm.

The absorbed energy density distribution in the case of a more complex topological structure, where the proximity effect cannot be ignored is shown in Fig.11. The results are presented for three inter-elemental spacings of d=0.5, 1.0, and 1.5 µm. This simulation result clarifies the effects of an adjacent element and its configuration on the energy distribution.

5. Absorbed energy approximation in the case of multilayers samples and heavy substrates

The proximity effect in the case of patterning the thin films of the most widely used HTS material, namely YBCO, deposited on two typical substrates (STO and MgO) is investigated (Gueorguiev et al., 1995; Gueorguiev et al., 1996; Gueorguiev et al., 1998; Olziersky et al., 2004; Vutova, 2007). HTS samples represent a more difficult case study since the substrate consists of bulk substrate STO or MgO and a very thin YBCO layer on top (multilayer substrate). The existence of the thin YBCO film between the bulk substrate and the resist changes the scattering phenomena and has to be carefully taken into account. This effect becomes more important as film thickness increases.

For substrates with larger mean atomic number and density the addition of one more function is necessary. In the case of YBCO films over STO or MgO substrates the addition of an exponential function in equation (19) is found to be adequate (Gueorguiev et al., 1996). Thus EDF(r) could be approximated as:

$$f(r) = \frac{k}{\pi(1+\eta+\nu)}\left[\frac{1}{\alpha^2}\exp(-\frac{r^2}{\alpha^2}) + \frac{\eta}{\beta^2}\exp(-\frac{r^2}{\beta^2}) + \frac{\nu}{2\gamma^2}\exp(-\frac{r}{\gamma})\right], \quad (29)$$

where the third term describes the energy deposition in the mid-lateral distances. The values of the parameters of this function (called proximity function) a, β, γ, η, ν and k are calculated using the MC technique (described above 3.).

5.1 Results and discussion

The variables studied in our work are the substrate material (STO and MO), the initial energy of accelerated electrons E_0 (25, 50 and 75 keV) and the HTS film thickness d (0, 100, 200, 300 and 1000 nm). The values of the proximity effect parameters are evaluated from the fitting of EDF(r) with a sum of suitable functions (Eq.29) and their dependence on all investigated variables are discussed (Gueorguiev et al., 1995; Gueorguiev et al., 1996; Gueorguiev et al., 1998; Olziersky et al., 2004). The absorbed energy distributions obtained and the calculated parameters of the proximity function can be used in a proper proximity effect correction algorithm as well as in a resist profile development model. The proximity effects are usually compensated for by applying proper correction algorithms which adjust the exposure dose and/or the shape and size of the exposed pattern. For the realization of such algorithms precise data are required about the spatial distribution of absorbed electron energy density in the resist. This distribution quantitatively describes the proximity effects.
In Fig.12 a comparison is made between the radial distributions of the absorbed energy density obtained by MC simulation for the structure 125 nm PMMA resist film /300 nm YBCO HTS film /MgO substrate at three beam energies -25, 50, and 75 keV, and the corresponding analytical fits. It is well seen that the combination of double Gaussian and exponential functions is a good approximation of these distributions. Although not shown here, the double Gaussian (Eq.19), as well as the triple Gaussian were also tested but they were found to be not adequate, especially in the intermediate regions (Fig.12).

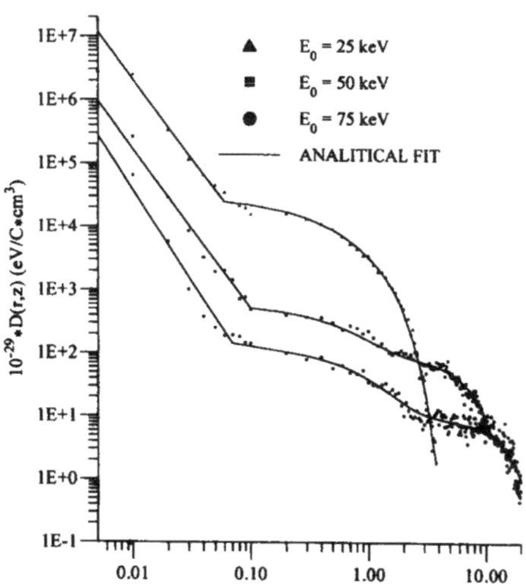

Fig. 12. Comparison between the exposure distributions obtained by MC simulation for the structure 125 nm PMMA resist film / 300 nm YBCO HTS film / MgO substrate at three beam energies - 25, 50, and 75 keV and the corresponding analytical fits.

Figure 13 shows the analytical fit to the radial distributions of absorbed energy density obtained by MC simulation for the structures 125 nm PMMA resist film /0, 100, or 300 nm

YBCO HTS films / STO or MgO substrates at 25, 50, and 75 keV, respectively. Since the aim is to investigate the proximity effects caused by YBCO film as well as by the substrate (STO, MgO) the backscattered exposure is of primary importance. For this reason here, in contrast to the Fig.12, a linear scale for the *x* axis is applied which although it compresses data points close to the beam axis (associated with the forward scattered electrons), it ensures a better distinction between the distributions in their intermediate and distant regions (associated with backscattered electrons).

The peaks of the distributions of the absorbed energy density are commonly attributed to the forward scattering of electrons or, in other words, to the single scattering of primary electrons into small angles in the resist. This scattering depends on the beam energy as well as on the material and thickness of the resist. In Figs.12-13 it is seen that the maximum values as well as the widths of the peaks decrease with increasing beam energy. This can be explained by both the more efficient scattering of primary electrons and the higher energy loss in the resist at lower energies.

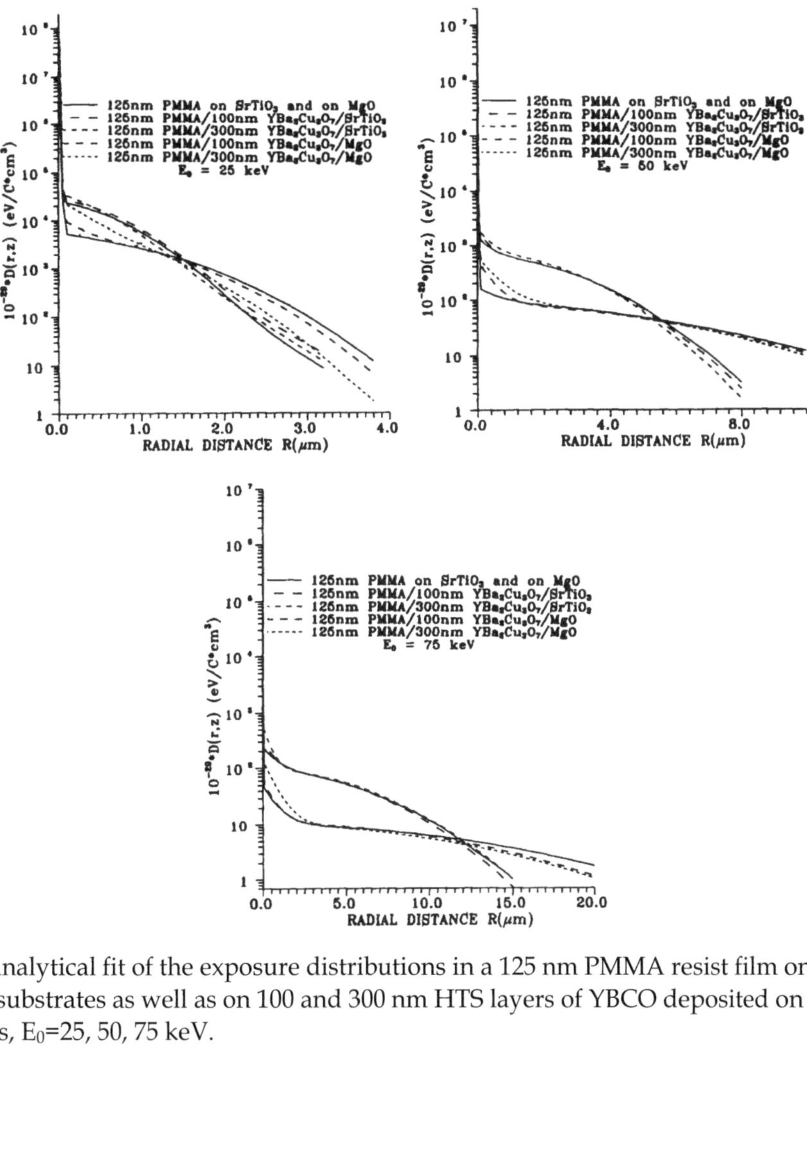

Fig. 13. Analytical fit of the exposure distributions in a 125 nm PMMA resist film on STO and MO substrates as well as on 100 and 300 nm HTS layers of YBCO deposited on the substrates, E_0=25, 50, 75 keV.

The calculated values of the parameters of the analytical function and their dependencies on the beam energy and on HTS film thickness are presented in the form of triangular diagrams as well as of 3D diagrams that can be used for easy approximate determination of the parameters at different beam energies and YBCO film thicknesses - Fig.14 and Fig.15.

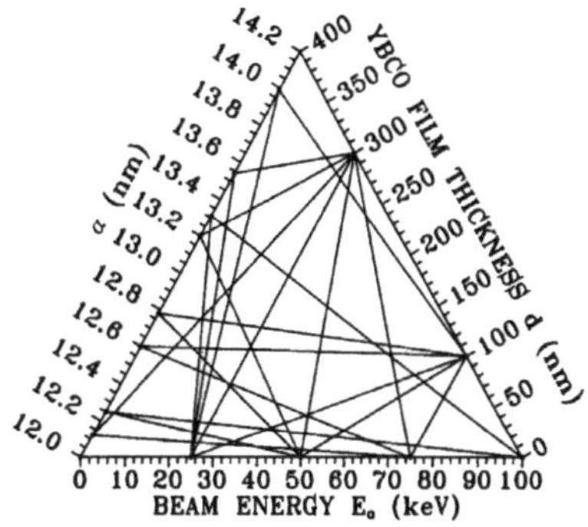

Fig. 14. Dependence of the parameter α of the proximity function on YBCO film thickness *d* and the beam energy E_0 for STO substrate.

The results show that the additional backscattering of primary electrons and, hence, the proximity effect, caused by the HTS film in the regions close to the incident point of the electron beam are not completely eliminated even at energies as high as 75 keV especially for the film thickness 300 nm as well as for the lighter substrate (MgO). The HTS thin film reduces the backscattering from the underlying substrate and this reduction is as greater as thicker is the film as well as lower is the beam energy.

In Fig.16, EDF(r) simulation results with the MC method in the case of YBCO/MgO substrates are presented. It is obvious that as d_{YBCO} increases, the overall energy deposition approaches the bulk YBCO case. When the YBCO layer is thin, due to the its relatively higher scattering parameters (density, mean atomic weight, mean atomic number) in comparison with those for the MgO substrate, the EDF(r) extends to regions far away from the point of incidence. On the other hand, when the YBCO layer is thick, the backscattered electrons come mainly from this layer rather than the MgO substrate. The external proximity effect obtained in regions far from the point of beam incidence (more than 4–5 μm) increases

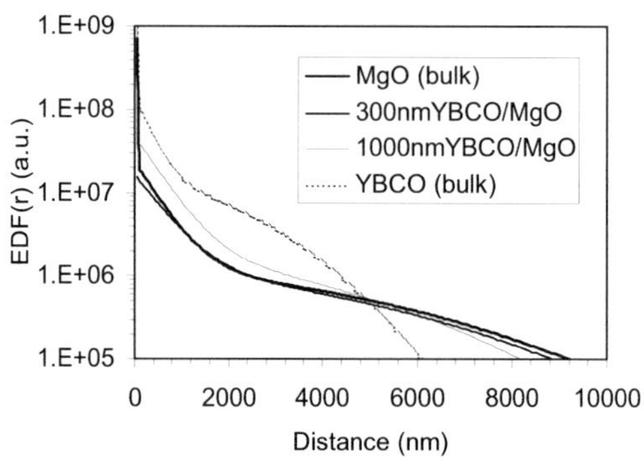

Fig. 16. EDF(r) simulations for various YBCO thicknesses over MgO bulk substrate using 50 keV e-beam energy.

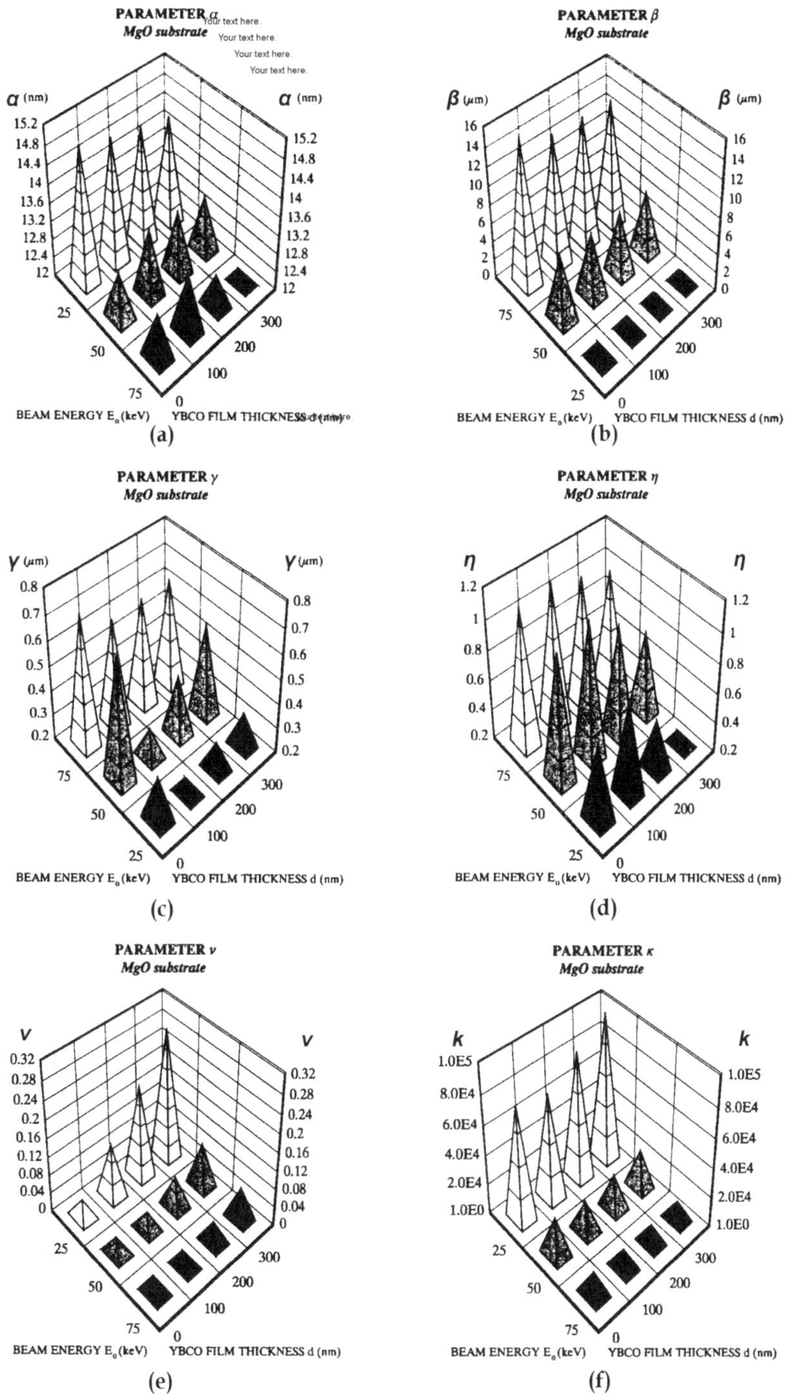

Fig. 15. Diagrams of the dependences of the parameters of the proximity function on the beam energy and YBCO film thickness for the MgO substrate: (a) α; (b) β; (c) γ; (d) η; (e) ν and (f) k.

when the $d_{YBCO} < T_{crit}$, while the proximity effect is lower in regions near to the point of beam incidence (T_{crit} is the YBCO thickness that causes energy deposition equal to the EDF(r) for bulk YBCO). The exponential function dominates in the intersection area between the two Gaussian functions which in the 50 keV case occurs in the ~ 40 nm to 2 μm range.

6. Comparison of simulated results with experimental data

The fundamental quantity in the simulation of EBL is the energy deposition due to point beam exposure (EDF(r)) or a single pixel exposure (SEDF(r)). When this quantity is in good agreement with the corresponding experimental data then it is possible to obtain good simulation results for the final resist profiles provided accurate dissolution rates are known.
Two different modules for the energy deposition calculation EDF(r,z) are applied in the case of Si and HTS films. The first module is based on the MC algorithm (Vutova & Mladenov, 1994; Gueorguiev et al., 1998) and the second based on an analytical solution of the Boltzmann transport equation (AS) (Raptis et al., 1998). Comparison between these approaches for various energies and substrates was performed in order to test the adequacy of calculated set of data. In addition, results from both modules were compared with experimental data for thin resist films on Si and HTS substrates (Gueorguiev et al., 1995; Olziersky et al., 2004; Vutova et al., 2007).
In Fig.17 the SEDF(r) in the case of 400nm thick epoxy resist on Si substrate is presented. The symbols represent the experimental data while the two lines represent the simulation results for MC and AS methods after convolution with the experimental beam diameter. Due to the 50 keV e-beam energy and the Si substrate used in this comparison, the backscattering contribution is spread out over a long distance from the point of incidence. It is obvious that both simulation methods provide results with good agreement with experimental data in the whole (forward scattered and backscattered) range.

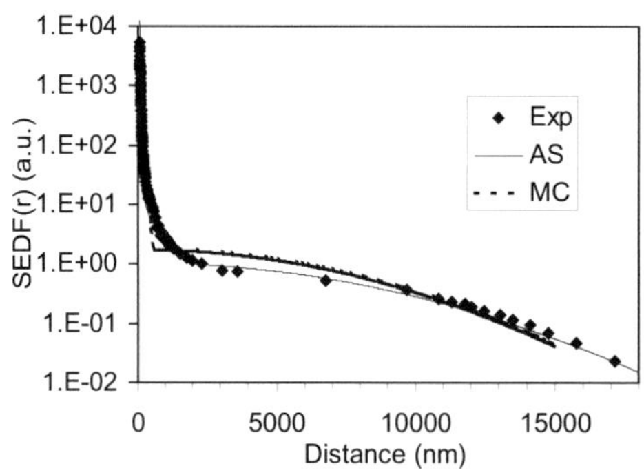

Fig. 17. Comparison of SEDF(r) between simulation and experiment in the case of Si substrate. The resist thickness was 0.4μm and the energy 50 keV.

In Table 2 the calculated β values in all cases (energy, substrate composition) is presented using both MC and AS modules, and the agreement between the calculated results is good.
In Fig.18 the simulation data (MC and AS) are presented along with the experimental SEDF(r) for STO and YBCO/STO substrates. The simulation results were prepared after convolution of the simulated EDF(r) (at the resist/substrate interface) from a point beam with a Gaussian beam diameter 30 nm. The SEDF(r) is obtained from exposure with 50 keV

electrons in the case of a 0.4 μm thick resist film. In all cases, the agreement between experiment and simulation is satisfactory. Even in the second case, where the substrate is not homogeneous but consists of two layers (a thin YBCO layer over bulk STO) the discrepancy is limited.

E_0 (keV)	d_{YBCO} (nm)	β (μm) (MC)	β (μm) (AS)
25	0	1.59	2.35
25	300	1.52	1.90
25	1000	1.14	1.34
25	bulk	0.98	1.34
50	0	6.05	7.20
50	300	5.92	6.50
50	1000	5.00	5.55
50	bulk	2.80	4.12
75	0	14.62	14.5
75	300	14.31	13.5
75	1000	10.77	12.00
75	bulk	5.60	8.20

Table 2. The standard deviation (β) of backscattering electron contribution for the 25-75 keV range in the case of YBCO/STO substrate. The data *d* in the second column represent the YBCO layer thickness.

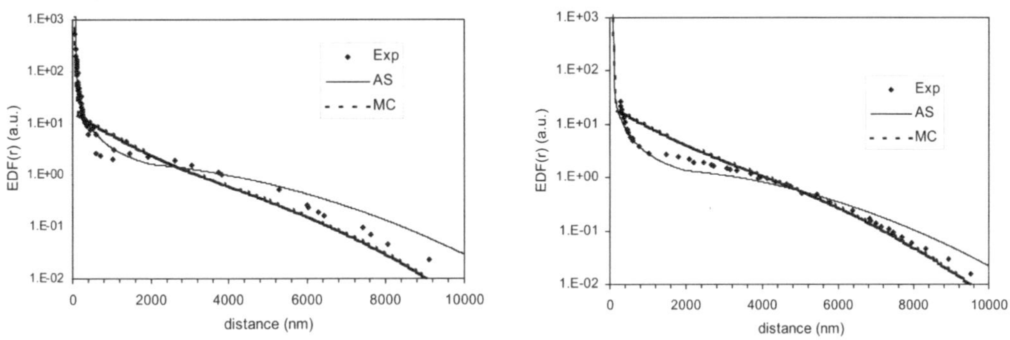

Fig. 18. EDF(r) in the case of 0.4 μm PMMA on: (a) $SrTiO_3$ substrate; (b) 100 nmYBCO/STO substrate. The e-beam energy is 50 keV.

7. Exposure process simulation at ion beam lithography (IBL)

A new look at previously obtained experimental results and simulation models applicable in IBL is a necessary task due to the progress in the development of IBL tools (Kaesmaier & Löschner, 2000) and their applications concerning exposure of one-component resist PMMA (Miller, 1989; Mladenov et al., 2001) as well as multi-component resists including chemically amplified resists (CARs) (Hirscher et al., 2001). The simulation tools for sub-150 nm ion beam lithography require more adequate models of the ion scattering, the energy losses and the development process. Accumulation and analysis of calculated results and physical parameters included in the models permit deeper understanding of the processes, comparison with experimental data and optimization of the simulation tools. Ion lithography of PMMA of thickness up to 400 nm is studied.

7.1 Model and computer program TRIM-MV for simulation of ion scattering and exposure intensity calculation in the case of ion bombardment

Stopping powers and trajectories of the penetrating atom particles in the materials under ion bombardment are quite well understood, due to the progress of ion implantation, ion sputtering and secondary ion mass spectrometry. Computer programs such as MARLOW, TRIM, PIBER, SASAMAL (Robinson & Torrens, 1974; Biersack & Haggmark, 1980; Adesida et al., 1982) were developed between 1970 and 1980. The program structure is the same as for electrons. The major difference is that the energy transfer is executed by two types of losses: nuclear stopping power, due to penetrating particle collisions with the sample atoms and electronic stopping power, due to penetrating projectile collisions with the electrons in the bombarded material. The first applications of these programs have been for PMMA (Adesida et al., 1982; Mladenov, 1985; Mladenov et al., 1985) to evaluate the ranges and energy losses.

The main computer program for ion scattering in amorphous materials using the Monte Carlo calculations is TRIM, first published by Biersack and Haggmark (Biersack & Haggmark, 1980), but later extended fully to cover collision cascades including the motion of recoils (Biersack & Eckstein, 1984). For the particular energy range in ion lithography the calculations are further speeded up by introducing an approximate but rather accurate treatment of the individual collision process. The program has spread in a large number of versions over the atomic collisions in solids. It is applied for the first time to the field of ion lithography, in (Mladenov et al., 1985).

We develop a program version, named TRIM-MV (Vutova & Mladenov, 1992), based on the famous Biersack model, to model the ion scattering in multilevel multicomponent amorphous targets. Depending on the different penetration velocities an accelerated ion changes its own charge state being neutralized at the sample surface or going to the highest charged state during its movement through the irradiated sample. The critical velocity for that change is the average orbital velocity of the electron in the statistical atom model of Thomas-Fermi:

$$V_{cr} = Z_1^{2/3} \cdot \frac{c}{137}, \tag{30}$$

where Z_1 is the atomic number of the penetrating particles. Ions penetrating in the solid with velocity less than V_{cr} , move as neutral atoms, taking an electron from the target material. In the case of $V>V_{cr}$, they loss own electrons. At these higher energies the interactions (and energy losses) due to collisions with the target electrons predominate while at the lower energies the interactions with the atom nucleus lead to more important type of energy losses. In the common case the total ion energy loss is calculated as a sum:

$$\left\langle -\frac{dE}{d\xi} \right\rangle_{\Sigma} = \left\langle -\frac{dE}{d\xi} \right\rangle_{n} + \left\langle -\frac{dE}{d\xi} \right\rangle_{el}. \tag{31}$$

The nuclear and electronic stopping powers are assumed to be independent of each other. The energy losses for different energies and combinations of types of penetrating ions and target atoms are known. Usually, these energy losses are presented as universal functions of dimensionless parameters ρ and ε (for the ranges and the energy, respectively). These parameters are connected with the real ion range R and the energy E by the relations:

$$\rho = 4\pi N a^2 \frac{M_1.M_2}{(M_1+M_2)^2}.R\ , \tag{32}$$

$$\varepsilon = 4\pi\varepsilon_0 a \frac{M_2}{Z_1 Z_2 e^2 (M_1+M_2)} E\ . \tag{33}$$

Here M_1 and M_2 are the atomic masses, Z_1 and Z_2 are the atomic numbers of the bombarding particle and the target atom; N is the density of scattering atoms, a is the shielding parameter of the interaction atomic potential. The Molier potential is appropriate approximation at low energies of the penetrating particle:

$$U(r)= \frac{Z_1 Z_2 e^2}{r}[0.35\exp(-0.3r/a)+0.55\exp(-1.2r/a)+0.1\exp(-6r/a)]\ , \tag{34}$$

where the shielding parameter a is presented by the Firsov distance:

$$a = \frac{a_o.0.885}{(Z_1^{1/2}+Z_2^{1/2})^{2/3}}\ . \tag{35}$$

The ranges through the penetrated particles' trajectories are defined by the total ion energy losses and are calculated using the relation:

$$R_L = \int_0^{E_o} \frac{dE}{(dE/d\xi)_\Sigma}\ . \tag{36}$$

In the case of low-energy the electronic energy losses are calculated on the basis of the Lindhard and Scharff formalism (Biersack & Haggmark, 1980):

$$(d\varepsilon/\ d\rho)_{el} = k.\varepsilon^{1/2}, \tag{37}$$

and κ is given by:

$$k = \frac{\varsigma_e.0.0793 Z_1^{1/2}.Z_2^{1/2}(M_1+M_2)^{3/2}}{(Z_1^{2/3}+Z_2^{2/3})^{3/4}.M_1^{3/2}.M_2^{1/2}}\ , \tag{38}$$

where ζ_e is a constant the value of which is about $Z^{1/6}$.
The electronic energy losses for high energies are calculated using the Bethe-Bloch equation as Z^2 is replaced by Z_1Z_2. The following interpolation expression is also used:

$$\frac{1}{(dE/d\xi)} = \frac{1}{(dE/d\xi)_{LSch}} - \frac{1}{(dE/d\xi)_{BB}}\ . \tag{39}$$

The nuclear losses are evaluated using the transferred (to the target atom) energy T at an elementary collision:

$$T = \frac{4M_1.M_2}{(M_1+M_2)^2}.E.\sin^2(\frac{\theta}{2})\ . \tag{40}$$

For calculation of the scattering angles θ at the nuclear collisions, using a "center of mass" coordinate system, the impact parameter value p is determined (the laboratory coordinate system and the "center of mass" system are identical in the case of collisions of the penetrating electrons with the target atoms, while these systems are different for penetration of ions in solids). The impact parameter is determined using random numbers:

$$p = [\frac{\overline{R_1}}{\pi . N^{2/3}}]^{1/2} . \quad (41)$$

Then the value of the scattering angle is calculated using the Biersack and Haggmark "scattering triangle" (Fig.19):

$$\cos(\theta/2) = \frac{R_1 + R_2 + p + \delta_1 + \delta_2}{R_1 + R_2 + r_o}, \quad (42)$$

where R_1 and R_2 are the trajectory curvature radiuses measured in the point of the minimum distance between the colliding particles M_1 and M_2; r_0 is the distance between the particles at the moment of minimum distance between them. The corrections δ_1 and δ_2 are low and have smaller values. The values of r_0 are determined using the value of the penetrating ion energy (in the center of mass system):

$$(\frac{p}{r_0})^2 = 1 - \frac{U(r_0)}{EM_2}(M_1 + M_2) . \quad (43)$$

The values of R_1 and R_2 are calculated using (Biersack & Haggmark, 1980):

$$R_1 + R_2 = 2[\frac{EM_2}{M_1 + M_2} - U(r_0)]/[-U'(r_0)], \text{ where:} \quad (44)$$

$$-U'(r_o) = \frac{M_1 . V_1^2}{R_1} = \frac{M_2 . V_2^2}{R_2} . \quad (45)$$

The values of δ_1 and δ_2 depend on the values of the ion energy E and of the target atomic mass (i.e. r_0). In (Biersack & Haggmark, 1980) their values are given in the case of the Coulomb potential using an appropriate analytical expression. For the Moliere potential the values of δ_1 and δ_2 are evaluated using the Robinson values of the scattering integral. The values of these corrections come to values of the scattering angle identical with the Rutherford scattering angle.

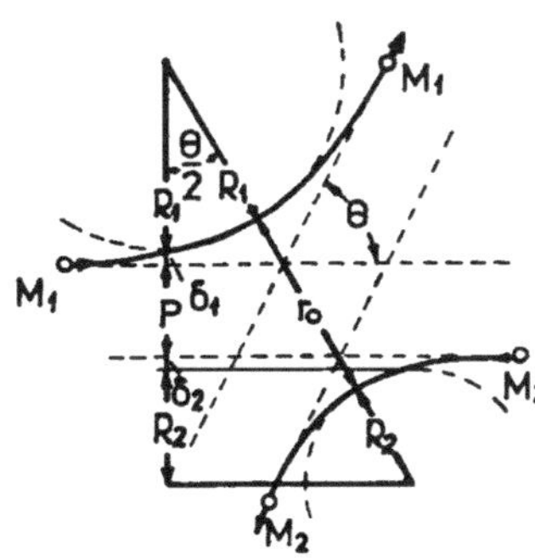

Fig. 19. Geometry of particles' scattering in a "center of mass" system.

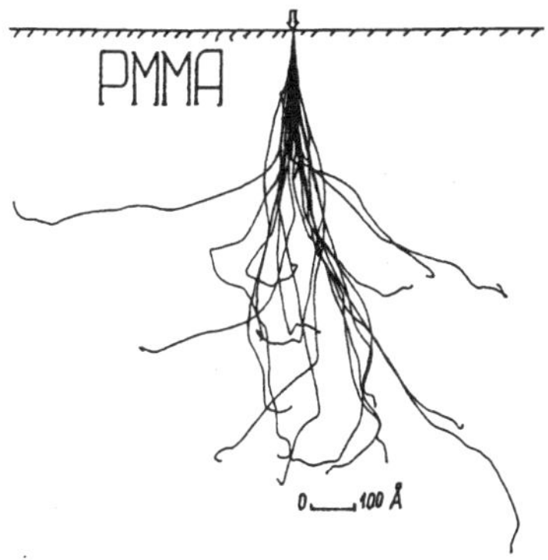

Fig. 20. 20 simulated trajectories in the case of 60 keV He ions in PMMA.

After evaluating the value of the scattering angle θ in the "center of mass" system, we use the following formula to calculate the value of θ_{ls} in the laboratory system:

$$\theta_{ls} = \mathrm{arctg}\left(\frac{\sin\theta}{\cos\theta + M_1 / M_2}\right) . \tag{46}$$

In order to achieve high computer efficiency, a suitable expanding of the distances between consecutive collisions at high energies is applied. The following expression is used instead of Eq.(41):

$$p = -\ln\frac{\overline{R_1}}{\pi NL} , \tag{47}$$

where the function $L(E,M_1,M_2)$ is evaluated as in (Wilson et al., 1977).
As in the case of simulation of electron scattering in the target, the penetration of each particle from one statistically significant extract (about 1000 – 10 000 particles) is considered as an elementary collision sequence with the target atoms. Characteristic changes in the particle motion for each collision, assuming a straight line trajectory between two collisions are calculated. The calculation continues until the cut-off conditions are not reached (the value of the penetrating particle is less than 100 eV or the particle is out of the sample). Bragg's rule (Biersack & Haggmark, 1980) is assumed to be valid. The role of the displaced secondary atoms in cascades is taken into account. The program involves the same MC technique to calculate the kind of the atom taking part in the collision (Eqs.(5-6)). The value of the free path between the collisions is also given by Eqs.(7-8). The difficulties concerning calculation of values of the different angles and the corresponding projection of the ranges (the straight line free paths between the sequential collisions of the projectile ion with two of the resist atoms) are overcome in the same way using Eqs.(11-18). The low-energy nuclear reactions (or anomalies) are not taken into account due to their minor influence on the ion stopping power.

7.2 Results and discussion

It is necessary to understand deeply the mechanisms of the processes at IBL in the critical dimension region below 150 nm and to predict by simulation more exactly the ion penetration, the absorbed energy dispersion and the resist modification as well as the subsequent process of the image development. For this aim more details of the calculated results can be experimentally confirmed or rejected. Accumulation and analysis of the calculation results after various simulation stages and of the physical parameters included in the mathematical modeling in nano-image region of IBL is also a question of present interest. The computer simulation allows us to obtain more numerical data and to reveal the dependence of these values on the incoming ion number or mass, as well as on ion energies. The values of projected and lateral ranges as well as the mean electronic and nuclear stopping powers have been calculated for various penetrating particles at different energies (Mladenov et al., 1985; Vutova & Mladenov, 1992; Vutova & Mladenov, 1994; (a) Vutova & Mladenov, 2001; Mladenov et al., 2001; Vutova, 2007), using our simulation tool TRIM-MV.

Fig.20 presents simulated trajectories for He ions in PMMA. The dependence of the electronic energy losses on the penetration depth in the PMMA resist is shown in Fig.21. It is evident that the contribution of the recoil particles to the total electronic losses is negligible. The nuclear energy losses are presented in Fig.22. We distinguish nuclear collisions accompanied with an energy transfer exceeding 25 eV (curve 1) and collisions with an energy transfer less than 25 eV (curve 2). The former is just the fraction of the nuclear energy losses which can be stored as defects in the material while the latter is the energy which can be converted directly into random (thermal) nuclear motion. The division line in the transferred energies, here chosen as 25 eV, for the creation of permanent defects on one side, and thermal vibrations on the other, is not well known for such polymers as PMMA. According to the Lindhard, Scharff and Schiott (LSS) theory, a comparison between Fig.21 and Fig.22 confirms that the electronic stopping is the dominant form of energy losses during most of the particle path through the solid, for the energies and light masses considered here. Only when the particle is about to come to rest, the nuclear energy loss will play a significant role. From Fig.22 one can also conclude that the nuclear energy losses, for which the transfer energies exceed the threshold value (here 25 eV), are more important than the energy losses due to the collisions with transfer energies less than 25 eV.

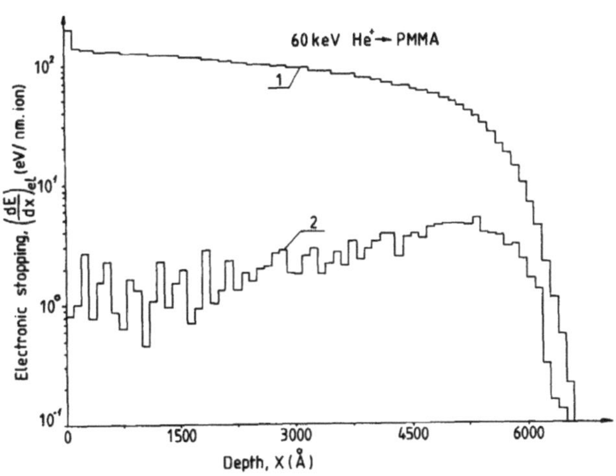

Fig. 21. Electronic energy losses vs penetration depth of 60 keV He^+ particles implanted in PMMA. The upper (1) and lower (2) curves correspond to the electronic stopping of primary and recoil particles, respectively.

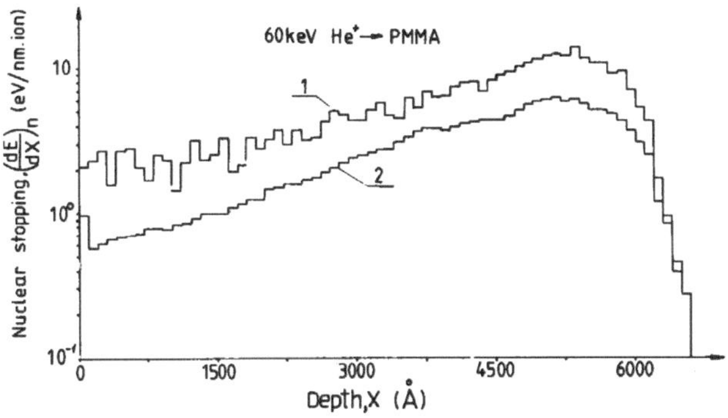

Fig. 22. Nuclear energy losses vs penetration depth of 60 keV He^+ particles implanted in PMMA. The upper (1) and lower (2) curves correspond to the nuclear collisions involving transfer energies higher and lower than the displacement threshold energy (here chosen to be 25 eV), respectively.

Depth distributions of the electronic and nuclear energy losses are presented in Fig.23. The lateral spread is low for the first few thousands Å in depth and then increases to a maximum value of about two thousands in the last part of the ion range - Fig.23. It can be seen that the lateral spread of the nuclear stopping is more pronounced than the electronic stopping. Figure 24 presents the two-dimensional range distribution of 60 keV implanted ^4He ions.

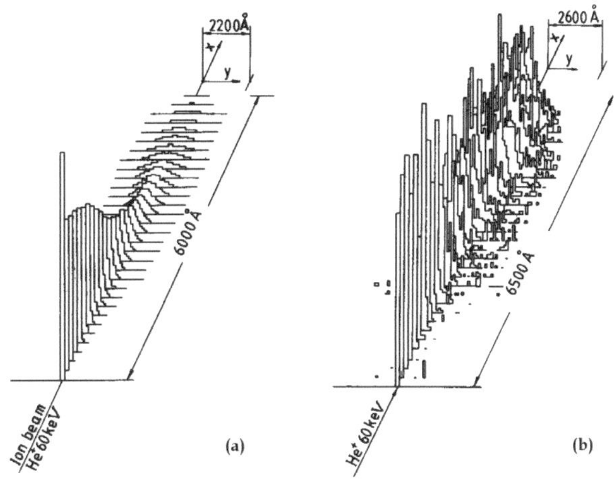

Fig. 23. Depth distributions of: (a) electronic energy losses, and (b) nuclear energy losses, calculated for 60 keV ^4He ions in PMMA. The vertical axis is normalized to an arbitrary unit.

Fig.25 shows the projected penetration ranges $R_p = \langle t_x \rangle$ (the mean value of the projections t_x of the ranges t on the axis x coincident with beam axis and with the sample depth; the brackets < > mark the average procedure) and the longitudinal (penetration) range deviation $\Delta R_p = \langle (t_x - \langle t_x \rangle)^2 \rangle^{1/2}$ of the ions (H^+, He^+, B^+, Ar^+ and Ga^+) in a wide energy region.
In Table 3 are shown the calculated projected penetration ranges $R_p = \langle t_x \rangle$ and $\Delta R_p = \langle (t_x - \langle t_x \rangle)^2 \rangle^{1/2}$ of the ions (H^+, He^+, Li^+ , Be^+, B^+, Ar^+ and Ga^+) with an energy of 100 keV in PMMA. Additionally, Table 3 shows the transverse ion range deviation $\Delta Y = \langle (t_y - \langle t_y \rangle)^2 \rangle^{1/2}$

$= \langle (t_y)^2 \rangle^{1/2}$ (the projected length t of the penetrated ion ranges between two collisions on the direction axis y parallel to the sample surface, characterizing the width of the irradiated line) in the case of mentioned irradiating ions.

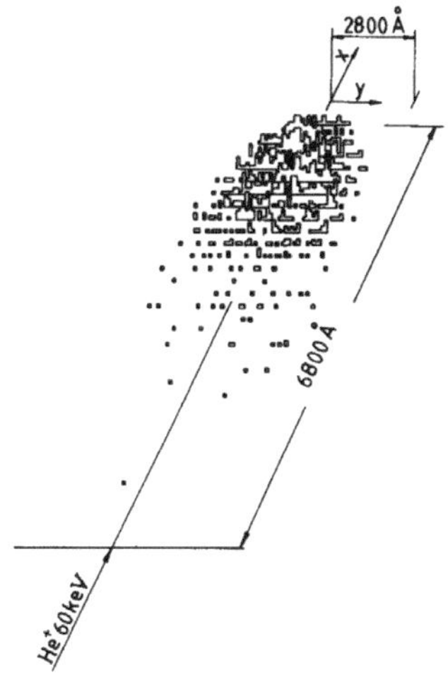

Fig. 24. Two-dimensional depth distribution of 60 keV ^4He particles implanted in PMMA.

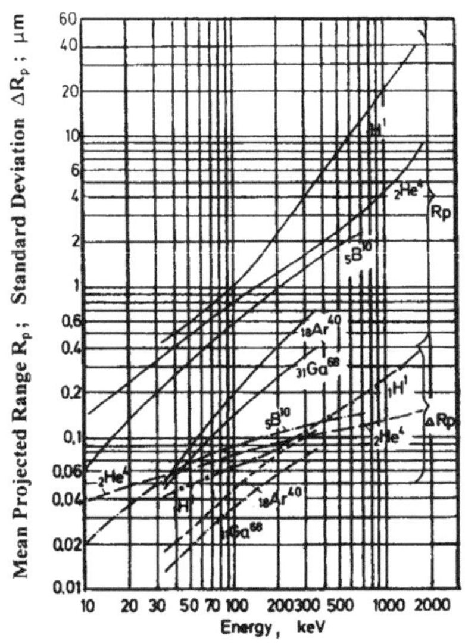

Fig. 25. Projected range R_p and mean quadratic deviation ΔR_p for different ions in PMMA over a wide energy region.

From these data it is concluded that only H^+, He^+ and B^+ cover the requirements of the present day IBL which aims to achieve 150 nm resolution patterns because the values of R_p are sufficient high at chosen resist thickness and these ions have appropriate values of ΔY for specified resolution. Among them He^+ ions are chemically inactive and lighter than the other ions, except than H^+, and therefore present some advantages. The ranges of heavier ions Ar^+ and Ga^+ at the mentioned accelerating voltages are satisfactory only for the resist thickness up to 100-150 nm. Note that heavier ions (Ar^+ and Ga^+) can be used for higher resolution. The issues are Coulomb interaction and the mentioned limited penetration depth. The latter issue can be solved with multilayer resist with top-surface imaging.

Ion	M_i	Z_i	R_p [nm]	ΔR_p [nm]	ΔY [nm]	$(dE/dx)_{el}$ [eV/nm]	$(dE/dx)_n$ [eV/nm]
H^+	2	1	994	58	107	120	0.13
He^+	4	2	796	78	115	185	1.96
Li^+	6	3	720	88	120	190	4.53
Be^+	9	4	562	80	108	195	10.16
B^+	11	5	514	77	91	205	16.64
Ar^+	40	18	183	47	30.9	371	158
Ga^+	70	31	140	35	22		

Table 3. Ranges and average energy losses (for the first 100 nm in depth) calculated for different ions and energy of 100 keV in a PMMA target.

One can see (Table 3) that the values of the transverse range deviation ΔY of the studied light ions are between 90 nm and 120 nm. At these values of the ion ranges ΔY, the patterning is affected by the ion scattering. The changes of the latent image dimensions in comparison with those of the irradiated image are considerable for high resolution layouts with critical dimensions of 50-150 nm. The situation is the same as in the case of electron lithography where the proximity effect and the related corrections of exposed patterns take part. In Table 3 calculated data of the energy losses for these light ions in PMMA are shown also where: $<dE/dx>_{el}$ stands for the mean electronic energy losses and $<dE/dx>_n$ stands for the mean nuclear energy losses. Data calculated for the all other moments characterising the spatial distribution of the He^+ with energies in the range of 30-300 keV penetrating in PMMA are shown in Fig.26-27. The upper curve (Fig.26) represents the calculated ranges through the penetrated particle trajectories $R_l = <t>$. The mean quadratic deviation $\Delta R_l = <(t - <t>)^2>^{1/2}$ and ΔY are also shown in Fig.26.

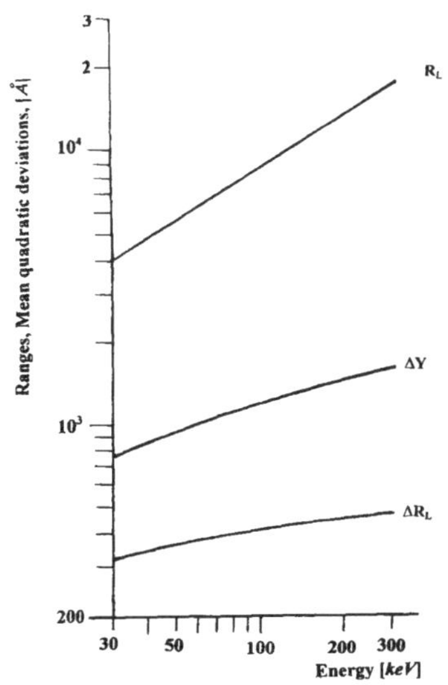

Fig. 26. Dependence of the mean range R_L, calculated through the bombarding ion trajectories, of ΔR_L, and of ΔY vs. the energy for He^+ ions and the resist PMMA.

The curve situated under the horizontal axis in Fig.27 represents the moment of the third order characterising the distribution symmetry i.e. the skewness $\gamma=<t^3{}_x/\Delta R_p{}^3>$. For example: $\gamma=0$ means that the distribution of the ranges is symmetrical. The next moment of the projected range distribution is the kurtosis $\beta=<t_x{}^4/\Delta R_p{}^4>$, which characterises the tail of the projected range distribution (curve 3 in Fig.27). For example: $\beta=3$ corresponds to the normal distribution; $\beta=6$ is in the case of an exponential distribution; the higher values of β mean that longer distribution tails exist. Curve 1 shows the transverse kurtosis concerning the distribution symmetry in a width. Curves 2 and 4 represent the moments $<t_x t_y{}^2/t_y{}^3>$ and $<t_x{}^2 t_y{}^2/t_y{}^4>$, respectively. By such calculations one can see that the scattering and lateral dispersion of the penetrating particles through a resist are small for the ions, which are three orders of magnitude heavier than the electrons. Therefore the proximity effects that are the major problem in the EBL are minimal in the IBL.

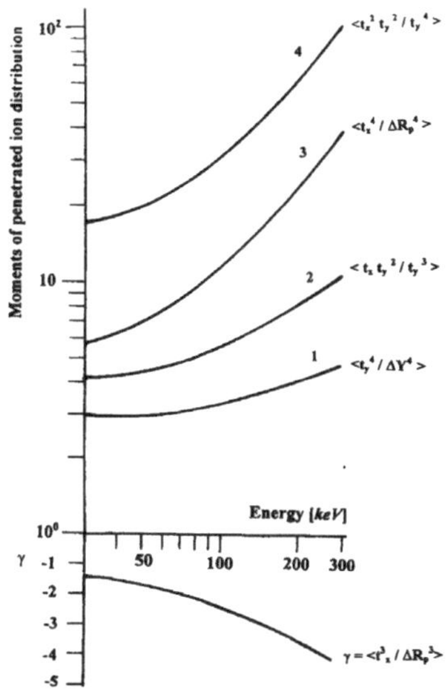

Fig. 27. Dependences of the calculated moments of the space distributions in the case of He^+ ions, penetrating PMMA: curve 1 is $<t_y{}^4/\Delta Y^4>$; curve 2 is $<t_x t_y{}^2/t_y{}^3>$; curve 3 is the kurtosis $\beta=<t_x{}^4/\Delta R_p{}^4>$; curve 4 is $<t_x{}^2 t_y{}^2/t_y{}^4>$. The curve under the horizontal axis is the skewness $\gamma= <t^3{}_x/\Delta R_p{}^3>$.

Fig.28 shows calculated energy losses in the case of He^+ bombardment of PMMA where $<dE/dx>$ is the mean electronic energy loss. Indexes 1 and 2 correspond to the primary and secondary ions, respectively. The upper curve represents the electronic energy losses of primary ions, whereas the lower curve describes these losses for secondary ions. The total electronic energy losses (the sum of these losses of primary and secondary ions) coincides with the upper curve (Fig.28) since secondary ions have a small contribution to the total electronic energy losses. The index f of the mean nuclear energy losses $<dE/dx>_{nf}$ means losses of energy through the nuclear collisions (Fig.29). The maximum value of these losses is assumed to be equal to 20 eV. It is assumed that these collisions create only phonons (i.e. heating). The index d of $<dE/dx>_{nd}$ indicates the higher portions of lost energy (>20 eV) through the nuclear type of collisions, which are able to change the resist structure completely by the creation of defects – curve $<dE/dx>_{nd}$, Fig.29. The total nuclear energy losses $<dE/dx>_{n\Sigma}$ are also shown in Fig.29. A comparision between the electronic and nuclear stopping power values in the case of He^+ in PMMA confirms the more general conclusion, based on the experimental data, that the electronic stopping losses determine the solubility changes in the resists (Mladenov & Emmoth, 1981). As an additional reason can be indicated the fact, that the values of $<dE/dx>_{el}$ are more than two orders of magnitude higher than $<dE/dx>_n$ at the studied kinds of ions and their energies.

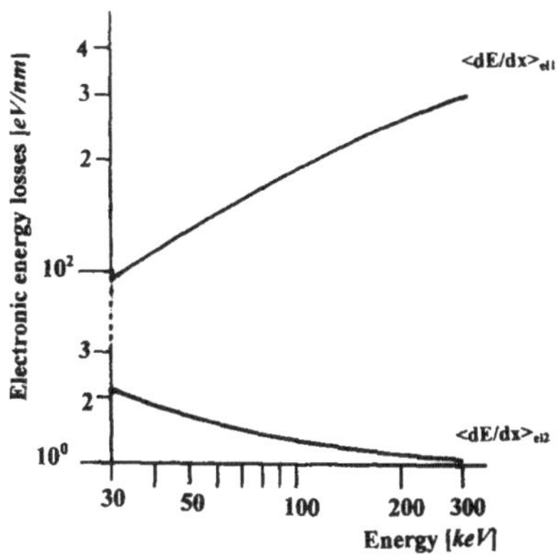

Fig. 28. Electronic stopping powers for He^+ penetrating PMMA. The lower curve - the electronic losses $<dE/dx>_{el2}$ of secondary particles, moved after collisions with the projectile ion. The upper curve - the electronic losses of primary ions as well as the total electronic losses: $<dE/dx>_{el1} \approx <dE/dx>_{el\Sigma}$.

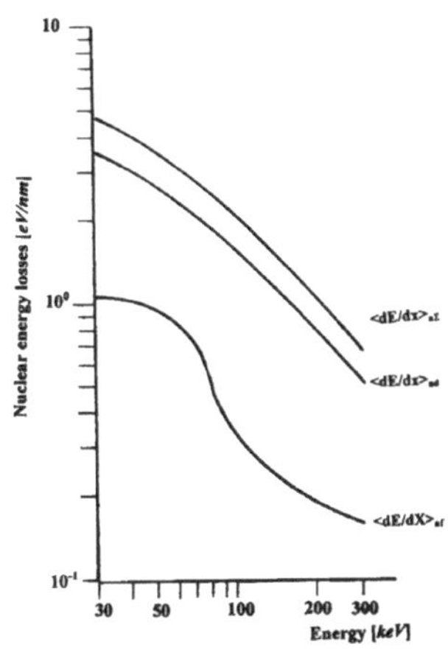

Fig. 29. Nuclear stopping power for He^+ in PMMA.

Using these data it can be concluded, that the radiation changes due to atom mixing of resist atoms at nuclear collisions of He projectiles with resist components in the studied energy range have a low probability. The situation is opposite to the case of 120 keV Ar^+ irradiation ((b) Vutova & Mladenov, 2001) where lower solubility rate at higher doses of ion exposure had been explained by ion irradiation defects. In the same time the electronic energy losses are not so high, that erosion, decomposition and change of physical properties of the irradiated resist take place (Braun et al., 1983; Emmoth & Mladenov, 1983).
To obtain the absorbed energy space distribution when exposing an arbitrary pattern, using an arbitrary exposure dose distribution, one must take into account the influence of a large

number of exposed points. Due to the large number of calculation, the simplified procedure (described above for EBL) is used to calculate the integral space distribution of the absorbed energy. To calculate the δ_f, δ_b and η_E parameter values (Eq.19), the linear approximation along the resist depth is used. In some ion lithography cases (heavy-ion bombardment, high-energy ion beams, etc) it is necessary to use more than two layers along the resist depth (Vutova & Mladenov, 1994). The values of δ_f, δ_b and η_E change linearly among them.

8. Conclusions

In this chapter we describe the basic steps when modeling the ion and electron exposure processes in multilevel multicomponent amorphous targets. The modeling of the exposure process in EIL involves the simulation of particle penetration into materials as well as the calculation of the absorbed energy distribution within the targets. The models presented for the exposure process simulation are realized by means of a software package. Using our simulation tools the effects of: irradiated particles (electrons and ions), exposure doses, substrates and dense patterns are discussed in order to extract the necessary values for high resolution patterning.

EBL simulation of the exposure process was carried out in the case of semiconducting and superconducting substrates using the developed models and the computer programs. For the calculation of energy deposition in the resist film due to a point beam (EDF(r,z)) a MC simulation was applied. The problem concerning the insufficient statistics of the discrete data for the absorbed energy in the case of large lateral distances is overcome. In the Si case, the EDF(r) is approximated by a sum of two Gaussian functions, while in the case of YBCO layers deposited on STO or MgO substrates, the EDF(r) could be approximated as a combination of a double Gaussian and an exponential function. In this case the division of scattering electrons into forward and backward is not the same as for Si samples since the backscattering from YBCO film is important in the middle region of the EDF(r) (from ~ 40 nm to 2 μm). Data for the scattering parameters, such as standard deviations of EDF(r), was obtained. A methodology and a computer program are proposed for the calculation of the absorbed energy integral space distribution when exposing real topological structures in EBL, using analytical expressions. The simulation results were compared to the experimental and other calculated data in the case of Si and YBCO/STO(MgO) substrates and the agreement was good.

A program version, named TRIM-MV, for exposure process modeling at IBL, based on the famous Biersack model, is developed. From the detailed simulation study, significant information on the energy losses, ranges and moments of the space distributions for different ions is obtained. It is concluded that the exposure with He ions with energies from 60 to 100 keV is favorable for sub-150 nm region. On the basis of our experimental and calculated results, it is assumed that the electronic energy losses are responsible for solubility modification of polymer resists in IBL. The nuclear stopping power has an effect on the solubility modification in the case of heavy low-energy ion irradiation at non-linear (multiciphered) development rates.

The obtained results and comparisons show that the models for exposure process modeling at EBL and IBL are adequate and have a very good potential for application in the case of multiplayer sub-quarter-micron patterns. The simulation results are very useful when optimizing particular technological processes for complex layouts in the sub-quarter-micron lithography. The ion lithography should be considered as one of the competitors for further improvement of the integration level of the produced integral circuits.

9. References

Adesida, I.; Everhart, T.E. & Shimizu, R. (1979). High resolution electron-beam lithography on thin film. *Journal of Vacuum Science & Technology*, 16(6), pp.1743-1748, ISSN 0022-5355

Adesida, I.; Ilesanmi; Karapiperis & Leonidas. (1982). Monte Carlo simulation of ion beam penetration in solids. *Radiation Effects*, 61 (3-4), pp.223-233, ISSN 0033-7579

Biersack, J. & Eckstein, W. (1984). Sputtering studies with the Monte Carlo Program TRIM.SP. *Applied Physics A Solids and Surfaces*, 34 (2), pp. 73-94, ISSN 0947-8396

Biersack, J. & Haggmark, L. (1980). A Monte Carlo computer program for the transport of energetic ions in amorphous targets, *Nucl. Instr. and Meth*, 174 (1-2), pp.257-269, ISSN 0168-9002

Braun, M.; Emmoth, B.; Mladenov, G.M. & Satherblom, H.E. (1983). Erosion of polymer thin films during ion bombardment. *J.Vac.Sci.Technol. A*, 1(3), 1383-1387, ISSN 0734-2101

Chang, T.H.P. (1975). Proximity effect in electron-beam lithography. *J. Vac. Sci. Technol.* 12 (6), pp.1271-1275, ISSN 0022-5355

Emmoth, B. & Mladenov, G.M. (1983). Sputtering of a polymer layer deposited on metal substrates. *Journal of Applied Physics*, 54 (12), pp.7119-7123, ISSN 0021-8979

Gavrilov, N.; Yakusheva, D. & Kondyurin, A. (1998). Structure of polyethylene after pulse ion beam treatment. *J. Applied Polymer Science*, 69 (6), pp.1071-1077, ISSN 0021-8995

Gueorguiev, Y.M. (1996). Monte Carlo simulation of inclined incidence of fast electrons to solids. *J.Vac.Sci.Technol. B*, 14 (4), pp.2462-2466, ISSN 1071-1023

Gueorguiev Y.; Mladenov, G. & Ivanov D. (1994). Monte Carlo simulation of electron-beam exposure distributions in the resist on structures with high-T_c superconducting thin films. *Thin Solid Films*, 251, pp. 67-71, ISSN 0040-6090

Gueorguiev, Y.M.; Vutova, K. & Mladenov, G. (1995). A Monte Carlo study of proximity effects in electron-beam patterning of high-Tc superconducting thin films. *Physica C*, 249 (1-2), pp.187-195, ISSN 0921-4534

Gueorguiev, Y.M.; Vutova, K. & Mladenov, G. (1996). Analysis of the proximity function in electron-beam lithography on high-Tc superconducting thin-films, *Supercond. Sci. Technol.*, 9 (7), pp.565-569, ISSN 0953-2048

Gueorguiev, Y.; Vutova, K. & Mladenov, G. (1998). Numerical modelling of the processes of exposure and development in electron beam lithography on high-temperature superconducting thin films,*Thin Solid Films*, 323 (1-2), pp.222-226, ISSN 0040-6090

Hatzakis, M.; Ting, C.H. & Viswanathan, N.S. (1974). Fundamental aspects of electron beam exposure of polymeric resist system. *Proceed. of 6-th Int.Conf. on Electron and Ion Beam Science and Technol. Electrochem Soc.*, pp.542-579, R.Bakish Ed., Princeton, N.J.

Hawryluk, R.J.; Hawryluk, A.M. & Smith, H.I. (1974). Energy dissipation in a thin polymer film by electron beam scattering. *Journal of Applied Physics*, 45 (6), pp.2551-2566, ISSN 0021-8979

Hirscher, S.; Kaesmaier, R.; Domke, W.-D.; Wolter, A.; Löschner, H.; Cekan, E.; Horner, C.; Zeininger, M. & Ochsenhirt, J. (2001). Resist process development for sub-100-nm ion projection lithography. *Microelectronic Eng.*, 57-58, pp.517-524, ISSN 0167-9317

Kaesmaier, R. & Löschner, H. (2000). Ion projection lithography: progress of European MEDEA & International Program. *Microel. Eng.*, 53 (1), pp.37-45, ISSN 0167-9317

Kyser, D.F. & Murata, K. (1974). Quantitative electron microprobe analysis of thin films on substrates. *IBM J. of Research and Development*, 18 (4), pp.352-363, ISSN 0018-8646

Miller, P. (1989). Image-projection ion-beam lithography. *J. Vac. Sci. Technol. B*, 7 (5), pp.1053-1063, ISSN 1071-1023

Mladenov, G. (1985). Ion implantation in semiconductors and other materials. *Proceedings of the 7th International Conference*, p. 230. Vilnus, Lithuania

Mladenov, G.M.; Braun, M.; Emmoth, B. & Biersack, J.P. (1985). Ion beam impact and penetration of polymethyl methacrylate. *Journal of Applied Physics*, 58 (7), pp.2534-2538, ISSN 0021-8979

Mladenov, G.M. & Emmoth, B. (1981). Polymethyl methacrylate sensitivity variation versus the electronic stopping power at ion lithography exposure. *Applied Physics Letters*, 38 (12), pp.1000-1002, ISSN 0003-6951

Mladenov, G. & Vutova, K. (2002). Computer simulation of exposure and development in EBL and IBL. *Proceedings of St.-Petersburg Electro-technical University, Solid State Physics and Electronics*, 1, pp.133-173, ISBN 020617, St.-Petersburg, Russia, April 2002, ed. B. A.Kalinikos, publisher SPbGETU LETI

Mladenov, G.; Vutova, K.; Raptis, I.; Argitis, P. & Rangelow, I. (2001). Simulation of latent image formation for ion beam projection lithography. *Microelectronic Engineering*, 57-58, pp.335-342, ISSN 0167-9317

Olziersky, A.; Vutova, K.; Mladenov, G.; Raptis, I. & Donchev, T. Electron beam lithography simulation on superconducting substrates. *Supercond. Sci. Technol.*, 17 (7), pp.881-890, ISSN 0953-2048

Raptis, I.; Meneghini, G.; Rosenbusch, A.; Glezos, N.; Palumbo, R.; Ardito, M.; Scopa, L.; Patsis, G.; Valamontes, E. & Argitis, P. (1998). Electron-beam lithography on multilayer substrates: experimental and theoretical study. *Proc. SPIE*, vol. 3331, pp.431-441, Santa Clara, CA, USA, February 1998

Robinson, M.T. & Torrens, I.M. (1974). Computer simulation of atomic-displacement cascades in solids in the binary-collision approximation. *Physical Review B*, 9 (12), pp.5008-5024, ISSN 0163-1829

Vutova, K. (2007). *DSc Thesis: Modeling of physical processes at material treatment by electron and ion beams*, Institute of Electronics, Bulgarian Academy of Sciences, Sofia, Bulgaria

Vutova, K. & Mladenov, G. (1991). Absorbed energy distribution in electron lithography of simple patterns. *J. Inform. Recording Materials*, 19 (4), pp.261-269, ISSN 1025-6008

Vutova, K. & Mladenov, G. (1992). Modelling of physical processes in ion lithography. *Thin Solid Films*, 214 (2), pp.144-149, ISSN 0040-6090

Vutova, K. & Mladenov, G. (1994). Modelling of exposure and development processes in electron and ion lithography. *Modelling and Simulation in Materials Science and Engineering*, 2, pp.239-254, ISSN 0965-0393

(a) Vutova, K. & Mladenov, G. (2001). Why light Ions in Future Ion Lithography. *Vacuum* 62 (2-3), pp.273-278, ISSN 0042-207X

(b) Vutova, K. & Mladenov, G. (2001). Sensitivity, contrast and development process in electron and ion lithography. *Microelectr. Eng.*, 57-58, pp.349-353, ISSN 0167-9317

Vutova, K.; Mladenov, G.; Raptis, I. & Olziersky, A. (2007). Process simulation at electron beam lithography on different substrates. *Journal of Materials Processing Technology*, 184 (1-3), pp.305-311, ISSN 0924-0136

Wilson W.D.; Haggmark L.G. & Biersack J.P. (1977). Calculations of nuclear stopping, ranges, and straggling in the low-energy region. *Physical Review B* 15 (5), pp.2458-2468, ISSN 0163-1829

18

Computer Simulation of Processes at Electron and Ion Beam Lithography, Part 2: Simulation of Resist Developed Images at Electron and Ion Beam Lithography

Katia Vutova, Elena Koleva and Georgy Mladenov
Institute of Electronics at Bulgarian Academy of Sciences
Bulgaria

1. Introduction

Polymer resists are recording and transfer media in electron, ion and photon lithography. The primary aim of lithography process is mastering of relief image in polymer resist layer. This developed and baked pattern is performed to be transferred to substrate trough either: a) plasma or ion (dry) processes or b) wet chemical etching. The transfer of resist relief image can be realised by ion implantation or by "lift off" of deposited on its top thin film (only deposited in open windows in the resist pattern remains after treatment on a strong solvent - namely acetone or methylene chloride). The use of these various microstructure technologies is to create optical, X-ray, electron or ion masks as well as for direct writing of the desired microstructure on the semiconductor wafers for production of integrated circuits. In the same way the resists are used at mastering the micro-structured elements of micro-devices.

The conventional polymer resist is dissolved in a liquid solvent. Usually the liquid resist is dropped onto the sample, which then spins at 1000 to 6000 rpm to form a smooth and uniform film. Immersion deposition of very thin resist film is also applicable. After that the deposited resist layer is pre-baked to 70-90°C to dry the prepared film.

During exposure with appropriate radiation scission and/or cross-linking of the polymer molecules are observed in the resists. An alternative for more complicated resist as chemically amplified resists (CAR's) is creation of an acid agent (from an acid generator component in the compound of the two- or three- component resists) that changes the dissolving protection of the base resin. In this case there are two bake steps: a) pre-bake performed after the spin-coating, aiming to dry the resist by removing the solvent from the deposited resist layer as well as b) post exposure bake (PEB) - the purpose of which is to cause a thermal catalysis of the chemical reactions amplifying the latent bulk image. The second baking step (PEB) is performed in an oven or on a hot plate. The PEB temperature and time are critical for the resist acid generator modification and for acid diffusion and reduction (degradation losses) as well as its evaporation.

At the end of the transfer process the resist is washed away in a solvent such as acetone or a special resist stripper. Often the strip of the resist is done by reactive ion etching in O_2.

At electron and ion beam lithography (EBL and IBL) the resist profile is created as a result of the development process of exposed latent images, containing resist segments with locally modified solubility. This relief image is produced due to the changes of resist thickness and local removal of resist material up to full opening of the irradiated or the non-irradiated areas (respectively in the case of positive tone or negative tone resists). The type of the used irradiation particles - electron or ion is not important when applying the simulation models to the development process. More detailed knowledge and/or comparison of predicted results with experimental data are needed to obtain correct simulation because of the distinct sensitivity behavior for various material compositions.

2. Sensitivity and contrast of electron and ion resists and their developers

The sensitivity and contrast are base parameters of the resist layers, used in the optical lithography, as well as in EBL and IBL. One can evaluate the sensitivity and the contrast of the resist using the experimental dependence d(D) of the resist thickness d at given initial thickness d_0 and at development time t, after irradiation with an average exposure dose D on an area (namely 100x100 μm) and after development of the exposed latent image. At this approach one evaluates change of the resist thickness d in that area during the development process in a chosen solvent called developer. This change is caused by the mentioned dissolution removal of the exposed or the un-exposed polymer. Variable value is the dose D at chosen energy E of the radiation projectiles as well as at constant development conditions (developer type, time and temperature of development). An example of such curve d(D) is given in Fig.1.

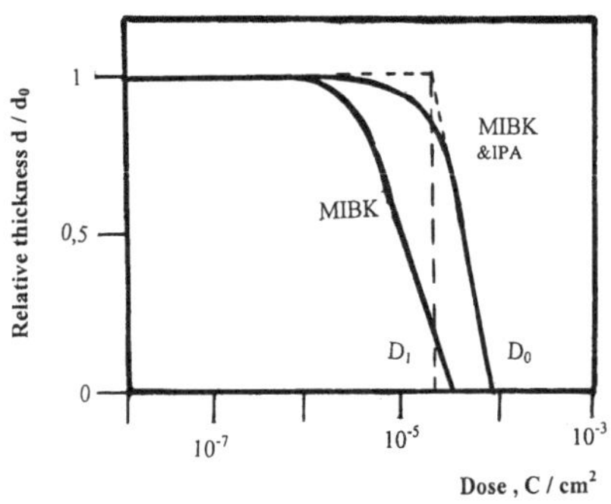

Fig. 1. Dependence of the relative thickness d/d_0 vs. the exposure dose for a positive-tone resist in two developers (MIBK and MIBK& IPA).

The sensitivity of the polymer resist to the radiation is measured as the minimal dose D_0, needed for the development process of the exposed image. This dose is also called dose-to-clear in contrast to the dose, optimized to obtain near to vertical side walls of the developed patterns, usually different than D_o.

In the field of high-resolution lithography it is very important to characterize the edges of the developed profiles in the resist. The contrast parameter γ_d is defined by the dose interval between the initial exposure dose D_1 (at which a start of development in the MIBK&IPA developer can be observed) and the mentioned dose D_0 (for full development of studied resist thickness). These doses are evaluated at increasing doses and chosen time and temperature conditions of the developing process using the bigger dose value of remaining

initial resist thickness d_0 of the utilised resist layer (in the case of positive one) and the minimal dose for full development. Often the curve shown in Fig.1 is called contrast curve.
The values of sensitivity and contrast characteristics in the case of EBL are estimated experimentally in many papers for some average molecular weight of the used resist (measured for example in a.u.) and for a chosen developer. The given data (Brault & Miller, 1980; Karapiperis et al., 1981) presents the sensitivity D_0 and the contrast γ_d in the case of some ion beam resists. Generally, the sensitivity of the polymer resists at the ion exposure is assumed to be higher than at the electron exposure (Brewer 1980; Ryssel et al.,1991). The comparison is not precise due to the different conditions and the peculiarities of the figures of merit used.
The sensitivity and the contrast are described usually as independent properties. Another observed experimental fact is that the solubility rate S of the irradiated spots depends on the used exposure dose D. So at selected (in a set of experiments) resist thickness and development time the curves d(D) are function of the concrete values of the solubility rate S at different exposure doses D.
A definition of another contrast parameter $\gamma_s = (\Delta S/\Delta D)$ as the slope of the solubility rate dependence on the exposure dose S(D) at a given exposure dose (see Fig.2) was accepted. Then can be seen that the parameters γ_s and D_0 for a resist-developer combination are mutually connected (Mladenov & Seyfarth, 1986; Vutova & Mladenov, 2001).
The sensitivity is determined by the radiation efficiency of the couple resist - irradiated particle. The radiation efficiency of the resist g at used kind of radiation can be measured by the mean number of chemical events (chain scission or destruction) per one unit (namely electron volt) of the absorbed energy.
The value of the contrast parameter γ_d for the positive-tone resists and definitions given in Fig.1 can be calculated using the relation:

$$\gamma_d = [\lg (D_1/D_0)]^{-1}, \tag{1}$$

when the removed normalised thickness (namely the ratio $\Delta d/d_0$, where Δd is the removed resist thickness) is equal to 1. In the case of negative-tone resists by analogy:

$$\gamma_d = [\lg (D_0/D_1)]^{-1}, \tag{2}$$

where D_1 and D_0 are the exposure doses at which a start of the development removal (D_1) and the full development, respectively, due to removal of non-irradiated adjacent to exposed pixels (D_0) occurs.
Due to the more smooth transition parts of the d(D) curves observable some times near to the initial dose (D_1) (or near 0% removed irradiated resist thickness for positive resists) and usually at 100% developed image - for negative resists, the rest - straight line sector of d(D) is continued. The doses corresponding to the crossing points of this straight line with the 0 % (or 100%) line is chosen as D_1 and D_0, respectively.

3. Solubility rate dependencies on the exposure dose

The goal of the computer simulation of the processes in EBL and IBL is prediction of the resist profile and the image dimensions of the developed exposed microstructure. During the exposure of the samples, covered by a sensitive to irradiation layer, the local solubility rate of this resist is modified and as a result a developed image as removed and non-removed areas of the treated resist layer could be observed. Respectively to solubility changes of irradiated areas one could distinguish positive-tone or negative-tone resists. The first important step of the modeling of the resist thickness changes is to transfer the absorbed energy distribution in the resist, caused by exposure, in a distribution of the local solubility (in the case that is also the development) rate. Due to different chemistry composition of utilized resists and to increasing requirements on the exactness of the predicted profiles in various dimension ranges of the developed images (for example- in micron or in sub-micron regions, in sub 0.250 μm or less than 50 nm of the image critical dimensions) the models used in computer simulation of profile dimensions and of its edge are different. The choice of a concrete model and fitting parameters used is the most critical step of this prediction. The integral-circuit-manufacturers use concrete and limited number of resists and any available simulation tool and keep the obtained fitting parameters and applied approximations as confidential as possible.

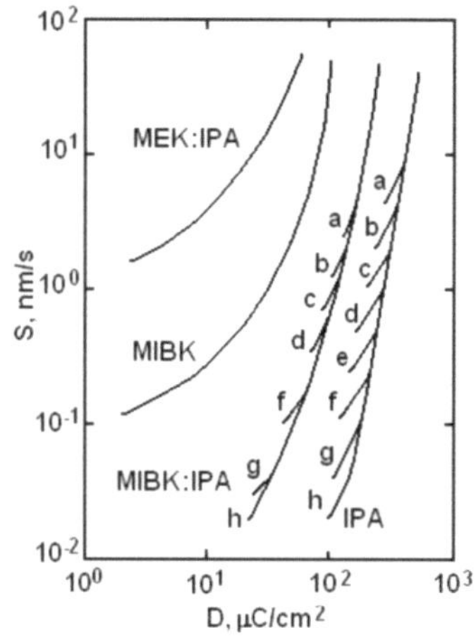

Fig. 2. Dependence of the solubility rate S on the dose D in the case of electron exposure for different developers. The resist thickness is 0.5 μm, the development time is: a - 15 s, b - 30 s, c - 60 s, d - 120 s, e - 240 s, f - 480 s, j - 960 s, h - 1920 s

Fig.2 represents the dependence S(D) in the case of electron exposure of polymethyl methacrylate (PMMA) and the following developers - pure methyl-isobutyl-ketone (MIBK), pure isopropyl alcohol (IPA), MIBK/IPA 1:1 solution and methyl-ethyl-ketone (MEK)/IPA 1:1 solution. This figure shows indirectly that the solubility rate S of the resist is a function of the absorbed energy in the resist for a developer. Experimental investigations also show that S depends on the radiation efficiency of the charged particles in the resist, the resist density and the molecular dispersion. The relation S(D) can be experimentally obtained for a given resist thickness and irradiated particles' energy. The calculated ranges and absorbed energy distribution of the irradiating species can easily convert this dependence in more physically important dependence S(E/V), where E/V is the absorbed energy per one unit resist volume.

In the case of positive-tone resists the solubility rate of the irradiated areas increases while in the case of negative-tone resists - the solubility rate decreases. The radiation affects the density or the number of the modified resist molecules. The changes are weak at lower exposure doses and one can develop the irradiated image at low contrast (and for a very long developing time). The contrast parameter γ_s can be defined by the slope of the curve S(D) or $S/S_0(D)$. Taking into account that $S=\Delta d/t$, where $\Delta d = (d_0-d)$ is the removed thickness and d_0 is the initial resist thickness, it can be found that $\gamma_s=\gamma_d/(t.d_0)$, i.e. γ_s is proportional to γ_d. In the cases of higher doses (that means - lower sensitivity of the resist) one can obtain a desirable irradiating image for shorter time, that is practical, and usually at higher contrast. This means that one can trade sensitivity for contrast.

The use of a mixture solvent/non-solvent as a developer of electron irradiated polymer (such weak developer is the mixture of MIBK and IPA in the case of PMMA) improves the contrast together with a loss of the sensitivity. In this case it was observed a non-linear behaviour of the solution process (Mladenov & Seyfarth, 1986). In Fig.2 one can see two cases of polymeric layer solution: (i) linear resist solution with a constant solubility rate during development process and (ii) non-linear resist solution with variable solubility rate during the development time. An initial dose from which development begins and an initial solubility rate at every chosen development time take place in the non-linear development of PMMA and a weak solvent is used as a developer. This phenomena is connected to the selectivity of the solution at different molecular weights of the fractions and to the diffusion processes of the developer in the polymer resists. The differences in the observed multi-ciphered dependencies are small and they could be neglected in numerical simulation of the development process in PMMA resist.

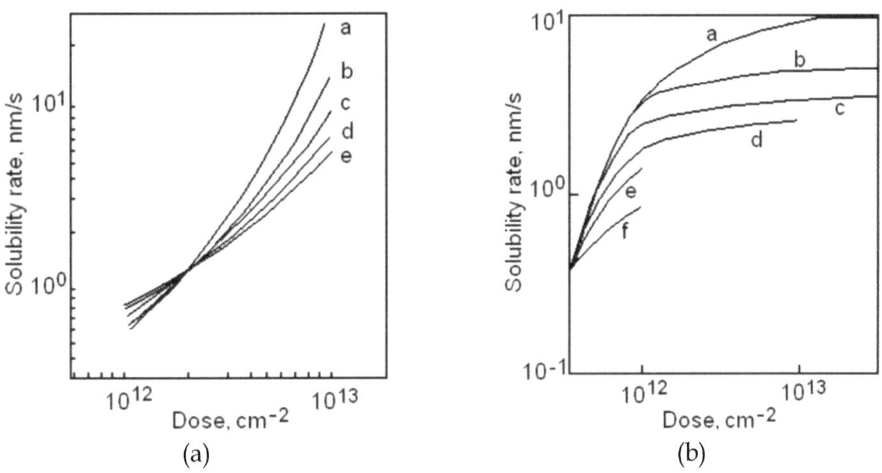

Fig. 3. Solubility rate of PMMA (M=675 000) using MIBK:IPA (1:1) at 20^0 C after exposure with: a) 80 keV He^+: a - 30 s, b - 60 s, c - 90 s, d - 120 s, e - 150 s; and b) 120 keV Ar^+: a - 30 s, b - 60 s, c - 90 s, d - 120 s, e - 150 s, f - 300 s.

In the case of ion exposure process a similar linear or non-linear solution processes were observed pertaining to ion mass (Brault & Miller, 1980; Karapiperis et al., 1981; Vutova & Mladenov, 2001) due to the different radiation ion efficiency to the energy transfer and subsequently, to different latent images. In Fig.3 one can see the non-linear (multi-valued) solubility rate obtained in the case of He^+ (Fig.3a) and Ar^+ (Fig.3b) irradiations of PMMA. A dependence of the contrast on the developing time (and the resist thickness) can be seen. The difference between the electron resist non-linear solution process and ion resist non-linear solution process is connected to the difference of the energy deposition processes in the polymer in both cases. In the case of EBL there is high probability that the electrons will

be more deflected from the initial path (up to back scattering) as well as the created secondary electrons will have high energies. As a result the resist layer up to a distance of a lot of microns from the irradiating point is changed. Differently, the electronic stopping power is responsible for the resist solubility modification in the case of ion exposure with low masses (Mladenov & Emmoth, 1981; Mladenov et al., 1985). This leads up to a distribution of the modified resist pixels (in an order of 100 nm) near and around the ion track. Ion irradiation defects in the polymer layer, caused by nuclear energy losses, can be a reason for the lower solubility rate at high doses as well as the more strong decrease of solubility rate at heavier irradiating ions. Note that in IBL development the increase of the exposure dose causes bigger difference between the development rates for various development times (opposite to PMMA and weak developer case, shown in Fig.2).
As a conclusion it could be noted, that non-linear behavior during the time of development could be expected to be a more general case and probably an important phenomena in the processes of development of the smallest images where more details of the development process are of considerable value. Note that at more complicated processes of solubility modification (in the case of three component resists, chemically amplified resists) the dependencies S(D) are functions of more parameters such as initial photo-acid generator concentration, times and temperature of post exposure bake, various diffusion coefficients etc. At robust process they are constants and S(D) is an approximation that is applicable for the computer simulation.

4. Data for the resists used in the electron beam nanolithography

4.1 PMMA in nanolithography

An organic one-component positive tone resist, which is extensively used in nanolithography, is special grades of polymethyl methacrylate (PMMA) with molecular formula $[CH_2C(CH_3)\ (CO_2CH_3)]$. The wide use of PMMA is due to the fact that it has high resolution, wide processing latitude, it is easy to handle and forms excellent film. In many papers, a resolution of developed PMMA patterns bellow 10 nm of isolated features and dense array of 30 nm parallel lines at a pitch of 30 nm (periodic grating) is demonstrated. For this aim thin membrane substrates, thick polymer sub-layer or high accelerating electron energy (80-200 keV), ultrasonic assisted development were used (Moreau et al., 1979; Cumming et al., 1996; Hatzakis, 1998; Yasin et al., 2002)
The predominant mechanism of polymer solubility modification in PMMA is chain scission between the carbon atoms changing the molecular weight distribution and decomposition of carbonyl radicals to volatile products. The required irradiation dose is high (as example 50 $\mu C/cm^2$ at 20 keV accelerating electron energy and developer methyl butyl ketone - MIBK). Some co-polymerizing PMMA compounds (with polymethacrylic acid or anhydride) could enhance 2-3 times the PMMA sensitivity (Hatzakis, 1998). The concentration change of the developer (to a mixture of strong dissolving agent and weak or non-solvent agents as iso-propanol (IPA) or water) is a way to improve the contrast (that is important for higher resolution) at decreasing of the solubility (Vutova & Mladenov, 2001; Yasin et al., 2002). Experiments were executed using 400-1000 nm thick PMMA films. In the sub-10 nm region the use of development assisted with ultrasonic agitation, instead the standard puddle development is important (Chen & Ahmed, 1993). This increases the development rate and in this way a lower dose of exposure could be applicable: 6.4 - 7.5 $\mu C/cm^2$ at 20 keV accelerating electron energy (Vutova & Mladenov, 2001) and 200 nm thick PMMA. Rinse in IPA decreases the roughness of the developed lines (Yasin et al., 2004).

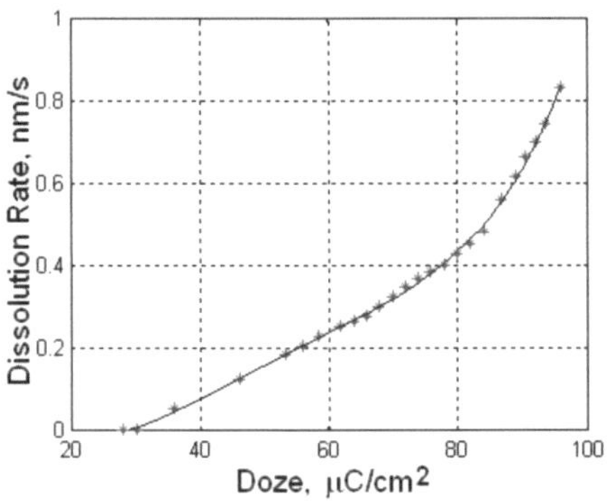

Exposure	electrons	
PMMA	A2	Microchem
Mol. weight	950 000	
Thickness	50 nm	
Substrate	Si	
Developer	MBK:IPA	1:3
Development Time	60 sec	
Developer Temperature	20° C	
PreBake	170° C	30 min oven
E-beam	30 keV	

Fig. 4. Dissolution rate, calculated for PMMA at the given parameters (50nm, 30keV)

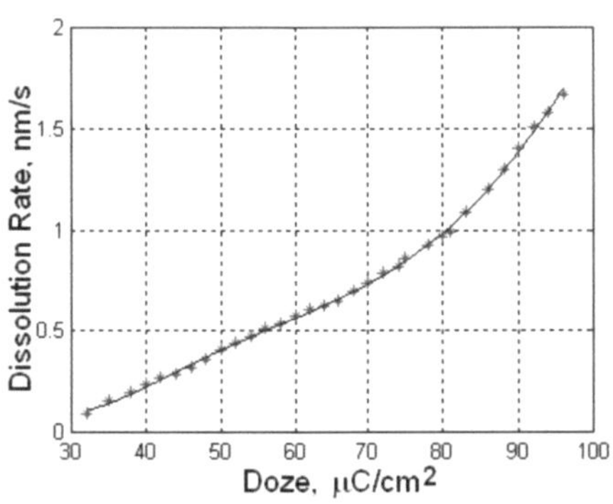

Exposure	electrons	
PMMA	A2	Microchem
Mol.weght	950 000	
Thickness	100 nm	
Substrate	Si	
Developer	MBK:IPA	1:3
Development Time	60 sec	
Developer Temperature	20° C	
PreBake	170° C	30 min oven
E-beam	20 keV	

Fig. 5. Dissolution rate, calculated for PMMA at the given parameters (100nm, 20keV)

Our experience concerning development of thin PMMA films - 100 nm and 50 nm thick films (PMMA with 950 000 Mw, development time 60s, temperature 20° C, developer MIBK:IPA 1:3, substrate Si, Pre-bake 170° C, electron energy from 20 keV to 30 keV) shows a lower contrast in the region of exposure doses from 30 $\mu C/cm^2$ to about 80 $\mu C/cm^2$ and an improved contrast parameter at exposure doses higher than 80 $\mu C/cm^2$ (see Fig.4 and Fig.5). The calculated dissolution rates, using experimentally measured relative changes of resist thickness during development process, at various doses of electron exposure are different for different resist thicknesses (in Fig4 and Fig.5 the energies of the exposure electrons are also different). There A2 means 2% dissolution in Anisole as a solvent.
To check the role of the changed electron energy we calculated the dependence of the absorbed electron energy distribution on the radial distance (from the direction of the initial movement of the beam electrons) using the approach discussed in the Part 1 of this Chapter (Fig.6 and Fig.7). One can see that the back scattered electrons are more widely distributed and have a lower amplitude in the case of a higher accelerated voltage (Fig.6). So the increase of the development rate in that case could be expected at higher exposure doses and this factor could not change the development rate at lower doses. The values of all development rates are higher in the case of thicker resist layer use (Fig.5). This could be explained by the more important role of the modified surface and interface sub-layers of the

resist film, when the used resist layers have thickness equals to or less than 100 nm. In that way, the characteristics of bulk PMMA become less important than modified boundary sub-layers of the resist at nanostructures lithography using PMMA.

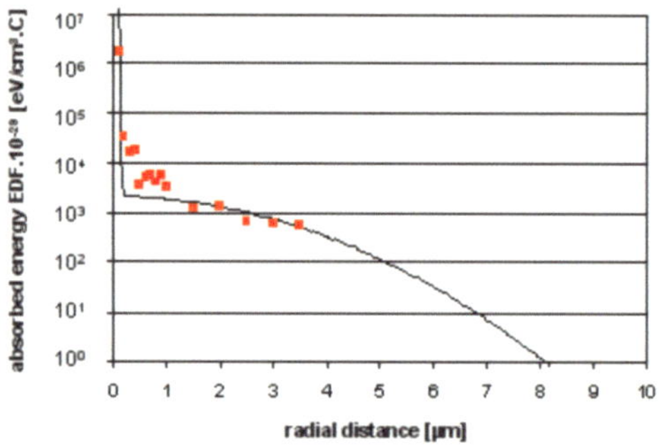

Fig. 6. Absorbed energy on a radial distance, calculated for PMMA 50 nm, 30 keV

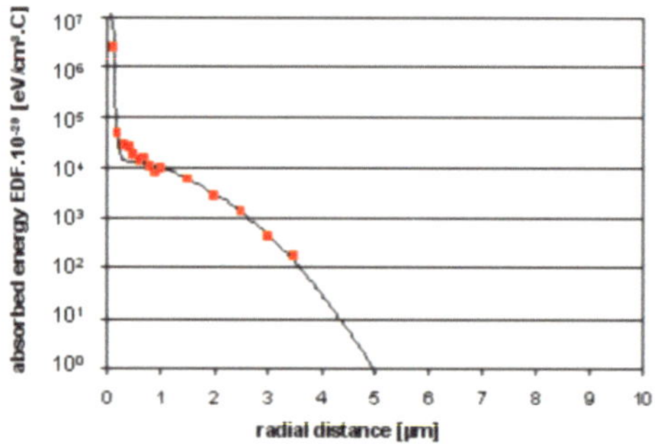

Fig. 7. Absorbed energy on a radial distance, calculated for PMMA 100 nm, 20 keV

For computer simulation of the developed EBL profiles the relations, shown in figures 4 and 5, are estimated using regression models and are given in Table 1.

parameters	equation
PMMA 50nm, 30keV, from 28.8 to 96 µC/cm^2	R=0.4060855351229032219441817 02805- -0.0453482943810538666922091450 06*D+ +0.00160897642509069575316387495 27031*(D^2)- -0.00002110576980639599621652039 6068051*(D^3) +0.00000010162114917206451072185 43839274*(D^4)
PMMA 100nm, 20keV, from 32 to 96 µC/cm^2	R = 3.4811250845045663734447886 11273- -0.3406885444772432294017684 60987*D+ +0.0124852802305000737142869156 19175*(D^2)- -0.000208081022063644416412397405 85732*(D^3)+ +0.00000164333655088755838571540 8896985*(D^4)- -0.00000000485841447194225023411 09588036606*(D^5);

Table 1. Dissolution rates R on the exposure doses D, evaluated using regression models.

4.2 Chemically amplified resists in electron beam nanolithography

To improve the throughput of the exposure process chemically amplified resists (CARs) are developed and used. Due to higher sensitivity the dwell time for imaging of these resists is shorter. Most CARs are a polymer matrix material, the solubility of which is blocked by a component. As an example poly (hydroxysturene) polymer could be blocked with t-butyl carbonyl. In the resist compound a reactive photo-acid generator (PAG) is added and an acid is produced during the electron exposure. After exposure a post exposure thermal bake step is performed to catalyze the reaction between the acid molecules and the protecting group. The de-protecting reaction switches the solubility of the polymer in an aqueous base while regenerating the acid further could amplify the de-protecting reaction. From Table 2 one could see a higher sensitivity at lower exposure doses (less than 100 μC/cm²). Due to that a resist platform for projection electron lithography in nanoelectronics manufacture exists.

Resist	Sensitivity μC/cm²	Resolution nm	Electron energy keV	Film Thickness nm	Paper
SAL605	1,6	50	20	450	Fedynishin et al., 1990
CAMP	60	250	50	300	Novembre&Munzel,1996
ARCH	10	100	50	350	Novembre&Munzel,1996

Table 2. Properties of some CARs in the nano-EBL

In Fig.8 contrast curves of CAMP6 and of two types ARCH CARs are compared with PMMA positive tone one-component nanoresist. The CAMP6 resist has nonlinear behavior during development (Fig.9 – (a) Vutova et al., 2009). The corresponding development rate vs. the dose (Fig.10) is a multi-ciphered dependence. As a result at computer simulation of developed profiles in the resist at the higher exposures doses the resist areas exhibit high dissolution rates and a central zone of the resist has rapid resist removal.

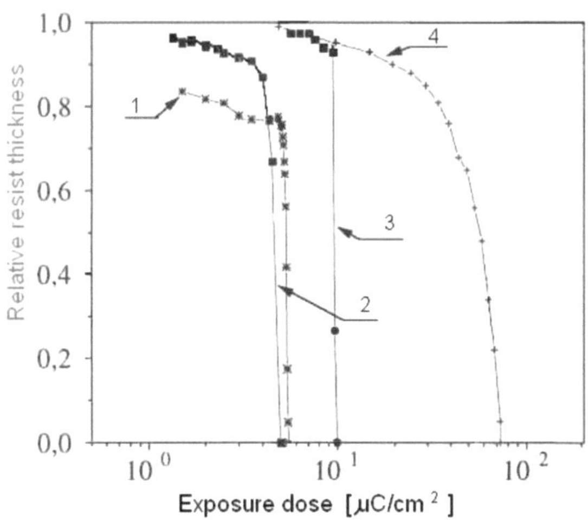

Fig. 8. Contrast curves for: 1- CAMP 6; 2- ARCH2; 3- ARCH; 4- PMMA (bulk) EBL resists

The regression equations for R(D) curves at times of development 15 s and 30 s are shown in Table 4.

Procedure	CAMP	ARCH/ARCH2
Wafer priming in air	HMDS / 60s	HMDS / 60s
Resist spin coating	1 μm, 2379 rpm 0.65 μm, 6000 rpm	1 μm, 2350 rpm 0.7 μm, 6000 rpm
Protective layer spin coating	0.065 μm, 4000 rpm	-
Pre-bake on hot plate in air	120°C, 60s	110°C- 130°C, 60s
EB exposure shaped beam	30keV	30keV shaped beam
Post exposure bake on hot plate in air (immediately after EB exposure)	120°C±2°C, 60s±5%	110°C, 60s
Development (immersion)	OPD262 Developer	OPD262 Developer/OPD4262
Rinse	DI water, 20s	DI water, 20s
Dry spinning in air	1000 rpm, 60s	1000 rpm, 60s
Hard bake on hot plate in air.	120°C, 60s	120°C, 60s

Table 3. Experimental procedures for EBL of CAMP6 and ARCH/ARCH2 resists

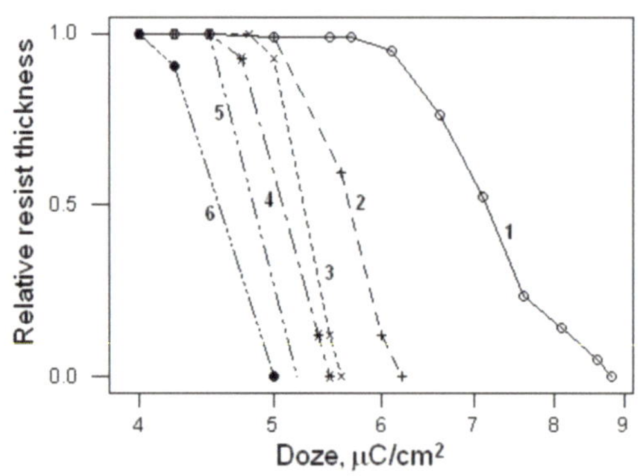

Fig. 9. Contrast curve for 1μm thick CAMP6 CAR at various development times: 1 - 15 s; 2 - 30 s; 3 - 60 s; 4 - 120 s; 5 - 240 s; 6 - 360 s. Thickness loss due to PEB is 23% and d_o = 770 nm

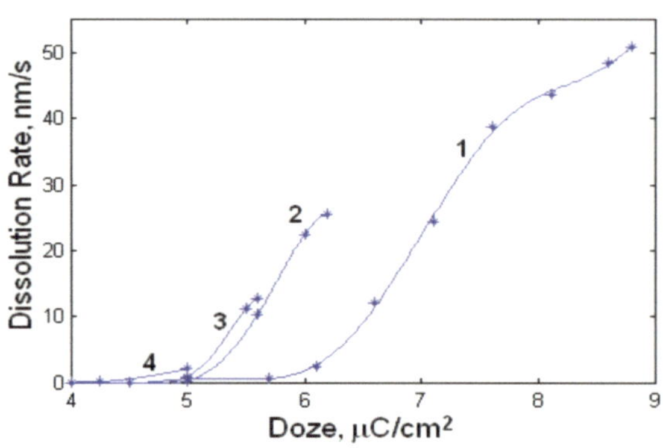

Fig. 10. Experimental characteristics of solubility rate vs dose of electron exposure at various times of development: 1 - 15 s; 2 - 30 s; 3 - 60 s; 4 - 360 s for CAMP6 CAR

Time, dose	Equation
15 s D= 4.5 ÷ 8.8 μC/cm²	R= -102813.40551880810852004267642035+142167.00666457094955037341868513.D-84825.995315595945739978661788758.D2+28510.530606400510592897424172967.D3-5901.5392059851775643956713915221.D4+770.16977950614902453894223441137.D5--61.876802541845874295076969973777.D6+2.798542091894069835009907026354.D7--0.054571727048585807708428593202256.D8
30 s D=4.95-6.2 μC/cm²	R= 16087.902418081173044563523355944-17028.97669942502685484722230845.D++7163.3636397166115081558015783809.D2-1495.8434483486597413241768135557.D3++154.9219377712650030114684168368.D4-6.3592278222426436346250566590707. D5

Table 4. Dissolute rates R on the exposure doses D, evaluated using regression models, for CAMP6 & OPD262 developer (non-linear development process)

4.3 Inorganic EBL nanoresists

After two decades of tests of vacuum deposited inorganic resists, aiming to reach higher resolution due to exhibition of very high contrast and to utilize excellent etch resistance, many inorganic candidates for EBL resists were avoided. The mainly low sensitivity and film quality (pinholes etc) put barriers to its practical development. Recently a spin coated inorganic resist was observed as more attractive - namely the hydrogen silsesquioxane (HSQ) (Namatsu et al., 1998; Namatsu, 2001). The stoichiometric formula of this negative tone resist is $[(HSiO_3)_{2n}]$. The resist is comprised of HSQ resin in a carrier solvent of methylisobutil-ketone. The first use of HSQ was low dielectric constant material for isolation layers in IC fabrication. Its dielectric constant is about 2.6 - 3.0. It provides high thermal stability, good gap-filling and crack free adhesion to metal film surfaces and multilevel interconnects. HSQ is oligomer composed of a caged silsesquioxane along with a linear Si-O network. A standard thermal cure is performed to convert the cages to the highly cross-linked network through reaction in silicon hydride (Si-H) parts of the chains. The same curing reaction could also be accomplished through electron irradiation. It was proposed that Si-H bonds are scissioned during exposure and being converted to silanol (Si-OH) groups in presence of absorbed moisture in the film. These silanol groups are unstable and modified the caged molecules into a linear network. This transition drastically decreases the dissolution rate of matrix in an aqueous solvent performing a negative tone relief pattern. Furthermore HSQ can also be utilized in a bi-layer lithography scheme due to its high etch resistance properties where the patterns are transferred through a sub-layer using a reactive ion etching. Etch resistance of HSQ is excellent. As alkaline developer of HSQ is used usually aqueous solution of tetra methyl ammonium hydroxide (TMAN) - for example 2.38% solution. Resist drying could be optimized effectively by replacing the rinsing liquid with a supercritical fluid (Wahlbrink et al., 2006). Using 30 nm resist and 50 keV electrons a resolution of 6 nm isolated lines and 27 nm period grating were demonstrated (Word et al., 2003). Typical contrast reported for HSQ ranges from 0.55 to 3.2 (Word et al., 2003). The not very good contrast often results in undesired bridging or footing between the developing closely spaced structures. Optimal doses depend upon the beam energy, desired resolution and film thickness. Instead using the mixture of salt and alkali, development in an aqueous mixture of NaOH alkali and NaCl salt enhances the contrast. Contrast values as high as 10 in a 115-nm-thick resist were achieved (Yang & Berggren, 2007). Increasing the temperature of development to 45°C also increases the contrast and resolution. After baking the resist surface roughness is better than 2 nm over a chip area measured by an atomic force-microscopy.

Thickness nm	Planarization sub-layer nm	Energy keV	Dose μC/cm²	Resolution nm	Paper
20	180	1-3	50	50	Millard et al., 2002
50	Novolack	30		20	Delft et al., 2000
90	-	50	1400	20	Jamieson et al., 2002

Table 5. HSQ characteristics as nano-EBL resist film

From Table 5 one can see the extremely high resolution of HSQ, obtained at cost of higher than in the CARs exposure doses. The contrast and line-wall roughness (line-wide fluctuations) are much lower than the contrast in traditional chain scission PMMA due to the oligomer nature of HSQ. In the references (Namatsu et al., 1998; Grigorescu et al., 2007) the resolution of sub-20 nm features has been also shown.

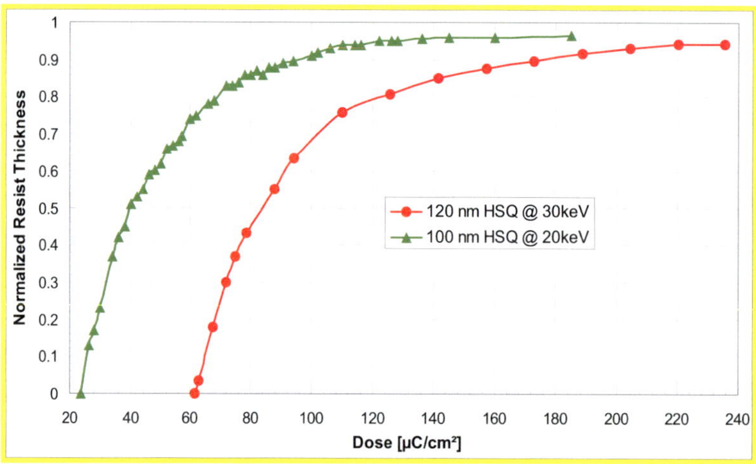

Fig. 11. Resist thickness changes on exposure electron doses at processing parameters - Table 6

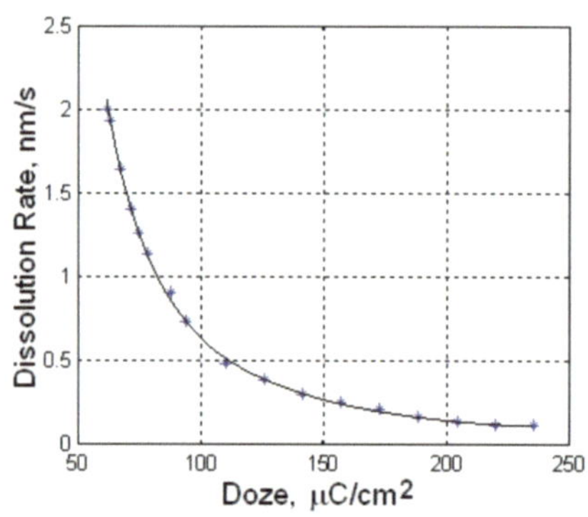

Fig. 12. Dissolution rate calculated for HSQ and parameters in Table 6

The experimental data for HSQ contrast curve (normalized resist thickness vs. exposure dose) is shown in Fig.11. The processing parameters used are given in Table 6. Fig.12 presents the dissolution rate dependencies on the exposure dose, evaluated from the contrast curves shown in Fig.11.

Resist	HSQ FOX-12,Dow Corning
Thickness	120 nm
Substrate	Si
Developer	MF322
Development Time	60 sec
Developer Temperature	20° C
PreBake	120° C, 120 sec oven
E-beam	30 keV
E-beam Current	0.393 A/cm^2

Table 6. Parameters and conditions of HSQ treatment

Evaluated dependencies using regression models of the development rate resist on the exposure dose D for HSQ are given in Table 7. The thickness of the resist film is 120 nm and the accelerating energy of electrons is 30 keV.

parameters	Equation
HSQ 120 nm, 30keV, D=61,5 to 235.8 μC/cm^2	R = exp(5.120539610557917031595849800305 8- -0.11206083135939619970953043371 2*(D) +0.000842589104462024658078794014084 91*(D^2)- -0.0000032874731904152225949993495776176*(D^3)+ +0.0000000049745270627310741552861561907296*(D^4));

Table 7. Development rate vs. the dose of exposure for HSQ resist and MF322 developer

4.4 Nanocomposite systems in nanolithography

Attempts for optimization of the resist etch characteristics inclusion of nano-particles were tested (Merhari et al., 2002; Ishii et al., 1997), but deeper understanding is still needed. The composition must contain low loading of nano-particles component, such that coagulation and phase segregation are unobserved. The distribution of nano-particles in nano-composite must be homogenous. The studied nanoparticles has been fullerenes and silica (SiO_2) included in CARs. In (Merhari et al., 2002) was been shown, that 7wt % of silica in ZEP520 resulted in a 20-fold reduction in etch rate in oxygen plasma. Resolution in that case was 50 nm. The mechanical stability of nano-images is higher in the case of a wet development. Radiation sensitive metal-organic precursors (Barstow et al., 2002) are tested as other potential high resolution nano-EBL resists.

Assuming the thickness of one-molecule layers as a limiting parameter for a thinnest resist, as an important task for future investigations the authors could point the need for studying the relations between the line width and the minimal resist thickness at various energies (radial energy spread and absorbed energy decrease at decreased resist thickness; there is a situation without losses from the forward penetrating electrons and only back-scattered electrons will modify the resist chains) as well as the role of nanopowder (or by fullerenes) loading of thin resist films at these optimal conditions.

5. Modeling of the resist development

5.1 Background

Various models are presented in the literature, describing the developed resist surface motion and its deformation. Several algorithms have been suggested in the former years (Greeneich, 1974; Dill et al., 1975; Jewett et al., 1977; Hatzakis et al., 1974; Neureuther et al., 1979; Jones & Paraszczak, 1981; Matsuzawa, 1985; Ferguson et al., 1990; Fukuda & Okazaki, 1990; Vutova & Mladenov, 1991; Weiβ et al., 1995; Glezsos et al., 1996) based on these models. The result of the computer simulation is a profile of the developed simple topological patterns, during their formation under particular processing conditions. These models can also be treated as particular cases, with respect to more general study of the profile evolution modeling, during ion etching, film growth and other similar situations, where the anisotropy, the surface tilt influence, the reflection and the secondary deposition make the process rather complex.

One of the first models for the developed contours time evolution described (Greeneich, 1974) is simple, but it is significantly inaccurate. In this model the required development time of each element is determined by summing up the time for reaching its depth and the time of dissolution of its radial distance.

Other authors (Dill et al., 1975; Hatzakis et al., 1974) describe a model, where the resist is partitioned in small elements (cells), coinciding with those, for which the absorbed energy is modeled. Within these elements the solubility rate value is taken to be a constant. Then the element dissolution time is calculated, taking into account the number of the cells' sides that suffer the development process. The profiles are obtained by a subsequent interpolation (on the base of solution times of the various cells).

Two models are described in (Jewett et al., 1977): ray-tracing model and string model. The ray-tracing model is based on the optical analogy of the profile evolution to the light propagation within a medium of variable refractive index. The corresponding value of this index is defined as the ratio of the maximum to the local value of the solubility rate. In the string model the simulated profile is presented by straight line segments, whose ends advance along the angle bisector of the two adjoining segments according to the local value of the solubility rate. During the modeling process, the segments are kept roughly equal in length by adding points in regions of expansion and removing points in regions of contractions. The majority of the mentioned published simulation models describe the solution and removal of the resist material as a surface process, that take place on the interface solid resist/liquid developer. We will see that when one tries to simulate resist material removal during nanolithography processing, this assumption becomes too rough.

A number of problems appear when trying to use in practice the described models. Some of them are listed bellow:

- most often, part of the model details (the calculation of the absorbed energy in the resist, the relation absorbed energy/solubility rate, the profile evolution algorithm etc.) are not presented in the papers and that's why the models are practically unusable;
- different experimental data has been used for the solubility rate characterization (usually it is not mentioned that the measured solubility rate is the average value and the range of calculating this mean is not defined);
- in order to obtain a final result of an appropriate accuracy, even the modeling of simple elements requires a proper choice of a fine grid, which leads to higher costs of the

simulation. This explains why only in refs. (Jones & Paraszczak, 1981; Matsuzawa, 1985) numerical calculations are carried out for 3D profiles of simple developed patterns.

There are few development models (Jones & Paraszczak, 1981; Matsuzawa, 1985; Ferguson et al., 1990; Fukuda & Okazaki, 1990; Weiβ et al., 1995) based on knowledge of 3D concentration distribution of a dissolution inhibitor in the resist layer after the exposure (including the delay and post exposure baking in the case of CAR's). The conversion of the resist material during electron irradiation (catalytic decomposition or creation of an acid from the acid generator) and the loss or diffusion of this inhibitor need a large number of input parameters and a simple model can be found in (Neureuther et al., 1979). But in this approach (solubility rate versus dissolution inhibitor concentration) the evaluation or experimental measuring of this inhibitor concentration is an open question.

Despite the common nature of the listed models, the development of a particular algorithm and the corresponding computer program need the solution of several problems such as: (i) the creation of developed resist profile with a slope in one of the two possible directions with respect to the initial surface normal; (ii) the particularities of the profile evolution at the points of smoothness discontinuity, due to the model assumptions; (iii) the graphical interpretation of the surfaces; (iv) the appropriate links of the program units, forming the whole working package.

In our model of development process (Mladenov et al., 1987; Vutova & Mladenov, 1991; Vutova & Mladenov, 1994; Mladenov & Vutova, 2002, Vutova & Mladenov, 2008; (a) Vutova et al., 2009; (b) Vutova et al., 2009) we assume that the resist is homogeneous enough and the development process is isotropic. To increase the accuracy of the end point position of the evolving point trajectory at a given time smaller time steps are used. The main features of the model are: (1) the motion of the evolving point takes place along the normal to the corresponding profile; (2) the space modification of the absorbed energy distribution is also taken into account; (3) in our model, we use a cubic spline in the 2D case and a bicubic spline in the 3D case to describe the developed profiles; (4) an original procedure is applied for the profile discretization to increase the contour accuracy. There is a reduction of the evolving points number (i.e. density) in the profile regions, where the normal tilt is kept close to the initial profile slope, and increasing the evolving points number in the profile regions, where a maximum change of the slope is observed.

5.2 Simulation of 2D developed contours of the resist profile at a linear development process

Our model (applied for a positive resist) can be described by the following steps:

Step 1. In order to trace the development contours in the resist plane a Cartesian coordinate system is introduced. The axis Ox is at the resist surface, and Oz is in its depth. A finite number of points on the resist surface is chosen, where the evolution of the time development contour will be traced.

Step 2. The spatial distribution of the adsorbed energy E_i per unit of the resist volume V and at used dose D is calculated (see part 2 and 3). The subscript i denotes the number of the points from the examined contour.

Step 3. The change of the resist solubility that is determined from the values of the absorbed energy, results in changes of the polymer molecular weight. That is why during this step, the value of the local deposited energy density $D_i.(dE/d\xi)_i$ is transformed after radiation into the new molecular weight values of the resist in vicinity of the point i, using the relation (Greeneich, 1974):

$$\frac{1}{\overline{M_{fi}}} = \frac{1}{\overline{M_n}} + \frac{g.D_i}{\rho.N_A} \cdot \left(\frac{dE}{d\xi} \right)_i , \tag{3}$$

where $\overline{M_n}$ is the original average molecular weight; $\overline{M_{fi}}$ is the fragment average molecular weight; g is the radiation efficiency value; ρ is the polymer mass density; N_A is the Avogadro's number and D is the average exposure dose of a single spot.

The movement of the point is traced along the normal to the initial contour (for instance z=0). Its velocity is calculated by converting the molecular weight values $\overline{M_n}$ into solubility rate values S_i , using the empirical formula:

$$S = R_0 + B / \overline{M_{fi}}^{A} , \tag{4}$$

and the relation (3). The development time (i.e. the time of movement of the evolving point along the normal) must be sufficiently small. Criteria for this are the number of the required profiles, the accuracy of their modelling as well as the missing intersection of adjacent evolving point trajectories. Using this way, one can obtain the coordinates of the points, defining the new contour.

Step 4. N-cubic splines are build. So the new developed profile is described on the basis of the obtained points set:

$$z = C.z_i + D.z_{i+1} + G.z_i'' + H.z_{i+1}'' , \tag{5}$$

where

$$C = \frac{x_{i+1} - x}{x_{i+1} - x_i} ; \qquad D = 1 - C ; \qquad G = \frac{1}{6}\left(C^3 - C\right).\left(x_{i+1} - x_i\right)^2 \tag{6}$$

and

$$H = \frac{1}{6}.\left(D^3 - D\right).\left(x_{i+1} - x_i\right)^2 . \tag{7}$$

Here z'' is the second derivative of the z function.

Step 5. The angle α between every two evolving points is calculated. Its value is compared to π/2. If the two values are close enough, then the modification of the development contour between the two points is insignificantly small. In this case the point (x_{i+1}, z_{i+1}) is not considered in the subsequent calculation. If this angle is small, i.e. there is a significant profile modification a new point $x = \frac{x_i + x_{i+1}}{2}$ between these two points is introduced, in order to estimate more precisely the profile change in this area. This test was done for all the evolving points. Then the second step is applied again until a given development time is reached.

In the case of negative resists in Eq.(4) the power index A is negative and one can use the relation:

$$S = S_0 . \left[C_0 + \frac{2g.D}{e.\rho.N_A} . \left(\frac{dE}{d\xi} \right) \right]^{-A} , \tag{8}$$

where S_0, g and A are parameters, the physical meaning of which are: the initial solubility rate at the resist surface ($S_0.C_n^{-A}$), g(cl) is the radiation efficiency for cross-linking defined as a number of the chemical joining per 1 eV absorbed energy, and A is the power index of the solubility rate decrease.
Here it is also assumed that there an associated molecule at bonding two polymer molecules by one cross linking is created. The constant C_n is connected to the average initial molecular weight of the resist $\overline{M_n}$. The exposure dose is in [C.cm^{-2}].

5.3 Peculiarities of ion lithography 2D development model

A formula that gives the relationship between the molecular weight of the resist after radiation $\overline{M_{fi}}$, the original molecular weight $\overline{M_n}$ and the exposure parameters can be found in a similar way to the case of electron exposure. In the case of ion exposure of the resist the $\overline{M_{fi}}$ can be represented as a function of the electronic and nuclear energies:

$$\frac{1}{\overline{M_{fi}}} = \frac{1}{\overline{M_n}} + \frac{\left[g_{el}.\left(dE/d\xi\right)_{el} + g_n.\left(dE/d\xi\right)_n\right].D}{\rho.N_A}, \tag{9}$$

where g_{el} and g_n are the corresponding radiation efficiencies for the two component of the absorbed energies. If we define:

$$g = g_{el} + g_n.\frac{(dE/d\xi)_n}{(dE/d\xi)_{el}}, \tag{10}$$

then Eq.(9) can be written as Eq.(3) and the solubility rate S is given by Eq.(4) that were used in the case of electron exposure.
In the studied case of ion beam exposure one can assume that the particular development characteristics R_o, B and A in Eq.(8) have the same values as in the case of electron lithography. This assumption is based on the dominance of the electronic energy losses. When the value of the ratio of the nuclear to electronic stopping powers in Eq.(10) is negligibly small, g=g_{el}. This is true in the case of light ions incident on a polymer in ion implantation energy region (tens hundreds keV) and for all ions at higher energies.

5.4 Three dimensional model for the simulation of the development profile

Introducing the coordinates of the exposed pattern (for example a rectangle) one can examine the evolution of the resist surface, using a finite set of points, as in the case of 2D case. Similarly, we obtain a new set of points and approximate the new surface using a bicubic spline. Let x_1, x_2, x_3....x_i....x_m and y_1, y_2, ...y_j....y_n are a set of points with corresponding function values $z_{ij} = f(x_i, y_j)$ where $i = 1, 2, ...m$, $j = 1, 2, ..., n$. The aim is to estimate the function f at some non-tabulated point (x^*, y^*) by interpolation procedure. More precisely, if:

$$x_i \le x^* \le x_{i+1} \qquad y_j \le y^* \le y_{j+1} \tag{11}$$

then the next expression gives the necessary values:

$$f(x',y') = \sum_{k=1}^{4}\sum_{l=1}^{4} C_{k,l}.t^{k-1}.u^{l-1}, \tag{12}$$

where $C_{k,l}$, k, l = 1, 2, 3, 4 are constant values. The formulas that obtain the sum from the functions and derivative values are just a complicated linear transformation with coefficients which, having been determined once, can be tabulated (Abramowitz & Stegun, 1964; Dahlquist & Bjorck, 1974). Then:

$$t = \frac{x' - x_i}{x_{i+1} - x_i} \quad \text{and} \quad u = \frac{y'' - y_j}{y_{j+1} - y_j}. \tag{13}$$

After the calculation of the new development profile, the step is repeated for the next time interval. The process stops when a given development time is reached.

5.5 Peculiarities of the simulation of the development profile in CARs - a simplification

As it was shown, in the case of PMMA, as typical example of the one-component resists, the conversion of the absorbed energy distribution in the resist to the solubility rate is based on the radiation modification of the polymer molecular weight in studied point of the resist. At chosen couple resist – developer, conditions of development, resist initial molecular weight, exposure dose and radiation yield, a definitive development rate could be estimated. The dissolution rate of the unexposed resist at these conditions is another parameter used in the simulation. Often the changes of the absorbed energy density distribution on the resist depth are neglected and only its radial (lateral) distribution is used at the simulation. Two types of the resist solubility were distinguished: linear and nonlinear solutions. Reasons for these phenomena are various: the role of an interface layer with solubility rate other than one of the bulk resist material, time dependence of the developer penetration into the solid resist, the role of the nuclear energy losses that deteriorate the polymer structure at ion modification of the resist material etc.

Then the resist removal in a linear development case is simulated as a movement of the interface between the solid polymer and the liquid developer. The developer – resist interface is a function of the time of development. For a short developing time each evaluating point from the moving development contour (or plane in the case of 3D simulation) advances along the normal to the profile with its local solubility rate. A procedure to increase the contour (plane) accuracy by increasing the evaluating point numbers in the parts with significant curvature is applied.

In the case of chemically amplified resists (CARs) there are two or three polymer components one of which is a radiation active component (RAC). During exposure of RAC the absorbed energy effect is a generation of a radiation product. Additionally a diffusion of that radiation product takes place during a thermal process, called post-exposure bake (PEB) applied in this case. In the CARs this radiation product (an acid) works as a catalytic agent and modifies the base component solubility rate. The models of the conversion of the solubility rate through the RAC concentration and the solubility rate modification products are described (Dill et al., 1975; Kim et al., 1984; Ferguson et al., 1990; Glezos et al., 1996).

In these models various descriptions of the diffusion and loss of the catalytic species (acid) and of the kinetics of the de-protection (switching resist solubility) reaction are used. The

surface changes of the development rate are the reason for other problems. Due to the complicated processing steps and the need of unknown numerical data that describes the reaction kinetics, the published models for simulation of CARs development are not exact. Some models for simulation of the developed resist profiles in CARs consist of models of coupling between diffusion and kinetic reactions; other models describe nonlinear or linear diffusion; there are also various development-rate models (Fedynyshyn et al., 1990; Ferguson et al., 1990). Surface changes (retardation or increase) of the development rate is evaluated using depth dependencies of the resist solubility rate during development (Kishkovich et al., 1999; Vutova & Mladenov, 2001). The processes of resist de-protection and resist removal show threshold or strongly nonlinear behaviour not understood exactly and described quantitatively. Due to that, the choice of a suitable overall lithography simulation model for the concrete resist-developer pair and alternating processing conditions is difficult. Due to many reasons, final experimental calibration of the model parameter set used in such a simulation is needed (Byers et al., 2002; Yen, 2002; Sekiguchi et al., 2001). Some reasons are the following: (i) variations of resist-developer pair and development conditions; (ii) different values of PEB times and temperatures; (iii) different procedures of the experimental parameter extraction; (iv) parameter values that depend on geometry factors; (v) batch-to-batch variations between the resist material characteristics.
A question ((a) Vutova et al., 2009) arises: whether to evaluate each reaction quantitatively, using experimental and theoretic values of too many factors, especially collected in an academic way, or to build one simplified model based on the existing simple experimental calibration procedure. In this model for simulation of development process only main steps could be used: (i) calculation of the absorbed energy distribution in the resist after exposure, (ii) application of experimental solubility rate as a function of the exposure dose and the time of development (multi-ciphered experimental dependencies of the solubility rate on the exposure dose) for approximation of the local development rates in the exposed resist, (iii) application of the procedures for simulation of the developed contour (plane) movement for small time intervals ((a) Vutova et al., 2009; (b) Vutova et al., 2009) in a time dependent calculation scheme. In this way the number of experiments for determining the quantities, relevant for simulation of the development kinetics of the resist profile at a concrete resist-developer system and chosen processing parameters and conditions is minimized.
This simplified model was applied for resist profile simulation in CAMP6 resist. During the development process, the line edge roughness is changed in various phases. Results are shown in Fig.23 below.

6. Calculated developed profiles at EBL and IBL

We assume that our exposed pattern is a 3.1 x 8.1μm^2 rectangle and that the electron beam has an energy of 20 keV and has Gaussian distribution. The results of the calculated contours, characterising the development profile, are obtained in the plane, that contains the line which bisects the longer sides of the rectangle and is perpendicular to the resist surface.
We investigate the modification of the time development contours (at every 1 minute of the solubility process), starting from the surface (Fig.13). Fig.13(a) shows the development profile contours in the chosen cross section. The dose is 1.4x10^{-4} C/cm^2 and the developer is a 1:3 solution of MIBK and IPA. The profile contours in the case of the more powerful developer MIBK are shown in Fig.13(b). In this case we observed an acceleration of the process and a very large difference between the exposed and the developed images.

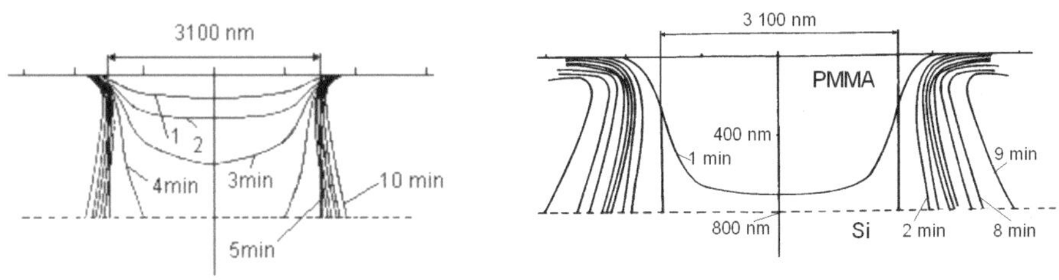

a b

Fig. 13. Time evolution of the cross sectional profiles for a rectangular exposure by electrons of 20 keV at a dose of $1.4x10^{-4}$ C/cm^2. (a) Development using standard developer which is 1:3 solution of MIBK and IPA; (b) development using MIBK solvent. Resist is PMMA of 800 nm thickness, on Si.

A general test pattern is shown in Fig.14. The structure consist 4 rectangles 1000x1000 nm^2 and a line between them, the width of which is 200 nm. The 3D profile obtained by computer simulation using this test pattern and shown in Fig.15 represents the developed gap surface at a dose of $2x10^{-7}$ C cm^{-2}. The beam is 60 keV H^+ and the development time is 0.25 min.

Fig.16 shows the shapes of the developed central lines of structure created by exposure of a layout, consisting many 250 nm parallel lines spaced by 250 nm at a dose of 250 μC/cm^2, yield - 0.01, development time – 45 s. The developed line width is 330 nm. The resist used is 400 nm PMMA on Si and the e-beam energy is 50 keV.

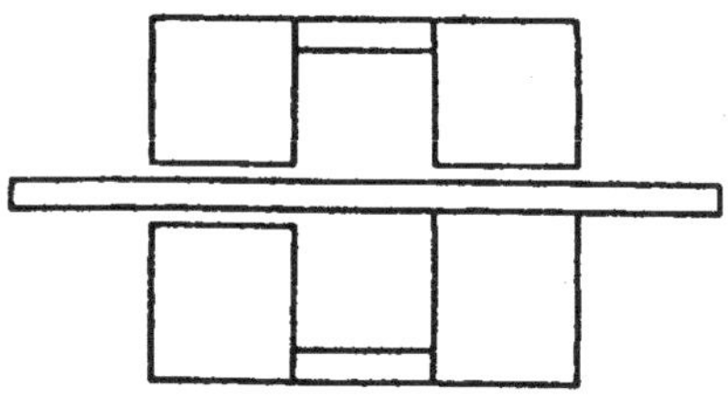

Fig. 14. Two-dimensional pattern for illustration purposes. All small gaps are 100 nm

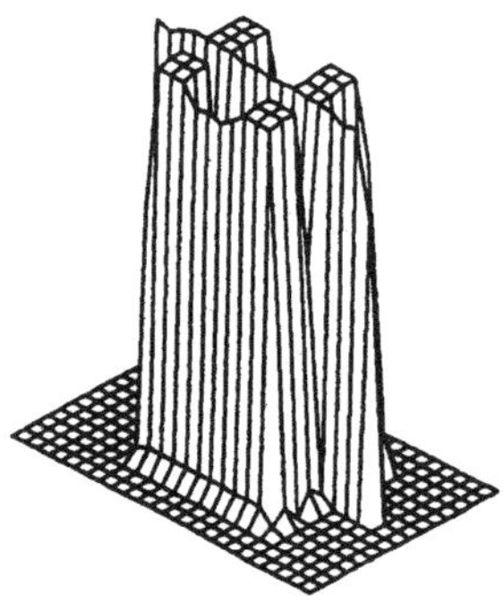

Fig. 15. Computer simulation results for a developed surface using 1:3 solution of MIBK and IPA, applying the test pattern shown in Fig.14.

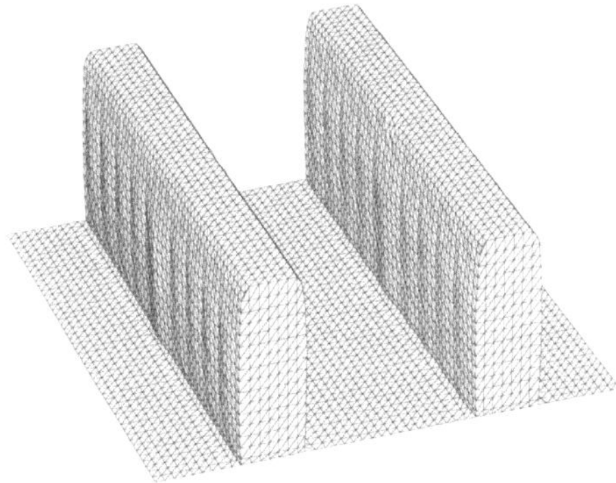

Fig. 16. Part of the 3D resist profile, created by EBL (only the central part of the two parallel lines, situated in the center of the exposed and developed structure, is shown)

Using our computer program for mathematical modeling in IBL, we are able to predict results that can be expected under chosen technology conditions and to learn more features concerning IBL in the region below 100 nm. Some simulation results in the case of ion beam exposure (He^+ ions with energy 100 keV) and development of ion sensitive resist (the exposed resist is 400 nm thick PMMA and the radiation yield is 0.05 eV^{-1}) are presented.

Fig.17 presents 2D simulated resist profile for a line developed in PMMA resist after exposure with 100 keV He^+ ion beam (the image width is 30 nm). In a series of simulations of IBL, the calculated EDF(r) function (see Part 1 of the Chapter) of the absorbed energy for concrete ions was incorporated into the SELID tool. The beam width is presented by the white band in the black mask on the top of the shown structure in Figs.17-19 (the substrate is not shown, the resist layer and the symbolic black mask, added virtually, are indicated on these figures). The developed line width, seen in Fig.17, is 101 nm at the top of the resist layer. This width is 106 nm at the Si substrate while it is 96 nm at the narrowest part of the developed line cross section. This extension of the developed line width is similar to the internal proximity effect obtainable at electron lithography. The resist thickness loss during the concrete development process is 7.1%.

In Fig.18, simulation results are provided for three 30 nm lines exposed with various ion doses. One can see the change of the developed profile shapes when changing the exposure dose. The calculated resist thickness loss is 7.3% at these development conditions. Central line is not developed enough and the optimal, from point of view of obtaining near to vertical side walls of the developed line (trench), is the regime at x= -1 μm.

In the case of IBL of narrow parallel lines, the difference between the width of the exposed and developed lines is also seen, nevertheless the absence of external proximity effect at ion exposure. This case is shown in Fig.19, where Fig.19a presents 2D simulation and Fig.19b presents the result of 3D simulated profile of the resist.

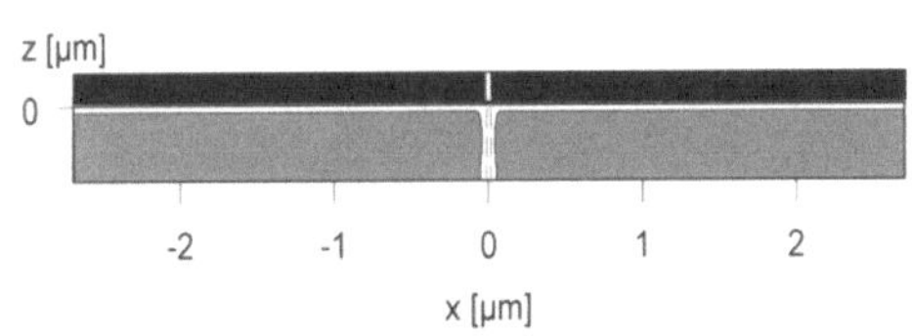

Fig. 17. Simulated resist profiles in the case of 30 nm wide isolated long line at a dose of 0.35 μC/cm², 400 nm thick PMMA on silicon, the development time is 5 min

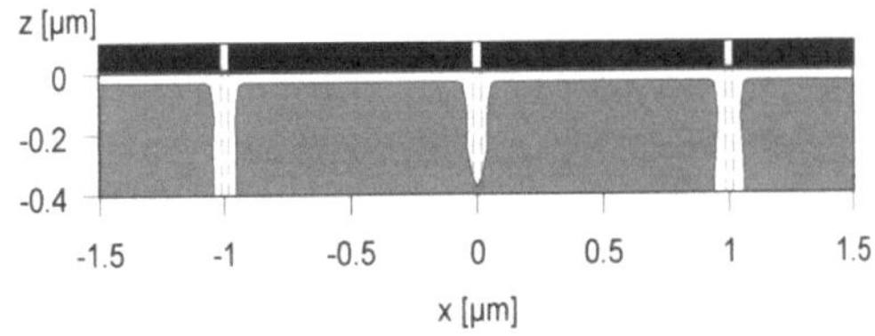

Fig. 18. Simulated resist profiles in the case of EBL of three lines for different exposure doses: 0.325 µC/cm² for the line at x=-1 µm; 0.25 µC/cm² for the line in the center of the figure and 0.4 µC/cm² for the line at x=1 µm

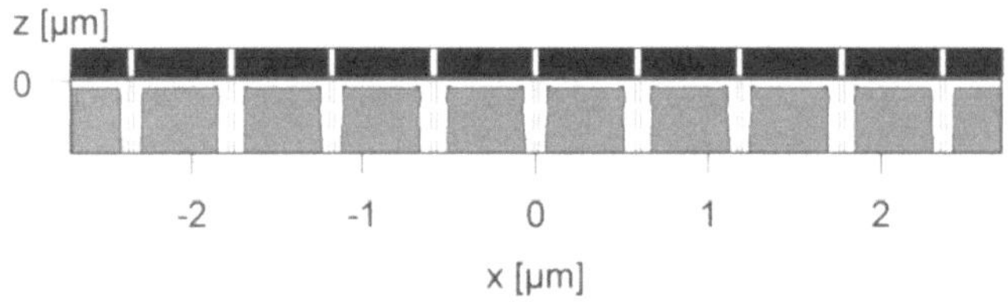

a)

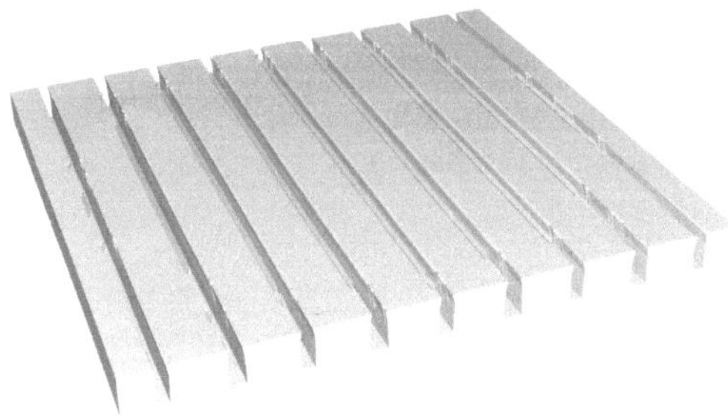

b)

Fig. 19. Simulated resist profiles for nine 30 nm wide long parallel lines spaced by 560 nm after exposure with 100 keV He^+ ion beam. The resist used is 400 nm thick PMMA on Si : a) 2D profile of developed lines; b) 3D resist developed profiles of the exposed parallel lines

A comparison of the same microstructure elements exposed by 100 keV He ions and by electrons is presented in Figs.20-21. The main difference, due to the proximity effect, observable in the case of electron lithography, is that the distance between the line and the squares is 635 nm (Fig.20), while it is 185 nm in the case of IBL (Fig.21). These distances were obtained aiming to minimize it, keeping elements of the structure well different from one another.

Fig. 20. Simulated 3D resist profiles for a layout consisting one 30 nm wide line and four 800 x 800 nm² squares at 50 keV e-beam

Fig. 21. Simulated 3D resist profiles for the layout in Fig.20 after exposure with 100 keV He^+ ion beam

In Fig.20 the influence of the proximity effect on the final result of the whole EB lithographic process is presented - after exposure and development of 30 nm wide central line situated between four 800 x 800 nm^2 squares, the developed line width is 410 nm due to the exposure of the line and absorbed energy added from the squares in the periphery of the exposed central line. In the case of IBL lack of proximity effect is demonstrated and it is important advantage over EBL.

In the case of CARs a time dependent (and multi-ciphered) characteristics were observed (Figs.9-10) due to more detailed study. The developed resist profiles are a result of the developer penetration, volume removal of polymer molecule segments and more complicated chemical and physical changes in the resist than ones in approximation of the developer-resist interface as a simple liquid to solid one.

Fig.22 shows experimentally observed shapes and profile evolution of the 0.5 µm exposed single line in CAR CAMP6. The central part of the line is removed at the beginning and then the resist removal of the line walls widens the trench and vertical side-walls of the developed line (trench) can be observed at higher values of the development time.

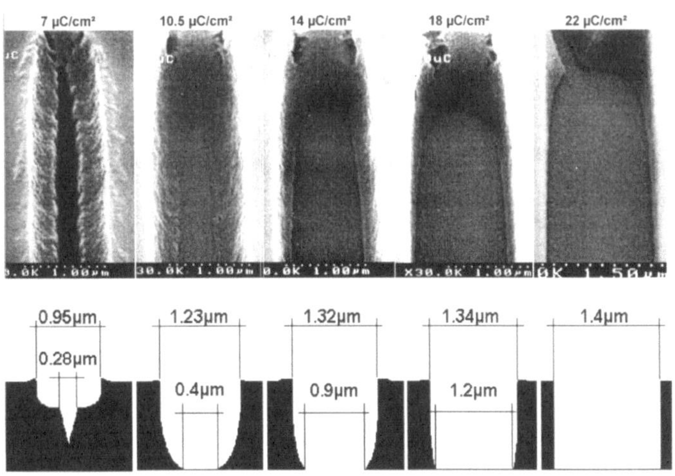

Fig. 22. SEM micrographs (first row) and the corresponding experimental cross-section shapes (second row) of the profile evolution at various exposure doses when 500 nm single lines in 1700-nm-thick (initial resist thickness) CAMP6 (OCG) are exposed.

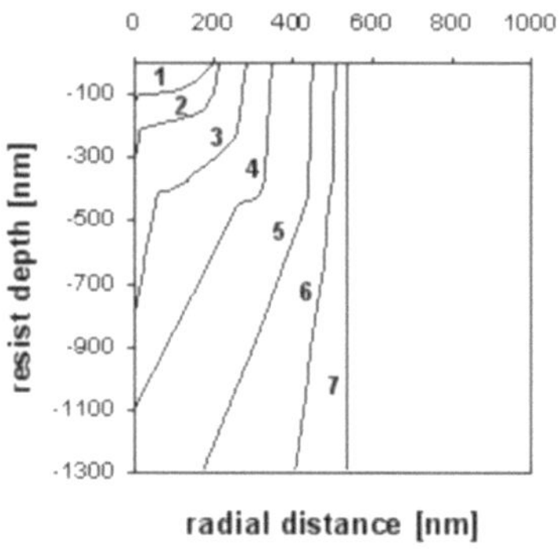

Fig. 23. Simulated development profiles of an isolated 500-nm line created in CAMP6 resist film on Si for different times of development: 1 – 17 s; 2 – 19 s; 3 – 21 s, 4 – 22, 5 – 23 s, 6 – 25 s and 7 – 27 s (half the developed profiles are shown). The developer is OPD-262.

Fig. 23 shows the time evolution of the simulated development profiles for a line in CAMP6 chemically amplified resist using OPD-262 developer. The developed profiles are obtained by computer simulation for the conditions in Fig.22: a single line with a 500 nm width in 1.7 µm thick CAMP6 (initial resist thickness) deposited on Si, the incident electron energy is 30 keV, the beam diameter is 500 nm. From the initial resist thickness of 1.7 µm the remaining thickness was 1.3 µm due to a resist-film thickness loss of non-exposed areas after PEB and development. Higher dissolution rates in the zones with applied higher exposure doses lead to creating a central region of the exposed structure with a rapid removal of the resist. The generated central profile fragment has nearly vertical walls and bottom, which come very quickly to the interface resist/silicon substrate. The resist profiles calculated using our model - curves 3 and 4 (Fig.23) - are in qualitative agreement with the first experimental resist profile, presented in Fig.22 (second row). This resist profile is then extended towards the periphery of the developing structure, obeying the necessary delay time at the boundaries of the different zones. The next profiles – curves 5, 6, and 7 (Fig.23) – with almost vertical walls correspond qualitatively to the experimentally obtained shapes of the rest of the developed resist profiles (Fig.22, second row).

7. Conclusion

In the current work the main steps of the development models for EBL and IBL and their peculiarities are shortly described and discussed. Using the simulation tools the effects of irradiated particles – electrons and ions, exposure dose, dense patterns and nature and characteristics of the development process are discussed in order to extract the necessary values of regime for high resolution patterning. From the obtained results it is seen that the developed algorithms and created new or improved commercial computer tools describe adequately the processes during the resist development at EBL and IBL. The created simulation tools permit to predict with good accuracy the critical dimensions and profiles of the developed pattern. The simulations of IBL structures demonstrate the potential of ion lithography to be considered as one of the main competitors for further improvement of fabrication performance of the new generations of nano-electronic devices.

8. References

Abramowitz, M. & Stegun, L.A. (1964). *Handbook of Mathematical Functions,* Applied Mathematics Series, vol.55

Barstow, S.J.; Jeyakumar, A. & Henderson, C.L. (2002). Direct photo-patterning of metal oxide structures using photo-sensitive metal-organics precursor films, *Proceed. SPIE-Emerging Lithographic Technologies* VI, 421-430, ISSN 0167-9317

Brault, R.G. & Miller, L.J. (1980). Sensitivity and contrast of some proton-beam resists. *Polymer Engineering and Science,* 20 (16), 1064-1068, ISSN 0032-3888

Brewer, G. (1980). *Electron Beam Technology in Microelectron. Fabrication*. Academic Press, N.Y

Byers, J.; Mack, C.; Huang, R. & Jug, S. (2002). Automatic calibration of lithography simulation parameters using multiple data sets, *Microelectronic Engineering,* 61-62, 89-95, ISSN 0167-9317

Chen, W. & Ahmed, H. (1993). Fabrication of 5–7 nm wide etched lines in silicon using 100 keV EB lithography and polymethylmethacrylate resist, *Applied Physics Letters,* 62 (12), 1499-1501, ISSN 1077-3118

Cumming, D.R.S.; Thomas, S.; Weaver J.M.R. & Beaumont, S.P. (1996). 3 nm NiCr wires made using EBL and PMMA resist, *Microel. Eng.* 30 (1-4), 423-425, ISSN 0167-9317

Dahlquist G. & Bjorck A. (1974). *Numerical methods,* v.7, 7.

Delft, F.C van.; Weterings, J.P.; Langen-Suurling, A.K. & Rumijn, H. (2000). HSQ novolac bilayer resist for high aspect rathio nanoscale EBL, *Journal Vac.Sci.Technol.*B, 18 (6) 3419-3423, ISSN 1071-1023

Dill, F.H.; Neureuther, A.R.; Tuttle, J.A. & Walker, E.J. (1975). Modelling of projection printing of positive photoresists, *IEEE Trans. on Electron Devices,* ED-22 , 456-464, ISSN 0018-9383

Fedynyshin, T.; Cronin, M.F.; Poly, L.C. & Konde, C. (1990). Process optimization of the advanced negative EB resist SAL 605, *J.Vac.Sci.Technol,*.B, 8(6), 1454-1460, ISSN 0734-211X

Ferguson, R.A, Hutchingson, J., Spence, C.A. & Neureuther A. R. (1990). Modeling and simulation of a deep UV acid hardening resist, *J.Vac.Sci.Technol.* B, 8 (6), 1423-1427, ISSN 0734-211X

Fukuda, H. & Okazaki S. (1990). Kinetic model and simulation for chemical amplification resists, *Electrochem.Soc.,* 137 (2), 675-679, ISSN 0013-4651

Glezos, N.; Patsis, G.P.; Raptis, I. & Argitis, P. (1996). Application of a reaction-diffusion model for negative chemically amplified resists to determine electron-beam proximity correction parameter, *J.Vac.Sci..Technol.* B, 14(6), 4252-4256, ISSN 1071-1023

Greeneich, J.S. (1974). Time evolution of developed contours in PMMA electron resist, *J. Appl. Phys.,* 45, 5264-5268, ISSN 0021-8979

Grigorescu, A.E.; Krogt, M.C.v.d.; Hagen, C.W. & Kruit, P. (2007). 10 nm lines and spaces written in HSQ using EBL, *Microel. Engineering,* 84 (5-8), 822-824, ISSN 0167-9317

Hatzakis M. (1998). PMMA co-polymers as high sensitivity electron resist, J.Vac. Sci. Technol. B, 16 (6), 1984-1988, ISSN 0022-5355

Hatzakis, M.; Ting, C.H. & Viswanathan, N.S. (1974). Fundamental aspects of electron beam exposure of polymeric resist system, *Proceed. of 6-th Int. Conf. on Electron and*

Ion Beam Science and Technol. Electrochem Soc., Princeton, N.J., 542-579, ISSN 0013-4651

Ishii, T.; Nozawa, H.; Tamamura, T. & Ozawa, A.. (1997). C_{60}-incorporated nanocomposite resist system for practical nanometer pattern fabrication, *Journal Vac.Sci.Technol.*B, 15 (6), 2570-2574, ISSN 1071-1023

Jamieson, A.T.; Willson, C.G.; Hsu, Y. & Brodie, A.D. (2002). HSQ bilayer resist process for low voltage EBL, *Proc. SPIE 4690*, 1171-1179, ISSN 0167-9317

Jewett, R.E.; Hagouel, P.I.; Neureuther A.R. & Duzer T.Van. (1977). Line profile resist development simulation techniques, *Polymer Engineering and Sciences*, 17(6), 381-384, ISSN 0032-3888

Jones, E. & Paraszczak, J. (1981). Computer simulation of resist development of 3D images, IEEE Trans. on El.Devices ED-28, 12, 1544-1552, ISSN 0018-9383

Karapiperis, L.; Adesida, I.; Lee, C.A. & Wolf, E.D. (1981). Ion beam exposure profiles in PMMA computer simulation, *J.Vac.Sci.Technol.* 19 (4), 1259 1263, ISSN 0022 5355

Kim, D.J.; Oldham W.G. & Neuteuther A.R. (1984). Development of positive resist, *IEEE Trans. Electr.Devices*, vol. ED-31, 12, 1730-1736, ISSN 0018-9383

Kishkovich, O.; Kinkead, D.; Higley, .; Kerwin, R. & Piatt, J. (1999). Real-time methodologies for monitoring airborne molecular contamination in modern DUV photolithography facilities, *Proc. SPIE*, 3677 (I), 348-376, ISSN 0277-786X

Mack, C. (1987). Development of positive photoresist, *J.Electrochemical Soc.*, 134 (1), 148-152, ISSN 0013-4651

Matsuzawa, T. (1985). 3D photoresist image simulation, *IEEE El.Devices Letters*, EDL-6, 8, 416-418, ISSN 0193-8576

Merhari, L.; Gonsalves, K.E.; Hu, Y.; He, W.; Huang, W.S.; Angelopoulous, M.; Bruenger, W.H.; Dziong, C. & Torkler, M. (2002). Nanocomposite resist system for next generation lithography, *Microel.Engineering*, 63 (4), 391-402, ISSN 0167-9317

Millard, L.; Cunde, G.; Tedeso, S.; Dal'zotto, B. & Voucher, J. (2002). HSQ hybrid lithography for 20 nm CMOS device development, *Microel. Eng.* 62-63, 755-761, ISSN 0167-9317

Mladenov, G.M. & Emmoth, B. (1981). PMMA Sensitivity Variation versus the Electronic Stopping Power at Ion Lithography Exposure, *Appl. Phys. Letters*, 38 (12), 1000-1002, ISSN 0003-6951

Mladenov, G.M.; Braun, M.; Emmoth, B. & Biersack, J.P. (1985). Ion Beam Impact and Penetration of Polymethil Methacrylate, *J.Appl.Physics*, 58(7), 2534-2538, ISSN 0021-8979

Mladenov G. & Seyfarth H. (1986). General problems of microlithography, *Vacuum*, 36, 649-653, ISSN 0042-207X

Mladenov, G.; Dimitrova, R. & Vutova, K. (1987). Numerical modeling of penetration of accelerated electrons in materials, *Bulg.Physical Journal*, 14 (5), 446-456, ISSN 0584-0279

Mladenov, G. & Vutova, K. (2002). Computer simulation of exposure and development in EBL and IBL. *Proceedings of St.-Petersburg Electro-technical University, Solid State Physics and Electronics*, 1, pp.133-173, ISBN 020617, St.-Petersburg, Russia, April 2002, ed. B. A.Kalinikos, publisher SPbGETU LETI

Moreau, W.; Merrit, D.; Moyer, W.; Hatzakis M.; Johnson, D. & Pederson, L. (1979). Speed enhancement of PMMA resist, *J.Vac.Sci.Technol.* B, 16(6), 1989-1991, ISSN 0734-211X

Namatsu, H. (2001). Supercritical resist drying for isolated nanoline formation, *Journal Vacuum Sci. Technology* B, 19 (6), 2709-2712, ISSN 1071-1023

Namatsu, H.; Yamaguchi, T.; Nagase, N.; Yamazaki, K. & Kurihara, K. (1998). Resist Materials Providing Small Line-Edge Roughness, *Microel. Engineering,* 41-42, 331-334, ISSN 0167-9317

Neureuther, A.R.; Kyser, D.F. & Ting, C.H. (1979). Electron beam resist edge profile simulation, *IEEE Trans. on electron devices* ED-26, (4), 686-693, ISSN 0018-9383

Novembre, A.E. & Munzel, N. (1996). The EB and X-ray lithographic performance of the high resolution CAMP and ARCH family of CARs, *Microel.Engineering,* 32, 229-236, ISSN 0167-9317

Ryssel, H.; Haberger, K. & Kranz, H. (1991). Ion beam sensitivity of polymer resists, *J.Vac. Sci. Technol.,* 19 (4), 1358-1362, ISSN 0734-2101

Sekiguchi, A.; Tajima, K.; Matsuzawa, T. & Kurihara, M. (2001). System for measurement of the development parameters of electron beam resists, *Electronics and Communications in Japan, Part II,* 84 (4), 16-25, ISSN 8756-663X

Vutova, K. & Mladenov, G. (1991). Absorbed energy distribution in electron lithography of simple patterns, *J. Inform. Recording Materials,* 19 (4), 271-282, ISSN 1025-6008

Vutova, K. & Mladenov, G. (1994). Modelling of exposure and development processes in electron and ion lithography. *Modelling and Simulation in Materials Science and Engineering,* 2, 239-254, ISSN 0965-0393

Vutova, K. & Mladenov, G. (2001). Sensitivity, contrast and development process in electron and ion lithography. *Microelectr. Eng.,* 57-58, 349-353, ISSN 0167-9317

Vutova, K. & Mladenov, G. (2008). Computer simulation of micro- and nano-structures at electron and ion lithography. *Journal of optoelectronics and advanced materials,* 10 (1), 91-97, ISSN 1454-4164

(a) Vutova, K.; Koleva, E.; Mladenov, G. & Kostic, I. (2009). Some peculiarities of resist-profile simulation for positive-tone chemically amplified resists in electron beam lithography. *J.Vac.Sci.Technol. B,* 27 (1), 52-57, ISSN 1071-1023

(b) Vutova, K.; Koleva, E.; Mladenov, G.; Kostic, I.; Tanaka, T. & Kawabata. K. (2009). A simulation model for CAR CAMP6. *Microel. Eng.,* 86 (4-6), 714-717, ISSN 0167-9317

Wahlbrink,T.; Kupper, Dan.; Georgiev, Y.; Boten, J.; Moeler, M.; Kupper, Dav.Lemme, M.C. & Kurz, H. (2006). Supercritical drying process for high-aspect-ratio HSQ nano-structure, *Microel. Engineering,* 83 (4-9), 1124-1127, ISSN 0167-9317

Weiβ M.; Binder, H. & Schwalm, R. (1995). Modeling and simulation of a chemically amplified DUV resists, *Microelectronic Engineering,* 27 (1-4), 405-408, ISSN 0167-9317

Word, M.J.; Adesida, I. & Berger, P.R. (2003). Nanometer-period gratings in HSQ fabricated by EBL, *Rapid commun. in Journ.Vac.Sci.Technol.* B, 21 (6), L12-L15, ISSN 1071-1023

Yang, J.K.W. & Berggren, K. (2007). Using high-contrast salty development of HSQ for sub-10-nm half-pitch lithography, *J.Vac.Sci.Technol.* B, 25 (6), 2025-2029, ISSN 1071-1023

Yasin, S.; Hasko, D.G. & Ahmed, H. (2002). Comparison of MIBK/IPA and water/IPA as PMMA developers for EB nanolithography, *Microel. Eng.*, 61-62, 745-753, ISSN 0167-9317

Yasin, S.; Khalid, M.N. & Hasko, D.G. (2004). Reduction in Roughness of Resist Features in PMMA due to the Absence of a Rinse, *Jpn.J.Appl.Phys.*,43,6984-6987, ISSN 0021-4922

Yen, A. (2002). Erratum: (Optical Microlithography XV), *Proc. SPIE*, v.4691 II, 1769-1770, ISSN 0167-9317

Soft Lithography

19

Soft Lithographic Fabrication of Micro Optic and Guided Wave Devices

Angel Flores and Michael R. Wang
University of Miami
U.S.A.

1. Introduction

Since the advent of the laser, guided wave and integrated optical devices have attracted significant research interest for use in advanced telecommunication and interconnection systems. Based on the device substrate or material used, i.e., silicon, LiNaO, LiTaO, GaAs, or polymer, different manufacturing techniques have been developed for fabrication of these optical devices. While various methods can effectively produce guided wave devices, none have been able to match the high-yield, low-cost, mass productivity schemes that define the photolithographic technique in the semiconductor industry.

For example, silicon and silicon dioxide waveguides (Bowers, et al., 2007) are normally produced through standard photolithographic methods; requiring customary thin film deposition (sputtering, chemical vapor deposition, or thermal oxidation), UV mask exposure, and post dry-etching procedures. Despite their high yields and exceptional cost performance, photolithography demands the use of a clean-room facility equipped with elaborate semiconductor manufacturing equipment (sputtering machine, e-beam evaporator, mask aligner, reactive-ion etcher, to name a few), leading to undesirable startup costs and prolonged lead times. Similarly, advanced manufacturing schemes derived from semiconductor production methods, including epitaxial growth waveguides (Brown, et al., 1987) experience comparable cost-prohibitive drawbacks.

Waveguides fabricated on glass substrates typically rely on an ion-exchange process (Ramaswamy & Srivastava, 1988) that may circumvent some of the equipment overhead required in photolithography. In the ion exchange process, the device substrate is placed in a molten cation bath causing the sodium ions in the glass substrate to exchange with one of the cations (ie., K^+, Li^+, Cs^+). The ion alteration raises the local refractive index of the substrate and creates a waveguiding region in the glass. Because of their low propagation losses, minimal production costs, and compatibility with optical fibers, the use of ion-exchange waveguides for integrated optical applications has been extensively researched. In spite of its advantages, issues regarding device yield and reproducibility still remain.

Consequently, polymers have become an attractive alternative to glass and Si/SiO_2 as materials for optical waveguide devices. Polymers are less fragile and less expensive than glass and silicon. Fittingly, polymer waveguides can be made flexible, accommodating non-planar approaches. On the other hand, waveguides fabricated on glass or semiconductor substrates are normally nonflexible and limited to static planar applications. Furthermore, fabrication of polymer devices is aided through mass-replication techniques. The fabrication methods generally used to create polymer devices are based on casting, embossing, or injection molding (Heckele & Schomburg, 2004) replication techniques that are normally faster and more cost effective than conventional photolithographic and ion exchange methods used on glass and Si/SiO_2 materials. More recently, soft-molding replication techniques known as soft lithography are being actively investigated for low-cost, rapid micro-device replication. To that end, we have been researching diverse soft lithographic techniques for guided wave device fabrication.

Soft lithography is a micro-fabrication technique that has been shown to generate high quality micro and nanostructures as small as 10 nm. It eliminates the use of costly and time-consuming lithographic techniques and equipment. Unlike photolithography which is expensive, has little flexibility in material selection, cannot be applied to non-planar surfaces, and provides little control over chemistry of patterned surfaces; soft lithography can circumvent many of these problems. Soft lithography can tolerate a wide selection of materials, can be used for non-planar and three-dimensional structure fabrication, and most importantly can reproduce high-resolution nano/microstructures at very low cost. As a result, soft lithography has generated considerable research interest over the past decade.

Similarly, microfluidic systems with a broad range of chemical and biological applications continue to be an active research area. Microfluidic based devices process or control small amounts of fluids through utilization of channels with micrometer dimensions (Whitesides, 2006). Particularly, a few of the widely reported microfluidic applications include forensics, gene expression assays (Liu et al., 1999), environmental tests (van der Berg et al., 1993), biomedical implantable devices (Santini et al., 1998), and clinical blood analysis (Lauks, 1998). To date, the majority of microfluidic systems have been fabricated using either photolithography, hard replica molding, or more recently, soft lithographic methods (Xia & Whitesides, 1998). Correspondingly, in this chapter we introduce and describe a novel soft lithographic fabrication technique; a vacuum assisted microfluidic (VAM) method that eliminates the polymer background residue inherent in traditional soft molding fabrication techniques. Incorporation of a microfluidic approach with soft lithography allows high-quality guided wave devices to be fabricated rapidly and inexpensively.

The VAM technique is used to develop guided wave devices including single mode and multimode channel waveguides, and array waveguide evanescent coupler (AWEC) ribbons for high speed optical interconnections. The fabrication of these devices demonstrates the cost effectiveness and promise of the proposed approach for the development of inexpensive, high-quality, and mass-produced polymer guided wave devices.

2. Soft lithography

Soft lithography represents a set of high-resolution patterning techniques in which an elastomeric stamp or mold is used for pattern definition. Once the replica stamp is created, multiple copies of the pattern can be defined through straightforward experimental methods. These non-lithographic techniques require minimal monetary investment (clean room not necessary), can be conducted under normal bench top laboratory conditions, and are conceptually simple to fabricate. Some of the diverse fabrication methods known collectively as soft lithography include: replica molding (Xia et al., 1997), micromolding in capillaries-MIMIC (Zhang et al., 2002), microcontact printing-μCP (Quist et al., 2005), and microtransfer molding-μTM (Zhao et al., 1996). Schematic illustrations of some these procedures are depicted in Fig. 1.

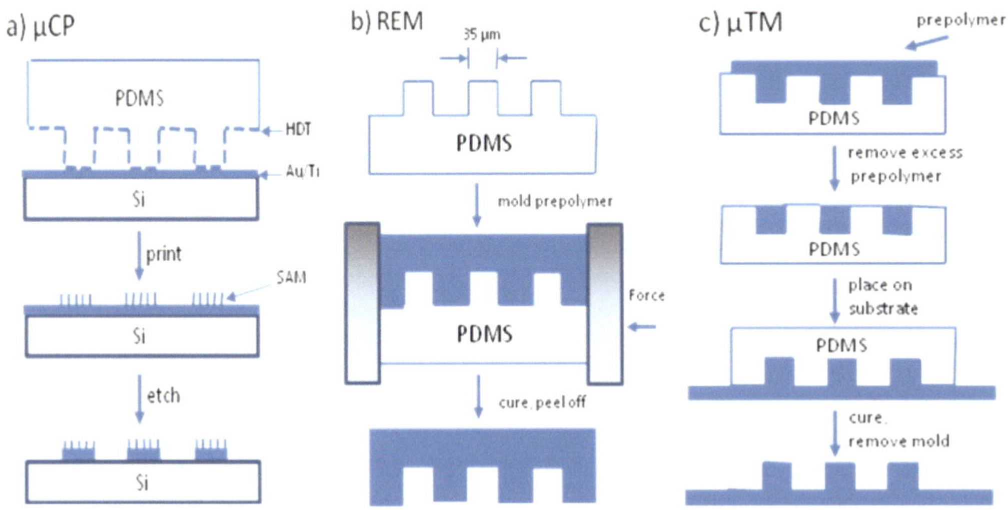

Fig. 1. Schematic illustrations of soft lithographic techniques for a) microcontact printing (μCP), b) replica molding (REM), and c) microtransfer molding (μTM).

Microcontact printing is a flexible, non-photolithographic method that forms patterned self assembled monolayers (SAM) with micron to nanometer scale dimensions. SAMs are surfaces consisting of a single layer of molecules which are prepared by adding a solution of the molecule to the substrate and washing off the excess mixture. Depending on the molecular structure and substrate surface, various molecules can be self assembled without the use of molecular beam epitaxy or vapor deposition. The procedure, demonstrated in Fig. 1a, is simple; an elastomeric polydimethylsiloxane (PDMS) stamp is used to transfer molecules of a hexadecanethiol (HDT) ink to the gold surface of the substrate by contact. After printing, any undesired gold material can be etched away to yield the desired pattern. The technique has been shown to be successful for device fabrication on non-planar surfaces and complex micro patterns.

In replica molding (REM), shown in Fig. 1b, an elastomeric mold rather than a rigid mold, is used to create replica patterns (Xia et al., 1997). Here the organic polymer is placed in contact with the PDMS while the mold is being deformed or compressed in a controlled manner. Deformation of the elastomer provides a method to fabricate structures that would be difficult or unpractical through other procedures.

Alike in several ways, μTM is based on the application of a liquid prepolymer against a patterned PDMS mold. After the excess liquid is removed (by scraping or blowing), the filled mold is placed in contact with a substrate, cured and then peeled to generate the patterned microstructure. Subsequently, soft lithography represents a collection of quick and convenient replication techniques suitable for the definition of both large core (> 100 μm) and nanometer scale devices as well as nanostructures. Through utilization of soft lithographic methods several optical and photonic components have already been successfully demonstrated, such as photonic bandgap structures (Schueller, et al., 1999), distributed feedback structures (Rogers et al., 1998), and microlens arrays (Kunnavakkam et al., 2003). Notably, the lower cost, ease of fabrication, rapid prototyping, and high resolution patterning capabilities are well suited for the replication of guided wave devices.

2.1 Master and PDMS stamp fabrication

The key elements in soft lithography are transparent elastomeric PDMS stamps with patterned relief structures on its surface. PDMS is a polymer having the elastic properties of natural rubber that is able to deform under the influence of force and regain its shape when the force is released. This enables PDMS to conform to substrate surfaces over a large area and adapt to form complex patterned structures. Accordingly, our PDMS molds are produced with Sylgard 184 from Dow Corning; a two-part elastomer that is commercially available at low cost. Once a replica stamp is created, multiple copies of the pattern can be defined through straightforward experimental methods, as illustrated in Fig. 1.

A schematic illustration of the PDMS stamp fabrication process is depicted in Fig. 2. A master silicon device (channel waveguide array) is developed in SU-8 photoresist through photolithography, as shown in Fig. 2a. To begin, SU-8 is spin coated and exposed to UV irradiation through a chromium photomask using a mask aligner. The mask, created via laser-direct writing (Wang & Su, 1998), is a positive replica of the desired channel waveguide arrays. After post exposure baking and photoresist development, the waveguide array master device is realized. Notably, SU-8 patterns processed on silicon wafers are robust, durable and can be used indefinitely (Saleh & Sohn, 2003).

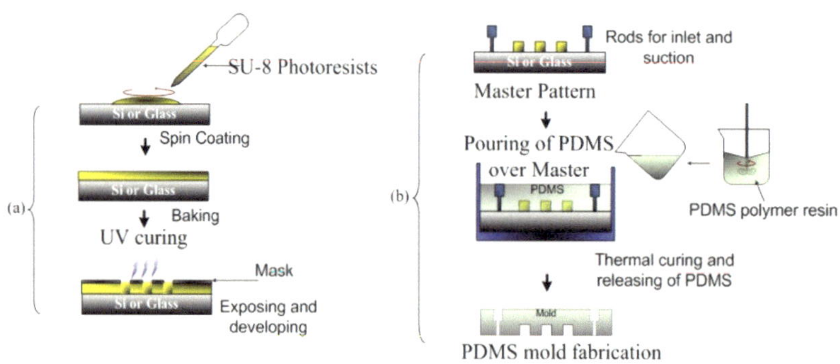

Fig. 2. a) Master pattern development process using SU-8 photoresist. (b) Subsequent generation of the PDMS replication stamp.

Once the master pattern is formed, casting the PDMS prepolymer against the desired surface profile generates a negative replica stamp. The prepolymer is left to settle for 8 hours to eliminate bubbles (and uniformly settle) and then baked for 1h at 60°C. After thermal curing, the solid prepolymer was peeled off to produce a PDMS replica stamp, as shown in Fig. 2b. The replica stamps can be used to create high-fidelity (nanometer scale) copies of the original master pattern. Additionally, the stamps can be reused multiple times (50~100 times) without degradation for mass replication. Such favorable traits have led to the exploration of soft lithography for low cost, mass prototyping device fabrication.

2.2 Microtransfer molding (μTM)

Subsequently, initial fabrication of our guided wave devices was based on microtransfer molding. μTM relies on conformal contact between the stamp and substrate surface to create the waveguide patterns. The approach represents the simplest and most cost-effective fabrication strategy. A schematic description of a standard μTM approach for polymeric waveguide fabrication is presented in Fig. 3a. To begin, the device substrate is coated with a low index buffer to act as the cladding layer. Then, a UV curable prepolymer resin is applied onto the PDMS stamp and placed in direct contact with the device substrate. Next, adequate force is uniformly applied to the stamp to assist in the pattern generation. Consequently, UV irradiation through the transparent PDMS mold creates a crosslinking reaction to solidify the waveguide core pattern. After the resin is cured to a solid, the mold is lifted-off (peeled) to leave a patterned structure on the substrate.

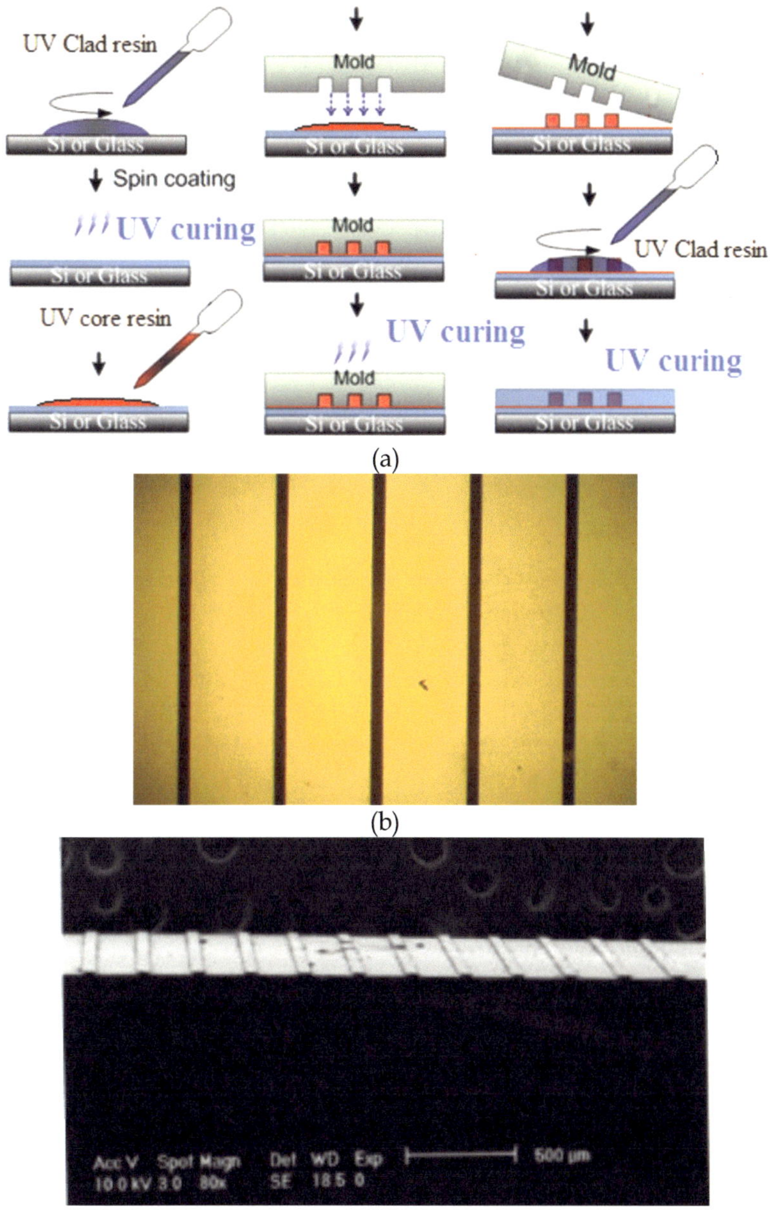

Fig. 3. Schematic illustration of polymeric waveguide fabrication via µTM (a). Micrographs (20× objective) of 35 µm wide master waveguide array (b) and replicated waveguide array (c).

Microscopic images of the master and replicated channel waveguide arrays are shown in Figs. 3b and 3c. Significantly, surface profile measurements exhibit near identical dimensions. We note that because the PDMS mold acts as a secondary master with no influence on the master fabrication, as long as a lower index cladding (or buffer) layer is processed on top of the substrate, a wide array of substrate materials such as glass, silicon wafer, or polymers can be employed. This will be advantageous as we begin to explore lithe substrates for flexible waveguide performance.

Microtransfer molding can generate microstructures over relatively large areas within a short period of time (<1 min). In addition, once the stamp is developed it can be reused many times for device replication. Due to its quick curing time and substantial working area the microtransfer molding technique can be used for fast and accurate prototyping. Nevertheless, it is important to mention that the elasticity of PDMS also leads to several drawbacks. For example, aspect ratios that are too high or too low cause the microstructures in PDMS to deform or distort. Gravity, adhesion and capillary forces exert stress on the elastomeric material causing it to collapse and generate defects in the pattern. Some of the common defects affecting PDMS generation including feature sagging, ineadequate aspect ratios and surface nonuniformity are a consequence of force applied during the soft molding pattern generation. Solutions to these and other common defects affecting PDMS replication including polymer residue and structure warping will be explored later.

3. Polymeric waveguide

The polymer material design is critical for the desired high-performance, high-resolution and low-loss guided wave device. As such, novel UV curable polymeric waveguide materials were developed (Song, S., et al., 2005). The waveguide materials are specifically suited for the fabrication of guided wave devices using soft lithography. The material adheres to the device substrate upon curing without bonding to the PDMS mold during lift-off (peel). Furthermore, we anticipate using both single mode and multimode waveguide structures so the material should be able to create small and large-core devices.

We designed and synthesized two types of photo curable oligomers; epoxy and acrylate oligomers. The epoxy type oligomer resins were prepared from commercially available dihydroxy (OH) monomers and epichlorohydrin. The acrylate oligomers were synthesized in two steps consisting of an initial reaction between the polyol and diisocyanate monomers, followed by the reaction between the first step byproduct and hydroxy-terminated methacrylate monomers. Their respective chemical schemes are shown in Figure 4.

Ultimately, the epoxy type resins outperformed the acrylate oligomers in terms of UV curing time, with the epoxy resins curing in about 30 seconds under 20,000 mW/cm^2 UV irradiation. The prepolymer resins were formulated from the synthetic oligomer, diverse photo curable monomers, additives, and catalytic amounts of photoinitiators. After all the reagents were discharged in a bottle they were dispersed and mixed in an ultrasonic bath for 15~30 min. The formulation study focused on reducing the curing time, shrinkage and determining the proper viscosity. The curing reactivity and viscosity of the resin can be controlled by addition of multifunctional monomers. The general formulation ratio utilized is given below:

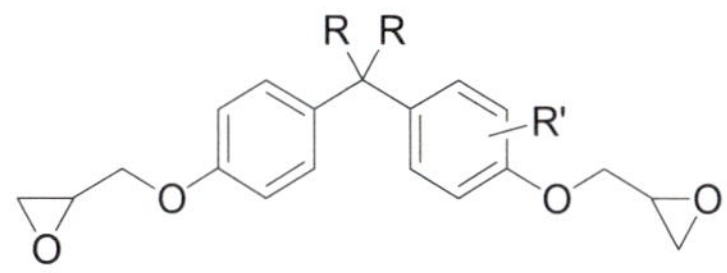

a) epoxy type oligomer

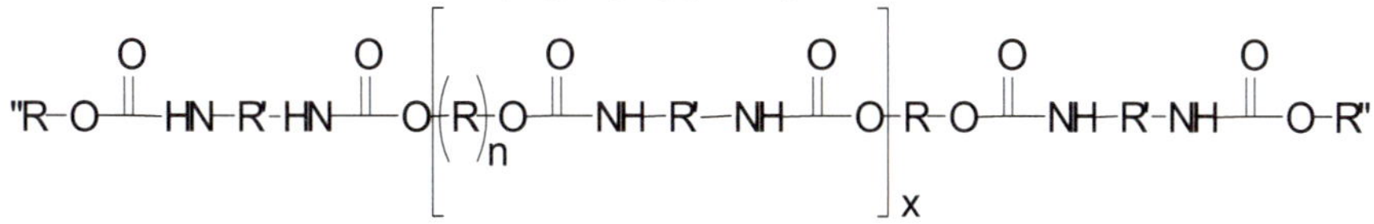

b) acrylate type oligomer

Fig. 4. Chemical structures of the UV curable oligomers for polymeric waveguides. a) The substituent R and R′ represent the fluorinated or non-fluorinated chemical groups, and b) the terminal substituent R″ is the acrylate group.

Fluorinated oligomers: 10 ~ 50%
Multifunctional monomers: 20 ~ 60%
Monofunctional additives: 5 ~ 20%
Photoinitiator: 1 ~ 5%
Other additives: 1 ~ 5 %

The roll of the fluorinated oligomer is an integral part of the composite resin and is very important for determining waveguide properties such as refractive index, optical loss, and hardness of the cured solid. Resins for both the core (CO-1) and cladding (C1-1) waveguide materials were synthesized, where the core material was designed with a marginally higher refractive index.

After synthesizing the waveguide material we analyzed some of its optical properties. The spectrum of both CO-1 and C1-1 (15 micron thick film samples) were measured with a UV-VIS-NIR spectrophotometer. The plot, shown in Fig. 5, discloses the excellent optical transparencies (> 90%) of the synthesized materials in the visible to near IR communication region. Significantly, the flat transparency curve allows for future flexibility in wavelength

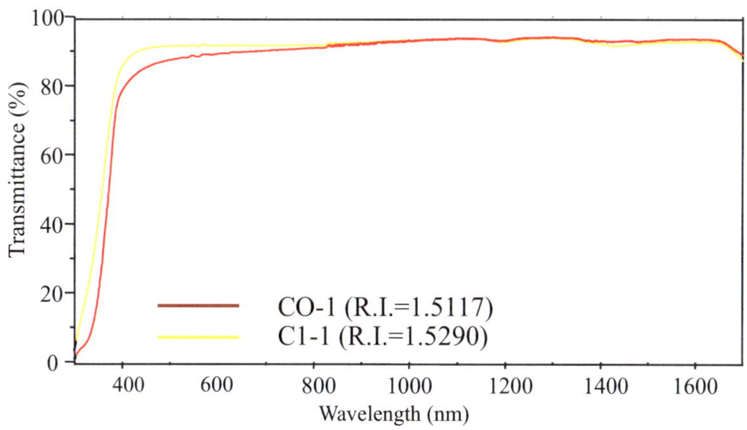

Fig. 5. Transmittance spectrum of formulated core and cladding material measured with a UV-VIS-NIR spectrophotometer.

selection. After establishing excellent optical transparency, the refractive indices of the core and cladding materials were carefully regulated. The refractive index can be explicitly controlled through alteration of the formulation ratio in the fluorinated oligomer portion. Specifically, the final waveguide core and cladding resins exhibited refractive indices of 1.5117 and 1.5290, respectively (Δn = 0.011).

In conclusion, polymeric waveguide resins based on fluorinated oligomers were developed. The material developed consists of a controlled mixture of fluorinated epoxy type oligomers, various photo curable additives, and photoinitiators. The polymer material exhibits excellent broadband (visible to near IR) optical transparency, tunable index control, rapid curing, and light guiding functionality. Moreover, the materials were specifically tailored to meet our soft lithographic fabrication technique which enables rapid device prototyping.

Array waveguide device replication

Once the prepolymer resins were developed, the feasibility of the proposed approach for guided wave device replication was assessed through production of 12 channel waveguide arrays using μTM. BeamPROP software from Rsoft Inc. was employed to design the waveguide array depicted in Fig. 6a. Accordingly, the electric field distribution of the AWEC device is shown on the right. The 12-channel waveguide array (each 10 mm long) has dimensions of 35 μm by 35 μm with a 250-micron pitch. Notably, the 250-micron pitch represents the standard pitch for optical transmitter/receiver arrays. A cross section schematic of the waveguide array device is also presented in Fig. 6b, where the large dimensions lead to a multimode structure. Once the simulation yielded satisfactory results, the waveguide pattern was transferred to our laser-writing machine for direct generation of the mask pattern.

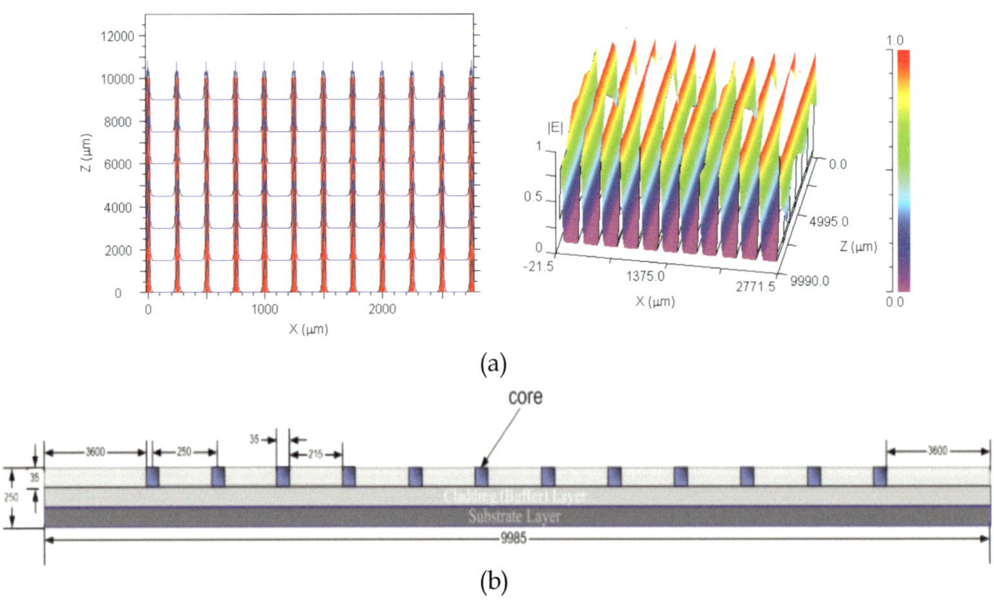

Fig. 6. a) 12 channel waveguide array designed using BeamProp software, and b) cross section schematic and dimensions of the waveguide array.

After generating the mask model, a silicon wafer with various waveguide array patterns was fabricated using conventional photolithographic techniques. Soft lithography was then employed to produce a high-resolution mold of the master pattern. A picture of the transparent mold generated is presented in Fig. 7a. Following mold generation, several channel waveguide array patterns were produced on a silicon substrate wafer, as shown in Fig. 7b. In addition, a micrograph of the 35 μm wide waveguide array is depicted in Fig. 7c.

(a) (b) (c)

Fig. 7. a) Soft lithographic mold of waveguide array master pattern and b) subsequent waveguide arrays fabricated on silicon substrate using soft lithography. c) Micrograph (20x objective) of 35 μm wide waveguide array.

The demonstration of channel waveguide arrays on a silicon substrate led to the investigation of similar waveguide fabrication on flexible substrates. We explored the use of different types of soft polymer substrates where the ideal material should be low cost, robust, impact resistant and durable. In addition, properties of high bending radius, UV curing compatibility, excellent thermal resistance, and optical transparency are desired.

Polymer sheets of PETG Vivak® co-polyester material were chosen for the waveguide device substrate. The Vivak® sheet is a low-cost, transparent, thermoplastic sheet widely used for assorted engineering applications. The co-polyester provides superior impact strength, durability and performed well under high-intensity UV illumination. A picture of the flexible waveguide array fabricated through μTM is shown in Fig. 8a. We observed no thermal shrinkage of the co-polyester sheet during UV illumination and the waveguide cladding and core materials bonded and adhered effortlessly to the flexible substrate. Furthermore, flexible waveguide arrays produced on substrates as thin as 200 μm were successfully demonstrated.

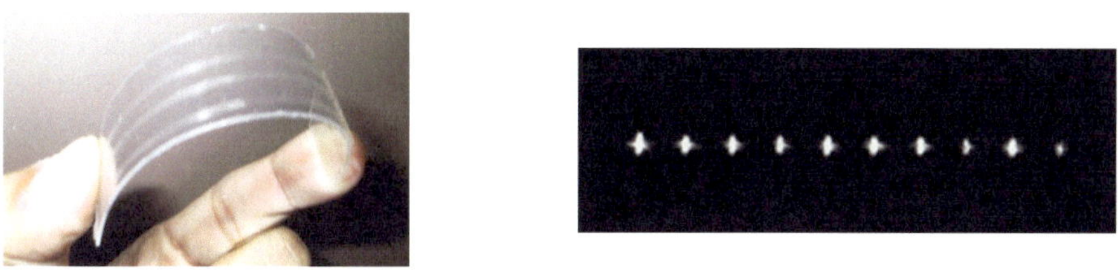

(a) (b)

Fig. 8. a) Flexible ribbon array waveguide on co-polyester substrate and b) resulting output mode spots emerging from waveguide array device.

Successful fabrication of our AWEC waveguides was followed by waveguide performance testing. Here elementary waveguide coupling into the channel array was examined. Resulting output beam spots emerging directly from the optical waveguide array device are shown in Fig. 8b. A VCSEL array source (850 nm wavelength) was used to simultaneously incite all of the array waveguides and light confinement in all channels was achieved. Mode profile non-uniformity can be attributed in part to the lack of optical preparation (and polishing) on waveguide edge facets. For improved edge quality and coupling performance advanced polishing machines for polymer substrates or end facet preparation through excimer laser micromachining has been demonstrated (Jiang, et al., 2004).

After stamp production, the rest of the processes are straightforward replication steps, appropriate for mass production, similar to the stamping of a compact or digital videodisk. While soft lithographic fabrication techniques provide a low-cost, mass production solution, our microtransfer molding approach has some drawbacks. Namely, in μTM the PDMS replica stamp is forcibly pressed against the device substrate, which leads to the creation of a small planar waveguide layer underneath the channel core. Such undesirable byproduct may hinder waveguide alignment, coupling efficiency and propagation loss performance. In the ensuing section we introduce a novel soft lithographic fabrication technique that eliminates the required pressing action and waveguide residue layer through a vacuum assisted microfluidic approach.

4. Vacuum assisted microfluidic waveguide fabrication

Currently, microfluidic guided wave devices based on liquid core waveguide structures are being investigated. Microfluidic optical waveguides have been constructed by inserting liquid into a rectangular channel, where light is guided when the index of the liquid exceeds the surrounding medium (Mach et al., 2002). As another example, liquid-core/liquid-cladding (L^2) optical waveguides have been demonstrated where an optically dense fluid flows in a microfluidic channel within an envelope of fluid with lower refractive index (Wolfe et al, 2004). While these optofluidic waveguides work well in certain biological or analytical instrumentations, their liquid state hinders their applicability in rugged optical communication and photonic applications. Hence, a majority of photonic components manufactured via soft-lithography have been developed through μTM or μCP techniques.

In the preceding section, we employed a μTM scheme to develop guided wave devices where multimode channel waveguide arrays were configured. Such fabrication, where an external force is applied to create the replicated device inevitably leads to the formation of polymer background residue; an unavoidable trait in contact based soft lithography. As the stamp is depressed, the solution on the substrate is forced into the waveguide structure and the excess solution escapes to the edges of the mold. The solution that does not escape forms pockets surrounding the waveguides, resulting in the creation of a planar film layer along the channel waveguide after curing.

When the layer is thick enough, it can become an undesirable planar waveguide, greatly affecting the overall waveguide performance or inducing channel-to-channel crosstalk. Even when the layer is thin, it can still affect the channel waveguide confinement resulting in unfavorable waveguide mode profile changes and modal effective index changes. The strong physical contact on the mold can also lead to pattern deformation and warped structures. Hence, the background polymer residue can alter the desired optical performance of the guided wave device.

Several possible solutions have been addressed to eliminate polymer background residue, including decreasing the applied force. Nonetheless, it was shown that the force with which the mold is depressed has little effect on the polymer background residue (Paloczi, et al.,

2004). In addition, post processing chemical or plasma etching for reducing the residue not only increases manufacturing costs but may affect overall device features such as smoothness and reduction of waveguide thickness. Recently, a new approach for residue reduction was proposed by altering the concentration of the polymer solution (Paloczi, et al., 2004). Since the residue thickness is dependent on the solution viscosity and density, the background residue can be marginally reduced. However, by thinning the solution, it may reduce the waveguide refractive index resulting in poor waveguide confinement and other device performance including electro-optic effects. In addition, the thickness of the waveguides is subsequently reduced because there is not enough solid density in the volume of the solution.
Towards that end, a vacuum assisted microfluidic (VAM) technique was developed that can circumvent such drawbacks. Although based on microfluid filling of the channel device, the microfluidic resin is cured for the creation of solid core structures. The master and PDMS stamp are produced in the same fashion as depicted in Fig. 2. The UV curable core and cladding resins remain the same as well. However, before mold replication, posts are placed on the master to generate holes for the microfluidic inlet reservoir and outlet suction, respectively. To further illustrate the concept, a schematic top view of the proposed VAM technique is presented in Fig. 9.

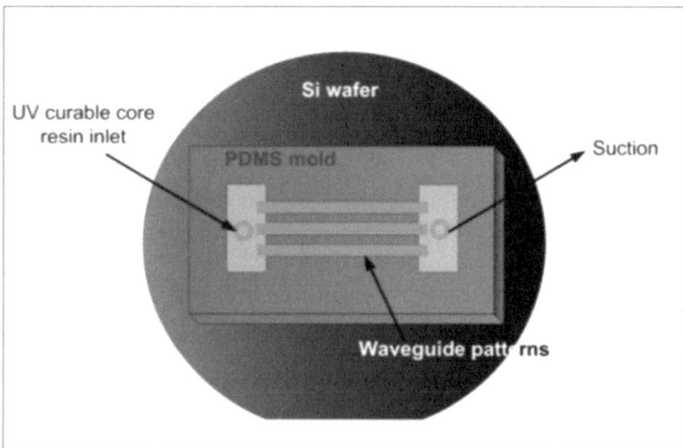

Fig. 9. Schematic of vacuum assisted microfluidic approach for waveguide fabrication.

The process design flow is depicted in Fig. 10. A UV curable cladding layer is spin coated and cured upon the desired substrate as seen in Fig. 10a. Then, the PDMS mold and substrate material are placed in conformal contact and a drop of UV curable core resin is placed through the inlet (see Fig. 10b). Next, rapid filling of the microfluidic channels is assisted by an aspirator (vacuum of 20 mbar) or syringe attached to the outlet opening. The moderate suction provided by an aspirator avoids bubble creation or channel deformation that can be normally caused by strong suction devices such as a rotary or diffusion vacuum pumps. Once filled, the mold is exposed to UV irradiation (20,000 mW/cm^2) and cured for 30 seconds as shown in Fig. 10c. After curing the mold is peeled off to reveal the replicated guided wave structure (Fig. 10d). Lastly, an upper cladding layer may be added, as shown in Fig. 10e. A top view picture of the PDMS microchannel structure along with the inlet and outlet orifices of the stamp is presented in Fig. 10f. Significantly, the quick filling time (1~2 min for 35 μm ×35 μm structure) and fast curing time (30 sec) leads to the rapid prototyping of optical waveguides.

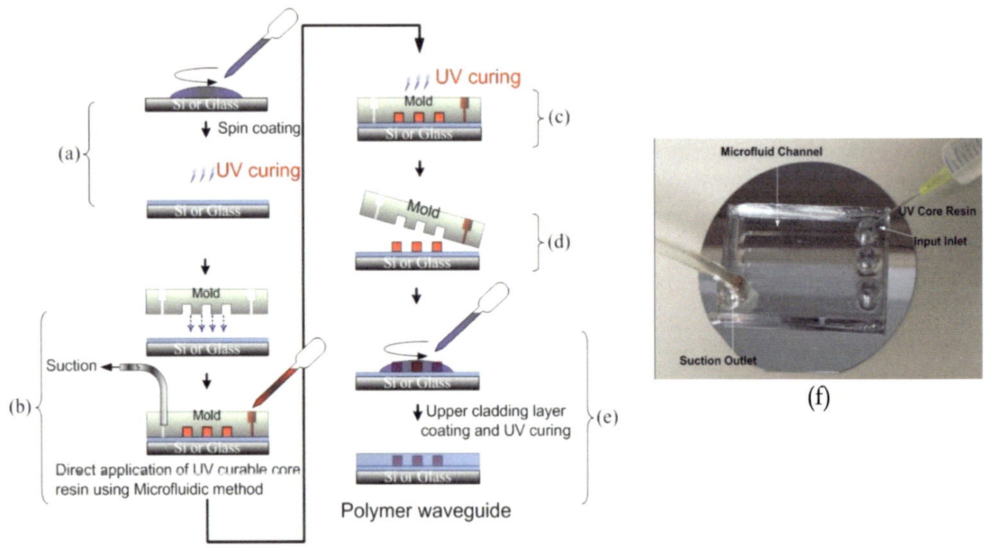

Fig. 10. Process design flow for device replication via VAM. a) UV curable cladding layer is spin coated upon desired substrate. b) After placing PDMS stamp over substrate UV curable core resin drops are placed through an inlet hole. c) Once channels are filled the resin is cured through UV irradiation. d) Stamp is peeled off to reveal the replicated master pattern. e) Upper cladding is coated and cured if desired. f) A top view of the PDMS microchannel structure along with the inlet and outlet orifices of the stamp.

In addition, the VAM method may also be applied to flexible substrates as thin as 200 μm. The ability to fabricate waveguides on rigid substrates (ie., glass, Si) and lithe polymer substrates (co-polyester, polycarbonate, and acrylic) alike, establish the versatility of the VAM approach for fabrication of a wide variety of guided wave devices.

Cross sections of identical waveguides fabricated via both μTM and the VAM technique are detailed in Fig. 11 The sample prepared through μTM contains a thin remnant layer (almost 3~5 microns) due to background residue along the channel structure. In contrast the waveguide prepared through the VAM approach was free of polymer background residue. We also note the improved channel waveguide structure and sidewall edge due to the absence of applied force during the microfluidic fabrication. To further compare both soft lithographic techniques, output mode profile spots of waveguides fabricated by both methods are depicted in the insets of Fig. 11. Once again, improved waveguide formation and elimination of background residue is evident. As a result, a superior mode profile spot is observed via the VAM approach. The images further attest to the elimination of the planar residue layer and satisfactory guided wave channel structure. We cite that the fabrication of the PDMS stamp, core and cladding resins, and UV curing procedures are identical to those previously employed in the μTM approach. Hence, the same low-cost, rapid prototyping of μTM were achieved by VAM while eliminating the cumbersome background residue.

Optical propagation loss is a vital parameter in communication system design. As such, the optical loss performance of the fabricated waveguide arrays was analyzed via the cutback method. As expected, improved waveguide formation through use of the microfluidic technique resulted in lower optical loss. The μTM waveguides exhibited an average loss of 1.1 dB/cm, while the microfluidic waveguide approach generated a loss of 0.68 dB/cm

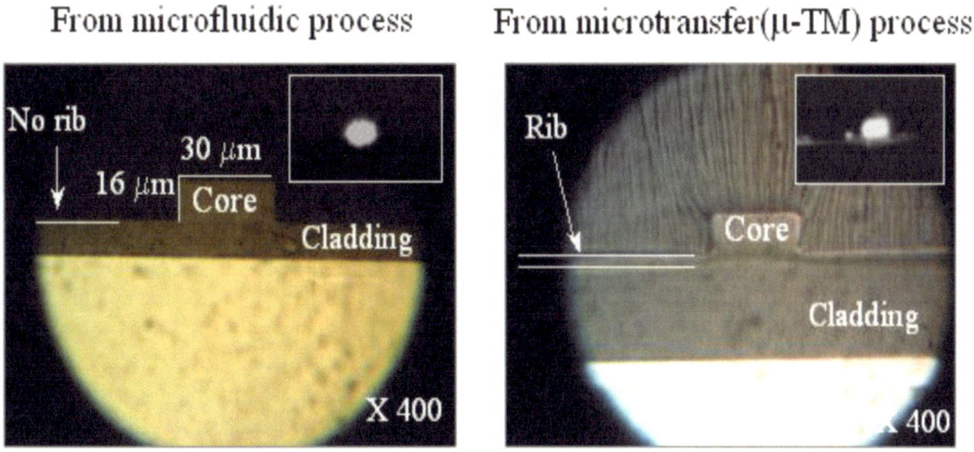

Fig. 11. Waveguide cross-section comparisons of μTM waveguide (right) and VAM waveguide (left). Image confirms elimination of polymer background residue and superior device formation through VAM.

using identical UV curable resins. A plot of the average transmitted power versus waveguide lengths for both methods is presented in Fig. 12 (measurement wavelength of 660 nm). The waveguide losses are attributed in part to the unpolished facet edges and optical preparation of the input and output facets should yield much improved loss performance. Nevertheless, propagation loss reduction (nearly half) is clearly demonstrated, most probably due to the improved waveguide core and reduced defect scattering. Likewise, the reduced waveguide losses further validate the VAM approach for inexpensive, mass fabrication of guided wave devices.

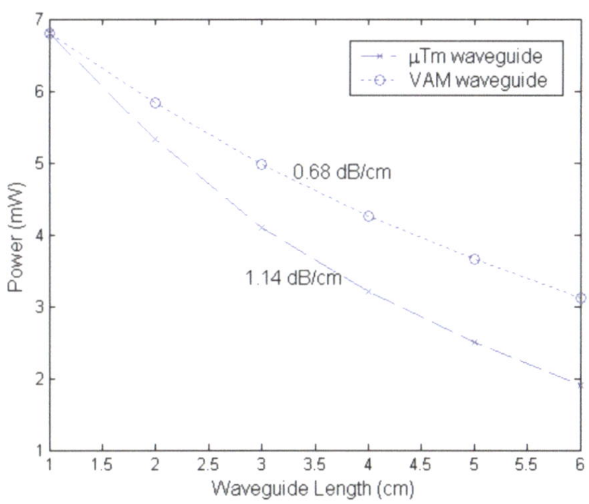

Fig. 12. Average transmitted power versus waveguide length for both μTM and VAM methods.

A lower cost, rapid prototyping, and high resolution patterning soft lithographic technique has been formulated. Furthermore, low-cost polymer materials exhibiting excellent broadband optical transparency, tunable index control, rapid curing, and light guiding functionality were developed in accordance with the fabrication method. More importantly through the VAM approach, microscopic and SEM analysis depicts improved waveguide structures with no bubbles, defects or planar rib layers. The VAM approach also results in lower propagation losses due to the improved sidewall edges and polymer background residue elimination.

5. VAM fabrication of guided wave devices

In this section we detail the fabrication of several integrated optic and guided wave devices. The VAM technique is used to develop single and multi-mode channel waveguides and array waveguide evanescent coupler (AWEC) ribbons for high-speed optical interconnection (Flores, et al., 2008). The fabrication of these devices demonstrates the cost effectiveness and promise of the proposed approach for the development of inexpensive, mass fabrication of polymer guided wave devices.

5.1 Single mode waveguide

Waveguides can be classified according to the total number of guided modes within the steering structure. Guided wave structures designed to carry only a single allowed propagation parameter are termed single-mode waveguides. Correspondingly, multimode waveguides are designed to accept multiple modes within a guided wave device. The light propagating in each mode has a distinct angle of incidence θ_m (m = 1, 2, 3…), and travels in the z direction with a phase velocity and propagation constant that is characteristic of that mode. In general, guided wave devices designed for long distance applications (> km) employ single-mode structures, while less-expensive and more efficient multimode configurations are used for communication over shorter distances (< km).

In the section 3.1, multimode waveguide arrays were devised through a μTM technique. Multimode guided wave devices offering excellent propagation path stability and lower production costs appear to have excellent potential for card-to-backplane optical interconnect applications. Specifically, the development of multimode guides for high speed optical interconnects will be discussed in a later section. Likewise, single mode waveguides guiding only the fundamental mode have a variety of applications for long distance telecommunications. Particularly, multimode waveguides supporting up to thousands of modes can lead to undesirable modal dispersion effects which are avoided in single mode structures.

Single mode waveguides were developed in accordance with the μTM and VAM techniques. The resins used for the cladding and core were Epotek OG 169 and Norland Optical Adhesive (NOA) 74, respectively. Epotek OG 169 and NOA 74 have viscosities of 200 and 80 cps, respectively and cured refractive indices of 1.5084 and 1.51 for the 1550 nm wavelength, respectively. In particular NOA 74 was chosen for its relatively low viscosity and its higher refractive index relative to the cladding. The low refractive index difference of 0.0016 yields a numerical aperture (NA) of 0.07. Subsequently, a channel waveguide with dimensions of 5 μm x 9 μm was devised to demonstrate single mode performance.

Referring to the channel waveguide dimensions and numerical aperture the number of modes in a channel structure can be approximated as (Saleh, B. & Teich, M., 1991)

$$M = \frac{\pi}{4}\left(\frac{2}{\lambda}\right)^2 d_x d_y NA_x NA_y \, , \tag{1}$$

where d_x and d_y represent the horizontal and vertical geometrical dimensions and NA the numerical aperture ($\left(n_{core}^2 - n_{clad}^2\right)^{1/2}$). Hence, the 5 µm × 9 µm rectangular waveguide (M<1) is single mode at the 1550 nm wavelength. A mode profile of the single mode waveguide fabricated through the VAM method is shown in Fig. 13a. For comparison, the multimode (35 µm x 35 µm) profile spot at the same wavelength is depicted in Fig. 13b. Significantly, both single and multimode waveguides with excellent structural quality were created through the VAM technique. Fig. 13c shows the mode field distribution of the single mode waveguide captured by a beam profiler. The Gaussian intensity distribution and subsequent single mode field profile is observed. In general, although single mode waveguides outperform their multimode counterparts they are deemed too costly and impractical for short distance applications. Through both the µTM and VAM methods, the low cost outlays for both single mode and multimode waveguides are identical. Notably, low cost and rapid prototyping production of single mode waveguides with tight fabrication and alignment tolerances has been accomplished.

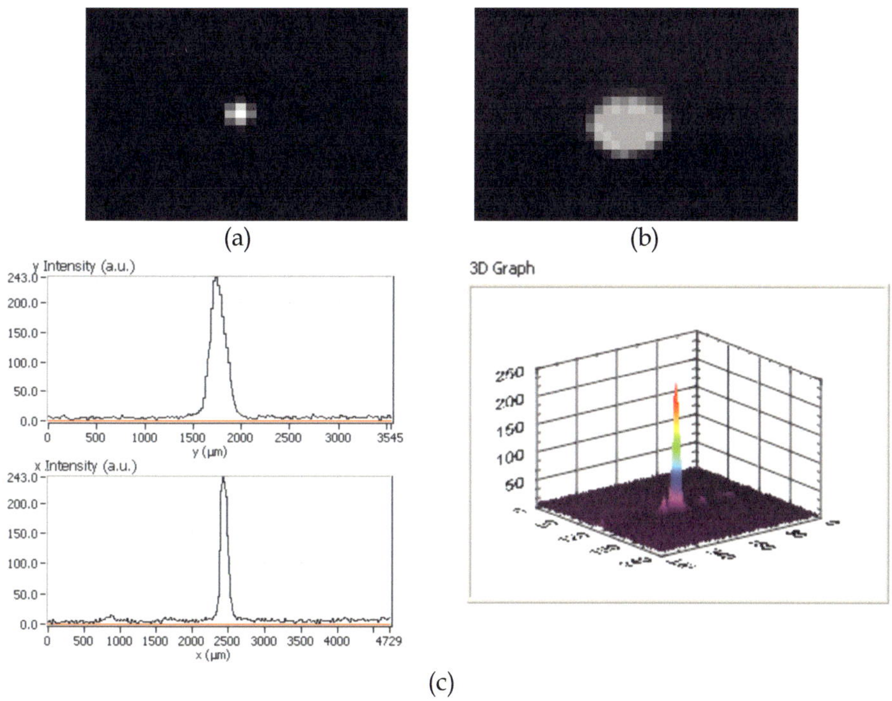

Fig. 13. Channel waveguide mode spots of a) single mode and b) multimode waveguides fabricated via VAM. c) Single mode VAM waveguide beam profile showing x and y axis transverse fields and 3D intensity plot.

An important consideration when using VAM is that unlike µTM (in general), the viscosity of the core resins plays a significant role in the realization of channel waveguide structures.

Low viscosity core resins allow for improved laminar flow, which in turn allows for a quicker distribution among the channels. The tradeoff though comes at the potential cost of intermittent formation of bubbles and a non-uniform density profile along both the length and the central axis (i.e. a dense central region relative to the sides) of the waveguide. On the other hand, high viscosity core resins do not have these associated problems (or at least not as prevalent), but must contend with a slower filling rate and the increased occurrence of partially complete channels. For our fabricated waveguides, low viscosity resins were used for the core material to achieve complete filling of the waveguide channels as the use of higher viscosity resins, especially those with viscosities over 500 cps, often resulted in partial, and/or sparsely filled channels. In particular, as the cross sectional feature size, through which the fluid propagates, is reduced, lower viscosity resins are necessary for successfully filling the channels. Specifically, formation of the smaller dimension single mode waveguides was only accomplished with low viscosity resins (under 200 cps).

Comparison of the cross sections of the single mode channel waveguides fabricated by both the μTM and VAM techniques, are shown in Fig. 14. Similar to the multimode waveguides, shown in Fig. 11, the sample prepared through μTM contains a thin remnant layer (~3 microns) due to background residue along the channel structure. In contrast the waveguide prepared through the microfluidic approach was free of polymer background residue. Once again, improved waveguide formation and elimination of background residue by VAM is evident.

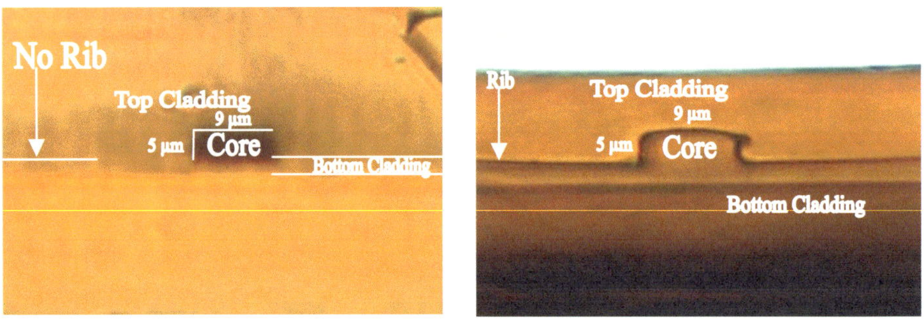

Fig. 14. Waveguide cross-section comparisons single mode waveguides fabricated by VAM technique and by μTM technique. Image reaffirms the elimination of polymer background residue and superior device formation through VAM.

5.2 Array waveguide evanescent coupler ribbon

Recent advances in computing technology have highlighted deficiencies with electrical interconnections at the motherboard and card-to-backplane levels. Specifically, the CPU speeds of computing systems are drastically increasing with on-chip local clock speeds expected to approach 6 GHz by 2010 (International SEMATECH, 2007), yet, card-to-backplane communication speed is unable to maintain the same pace. Due to severe frequency dependant physical factors such as crosstalk, power dissipation, packaging density, and electromagnetic interference (EMI); copper interconnections used on existing motherboards are expected to cause drastic bottleneck problems for board-to-board or off-chip data bus transfers.

Consequently, optical links have been extensively researched for high-speed backplane applications (Glebov, et al., 2005). The most significant benefit that optical interconnects provide over electrical links is the tremendous gain in bandwidth capacity. For example, the bandwidth capacity of a single optical interconnect (guided wave) line was experimentally characterized to be 2.5 THz (Kim, G. & Chen, R. T., 1998), and the bandwidth capacity of optical silica fibers can theoretically reach speeds of up to 50 THz (Kogelnik, 2000). In addition, attenuation losses in optical interconnects are data rate and EMI independent, and offer improved packaging densities. Actively researched solutions at the board-to-board level include free-space optical interconnects, guided-wave interconnects, and fiber based links. While each approach appears promising for future card-to-backplane applications various drawbacks must be addressed before optical links replace copper interconnections inside computers.

For example, the guided-wave approach offers excellent interconnect path stability, low cost processing (ie., polymer waveguides) and are suitable for multi-drop interconnect architectures. However, a major obstacle in guided-wave techniques is the need of 90° out of plane optical deflectors to couple light into or out of the interconnecting waveguides (Glebov, et al., 2007). Such micro-mirrors suffer from reflection losses (i.e., 0.5 dB) due to roughness of diced surface and absorption and scattering of the metal film. Additional microlenses or fiber-coupler adaptors are also regularly used to assist the out-of-plane deflector. These constraints increase manufacturing complexity and cost, degrades backplane reliability, and results in local waveguide terminations (deflecting mirror reflects all light preventing further waveguide transmission for multi-drop interconnects).

Moreover, the out of plane deflectors are not energy efficient since they consume optical power even when cards are not plugged into the backplane. Previously, we reported on the concept of array waveguide evanescent couplers (AWEC) for card-to-backplane optical interconnections (Yang, et al., 2007). By evanescently tapping optical signal power from a backplane bus to a flexible optical bus on the daughter card, the proposed concept eliminates local waveguide termination and the use of 45° micro-mirrors or prisms for the 90° out of plane turns. An initial AWEC optical ribbon link was successfully demonstrated. Nevertheless, the initial AWEC ribbons were limited by excessive manufacturing and fabrication costs (photolithography) and the reported operating speed was limited to 2.5 GHz.

As a result, a VAM approach is adopted for AWEC fabrication. The high-resolution rapid AWEC prototyping technique can result in overall lower coupler fabrication and system cost. A schematic of the proposed AWEC technique and its use for multicard backplane optical interconnects is shown in Fig. 15. The interconnection scheme is based on the exposed core evanescent coupling between a backplane waveguide bus and a flexible bus connected to the plug-in card (or daughter board). Comparable to electronic backplanes, a plug-in card can simply be plugged into the designated backplane AWEC connector to gain access to shared media bus.

Our AWEC interconnect technique is able to efficiently tap optical signal power from the backplane waveguide to card waveguide without any local waveguide terminations. The diagram demonstrates high-speed signals exiting the backplane waveguide without the use of 45° micro-mirrors or prisms. In this particular case, the backplane waveguide signal is evanescently routed to an identical waveguide in the plug-in card, through a flexible AWEC ribbon. A locking mechanism is used to control the interaction length and allow a relatively uniform pressure for consistent power coupling from the backplane waveguide to the AWEC card ribbon. The locking device can be made automatic using spring loaded mechanism not presented herein.

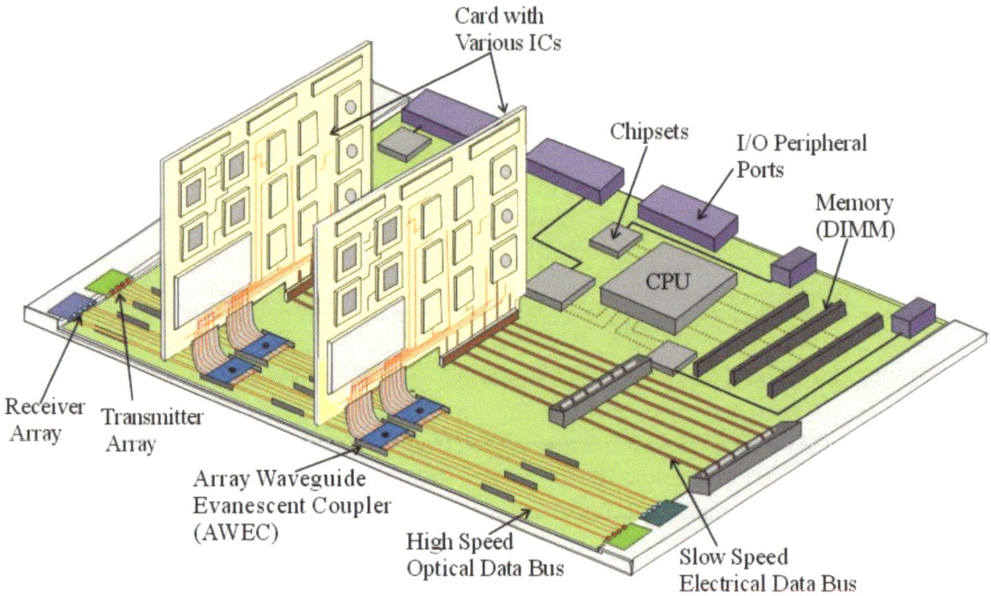

Fig. 15. Schematic detailing AWEC optical interconnect technique for card-to-backplane motherboard applications.

The operational principle of the AWEC technique can be explained by the directional coupling (Yariv, 1973) between the waveguide on the AWEC ribbon and its counterpart on the backplane surface. We consider a general AWEC waveguide structure with a refractive index distribution given by

$$n^2(x,y) = \begin{cases} 1, & |y| < d, \quad \text{all } x \\ n_{core}, & d \le |y| \le 2a + d, \quad |x| \le a \\ n_{clad}, & 2a + d < |y| \le 2a + 2b + d, \quad |x| \le c \end{cases}, \quad (2)$$

where all dimensions are defined in Fig. 16 except that the coordinate origin is at the center of the waveguide gap. Two waveguides, labeled W_1 and W_2, which are initially separated, are brought close to one another over some interaction length L. When the two waveguides are closely spaced and aligned in the lateral direction, the evanescent fields of the guided modes in the two waveguides overlap causing an alteration of the optical mode and field distributions of the waveguide system.

Due to the weakly guided nature of the AWEC system (Δn = 0.011), coupled mode analysis (Yariv, 1973) can be used to analyze the electromagnetic behavior of the complete structure. In this regard, the goal of the coupled-mode theory is to express the electromagnetic fields of the complete structure as a superposition of the unperturbed waveguide electric fields. Most importantly, the coupled-mode theory assumes that the waveguide modes remain approximately the same and the coupling interaction modifies the amplitudes of the modes without affecting the modal transverse field distributions or propagation constants of the waveguides. Thus, in the presence of waveguide coupling the modal amplitudes in the two waveguides become functions of the propagation path z.

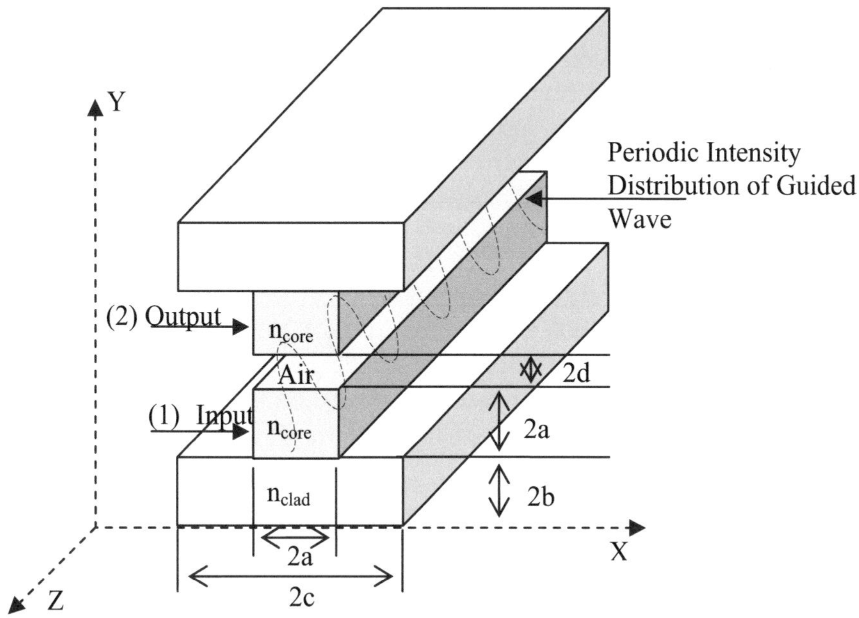

Fig. 16. Diagram depicting AWEC coupling structure and directional coupling between two parallel channel waveguides.

The overlapping fields result in the modification of the waveguide mode solutions and the introduction of an energy exchange coupling constant between the two waveguide modes. Through the coupled-mode theory, the optical powers of the propagating modes in the two channel waveguides, $P_1(z)$ and $P_2(z)$, can be approximated as

$$P_1(z) = P_1(0)\cos^2 \kappa z$$

$$P_2(z) = P_2(0)\sin^2 \kappa z \quad , \tag{3}$$

when the two waveguide channels are phase matched. Here, κ is the propagation constant of the waveguide mode. $P_1(0)$ and $P_2(0)$ are the initial light power of the two waveguide channels, respectively.

Notably, the coupling constant κ is proportional to channel separation. Although the coupling coefficient can be controlled through pressure regulation between the two waveguides, - which in turn influences waveguide separation - precise pressure control is both impractical and difficult to achieve. For the AWEC case, the waveguide interaction length will be employed to deliver the desired coupling coefficient, as the variation of the interaction length on the millimeter scale is much easier to resolve and repeat. By adjusting the interaction length between waveguides, the amount of energy transferred and coupling efficiency between the guides can be regulated.

It is important to mention that the coupled-mode analysis discussed above pertains to single-mode channel waveguides. The more pertinent case of multimode waveguides can be analyzed by noting that the complex optical fields are a superposition of all the modes excited within the channel. As such CAD simulations modeled via BeamPROP were executed to examine multimode coupler behavior between the AWEC ribbons (Flores, et al, 2008). CAD simulation results demonstrated strong evanescent coupling between the AWEC ribbons and dominant energy distribution (> 90%) within the fundamental mode and agreed well with the experimental coupling results.

To demonstrate effective AWEC coupling, flexible ribbon arrays identical to the multimode arrays presented in Fig. 6 were fabricated via a VAM approach. The flexible ribbon arrays consist of 12 channels with a 250 μm pitch for integration into commercial parallel optical transmitter/receiver array modules. By incorporating a soft lithographic method we were able to generate the waveguide arrays essential for AWEC fabrication without the intricate use of conventional photolithography. Moreover, a scanning electron micrograph (SEM) of a multi-channel AWEC device fabricated on a silicon substrate is shown in Fig. 17. The image once again attests to the elimination of the planar rib layer and satisfactory guided wave channel structure.

Fig. 17. SEM photograph of the replicated AWEC device fabricated via soft lithography. Inset shows a cross-section micrograph of an AWEC channel.

For the optical interconnection, the flexible waveguide ribbons were integrated into parallel optical transmitter array and receiver array modules. One flexible ribbon was interfaced to the VCSEL transmitter board, while another ribbon was connected to the PIN receiver board and the two AWEC ribbons displaced at 90° were then evanescently coupled at an interaction length of 11 mm. Index matching fluid (n = 1.515) was inserted between the ribbons to facilitate coupling as predicted by the simulation results. Successfully, at speeds of up 10 GHz we were able to easily demonstrate evanescent coupling of pulse data information. We note that our high speed link was solely limited by our electronic signal generator and digitizing oscilloscope which cannot operate beyond 10 GHz.

Figure 18 shows the 10 Gbps eye diagram for the AWEC optical interconnect link. Optical interconnections at 10 Gbps have been successfully achieved for each of the 12 channels on the AWEC ribbon, resulting in an aggregate data rate of over 100 Gbps. The modal dispersion analysis for the multimode waveguide predicts speeds of up to 40 Gbps can be accommodated for each interconnect channel (Flores, et al., 2008).

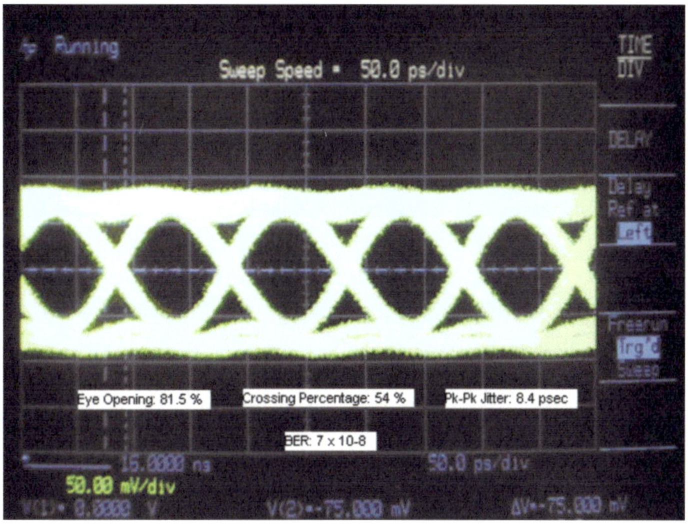

Fig. 18. Recovered eye diagram for our AWEC link at 10 Gbps.

6. Conclusions

The design and development of guided wave devices based on soft lithographic fabrication techniques was examined. Notably, a novel vacuum assisted microfluidic technique for fabrication of guided wave and integrated optical devices was described. Importantly, the technique eliminates the polymer background residue inherent to traditional soft molding fabrication techniques. In addition, UV curable resins with tunable index control specifically tailored for soft lithography were developed.

Comparisons to conventional soft lithography demonstrate that the VAM approach results in lower propagation losses, lower crosstalk, and improved waveguide structures. More importantly, microscope analysis portrays improved device formation, sidewall edges and the elimination of the polymer background residue intrinsic to conventional soft lithography. As a low-cost rapid prototyping technique the VAM soft lithographic method allows guided wave devices to be implemented rapidly and inexpensively.

Moreover, through adoption of our VAM technique we were able to successfully fabricate several integrated optic and guided wave devices. The VAM technique is used to develop single and multi-mode channel waveguides, and array waveguide evanescent coupler (AWEC) ribbons for high-speed optical interconnection. Notably, through a soft lithographic approach the overall fabrication costs were reduced (without sacrificing ribbon quality and performance) and data rates of up to 10 Gbps per channel were demonstrated. We expect that the AWEC scheme will be significant for high-speed optical interconnects in advanced computing and backplane systems. Overall, we believe that the novel VAM technique can yield lower production costs and manufacturing complexity for polymer based photonic integrated circuits.

7. Acknowledgements

The projects were supported in part by the Department of Defense and the American Society for Engineering Education.

8. References

Bowers, J. E.; Park, H; Kuo, Y. -H; Fang, A. W.; Jones, R.; Paniccia, M. J.; Cohen, O. & Raday, O. (2007). Integrated Optical Amplifiers on Silicon Waveguides. *Integrated Photonics and NanoPhotonics Research and Applications (IPNRA)* 2007, ITuG1.

Brown, T. G.; Bradfield P. L.; Hall D. G. & Soref R. A. (1987). Optical emission from impurities within an epitaxial-silicon optical waveguide. *Optics Letters*, Vol. 12, No. 9, 753-755.

Flores, A.; Song, S.; Yang, J. J.; Liu, Z. & Wang, M. R. (2008). High-speed optical interconnect coupler based on soft lithography ribbons. *IEEE Journal of Lightwave Technology*, Vol. 26, No. 13, 1956-1963.

Glebov, A. L.; Roman, J.; Lee, M. G. & Yokouchi, K. (2005). Optical interconnect modules with fully integrated reflector mirrors. *IEEE Photonic Technology Letters*, Vol. 17, No. 7, 1540-1542.

Glebov, A. L.; Lee, M. G. & Yokouchi, K. (2007). Integration technologies for pluggable backplane optical interconnect systems. *Optical Engineering*, Vol. 46, No. 1, 015403-015410.

Grundy, K.; Liaw, H.; Otonari, G. & Resso, M. (2006). Designing scalable 10G backplane interconnect systems utilizing advanced verification methodologies. *DesignCon 2006*, Paper 8-WP2.

Heckele, M. & Schomburg, W. K. (2004). Review on micromolding of thermoplastic polymers. *Journal of Micromechanics And Microengineering*, Vol. 14, No. 3, R1-R14.

Heinrich, J.; Zeeb, E. & Ebeling, K. J. (1997). Butt-coupling efficiency of VCSELs into multimode fibers. *IEEE Photonics Technology Letters*, Vol. 9, No. 12, 1555-1557.

International SEMATECH (2007). The National Technology Roadmap for Semiconductors (ITRS)- Technology. Semiconductor Industry Association.

Jiang, J.; Callender, C. L.; Noad, J. P.; Walker, R. B.; Mihailov, S. J.; Ding, J. & Day. M (2004). All-polymer photonic devices using excimer laser micromachining. *IEEE Photonics Technology Letters*, Vol. 16, No. 2, 509-511.

Kim, G. & Chen, R. T. (1998). Three-dimensionally interconnected bidirectional optical backplane. *IEEE Photonic Technology Letters*. Vol. 11, No. 7, 880-882.

Kogelnik, H. (2000). High-capacity optical communication. *IEEE Journal of Selected Topics in Quantum Electronics*. Vol. 6, No. 6, 1279-1286.

Kunnavakkam, M. V.; Houlihan, F. M.; Schlax, M.; Liddle, J. A.; Kolodner, P.; Nalamasu, O. & Rogers, J. A. (2003). Low-cost, low-loss microlens arrays fabricated by soft-lithography replication process. *Applied Physics Letters*, Vol. 82, No. 8, 1152-1154.

Lauks, I. R. (1998). Microfabricated biosensors and microanalytical systems for blood analysis. *Accounts of Chemical Research*, Vol. 31, No. 5, 317-324.

Liu, S.; Shi, Y.; Ja, W. W.; & Mathies, R. (1999). Optimization of high-speed DNA sequencing on microfabricated capillary electrophoresis channels. *Analytical Chemistry*, Vol. 71, No. 3, 566-573.

Mach, P.; Dolinski, M.; Baldwin, K. W.; Rogers, J. A.; Kerbage, C.; Windeler, R. S. & Eggleton, B. J. (2002). Tunable microfluidic optical fiber. *Applied Physics Letters*, Vol. 80, No. 9, 4294-4296.

Paloczi, G. T.; Huang, Y.Y.; Scheuer, J. & Yariv, A. (2004). Soft lithography molding of polymer integrated optical devices: reduction of background residue. *Journal of Vacuum Science and Technology B*, Vol. 22, No. 4, 1764-1769.

Ramaswamy R. V. & Srivastava R. (1988). Ion-exchanged glass waveguides: a review. *IEEE Journal of Lightwave Technology*, Vol. 6, No. 6, 984-1000.

Rogers, J. A.; Meier, M. & Dodabalapur, A. (1998). Using printing and molding techniques to produce distributed feedback and Bragg reflector resonators for plastic lasers. *Applied Physics Letters*, Vol. 73, No. 13, 1766-1768.

Saleh, B. E. & Teich, M. C. (1991). *Fundamentals of Photonics*. John Wiley & Sons, ISBN: 0471839655, New York.

Saleh, O. A. & Sohn, L. L. (2003). An artificial nanopore for molecular sensing. *Nano Letters*, Vol. 3, No. 1, 37-38.

Santini, J.T.; Cima, M. J. & Langer, R. (1998). A controlled-release microchip. Nature, Vol. 397, No. 6717, 335-338.

Schueller, O. J.; Whitesides, G.; Rogers, J. A.; Meier, M. & Dodabalapur, A. (1999). Fabrication of photonic crystal lasers by nanomolding solgel glasses. *Applied Optics*, Vol. 38, No. 27, 5799-5802.

Song, S.; Flores, A.; Yang, J. J. & Wang, M. R. (2005). Fabrication of multimode polymer waveguide by using soft lithographic technique. *OSA Annual Meeting 2005*, Tuscon, Az., FThQ7.

Quist, A. P.; Pavlovic, E. & Oscarsson, S. (2005). Recent Advances in microcontact printing: Analysis of biomaterials. *Analytical and bioanalytical chemistry*, Vol. 381, No. 3, 591-600.

van der Berg, A.; Grisel, A.; Verney-Norberg, E.; van der Schoot, B. H.; Koudellka-Her, M. & de Rooij, N. F. (1993). On-wafer fabricated free-chlorine sensor with ppb detection limit for drinking-water monitoring. *Sensors and Actuators B*, Vol. 13, No. 1-3, 396-399.

Wang, M. R. & Su, H. (1998). Laser direct-write gray-level mask and one-step etching for diffractive microlens fabrication. *Applied Optics*, Vol. 37, No. 32, 7568-7576.

Whitesides, G. M. (2006). The origins and the futures of microfluidics. *Nature*, Vol. 442, No. 7101, 368-373.

Wolfe, D. B.; Conroy, R. S.; Garstecki, P.; Mayers, B. T.; Fischbach, M. A.; Paul, K. E.; Prentiss, M. & Whitesides, G. M. (2004). Dynamic control of liquid-core/liquid-cladding optical waveguides. *Proceedings of the National Academy of Sciences USA*, Vol. 101, No. 34, 12434-12438.

Xia, Y.; Kim, E.; Zhao, X.; Rogers, J. A.; Prentiss, G. M. Whitesides, G. M. (1997). Complex optical surfaces formed by replica molding against elastomeric masters. *Science*, Vol. 273, No. 5273, 347-349.

Xia Y. & Whitesides, G. M. (1998). Soft lithography. *Angewandte Chemie International Edition*, Vol. 37, No. 5, 570-575.

Yang, J. J.; Flores, A. & Wang, M. R. (2007). Array Waveguide Evanescent Ribbon for Card-to-Backplane Interconnects. *Optics Letters*, Vol. 32, No. 1, 14-16.

Yariv, A. (1973). Coupled-mode theory for guided-wave optics. *IEEE Journal of Quantum Electronics*, Vol. 9, No. 9, 919-933.

Zhang, F.; Nyberg, T. & Inganas, O. (2002). Conducting polymer nanowires and nanodots made with soft lithography. *Nano Letters*. Vol. 2, No. 12, 1373-1377.

Zhao, X.; Xia, Y. & Whitesides, G. M. (1996) Fabrication of three-dimensional micro-structures: Microtransfer molding. *Advanced Materials*, Vol. 8, No. 10, 837-840.

20

Application of Soft Lithography for Nano Functional Devices

Shin-Won Kang
School of Electrical Engineering and Computer Science, Kyungpook National University
1370 Sankyuk-dong, Bukgu, Daegu,
Republic of Korea

1. Introduction

Since photolithography method that is for producing minuteness electronic devices was developed, attempts against the new methods are progressing constantly as increasing demands for the electronic devices. Currently, because of the resolution required for the integrated element is decreased to below 100 nm, the lithography method using ultraviolet rays is developed as various methods like EUV(Extreme UV), X-ray, E-beam lithography. However, the methods mentioned previously are basically non-environmental friendly, and the development of photoresist that reacted on the source should be preceded. Also, it has limitations such as substrate and material selections, low throughput and high cost of methods.

To overcome these limitations and guarantee the high throughput, the soft lithography method is a new counter plan, so a lot of researches are executed. This indicates the producing technique that making patterns with mechanical method by using the master of polydimethylsiloxane (PDMS) stamp, so it has advantages to micro and nano structure patterning on the substrate that is not uniform than photolithography producing technique. Specially, it is useful to produce of optics, mechanics and heat fluid structure of MEMS/NEMS. The detail methods related to nano imprint lithography (NIL) and nano moldings, and each of them are effected on the producing structure that size of between 25 ~ 100 nm, 10 ~ 100 nm, respectively.

NIL is the technique that can effectively produce nano pattern that line width below 100 nm, the limitation of UV lithography, and nano contact printing method produces and uses the stamp with polymer such as PDMS by using patterned master by electron beam, and after transfer to the self-assembled monolayer (SAM) substrate that created by contacting of the stamp that has ink by arranging ink element on the stamp and substrate, use them in the wet etch mask to produce the structure.

So, we fabricated gas chamber that is for collecting gas diffused on the skin, optical waveguide, and pixel definition for polymer light-emitting diode (PLED) by using above mentioned methods and evaluated the possibilities.

Soft Lithography mentioned above can overcome the resolution limitation that photolithography method has, and the method is simple and it has advantages on cost saving. Also, like lens and optical fiber, it is available on the method in large area like non planar surface, so it can be applied to the not only cell biology industry but microelectronics, optics and display areas.

2. Trend of research and development

In late 1960's and early 1970's, Gorden Moore, a founder of Fairchild Semiconductor and Intel, argued that the circuit integration of semiconductor is estimated to double its degree every eighteen months (Younan Xia & G.M. Whitesides, 1998). His prediction later becomes Moore's law (R.W. Keyes et al., 1992). Dr. Hwang in Korea published a "new memory growth theory" in 2002 asserting a Hwang's law, which argues that degree of integration of semiconductor doubles every twelve months. Samsung company demonstrated the doubling growth of integration degree that continued over a period of seven years, beginning from 256-mega in 1999 to 32-giga in 2006. Such accomplishment was possible thanks to the continuous advancement in photolithography technology that doubled its resolution every three years over the past thirty years, as many trends in semiconductor industry followed those laws. (Figure 1)

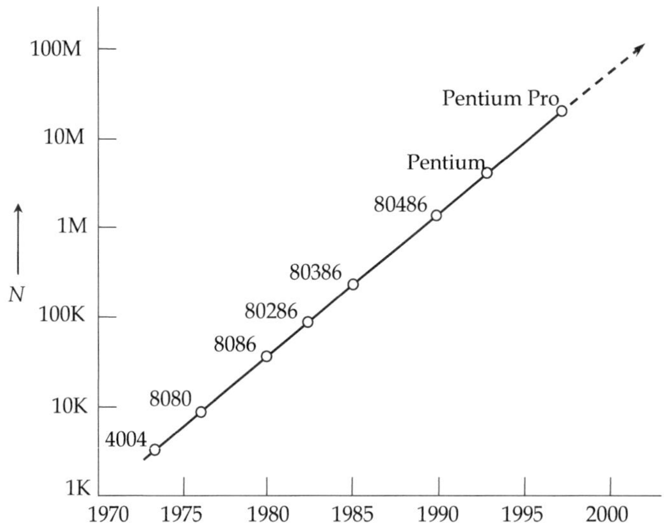

Fig. 1. The integration trend given by Moore's law, and how microprocessors manufactured by Intel have followed this law since 1973. N is number of transistors per chip. (C.R. Barret, 1993; R.F. Service, 1996).

This curve reflects the general trend of scaling technology that was made possible due to micro lithography which can also be applied to RAM, DRAM, micro processor, etc. Assuming that development of new short wavelength light source and photosensitive film continues following this trend, creating semiconductor with a so-called minimum line width of 100 nm can become possible.

However, creating chips that are small than 100 nm are extremely limited due to light diffraction, problems in creating light masks, lens resolution, etc. In fine processing technology, reducing the line width under 100 nm necessarily requires new approach.

The current photolithography technologies include EUV, soft X-ray lithography, e-beam writing, focused ion beam writing, proximal-probe lithography, etc (W. M. Moreau, 1998; R.F.W. Pease, 1992).

Even though these technologies can realize a very small chips, originalities are critical matters when is required when actually applying it to mass-producing technology that require mass production at low cost.

The photo lithography field for creating chips smaller than 100 nm is facing new technological challenge, and there is no guarantee that photolithography technology is the optimal technology. For example, in chemistry, biology, and materials science, reducing the size of an object requires high cost both in capital and operational manner. Moreover, patterning the non-uniform surface requires difficult technology and using in case of glass, plastic, ceramic, carbon-based materials that have great potential for next-generation technology is very limited.

Currently, a development of practical technologies that can produce structures smaller than100 nm is one of the most critical matters and the most challenging issues in micro-integration technology field. As a result, a variety of non-photolithography technologies are being introduced in creating high-quality micro structure and nanostructure chip, as is shown in Table 1.

Method	Resolution
Injection molding	10 nm
Embossing (imprinting)	25 nm
Cast molding	50 nm
Laser ablation	70 nm
Micromachining with a sharp stylus	100 nm
Laser-induced deposition	1 um
Electrochemical micromachining	1 um
Silver halide photography	5 um
Pad printing	20 um
Screen printing	20 um
Ink-jet printing	50 um
Electrophotography	50 um
Stereolithography	100 um
Soft Lithography	
Microcontact printing	35 nm
Replica molding	30 nm
Microtransfer molding	1 um
Micromolding in capillaries	1 um
Solvent-assisted micromolding	60 nm

Table 1. Non-photolithographic methods for micro- and nanofabrication (Younan Xia & G.M. Whitesides, 1998).

This chapter focuses on soft lithography technologies that are currently under research, such as microcontact printing (A. Kumar & G.M. Whitesides,1993), replica molding (Y. Xia et al, 1996), embossing, elastomeric stamp (X.-M. Zhao et al., 1996), mold, and micromolding in capillaries (MIMIC) (E. Kim et al., 1995). The name of soft lithography originates from the following facts. First, different from the photolithography, elastomeric stamp and mold play an important role as a board in transferring patterns. Second, it uses flexible organic device instead of rigid minerals.

Such soft lithography creates SAM thin film type fine patterns using contact printing or builds fine structure using embossing (imprinting) or replica molding. Figure 2 describes soft lithography in a technological general procedure that we call as "rapid prototyping. (Younan Xia & G.M. Whitesides, 1998)" The biggest strength of soft lithograph is that

cloning process is possible through creating master or mold without complicated process such as photolithography. Other advantages of the technology include relatively low investment cost and simple procedure which does not require special environment such as the clean room. Hence the research can be generally conducted in a normal lab and it is not affected by the diffraction of light or transparency. With these merits, the soft lithography is receiving increasing amount of attention as an alternative to the photolithography technology in creating structures smaller than 100 nm. Moreover, it opens door to a new approach to creating those that are hard to be created using the photolithography technology, such as a surface, optical structure, sensor, etc.
Therefore, this chapter explains the fundamental theory of soft lithography and patterning technology and presents the application research results.

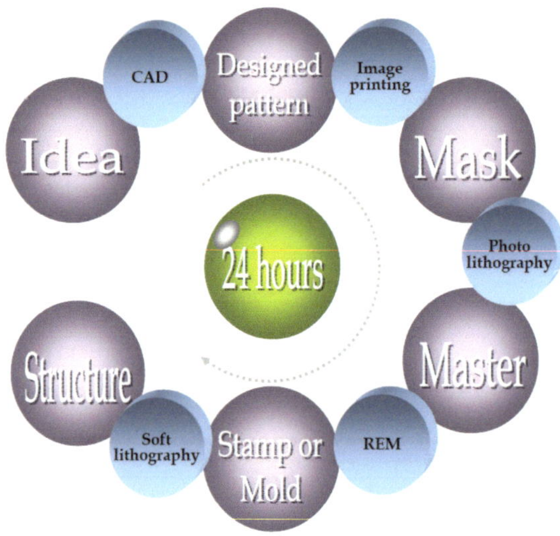

Fig. 2. The rapid prototyping procedure for soft lithography.

3. Method of soft lithography

3.1 Self-Assembly

Photolithography has been regarded as an extremely new approach in micro-integration technology as a technical challenge for 100 nm and lower resolutions. Through amazingly extensive contributions to the practical and conceptual aspects in chemistry and biology, it provided a new methodology in micro-integration area and many means for achieving smaller size and lower cost with conceptually new strategies. A representative example is self-assembly which has been most perfectly studied and actually implemented.
In a self-assembly, molecules or objects form continuous structures in stable form which are very well defined by non-covalent forces (J.-M. Lehn, 1990).
One of the key concepts of self-assembly is that the final structure is almost thermodynamically stable and often has a better system than non-self-assembly structure.
Studies on the technology of self-assembly have steadily developed and it has been applied to the integration of structures of two and three dimensions that include various levels from molecules to middle structures and large structures (J.-M. Lehn, 1988; C.A. Mirkin et al., 1996; A.S. Dimitov & K. Nagayama, 1996; A. Terfort et al., 1997).

3.2 Self-Assembly monolayers

SAM has been studied in most extensive areas and many developments have been achieved in the self-assembly systems of non-biological areas (C.D. Bain & G.M. Whitesides, 1989; J. Xu et al., 1995). It refers to the self-organization in a continuous form of functionalized organic molecules with chemical adsorption and long chains on the surface of an appropriate substrate. It is realized by soaking the substrate in a solution that contains ligands or exposing the substrate to a gas that contains reactive species. Table 2 lists various mechanisms known as SAM, and many studies are being conducted in new areas in addition to them (Younan Xia & G.M. Whitesides, 1998; P. Fenter et al., 1994).

Substrate	Ligand or Precusor	Binding
Au	RSH, ArSH (thiols)	RS-Au
	RSSR′ (disulfides)	RS-Au
	RSR′ (sulfides)	RS-Au
	RSO_2H	RSO_2-Au
	R_3P	R_3P-Au
Ag	RSH, ArSH	RS-Ag
Cu	RSH, ArSH	RS-Cu
Pd	RSH, ArSH	RS-Pd
Pt	RNC	RNC-Pt
GaAs	RSH	RS-GaAs
InP	RSH	RS-InP
SiO_2, glass	$RsiCl_3$, $Rsi(OR')_3$	Siloxane
Si/Si-H	$(RCOO)_2$ (neat)	R-Si
	$RCH=CH_2$	RCH_2CH_2Si
Si/Si-Cl	Rli. RMgX	R-Si
Metal oxides	RCOOH	RCOO-...Mon
	RCONHOH	RCONHOH...Mon
ZrO_2	RPO_3H_2	RPO_3^{2-}...Zr^{IV}
In_2O_3/SnO_2 (ITO)	RPO_3H_2	RPO_3^{2-}-...M^{n+}

Table 2. Substrates and ligands that form SAMs.

One example that best represents the characteristics of SAM is the reaction of Au and alkanethiolates $CH_3(CH_2)nS\ddot{y}$ (Figure 3) (P. Fenter et al., 1994). From liquid state, alkanethiols react with gold surface in continuous chemical adsorption and alkanethiolates are adsorbed as a result. Although there is no established theory related to the fracture of hydrogen atoms, it is assumed that this process occurs together with the loss process of dihydrogen. Sulfur atoms combine with gold by bringing alkali atoms near them to the gold surface. This approach of atoms is characterized by stabilized structural entropy and attainment of orderly structure.

In the case of about 20 carbon combinations, the degree of interaction of molecules in SAM increases in accordance with the molecular density on surface and the length of alkali backbone. Only alkanethiolates with n>11 form a close, solid structure, and two-

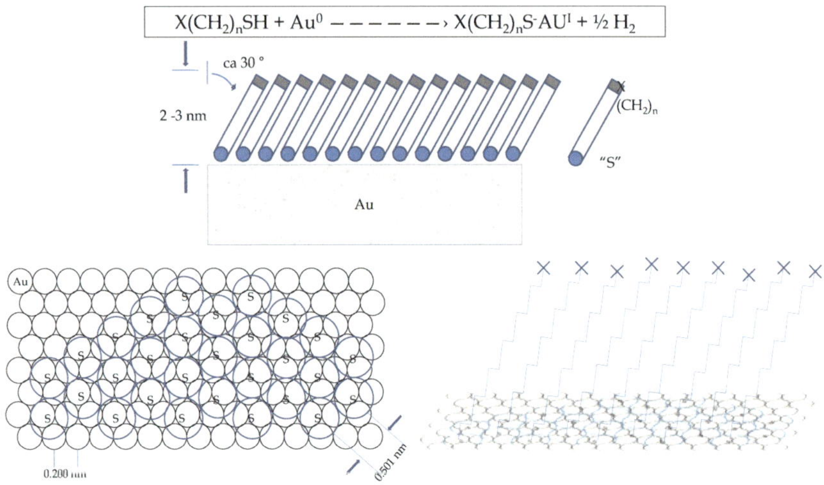

Fig. 3. Representation of a highly ordered monolayer of alkanethiolate formed on a gold surface. The sulfur atoms form a commensurate overlayer on Au(111) with a (√3×√3)R30 degree structure, whose thickness is determined by the number of methylene groups (n) in the alkyl chain. The surface properties of the monolayer can be easily modified by changing the head group X. The alkyl chains $(CH_2)_n$ extend from the surface in a nearly alltrans configuration. On average they are tilted approximately 30 degrees from the normal to the surface to maximize the van der Waals interactions between adjacent methylene groups.

dimensional organic quasi-crystals necessarily sustain the form supported by gold, which is the most useful case of the application of soft lithography in SAM (E. Delamarche et al., 1996).

The orderly structure formed on gold starting from alkanethiols exhibit relatively fast progressing speed. In this way, the structure in which hexadecanethiolates are very orderly aligned on gold can be fabricated by soaking a gold substrate in ethanol solution containing hexadecanethiol for a few minutes. It is formed for a few seconds during mCP. The ability to form an orderly structure in a short period of time is one of the factors that mCP can be implemented as a successful process.

As can be seen from the before mentioned alignment of alkanethiolates on a gold substrate, the structure and characteristics of SAMs have been experimented using various techniques (Table 3). (G.E. Polner, 1997; C.A. Alves et al. 1992; M.R. Anderson et al., 1996; N. Camillone et al., 1996; W.B. Caldwell et al., 1995; L. Strong & G.M. Whitesides, 1988; M.A. Bryant & J.E. Pemberton, 1991; Q.Du et al., 1994; J.P. Folkers et al., 1992; L.H. Dubois et al., 1990; Y. Li et al., 1992; C.D. Brain & G.M. Whitesides, 1988; C.D. Brain et al., 1989; T.W. Schneider & D.A. Buttry, 1993; M.D. Ward & D.A. Buttry, 1990; S. Li & R.M. Crooks, 1993; X.-M. Zhao et al., 1996)

In general, sulfur atoms have been known to form R30 degree overlayer on the Au(111) surface (Figure 3), and recent STM studies revealed that these systems consist of heterogeneous, complex structures. Alkyl chain forms a superlattice on single film surface which is different from the symmetrical hexagonal lattice formed by sulfur atoms at the bottom. This result indicates that the top part of SAM is not affected by sulfur atoms which are directly attached to the gold surface and strongly depends on the intra-molecular interactions between alkyl backbones.

Alkanethiolates SAM on gold explains the reasons that self-assembly system is an excellent technology: easy fabrication, low defects for wide applications, stable characteristics in laboratory environment, technical applicability, and the possibility of variation of characteristics by the adjustment of the system interface characteristics (physical, chemical, electrochemical, biological).

Consequently, SAM provides excellent models for studies in various areas such as wet, adhesive, lubricating, and erosive, nuclear-structural methods, usage of protein absorption, and cell attachment method. Furthermore, it is also an appropriate technique and basis for horizontal unit pattern in the range from nanometer to micrometer, as well as structural and integrated devices.

Patterning SAMs in the plane of the surface has been achieved by a wide variety of techniques (Table 4) (J.L. Wilbur et al., 1994; Y. Xia et al., 1996; T.P. Moffat & H. Yang, 1995; Y. Xia et al., 1995; P.M. St. John & H.G. Craighead, 1996; J. Huang & J.C. Hemminger, 1993; J. Huang et al., 1994; K.C. Chan et al., 1995; E.W. Wollman et al., 1993; A.C. Pease et al., 1994; W.J. Dressick & J.M. Calvert, 1993; J.A.M. Sondag-Huethorst et al., 1994; M. Lercel et al., 1993; M.J. Lercel et al., 1996; G. Gillen et al., 1994; K.K. Berggren et al., 1995; K.S. Johnson et al., 1996; C.B. Ross et al., 1993; N.L. Abbott et al., 1992; A. Kumar et al., 1992). Each technique has advantages and disadvantages. Only micro-contact printing will be discussed in this review since it is the one that seems to offer the most interesting combination of convenience and new capability.

3.3 Contact printing, replica molding and embossing

Contact printing is the most efficient pattern transfer method. The biggest benefits of this printing are simplicity and convenience (A. Voet, 1952). Once a stamp is available, it is possible to produce repeated patterns. Moreover, it minimizes the waste of materials and

Property of SAM	Technique
Structure and order	Scanning probe microscopy
	STM, AFM, LFM
	Infrared spectroscopy
	Low-energy helium diffraction
	X-ray diffraction
	Transmission electron diffraction
	Surface Raman scattering
	Sum frequency spectroscopy
Composition	X-ray photoelectron spectroscopy (XPS)
	Temperature programmed desorption (TPD)
	Mass spectrometry (MS)
Wettability	Contact angle
Thickness	Ellipsometry
Coverage	Quartz crystal microbalance (QCM)
Degree of perfection	SAW device
	Electrochemical methods
Defects	STM and AFM
	Wet etching

Table 3. Techniques for characterizing SAM of alkanethiolates on gold.

Technique	SAM	Resolution
Microcontact printing (uCP)	RSH/Au	35 nm
	RSH/Ag	100 nm
	RSH/Cu	500 nm
	RSH/Pd	500 nm
	RPO_3H_2/Al	500 nm
	Siloxane/SiO_2	500 nm
Photooxidation	RHS/Au	10 um
Photo-cross linking	RHS/Au	10 um
photoactivation	RHS/Au	10 um
	Siloxane/glass	10 um
Photolithography /plating	Siloxane/SiO_2	500 nm
E-beam writing	RHS/Au	75 nm
	RHS/GaAs	25 nm
	Siloxane/SiO_2	5 nm
FIB writing	RSH/Ag	10 um
Neutral metastable atom writing	RSH/Au	70 nm
	Siloxane/SiO_2	70 nm
SPM lithography	RSH/Au	10 nm
Micromachining	RSH/Au	100 nm
Micropen writing	RSH/Au	10 um

Table 4. Techniques that have been used for patterning SAM.

has potential for large-area patterning. Contact printing is optimized for the production of two-dimensional devices and provides the advantage of extending its application to three-dimensional structures through a process that uses metal plates, etc (P.O. Hidber et al., 1996).

Replica molding is to replicate the shape, form, structure and other information of the master, and can accept formative information of materials in a wider range than photolithography. Furthermore, it allows the replication of three-dimensional morphology through only one processing step, which is impossible in photolithography. Replica molding has been used for mass production of objects that have stable surface structures such as diffraction grating (B.L. Ramos & S.J. choquette, 1996), holograms (M. Nakano, 1979), CD [(H.C. Haverkorn et al., 1982), and microtools (D.A. Kiewit, 1973). Replica molding that uses appropriate materials can replicate reliably down to nanometer unit even materials with very complex structures in a simple, cheap method. The excellent replication property of replica molding is determined by Van der waals interaction, wet method, and dynamic factors used for filling the mold. Due to this physical interaction, replica molding enables more accurate replication in the smaller sizes than 100 nm which cannot be done with photolithography because of its limitation by diffraction.

Embossing is another technique for stamping thermoplastic materials and has advantages in terms of price to performance ratio and high yield. For example, the technique for stamping polycarbonate using Ni master is used as a basic technique for CD production, and the technique for stamping SURPHEX photopolymer (Du-Pont) using a master with melted quartz is used as a basic technique for producing holograms (Sing H. Lee, 1993). In recent years, embossing technique has rapidly developed as it is used in semiconductors, metals, and micro electronic circuits. Chou group demonstrated the possibility of forming 25 nm-class patterns on silicon with embossing technique, and reported on its potential. This potential indicates that patterning techniques can develop through new materials and technical approaches. In particular, merging self-assembly technique with various soft lithography techniques such as elastic stamp, mold, mask, etc. will enable more innovative developments than any others (S.Y. Chou et al., 1995).

These technical fusions can complement the limits of photolithography, and provide new opportunities for micro- and nano-unit structures or integrated devices. We are extending the capability of these patterning techniques by bringing new approaches and new materials into these areas. In particular, a combination of self-assembly (especially of self-assembled monolayers) and pattern transfer using elastomeric stamps, molds, or masks constitutes the basis of soft lithographic methods. It complements photolithography in a number of aspects and provides a wide range of new opportunities for micro- and nanofabrication (X.-M. Zhao et al., 1997)

3.4 Elastomeric stamps and molds

The technique for separating after contacting of elastomeric stamp, mold, and mask with surface is a core technique in soft lithography (X.-M. Zhao et al., 1997). The use of elastomeric stamp and mold is based on the technique for forming a pattern by applying liquid prepolymer that is contrary to the characteristics of mater to the surface and removing it which is used in replica molding (Figure 4). Typical materials used for this purpose include PDMS Sylgard 184 series from Dow Corning, polyurethanes, polyimides, and cross-linked Novolac resin (a phenol formaldehyde polymer) (J.L. Wilbur et al., 1994; 1996; A. Kumar et al., 1994; Y. Xia et al., 1998).

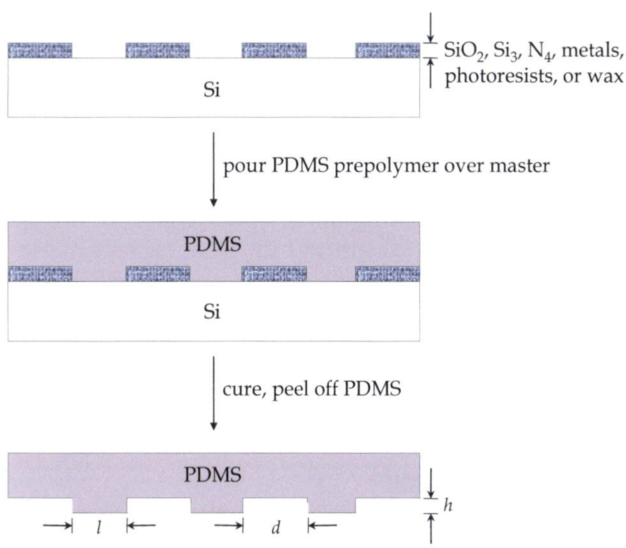

Fig. 4. Schematic illustration of the procedure for casting PDMS replicas from a master having relief structures on its surface.

The reason that soft lithography can produce high quality patterns and structures is because it has many excellent characteristics of PDMS.
Firstly, PDMS is an elastomer, and highly adhesive to substrates in relatively wide area of the surface even in micrometer unit, which allows conformal contact. Furthermore, its elastic property facilitates attachment to and detachment from even the surface of complex, brittle structures. Secondly, PDMS is free from the surface in terms of energy and chemically inactive; so polymers can be easily attached to or detached from the surface of mold-shaped PDMS. Thirdly, PDMS is homogeneous, isotropic, and optically transparent to the wavelength range of 300 nm, so it allows the UV cross-linking of prepolymers even in mold form (J.L. Wilbur et al., 1996).
Therefore, it is used in photomasks which are used in UV photolithography and contact phase-shift photolithography and in elastic optical instruments which are used in adaptive optics. Fourthly, PDMS has excellent durability and its functions do not degrade even after 100 or more repeated works for several months. Fifthly, the surface characteristics of PDMS can be easily changed through plasma treatment using a SAM method. Various surface interactions can be generated by freeing the surface energy through this treatment (Figure 5). (G.S. Ferguson et al., 1991)

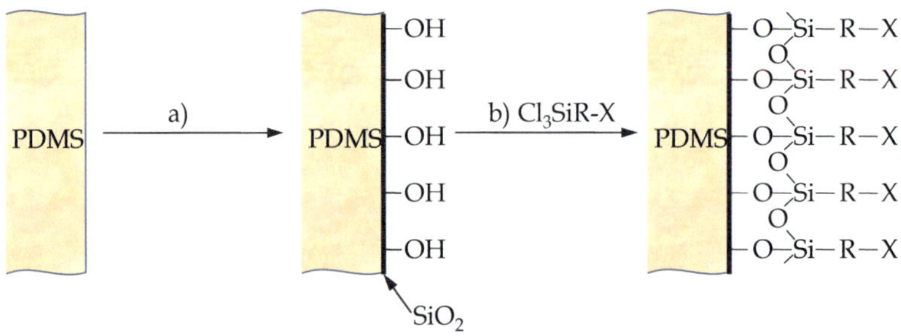

Fig. 5. Schematic procedure for the modification of the PDMS surface. (a) Treatment with an O_2 plasma, (b) reaction with silyl chloride vapor.

The PDMS has most serious technical problems that must be solved before soft lithography can be used in forming complex patterned structures (Figure 6). First, gravity, adhesion and capillary forces (T. Tanaka et al., 1993) exert stress on the elastomeric features and cause them to collapse and generate defects in the pattern that is formed (E. Delamarche et al., 1977). If the aspect ratio of the relief features is too large, the PDMS microstructures fall under their own weight or collapse owing to the forces typical of inking or printing of the stamp. Second, when the aspect ratios are too low, the relief structures are not able to withstand the compressive forces typical of printing and the adhesion between the stamp and the substrate; these interactions result in sagging. Third, achieving accurate registration without distorting the multilayer fabrication process is substantially more difficult with a flexible elastomer than with a rigid material. Therefore, these problems must be improved to technique by material, design and configuration for nano/micro structure application.

3.5 Micromolding in capillaries (MIMIC)

There was the trial that a capillary phenomenon applies to lithography of nano-scale 15 years ago. In 1995 Prof. George Whitesides in Havard university reported MIMIC process which is representative lithography method using a capillary phenomenon (E. Kim et al., 1995; D. Myers, 1991).

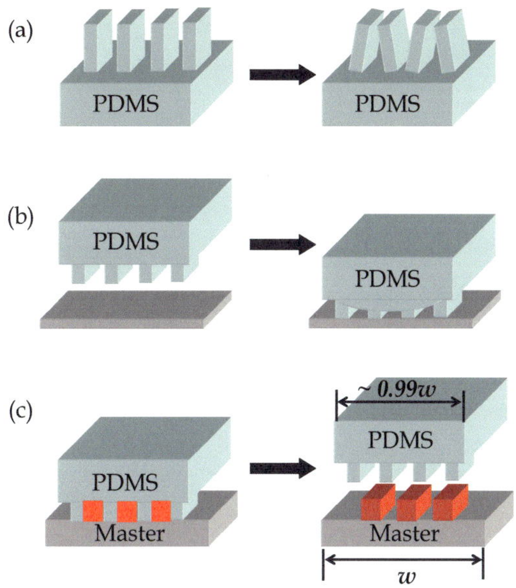

Fig. 6. Schematic diagram of possible deformations and distortions of microstructures in the surfaces of elastomers such as PDMS. (a) pairing, (b) sagging, (c) shrinking.

The capillary is natural phenomenon we can see easily when liquid like water go through a narrow tube, it goes up or down because of the Laplace pressure. The pressure is canceled out by the gravity, we can predict a rising or falling height of liquid based on Young-Laplace equation. When a glass tube is submerged under water or mercury, water causes the capillary rising because a contact angle is smaller than 90 degree and mercury causes the capillary falling because a contact angle is more than 90 degree. In these cases, Young-Laplace equation is

$$\Delta P = \frac{2\gamma}{r}\cos\theta,\, h = \frac{2\gamma}{\rho g r}\cos\theta \tag{1}$$

in which ΔP is Laplace pressure because of the curvature, γ is the surface tension, r is the radius of tube, θ is the contact angle, ρ is the density of liquid, g is the acceleration of gravity. If a tube is tetragonal not round, the curvature decreases and so a numerator in the Laplace pressure changes 2 to 1. To explain a MIMIC phenomenon physically, we use the mathematical modeling shown in eq. 2

$$\frac{dz}{dt} = \frac{R\gamma_{LV}\cos\theta}{4\eta z} = \frac{R(\gamma_{SV} - \gamma_{SL})}{4\eta z} \tag{2}$$

Here, R is value that the area of fluid flowing into hydraulic radius divides into the parameter of area, η is the viscosity of liquid, z is the path fluid flowed in. Three surface tensions are values that affect on the surface between liquid and vapor, solid and vapor, solid and liquid. It can be easily acquired that the length of channel is proportional to the square root of time, and reported it is accorded with the experimental results.

The trial that a capillary applies to the photolithography is reported as a form of MIMIC process in 1995, it is shown in figure 7.

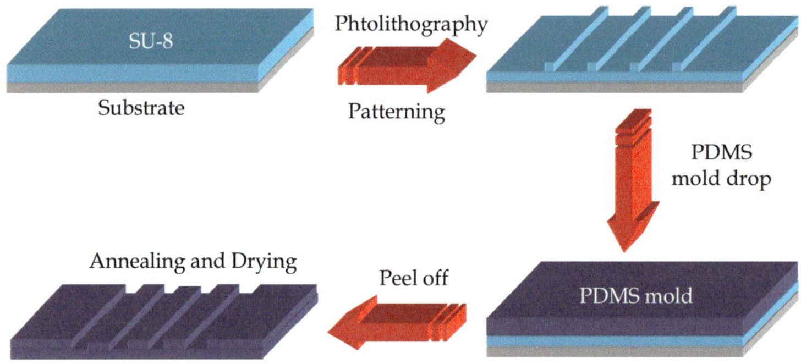

Fig. 7. Fabrication process of PDMS channel.

When we actually perform MIMIC process, patterns are made in the elastic mold PDMS using soft-lithography and contact with the Si wafer or glass surface. When this time, general contact can be achieved without external pressure because PDMS is an elastomer. And due to the small surface tension of about 21 mJ/m^2, stable surface can be achieved. From a simple line to complicated structure can be fabricated easily using the MIMIC process and applicable to polyurethane, polyacrylate, poly(methylacrylate) etc, and if materials can be hardening by heats or ultraviolet, MIMIC can be applied to almost all materials (Figure 7).

It is very encouraging the MIMIC process is in the spotlight and applies to display devices or optic device fabrication but there are some limits because of intrinsic characteristics.

First of all, because making the network-structure connected each other is necessary to fill the vacant space of a channel, so it is impossible to make the dot-type structure isolated each other. Secondly, it is difficult for bio-fluid and water to flow in the channel because PDMS is hydrophobic. Lastly, it is often seen a fluid flowing in the vacant space of the channel stops its flowing. It can be a reason that one side of channel is closed, but mostly roughness of surface and other external factor is main reason.

To overcome these problems, applying the vacuum condition to the channel is tried and it showed better characteristics. But when length of the channel is shorter than 1 μm, the resistance increases and capillary movement of flowing in the side direction shows limit. Especially the movement has an unusual sensibility to molecular weight of materials melted in the fluid, limits of the structure using MIMIC is mostly micro-level.

The next section will describe the studies on devices manufactured using various methods mentioned above and examine their characteristics.

4. Research and application

4.1 OLED device

4.1.1 Stamp method

We proposed the stamp method using soft lithography method to define the PLED's color pixel (W.J. Cho et al., 2006). This is subjected to using the merits of a spin-coating method or an Ink-jet printing method applying the roll-to-roll method (T. Zyung et al., 2005). This method requires a very simple process compared with the current spreading method and has a lot of merits can easily fabricate the uniform thickness of the respective pixel materials, surface uniformity and pattern's shape that brings on problems when we fabricate the fine patterns according to the form of the stamp (P.W.M. Blom & M.J.M. de Jong, 1998). On the other hand, the surface uniformity of the patterns is a critical point when the polymer ink is hardened. To solve it, we used the PDMS, which is elastomeric material.

As shown in Figure 8 (a), this study first forms a mold using the soft lithography process, and then uses this mold to manufacture a stamp to define polymer light emitting pixels. The master used as the mold in this study is formed by laying a highly viscous SU-8 with a negative PR on a glass 100 um thick (C. Thibault et al., 2006). Then the pixel pattern is defined through the photolithography method. After forming the master, Sylgard 184A PDMS and 184B (Dow Corning Company, USA) hardener are diluted in a ratio of 10:1, and sprayed on the upper part of the master. To remove the bubbles which promote unevenness in the lower surface, the study uses a vacuum processing method under an atmosphere of 25 mmHg while manufacturing the stamp. Then the stamp goes through a heating process for 40 minutes at 120°C. In order to make the stamp easily separate from the polymer substances, an O_2 plasma process is also applied (P. Yimsiri & M.R. Mackley, 2006; W.P. Hsu, 2005).

The device fabricated in this study has a four-layer structure of anode, HIL, EML and cathode. A 170 nm thick ITO sputtering with a sheet resistance of 15 Ω/□ is used to pattern the anode. EML, polymer ink, is stamped by using soft-lithography after making an easy hole-injection with spin-coated PEDOT (Poly(3,4-ethylenedioxythiophene)) on the ITO patterned by photolithography. Finally, the device is completed with aluminum deposition (100 nm) by using a thermal evaporator. Figure 8 (b) shows the process of manufacturing the PLED device.

In order to define the EML layer of the PLED device with a four-layer structure, a stamp patterning system was designed as shown in Figure 9. This system can 1) simplify the overall process, 2) run the process at room temperature, 3) define the pattern consecutively and 4) resolve the shortcomings of the previous methods of defining PLED. The designed system is divided mainly into a device JIG part, a stamp location coordination part and a stamping controller. In order to move the light emitting device into an accurate position, an x-z stepping motor (Sigma Koki Co., Ltd., Japan) is used in the stamp location coordination part. In addition, the manufactured stamp is installed in stage z to adjust the stamping pressure and to define the light emitting pattern. In order to control the location accurately, software (SGTERM Ver. 1.20) from Sigma Koki Co., Ltd. is used.

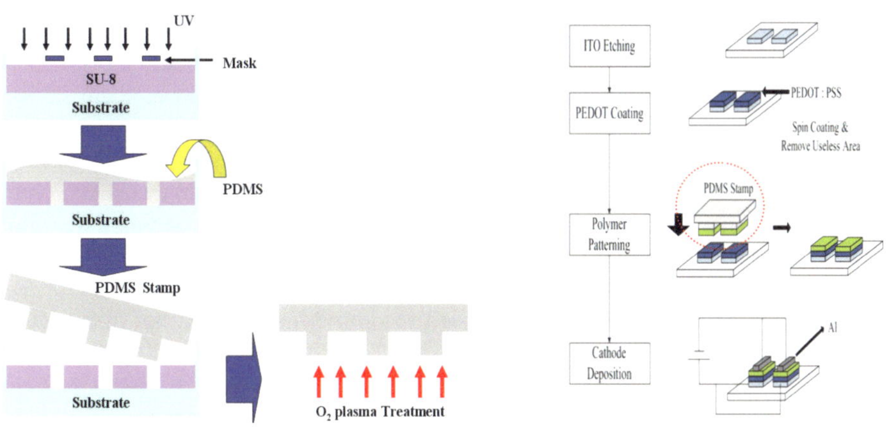

(a) (b)

Fig. 8. Schematic diagram of stamp fabrication for pixel definition.

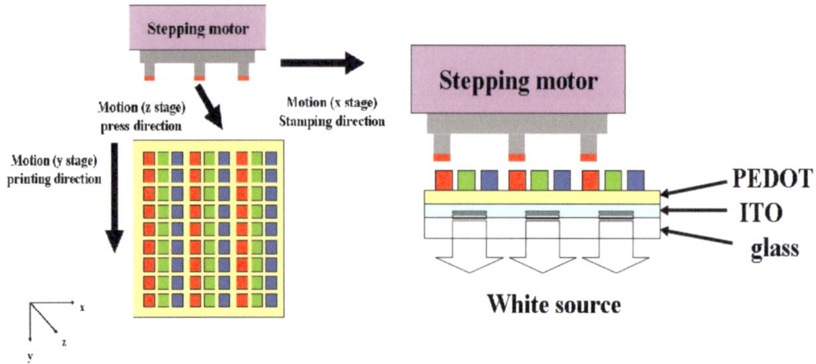

Fig. 9. Process of unit pixel definition using fabricated stamp.

4.1.2 MIMIC process

It can be a simple process to fabricate the PLED using the MIMIC process because we can form the emitting layer just drop the polymer solution to the channel (E. Kim et al., 1995). We used masks for MIMIC process in figure 10. There are cathode and organic channel, respectively.

Fig. 10. Masks of device fabrication by MIMIC process.

To make the 4 inch master structure, we formed the SU-8 2007 (Microchem Inc.) on the silicon wafer by spin-coating and defined channel pattern using photolithography. The height and width of fabricated channel of master structure are 7 μm and 600 μm, respectively. We fabricated the PDMS channel using made master structure, formed emitting layer by dropping the polymer solution to the substrate which anode has formed on the surface and finally confirm characteristics of the polymer light emitting diodes. The fabricated PLED showed about 50 cd/m^2 luminance and 0.2 cd/A efficiency characteristics.
In the experimental process, when the polymer material which has over 45,000 molecular weights is dissolved with over 0.7 wt% concentration, it is not flown in the PDMS channel.
Both molecular weight and concentration affect to viscosity of the polymer material, so it interrupted the capillary phenomenon of polymer material and eventually polymer material cannot be injected to the channel.
But we confirmed MIMIC process can be used to define line pattern but also unit pixel of PLED.

4.2 Optical transcutaneous pCO$_2$ gas sensor

In this study, the proposed system is an optical system that detects the selected wavelength in the range of mid-IR radiated from a light source without using prism or diffraction

grating by NDIR method (A. Verdin, 1973). This optical system is composed of a light source which provides a mid-IR and an optical reaction chamber to make a vibration energy level of the sample gas changed in this part, a detector which detects an intensity of light and a signal processing circuit to amplify a micro signal. Then, the optical reaction chamber is designed by a 1 mm optical path length and 64 μℓ volumes as considering that a very small amount of CO_2 gas is exhausted from a skin (D.E. Kim et al., 2005).
Figure 11(a) and 11(b) show a fabricating process of the optical reaction chamber, respectively. Si-based optical reaction chamber is fabricated by photolithography with etching and bonding process. To make 1 mm optical path length, two Si substrates are etched up to 300 μm through wet etching. Then, Si supporting layer is bond between two Si substrates. In this case, however, this process makes a surface of Si substrate roughen and results in decrease of IR transmittance rate. Therefore, LiF glass, which is used for a material of a prism is used to fabricate optical reaction chamber because it has very high optical transmittance compared to Si substrate. This LiF-based optical reaction chamber has 1 mm optical path length by soft-lithography method which results in short fabricating process time. Also, the surface of LiF glass isn't rough because this soft-lithography method doesn't need etching process. It means that LiF-based optical reaction chamber is used to increase process efficiency and detecting efficiency. Also, LiF-based optical reaction chamber is used by thermally and chemically stable acetal stamp master to define the 1 mm optical path length by soft-lithography method. These two types of the fabricated optical reaction chamber and unification style are shown in Figure 11(inset image) (D.E. Kim et al., 2005; H.Y. Bang, 2007).

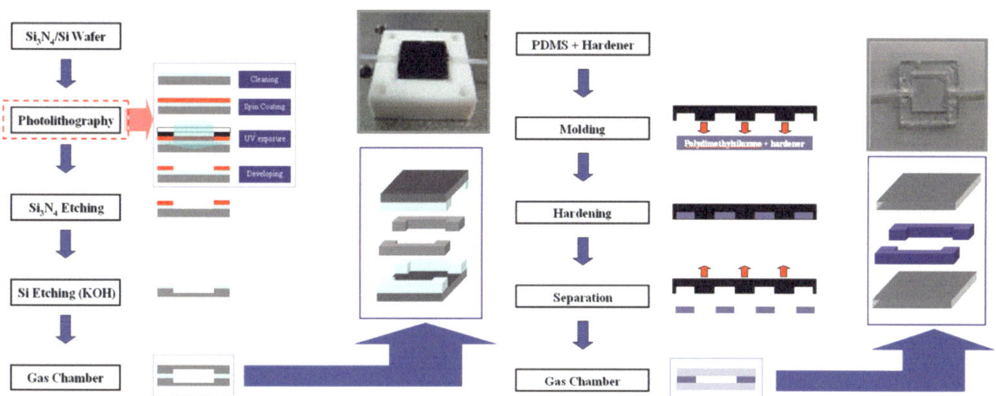

(a) (b)

Fig. 11. Fabrication of optical reaction chamber. (a) is the Si optical reaction chamber and (b) is the LiF optical reaction chamber.

The IR transmittance of the two types of the fabricated optical reaction chamber was measured by using FT-Infrared Spectrophotometer (Mattson Instruments, Inc., Galaxy 7020A) and appeared to Figure 12. This figure shows that LiF-based optical reaction chamber has approximately 70 % IR transmittance comparing with Si-based optical reaction chamber of 35 % IR transmittance in the 4.26 μm ranges. Figure 13 shows the absorbance according to the concentration of CO_2 gas. This figure exhibits the absorbance in the arterial pCO_2 concentration region, 0 ppm ~ 5,000 ppm, after blowing CO_2 gas into each optical reaction chamber by using MFC (Mass Flow Controller, P.J KODIVAC, Japan). The variation result of the absorbance is 3.78 × 10^{-6} absorbance/ppm and 6.50 × 10^{-6} absorbance/ppm in the arterial pCO_2 concentration region. As a result, LiF-based optical reaction chamber has more efficiency increased by 65 % comparing with Si-based optical reaction chamber.

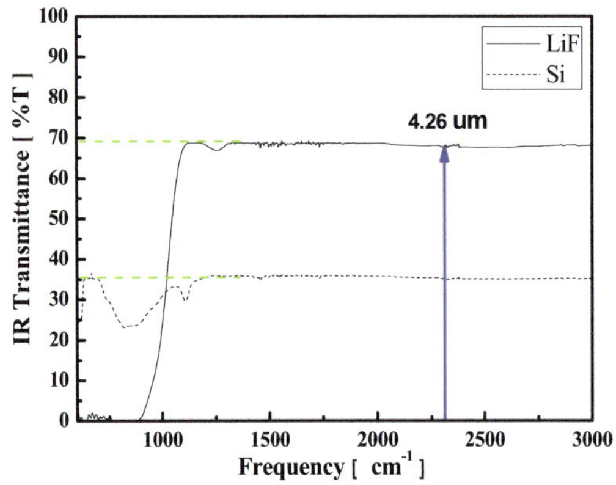

Fig. 12. IR transmittance of the fabricated optical reaction chamber.

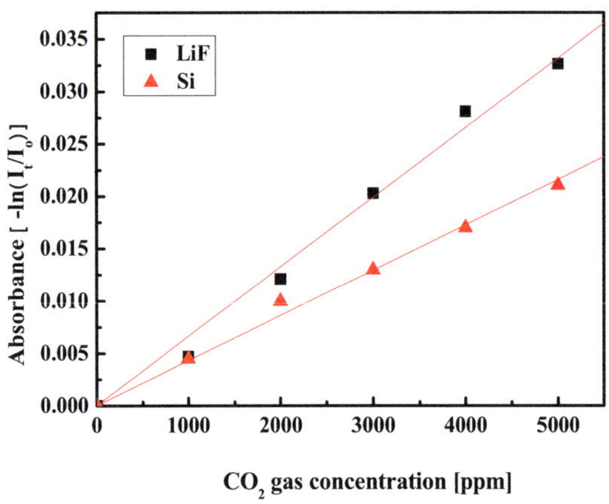

Fig. 13. Absorbance according to the concentration of CO_2 gas in each optical reaction chamber.

4.3 Optical waveguide

The polymer waveguide presents several advantages including easy fabrication process, wide bandwidth and control of refractive index, high optical coefficient, and low cost compared to inorganic waveguide such as LiNbO (C. Bulmer & W. Burns, 1983), III–V compound semiconductor (C. Glingener et al., 1995), and glass substrate (C.C. Lee & R.W. Chuang, 2004). For these reasons, the polymeric waveguide has been widely applied to optical switch (K. Chen et al., 2005), modulator, optical interconnection device (H.-D. Bauer et al., 2000), etc. Recently, optic devices using polymers such as PMMA, polyimide, SU-8 and PDMS have been reported. Especially, PDMS is an optically clear, biocompatible and easily sealed so as to produce waveguides and micro/nano fluidic channels. In this paragraph, optical waveguide fabrication with soft-lithography method is presented.

In our experiment, SU-8 photoresist (Microchem Inc.) and PDMS were used as the mold and waveguide, respectively. SU-8 photoresist, is commonly used to the MEMS application, has advantages such as optical and thermally stability, high solidity, and simple fabrication with high resolution so it suitable to apply the mold formation in soft-lithography. Procedure of fabricating SU-8 is similar to the normal process of photolithography in semi-conductor fabrication. First, the SU-8 polymer was spin-coated on the substrate - glass or Si wafer. This step can define the thickness of the mold. Specification, was stated at Microchem Inc., of the spin speed versus thickness was shown in Figure 14 (http://www.microchem.com). Second, soft baking and exposure process was performed. Then, post exposure baking was performed before the develop process. This process which is one of the important fabrications to form the SU-8 mold functions the acid-initiated, thermally driven epoxy cross-linking. Finally, SU-8 master was formed after develop process. The example of the SU-8 mold fabrication is shown in Figure 15.

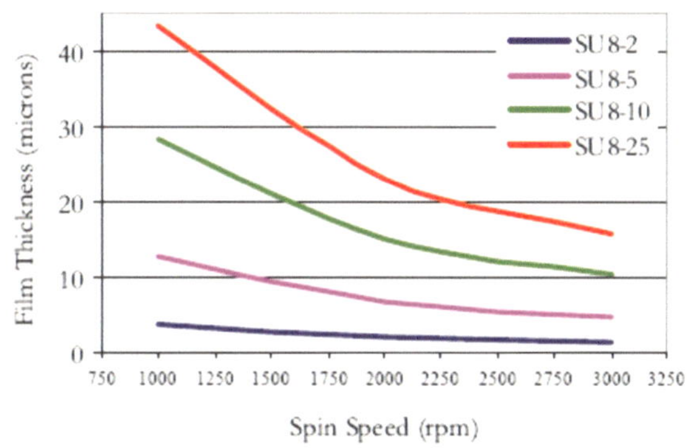

Fig. 14. Spin speed versus thickness

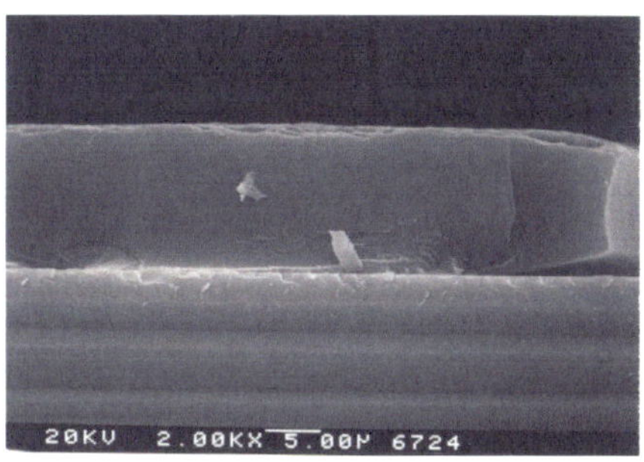

Fig. 15. SEM image of fabricated SU-8 master.

Next step is forming the PDMS waveguide. First, PDMS elastomer and curing agent were mixed in proper ratio. When mixing elastomer and hardner, refractive index of the PDMS waveguide could be changed with mixing an additive such as hexane (Jack Sheng Kee et al., 2008). Then, premixed PDMS solutions were poured on the pre-fabricated SU-8 master. After that, excess surface PDMS was scraped off with razor edge to form waveguides and PDMS was baked at 150 °C for 60 min. Then, PDMS was poured over cooled water to form optical substrate. Finally, substrate and waveguides peeled from SU-8 master. Waveguided photograph of fabricated PDMS waveguide was shown in Figure 16 (Su-Won Jang et al, 2006).

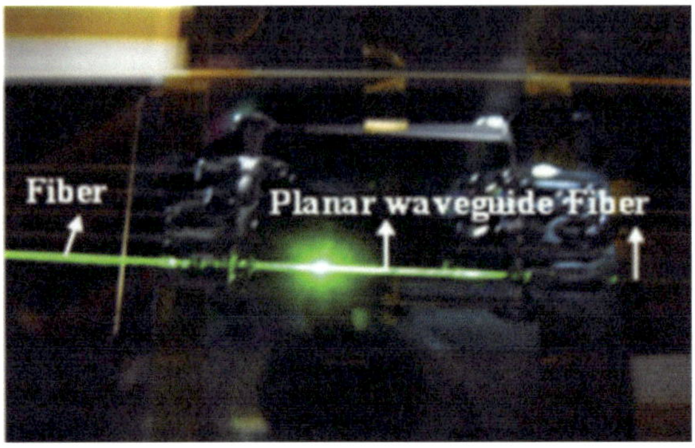

Fig. 16. Photograph of planar waveguide using PDMS.

5. Conclusion

The continued growth of the semiconductor industry is a direct result of the capability to transfer smaller and smaller circuit patterns onto semiconductor wafers or other substrate. Currently, the photolithography field for creating chips smaller than 100 nm is facing new technological challenge, and there is no guarantee that photolithography technology is the optimal technology. To create devices that are small than 100 nm is extremely limited due to light diffraction, problems in creating light masks, lens resolution, etc. So, vast majority of photolithographic equipment is optical system.

To overcome these limitations and guarantee the smaller electric devices, the soft lithography method is a new alternative plan, so a lot of researches are executed. Soft Lithography mentioned above can overcome the resolution limitation that photolithography method has, and the method is simple and it has advantages on cost saving. Also, like lens and optical fiber, it is available on the method in large area like non planar surface, so it can be applied to the not only cell biology industry but microelectronics, optics and display areas.

6. Acknowledgement

This work was supported by Korea Science and Engineering Foundation (KOSEF) grant funded by the Korea government (MEST) (No. 2009-0063405)

7. References

Younan Xia, George M. Whitesides (1998). Soft Lithography. *Angew. Chem.*, Vol. 37, (550-575), 1433-7851

R. W. Keyes (1992), The Future of Solid-State Electronics. *Phys. Today,* Vol. 45, No. 8, (42-48), 0031-9228

C. R. Barrett (1993), General accounts on the development of microelectronics. *MRS Bull.* Vol. 28, No. 7, (3-10)

R. F. Service (1996), Can Chip Devices Keep Shrinking?. *Science,* Vol. 274, (1834-1836), 0036-8075

W.M. Moreau (1988), Semiconductor Lithography: Principles and Materials, Microdevices, Plenum Press, 0306421852, New York

R. F. W. Pease (1992), Nanolithography and its prospects as a manufacturing technology, *J. Vac. Sci. Technol. B*, Vol. 10, (278-285), 1071-1023

A. Kumar, G. M. Whitesides (1993), Features of gold having micrometer to centimeter dimensions can be formed through a combination of stamping with an elastomeric stamp and an alkanethiol "ink" followed by chemical etching, *Appl. Phys. Lett.*, Vol. 63, (2002–2004), 0003-6951

Y. Xia, E. Kim, X.-M. Zhao, J. A. Rogers, M. Prentiss, G. M. Whitesides (1996), Complex Optical Surfaces Formed by Replica Molding Against Elastomeric Masters, *Science*, Vol. 273, (347-349), 0036-8075.

X.-M. Zhao, Y. Xia, G. M. Whitesides (1996), Fabrication of three-dimensional micro-structures: Microtransfer molding, *Adv. Mater.*, Vol. 8, No. 10, (837-840), 0935-9648

E. Kim, Y. Xia, G. M. Whitesides (1995), Polymer Microstructures Formed by Molding in Capillaries, *Nature*, Vol. 376, (581-584), 0028-0836

J.-M. Lehn (1990), Perspektiven der Supramolekularen Chemie - von der molekularen Erkennung zur molekularen Informationsverarbeitung und Selbstorganisation, *Angew. Chem.*, Vol. 102, (1347-1362), 1433-7851

J.-M. Lehn (1988), Supramolekulare Chemie - Moleküle, Übermoleküle und molekulare Funktionseinheiten (Nobel-Vortrag), *Angew. Chem.*, Vol. 100, (91-116), 1433-7851

C. A. Mirkin, R. L. Letsinger, R. C. Mucic, J. J. Storhuff (1996), A DAN-based method for rationally assembling nanoparticles into macroscopic materials, *Nature*, Vol. 382, (607-609), 0028-0836

A. S. Dimitov, K. Nagayama (1996), Continuous Convective Assembling of Fine Particles into 2-Dimensional Arrays on Solid-Surfaces, *Langmuir*, Vol. 12, No. 5, (1303-1311), 0743-7463

A. Terfort, N. Bowden, G. M. Whitesides (1997), Three-dimensional self-assembly of millimeter-scale components, *Nature*, Vol. 386, (162 ± 164), 0028-0836

C. D. Bain, G. M. Whitesides (1989), Modeling organic surfaces with self-assembled monolayers, *Angew. Chem.*, Vol. 101, (522-528), 1433-7851

J. Xu, H.-L. Li (1995), The Chemistry of Self-Assembled Long-Chain Alkanethiol Monolayers on Gold, *J. Colloid Interface Sci.*, Vol. 176, No. 1, (138-149), 0021-9797

P. Fenter, A. Eberhardt, P. Eisenberger (1994), Self-Assembly of n-Alkyl Thiols as Disulfides on Au(111), *Science*, Vol. 266, (1216-1218), 0036-8075

E. Delamarche, B. Michel, H. A. Biebuyck, C. Gerber (1996), Golden interfaces: The surface of self-assembled monolayers, *Adv. Mater.*, Vol. 8, (719-729), 0935-9648

G. E. Polner (1997), Characterization of Organosulfur Molecular Monolayers on Au(111) using Scanning Tunneling Microscopy, *Chem. Rev.*, Vol. 97, (1117-1127), 0009-2665

C. A. Alves, E. L. Smith, M. D. Porter (1992), Atomic scale imaging of alkanethiolate monolayers at gold surfaces with atomic force microscopy, *J. Am. Chem. Soc.*, Vol. 114, (1222-1227), 0002-7863

M. R. Anderson, M. N. Evaniak, M. Zhang (1996), Influence of Solvent on the Interfacial Structure of Self-Assembled Alkanethiol Monolayer, *Langmuir*, Vol. 12, (2327-2331), 0743-7463

N. Camillone III, T. Y. B. Leung, P. Schwartz, P. Eisenberger, G. Scoles, Chain Length Dependence of the Striped Phases of Alkanethiol Monolayers Self-Assembled on Au(111): An Atomic Beam Diffraction Study, *Langmuir*, Vol. 12, (2737-2746), 0743-7463

W. B. Caldwell, D. J. Campbell, K. Chen, B. R. Herr, C. A. Mirkin, A. Malik, M. K. Durbin, P. Dutta, K. G. Huang (1995), A Highly Ordered Self-Assembled Monolayer Film of an Azobenzenealkanethiol on Au(111): Electrochemical Properties and Structural Characterization by Synchrotron in-Plane X-ray Diffraction, Atomic Force Microscopy, and Surface-Enhanced Raman Spectroscopy, *J. Am. Chem. Soc.*, Vol. 117, (6071-6082), 0002-7863

L. Strong, G. M. Whitesides (1988),Structures of self-assembled monolayer films of organosulfur compounds adsorbed on gold single crystals: electron diffraction studies, *Langmuir*, Vol. 4, (546-558), 0743-7463

M. A. Bryant, J. E. Pemberton (1991), Surface Raman scattering of self-assembled monolayers formed from 1-alkanethiols at silver [electrodes], *J. Am. Chem. Soc.*, Vol. 113, (3629-3637), 0002-7863

Q. Du, E. Freysz, Y. R. Shen (1994), Surface Vibrational Spectroscopic Studies of Hydrogen Bonding and Hydrophobicity, *Science*, Vol. 264, (826-828), 0036-8075

J. P. Folkers, P. E. Laibinis, G. M. Whitesides (1992), Self-assembled monolayers of alkanethiols on gold : comparisons of monolayers containing mixtures of short-and long-chain constituents with methyl and hydroxymethyl terminal groups, *Langmuir*, Vol. 8, (1330-1341), 0743-7463

L. H. Dubois, B. R. Zegarski, R. G. Nuzzo (1990), Fundamental studies of microscopic wetting on organic surface. 2. Interaction of secondary adsorbates with chemically textured organic monolayers, *J. Am. Chem. Soc.*, Vol. 112, (570-579), 0002-7863

Y. Li, J. Huang, R. T. McIver, Jr., J. C. Hemminger (1992), Characterization of thiol self-assembled films by laser desorption Fourier transform mass spectrometry, *J. Am. Chem. Soc.*, Vol. 114, (2428-2432), 0002-7863

C. D. Bain, G. M. Whitesides (1988), Correlation between wettability and structure in monolayers of alkanethiols adsorbed on gold, *J. Am. Chem. Soc.*, Vol. 110, (3665-3666), 0002-7863

C. D. Bain, E. B. Throughton, Y.-T. Tao, J. Evall, G. M.Whitesides, R. G. Nuzzo (1989), Formation of monolayer films by the spontaneous assembly of organic thiols from solution onto gold, *J. Am. Chem. Soc.*, Vol. 111, (321-335), 0002-7863

T. W. Schneider, D. A. Buttry (1993), Electrochemical quartz crystal microbalance studies of adsorption and desorption of self-assembled monolayers of alkyl thiols on gold, *J. Am. Chem. Soc.*, Vol. 115, (12391-12397), 0002-7863

M. D. Ward, D. A. Buttry (1990), In Situ Interfacial Mass Detection with Piezoelectric Transducers, *Science*, Vol. 249, (1000-1007), 0036-8075

S. Li, R. M. Crooks (1993), Indirect visualization of defect structures contained within self-assembled organomercaptan monolayers: combined use of electrochemistry and scanning tunneling microscopy, *Langmuir*, Vol. 9, (1951-1954), 0743-7463

X.-M. Zhao, J. L. Wilbur, G. M. Whitesides (1996), Using Two-Stage Chemical Amplification To Determine the Density of Defects in Self-Assembled Monolayers of Alkanethiolates on Gold, *Langmuir*, Vol. 12, (3257-3264), 0743-7463

J. L. Wilbur, A. Kumar, E. Kim, G. M. Whitesides (1994), Microfabrication by microcontact printing of self-assembled monolayers, *Adv. Mater.*, Vol. 6, (600-604), 0935-9648

Y. Xia, E. Kim, G. M. Whitesides (1996), Microcontact Printing of Alkanethiols on Silver and Its Application in Microfabrication, *J. Electrochem. Soc.* Vol. 143, (1070-1079), 0013-4651

T. P. Moffat, H. Yang (1995), Patterned Metal Electrodeposition Using an Alkanethiolate Mask, *J. Electrochem. Soc.*, Vol. 142, (L220-L222) , 0013-4651

Y. Xia, M. Mrksich, E. Kim, G. M. Whitesides (1995), Micorcontact Printing of Octadecylsiloxane on the Surface of Silicon Dioxide and Its Application in Microfabrication, *J. Am. Chem. Soc.*, Vol. 117, (9576-9577), 0002-7863

P. M. St. John, H. G. Craighead (1996), Microcontact printing and pattern transfer using trichlorosilanes on oxide substrates, *Appl. Phys. Lett.*, Vol. 68, (1022-1024), 0003-6951

J. Huang, J. C. Hemminger (1993), Photooxiation of thiols in self-assembled monolayers on gold, *J. Am. Chem. Soc.*, Vol. 115, (3342-3343), 0002-7863

J. Huang, D. A. Dahlgren, J. C. Hemminger (1994), Photopatterning of Self-Assembled Alkanethiolate Monolayers on Gold: A Simple Monolayer Photoresist Utilizing Aqueous Chemistry, *Langmuir*, Vol. 10, (626-628), 0743-7463

K. C. Chan, T. Kim, J. K. Schoer, R. M. Crooks (1995), Polymeric Self-Assembled Monolayers. 3. Pattern Transfer by Use of Photolithography, Electrochemical Methods, and an Ultrathin, Self-Assembled Diacetylenic Resist, *J. Am. Chem. Soc.*, Vol. 117, (5875-5876), 0002-7863

E.W. Wollman, C. D. Frisbie, M. S. Wrighton (1993), Scanning electron microscopy for imaging photopatterned self-assembled monolayers on gold, *Langmuir,* Vol. 9, (1517-1520), 0743-7463

A. C. Pease, D. Solas, E. J. Sullivan, M. T. Cronin, C. P. Holmes, S. P. A. Fodor (1994), Light-generated oligonucleotide arrays for rapid DNA sequence analysis, Proc. Natl. Acad. Sci. USA, Vol. 91, pp. 5022-5026, 0027-8424

W. J. Dressick, J. M. Calvert (1993), Patterning of Self-Assembled Films Using Lithographic Exposure Tools, *Jpn. J. Appl. Phys.*, Vol. 32, (5829-5839), 0021-4922

J. A. M. Sondag-Huethorst, H. R. J. van Helleputte, L. G. Fokkink (1994), Generation of electrochemically deposited metal patterns by means of electron beam (nano)lithography of self-assembled monolayer resists, *Appl. Phys. Lett.*, Vol. 64, (285-287), 0003-6951

M. Lercel, R. C. Tiberio, P. F. Chapman, H. G. Craighead, C. W. Sheen, A. N. Parikh, D. L. Allara, J. Vac. Sci. Technol. B 1993, 11, 2823 ± 2828.

M. J. Lercel, H. G. Craighead, A. N. Parikh, K. Seshadri, A. L. Allara (1996), Sub-10 nm lithography with self-assembled monolayers, *Appl. Phys. Lett.*, Vol. 68, (1504 ± 1506), 0003-6951

G. Gillen, S. Wight, J. Bennett, M. J. Tarlov (1994), Patterning of self-assembled alkanethiol monolayers on silver by microfocus ion and electron beam bombardment, *Appl. Phys. Lett.*, Vol. 65, (534-536), 0003-6951

K. K. Berggren, A. Bard, J. L. Wilbur, J. D. Gillaspy, A. G. Heig, J. J. McClelland, S. L. Rolston, W. D. Phillips, M. Prentiss, G. M. Whitesides (1995), Microlithography by using neutral metastable atoms and self-assembled monolayers, *Science,* Vol. 269, (1255-1257), 0036-8075

K. S. Johnson, K. K. Berggren, A. J. Black, A. P. Chu, N. H. Dekker, D. C. Ralph, J. H. Thywissen, R. Youkin, M. Prentiss, M. Tinkham, G. M. Whitesides (1996), Using neutral metastable argon atoms and contamination lithography to form nanostructures in silicon, silicon dioxide, and gold, *Appl. Phys. Lett.*, Vol. 69, (2773-2775), 0003-6951

C. B. Ross, L. Sun, R. M. Crooks (1993), Scanning probe lithography. 1. Scanning tunneling microscope induced lithography of self-assembled n-alkanethiol monolayer resists, *Langmuir*, Vol. 9, (632-636), 0743-7463

N. L. Abbott, J. P. Folkers, G. M. Whitesides (1992), Manipulation of the Wettability of Surfaces on the 0.1- to 1 -Micrometer Scale Through Micromachining and Molecular Self-Assembly, *Science*, Vol. 257, (1380-1382), 0036-8075

A. Kumar, H. Biebuyck, N. L. Abbott, G. M. Whitesides (1992), The use of self-assembled monolayers and a selective etch to generate patterned gold features, *J. Am. Chem. Soc.*, Vol. 114, (9188-9189), 0002-7863

A. Voet (1952), *Ink and Paper in the Printing Process*, John Wiley & Sons - Interscience, 0-471-75721-7, New York

P. O. Hidber, W. Helbig, E. Kim, G. M. Whitesides (1996), Microcontact Printing of Palladium Colloids: Micron-Scale Patterning by Electroless Deposition of Copper, *Langmuir*, Vol. 12, (1375-1380)

B. L. Ramos, S. J. Choquette (1996), Embossable grating couplers for planar waveguide optical sensors, *Anal. Chem.*, Vol. 68, (1245-1249), 0003-2700

M. Nakano, N. Nishida (1979), Holographic laser scanners using generalized zone plates, *Appl. Opt.*, Vol. 18, (3073-3074), 0003-6935

H. C. Haverkorn van Rijsewijk, P. E. J. Legierse, G. E. Thomas (1982), Manufacture of Laservision Discs by a Photopolymerization Process, *Philips Tech. Rev.*, Vol. 40, (287-297), 0031-7926

D. A. Kiewit (1973), Microtool fabrication by etch pit replication, *Rev. Sci. Instrum.* Vol. 44, (1741-1742), 0034-6748

Sing H. Lee (1993), *Diffractive and Miniaturized Optics*, SPIE-International Society for Optical Engineering, 0819412910, Bellingham

S. Y. Chou, P. R. Krauss, P. J. Renstrom (1995), Imprint of sub-25 nm vias and trenches in polymers, *Appl. Phys. Lett.*, Vol. 67, (3114-3116), 0003-6951

X.-M. Zhao, Y. Xia, G. M. Whitesides (1997), Soft lithographic methods for nano-fabrication, *J.Mater. Chem.*, Vol. 7, (1069-1074), 0959-9428

J. L. Wilbur, A.Kumar, E. Kim, G. M. Whitesides (1994), Microfabrication by microcontact printing of self-assembled monolayers, *Adv. Mater.*, Vol. 6, (600-604), 0935-9648

J. L. Wilbur, A. Kumar, H. A. Biebuyck, E. Kim, G. M. Whitesides (1996), Microcontact printing of self-assembled monolayers: application in microfabrication, *Nanotechnology*, Vol. 7, (452-457), 0957-4484

A. Kumar, H. Biebuyck, G. M. Whitesides, Patterning Self-Assembled Monolayers: Applications in Materials Science (1994), *Langmuir*, Vol. 10, (1498-1511), 0743-7463

Y. Xia, N. Venkateswaran, D. Qin, J. Tien, G. M. Whitesides, Use of Electroless Silver as the Substrate in Microcontact Printing of Alkanethiols and Its Application in Microfabrication, *Langmuir*, Vol. 14, (363-371), 0743-7463

J. L. Wilbur, R. J. Jackman, G. M. Whitesides, E. L. Cheung, L. K. Lee, M. G. Prentiss (1996), *Chem. Mater.*, Vol. 8, (1380-1385), 0897-4756

G. S. Ferguson, M. K. Chaudhury, G. B. Sigal, G. M. Whitesides (1991), Contact Adhesion of Thin Gold Films on Elastomeric Suppots: Cold Welding Under Ambient Conditions, *Science*, Vol. 253, (776-778), 0036-8075

T. Tanaka, M. Morigami, N. Atoda (1993), Mechanism of Resist Pattern Collapse during Development Process, *Jpn. J. Appl. Phys.*, Vol. 32, (6059-6061), 0021-4922

E. Delamarche, H. Schmid, H. A. Biebuyck, B. Michel (1997), Stability of molded polydimethylsiloxane microstructure, *Adv. Mater.*, Vol. 9, (741-746), 0935-9648

D. Myers (1991), *Surfaces, Interfaces, and Colloids*, Wiley-VCH, 0-471-75721-7, New York

W.J. Cho, S.W. Kang et al (2006), Development of Patterning Technique Using a Stamp Method and Evaluation of Characteristics for Polymer OLED, Proc. of ASID 2006, pp. 224-225

Taehyoung Zyung, Seong Hyun Kim, Sang Chul Lim, Jung Hun Lee, Hye Yong Chu, Jeong-Ik Lee, Ji-Young Oh (2005), Novel Method for Combining Flexible Organic Light-Emitting Diodes with Organic Thin-Film Transistors, *J. Korean Phys. Soc.*, Vol. 48, (S111-S114), 0374-4884

Paul W. M. Blom, Marc J. M. de Jong (1998), Electrical Characterization of Polymer Light-Emitting Diodes, *IEEE Journal of selected topics in quantum electronics,* Vol. 4, No.1, (105-112), 1077-260X

C. Thibault, C. Severac, E. Trevisiol, C. Vieu (2006), Microtransfer molding of hydrophobic dendrimer, *Microelectron. Eng.*, Vol. 83, (1513-1516), 0617-9317

P. Yimsiri, M. R. Mackley (2006), Spin and dip coating of light-emitting polymer solutions: Matching experiment with modeling, *Chem. Eng. Sci.*, Vol. 61, (3496-3505), 0009-2509

W. P. Hsu (2005), Soft lithography contacts to organics, *Mater.today*, Vol. 8, (42-54), 1369-7021

A. Verdin (1973), *Gas Analysis Instrumentation*, John Wiley & Sons - A Halsted Press Book, 0470906154, New York

Do Eok Kim, Shin Won Kang et al (2005), Noninvasive Optical Transcutaneous pCO_2 Gas Sensor, *Sensors and Materials*, Vol. 17, No. 5, pp. 249-257, 0914-4935.

Hyang-Yi Bang, Shin-Won Kang et al (2007), Characteristics of Transcutaneous pCO_2 Gas Sensor Based on LiF Glass using Soft Lithography, *Sensors and Materials*, Vol. 19, No. 8, (465-476), 0914-4935.

C. Bulmer and W. Burns (1983), Polarization characteristics of LiNbO channel waveguide directional couplers. *J. Lightw. Technol.* , Vol. 1, No. 1, (227–236), 0733-8724.

C. Glingener, D. Schulz, and E. Voges (1995), Modeling of optical waveguide modulators on III-V semiconductors. *IEEE J. Quantum Electron.*, Vol. 31, No. 1, (101–112), 0018-9197

C. C. Lee and R. W. Chuang (2004), A dry electromigration process for fabricating deep optical channel waveguides on glass their characterization. *Mater. Sci. Eng. B*, Vol. 111, (40–48), 0921-5107.

K. Chen, P. L. Chu, and H. P. Chan (2005), A vertically coupled polymer optical waveguide switch. *Opt. Commun.*, Vol. 244, (153–158), 0030-4018.

H.-D. Bauer, W. Ehrfeld, and M. Harder (2000), Polymer waveguide devices with passive pigtailing: And application of LIGA technology, *Synthetic Metals*, Vol. 115, (13–20), 0379-6779.

http://www.microchem.com, Microchem Inc.

Jack Sheng Kee, Daniel Puiu Poenar, Pavel Neuzil and Levent Yobas (2008), Design and fabrication of Poly(dimethylsiloxane) single-mode rib waveguide, *Optics Express*, Vol. 17, No. 14, (11739-11746), 1094-4087.

Su-Won Jang, Shin-Won Kang et al. (2006), UV-Sensitive Photofunctional Device Using Evanescent Field Absorption Between SU-8 Polymer Optical Waveguide and Photochromic Dye, *IEEE Photonic. Tech. Lett.*, Vol. 18, No. 1, (82-84), 1041-1135

21

Fabrication of SiC-based Ceramic Microstructures from Preceramic Polymers with Sacrificial Templates and Softlithography Techniques

Tae-Ho Yoon[1], Lan-Young Hong[1] and Dong-Pyo Kim[1,2]
[1]*Center of Applied Microfluidic Chemistry, Chungnam National University*
[2]*Graduate School of Analytical Science and Technology, Chungnam National University*
Korea

1. Introduction

Silicon derived polymers containing nitrogen, carbon and boron have been considered as precursors for various non-oxide ceramics such as SiC, SiCN and SiBCN (Madou, 2002, Nguyen & Wereley, 2002, Liew et al., 2003). These ceramics can be easily shaped using various forming processes and then crosslinked by exposure to heat or UV radiation to form an infusible solid. The consolidated preceramic polymers are finally pyrolyzed at high temperatures to transform into the dense ceramic phases. These materials can be used for high temperature applications in areas such as structural composites (Kim et al., 1996), electronic devices (Xia & Whitesides, 1998) and catalytic chemical reactions (Xia et al., 1999). Table 1 shows some selected important preceramic polymers that have been studied in various aspects. In particular, silicon carbide (SiC) is a typical non-oxide ceramic that has attracted the most interest on account of its unique physical and chemical properties such as high thermal conductivity, excellent thermal stability, superior stability towards oxidation compared with carbon, high mechanical strength and chemical inertness. Commercially available polysilazane (VL-20, KiON Corp. USA) and two types of polycarbosilanes, Polymethylsilane (PMS) and Allyhydridopolycarbosilane (Starfire System, USA) are readily used as preceramic polymers for SiCN and SiC ceramics, respectively.

Preceramic polymers	Polymeric unit	Ceramic	Ceramic yield (%)
Polycarbosilane	$(\text{-}R_2SiCH_2\text{-})_n$	SiC	65
Polysiloxane	$(\text{-}R_2SiO\text{-})_n$	Si-O-C	30-60
Polysilazane	$(\text{-}R_2SiNR\text{-})_n$	Si_3N_4	20-90
Aluminum amide	$(=AlNR\text{-})_n$	AlN	20-50
Polyborazine	$(\text{-}B_3N_3H_x\text{-})_n$	BN	85
Polytitanium imide	$(=Ti(NR)_2\text{-})_n$	TiN	50-70

Table 1. A list of typical preceramic polymers

A variety of synthetic approaches have been proposed for the development of porous materials with a high surface area and a controlled pore size distribution due to their many potential applications. There are many reports on the variety of porous carbon, oxides, sulfides and metals prepared from various hard and soft templates. Moreover, porous ceramics with a different porous morphology and size distribution have been fabricated via different routes, such as burning out a polymeric sponge impregnated with a solid-state sintering (Kwon et al., 1994), replica of a polymer foam by impregnation (Peng et al., 2000), ceramic slurry (Zhu et al., 2002), sol-gel process (Geis et al., 2002) and gel casting methods (Zhang et al., 2006). Because of the low oxidation resistance of carbon and the poor hydrothermal stability of porous silica materials, SiC with a high surface area has attracted considerable attention as a support material in the catalysis. Therefore, several attempts have been made to prepare various SiC macroporous and mesoporous materials.

On the other hand, the fabrication of 3D microstructures has been recently developed for use in photonic crystals, biochips, micro/nanofluidic devices and nano/micro-electromechanical systems (N/MEMS) (Yamazaki & Namatsu, 2004, Lee & Seung, 2004). In addition, it is expected that there will be considerable demand for ceramic devices that can be used in harsh environments in the fields of aerospace, military and energy industry. In terms of practical fabrication techniques, the mechanical method of machining has mainly been used for various materials and has played an important role in the fabrication of ceramic microstructures. However, this method has shortcomings when it comes to fabricating controlled 3D ceramic microstructures. As an alternative to the machining process, lithographic techniques have been investigated for producing 3D ceramic microstructures with a nanoscale resolution using preceramic polymers or ceramic-powder mixed polymers.

This chapter reviews the recent development of porous SiC materials from templated preceramic polymers and the fabrication of small and complicated SiC ceramic features using near-net shape processing techniques such as soft lithography. There is a need to summarize these types of SiC structural materials on the nanoscale in order to extend their utility into nanotechnology devices. Besides, it is obvious that one of the challenging strategies in ceramic applications is the integration of preceramic polymers into existing manufacturing processes to achieve nano-level process control and the ability to produce useful architectures. In this context, it is meaningful to introduce several preliminary results from our own laboratory in this area.

2. Porous ceramic structure form sacrificial template

This section defines the scope for the preparation of various SiC porous materials using different types of sacrificial hard templates. The main concern is on macroporous SiC with pores larger than 50 nm, mesoporous SiC with pores ranging from 2 to 50 nm, and SiC nanotubes. These porous ceramics have a wide variety of applications including filters, membranes, sensors, catalyst supports, as well as biomedical and construction materials (Sepulveda, 1997).

2.1 Macroporous SiC-based ceramics

There have been many studies on macroporous structures using oxide and carbon materials, but there are only few on SiC for making macroporous structures. The macroporous structure has the advantage of a lower pressure drop than that of a mesoporous structure when used as a catalyst support. Table 2 shows the various precursors and templates used to prepare macroporous SiC and SiCN ceramic materials. According to the Quin et al., a SiC based macropore structure 'wood ceramic' was prepared from carbonized wood powder and phenol resin via a direct reaction with Si powder (Quin, 2003). But, the wood ceramic product showed disordered porosity with broad range of pore size distribution. Accordingly, the sacrificial template method has been used in the manufacture of highly ordered macroporous materials with a narrow pore size distribution. Firstly, homogeneous colloidal silica spheres ranging in size from 137 to 700 nm, as shown in figure 1(a) and (b), were gently precipitated to form a closed packed crystal template (Sung et al., 2002). A low molecular weight polymer precursor, polymethylsilane (PMS), was then infiltrated into the sacrificial colloidal silica crystalline arrays, which were subsequently etched with HF after pyrolysis in an argon atmosphere (Wang et al., 2004). Pore sizes of approximately 84~658 nm and a BET surface of approximately 585 m^2g^{-1} ~ 300 m^2g^{-1} of the obtained porous products in proportion to the sizes of the sacrificial templates were obtained. It is believed that the high surface area was due to the interfacial area between the sphere and the infiltrated polymer as well as to the formation of micropores at the ceramic wall during pyrolysis. In addition, 3-dimensionally ordered macroporous (3DOM) SiC ceramics were prepared using polysilazane and silica spheres ranging in size from 112 to 650 nm. This was followed by a thermal curing step, pyrolysis at 1250 °C in a N_2 atmosphere, and an identical etching process (Wang et al., 2005). Table 3 summarizes the comparative pore characteristics using silica sphere templates with various sizes (112 ~ 500 nm) and different types of preceramic polymers.

On the other hand, porous carbon was used as an alternative sacrificial template to prepare a different type of macroporous SiC ceramic with a unique morphology. A 3DOM carbon template was prepared by infiltrating sugar or phenolic resin into a closed packed silica

Precursor	Template	Final product	Pore sizes and types	BET surface area (m^2g^{-1})	Ref.
phenol resin, Si	wood powder	SiC	10-30μm Irregular channel	--	Quin et al., 2003
PMS PCS	monolayered silica sphere (500nm)	SiC	340 nm ordered macropore	150~172	Sung et al., 2002
PMS, polysilazane	macroporous carbon (150-1000 nm)	SiC SiCN	135-896 nm hollow spheres	50.8-5.0	Wang et al., 2004
PMS, PCS PMS&PCS hybrid	alumina membrane (100-400 nm)	SiC	200 nm hollowed tube	339	Wang et al., 2005
polysilazane	silica spheres (112-650 nm)	SiCN	98-578 nm ordered macropore	455.6-250.3	Wang et al., 2005

Table 2. Summarized characteristics of the macropores originating from different precursors and templates.

PMS: polymethylsilane, PCS: polycarbosilane

Fig. 1. The SEM images of a representative silica template and porous SiC with different diameters; (a) 137 nm silica template (b) 300 nm silica template, (c) porous SiC from 192 nm template and (d) Porous SiC from 700 nm silica template (Sung et al., 2002).

Precursor	SiO_2 sphere (nm)	BET Surface area ($m^2 g^{-1}$)	Average pore size (nm)	Pore volume ($cm^3 g^{-1}$)	Ref.
SiC	112	584.6	4.9	0.68	Wang et al., 2004
	300	387.1	4.9	0.35	
	500	362.3	2.8	0.26	
SiCN	112	455.6	4.5	0.31	Wang et al., 2005
	243	367.7	3.8	0.23	
	500	316.4	3.1	0.18	
BCN	145	412.9	2.5	0.40	Hong et al., 2005
	300	387.1	2.4	0.35	
	500	315.3	4.9	0.27	
SiBCN	145	220.5	2.5	0.14	Hong et al., 2005
	300	211.7	3.9	0.12	
	500	203.2	5.0	0.11	
SiC-$MoSi_2$	500	232.4	3-5	0.46	Wang et al., 2005

Table 3. Summarized pore characteristics of various ordered macroporous ceramics obtained from silica sphere templates and different preceramic polymers.

sphere assembly, followed by an oxidation or curing step and subsequent carbonization at 900 °C (Wang et al., 2004). The 3DOM carbon, as a sacrificial template, was gently infiltrated by low molecular weight preceramic polymers. In order to obtain the hollow nanosphere assembly, a polymeric precursor diluted to 25 mass% in THF was used to induce polymer adsorption on the inner wall of the carbon template during solvent evaporation. The carbon-precursor composites were cured at 160 °C for 6 hr, and then pyrolyzed at 1250 °C. Finally, the carbon was oxidized at 650 °C in air to obtain an interconnected SiC sphere assembly, as shown in figure 2(a). The TEM image (figure 2(b)) clearly shows a regular ordered array of hollow spheres with dense shells (Wang et al., 2005). It should be noted that a hollow nanosphere with an empty core and inter-connections might have applications as capsules for drug delivery systems (DDS), pigment stabilizers in paints, photonic materials, chemical and biological sensors, and catalysts (Caruso etal., 2001). On the other hand, an ordered assembly of SiCN ceramic spheres with filled cores was produced when low viscous polysilazane with no dilution was inserted into a carbon template, as shown in the SEM and TEM image in figure 2(c) and (d), respectively. The filled SiCN sphere nanostructures with diameters ranging from 142 to 944 nm were proportional to the initial pore sizes of the sacrificial carbon templates used (approx. 150 ~ 1000 nm).

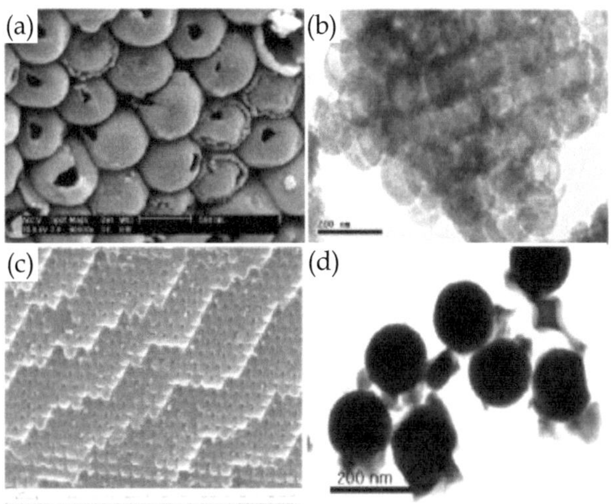

Fig. 2. Three representative SEM(a) and TEM(b) images of 500nm and 135nm hollow SiC sphere assemblies, and SEM(c) and TEM(d) images of 613nm and 142nm filled SiCN sphere assemblies, respectively (Wang et al., 2005).

2.2 Mesoporous SiC-based ceramics

Since mesoporous silicates (M41S) were first discovered in the early 1990s, many efforts have been devoted to producing various mesostructure materials including mesoporous carbon CMK-I, oxides and metal. The use of a hard sacrificial template for the replication of nanoscale structures using a direct-templating process has sparked excellent contributions in this field. According to this strategy, some disordered mesoporous SiC materials were originally prepared using gas-phase infiltration techniques. For example, Pham-Huu et al. prepared high surface area SiC by a reaction between SiO vapor and active charcoal at temperatures ranging from 1200 to 1500 °C, which is known as the shape-memory synthesis (SMS) method (Pham-Huu et al., 1999). Parmentier et al. synthesized mesoporous SiC with a surface area of 120 m^2g^{-1} via a carbothermal reduction reaction between mesoporous MCM-48 silica with pyrolytic carbon filled using chemical vapor infiltration (CVI) with propylene as the carbon precursor at temperatures ranging from 1250 to 1450 °C (Parmentier et al., 2002). Krawiec et al. produced disordered mesoporous SiC with a high surface area (508 m^2g^{-1}) using a CVI process involving the introduction of a gaseous SiC precursor, dimethyldichlorosilane (DDS), into nanoporous SBA-15 silica as summarized in Table 4 (Krawiec et al., 2004).

Template	Structure	SiC precursor	Space group	$S_{BET}(m^2g^{-1})$
Active Charcoal	Disordered	SiO vapor, Carbon	-	~100
MCM-48	Disordered	C, SiO_2	-	120
SBA-15	Disordered	DDS vapor	-	508
Silica nanosphere	Disordered	AHPCS	-	612
SBA-15	Ordered	AHPCS	*P6mm*	260
MCF	Ordered	AHPCS	Unknown	250
SBA-15	Ordered	PCS	*P6mm*	720
KIT 6	Ordered	PCS	*Ia3d*	590

Table 4. Comparative summary of the reported mesoporous SiC products and corresponding templates.

The first report on the production of mesoporous SiC using a preceramic polymer showed a simple method, similar to that used to produce macroporous SiC, involving the infiltration of low viscous allyhydridopolycarbosilane (AHPCS, SP matrix, Starfire sys., USA) into a randomly packed silica colloidal sphere template with a diameter of 20 ~ 30 nm. The disordered mesoporous SiC exhibited an amorphous foam-like SiC with a high surface area of 612 m^2g^{-1} and a total pore volume of 0.81 cm^3g^{-1} (Park et al., 2004). The above mesoporous SiC showed no long-range order of porosity because the silica nanosphere could not be precipitated into a closed-packed mode as a result of the strong electrostatic interactions between spheres. Because ordered mesoporous carbon such as CMK-3 have been formed from the use of ordered mesoporous silica templates, highly ordered mesoporous SiC materials were also prepared using trimethylsilyated SBA-15 and mesocellular siliceous foam as sacrificial hard templates. It is well known that SBA-15, which is prepared using triblock copolymers as a structure directing agent, is a two-dimensional hexagonally ordered mesoporous silica with channel-interconnecting micropores (6.5 nm) within the wall (Zhao et al., 1998). Mesocellular siliceous foam was also composed of uniform and large spherical cells (~20 nm) and connecting windows (Schmidt-Winkel et al., 1999). The diluted allylhydridopolycarbosilane was infiltrated into two types of surface modified nanoporous silica templates. The silica templates were subsequently etched off after pyrolysis at 1000 °C under a nitrogen atmosphere to leave an ordered mesoporous structure. Both synthesized mesoporous SiC materials had a high BET surface area in the range of 250~260 m^2g^{-1} with a pore size of 3.4~3.6 nm. The mesoporous SiC materials prepared from the two types of silica templates were exact inverse replicas of their templates, as shown in figure 3 (Yan et al., 2006).

A similar study was also carried out by Zhao's group as listed in Table 4. Highly ordered mesoporous SiC ceramics were synthesized via a one-step nanocasting process using commercial polycarbosilane (PCS) as a precursor and mesoporous silica materials, SBA-15 and KIT-6, as hard templates (Shi et al., 2006). The obtained mesoporous SiC ceramics with 12% excess of carbon were amorphous below 1200 °C, and were composed of randomly oriented β-SiC crystallites after being heated to 1400 °C. These ordered mesoporous SiC ceramics had very high BET specific surface areas up to 720 m^2g^{-1}, large pore volumes (ab. 0.8 cm^3g^{-1}) and a narrow pore-size distribution (2.0~3.7 nm). It is expected that these novel techniques will be suitable for synthesizing many other types of ordered mesoporous non-oxide ceramic materials with interesting pore topologies.

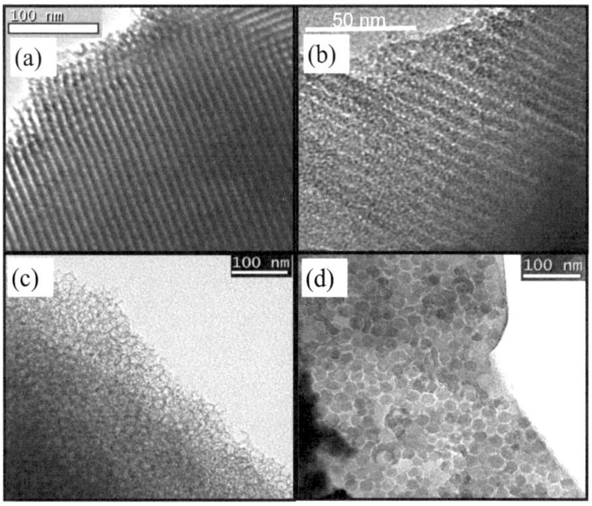

Fig. 3. TEM images of the mesoporous SiC products and corresponding nanoporous silica templates. (a) SBA-15 template, (b) SiC product from SBA-15, (c) MCF template and (d) SiC product from MCF (Zhao et al., 1998).

It should be noted that the porous SiC products prepared from silica templates had severe oxygen contamination as a result of oxygen diffusion at the interface during the pyrolysis of the infiltrated preceramic polymers. The mesoporous SiC obtained had a surface severely contaminated with SiC_xO_y impurities, which is detrimental to high temperature applications. Therefore, it is desirable to use a sacrificial template containing no oxygen, which can avoid the formation of silicon oxycarbide species in the produced mesoporous SiC. In this context, it is deserved to introduce that mesoporous boron nitride (BN) with a specific surface area of 540 m^2g-1, a mesoporous volume of 0.27 cm^3g^{-1}, and a narrow pore size distribution (4.4 nm), was obtained from tri(methylamino)borazine as a precursor using CMK-3 mesoporous carbon as a non-oxygen template (Dibandjo et al., 2005). The mesoporous carbon template route appears to be a promising method for fabricating mesoporous ceramics from polymeric precursors. BN and BCN nanostructures were alternatively prepared via a substitution reaction using carbon templates (Vinu et al., 2005).

2.3 SiC nanotube structure

Since the discovery of carbon nanotubes in 1991, there has been considerable interest in fabricating one-dimensional tubular structures for their potential applications as electric devices and sensors (Iijima, 1991). Recently, many types of organic materials (peptide, polypyrrole) and inorganic materials (nitride, sulfide, oxide, carbide) have been considered in the preparation of tubular structures (Wu et al., 2004). Different types of tubular SiC nanostructures were synthesized since Dai et al. first reported the preparation of SiC nanotubes using a shape memory synthesis method (Dai et al., 1995, Keller et al., 2003). Most preparation methods are based on a carbothermal reduction and/or chemical vapor deposition, resulting in randomly dispersed nanotube structures. An alumina (Al_2O_3) membrane with a 200 nm diameter was used as a template for making SiC arrays with a well-aligned tubular structure and a tailored diameter and wall thickness. A polymethylsilane solution was infiltrated into the dried alumina membrane at room temperature under a nitrogen atmosphere. After vacuum evaporation, the infiltrated polymer was cured and the polymer was heated to 1250 °C in an argon atmosphere. Figure 4 shows SEM and TEM images of well-aligned array of SiC tubes with a uniform wall thickness 35 nm. The SiC nanotube had an electrical resistance of 6.9 × 103 to 4.85 × 103 Ωm at temperatures ranging from 20 to 300 °K with a negative temperature dependence, which is similar to a semiconductor-like behavior. In addition, Pt/Ru alloy nanoparticles could be selectively deposited on the inner wall of the nanotube. This material might be useful in the fields of heat-resistant nanodevices, fuel cells and nanofluidic devices.

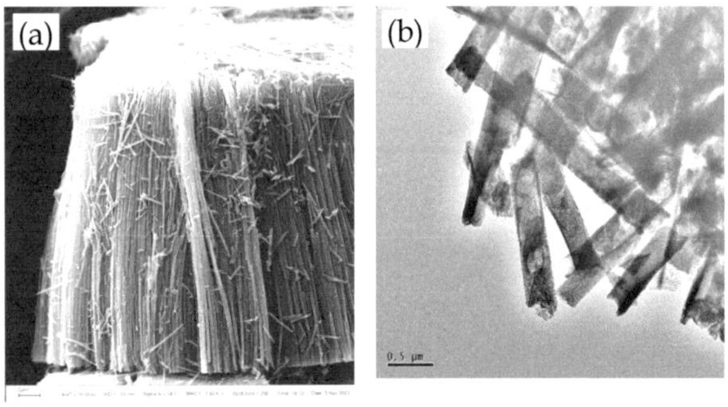

Fig. 4. The representative SEM(a) and TEM(b) image of 100nm tubular SiC derived from PMS (Wang et al., 2005).

3. Ceramic nanostructure via lithographic techniques

Whitesides suggested the use of elastomeric silicone rubber, mainly polydimethylsiloxane (PDMS), for micropattern transfer known as soft lithography (Whitesides et al., 1999). Soft lithography is a valuable tool for the low-cost microstructuring of liquid materials (Heule & Gauckler, 2001, Heule & Gauckler, 2003, Martin & Aksay, 2005). There are several variations available that are suited for the direct patterning of polymeric liquids. Microcontact printing (μ-CP) is used for stamping self-assembled monolayers serving as a resist or as functional layers. Replica molding is an efficient method for duplicating the information (i.e. shape, morphology and structure) present on the surface of a mold. In microtransfer molding (μ-TM), a thin layer of a liquid prepolymer is applied to a patterned surface of a PDMS mold and the excess liquid is removed by scraping it with a flat PDMS block or by blowing it off with a nitrogen stream. A low-viscosity fluid was patterned through the spontaneous filling of PDMS microchannels by capillary action in micromolding in the capillaries (MIMIC) (Heule et al., 2003). After curing the prepolymer into a solid, the PDMS mold was removed to reveal patterned microstructures of the polymer. In the imprinting method, a mold was pressed into a layer of a viscous prepolymer film on a substrate, which can flow under pressure to conform to the mold (Guo et al., 2004, Donsel et al., 2001, Chou et al, 1995).

It should be emphasized that synthetic routes using preceramic polymers are promising for producing small and complicated SiC ceramic features using soft lithography and a modified version of near-net shape processing techniques. Furthermore, recent developments for fabricating a porous channel structures have been introduced as a preliminary work for ceramic microreactor applications.

3.1 Non-porous ceramic patterning via softlithography

Most currently used MEMS devices in the silicon semiconductor industry are fabricated using photolithography coupled with surface machining and wet etching, which is the most common method for obtaining the micrometer sized surface features needed for sensors and actuators. Recently, the use of preceramic polymers offers a simple route for fabricating 2- or 3-dimensional ceramic microstructures using soft lithography techniques. SiC ceramic line patterns on the micron scale were fabricated using MIMIC method, which involved filling PDMS channels that had been formed by conformal contact of a low viscosity preceramic polymer to a silicon wafer, followed by curing and pyrolysis at 800 °C (Hong & Kim, 2005). Moreover, fine ceramic line patterns were also made by applying PDMS mold transfer techniques. Figure 5 shows SEM images of the dense SiC ceramic line patterns, which were exact replicas of the CD and DVD relief structure as an economic master, respectively (Dat etal., 2006). This suggests that preceramic polymers have excellent patterning processibility even on the nanoscale level by efficiently filling a narrow gap. This preliminary study highlights the feasibility of developing high temperature resistant nanoscale ceramic components including MEMS as well as NEMS (nano electromechanical system).

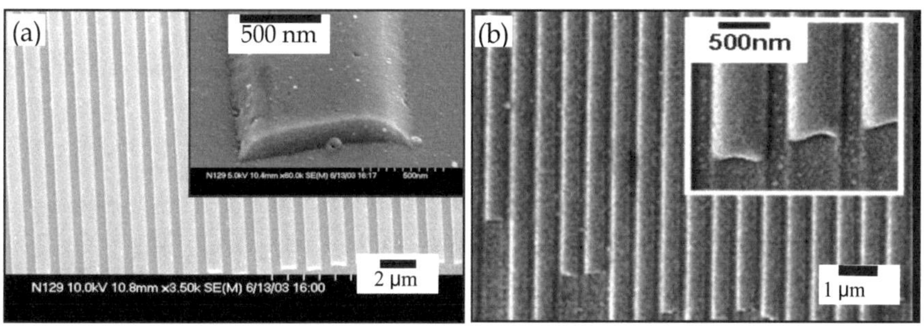

Fig. 5. SEM images of the imprinted SiC ceramic precursor pyrolyzed at 800 oC in an argon atmosphere; (a) SiC line pattern from CD master, (b) SiCN line pattern from DVD master.

R. Raj's group reported the very meaningful achievement by preparing SiCN ceramic MEMS devices using polyureamethylvinylsilazane as a precursor (Liew et al., 2001, Liew etal., 2002, Shah & Raj, 2005). Even primitive types of high-temperature MEMS, i.e. electrostatic actuators, a pressure transducers, and combustion chambers were developed mainly using preceramic polymers that forms SiCN ceramics by pyrolysis via a temperature or radiation induced transformation of a processable liquid state to an infusible solid state (cured polymer). This suggests that multi-layered ceramic MEMS can be fabricated by adding and curing successive layers of liquid polymers on top of each other using multi-level photopolymerization.

3.2 Porous SiC-based ceramic channels for microreactor

This section summarized the preparation of tailored, highly uniform SiC and SiCN porous structures by filling the void space in packed beds of silica spheres with a low viscous preceramic polymer. However, these products are a powdery type, which limits their utility

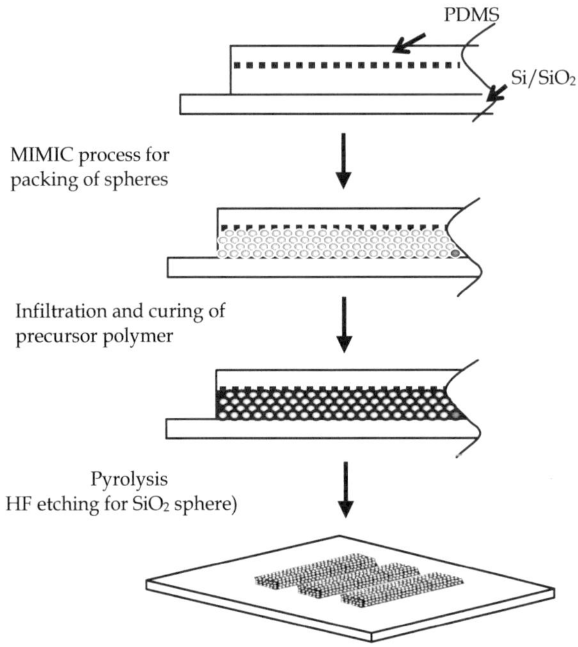

Fig. 6. Schematic diagram of the fabrication steps used to prepare microchannels.

to existing applications. Recently, we reported that the integration of templated preceramic polymers into a new fabrication technique such as soft lithography can produce useful products with new architectures. For the fabrication of tailored porous SiC and SiCN microchannels, as shown in figure 6, a PDMS mold was placed onto the flat surface of a silicon wafer, forming open channels at both ends. A solution containing colloidal silica or polystyrene spheres was allowed to flow slowly into the channels from one and end via capillary forces. The void space between the spheres was filled with the preceramic polymer through capillary action. After curing the preceramic polymer, the colloidal polystyrene spheres were decomposed during the early stages of the pyrolysis process, as shown in figure 7 (Sung et al., 2005). The inverted beaded SiC porous monoliths showed a crack-free ceramic microchannel replica with 150 ~ 200 nm of interconnecting windows for the 1 μm spheres used. The pore size could be tailored independently according to the bead size, allowing for the easy integration of porous monoliths into a microreactor. The SiC ceramic monoliths obtained were used in the decomposition of ammonia after depositing a ruthenium catalyst via wet impregnation and calcinations. The efficient conversion of NH_3

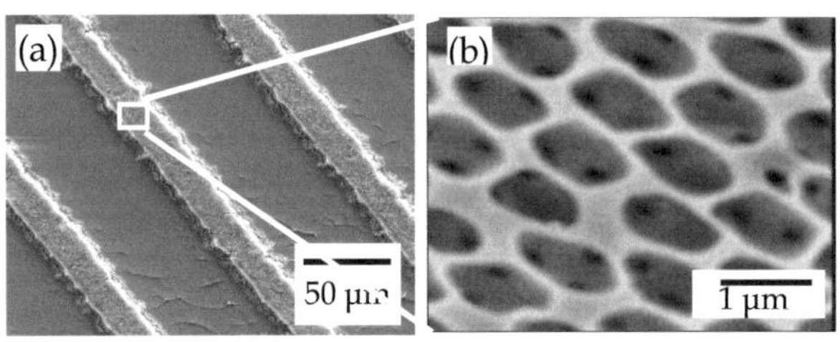

Fig. 7. SEM micrographs of (a) SiCN microchannel replica and (b) its 3-dimensionally interconnected pore structure containing 1 μm pores formed by pyrolysis (Sung et al., 2005).

to H_2 with increasing reaction temperature demonstrated its successful performance as a hydrogen reformer for fuel cells, as shown in figure 8. These novel porous materials show great promises for use in high temperature micro-reactors possibly for the on-demand reforming of higher hydrocarbons into hydrogen for portable power sources.

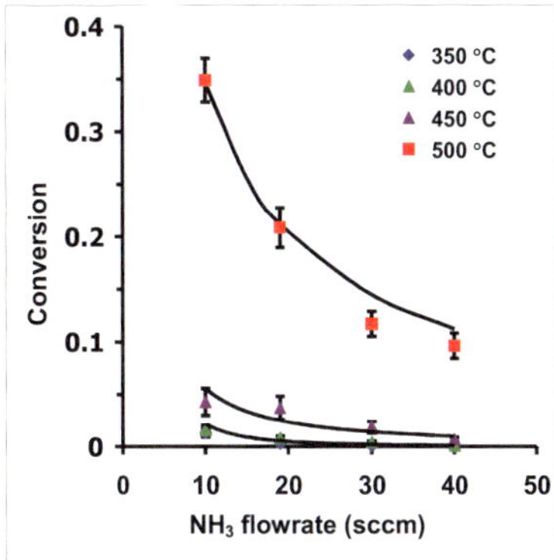

Fig. 8. Ammonia (NH_3) conversion as a function of temperature for different flow rates measured at different temperatures.

4. Advanced ceramic derived microstructure via softlithography

4.1 Fabrication of three-dimensional SiC ceramic microstructures with near-zero shrinkage via dual crosslinking induced stereolithography

3D ceramic microstructures with a submicron resolution will be very useful for a wide variety of applications to ceramic nanodevices. However, the general properties of organic polymers are not sufficient for the devices applicable to harsh environments requiring a tolerance to high temperatures, a resistance to corrosion, as well as tribological properties. Therefore, it is clear that there is a continuous demand for the development of a fabrication process of ceramic structures on the micro- or nano-scale. In terms of feasible fabrication techniques, the mechanical method of machining process has been widely utilized for various materials and has played an important role in fabricating ceramic microstructures. However, this has shortcomings when it comes to the fabrication of arbitrary 3D ceramic microstructures. For the alternative of machining process, the soft lithography and micro-stereolithography have been utilized to create 3D ceramic microstructures with a resolution of several micrometers using preceramic polymers or ceramic-powder mixed polymers (Kawata et al., 2001). At this point, stereolithography via two-photon absorption polymerization is a very interesting technique for fabricating 3D ceramic patterns. This section proposes a new chemical approach for the fabrication of SiC ceramic microstructures with near-zero shrinkage from a new photosensitive precursor system using bifunctional inorganic polymer allylhydridopolycarbosilane (AHPCS) incorporated with organometallic (cyclopentadienylmethyl)-trimethylplatinum ($CpPtMe_3$) as a versatile additive. AHPCS is a well known precursor forstoichiometric SiC ceramics (Park etal., 2004), and $CpPtMe_3$ plays a versatile triple role as a photo-hydrosilylation catalyst upon 365 nm UV irradiation, a thermal hydrosilylation catalyst at elevated temperatures, and a two-photon absorbing initiator when exposed to a 710–800 nm laser (Boardman, 1992, Coenjarts & Ober, 2004). Therefore, it is expected that a simple AHPCS–CpPtMe3 mixture would form elaborate dense networks via multiple curing routes between bifunctional groups, Si-H and the allyl group of the AHPCS, in sequential or/and coincidental reactions, which would result in little shrinkage with high ceramic yield during pyrolysis.

2D nanoscale line patterns were attempted with AHPCS mixed with 1 wt% of $CpPtMe_3$ (designated AHPCS-Pt) using a stereolithography process as an alternative terminology for the two-photon absorption fabrication technique. The AHPCS-Pt mixture was consolidated selectively using a laser beam with 780 nm wavelength for stereolithography, while AHPCS alone was not photocured. The patterned line width could be tailored by controlling the laser power and exposure time. The smallest line width of 320 nm was achieved at a laser power of 150 mW for a duration of 1 ms, which is a slightly lower resolution than the 210 nm reported previously for the acrylated polysilazane photoresist(Tuan et al., 2006).

Figure 9 shows various 3D SiC functional microstructures with dimensions of 1-5 μm obtained by pyrolysis at 600 °C in a nitrogen atmosphere from the AHPCS-Pt preceramic polymer features fabricated by a two-photon cross-linking process. There was no distortion or fracture of the structures, which often occurs through severe shrinkage. In contrast AHPCS alone or AHPCS mixed with two-photon absorbing (TPA) dyes could not form even 3D polymeric microstructures.

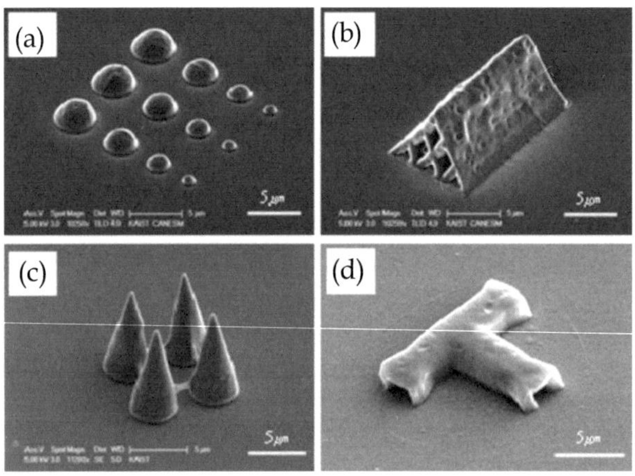

Fig. 9. Three-dimensional SiC ceramic microstructures obtained from pyrolysis at 600 1C of the AHPCS-Pt polymeric structure fabricated by stereolithography: (a) multi-scale hemispheres; (b) microchannel with multi-holes; (c) cones and (d) microscale multichannel

In the mushroom-shaped bolt shown in figure 10, the pyrolysed ceramic product (d,e) showed an 8.15 μm head dimension while the polymer structure (b,c) had a head size of 8.4 μm, as shown in figure 10(a). This clearly shows only 3% shrinkage in the top lateral direction during pyrolysis.

In order to confirm the extremely low shrinkage behaviour, spin-coated films on a Si wafer were prepared from AHPCS-Pt and AHPCS samples by UV exposure at 365 nm, post-curing at 160 °C, and pyrolysis at 600 °C. The effect of the $CpPtMe_3$ catalyst on the pyrolytic shrinkage was determined by comparing the behaviour of the AHPCS-Pt film with that of the alternative AHPCS film. The AHPCS-Pt derived ceramic film exhibited only 3% shrinkage; the thickness changed from 1.69 μm for the cured polymer to 1.64 μm after pyrolysis at 600 °C. On the other hand, AHPCS in the absence of a Pt additive exhibited 12% shrinkage with a change in film thickness from 1.32 μm to 1.16 μm. This is morphological

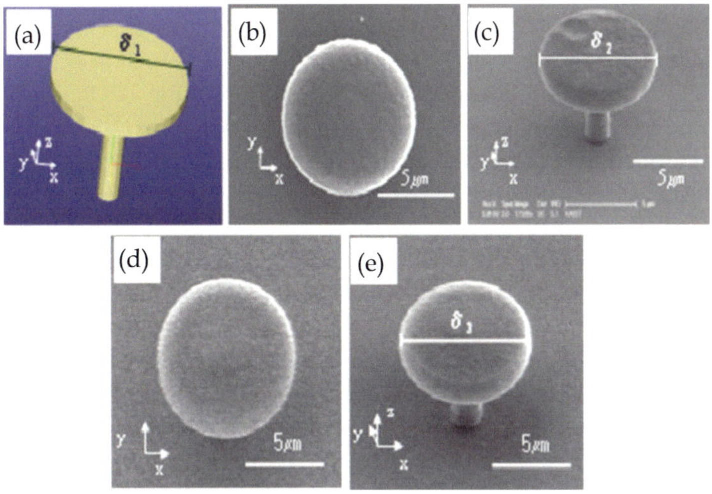

Fig. 10. Mushroom head bolt shape of SiC structure; (a) schematically designed structure (δ_1 = 8.4μm) and (b) top view of polymer state, (c) inclined view of the polymer state (δ_2 = 8.4 μm), (d) top view of the ceramic state, and (e) inclined view of the ceramic state (δ_3 = 8.15 μm).

evidence of the lower shrinkage of the AHPCS–Pt system during conversion from the polymer to ceramic phase, presumably due to the effective curing chemistry in the presence of a versatile $CpPtMe_3$ additive. In addition, the SiC film obtained from AHPCS without the Pt additive showed 12% shrinkage, which is significantly lower than the 40% in the case of SiCN from polyvinylsilazan being used as a precursor. This means that the AHPCS precursor also makes a significant contribution to decreasing the level of structural shrinkage.

4.2 Three-dimensional SiCN ceramic microstructures via nano-stereolithography of inorganic polymer photoresist

In this section, it is reported that a newly synthesized photosensitive preceramic polymer, a negative type of inorganic photoresist resin, is a suitable candidate for the fabrication of complex 3D sub-micrometer-sized structures via a two-photon absorbed crosslinking process followed by pyrolysis to form a ceramic phase which had been explained in previous section. The two-photon process including polymerization or/and crosslinking is recognized to be a promising technique for the fabrication of real 3D microstructures with a sub-micron resolution. This is shown for organic-based materials such as urethane acrylate, SU-8, and PDMS (Seet et al., 2005, Coenjarts & Ober, 2004). On the other hand, it is believed that an inorganic polymer photoresist is a promising material that could pave the way for a near-direct ceramic structuring process. In this work, a two-photon curable inorganic polymer (by mixing two-photon chromophore into a matrix of photosensitive inorganic polymer) was developed for the fabrication of 3D ceramic patterns.

In this work, the fabrication of two-dimensional (2D) and real three-dimensional (3D) nanoscale SiCN ceramic structures was attempted by a nano-stereolithography (NSL) process with a resolution of less than the diffraction limit, which can be difficult to obtain using the conventional photolithographic technologies. The developed inorganic polymer photoresist containing 0.4 wt% of [1,4-bis(2-ethylhexyloxy)-2,5-bis(2-(4-(bis(4-bromophenyl)amino))-vinyl)benzene] (EA4BPA-VB) photosensitizer was selectively consolidated via a two-photon absorbed photocross-linking route, while the initial polyvinylsilazane was not photocuable. It is generally known that voxels, i.e. unit-volume-pixels, were controlled by altering the processing parameters of the NSL such as exposure time, laser power, and truncation amount of voxels under the substrates. As shown in figure 11 (a), the patterned line width can be tailored easily by controlling the laser power and the exposure time. The smallest line width of 210 nm was achieved under the conditions of a laser power of 100 mW and duration of 1 ms, whose resolution is comparable to the cases of reported two-photon materials. Moreover, the diameter and length of an elliptical voxel were both steeply broadened with the increase of laser power and exposure time close to the field of threshold energy for crosslinking. It was also determined from the experimental results that the higher power and longer exposure time enabled the fabrication of thick patterns.

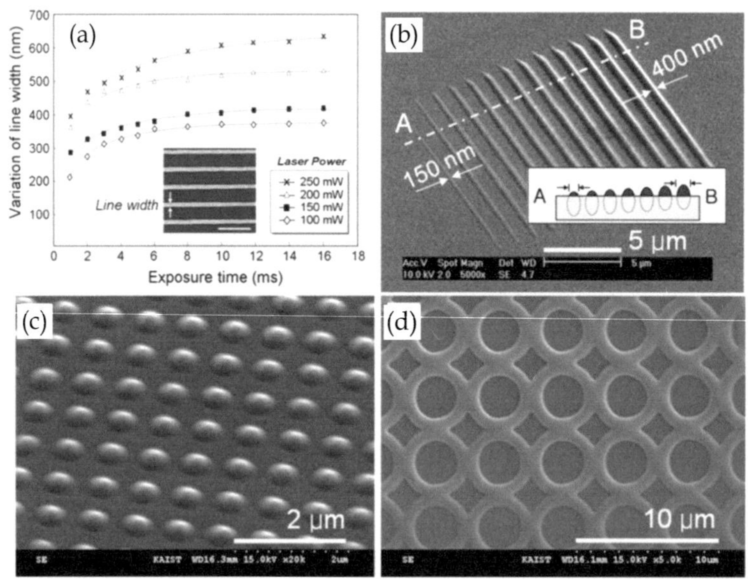

Fig. 11. (a) The dependence of line width at polymeric phase on laser power and exposure time as processing parameter study of nano-stereolithography. The inset shows an example of fabricated lines, and the scale bar is 2 μm. (b) to (d) show two-dimensional ceramic patterns pyrolyzed at 600°C; (b) multi-layered line patterns (c) ceramic nano-dots array with diameter 530 nm and (d) circle pattern.

Various 2D ceramic micropatterns were also fabricated by NSL and subsequent pyrolysis at 600°C (figure 11(b)~11(d)). Figure 11(b) shows line patterns fabricated by moving the beam focal position from the left-end to the right-end with 30 nm per line. It indicates that the line width can be controlled by the truncation amount under the glass plate, as illustrated in the inset of figure 11(b); actually reduced to 150 nm, that is lower than the minimum voxel diameter. Alternative 2D ceramic patterns such as nano-dot array with 530 nm diameter and circle patterns were fabricated by adapting the aforementioned process parameters (figure 11(c) and 11(d)). Here, it is noteworthy to point out that the obtained ceramic patterns are useful as a tribological ceramic stamp (mold) with a nanoscale resolution suitable for various lithographic techniques such as hot embossing process, which is not achievable through conventional molds.

Eventually, the fabrication of a real 3D ceramic woodpile structure was performed by piling up the line patterns by layer-by-layer technique. The rectangular shape with a 9μm x 9μm x 9μm dimension was designed as shown by the schematic diagram in figure 12(a). However, after pyrolysis at 600°C, the photocured woodpile structure (figure 12(b)) was significantly deformed into a pyramid-like structure that nonlinearly tapered in a perpendicular direction (figure 12(c)). The lateral length of the top surface (W_T) was reduced to 5.3 μm with no change of bottom surface length (W_B). A dimensional change with a 41% shrinkage in the top lateral direction during pyrolysis occurred anisotropically with a different extent of shrinkage along the normal direction to the substrate. It can be interpreted that the bottom section adhered strongly to the glass substrate and was pyrolyzed under the constraint conditions, while the top section has a nearly free-standing condition. Here, it is generally reported that linear shrinkage in the range of 20~30% intrinsically occurs due to the thermal conversion from low dense polymer to a highly dense ceramic phase, accompanied with the weight loss. However, it is much less severe than in the case of sol-gel process, thus the preceramic polymers have been widely utilized via thermal curing and pyrolysis steps in a variety of high temperature material applications such as fibers and composites. (Liu et al., 2002)

The observed anisotropic shrinkage behavior must be a severe detrimental factor for a precise fabrication of 3D ceramic microstructures. In an attempt to rectify this shrinkage problem, silica particles with approximately 10 nm of diameter was introduced as filler into the inorganic photoresist resin with various solid loading portions. It was observed that the polymer-nanoparticle mixtures were kept transparent against laser beam due to homogeneous dispersion. When the identical structuring process and subsequent pyrolysis were conducted, the total amount of shrinkage in the lateral direction was considerably reduced to 33%, 28%, and 24% in each case of 20 wt%, 30 wt%, and 40 wt% silica particle loading samples, respectively (figure 12(d), 12(e), and 12(f)). Interestingly, through this, it was verified that the shrinkage can be reduced with a linear relation of 41(1-x)% for the silica particle percentage (x). In addition, the ceramic structure containing 40 wt% silica particle exhibited relatively isotropic shrinkage owing to its sliding free from the substrate during pyrolysis, presumably due to its weak interfacial adhesion to the substrate. Furthermore, other ceramic examples for real 3D structures of the spiral micro tube and cruciform were also fabricated with the developed resin polymer mixed with 40 wt% silica particles (figure 12(g) and 12(h)), suggesting novel applicability for chemical channels, or even for use in mechanical devices. The both structures with 90° twisting angles were originally designed with a side length 7.6 μm squared. After pyrolysis, the nearly isotropic shrinkage has occurred as the side lengths at the top and the bottom showed smaller discrepancy, with 5.81 μm and 6.23 μm in figure 12(g), 5.75 μm and 6.39 μm in figure 12(h), in contrast to these parameters of the non-free-standing structures. Here, it is believed that the pyrolytic shrinkage can be improved by controlling the resin compositions as well as the geometric constraints, including the design of the compensated structure, with a prediction model simulation.

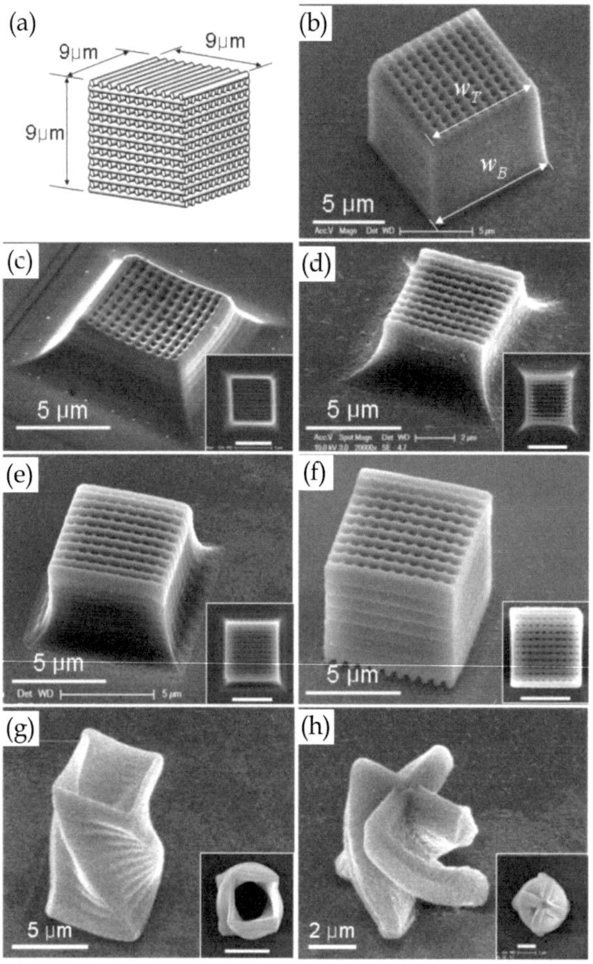

Fig. 12. Three-dimensional ceramic microstructures fabricated by nano-stereolithography or/and subsequent pyrolysis at 600°C; (a) schematically designed woodpile structure, (b) polymeric structure with no filler and (c) ceramic woodpile structure with no filler. Ceramic woodpile structure obtained from the mixed resin containing various amount of silica filler for reduced shrinkage; (d) 20 wt% silica, (e) 30 wt%, and (f) 40 wt%. Other 3D ceramic microstructures of spiral; (g) microtube and (h) microcruciform with twisting angle of 90° between their bottom and top; they are fabricated using the 40 wt% particle containing resin (each insert is top-views of the structure.).

5. Extended applications in the future

There are many possibilities for the adoption of micro ceramics structures in the fields of micro-reactors, diagnosis and sensors (e.g. microfludics, protein patterning and 3D ceramic structuring). For example, a microfluidic laboratory-on-a-chip have a wide variety of applications in liquid handling in biology and chemistry due to their ability to manipulate small sample quantities for rapid, high-resolution and low-cost analysis and synthesis (Guo et al., 2004, Sun & Kawata et al., 2004, Takada etal., 2005). Currently, microfluidics is mainly fabricated using polydimethylsiloxane (PDMS) and other organic polymers, owing to the flexibility and easy handling of the manufacturing microstructure. However, the general properties of organic polymers are not suitable for devices applicable to harsh environments requiring a high temperature tolerance, corrosion resistance and tribological properties. Therefore, there is a demand for the development of a process fabricating ceramic structures on the micro- or nano-scale. However, mechanical machining techniques in the fabricating the silicon, glass and metal microstructures have disadvantages such the requirement of high cost processing equipment as well as the limited resolution and geometry. It is believed that the integration of ceramic polymers into new economic manufacturing strategies can provide a break-through technology for the creation of complex ceramic microstructures with a nanoscale resolution. In addition, shrinkage control during pyrolysis is a critical issue in most applications of polymer-derived ceramics, which have been in strong demand for the development of an appropriate process for achieving the high dimensional stability of the final ceramic products.

6. Conclusion

Processes for fabricating the ceramic microstructures formed with SiC, SiCN and SiCBN, are in demand for use in areas such as microfluidics and MEMS that can be tolerated in harsh environments requiring a high temperatures tolerance, corrosion resistance, as well as tribological properties. This chapter describes the manufacturing techniques of SiC-based ceramic microstructrues by applying preceramic polymers with various fabrication techniques and subsequent pyrolysis to convert into ceramic phase. Firstly, macroporous and mesoporous SiC ceramic structures have been fabricated using various templates including packed silica sphere assemblies, porous carbon templates, and nanoporous silica structures. In addition, SiC nanotubes have also been obtained from an alumina membrane used as a template. These porous structures were generally obtained using a series of infiltration, curing and pyrolysis, and chemical or oxidative etching steps. Moreover, the feasible application of the microporous structures was attempted for high temperature ceramic microreactor to perform the on-demand reforming of higher hydrocarbons into hydrogen in portable power sources. Secondly, complex 3 dimensional SiC-based ceramic microfeatures were fabricated using near-net shape lithographic techniques with preceramic polymers. Line patterns and porous channels were produced by soft lithography, whereas the 3D woodpile structures were fabricated using stereolithography. These novel structures show great promises for future applications in tribological ceramic devices. It is expected that the integration of preceramic polymers into new economic manufacturing strategies such as precursor casting in templates and lithography utilization will be a break-through technology for the creation of complex ceramic nanostructures.

7. References

Boardman, L. D. (1992). *Organometallics*, Vol. 11, (4194-4201), 1099-0739

Caruso, F.; Shi, X. Y.; Caruso, R. A. & Susha, A. (2001). *Adv. Mater.*, Vol. 13, (740-744), 0935-9648

Chou, S.Y.; Krauss, P.R. & Renstrom, P. J. (1995). *Appl. Phys. Lett.*, Vol. 67, (3114-3116), 0034-6748

Coenjarts, C. A. & Ober, C. K. (2004). *Chem. Mater.* Vol. 16, (5556-5558), 0897-4756

Dai, H.; Wang, E. W.; Lu, Y. Z.; Fan, S.S. & Lieber, C. M. (1995). *Nature*, Vol. 375, (769-772), 0028-0836

Dibandjo, P.; Bois, L.; Chassagneux, F; Cornu, D.; Letoffe, J. M.; Toury, B.; Babonneau, F. & Miele, P. (2005). *Adv. Mat.*, Vol. 17, (571-574), 0935-9648

Donzel, C.; Geissler, M.; Berard, A.; Wolf, H.; Michel, B.; Hilborn, J. & Delamarche, E. (2001). *Adv. Mat.*, Vol. 13, (1164-1167), 0935-9648

Geis, S.; Fricke, J. & Löbmann, P. (2002). *J. Eur. Ceram. Soc.*, Vol. 22, 1155-1161, 0955-2219

Guo, L. J.; Cheng, X. & Chou, C. F. (2004). *Nano Lett.*, Vol. 4, 69-73, 1530-6992

Heule, M. & Gauckler, L. J. (2001). *Adv. Mat.*, Vol. 13, (1790-1793), 0935-9648 , (2003). *Sensor Actuat. B*, Vol. 93, (100-106), 0935-9648

Heule, M.; Vuillemin, S. & Gauckler, L.J. (2003). *Adv. Mat.*, Vol. 15, (1237-1245), 0935-9648

Hong, L. Y. & Kim, D. P. (2005). *Key Eng. Mat.* Vol. 287, (323-328), 1013-9826

Iijima, S. (1991). *Nature*, Vol. 354, (56-58), 0028-0836

Lee, K. C. ; Seung S. & Lee, S. S. (2004). *Sens. Act. A*, Vol. 111, (37-43), 0924-4247

Liew, L.; Zhang, W.; Bright, V. M.; An, L.; Dunn, M. L. & Raj, R. (2001). *Sensor Actuat. A*, Vol. 89, (64-70), 0924-4247

Liew L.; Liu, Y.; Luo, R.; Cross, T.; An, L.; Bright, V. M.; Dunn, M. L.; Daily, J. W. & Raj, R. (2002). *Sensor Actuat. A*, Vol. 95, (120-134), 0924-4247

Liew, L. ; Bright, V. M. & Raj, R. (2003). *Sensor Actuat. A*, Vol. 104, (246-262), 0924-4247

Liu, Y.; Liew, L. A.; Liu, Y.; Luo, R.; An, L.; Dunn, M. L.; Bright, V. M.; Daily, J. W. & Raj, R. (2002). *Sensor. Actuat. A*. Vol. 95, (143-151), 0924-4247

Madou, M. J. (2002). *Fundamentals of Microfabrication - The Science of Miniaturization, 2nd ed.*, pp. 519-521, 978-0849308260, CRC Press, Boca Raton

Martin, C. R. & Aksay, I. A. (2005). *J. Mater. Res.*, Vol. 20, (1995-2003), 0884-2914

Nguyen, N. T. & Wereley, S. T. (2002). *Fundamentals and Applications of Microfluidics*, pp. 82-97, 978-0-387-25564-4, Artech House, Boston

Kawata, S.; Sun, H. B.; Tanaka, T & Takada, K. (2001). *Nature,* Vol. 412, (697-698), 0028-0836

Keller, N.; Pham-Huu, C.; Ehret, G.; Keller, V. & Ledoux, M. J. (2003). *Carbon*, Vol. 41, (2132-2139), 0008-6223

Kim, E., Xia, Y. & Whitesides, G. M. (1996). *J. Am. Chem. Soc.*, Vol. 118, (5722-5731), 0002-7863

Krawiec, P.; Weidenthaler, C. & Kaskel, S. (2004). *Chem. Mater.* Vol. 16, (2869-2880), 0959-9428

Kwon, S.; Son, G.; Suh, J. & Kim, K. T. (1994). *J. Am. Ceram. Soc.*, Vol. 77, (3137-3141), 0002-7863

Parmentier, J.; Patarin, J.; Dentzer, J. & Vix-Guterl, C. (2002). *Ceram. Inter.*, Vol. 28, (1-7), 0272-8842

Park, K. H.; Sung, I. K. & Kim. D. P. (2004). *J. Mater. Chem.*, Vol. 14, (1-4), 0959-9428

Park, K. H.; Sung, I. K. & Kim, D. P. (2004). *J. Mater. Chem.* Vol. 14, (3436-3439), 0959-9428

Peng, H. X.; Fan, Z.; Evans, J. R. G. & Busfield, J. J. C. (2000). *J. Eur. Ceram. Soc.* Vol. 20, (807-813), 0955-2219

Pham, T. A.; Lim, T. W.; Park, S. H.; Yang D. Y. & Kim, D. P. (2006). *Adv. Funct. Mater.*, Vol. 16, (1235-1241), 1616-301X

Pham-Huu, C.; Bouchy, C.; Dintzer, T.; Ehret, G.; Estournes, M. & Ledoux, M. J. (1999). *Appl. Catal. A*, Vol. 180, (385-397), 0926-860X

Qin, D.; Xia, Y.; Rogers, J. A.; Jackman, R. J.; Zhao, X. M. & Whitesides, G. M. (1999). *Microsystem Technology in Chemistry and Life Science*, Ed. by Manz, A. and Becker, H., pp. 1-20, 10: 354063424X, Springer, Berlin

Quin, J.; Wang, J.; Zhihao, J. & Qiao, G. (2003). *Mat. Sci. Eng. A*, Vol. 358, (304-309), 0965-0393

Schmidt-Winkel, P.; Lukens, W. W. Jr.; Zhao, D. Y.; Yang, P. D.; Chmelka, B. F. & Stucky, G. D. (1999). *J. Am. Chem. Soc.* Vol. 121, (254-255), 0002-7863

Seet, K. K.; Mizeikis, V.; Matsuo, S.; Juodkazis, S. & Misawa, H. (2005). *Adv. Mater.* Vol. 17, (541-545), 0935-9648

Sepulveda, P. (1997). *Am. Ceram. Soc. Bull*, Vol. 76, (61-65), 1226-086X

Shah, S. R. & Raj, R., J. (2005). *J. Eur Ceramic Soc.*, Vol. 25, (243-249), 0955-2219

Shi, Y. F.; Meng, Y.; Chen, D. H.; Cheng, S. J.; Chen, P.; Yang, H. F.; Wan, Y. & Zhao, D. Y. (2006). *Adv. Funct. Mater.*, Vol. 16, (561-567), 1616-301X

Sun, H. B. & Kawata, S. (2004) *Adv. Polym. Sci.*, Vol. 170, (169-273), 0065-3195

Sung, I. K.; Yoon, S. B.; Yu, J. S. & Kim. D. P. (2002). *Chem. Comm.*, Vol. 2002, (1480-1481), 1359-7345

Takada, K.; Sun, H. B. & Kawata, S. (2005). *Appl. Phys. Lett.*, Vol. 86, (071122-1 - 071122-3, 0003-6951

Vinu, A.; Terrones, M.; Golberg, D.; Hishita, S.; Ariga, K. & Mori, T. (2005). *Chem. Mater.*, Vol. 17, (5887-5890), 0897-4756

Wang, H.; Yu, J. S.; Li, X. D. & Kim, D. P. (2004). *Chem. Comm.*, Vol. 2004, (2352-2353), 1359-7345

Wang, H.; Li, X. D.; Yu, J. S & Kim, D. P. (2004). *J. Mat. Chem.*, Vol. 14, (1383-1386), 0959-9428

Wang, H.; Li, X. D. & Kim, D. P. (2005). *Appl. Organomet. Chem.* Vol. 19, (742-749), 0268-2605

Wang, H; Li, X. D.; Kim, T. S. & Kim, D. P. (2005). *Applied Phy. Lett.*, Vol. 86, (1-3), 0003-6951

Wang, H.; Zheng, S. Y.; Li, X. D. & Kim. D. P. (2005). *Micropor. Mesopor. Mat.*, Vol. 80, (357-362), 1387-1811

Wu, G.; Zhang, L.; Cheng, B.; Xie, T. & Yuan, X. (2004). *J. Am. Chem. Soc.*, Vol. 126, (5976-5977), 0002-7863

Yamazaki, K. & Namatsu, H. (2004). *Microelectron. Eng.*, Vol. 73-74, (85-89), 0167-9317

Yan, J.; Wang, A. & Kim. D. P. (2006). *J. Phys. Chem. B*, Vol. 110, (5429-5433), 1089-5647

Xia, Y. & Whitesides, G. M. (1998). *Annu. Rev. Mater. Sci.*, Vol. 28, (153-184), 1225-0112

Xia, Y.; Rogers, J. A.; Paul, K. E. & Whitesides, G. M., (1999). *Chem. Rev.*, Vol. 99, (1823-1848), 0009-2665

Zhang, F. Z.; Kato, T.; Fuji, M. & Takahashi M. (2006). *J. Eur. Ceram. Soc.*, Vol. 26, (667-671), 0955-2219

Zhao, D. Y.; Feng, J. L.; Huo, Q. S.; Melosh, N.; Fredrickson, G. H.; Chmelka, B. F. & Stucky, G. D. (1998). *Science*, Vol. 279, (548-550), 0036-8075

Zhu, X.; Jiang, D. & Tan, S. (2002). *Mater. Res. Bull.*, Vol. 37, (541-553), 0025-5408

22

Soft Lithography, a Tool to Address Single-Objects Investigations

Aline Cerf and Christophe Vieu
CNRS; LAAS
Université de Toulouse; UPS, INSA, INP, ISAE; LAAS
France

1. Introduction

Nowadays, research is following a new trend where nanotechnology meets biology and vice-versa. In fact, nanotechnology offers remarkable tools to improve our understanding of biology's mechanisms but at the same time, biology shields precious clues that could inspire nanotechnology to answer to today's and tomorrow's issues and needs.

The study of cells and molecules at the single-object level has become essential in this complementary exchange and multidisciplinary effort. But, to track single proteins at work, watch nanoscale biomachines or control cell adhesion and differentiation, one of the main challenges is still the integration of these bioentities onto desired locations on a substrate. In other terms, one has to be able to single-out the object of interest from a given media, organize it and fix it on a surface while preserving its functionality. This integration of objects further implies being efficient and having a high-throughput capacity in order to conciliate single event recording with statistical analysis. Thus, efforts on creating defect-free and well-ordered assemblies are crucial for the study of single objects.

To define the number, position, and distances of the objects on a surface, it is often necessary to direct their assembly by means of suitable prepatterned substrates or stamps. Nanopatterned surfaces have been fabricated by nanoimprinting (Wang et al., 1999), contact printing (Kumar & Whitesides, 1993; Xu et al., 2007), microtransfer molding (Zhao et al., 1996), pattern replication induced by an electric field (Morariu et al., 2003), lithographically induced self-construction (Chou et al., 1999), grid-assisted self-organization (Cavallini et al, 2002), and self-assembly (Chou & Zhuang, 1999) among others. Most of these techniques rely on a soft mould or stamp (Quake & Sherer, 2000).

Soft-lithography emerged in the early 90's as a technique based on the use of an elastomer material named polydimethylsiloxane (PDMS). This polymer can be poured onto a silicon-based master with micro- or nanostructures made by photolithography or electron beam lithography respectively, to obtain a PDMS replication of the master's structures. The corresponding PDMS stamp can then be inked with a molecule and put in contact with a target surface to transfer the molecules into a patterned layer. This is what we call microcontact printing (Kumar & Whitesides, 1993; Xia & Whitesides, 1998). The major advantages of microcontact printing are that the contact with the surface is made in a very gentle manner, so the bioentities are transferred without damaging them. The flexibility of the

PDMS stamp and the ability to achieve conformal contact between the stamp and the surface enables high-resolution patterning on both planar and non-planar surfaces. Areas greater than 50 cm^2 have been patterned by microcontact printing (Xia et al., 1996).
So this technique is suitable for large-area micro- or nanopatterning, and is attractive because of its simplicity, biocompatibility, speed, and low cost.
Directed Capillary Assembly is another interesting technique that has received constant interest throughout the years. This technique was first introduced by Denkov et al. in the 90's. At that time, their objective was to produce photonic band gap materials based on the compact assembly of dielectric colloids (silica beads). The method relied on the direct evaporation of a droplet of a colloidal solution on a surface and was called Capillary Force Assembly (CFA). The nanoparticles dragged by the convective flux in the droplet were forced to assemble on the surface close to the liquid meniscus. A few years later, the group of Alivisatos (Klein et al., 1996) showed how this method could be combined with topographic patterns in order to control the position of the assembly as well as the shape of the nanoparticles' patterns and this was called "Lithography Guided Capillary Force Assembly". Kraus et al. further demonstrated that this high-precision inking could be performed by dragging the meniscus of the objects in suspension over a structured stamp with a controlled speed. As a result, the objects are trapped and assembled inside each cavity and the obtained assembly can be subsequently transferred onto a target surface by simple contact (Malaquin et al., 2007; Kraus et al., 2007).
Soft-lithography and directed capillary assembly thus remain very attractive and straightforward techniques capable of patterning and nano-ordering over large areas at low cost. Therefore, to study single objects, the methodology we developed is declined following two strategies: either we combine soft-lithography with a simple incubation technique, or we combine soft-lithography with directed capillary assembly. Both approaches allow us to tightly control the assembly process to direct the precise ordering of the objects of interest. With these two strategies, we have been able for example to generate arrays of single living bacteria for fundamental studies and regular arrays of single stretched DNA molecules for genetic analysis or medical diagnostics. We have also been able to produce periodic bi dimensional matrixes of gold nanoparticles for different applications in the field of plasmonics.

2. First strategy: chemical patterning

In the case of cells, we developed a fast, simple and reproducible procedure for isolating and depositing individual living cells onto a surface along well-defined and precisely registered micrometric patterns. During the process, the species are not extracted from their natural media so there is no risk of damaging them. The approach consists in chemically engineering the surface with hydrophobic/hydrophilic contrast to induce and localize single cell immobilization events in a deterministic manner. The principle of the methodology we perform for assembling cells is depicted in Figure 1. It is based in the combination of conventional microcontact printing with a simple incubation technique. In pratice, the PDMS stamp with topographic structures is used to print a first ink, in this case a hydrophobic molecule, on the surface. Then, by incubating a second ink, this time a hydrophilic molecule, the hydrophilic molecules fill out the empty patterns. After incubating a droplet of cells in suspension on this engineered surface, we obtain a selective adsorption of cells on the local chemical patterns producing highly ordered arrays of single living cells with a success rate close to 100% and over a 1cm² area. By tuning the chemistry of the surface we are able to modulate the binding strength of the cells onto the patterns. This allows the technique to be flexible as concerning the choice of the molecules used to graft the cells by using untreated or chemically treated cells. The molecule used mainly influences the strength of cell's binding onto the surface rather than their localization. In fact, only the surface microstructuration is responsible for the cells trapping, so this trapping occurs whatever the affinity of the cells with the surface is. This methodology has allowed us to realize more fundamental studies on single cells with the possibility to perform statistics on them as we obtain chips with high-throughput. By correlating AFM spectroscopy with staining techniques we have further proved that the grafted cells remain alive after patterning (Cerf et al., 2008; Cerf et al., 2009).

By extension, this methodology could be a basic building block for interfacing living cells with artificial microsystems or for generating arrays of any kind of microorganisms by simply using the appropriate pattern functionalization.

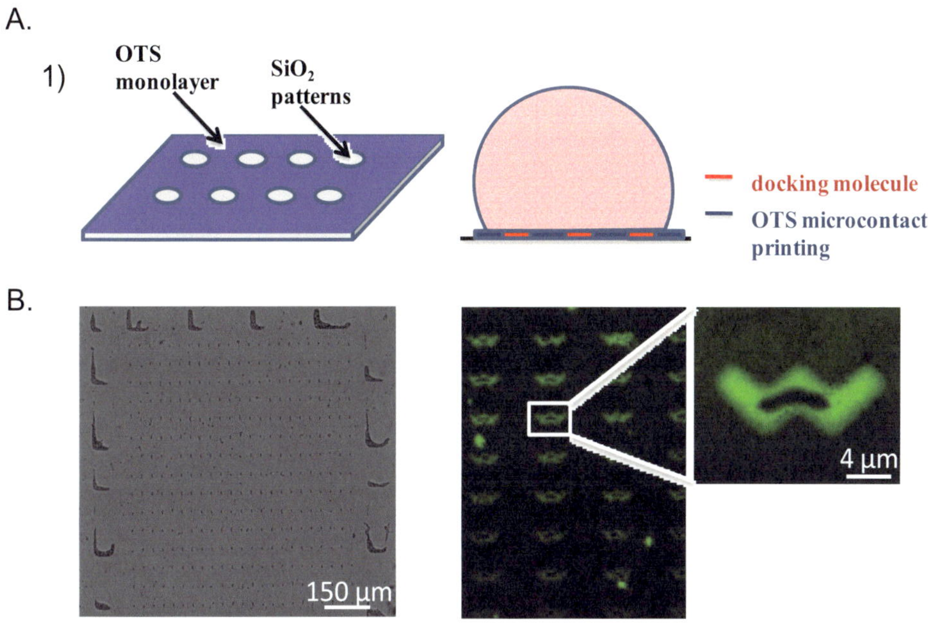

Fig. 1. Chemically engineered surfaces used to trap individual bacterial cells along the chemical patterns so-defined (biopatterning). A) OTS microcontact printing (1) followed by a functionalization by incubation of the bare patterns with docking molecules (2). B) On the left, bright field image of highly ordered arrays of single living bacteria and on the right, fluorescence image of the same array. Here the patterns are functionalized with streptavidin molecules and bacteria are biotinylated so that they strongly bind to the surface.

3. Second strategy: use of topographical structures

If we travel now from the micrometric scale down to the nanometric scale, we find other interesting objects to study such as DNA molecules. Our purpose in this study was to single out individual DNA molecules and stretch them into controlled arrays for fundamental investigations but also medical applications. This time, we did not use chemically

engineered surfaces but structured PDMS stamps as templates with cavities that would allow the trapping of the objects of interest inside each cavity. In fact, in this case, we combine two techniques: soft-lithography with directed capillary assembly.
The technique of directed capillary assembly that we perform is carried out using a dedicated setup. The so-called directed assembly setup consists in a motorized translation stage where we put a PDMS stamp with micro- or nanostructures. Above this stamp we place a fixed glass slide at a distance of 1mm and we put in between a certain volume of the colloidal suspension of the object to deposit. The liquid meniscus is then moved over the structured PDMS stamp at a controlled velocity (Scheme 1). As the meniscus encounters the stamp's topographic structures, it gets pinned during a certain time. Meanwhile, the objects undergo a convective flux and are directed towards the liquid front line. There, they are trapped by the capillary forces exerted inside each topographic cavity or well and when the elastic energy stored exceeds the pinning energy, the meniscus disrupts and releases the objects.
By adjusting the displacement speed of the stage, the concentration of the colloidal suspension and the temperature of the PDMS substrate thus the evaporation rate of the solvent, we are able to control the number of objects to be trapped inside each micro- or nanowell of the structured PDMS stamp with high placement accuracy. Once the objects are properly assembled under optimized experimental conditions, the objects can be transferred onto a target surface by simple contact printing.
In the case of DNA molecules, we use this particular methodology: the assembly of coiled DNA molecules is performed on a microstructured PDMS stamp at a relatively high velocity of 1mm/sec inducing almost simultaneously the selective trapping of single molecules at each well of the stamp and their stretching by the meniscus disrupting. The resulting assembly of single stretched molecules is then transferred by contact printing onto a freshly cleaved mica sheet or onto a glass slide coated by vapor deposition of 3-aminopropyltriethoxysilane (APTES) molecules (Scheme 1).

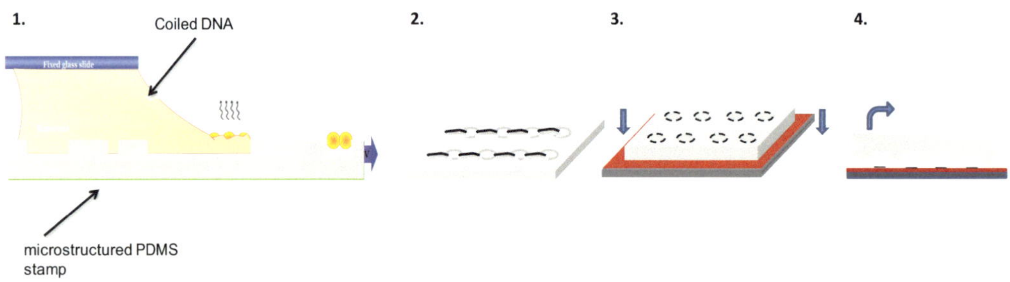

Scheme 1. Generation of single stretched DNA molecules arrays by directed assembly. 1) Capillary assembly of DNA molecules on a microstructured PDMS stamp. 2) Microstructured PDMS stamp with assembled and stretched DNA molecules. 3) The assembly is then transfer-printed on an APTES-coated surface. 4) The PDMS stamp is finally peeled away.

Figure 2A depicts the nice and regular arrays of single stretched DNA molecules we obtain with a success rate close to 100% (Cerf et al., 2009). We observe through AFM cross-section measurements that the stretched DNA molecules are 1.7 nm high which corresponds to the theoretical height of a single DNA molecule. So the generation of these chips with single DNA molecules could be interesting for genome mapping or medical diagnostics through a fast and more efficient optical readout. Another advantage of this methodology is that the experimenter is able to tune the size of the PDMS wells and their periodicity so as to obtain continuous lines of DNA molecules or much more complex DNA patterns. In fact, the size, organization and center-to-center distance between the wells also play an important role in the number of molecules to be anchored and in the arrangement we want to give to the stretched DNA molecules. In Figure 2B, we notice that with a shorter center-to-center distance between the wells, the combing of the molecule can continue over the next well, where the formation of another stretched molecule is initiated. As a result, the two consecutive stretched molecules bundle together in the knot that appears brighter in the fluorescence micrograph. Figure 2C shows the influence of the wells' arrangement and the moving direction of the meniscus on the DNA molecules' stretching. In the case of wells in a quincunx arrangement, the molecules tend to cling to the shape of the next microwell while stretching when the meniscus breaks. The orientation of the DNA molecule will be the same as the moving direction of the receding meniscus. Thus, by controlling the size, period and arrangement of the PDMS wells as well as the moving direction of the meniscus, we can further create more complex DNA templates. Thus, in a broader sense, this methodology could also be very useful for microelectronics as by metalizing these robust 1D nanostructures or by functionalizing them with nanoparticles (Braun et al., 1998; Keren et al., 2002; Deng & Mao, 2003; Monson & Woolley, 2003; Nguyen et al., 2008; Köhler et al., 2001), one is able to generate very appealing conducting wires and complex networks over a large area and at low cost.

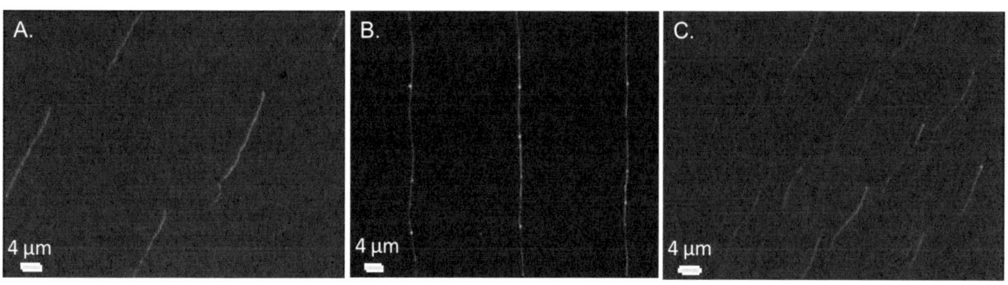

Fig. 2. Highly organized chips of stretched DNA molecules (A and C) and continous lines of stretched DNA molecules (B) on APTES-coated surfaces.

Another example of direct application is the integration of nanoparticles onto a target surface.

Recently, we explored the possibility to generate bi dimensional matrixes of periodic sub-100-nm gold nanoparticles patterns by directed assembly. Small nanoparticles are the target of intense research due to their astonishing properties and their large window of applications in biology, nanoelectronics, spintronics, optics, chemical sensing, etc. Assembling nanoparticles into well-ordered structures provides new nanostructures with remarkable and very distinct collective properties. In fact, by controlling their type, number, and precise arrangement, one can benefit from the unique properties they offer at that scale in the field of plasmonics in particular (Surface Enhanced Raman Scattering, Metal Enhanced Fluorescence).

By using the same technique as previously described, but this time by using a nanostructured PDMS stamp and by placing between the fixed glass slide and the mobile stamp a colloidal suspension of gold nanoparticles, we are able to generate matrixes of gold nanoparticles with different numbers of particles and different arrangements. In this case, the speed of the translation stage needs to be very low (1µm/sec). This time, accounting from the size of the nanoparticles, their irregular shape and the fact that the resulting assembly is dry, the formed nanoparticles arrays were extremely difficult to transfer onto a solid surface. So we investigated different strategies to transfer print the assembled nanoparticles without the need for an adhesion layer. Generally what is commonly used in the literature is an intermediate polymer layer to emboss so to speak the particles onto the target surface (Kraus et al., 2007; Yan et al., 2004). But for further technological steps, this polymer layer has to be removed from the surface by hydrogen plasma and this additional processing is not always feasible depending on the object or the application sought. Thus we explored other possibilities and we found an alternative approach to transfer print an assembly from a PDMS stamp onto a surface by performing the contact through a thin liquid film. In fact we deposit a droplet of solvent (deionized water or ethanol) on the target surface coated by vapour deposition of 3-aminopropyltriethoxysilane (APTES) molecules and we put the PDMS stamp with the assembled nanoparticles in contact. We place the overall in an oven at 100°C for a few minutes to evaporate all the solvent and we peel the stamp off (Scheme 2). Figures 3A and 3B show the results obtained after directed assembly of gold nanoparticles on a nanostructured PDMS stamp and transfer printing onto a silicon surface.

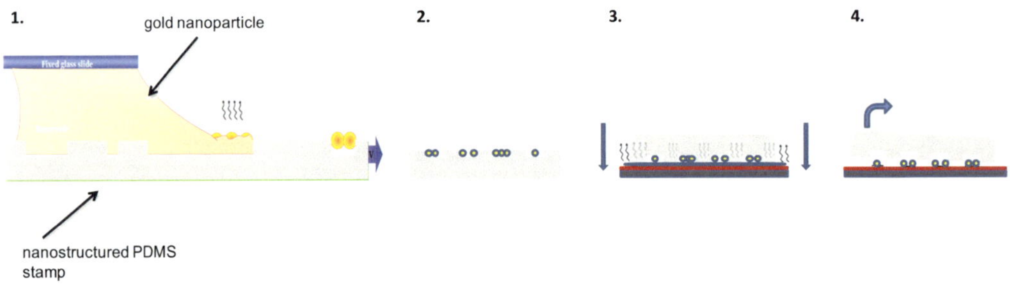

Scheme 2. Generation of 100nm gold nanoparticles arrays by directed assembly. 1) Capillary assembly of gold nanoparticles on a nanostructured PDMS stamp. 2) Nanostructured PDMS stamp with assembled gold nanoparticles. 3) The assembly is then transfer-printed with solvent mediation. 4) The PDMS stamp is finally peeled away.

We obtain nice and regular nanoparticle patterns and the totality of the particles is transferred. Furthermore, we can observe from the dark field micrographs that the assembled particles retain their optical activity (figure 3A), so in a broader sense, we could say that the assembly process does not alter the intrinsic properties of the objects (Cerf & Vieu, 2009).
These encouraging results allowed us to investigate the possibility to detect single objects or molecules at low concentrations via optical techniques. Optical techniques are very interesting indeed because they have the advantage of being non-invasive, non-destructive and non-contact, but are in principle not sensitive enough to study a single nano-object or isolated molecule. Different approaches have been explored in our group among which, those relying on diffractive sensing or more recently on surface enhanced Raman scattering. In this last case, a way to enhance the signal is to position the objects of interest in the close vicinity of noble metal nanocrystals (silver, gold). Such nanocrystals offer remarkable properties due to the localized surface plasmon resonances (SPR) that induce, under the

effect of an optical excitation with a frequency close to the SPR, a very intense local electric field enhancement nearby the surface. The interest of surface-enhanced Raman scattering (SERS) effect is that the enhancement can reach 10^{14}, sufficient enough to detect a single molecule (Nie & Emory, 1997). The conception of ordered nanometer-sized metallic structures that allow controlling the position of these "hot spots" where the electromagnetic field is locally amplified is the heart of intense research, both theoretical and experimental (Käll et al., 2005). What is known, is that some of these properties only appear with very precise arrangements such as regularly spaced coupled particles with short interdistance between them for instance (Haynes et al., 2005a). In this perspective, we set the experimental conditions of our directed assembly setup to generate gold nanoparticle arrays as previously described but aiming at investigating the exact number of particles giving the maximum local electric field enhancement. We deposited a Raman label onto the substrate and we performed some Raman measurements. We evidenced that, effectively, the electromagnetic field is locally amplified at the sites corresponding to the nanoparticle patterns. As we can further observe in figure 3D and 3E, the intensity varies among the patterns so the intensity of the enhanced Raman signal depends on several parameters, namely the number of particles, their arrangement, their size, and their orientation... This substrate showed its great efficiency as a SERS but also as a metal-enhanced fluorescence (MEF) substrate. In fact, figure 3C shows that when a fluorophore is deposited at low concentrations onto the substrate, fluorescence is locally amplified at the sites corresponding to the nanoparticles patterns only (Cerf et al., 2009). So we could later imagine including these imaging-enhancing structures colocated with DNA tethers for single molecule fluorescence.

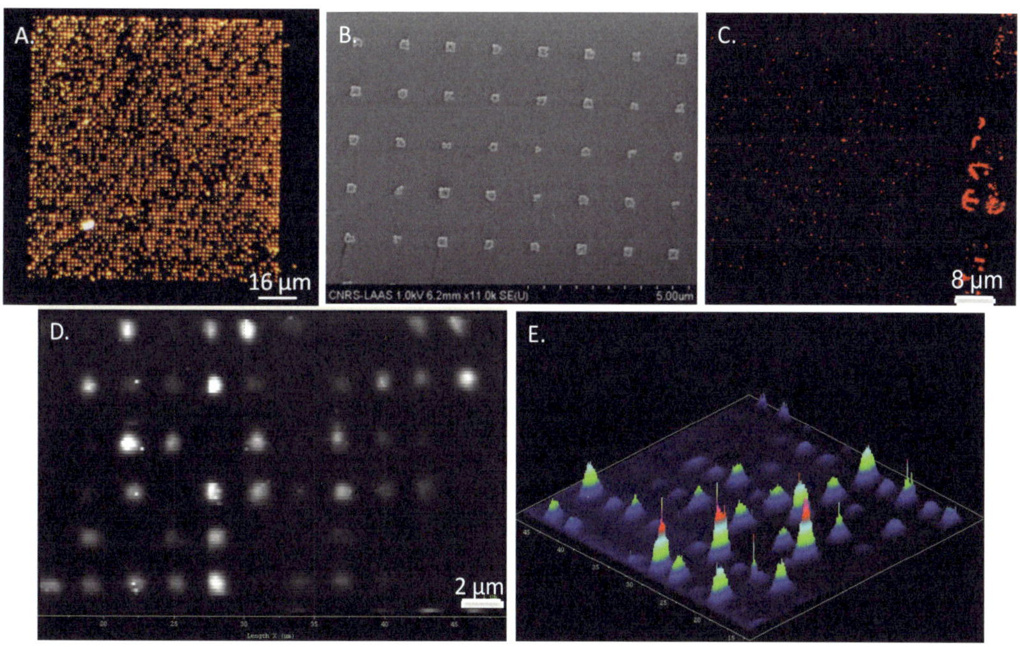

Fig. 3. A) Dark field image of a periodic bi dimensional matrix of 100nm gold nanoparticles. B) Scanning Electron Microscope image of a nanoparticle matrix. C) Confocal image of a controlled array of gold nanoparticles revealing a metal- enhanced fluorescence effect. D) Raman mapping of a controlled array of gold nanoparticles and E) its equivalent in intensity levels revealing a SERS effect.

So we could later imagine including these imaging-enhancing structures colocated with DNA tethers for single molecule fluorescence.

4. Conclusion

Therefore, soft-lithography has become an invaluable technique at the service of single objects studies. This process is essential in the creation of chips with single cell, molecule or particle resolution. We have set a versatile toolbox based on soft-lithography combined with capillary assembly and incubation techniques to generate controlled arrays of single objects at the micrometer and the nanometer scale for different applications. We have shown that the strategies developed are very efficient to generate arrays of all kinds of objects ranging from cells, molecules to metallic nanoparticles with a high success rate in a precise, large-scaled and cost-efficient manner. We have demonstrated that our methodology is capable of high-throughput, and is fully compatible with different types of supports including glass or silicon. The objects assembled preserve their functionality and intrinsic properties and exhibit remarkable potential for a great number of applications in the fields of medical analysis, gas sensing, optics, but also in the construction of tomorrow's nanoscale devices.
The next step in this development would be to couple these techniques and results to build more complex scaffolds and monitor specific interactions at the single object level. For instance, one should later imagine evolving towards a label-free biochip with optical readout by coupling SERS substrates with biological molecules or cells.

5. References

Braun, E.; Eichen, Y.; Sivan, U. & Ben-Yoseph, G. (1998). DNA-templated assembly and electrode attachment of a conducting silver wire. *Nature*, 391, 6669, 775, ISSN 0028-0836

Cavallini, M.; Biscarini, F.; Farran-Morales, A.; Massi M.; Leigh, D. A. & Zerbetto, F. (2002). Spontaneous fabrication of microscopic arrays of molecular structures with submicron length scales. *Nano Lett.*, 2, 635, ISSN 1530-6992

Cerf, A.; Cau, J.C. & Vieu, C. (2008). Controlled assembly of bacteria on chemical patterns using soft lithography. *Coll. Surf. B.*, 65, 2, (September 2008) 285-291, ISSN 0927-7765

Cerf, A.; Cau, J. C.; Vieu, C. & Dague, E. (2009). Nanomechanical properties of dead or alive single patterned bacteria. *Langmuir*, 25, 10, (May 2009) 5731-5736, ISSN 0743-7463

Cerf, A.; Thibault, C.; Geneviève, M. & Vieu, C. (2009). Ordered arrays of single DNA molecules by a combination of capillary assembly, molecular combing and soft-lithography. *Microelec. Eng.*, 86, 4-6, (April-June 2009) 1419-1423, ISSN 0167-9317

Cerf, A. & Vieu, C. (2009). Transfer printing of sub-100 nm nanoparticles by soft lithography with solvent mediation. *Coll. Sur. A.*, 342, 1-3, (June 2009) 136-140, ISSN 0927-7757

Cerf, A.; Molnar, G.; Vieu, C. (2009). Novel approach for the assembly of highly efficient SERS substrates. *ACS Appl. Mater. Interfaces*, 1, 11, 2544-2550, ISSN 1944-8244

Chou, S. Y.; Zhuang, L. & Guo, L. (1999). Lithographically induced self construction of polymer microstructures for resistless patterning. *Appl. Phys. Lett.*, 75, 1004, ISSN 1077-3118

Chou, S. Y. & Zhuang, L. *J.* (1999). Lithographically induced self-assembly of periodic polymer micropillar arrays. *Vac. Sci. Technol. B*, 17, 3197, ISSN 1520-8567

Deng, Z.X. & Mao, C.D. (2003). DNA-templated fabrication of 1D parallel and 2D crossed metallic nanowire arrays. *Nano Lett.*, 3, 1545, ISSN 1530-6992

Denkov, N. D.; Velev, O.D.; Kralchevsky, P.A.; Ivanov, I.B.; Yoshimura, H. & Nagayama, K. (1992). Mechanizm of formation of two-dimensional crystals from latex particles on substrates. *Langmuir*, 8, 3183, ISSN 0743-7463

Haynes, C. L.; McFarland, A. D. & Duyne, R. P. V., (2005a). Surface-enhanced Raman spectroscopy. *Anal. Chem.*, 77, 338A-346A, ISSN 0003-2700

Käll, M.; Xu, H. X. & Johansson, P. (2005). Field enhancement and molecular response in surface-enhanced Raman scattering and fluorescence spectroscopy. *J. Raman spectr.*, 36, 510-514, ISSN 1097-4555

Keren, K.; Krueger, M.; Gilad, R.; Ben-Yoseph, G.; Sivan, U. & Braun, E. (2002). Sequence-Specific Molecular Lithography on Single DNA Molecules. *Science*, 297, 72, ISSN 1095-9203

Kinge, S.; Crego-Calama, M. & Reinhoudt D. N. (2008). Self-assembling nanoparticles at surfaces and interfaces. *ChemPhysChem.*, 9, 20-42, ISSN 1439-7641

Klein, D.L.; McEuen, P.L.; Katari, J. E. B.; Rothand, R. & Alivisatos, A. P. (1996). An approach to electrical studies of single nanocrystals. *Appl. Phys. Lett.*, 68, 2574, ISSN 1077-3118

Köhler, J. M.; Csáki, A.; Reichert, J.; Möller, R.; Straube, W. & Fritzsche, W. (2001). Selective labeling of oligonucleotide monolayers by metallic nanobeads for fast optical readout of DNA-chips. *Sens. Actuators B.*, 76, 166–172, ISSN 0925-4005

Kraus, T.; Malaquin, L.; Schmid, H.; Riess, W.; Spencer, N. D. & Wolf, H. (2007). Nanoparticle printing with single-particle resolution. *Nat. Nanotechnol.*, 2, 570-576, ISSN 1748-3387

Kumar, A. & Whitesides, G. M. (1993). Features of gold having micrometer to centimeter dimensions can be formed through a combination of stamping with an elastomeric stamp and an alkanethiol ink followed by chemical etching. *Appl. Phys. Lett.* 63, 2002-2004, ISSN 0003-6951

Malaquin, L.; Kraus, T.; Schmid, H.; Delamarche, E. & Wolf, H. (2007). Controlled particle placement through convective and capillary assembly. *Langmuir*, 23, 11513-11521, ISSN 0743-7463

Monson, C.F. & Woolley, A.T. (2003). DNA- templated construction of copper nanowires. *Nano Lett.*, 3, 359, ISSN 1530-6992

Morariu, M. D.; Voicu, N. E.; Scha¨ffer, E.; Lin, Z.; Russel, T. P. & Steiner, U. (2003). Hierarchical structure formation and pattern replication induced by an electric field. *Nat. Mater.*, *2*, 48, ISSN 1476-4660

Nguyen, K.; Monteverde, M.; Filoramo, A.; Goux-Capes, L.; Lyonnais, S.; Jegou, P.; Viel, P.; Goffman, M. & Bourgoin, J. P. (2008). Synthesis of thin and highly conductive DNA-based palladium nanowires. *Adv. Mat.*, 20, 1099–1104, ISSN 1521-4095

Nie, S. M. & Emory, S. R. (1997). Probing single molecules and single nanoparticles by Surface-Enhanced Raman Scattering. *Science*, 275, 1102, ISSN 1095-9203

Quake, S. R. & Scherer, A. (2000). From micro to nano fabrication with soft materials. *Science*, 290, 1536, ISSN 1095-9203

Wang, J.; Sun, X.; Chen, L. & Chou, S. Y. (1999). Direct nanoimprint lithography of submicron organic light-emitting structures. *Appl. Phys. Lett.*, 75, 2767, ISSN 1077-3118

Xia, Y.; Qin, D. & Whitesides, G. M. (1996). Microcontact printing with a cylindrical rolling stamp: A practical step toward automatic manufacturing of patterns with submicrometer-sized features. *Adv. Mater.*, 8, 1015–1017, ISSN 1521-4095

Xia, Y. & Whitesides, G. M. (1998). Soft lithography. *Angew. Chem. Int. Ed.*, 37, 550–575, ISSN 1521-3773

Xu, L.; Robert, L.; Ouyang, Q.; Taddei, F.; Chen, Y.; Lindner, A. B. & Baigl, D. (2007). Microcontact printing of living bacteria arrays with cellular resolution. *Nano Lett.*, 7, 2068-2072, ISSN 1530-6992x. 0,

Yan, X.; Yao, J.; Lu, G.; Chen, X.; Zhang, K. & Yang, B. (2004). Macrocontact printing of colloidal crystals. *J. Am. Chem. Soc.*, 126, 34, 10510–10511, ISSN 0002-7863

Zhao, X.-M.; Xia, Y. & Whitesides, G. M. (1996). Fabrication of three-dimensional microstructures: microtransfer molding. *Adv. Mater.*, 8, 837, ISSN 1521-4095

Nanoimprint Lithography

23

Nanoimprint Lithography

Hongbo Lan[1,2] and Yucheng Ding[2]
[1]*School of Mechanical Engineering, Shandong University,*
[2]*State Key Laboratory for Manufacturing System Engineering, Xi'an Jiaotong University*
China

1. Introduction

The Nanoimprint lithography (NIL) is a novel method of fabricating micro/nanometer scale patterns with low cost, high throughput and high resolution (Chou et al., 1996). Unlike traditionally optical lithographic approaches, which create pattern through the use of photons or electrons to modify the chemical and physical properties of the resist, NIL relies on direct mechanical deformation of the resist and can therefore achieve resolutions beyond the limitations set by light diffraction or beam scattering that are encountered in conventional lithographic techniques (Guo, 2007). The resolution of NIL mainly depends on the minimum template feature size that can be fabricated. Compare with optical lithography and next generation lithography (NGL), the difference in principles makes NIL capable of producing sub-10 nm features over a large area with a high throughput and low cost (Chou et al., 1997). Therefore, the charm of NIL largely comes from its capability for patterning with high resolution, high fidelity, high throughput, and low cost. In addition, nanometer sized patterns can easily be formed on various substrates, e.g., silicon wafers, glass plates, flexible polymer films, and even nonplanar substrates. The process has been added to the International Technology Roadmap for Semiconductors (ITRS) for the 32 and 22 nm nodes. Toshiba, moreover, has validated it for 22 nm and beyond. What is more significant is that NIL is the first sub-30 nm lithography to be validated by an industrial user (Yoneda et al., 1997).

Nanoimprint lithography was first invented by Chou and his students in 1995 as a low-cost and high throughput alternative to photolithography and e-beam lithography (EBL) for researchers who need high resolution patterning, motivated by the high expense and limited resolution of optical lithography. Due to historical reasons, the term NIL initially refers to a hot embossing lithography (HEL) process, and was also used as a synonym for thermal NIL (Chou et al., 1995). However, NIL has now an extended meaning which includes not only two fundamental types (Hot Embossing Lithography and UV-based Nanoimprint Lithography, UV-NIL) but also many different variations developed such as roll imprint process, laser-assisted direct imprint, reverse imprint lithography, substrate conformal imprint lithography, ultrasonic NIL, etc. Compared to other lithography processes and next generation lithography with nanoscale resolution, such as e-beam lithography and extreme ultraviolet lithography (EUVL), the most prominent advantage of NIL is its ability to pattern 3D and large-area structures from micron to nanometer scale and its potential to do so at a high throughput and low cost. This paves the way for many applications in the area of data storage, nano-optoelectronic, optical elements, NEMS and NOEMS, etc (Balla et al., 2008). Furthermore, NIL has currently demonstrated great potential and commercial prospects in many application fields, such as Hard Disk Drive (HDD), LED, OLED, flexible display, optical and biological devices, etc. After more than ten years of NIL process development, a range of standard tools and materials is available from various industrial providers.

NIL technology involves two fundamental aspects: the basic research and the application research, as shown in Fig.1. The basic research consists of the process, tool, template (mold), material (resist, functional material, etc) which aim to meet the different application requirements, namely the micro/nano structures or devices fabrication. NIL applications mainly cover nanoelectronics, nano-optoelectronics, nanophtonic, nano-biology, optical components, etc.

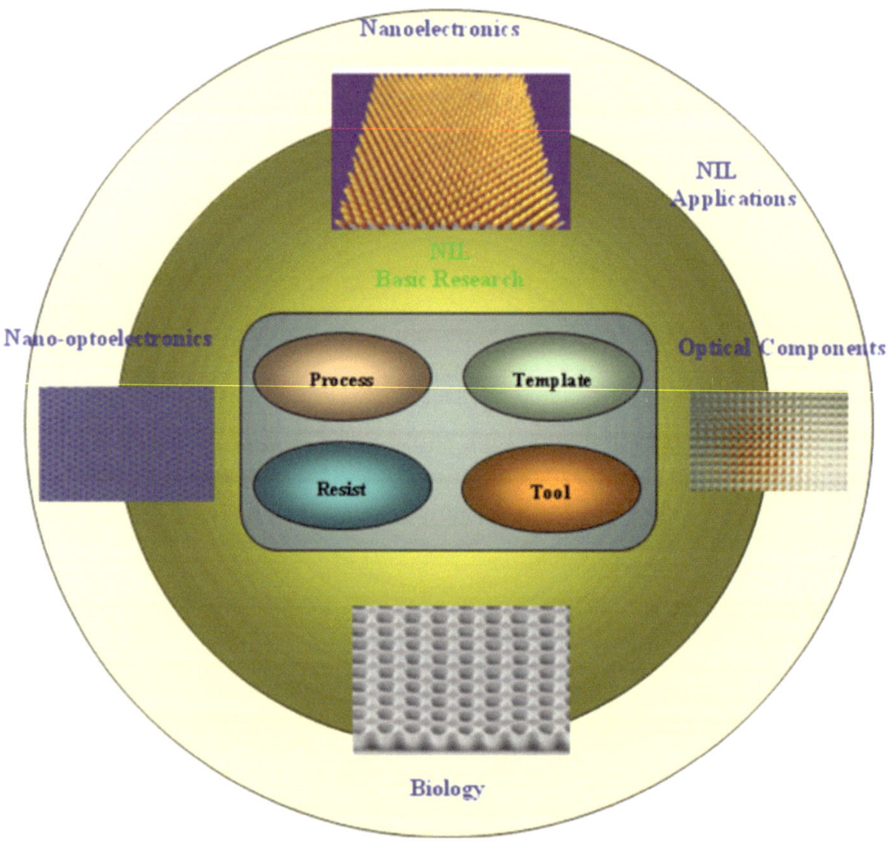

Fig. 1. Overview of the NIL

The chapter systematically presents the NIL technique and various applications. Moreover, some key issues and recent progresses for NIL are discussed in detail. Finally, the prospect and challenges in NIL are also addressed. The aim of this chapter is to make readers to systematically and deeply understand NIL, and better apply the new technique.

2. NIL process

2.1 Principle of NIL

NIL is based on the principle of mechanically modifying a thin polymer film (mechanical deformation of the resist) using a template (mold, stamp) containing the micro/nanopattern, in a thermo-mechanical or UV curing process. In other words, NIL uses the direct contact between the mold (template) and the thermoplastic or UV-curable resist to imprint (or replicate) the pattern, unlike optical lithography, does not require expensive and complex optics and light sources for creating images. The switch from using light to using contact to pattern brings some advantages. For instance, it can therefore achieve resolutions beyond the limitations set by light diffraction or beam scattering that are encountered in conventional techniques, simplifies process and largely reduces cost. However, that will also indeed bring new challenges and issues, the most important of which are alignment and the 1x mask/template fabrication. Since NIL can be considered as such a process based on squeeze flow of a sandwiched viscoelastic material between a mold and a substrate, the property of interface between the two materials has to be considered throughout the entire process, both from topographical, chemical, and mechanical points of view. Furthermore, the characteristics of the interface and surface have a great impact on the demolding capability and filling behavior which can strongly influenced pattern quality and throughput (Schift, 2008, Bhushan, 2007, Guo, 2004).

The patterned polymer can even act as a functional device, e.g. lens for imaging sensors, micro fluidic chip, biomedical array etc. It can also be used as a high resolution mask for subsequent steps of the process (metal deposition, electroplating, etching and lift-off process). Moreover, various substrates, including silicon wafers, glass plates, flexible polymer films, polyethylene terephtalate (PET) polymer film, and even nonplanar substrates can be utilized for NIL (Costner et al., 2009). The ultimate resolution of the patterns fabricated by NIL is primarily determined by the resolution of the features on the surface of the mold. Because of the 1X nature of NIL compared with 4 X for photolithography, the 1X template must be more accurate than conventional masks.

As a result, distinct features for NIL involve two points: (1) the contact nature of the process; (2) direct mechanical deformation of the resist. Two crucial steps, namely the resist filling rheology behavior and demold capabilities, have decisive influence on transferred pattern quality and throughput for NIL. The particular advantage of NIL compared to other lithography techniques and NGL is the ability to fabricate large-area and complex three-dimensional (3D) micro/nanostructures with low cost and high throughput.

2.2 Two fundamental processes for NIL

Currently, there are a great variety of NIL process types, but two of them are most important and fundamental: Hot Embossing Lithography (HEL) or thermal nanoimprint lithography (T-NIL), UV-based Nanoimprint Lithography (UV-NIL), as shown in Fig. 2 (Steward & Willson, 2005). Both thermal and UV-NIL have demonstrated a sub-10 nm resolution.

T-NIL is the earliest NIL developed by Stephen Chou's group. In a standard T-NIL process, a thin layer of imprint resist (thermoplastic polymer) is spin-coated onto the substrate. Then the mold, which has predefined topological patterns, is brought into contact with the substrate and they are pressed together under certain pressure. When heated up above the glass transition temperature (Tg) of the polymer, the feature pattern on the mold is pressed into the melt polymer film. After being cooled down, the mold is separated from the

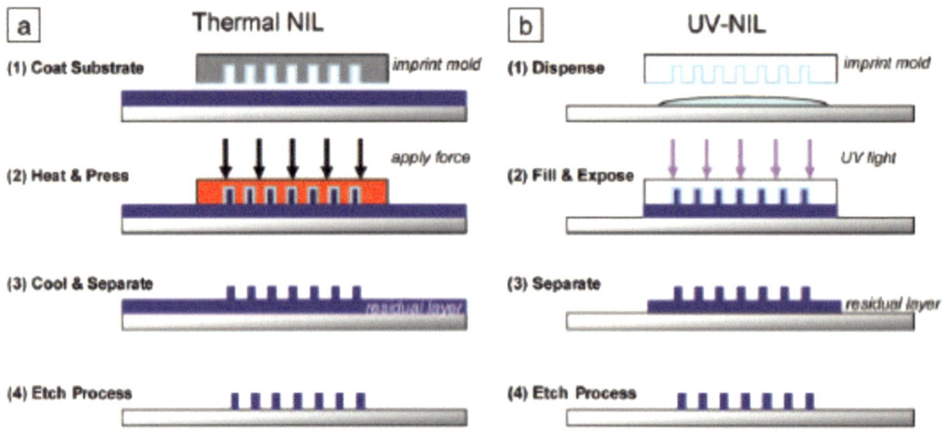

Fig. 2. Two fundamental process types for NIL (Steward & Willson, 2005)

substrate and the pattern resist is left on the substrate. A subsequent pattern transfer process (e.g. reactive ion etching) can be used to transfer the pattern in the resist to the underneath substrate. In UV-NIL, a UV-curable liquid photopolymer instead of thermoplastic as resist is applied to the substrate and the mold is normally made of transparent material like fused silica, quartz mold. After the mold and the substrate are pressed together and the cavities (trenches) are fully filled by resist, then the resist is cured in UV light and becomes solid. After demolding, a similar pattern transfer process can be used to transfer the pattern in resist onto the underneath material. The polymer residual layer is removed (Nanoimprint lithography, 2009). The basic difference between UV-NIL and T-NIL is that a resin, which is liquid at room temperature, is shaped by a moderate pressure, which is then crosslinked and hardened by curing. Each process has of its own prominent advantages, e.g. while UV-NIL can be performed at room temperature and low pressure, hot embossing is low-cost since nontransparent molds can be used (Less restrictions on mold). Schift and Kristensen provide a comparison of T-NIL and UV-NIL, with typical parameters of current processes (Bhushan, 2007). However, UV-NIL has established itself as a promising alternative to NIL in which imprint lithography is conducted at room temperature under low pressure conditions. UV-NIL is one of the most important NIL technologies for structuring of large wafer areas up to 300 mm in diameter. The process offers several decisive technical advantages concerning overlay alignment accuracy, simultaneous imprinting of micro- and nanostructures and tool design due to the absence of high imprint pressures and thermal heating cycles (Fuchs et al., 2008, Bender et al., 2006).

UV-NIL offers two approaches for patterning using either rigid quartz glass molds (Hard UV-NIL) or soft molds (Soft UV-NIL) for structuring of UV sensitive resists resulting in an etching mask for the substrate to be patterned (Glinsner et al., 2007, Plachetka et al., 2004). For UV-NIL using a rigid mold, the hard mold brings about two weaknesses, one is the sticking characteristic which can lead to the following shortcomings that a release agent or surfactant is necessary, and the demolding force is especially large. Another is the limitation in imprint area due to having the surface waviness onto the mold and the substrate surfaces. Furthermore, it is rather difficult to ensure uniform and parallel surface contact between a template and a wafer during imprinting process. Compared with the hard mold, using a soft or flexible mold can avoid the conglutinating of resist, acquire high precision feature, enlarge the pattern transferring area, reduce the parallelism error between the mold and the substrate and lengthen the life-time of the master. In particular, high flexibility of the mold enabled conformal contact as well as imprinting at significantly reduced imprint pressure. However, the elastomeric behavior of the soft mold has both, positive and negative attributes. On one side it offers numerous advantages, but one the other hand some defects such as resolution limitations and non-uniformity of the transferred patterns, etc., will have to be considered and resolved. In addition, Swelling is a commonplace issue with PDMS based molds since most organic liquids will swell PDMS. In the case of flexible mold local deformations limit the resolution of soft UV-NIL principally. Therefore, compared to HEL and UV-NIL used rigid molds, it is particularly important for the soft UV-NIL to understand and reduce as much as possible the mold deformation for the practical application of the technique. The current capability for the process from AMO and Süss can ensure nanoscale resolution down to sub 50 nm and perfect pattern stability (SCIL, 2009). However, the deformations of the soft mold during imprinting process which can cause serious consequences have to be considered for the practical application of the process and further improving the pattern resolution (Lan et al., 2009).

The Jet and Flash™ Imprint Lithography (J-FIL®) process, formerly called Step and Flash® Imprint Lithography, (S-FIL®) developed at the University of Texas at Austin, is a typical and fundamental UV-NIL process. The fused silica surface, coated with a release layer, is gently pressed into a thin layer of low-viscosity resist. The resist is deposited in a customized pattern matching the template using the IntelliJet™ Drop Pattern Generator. When illuminated by a UV lamp, the surface is polymerized into a solid layer. Upon separation of the fused silica template, the pattern is left on the substrate surface. A residual layer of polymer between features is removed by an etch process, and a perfect replica of the pattern is ready to be used in subsequent processing for etch or deposition. Fig.3. demonstrated the Jet and Flash™ imprint lithography process. Sub-10nm features have been made that exceed the present requirements outlined in the ITRS, as well as most patterned media roadmaps (J-FIL, 2009).

In addition, the NIL process can also be divided into two broad categories: single-step imprinting and multi-step imprinting based on the size of a template and imprinting time. The former is to imprint resist on a wafer using a wafer-sized template at a time, whereas the latter is to use a chip-sized template by the step and repeat process. For the single-step NIL, the wafer-sized template consists of multiple groups of chip-sized patterns that are uniformly distributed with the equal in between space. According to the layer number of an imprinting pattern, it can be further divided into the monolayer patterning process (single level) and multilayer patterning process (multilevel). High-resolution overlay is considered to be an important challenge for multilayer imprinting processes. Therefore, there are four types of corresponding imprinting processes: (i) single-step and monolayer, (ii) single-step and multi-level, (iii) multi-step and monolayer, (iv) multi-step and multi-level (a real commercialization process). The current NIL machines mainly involve first these forms of (i), (ii) and (iii).

2.3 Variants of NIL

In the recent years, a variety of new processes have been proposed and investigated, such as reverse NIL, soft UV-NIL, Laser assisted direct imprint (LADI), Sub-10 nm NIL, chemical nanoimprint, electrical field-assisted NIL, etc., which aim to implement the micro/nano structures fabrication with large area, 3-Dimension, high throughput, high resolution, free

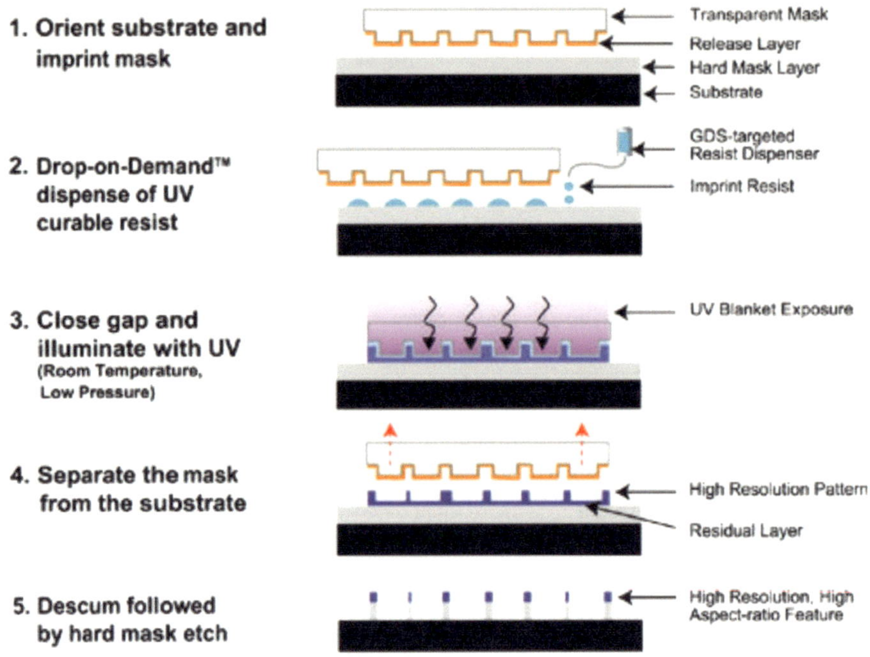

Fig. 3. Jet and Flash™ imprint lithography process (J-FIL, 2009)

defect, and to directly make various functional structures, as well as to improve the throughput and pattern quality.

2.3.1 Combined thermal and UV-NIL

STU® (Simultaneous Thermal and UV) technology form Obducat enables simultaneously combined thermal and UV-NIL, as shown in Fig.4, allowing the complete imprint sequence into UV-curable thermoplastic pre-polymers to be performed at a constant temperature. By using the unique STU® technology, problems related to thermal expansion mismatch between stamp and substrate are avoided. The method allows the use of spin-coated UV-curable polymers with a homogeneous thickness distribution on wafer scale, crucial for CD control and enabling pattern transfer to an underlying substrate. Obducat has proved the ability to imprint 17nm features with its proprietary IPS (intermediate polymer stamp)-STU process. Furthermore, 17 nm dots have been printed uniformly with a residual layer below 7 nm (Key Technologies, 2009).

2.3.2 Reverse imprint process

For the reverse NIL process, a polymer film is firstly spin-coated onto the mold (rather than substrate), the polymer will fill up the trench regions of the surface relief patterns. This means that a replica of the mold pattern is formed in the polymer film simply by spin coating. Subsequent, this film can be transferred from the mold to a substrate, patterned structures are obtained. The key to the successful film transfer lies in the fact that the mold has a lower surface energy than does the substrate, and so the polymer film has better adhesion to the substrate and therefore can be detached from the mold. The process has the ability to construct the three-dimensional and multilayer micro/nanostructures.

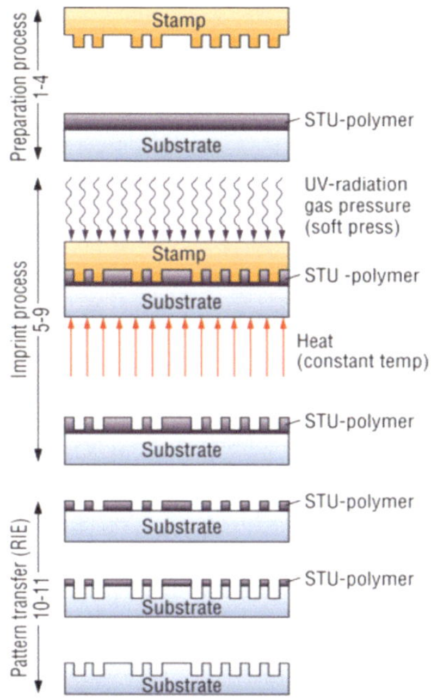

Fig. 4. A simultaneous thermal and UV (SUV®) imprint process (Key Technologies, 2009)

Furthermore, the crucial advantage of this technique is the possibility to construct three-dimensional device-like structures without having to etch polymer residual layer at any intermediate step. Some devices such as three-dimensional photonic crystals, multi-layered nano-channels, polymer optical devices, gold gratings (metallic nanostructures) have been fabricated using the reverse NIL process or the combination of the reverse imprint process and other micro-fabrication technologies (Guo, 2004, Kehagias et al., 2006, Han et al., 2007). Fig.5 shows the schematics of a reverse UV contact NIL for 3D Nanofabrication (Kehagias et

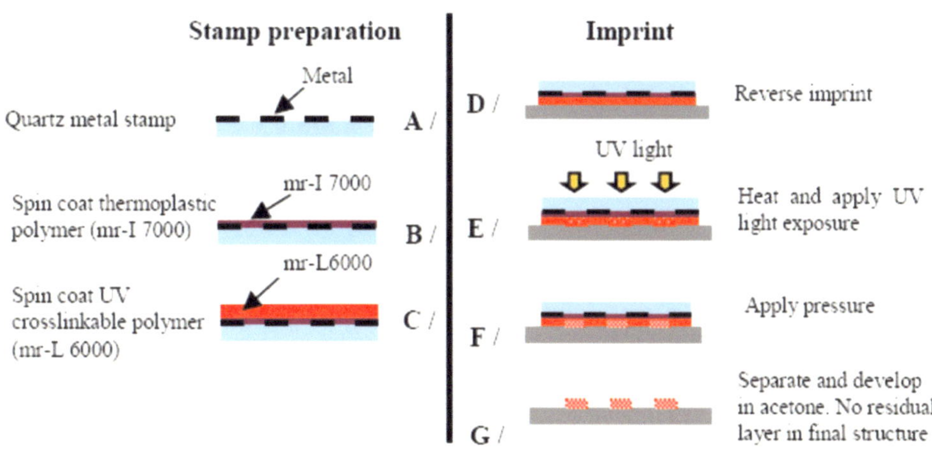

Fig. 5. Reverse UV contact NIL process (Kehagias et al., 2006)

al., 2006). Combined the UV-curable reverse NIL process with a water soluable PVA (polyvinyl alcohol) based removable template and home made UV-curable glue, Lee's team implement the successful fabrication of multi-stacked 2D nano patterned slabs on various substrates including flexible polymer film. The highlight for the process is to develop a PVA mold which be dissolved by water (Han et al., 2007).

2.3.3 Laser-assisted direct imprint

Laser assisted direct imprint (LADI) is a rapid technique for patterning nanostructures that does not require etching. LADI is based on the following principle: a single excimer laser pulse melts a thin surface layer of the functional materials, and a mold is embossed into the resulting liquid layer, as shown in Fig. 6. It has been used for making nanostructures in silicon and metals with a resolution better than 10 nm. LADI offers direct patterning without etching for compound semiconductors which are hard to be etched. Using this method, applicants have directly imprinted into silicon large area patterns with sub-10 nanometer resolution in sub-250 nanosecond processing time. The method can also be used with a flat molding surface to planarize the substrate. The high resolution and speed of LADI could open up a variety of applications and be extended to other materials and processing techniques (Chou al., 2002).

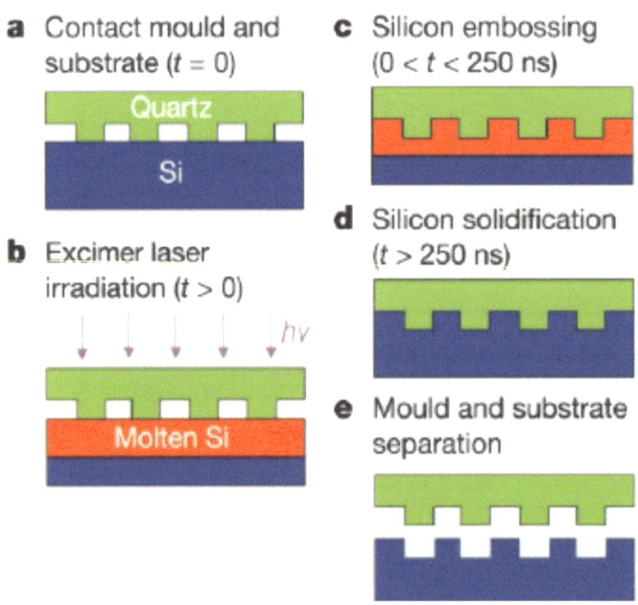

Fig. 6. Laser assisted direct imprint (LADI) (Chou al., 2002)

2.3.4 Roll imprint process

For conventional NIL processes, one of the most important problem is that it cannot significantly improve the throughput in the patterning of large area product with low cost because it is not a continuous process. To overcome this limitation, roller-type nanoimprint lithography (RNIL) has been developed and is becoming the most potential manufacturing method for industrialization of nanoimprinting process, due to its prominent advantage of continuous process, simple system construction, high-throughput, low-cost and low energy consuming. Compared to other NIL processes, the unique advantage for the RNIL is only a continuous process with a high throughput to fabricate the large-area patterns. The RNIL involves three essential steps: deposition, patterning and packaging (Fig.7 (II)). Two molds (roller mold and flat mold, Fig. 7 (III)) and two substrates (flexible substrate and rigid substrate, Fig. 7 (I)) can be used for the RNIL (Lan et al., 2008, Ahn et al., 2006, Ahn & Guo, 2008, Youn et al., 2008, Kao et al., 2005, Lee et al., 2008, Chang et al., 2006). Lan et al., have presented a general literature review on the RNIL (Lan et al., 2008). Guo et al., demonstrated a Roll-to-Roll NIL process in which polymer patterns down to 70nm feature size were continuously imprinted on a flexible web (Ahn & Guo, 2008). A thermal roller imprint lithography (RIL) system was developed and applied to RIL tests to evaluate its feasibility for the large area replication of an optical micro device. The system has the capacity to replicate ultra-precision structures on an area of 100mm×100mm at the scanning speed range of 0.1–10 mm/s. A light guide plate (LGP) for a back light panel was fabricated. The system is suitable for the fabrication of various optical micro devices such as flat panel displays, electronic papers, functional films, and others (Youn et al., 2008). A combination method of the roller-type imprinting lithography and photolithography (CRIP), followed by wet chemical etching was used to fabricate the patterned organic light emitting devices (OLEDs) with pixels of 500 μm × 300 μm on the flexible PET substrates. Compared with the conventional imprint lithography or photolithography, CRIP using the hybrid mold has the advantages of better uniformity, less force, less time-consuming, lower cost and higher aspect ratio. This technique is potentially cost-effective, offers high throughput, less time-consuming and is suitable for fabrication on flexible substrate (Chang et al., 2006).

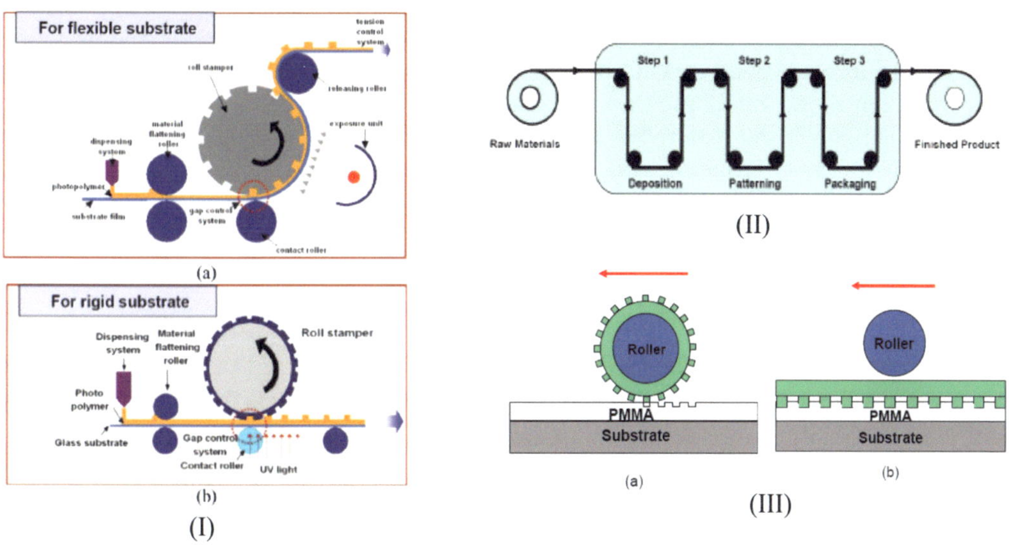

Fig. 7. Roller-type Nanoimprint Lithography (Lan et al., 2008)

2.3.5 Large area imprint

Large area imprint over the full wafer surface is needed often in single layer imprinting when high throughput is needed. In a full wafer nanoimprint scheme, all the patterns are contained in a single nanoimprint field and will be transferred in a single imprint step. This allows a high throughput and uniformity. To ensure the pressure and pattern uniformities

of full wafer nanoimprint processes and prolong the mold lifetime, a pressing method utilizing isotropic fluid pressure, named Air Cushion Press (ACP) by its inventors, was developed and being used by commercial nanoimprint systems. ACP applied air pressure for conformal contact and imprint, therefore achieved ultra-uniformity over the whole imprint field. It has advantages over traditional parallel plate press: capable of handling the substrate with uneven back and pattering on curved surface. Presently, Nanonex provides three series of NIL tools (NX-1000, NX-2000, and NX-3000) for T-NIL and P-NIL, with and without alignment. All of them use the ACP to achieve excellent pattern uniformity (Tan et al., 2004, Gao et al., 2006). Pelzer et al., studied full wafer replication of nanometer features, and presented results on full wafer imprints up to 200mm with high-resolution patterns for microelectronic applications. There are no physical limitations encountered with imprinting techniques for fully replicated structures, in the sub-10nm range. The real challenge for the technique is its utilization for dense structured full wafer imprints up to 200mm (Pelzer et al., 2005).
The step and repeat process is another approach to pattern on large areas (e.g. 300 mm wafers scale). The imprint field (die) is typically much smaller than the full wafer nanoimprint field. The die is repeatedly imprinted to the substrate with certain step size. This scheme is good for nanoimprint mold creation. It is currently limited by the throughput, alignment and street width issues (Jeong et al., 2005).

2.3.6 Substrate conformal imprint lithography

SCIL (Substrate conformal imprint lithography), that was developed based on a cooperation between Philips Research and SUSS MicroTec in 2008, is an enabling technology offering large-area soft stamps with repeatable sub-50nm printing capability, while avoiding stamp deformation as no contact force is applied, non-UV based curing at room temperature and allowing high aspect ratios even up to 1:5 and more. The new SCIL technology has been designed for sub-50nm patterning and is bridging the gap between small rigid stamp application for best resolution and large-area soft stamp usage with the usual limited printing resolution below 200nm. Süss believes that the SCIL represents an enabling new technology that paves the way for further commercialization of NIL (SCIL, 2006).

2.3.7 Nanoelectrode lithography

Nanoelectrode lithography, which is a pattern duplication method that combines nanoimprint with an electrochemical reaction. The conductive mold pattern undergoes an electrochemical reaction that enables an oxide pattern to be fabricated directly on the surface of a semiconductor or metal layer. Since this technique transfers the mold pattern to a target surface chemically, it is categorized as chemical nanoimprint, while conventional nanoimprint physically transfers a mold pattern having peaks and valleys to the target (Fig.8). This patterning phenomenon gives nanoelectrode lithography some advantages such as resistless patterning and multiple patterning, which will improve the accuracy and flexibility of nanoimprint (Yokoo & Namatsu, 2009).

2.3.8 Hybird NIL process

Mix-and-match approaches are used to combine the advantages of two or more lithographic processes or simply to avoid their mutual disadvantages. This is also a way to improve throughput and reliability, e.g. since the fabrication of large-area nanostructures is often costly, the definition of microstructures can be done with optical lithography, while the

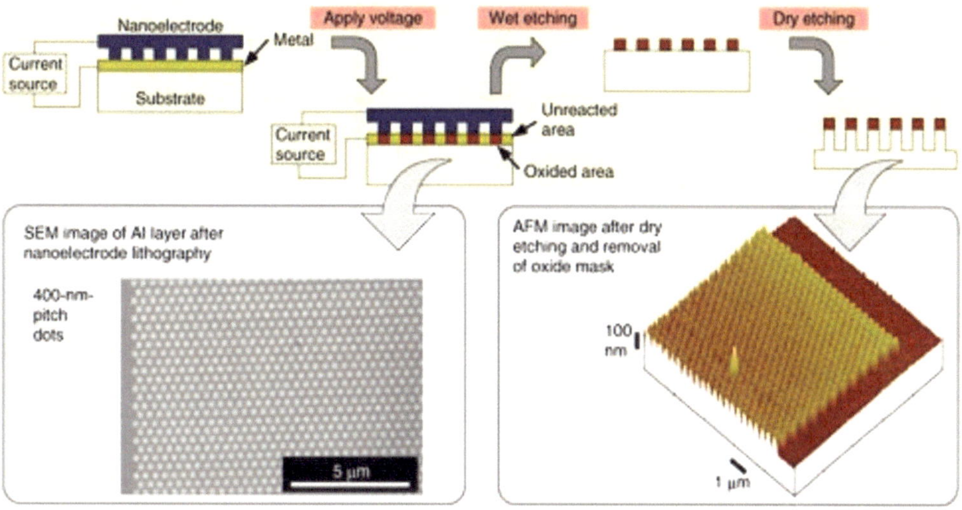

Fig. 8. Chemical nanoimprint (nanoelectrode lithography) (Yokoo & Namatsu, 2009)

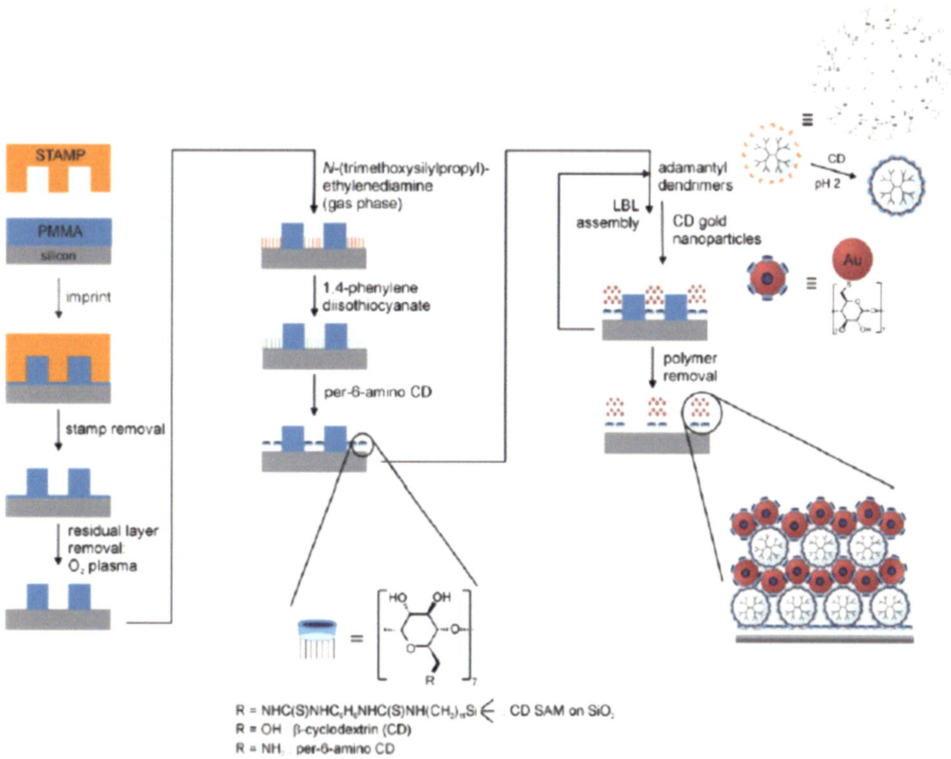

Fig. 9. Integrated nanofabrication scheme on nanoimprinted pattern (Plachetka et al., 2008)

nanopatterning of critical structures in small areas can be done by NIL. An integrated process combining top-down nanoimprint lithography and bottom-up layer-by-layer (LBL) self-assembly was applied to the fabrication of 3D hybrid nanostructures, as shown in Fig.9.

The method can implement the fabrication of 3D nanoobjects of arbitrary shapes on substrates, where the x,y dimensions are determined by NIL and the z dimension by the LBL assembly. The process can be used to fabricate the photonic components. Passive photonic devices in silicon waveguide technology have been fabricated with quite acceptable results in comparison with other lithography methods (Plachetka et al., 2008). Cheng and Guo proposed a combined-nanoimprint-and-photolithography (CNP) technique which introduced a hybrid mask mold made from UV transparent material and with a light-blocking metal layer placed on top of the mold protrusions. The CNP method using such a hybrid mold can achieve resist patterns without residual layer, and the resist patterns can have higher aspect ratio than the feature on the mold. In addition, the photoresist used in the CNP technique can provide higher etching durability compared with thermal plastic polymers that are commonly used in NIL. Compared with contact photolithography techniques, the CNP can achieve much higher resolution by reducing the effective resist thickness down to tens of nanometers (Cheng & Guo, 2004).

2.3.9 High resolution NIL

HP and MIT developed a sub-10 nm NIL by wafer bowing, introduce the concept of wafer bowing to affect nanoimprinting, as shown in Fig. 10. In the scheme, the imprint force is applied uniformly and systematically from center to edge, preventing air from being trapped. More importantly, it shortens the mechanical path between the mold and wafer; this makes the imprinter less susceptible to ambient vibration and helps to preserve the alignment during mold-wafer approach. After an UV exposure step, air can be let into the module to effect mold-wafer detachment. These will enable achieving excellent patterning and overlay at much lower cost (Wu et al., 2008).

In order to meet the manufacturing requirements of a variety of micro/nano devices and structures, lots of new NIL processes are being proposed and developed in recent year. Here only presents some principal and typical NIL processes. As the rapid development of the NIL technique and micro/nano fabrication technologies, much more innovative processes or methods concerning NIL will emerge in the future.

2.4 Crucial process issues for NIL

2.4.1 Thickness and uniformity of residual layer

A key characteristic of NIL is the residual layer following the imprint process. It is preferable to have thick enough residual layers to support alignment and throughput and low defects. However, this renders the NIL step less critical for critical dimension (CD) control than the etch step used to remove the residual layer (Nanoimprint lithography, 2009). In addition, as a practical technology for the mass production of nanosized patterns, NIL must have an ability to produce a uniform layer with minimal residual layer thickness. Reducing the residual layer is important as it limits the effect RIE has on the resist mask improving tolerance control. Therefore, having a thin and uniform residual layer plays an important role for various NIL processes. In order to satisfy the requirements, various methods have been proposed. Among these include (1) the addition of some aspects of photolithography such as selective UV-curing through a hybrid mask- mold followed by a development step, (2) contrast-modified exposure followed by development, (3) reducing the initial volume of resist to induce incomplete filling of the mold, (4) using high pressure to squeeze excess resin out from between the mold and the substrate, and (5) optimization of droplet positioning in the case of a liquid resin system (Jun et al., 2005, Dumond & Low,

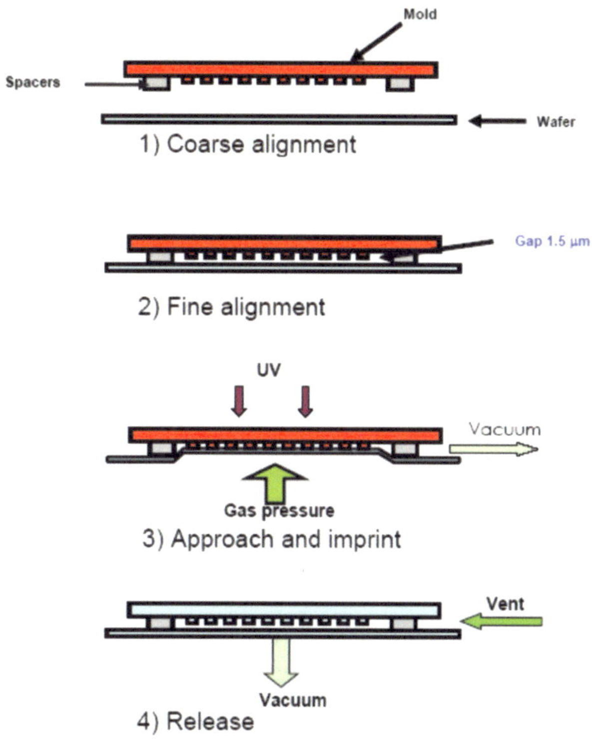

Fig. 10. Sub-10 nm NIL by wafer bowing (Wu et al., 2008)

2008, Lee & Jung, 2004, Bogdanski et al., 2007, Hiroshima, 2008). Jun et al., from HP Lab, devised a novel technique for dispensing resist that takes advantage of the opposing surface energies of the mold and the substrate to produce a uniform and air-free resist film (Jun et al., 2005). Dumond and low developed a method of imprinting resist structures wherein the residual layer is self-removed via failure while desirable resist features are transferred to an external substrate. The uniqueness of this technique is further enhanced by the relative ease with which it can be used to fabricate overhang structures useful for shadowing evaporated materials (Dumond & Low, 2008). Balla et al., discussed the relationship between the residual layer and initial resist thickness. Lee and Jung investigated the factors affecting residual layer thickness in UV-NIL. To obtain imprinted patterns with no residual layer, a thin and uniform layer of imprint resin must initially be applied. Effective rearrangement of the resin during imprinting is also essential to obtain no residual layer. The use of pressurized imprinting and a more fluidic imprint resin are very helpful. The high fidelity transfer of various patterns as small as 150 nm with no residual layer was successfully demonstrated by controlling the initial resin thickness at an imprint pressure of 15 atm (Lee & Jung, 2004). Investigators at University of California at Berkeley have developed a zero residual layer NIL process. This technology is applicable to large-scale nanoimprinting, flexible electronics, printing of nanoscale metal electrodes, patterning photoresist for subsequent processing, among other applications. The method provides a new way to change the fluid and solid interaction so as to exclude more fluid and eliminate the residual layer. Preliminary results show that the gold nanoparticles are confined to patterned features, and excluded from all other areas. The ability to pattern micro/nanoscale features without residual layer formation makes NIL significantly simpler and reduces processing time as compared to prior methods. This technology is application to all NIL n needs (Zero residual layer nanoimprint lithography, 2009).

2.4.2 Pattern fidelity

In most cases, NIL involves two transferring steps for whole imprint course. Firstly, the patterns in the template is replicated to the resist, subsequently, the resist pattern is to be further process for transferring the substrate applying the etching process or functional materials using combination of deposition process and the lift-off technique. Therefore, it is very important to ensure pattern fidelity in NIL process for achieving high-quality pattern. There are two problems to overcome to preserve the pattern fidelity in transferring the polymer pattern to the underlying substrate. One of them involves removing the residual resist from the recessed parts (windows) of the pattern by RIE, depositing a metal layer and applying a lift-off technique such that the metal layer left only in the windows can be used as the etch mask for further processing. Another approach is the use of a multilayer, typically a bilayer of one polymer on another polymer layer or a trilayer in which a layer easily deformable at relatively low temperature is added to a metal–polymer bilayer. This bilayer reversal imprint lithography offers a distinct advantage over other imprint techniques in allowing for a high aspect ratio of the pattern transferred onto a substrate, which has been difficult to obtain for small feature sizes. The method requires only one etching step as opposed to the two etching steps typically needed in the imprint lithography, which can degrade the pattern fidelity (Suh et al., 2004). Li et al., studied the pattern transfer fidelity of NIL by patterning sub-micron MESFET gates on six-inch wafers. The CDs of gate patterns in resist are 5.2% (or 37 nm) on average larger than those on the mould with a standard deviation of 1.2% (or 8 nm), and the CDs after oxygen RIE and metal lift-off are 42% (or 296 nm) on average larger than those on the mould with a standard deviation of 8% (or 30 nm). Compared with conventional photolithography, NIL has higher resolution and better pattern transfer fidelity with CD controls about four times smaller (Li et al., 2003). The primary measure of process quality in NIL is the fidelity of pattern transfer, comparing the dimensions of the imprinted pattern to those of the mold. A rapid, nondestructive technique termed critical dimension small angle X-ray scattering (CD-SAXS) is used to measure the cross sectional shape of both a pattern master, or mold, and the resulting imprinted films. CD-SAXS data are used to extract periodicity as well as pattern height, width, and sidewall angles. Films of varying materials are molded by thermal embossed NIL at temperatures both near and far from the bulk glass transition (Jones et al., 2006).

2.4.3 Defect control

In NIL, defects roughly could be divided into two groups: random distributed and repeated defects. Random distributed defects include particle-associated defect, gap (or void) associated defect, and separation related defects, and the residual after imprint, which are not repeatable in terms of location and amount. Repeated defects include those existing defects on mold and substrate, which are repeated in the process. The gap-associated defect is a unique phenomenon in NIL, which is induced by the incomplete contact between mold and substrate. Defects have been one of the biggest obstacles for NIL to be the real nanofabrication process in the industry. To the nanodevice manufacturer, defects are one of the key issues to product quality and yield (Chen et al., 2005). Unlike in other lithography, a particle induced defect in NIL is larger than the particle itself. To remove the particles, a dry clean process for the nanostructure-patterned surface can be used (Chen et al., 2005). When vacuum is not used during the imprint process, air can get trapped, resulting in bubble defects. This is because the imprint resist layer and the template features are not perfectly flat. There is an elevated risk when the intermediate or master stamp contains depressions (which are especially easy air traps), or when the imprint resist is dispensed as droplets just before imprinting, rather than pre-spun onto the substrate. Sufficient time must be allowed for the air to escape. Bubble defects are unavoidable when UV-NIL is carried out in air. Hiroshima et al.; proposed two methods to eliminate bubbles, i.e., resin squeezing by which bubbles are transferred out of the mold, and gas condensation by which bubbles are considerably decreased in volume. The resin squeezing method, which requires no additional cost, is effective for a thick initial film, but not for a thin initial film because resin flow transporting bubbles out of the mold is insufficient. The gas condensation method does not involve such restrictions. Due some limitations of the gas condensation method, they studied the elimination of bubble defects by gas condensation method using pentafluoropropane. When UV nanoimprint is carried out in an environment of a pentafluoropropane flow higher than 150 sccm with a hold time longer than 20 s, no bubble defects are generated in the entire imprint area. Bubbles are eliminated within a few seconds under an imprint pressure of 0.5 MPa. They consider that bubble elimination using pentafluoropropane is very useful and realizes UV nanoimprint of the same quality as that using vacuum but at a much lower cost (Hiroshima & Komuro, 2007). Liang et al., reported an experimental and theoretical study of two most critical yet still to-be-answered issues in dispensing-based nanoimprint lithography (D-NIL): air bubble formation and absorption, and discuss their impact on NIL yield and throughput. Their study shows that the key factors that affect the air dissolution time (and hence the air bubble shrinking time) are air bubble initial size, imprinting pressure, air solubility, and resist residue layer thickness. One of the key conclusions from the study, which has significant practical importance, is that although the air in a bubble can be completely dissolved in a resist liquid as long as the bubble is smaller than a certain size, the air absorption time might be too long for the dispensing-NIL operating in atmosphere or poor vacuum to have a necessary throughput in mass manufacturing. In addition, when the residual layer thickness is close to zero or a critical thickness, the bubble dissolution process can be significantly slowed down (Liang et al., 2007).

2.4.4 Filling process

The NIL involves two crucial process steps including the resist filling rheology behavior and demold characteristics which have decisive influence on pattern quality and throughput. In order to better understand the NIL mechanism and obtain the optimal imprint conditions (e.g., pressure, temperature, pattern layout of the mold, and time), it is necessary and important to investigate the flow and filling rheology behaviors of the resist. Two fill mechanisms have been observed: simple flow of the PMMA from the borders and formation of polymer mounds. A simple theory was used to estimate the embossing time required to fill a given stamp geometry. In both cases there is evidence of compression causing buckling of the polymer and also capillary action drawing the polymer up to the top of the stamp cavity (Scheer & Schulz, 2001). Heyderman et al., thought that the time to fill a nanostructure array with a large unstructured surrounding area is the same for a microcavity with the same surrounding area and cavity volume. The fastest embossing times for complete fill of the cavities were obtained at temperatures greater than 100 °C above T_g with PMMA viscosities in the range 300 to 3000 Pas (Heyderman et al., 2000). Jeong et al., study simulated hot embossing micro/nano-manufacturing with thin films, including the effects of capillary force and width of stamp groove on the flow behavior during embossing. Simulations showed double peak flow for regions of large width, slow printing speeds and with surface tension as a significant factor (Jeong et al., 2002). Macintyre and Thoms developed a method to directly observe the flow of resist arising from NIL. A fiducial grid was embedded in the polymer layer being imprinted, and they observed its distortions after imprinting (Macintyre & Thoms, 2005). Scheer and Schultz recorded wave-like fronts of resist moving between the stamp and the substrate during imprinting processes operating at more than 100 bar (Scheer & Schulz, 2001). The mechanism in the low-pressure/high-temperature imprinting depends on the type of mold used. When a rigid mold is used, capillarity is the mechanism. In the case of a flexible film mold, however, both capillarity and viscous flow are responsible for the imprinting. As time increases, only capillarity is operative. An understanding of the mechanisms involved in various imprinting methods is valuable on its own. More important is the insight it can provide into possible modifications of the imprinting process that can be tailored for specific applications (Khang & Lee, 2008).

3. NIL mold (Template)

The mold (template, stamp) is one of the most critical elements for the NIL process. The ultimate resolution of the patterns fabricated by NIL is primarily determined by the resolution of the features on the surface of the mold. Because of the 1X nature of NIL compared with 4 X for photolithography, the 1X template fabrication has now been considered the greatest challenge for NIL process. This section will mainly discuss three issues regarding NIL molds: material, fabrication method and crucial process issues.

3.1 Mold material

A variety of materials such as Silicon, SiO_2, Fused Silica (bulk), Quartz (fused), Glass, Silicon Nitride (Si_3N_4), Diamond, Nickel, PDMS, etc., have been utilized to make molds for NIL. The material chosen affects the mold lifespan and reliability. Harder materials provide better wear characteristics, while soft moulds may have a limited lifespan, but can simplify stamp creation (Pfeiffer et al., 2002). Not only the mechanical characteristics, but also optical and chemical properties are important when choosing a mold material for NIL. Critical mechanical parameters and their implications for NIL are hardness and thermal stability (lifetime and wear), thermal expansion coefficients and Poisson's ratio (dimension mismatch leading to distortions during demolding), roughness (higher demolding force and damage), Young's modulus (bending), and notch resistance (lifetime and handling). Issues related to fabrication are processability (etching processes, selectivity, clean room environment), and surface quality (resolution) (Bhushan, 2007). The handbook gives a brief overview of the mechanical and thermal properties of materials used for molds (Bhushan, 2007). The use in a NIL process is also determined by additional properties such as transparency, conductivity, anti-sticking properties (with/without anti-adhesive coating, e.g. by covalent coating), availability and cost (standard materials and sizes, tolerances, processing equipment and time), and how easy it is to employ in NIL (e.g. fixing by clamping, thermobonding, gluing). Currently, Silicon, Quartz, Nickel and Silicon Nitride (Si_3N_4) are typical materials frequently used for hard molds. Various polymeric materials, including polydimethylsiloxane (PDMS), polyurethane acrylate (PUA), polyvinyl alcohol (PVA) and polyvinyl chloride (PVC), have UV transparency, mechanical hardness and formability and thus can be used as the material for soft UV-NIL templates. Among these polymeric materials, PDMS is highly UV-transparent and has a very low Young's modulus which gives it the flexibility required for conformal contact. It has a very low reactivity and interfacial energy toward the polymeric materials and is sufficiently elastic that it can be separated from the polymeric structure without destruction or distortion. In addition, PDMS mold has a low surface energy at the polymer interface, eliminates the problem of the polymer sticking to the surface of the mold during detachment, which has proved a critical defect of NIL. Currently, PDMS has been considered as standard soft mold material due to its favourable properties concerning flexibility, UV-transparency and low surface energy. However, the main drawback of PDMS materials is the high viscosity and swelling. It is necessary to develop much more new mold materials with better performances to meet new NIL requirements (e.g. conductive mold for electrical field assisted NIL, release agent-free mold) (Guo, 2007, Costner et al., 2009, Yokoo & Namatsu, 2009, Pfeiffer et al., 2002, Bender et al., 2004, Choi & Park, 2004).

3.2 Mold fabrication

Mold fabrication is one of the biggest hurdles to imprint acceptance due to the 1X pattern resolution required. The three major challenges for mold manufacturing are resolution, quartz etch uniformity, and defect inspection. For many of the early adopters of this technology, the first two items are of greatest interest, as their designs are not defect sensitive. The minimization of defects and the ability to verify template quality are ultimately necessary for CMOS type applications. Patterning of most templates in a commercial mask making facility is done using e-beam lithography. There are two types of e-beam writers in use today, shaped beam tools and spot beam tools. Shaped beam tools are used predominately by mask makers due to their higher throughput and superior image placement accuracy. Spot or Gaussian beam tools offer finer resolution, but write times tend to be significantly longer. This requires a mask maker to determine the requirements of each template job and choose the appropriate e-beam writing strategy (Maltabes & Mackay, 2006, Maltabes et al., 2005).

For NIL to be accepted across the industry, a new infrastructure for 1X template fabrication, inspection, and repair needs to be established. This is a challenge. Obducat uses its own e-beam lithography technology to support and develop its template technology. Molecular Imprint has been working with many industrial partners, including Dupont Photronics, Toppan Photomasks (template fabrication), Motorola, KLA-Tencor (template inspection), and Carl Zeiss (template repair), to establish a template infrastructure. Recently, BenchMark Technologies, Inc. started offering e-beam written standard nano-imprint test templates. Motorola Labs has been focusing on developing the template and wafer-level processes while optimizing the imprinting process and collaborating with external partners to optimize both the inspection and repair of imprint templates (Hussain et al., 2007). Dauksher et al., reviewed recent results of template fabrication (including template repair and inspection), the imprinting process, the wafer-level pattern transfer processes, and applications (Dauksher et al., 2006).

A variety of methods and processes (e.g. focused ion beam lithography, gray-scale lithography, atomic layer deposition, DLC coating, glass wet etching, et al.) have been proposed and utilized to fabricate the NIL molds for meeting various application demands. The current challenges for NIL templates focus on developing new processes and materials to implement the low cost and high throughput fabrication for sub-50nm soft mold, 3-D mold, large-area, sub-10nm rigid mold, and propose better solutions to solve the anti-adhesion, defect inspection, and mold lifetime issues.

3.2.1 3D template

NIL with 3D templates can imprint advanced structures known as diffractive optical elements (DOEs), optical grating couplers, holograms, T-gates, and micro lens array (MLA), and the realization of 3D templates have been widely studied and reported. Currently, the fabrication process of 3D structured templates for UV-NIL applications by means of electron beam lithography (EBL) faces three major challenges. (1) E-beam exposures on UV transparent and therefore non-conductive substrates like quartz, (2) usage of thick resists layers with a low contrast for three dimensional patterning, and (3) reliable transfer of 3D structures into the template. Piaszenski et al., described a possible approach to solve all these different tasks. Through the use of low contrast PMMA resist, grey scale lithography could be performed for the definition of the molds pattern using a RAITH150 EBL system. The features could be transferred into the SiO_2 substrate via RIE while the addition of O_2 enables a modulation of the selectivity between the resist mask and the substrate. The mold was replicated via UV-NIL in one single step (Piaszenski et al., 2007). Kurihara et al., proposed two new techniques for practical 3D template manufacturing, which can conceptually make 3D templates faster than in the conventional manufacturing methods. The first one is to use a photolithography process with a dot modulated chrome mask, and the second one is a direct writing process with dot modulated pattern data, which is referred to as the dot modulated direct writing (DMDW). In both techniques, advantages of employing the dot modulation are to reduce the process cycles by using a single mask or single-exposure, and a single etching, and to improve the performance of the devices due to the relatively optimal analog relief surface, compared to the multi-level phase relief structures fabricated by multi-process technique (Kurihara et al., 2007). Taniguchi et al., described rapid and three-dimensional nanoimprint template fabrication technology using focused ion beam lithography (FIB). FIB lithography working with SOG resists can be used for speedy fabrication of three-dimensional nanoimprint template. Using fabricated SOG template, UV-NIL was carried out and replicated pattern with 59.8 nm gap was obtained. Because of its fast writing speed and good transfer properties, this combination of FIB lithography and SOG resist has the potential for alternative to EB cutting which is the next generation media mastering technique (Taniguchi et al., 2006). Ishii and Taniguchi proposed a fabrication method of three-dimensional mold using inorganic resist in low accelerating voltage electron beam lithography. Three dimensional patterns with a few hundred nanometer line width were fabricated with a control of acceleration voltage electron beam lithography (CAV-EBL). The pattern depths on inorganic resist were gradated with 5 nm depth-resolution per 30 V. By controlling the pattern depth, a seven stairs blade-shaped binary optics mold was fabricated, and then a replica pattern of the mold was made by using UV-NIL (Ishii & Taniguchi, 2007). High-quality 3D Ni stamps and molds featuring high aspect ratios as well as smooth vertical sidewalls and sharp straight edges were fabricated by PBW and nickel sulfamate electroplating. The stamp quality was improved by optimizing the plating process and utilizing pulse-reverse electroplating and backside polishing. To protect the original master, a process of metal-on-metal plating was employed (the so-called father–mother–son replication) to copy the master to secondary molds for imprinting applications (Ansari et al., 2006).

3.2.2 Large-area template

Full-wafer imprinting requires a full-wafer template; however, creation of a wafer-scale imprint template with sub-100 nm structures is not feasible with direct-writing approaches. Miller et al., described a practical methodology for creating wafer-scale templates suitable for full-wafer imprinting of sub-100 nm structures. The wafer-scale template is replicated from a smaller area master template using the S-FIL step-and-repeat process. A reverse tone process was used to maintain the original tone of the master template (Miller et al., 2007). Hussain et al., used atomic layer deposition (ALD) technique to fabricate nano-imprint templates with sub-15 nm template line width and 10 nm pitch length throughout the entire 200 mm wafer, varying the deposition thickness of multiple alternate films. ALD has the inherent advantage of being able to deposit ultra-thin conformal films even at atomic scale with excellent surface smoothness (Hussain et al., 2007). Resnick et al., addressed a methodology for creating high resolution thin templates for full wafer (or disk) imprinting. Half pitches of 21nm, using a PMMA lift-off process were formed on a template and imaged using an Imprio-1100 step and flash imprint lithography full wafer production tool. The templates have been applied toward the imprinting of both photonic crystal and patterned media devices using a large area printing tool developed around the S-FIL process (Resnick et al., 2007). Haatainen et al., proposed a step & stamp imprint lithography technique (SSIL) to produce large area metal stamps with wafer size up to 100 mm with 100 nm patterns. Electron beam patterned silicon stamps with size of a few square millimeters were used in patterning. The pattern of the stamp was transferred into a 100 mm silicon wafer coated with mr-I 7030 thermoplastic resist using SSIL sequential imprinting. The TiW/Cu field metallization was then sputtered onto the wafer. A 40–100 μm nickel layer was electroplated using a commercial plating bath and pulsed current source. The nickel stamp was detached from the substrate in a solvent. Compared to electron beam lithography the SSIL method is simple and fast leading to substantial saving in time (Haatainen et al., 2006).

3.2.3 High resolution template

Up to now, the resolution is not limited by the imprint process or the resist material applied but by the mold fabrication process. Recently, the small line edge roughness (LER) as well as the potentially high resolution that will ensure no-OPC mask feature is attracting many researchers. Since imprint lithography is a 1x lithography technology, sub-10 nm imprint mask with good LER over a large area is required for high throughput, low cost sub-10 nm imprint lithography. Currently, none of conventional lithography technologies are capable for the mask fabrication in this regime. Gu et al., showed experiments of using unconventional technology to fabricate imprint mask towards sub-10 nm regime with low LER over a large area. They reported for the first time the use of orientation dependent etching (ODE) of (110) c-Si in sidewall thin film technology for imprint mask fabrication with LER over a large area. Oxidation is used for sidewall thin film formation with a good critical dimension control. 2-dimensional oxidation effects are discussed. Features down to

12 nm have been fabricated successfully. Simulation showed that the fabricated oxide line is strong enough to imprint both thermoplastic and photo-curable imprint resists (Gu et al., 2005). Voisin et al., described a fabrication process developed to pattern nano scale features on thin silica wafers. The optimization of e-beam exposure dose was presented. Higher exposure doses are required on silica wafers coated with a chrome layer, compared to the case of silicon, but a larger process window is an advantage for the control of the exposed patterns. The etchings of the chrome layer and the silica wafer have been optimized and successfully performed. The optimization of both the exposure dose during e-beam lithography and the plasma dry etch steps, features with a resolution as low as 30 nm have been achieved (Voisin et al., 2007). Yoshida et al., investigated the possibility of 1X mold making with the extension of current photomask making, and found that there was a resolution limit of approximately 30nm for isolated spaces and approximately 50nm for dense features. By optimizing the resist material and quartz dry-etch process, 10nm isolated space and 35nm dense features are achieved using the spot beam writer. The resolution was improved by the non-CAR and the spot beam writer, instead of the CAR and the VSB writer. The quartz pattern side wall angle and the quartz pattern depth linearity on the mask were improved by modification of the quartz etching condition (Yoshida et al., 2006). Compare to silicon stamp, nickel stamp has high mechanical strength and durability, and can easily be duplicated using electroforming process. Hong et al., presented fabrication of 50 nm patterned nickel imprint stamp by combining hot embossing lithography and electroforming processes. Using hot embossing, the 50 nm features of original silicon stamp were transferred to PVC film without defects, nickel stamp, which has patterns as same as silicon master fabricated by nickel layer deposition on hot embossed PVC films (Hong et al., 2007). Yusa et al., achieved for the first time an hp22nm resolution on the 65nm template format. Both line and space patterns and hole patterns were well resolved. Regarding dot patterns, they still need improvement, but they have achieved resolution down to hp28nm (Yusa et al., 2007).

3.2.4 Flexible (soft) mold

Soft UV-NIL process enables the imprint of nanostructure on wafer scale but the resolution is limited by the fabrication processes of the flexible molds. One critical disadvantage of PDMS is its tendency to absorb organic solvents. To overcome the swelling of PDMS for organic solvent, Choi and Park applied the surface-modified PDMS mold by Teflon AF coating. To prevent the swelling of the PDMS mold by the monomer during the polymerization procedure, the mold surface is coated with an amorphous fluoropolymer, Teflon AF. The Teflon AF coating prevents the swelling and deformation of the PDMS mold and enables the creation of large area patterns with good fidelity. By using the surface-modified PDMS mold, the submicron scale polymer structures are successfully generated with good pattern fidelity (Choi & Park, 2003). The traditional fabrication method of the flexible molds is carried out by a cast molding process where an appropriate liquid mold material is deposited on a structured master, followed by thermal curing of the material. The main drawback of PDMS materials is the high viscosity. As a consequence, the profiles of nanopatterns on the mold are not completely defined, resulting in a loss of pattern height. Thus, the resolution of the mold fabrication process is limited by an inappropriate material flow for pattern geometries within the sub-100 nm regime. To overcome this problem previous literatures have reported the lowering of the viscosity of h-PDMS using triethylamine, toluene and hexane as solvent and demonstrated the imprinting of 75 nm lines with a pitch of 150 nm. However, due to the fragility of h-PDMS the usage for large scale imprint process and the reproducibility seems to be limited. Koo et al., presented an improved mold fabrication process using toluene diluted PDMS as mold material. They clearly demonstrated the capability of the technique to define patterns with 50 nm resolution over 4 in. with one imprint step only. The excellent homogeneity on wafer scale as well as the high degree of dimensional stability of both, mold fabrication and the imprint process was demonstrated (Koo et al., 2007). Chen et al., discussed a tri-layer soft mold, which consists of a rigid carrier, a low Young's module buffer and a top layer supporting nanostructure patterns to be replicated. Typically, the buffer and the top layer are made of PDMS of 5 mm thickness and PMMA of 10-50 μm thickness respectively. Patterning of the mold top layer can be done in three different ways, i.e., spin coating, nano-compression and direct writing, all resulting in 100 nm features over a large wafer area. The fabricated soft molds have been used for the fabrication of high resolution and high aspect ratio gratings and pillar arrays (Chen et al., 2005).

3.3 Key concern for NIL templates

3.3.1 Template surface treatment

The surface treatment, coating a low surface energy release layer or anti-adhesive layer, has been always regarded as the most important task for generating NIL templates. The surface treatment for NIL templates not only avoid the adhesion between template materials and resists which can improve imprint qualities, but it also increases the template lifetime significantly by preventing surface contamination. The template surface should allow the molded surfaces to detach easily from the mold, and once released, provide low friction, resulting in a continuous vertical slipping movement without sticking. Therefore, the templates need be treated using a low surface energy, self-assembled monolayer to ensure selective release at the template-etch barrier interface. This surface treatment is very durable and survives repeated imprints and multiple aggressive physical and chemical cleanings (Bhushan, 2007, Costner et al., 2009, DiBiase et al., 2006, Bailey et al., 2000). A common solution of this problem is to deposit a low-surface-energy release layer directly onto the mold. To accomplish this, self-assembled monolayers with low surface energies such as Teflon, $CF_3CH_2CH_2SiCL_3$(OTS) and $CF_3(CF_2)_5(CH_2)_2SiCL_3$(FDTS) have recently been adopted as an anti-adhesive layer to improve the demolding capabilities and to increase the lifetime of the mold. Fluorinated self-assembled monolayers (FSAMs) are also especially attractive as release agents because of their low surface energies, which in theory minimize the interaction between the polymer and the template. In addition, significant progress has been done to better improve the performance of antiadhesion surface treatment. Houle et al., discussed the adhesion between template materials and UV-cured nanoimprint resists. The process of interfacial separation of several model acrylate nanoimprint resist formulations cured in contact with clean (no release treatment) and release-treated silica glass surfaces has been quantitatively investigated. The results show that interfacial fracture energies are resist formulation dependent, that the resist-release layer systems studied are not chemically stable and that the release process is more complex than simple fracture at a glass-organic interface. Furthermore, they conclude that a criterion of low surface energy for a durable release coating in a UV curable nanoimprint system is not useful—it is low reactivity that is required (Houle et al., 2007). A thin film of PDMS is used as an anti-adhesion release layer because PDMS has a fairly low surface energy and allows for the easy release of the mold from the patterned polymer on the substrate. Lee et al., demonstrated the capability of the PDMS-based anti-adhesion surface treatment strategy for high resolution unconventional lithography using hard or soft molds as representatives of imprint lithography or soft lithography, respectively. The PDMS-based coating strategy used to achieve an anti-adhesion surface can be applied to various molds and polymeric materials on a substrate by adopting silane chemistry and the intrinsically low surface energy of PDMS (Lee et al., 2006). Surface treating imprint templates with FSAM layers is currently the dominant technique used to reduce interfacial adhesion in the S-FIL process. However, regardless of how these layers are applied, the durability of the FSAM over multiple imprints remains questionable (Costner et al., 2009). Park et al.; developed a new method of anti-adhesive coatings for Ni stamps which uses silane chemistry by introducing oxide surfaces. The (tridecafluoro-1,1,2,2-tetrahydrooctyl) trichlorosilane coating on SiO_2/Ni substrates shows lowest surface energy, lowest friction constant, and best imprint results among the different coatings investigated (Park et al., 2004). Wu et al., investigated anti-adhesive effects of diverse self-assembled monolayers in NIL. Self-assembled monolayers such as $CH_3(CH_2)_{17}SiCl_3$ (OTS) and $CF_3(CF_2)_7(CH_2)_2SiCl_3$ (FDTS) are adopted as an anti-adhesive layer to increase the lifetime of the mold. This anti-adhesive monolayer may be the best choice for high-temperature nanoimprinting (Wu et al., 2007).

3.3.2 Defect inspection

Actually, the most challenge for NIL (particularly for imprinting high-aspect-ratio 3D patterns) is the defect inspection of molds and replicated patterns. As Hector, manager of advanced lithography at Freescale Semiconductor Inc. said, NIL is dominated by inspection and overall yield. Compared with the 4 X masks used by traditional lithography, NIL, which is a 1X lithography technology, the defect inspection of 1X templates (molds) may be much more difficult (Actually, mask inspection and repair are not available for NIL templates with sufficiently small defect sensitivity). Moreover, since most masks can only involve 2D patterns, it is not viewed as an ideal solution for using current metrologies and tools to implement the defect inspection of molds and replicated patterns with 3D features. Therefore, it is especially important to understand defect reason and mechanism as well as establish effective inspection methodologies for replicating the high-quality 3D nano-scale patterns using the NIL process. He emphasized the need for an adequate inspection tool with very small pixel size and very high speed in order to make the templates (Hand, 2009). Due to the importance of defect inspection for NIL templates, a number of studies from both the academia and industrial bodies have been conducted to explore the issue. Some typical investigations are presented as follows. Yoshida et al., presented an optimization of measuring conditions for the templates of UV-NIL, and discussed some measuring techniques for CD precision and repeatability using a CD-SEM and a scanning probe microscope (Yoshida et al., 2008). JeffMyron et al., demonstrated that an advanced commercial photomask facility can fabricate templates with sub-100 nm critical dimensions, and that the CDs can be characterized using a commercially available CD-SEM metrology tool (Jeff Myron et al., 2005). Hess et al., developed a methodology to inspect the imprint templates with an existing commercial DUV inspection tool and discussed how e-beam inspection systems can extend the resolution beyond that of the DUV optical capability. They also reported the first systematic study of die-to-database electron beam inspection of patterns that were imprinted using an Imprio 250 system. The die-to-database inspection of the wafers was performed on an NGR2100 inspection system (Hess et al., 2004). McMackin et al., presented the study on high resolution inspection of step and flash imprint lithography for the 32 nm node and beyond. DiBiase et al., described a few non-traditional approaches to working with the extreme dynamic range of defect types found in the step and repeat NIL process. They examined the process of template defect inspection using a KLA-Tencor reticle inspection tool; Terascan 576. Inspection of the resulting wafers was performed on a KLA-Tencor 2365 wafer inspection tool. In addition to the conventional defect inspection tools, Archer Aim and KT 8450 CD SEM metrology tools were used to identify and characterize defect mechanisms (DiBiase et al., 2006).

3.3.3 Lifetime

The use of substantial pressure to not only contact but also penetrate a layer during imprinting accelerates the wear of imprint templates compared to other types of lithographic masks. That will lead to template wear, and shorten the lifetime of the template., in particular, for soft molds. How to extend template lifetime is another important issue for the real commercialization or industrial applications for NIL.

4. Imprint material

The progress and performance of NIL is closely linked with the availability of resists or imprint materials, which meet all the specific requirements of this technology. Imprint materials used for NIL are either utilized as an intermediate masking layer for the substrate or as a functional layer for a specific application. According to the process type, the materials used in NIL can be categorized into two main types: thermoplastic materials generally used in hot embossing lithography, and thermosetting (or curable) materials with either ultraviolet (UV) initiated precursors used in UV-NIL, or thermal initiation in the case of thermal curing imprint lithography. The unique advantage of a thermoplastic material is that the viscosity can be changed to a large extent by simply varying the temperature (Guo, 2007, Bhushan, 2007, Costner et al., 2009).

Guo (Guo, 2007) and Willson (Costner et al., 2009) recently presented a comprehensive reviews for NIL materials used. One wants to know much more details, please refer them. For the section, we only focused on some significant progresses in NIL materials. PMMA has been widely used as the thermoplastic polymer in NIL despite its requirements for high operating temperature (~ 200 °C) and pressure (~2000 psi). In order to low operating temperature and pressure, Jun et al., from HP Lab, developed an alternative resist that is a viscous fluid at room temperature and cures at a lower temperature of 70 °C than the operating temperature of the conventional thermal NIL (~ 200 °C) (Jun et al., 2005). Different enterprises and institutes have made UV-curable materials. Nanonex, MII, AMO, and Obducat offer these materials adapted to their proprietary process and tool. UV-curable NIL materials are composed of a mixture of monomers (or prepolymers) and a suitable photo initiator, and often chemicals are added which decrease the effect of radical scavengers on photo polymerization. *Micro resist technology* presented a novel, spin-coatable UV-NIL polymer, mr-UVCur A, which fulfills the specific requirements of an UV-NIL process. These are film thicknesses down to 100 nm, low viscosity, good film quality, fast UV-curing at low UV doses, good adhesion properties, high pattern fidelity and high plasma etch resistance (Kim et al., 2006).There is a demand for new materials with properties more appropriate for the particular requirements of nanoimprinting. One critical requirement is to provide mold

releasing properties during the de-molding process while not compromising the adhesion of the mold to the substrate. When imprinting high density patterns, the imprinted polymer tends to adhere to the mold, creating pattern defects that are not acceptable for many applications. Therefore, a material with low surface energy is desirable. Another critical property of resists is that they must have high etching resistance to allow pattern definition to be completed (Froelich, 2009). An electrically curable resist has been recently developed that can make electric imprint lithography (EIL) a reality. The resist is composed of a diaryliodonium salt photo acid generator and a cycloaliphatic epoxy monomer. Its polymerization takes place when an electric potential is applied between a conductive imprint mold and a substrate which sandwich the resist. A proof-of-concept pattern transfer by EIL with a micron-scale resolution has been demonstrated (Ahn et al., 2009).
A functional material may be imprinted directly to form a layer in a chip with no need for pattern transfer into underlying materials. The successful implementation of a functional imprint material would result in significant cost reductions and increased throughput by eliminating many difficult chip fabrication processing steps. Therefore, there is a strong need for the development of additional imprint material formulations that can become functional materials that remain on the imprinted structure and are used for other patterning applications (Kettle et al., 2008, Reuther, 2005).

5. NIL tool

This section presented recent progress in NIL equipments and key components. NIL can be carried out using three different types of machines: single, step-and-repeat, and roller imprint, as shown in Fig.11. Single-imprint machines pattern the surface of an entire wafer in one step. Thus the mold must have the same size as the wafer to be patterned. Step-and-repeat NIL machines pattern a smaller area of a wafer at a time, and then move to an unpatterned area, where the process is repeated. The process is continued until the whole wafer is patterned. Roller-type NIL tools continuously make large-area patterns based on roll-to-roll or roll-to-plate methods (Bhushan, 2007).

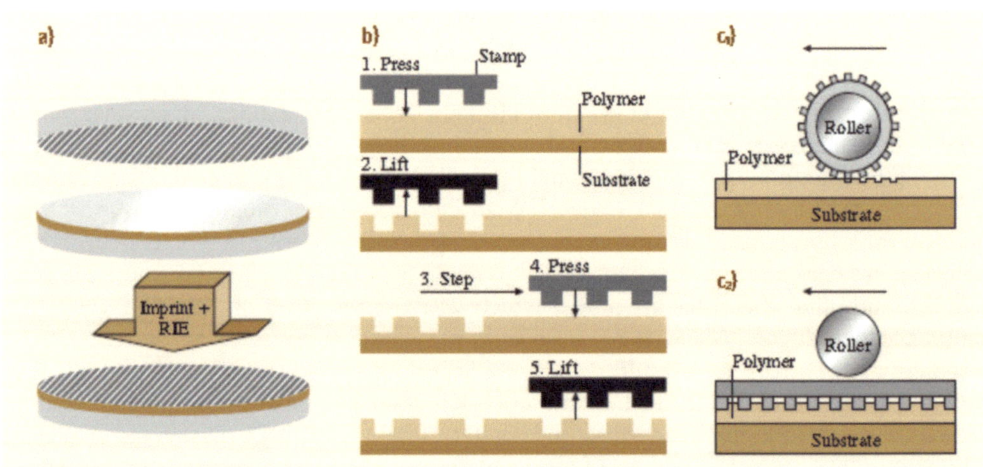

Fig. 11. Outline of the three most common types of NIL tools (Bhushan, 2007)

5.1 Overview of the typical NIL tools

Currently, there are five leading suppliers for the NIL tool and process in the world which include Molecular Imprints, Obducat, EVG, Süss, Nanonex. They have now developed and held their own proprietary NIL processes or patents respectively, for instance, Simultaneous Thermal and UV (STU®) technology and Intermediate Polymer Stamp (IPS®) technology from Obducat; J-FIL® (formerly named S-FIL®) and Double-sided Patterning from Molecular Imprints; Substrate Conformal Imprint Lithography (SCIL) for Süss from Philips patent; Air Cushion Press from Nanonex, and Soft UV-NIL and Step and Repeat Process for EVG from AMO. Recently, they have released new product families.

Obducat, a global leader in NIL equipment and processes, presented recently their High Volume Manufacturing (HVM) NIL system, featuring a throughput of 90 wafers per hour and cassette to cassette handling of up to 8-inch diameter substrates. The machine is designed for full area imprinting using the company's intermediate polymer stamp (IPS) process, which lowers the risk of substrate contamination by replicating the master stamp on an intermediate polymer film, which is used once for imprinting and then discarded (Obducat, 2009). The Imprio® HD2200 system represents the next generation in fully automated nano-imprint lithography for the hard disk drive (HDD) market. Its precision patterning fidelity, automated double-sided disk patterning capability and high throughput of 180 disks per hour all contribute to significant advantages in performance and cost of ownership (CoO) over alternative technologies such as spin-on UV imprint and thermal imprint. The Imprio HD2200 uses a liquid, room-temperature process and does not encounter the thermal distortions associated with the baking step or the defect issues associated with the lamination pressure of thermal imprint processes. Due to these and other advantages, HDD manufacturers are quickly standardizing on Molecular Imprints' imprint technology for their commercial discrete track recording (DTR) and bit patterned media (BPM) roadmaps. The Imprio® HD2200 is ideally suited for both development and pilot-production of patterned media applications. In Feb., 2008, Molecular Imprints, Inc. introduced the latest addition to its family of imprint lithography tools for semiconductor applications. Incorporating improvements in automation, tool throughput and overlay performance, the Imprio® 300 system represents the industry's highest resolution and lowest cost-of-ownership patterning solution for IC prototyping and process development at the 32nm node and beyond. The Imprio 300 offers sub-32nm half pitch resolution, sub-10nm overlay, integrated magnification control, and fully automated wafer and template loading capability. The Imprio 300 is ideal for device and process prototyping and pre-production for 32nm resolution, 22nm resolution, and beyond. The Imprio® 1100 system represents the next generation in fully automated nano-imprint lithography, combining the resolution and CD control of e-beam lithography with the throughput, overlay, and operating simplicity of a mask aligner. The Imprio 1100 is ideally suited for applications in LEDs, laser diodes, optical components, and patterned media development (MII, 2009). EVG 620 is a full-wafer Soft UV Nanoimprint tool which has high throughput mask aligner with >130 wafers per hour (first print, 1s exposure time), Wafer sizes up to 150 mm. IQ Aligner allows for customization with a special tooling to perform imprint processes with stamps and wafers from 100 mm to 300 mm in diameter. In 2009, EV Group unveiled next-generation UV-NIL Step and Repeat System-the EVG770 Gen II NIL Stepper. Some of the key new features the EVG770 touts include vacuum imprinting on a spun-on polymer layer, which eliminates defect issues caused by trapped air bubbles--ultimately resulting in superior pattern fidelity. With these new capabilities, the EVG NIL stepper tackles issues faced by other UV-NIL approaches in which vacuum-less environments and resists are dispensed as droplets rather than pre-spun, which cause easy air traps and impact structural integrity (EVG, 2009, AMO, 2009). A new generation of NIL production equipment has been developed by SUSS MicroTec in partnership with VTT Microelectronics, and introduced at Semicon Europa. The Nano-imPrinting Stepper NPS300 is the first NIL equipment which offers both hot and cold embossing of wafers in a single step-and-repeat system. The NPS300 provides a sub-20 nm imprinting resolution in a system that is optimized for the cost-effective volume production replication of micro or nanometer scale devices. Applications include integrated optical devices such as passive devices on diodes, gratings, photo refractive polymers, back side illumination devices; smart materials for microelectronics, such as sensors, resonators and transducers; sensors for temperature, light, molecules, and biomedical sciences; replication of three-dimensional objects such as MEMs components. When equipped with the automatic alignment option, the NPS300 has 250nm overlay accuracy and accepts stamps with sizes up to 100 mm and with thicknesses up to 6.5 mm. The imprinting force ranges from 5 newtons to 4,000 newtons. The active area of the stamp can be up to 40 x 40 mm for cold embossing and 100 x 100 mm for hot embossing (SUSS, 2009).

5.2 Key components for NIL tools

Generally, a NIL tool mainly consists of the following components: the imprint module, the substrate (wafer) stage, the overlay alignment module, the measurement module, and the environment control module. Among these basic elements, both the overlay alignment and the substrate (wafer) stage are considered as the core units.

5.2.1 Overlay alignment

Accurate multi-level overlay capability for NIL is essential to integrated circuit manufacturing and other multilayer imprint applications. The overlay alignment has been considered as one of the most challenges for NIL tools.

Overlay in NIL is different from that in conventional optical lithography and next generation lithography such as EUV. In optical lithography, one can fix the gap between the mask and the wafer and then perform the alignment and the exposure. In NIL, one needs to do alignment while bringing the template in contact with the resist on the wafer and then imprint into it. Accurate alignment is difficult with the mold and the substrate in contact, or nearly in contact, both because of mechanical interference as well as because of viscous forces arising from the polymer. This is of course especially true if heat and pressure are needed to mold the polymer. On the other hand, if the mold template and the substrate are separated by a large gap, the accuracy is limited by the difficulty of simultaneously imaging fine features on both of them (Jiang & Feldman, 2005). In addition they must be brought into contact without losing the alignment. Mühlberger et al., demonstrated alignment accuracies in the range of 100 nm and below in UV-NIL using a simple optical technique (a Moiré method). The advantages of this technique are the relative simplicity of the marker-design and the whole setup combined with the possibility of an upgrade of existing equipment and still ultra-high precision alignment capabilities (Muhlberger et al., 2007). Jiang and Feldman proposed a solution to the problem of obtaining accurate alignment over an extended imprinted area. They described how an alignment technique that was previously reported to have 18 nm 3σ accuracy can be combined with a stage that has low stiffness in the direction of motion and extremely high stiffness in the transverse directions. Such a combination would go a long way towards making imprint lithography a practical tool for integrated circuit fabrication (Jiang & Feldman, 2005). Shao et al., presented a novel overlay process for imprint lithography using load release and alignment error pre-compensation method. The experimental results demonstrated that the proposed process is an effective method to correct force-induced distortion and position shift, and the overlay precision is excel to 30 nm (3σ) as a soft mold is used (Shao et al., 2009). Li et al., reported sub-20-nm alignment in NIL using Moire´ fringe. Using the "beat" grating image (Moire´ fringe) generated by overlaying two sets of gratings that have slightly different periods, they obtained an alignment signal with a sensitivity better than 10 nm in NIL. The alignment signal is, as expected, independent of the size of the gap between the wafer and the imprint mold. A single-point overlay accuracy (error distribution) of sub-20 nm between the first and second imprinted layers by using two sets of Moire´ fringes was achieved. With higher precision nanopositioning stages, better single-point alignment accuracy is expected. Furthermore, they achieved sub-150 nm alignment over an area of 1 sq in and sub-250 nm over the entire area of a 4 in wafer using simple low-resolution stages without temperature control or wafer-mold mismatch compensation. With better stages, precision temperature control, and wafer-mold mismatch compensation, they believe that much higher overlay alignment accuracy over large areas (either in a 1 sq in die or a full wafer) is feasible (Li et al., 2006). Pan and Chen developed a self-built multilayer aligner for multilayer-imprint process to fabricate thin film transistor-liquid crystal display (TFT-LCD). Two stereo-microscopes were set up to examine angle deviation errors. The resolutionsof X-Y-Z and Ø axis are 1μm and 0.5 degree, respectively (Pan & Cheng, 2008). Shao et al., proposed an alignment measurement method using a moiré fringe pattern which involves relative rotation positioning and relative linear displacement measurement. Taking full advantage of the magnification effect of moiré fringe in angular and linear displacement, the relative rotation between the template and the wafer is determined first by measuring the inclination of the moiré fringe, and then the relative linear displacement between them is acquired by evaluating the spatial phase shift of two matched moiré fringes. The repeatability for the misalignment measurement has reached 4.8 nm 3σ (Shao et al., 2008).

5.2.2 Wafer stage

The prominent feature for the NIL is a contact imprinting process over the conventional optical lithography and NGL, therefore, the requirements for the NIL substrate stages are much different from and much more difficult than current wafer stages used. The substrate stage plays a particularly crucial role for the NIL machines. The substrate stage not only need carry out the precision position and an overlay alignment, but also ensure uniform and parallel surface contact between a template and a wafer while minimizing relative lateral motions during imprinting process. In other words, the wafer stage should implement the following functions: two translations along *X* and *Y*, and one tilting motion about *Z* (Θz or yaw) to carry out the step and repeat positioning and an overlay; one translation along *Z* and an two rotations about *X* and *Y* (Θx and Θy, or Roll and Pitch) to fulfill the parallel surface contact between a wafer and a template. It could be accomplished by either an actively controlled stage or a passive compliant mechanism. Passive compliant stages may have a simpler and more compact structure than actively controlled stages, and offer very high repeatability for lower cost and complexity. However, a passive stage cannot incorporate error correction and feedback into an operating scheme. It is also especially difficult for a passive compliant stage to carry out a high resolution overlay function.

Actively controlled stages may have a complicated structure and control system that consists of actuators, high resolution sensors, control algorithms, etc. But, an active stage with feedback control is one step toward a stage capable of performing high resolution layer-to-layer alignment and would provide a controlled mechanism of separation between the substrate and the template during imprinting process. This stage also enables the qualification of separation forces in a more rigorously controlled manner than is possible with the current configuration. For a real commercialization NIL machine, actively controlled stage may be an ideal and better scheme over passive compliant stage. Although a number of challenges and difficult works may have to be faced and resolved.
In order to further improve the performance of NIL machines and implement the multi-step and multi-level imprint process, it is necessary to develop a 6-DOF actively controlled stage. Two preliminary schemes for implementing 6-DOF actively controlled stage for NIL may be feasible. The first is a combination of two linear motors and an air bearing stage capable of entailing the dynamic load from zero to hundred Newton. Another is a combination of two linear motors and a flexure-based parallel kinematics stage which may be a more feasible and simpler scenario. In order to develop an actually actively controlled stage for NIL, a number of challenges and hard issues may still have to be faced and resolved. The mechanical stage generally suffers from poor dynamics, a result of the compound flexures used. There is increasing interest in the semiconductor industry to utilize magnetic levitation stage as the techniques become more and more feasible and lower cost. Since planar motors have light weight and compact structure, it is very attractive as an actuator in the flexure based motion stage. The future actively controlled stage for the NIL may a combination of a planar motor (coarse stage) and a magnetic levitation stage (fine stage) (Lan et al., 2007).

6. NIL applications

6.1 Introduction to NIL applications

NIL has been used to fabricate various devices for nanoelectronics, nano-optoelectronics, nanophtonic, optical components, biological applications, etc. The patterns can take many forms, from simple rectilinear patterns for nanowire development to complex diffractive optical elements for LED general lighting applications. The applications of NIL can be divided into two main categories: pattern-transfer applications and polymer devices. In the first category, pattern-transfer applications, the nanoimprinted resist structure is used as a temporary masking layer for a subsequent pattern-transfer step. In the second category, polymer devices, the imprinted pattern adds functionality to the polymer film, which is the end product (Bhushan, 2007). The NIL process offers new possibilities to form polymer devices with microscale to nanoscale features. Nanoscale patterned polymer films find a wide range of applications within optics, electronics and nano-biotechnology. The capability to form 3D polymer structures, with curved surfaces and high aspect ratios, paves the way for new classes of polymer based passive optical devices, such as lenses and zone plates, photonic crystals and integrated polymer optics. The NIL technique allows for choosing a wide range of polymers with optimized optical properties, and allows for the patterning of thin films of organic light-emitting materials and polymers doped with laser dyes to create organic light-emitting devices (OLED) and lasers. NIL is also suitable for nanoscale patterning of conducting organic films for cost-effective organic electronics (Bhushan, 2007, Guo, 2004).

According to current reports, NIL current applications mainly involve the following fields: magnetic storage media (hard disk media, NAND flash memory), optical storage media (HD-DVD, Blu-Ray), photonic (opto) electronics (high brightness LEDs, OLED, LCD, field emission display, flat panel display, organic light emitting display , flexible macro-electronic,), optical elements (microlens, diffractive grating, waveguide, tunable optical filters, nano wire grid polarizer), biological devices (biosensors, Nanofluidic devices, microarrays for genomics, proteomics and tissue engineering, nanoscale protein patterning), nanoelectronics (molecular electronics, AFM tips,), high-end semiconductors and high density interconnects, other NEMS/MEMS applications (solar cell, fuse cell, CNT sensor, etc), etc. In particular, NIL techniques currently have demonstrated great commercial prospects in several market segments, hard disk drives (HDDs), high-brightness light-emitting diodes (LEDs), flat panel displays, flexible macro-electronics devices, optical components and functional polymer devices. Some specific equipments have been developed by the key NIL providers, such as Imprio® HD2200 and Sindre® 800 for HDD, Imprio 1100 for LED. Table 1 shows the industrialization application roadmap for NIL from Obducat (Obducat, 2009).

Application areas – industrialisation status

Applications	Pre-production	Mass production
Magnetic storage	2005-2006	2006-2007
Optical storage (EBR)	2005-2006	2006-2007
Opto electronics	2005-2006	2006-2007
Displays	2005-2006	2006-2008
Polymer electronics	2005-2007	2007-2009
Bio applications	2005-2007	2007-2009
Molecular electronics	2006-2008	2007-2009
Semiconductors – Low End	2006-2008	2007-2009
Semiconductors – High End	2010-2012	2012-2013

Table 1. Industrialization application roadmap for NIL from Obducat (Obducat, 2009)

6.2 Typical applications

6.2.1 Hard-disk drives (HDDs)

Since the first demonstration of NIL in the mid 1990s, patterned magnetic media for hard-disk drives (HDDs) has been a key application, driving the development of NIL technology (Bhushan, 2007). Obducat researches performed a comparison among the electron beam lithography, EUV, scanning probe lithography, and NIL, as shown in Fig. 12, the results indicated that NIL is much more suitable for the fabrication process than other processes (Obducat, 2009). Now, HDD is the most successful case in NIL applications. NIL has emerged as a likely candidate for high-volume fabrication of patterned media disks, and electron beam lithography as the method required to produce nanoimprinting master stamps. Toshiba Corporation recently announced a prototype hard disk drive (HDD) that uses discrete track recording (DTR) technology, which enabled a capacity of a record-breaking 120GB on a single 1.8-in. disk. Toshiba plans to release product to the market with DTR at the end of 2009 (Toshiba, 2009). Nanoimprint providers and their users are also focusing on the high-density memory – such as NAND flash. Furthermore, some specific equipments such as, Imprio® HD2200 and Sindre® 800, have been developed for HDDs. However, many challenges in implementation of nanoimprinting technology into patterned media disk fabrication process still abound, such as choice of nanoimprinting resist, pattern transfer fidelity and uniformity, and lifetime of master stamp. System-level integration issues also create additional challenges in track-following, head-disk interface, and signal processing.

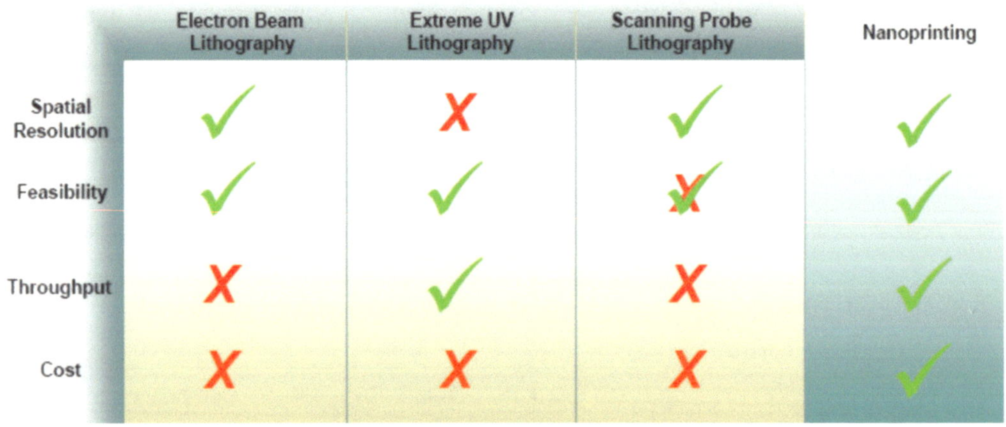

Fig. 12. Comparison of NIL and other processes (Obducat, 2009)

6.2.2 LED

NIL can be used as a most effective patterning tool to fabricate photonic crystal for high efficiency and low cost LED devices. Rapid progress is being made to reach manufacturing quality and cost targets suitable for high volume LED manufacturing. Obducat launched high volume LED NIL system in 2009. The system, named Sindre 400, is the world's first fully automated system for high volume manufacturing of LEDs. The first commercial system has already been ordered and the buyer is Luxtaltek, the Taiwan-based manufacturer of LEDs (Obducat, 2009). The Imprio® 1100 system developed by Molecular Imprints is ideally suited for applications in LEDs, laser diodes, optical components, and patterned media development (MII, 2009). Current throughput, cost and defect densities are acceptable for initial production.

NIL is being used by a number of leading edge device manufacturers and universities to develop photonic crystal and other light extraction technologies for LEDs. Scientists at Glasgow University along with the Institute of Photonics at the University of Strathclyde devise brighter LEDs via NIL. LEDs have not been introduced as the standard lighting in homes because the process of making the holes is very time consuming and expensive. Using NIL makes millions of microscopic holes on the surface of a LED bulb, which increase the amount of light the LEDs give off. They believe they have found a way of imprinting the holes into billions of LEDs at a far greater speed, but at a much lower cost (Glasgow, 2009).

Byeon et al., presented a fabrication process of photonic crystals in a p-GaN layer was established to improve the light extraction efficiency of LEDs by using NIL and inductively coupled plasma etching process. Two-dimensional pillar array patterns were uniformly fabricated on the p-GaN layer and the photoluminescence intensity of the photonic crystal patterned LED was increased by 2.6 fold compared to that of the same LED sample without photonic crystal patterns (Byeon et al., 2007). Ono et al., presented the fabrication of high-intensity LEDs using nanostructures by UV-NIL and electrodeposition. Nanostructures of 600 nm in depth, 300 nm in diameter, and in 500 nm pitch were fabricated on a GaN substrate. The radiant intensity of the LEDs was increased 1.5 times compared to that of a conventional LED (Fukuhara et al., 2007). Kao et al., reported a new patterning method to fabricate the patterned organic light emitting devices (OLEDs) with pixels of 500 μm × 300 μm on the flexible PET substrates using a combined roller imprinting and photolithography. This technique is potentially cost-effective, offers high throughput, less time-consuming and is suitable for fabrication on flexible substrate (Kao et al., 2005). However, there are several challenges that must be overcome before NIL can be applied to typical LED substrates (Hershey et al., 2007).

6.2.3 Optical elements

NIL is able to achieve aspect ratios greater than 10. A fine grating with high aspect rate pattern is one of the essential elements for advanced nano optical devices such as a quarter wave plate. The fabrication of high aspect ratio pattern as high as 8.75 has been successfully demonstrated using PMMA thin film on quartz substrate and Si based mold by NIL. This proposed method is a promising technology for industrial production of advanced nano optical elements having high aspect ratio structure (Hirar et al., 2004). Glinsner et al., presented soft UV-NIL for large area imprinting applications such as the production of optical components (e.g. micro lenses). A resolution well below 50 nm was demonstrated with soft stamp materials over areas as large as 150 mm wafers. Although the resolution of the patterning process is in the sub-50nm regime the currently most likely applications fields play in the resolution regime between 20μm and sub-200nm. This is the field for micro-optics, photonics, microfluidics and a vast number of sensory devices (Glinsner et al., 2007). Chaix et al., described how NIL was developed on 200 mm wafers for the fabrication of two optical devices: parts for optical encoders and organic light emitting diodes with enhanced light extraction efficiency (Chaix et al., 2007). NIL is also ideally suited for the fabrication of polymer nanophotonics and waveguide devices with submicron critical dimensions, defined over large areas. It is also compatible with many polymer materials, giving large freedom to choose a material with specific optical properties

As a result, data storage and optical displays will most likely be the first industrial application fields where replication techniques will be able to replace standard methods of lithography. Imprint structures and devices in polymer and functional polymers may become the next promising application field.

7. Prospects and challenges in NIL

NIL has now been considered as an enabling, cost-effective, simple pattern transfer process for various micro/nano devices and structures fabrications. Two unique benefits of NIL is the ability to pattern 3-D and large-area micro/nano structures with low cost and high throughput. In addition, NIL is able to achieve aspect ratios greater than 10. The progress made in the recent years enabled NIL not only a serious NGL candidate but also to a platform for one of the ten technologies in MIT Technology Review being evaluated to change the world (Fuchs et al., 2008). It has been added to the ITRS for the 32 and 22 nm nodes. Toshiba has shown impressive nanoimprint data connected to 18 nm feature size work: &1 nm CD uniformity, &2 nm LER, and the chipmaker confirms down to 20 nm overlay for the Molecular Imprints tool. Compared with the hot embossing or thermal nanoimprint lithography, UV-NIL offers several decisive technical advantages concerning overlay alignment accuracy, simultaneous imprinting of micro- and nanostructures and tool design due to the absence of high imprint pressures and thermal heating cycles. As a high-resolution patterning technique that has been used to print patterns as small as 2.4 nm, the capacity for high-resolution patterning makes NIL an attractive NGL technique for sub-50-nm device fabrication. Currently, NIL techniques have demonstrated great commercial prospects in several market segments, hard disk drives (HDDs), high-brightness light-emitting diodes (LEDs), flat panel displays, flexible macro-electronics devices, and polymer and functional imprint materials.

However, NIL is still facing many serious challenges. The currently crucial challenges for NIL include overlay alignment, template fabrication, defect control, high yield, and seeking especially suitable application fields. In particular, two key challenges remain for NIL before it can be adopted for semiconductor manufacturing: alignment (overlay) and template/mold fabrication. For the NIL process, the future development will focus on the following several aspects: external fields (e.g. electric field) assisted NIL, RNIL, combined NIL with other nanofabrication processes, as well as developing novel NIL processes. One exciting opportunity is the development of industrial-level roll-to-roll imprinting tools and processes, which could provide unprecedented throughput for many practical applications. In the future several years, the roller-type imprinting process will be commercially applied to OLED, flat panel display, and flexible macro-electronic fields. In addition, complex 3-D nanostructures fabrication based on NIL is also a crucial direction which be further studied. Furthermore, it is considered as a promising method to product optical elements, solar cell, 3D-photonic crystals, etc. The current challenges for NIL templates focus on developing new processes and materials to implement the low cost and high throughput fabrication for sub-50nm soft mold, 3-D mold, large-area, sub-10nm rigid mold, and propose better solutions to solve the anti-adhesion, defect inspection, and mold lifetime issues. In addition, new making mold methods also need be further developed to meet the requirements from emerging market applications. It is possible that self-assembled structures will provide the ultimate solution for templates of periodic patterns at scales of 10 nm and less. It is also possible to resolve the template generation issue by using a programmable template in a scheme based on double patterning. Compared to optical lithography, the alignment in NIL process is especially difficult to perform since the template and the substrate are in contact, or nearly in contact. For alignment, furthermore, a NIL tool lacks the expensive optics and extremely precise stages, which are two of the cost-drivers for a photolithography stepper. Currently, the highest alignment accuracy is sub-10nm (3 sigma, single point, X,Y) from the Imprio® 300. It is still a severe challenge for commercial NIL tools to develop new alignment solutions with low cost and high accuracy so far. In addition, it is necessary to exploit and develop the six-Degree-of-Freedom active control substrate stage with multi-step and multi-level functions and the new overlay solution with sub-10nm alignment accuracy. For resists, there are still significant challenges to develop low viscosity resists and new functional materials to meet the stringent requirements of new NIL processes and

NEMS/MEMS applications. Therefore, the future investigations for NIL materials should focus on the following three aspects; low viscosity resists, various functional materials and nanoimprint resist materials with high etching resistance. For the NIL application, apart from the current applications, it is the most critical challenges for NIL to find promising application fields and devices, for which NIL is considered as the most suitable fabrication process. Direct printing of functionalized resists finally will enter the extremely hot field of future device generations through a massive reduction of process steps in the fabrication sequence of complex systems. If the bridge between bottom up and top down is established the fabrication of highly dense complex structures with new functionalities will certainly dominate the future of UV-NIL (Fuchs et al., 2008). A crucial requirement for being able to further extend the success of the NIL technique is to improve resolution and throughput of conventional NIL schemes as well as to explore unconventional ways to utilize NIL processes for fabrication of novel nanodevices and nanomaterials.
There is still a long way to go for NIL technology and various industrial applications. Further potentials of NIL need still to be explored and developed.

8. Acknowledgements

This work was partly supported by the 973 Basics Science Research Program of China (Grant No. 2009CB 724202), National Science Foundation of China (Grant No. 50775176) and National Science Foundation of Shandong Province (Grant No.Y2007F49).

9. References

Ahn, S.; Cha, J. & Myung, H. (2006). Continuous ultraviolet roll nanoimprinting process for replicating large-scale nano- and micropatterns, *Appl. Phys. Lett*, 89, 213101

Ahn, S. & Guo, J. (2008). High-speed roll-to-roll nanoimprint lithography on flexible plastic substrates, *Adv. Mater.*, 20, 2044–2049

Ahn, Y.; Chen Y & Hahn, T. (2009). A resist for electric imprint lithography, *Microelectronic Engineering*, 86,392–396

AMO, *http://www.amo.de/imprint_process.0.html*, 2009.

Ansari, K.; Kan, J.; Bettiol, A.; et al. (2006). Stamps for nanoimprint lithography fabricated by proton beam writing and nickel electroplating, *Journal of Micromechanics and Microengineering*, 16, 1967–1974

Bailey, T.; Choi, B.; Colburn, M.; et al. (2000). Step and flash imprint lithography: Template surface treatment and defect analysis, *J. Vac. Sci. Technol. B*, 18, 6, 3572-3577

Balla, T.; Spearing, S. & Monk, A. (2008). An assessment of the process capabilities of nanoimprint lithography, *J. Phys. D: Appl. Phys.*, 41, 174001

Bender, M.; Plachetka, U.; Ran, J.; et al. (2004). High resolution lithography with PDMS molds, *J. Vac. Sci. Technol. B*, 22, 6, 3229-3232

Bender, M.; Fuchs,A.; Plachetka, U.; et al. (2006). Status and prospects of UV-nanoimprint technology, *Microelectronic Engineering*, 84, 4-9, 827-830

Bhushan, B. (2007). *Handbook of Nanotechnology*, Springer, ISBN: 978-3-540-29855-7, Berlin Heidelberg

Bogdanski, N.; Wissen, M.; Möllenbeck, S.; et al. (2007). Challenges of residual layer minimisation in thermal nanoimprint lithography, *Proceedings of the SPIE*, 6533, pp.65330Q

Busnaina, A. (2006). *Nanomanufacturing handbook*, Talor & Francis, ISBN: 978-0-8493-3326-2, Boca Raton, London, New York

Byeon, K.; Hwang, S.; Lee, H. (2007). Fabrication of two-dimensional photonic crystal patterns on GaN-based light-emitting diodes using thermally curable monomer-based nanoimprint lithography, *Applied Physics Letters*, 91, 091106

Chaix, N.; Landis, S.; Gourgon, C.; et al. (2007). Nanoimprinting lithography on 200 mm wafers for optical applications, *Microelectronic Engineering*, 84, 5-8, 880-884.

Chang, C., Yang, S. & Sheh, J. (2006). A roller embossing process for rapid fabrication of microlens arrays on glass substrates, *Microsystem Technologies.*, 12, 8, 754-759

Chen, L.; Deng, X.; Wang, J.; et al. (2005). Defect control in nanoimprint lithography. *J. Vac. Sci. Technol. B.* , 23, 6, 2933-2938

Chen, Y.; Roy, E.; Kanamori, Y.; et al. (2005). Soft nanoimprint lithography, *Proceedings of SPIE*, 5645, pp.283-288

Cheng, X. & Guo, J. (2004). A combined-nanoimprint-and-photolithography patterning technique, *Microelectronic Engineering*, 71, 3-4, 277-282

Chou, Y.; Krauss, R. & Renstrom, J. (1995). Imprint of sub 25 nm vias and trenches in polymers, *Appl Phys Lett.*, 67, 21, 3114-6

Chou, Y.; Krauss, P.; Renstrom P. (1996). Imprint lithography with 25-nanometer resolution, *Science*, 272, 5258, 85-87

Chou, Y.; Krauss, P.; Zhang, W.; et al. (1997). Sub-10 nm imprint lithography and applications, *J. Vac. Sci. Technol. B.*, 15, 6, 2897-2904

Chou, Y.; Keimel, C. & Gu, J. (2002). Ultrafast and direct imprint of nanostructures in silicon, *Nature*, 417, 835-837

Choi, W. & Park, O. (2003). A soft-imprint technique for direct fabrication of submicron scale patterns using a surface-modified PDMS mold, *Microelectronic Engineering*, 70, 131–136

Choi, W. & Park, O. (2004). A soft-imprint technique for submicron-scale patterns using a PDMS mold, *Microelectronic Engineering*, 73–74, 178–183

Costner, E; Lin, M.; Jen, W.; et al. (2009). Nanoimprint lithography materials development for semiconductor device fabrication, *Annu. Rev. Mater. Res.* 39, 155–80

Dauksher, J.; Le, N.; Ainley, E.; et al. (2006). Nano-imprint lithography: templates, imprinting and wafer pattern transfer, *Microelectronic Engineering*, 83, 929–932

DiBiase, T.; Maltabes, J.; Reese, B.; et al. (2006). Building 1X NIL templates: challenges and requirements, *Proc. of SPIE*, 6151, 61511E

Dumond, J. & Low, H. (2008). Residual layer self-removal in imprint lithography, *Adv. Mater.* , 20, 1291–1297

EVG, http://www.evgroup.com/, 2009

Froelich, S. New materials and applications of nanoimprint technology. http://www.nnin.org/doc/2004NNINreuFroelich.pdf, 2009

Fuchs, A.; Bender, M.; Plachetka, U.; et al. (2008). Lithography potentials of UV-nanoimprint, *Current Applied Physics*, 8, 669–674

Fukuhara, M.; Ono, H.; Hirasawa, T.; et al (2007). UV nanoimprint lithography and its application for nanodevices, *Journal of Photopolymer Science and Technology*, 20, 549-554

Gao, H.; Tan, H.; Zhang, W.; et al. (2006). Air cushion press for excellent uniformity, high yield, and fast nanoimprint across a 100 mm field, *Nano Lett.*, 6,11, 2438-2441

Glasgow, Brighter LED lights could replace household light Bulbs within three years, http://www.sciencedaily.com/releases/2008/01/080109083914.htm, 2009
Glinsner,T.; Plachetka,U.; Matthias, T.; et al. (2007). Soft UV-based nanoimprint lithography for large area imprinting applications, *Proc. of SPIE*, 6517, pp.651718
Gu, J.; Jen, C.; Wei, Q.; et al. (2005). Mask fabrication towards sub-10 nm imprint lithography, *Proc. of SPIE*, 5751, 382-391
Guo, J. (2004). Recent progress in nanoimprint technology and its applications, *Journal of Physics D: Applied Physics.*, 37, R123–R141
Guo, J. (2007). Nanoimprint lithography: methods and material requirements, *Advanced Materials*, 19, 4, 495 - 513
Haatainen, T.; Majander, P.; Riekkinen, T.; et al. (2006). Nickel stamp fabrication using step & stamp imprint lithography, *Microelectronic Engineering*, 83, 948–950
Han, K,; Hong, S.; & Lee, H. (2007). Fabrication of complex nanoscale structures on various substrates, *Applied Physics Letters*, 91, 2, 123118
Hand, A. Nanoimprint templates need high-quality inspection, http://www.semiconductor.net/article/CA6303358.html, 2009
Hershey, R.; Doyle, G.; Jones, C.; et al. (2007). Imprint lithography advances in LED manufacturing, *Physica status solidi. C. Current topics in solid state physics*, 4,1, 21-24
Hess, H.; Pettibone, D.; Adler, D.; et al. (2004). Inspection of templates for imprint lithography, *Journal of Vacuum Science and Technology B*, 22, 6, 3300-3305
Heyderman, L.; Schift, H.; David, C.; et al. (2000). Flow behaviour of thin polymer films used for hot embossing lithography, *Microelectronic Engineering*, 54, 229–245
Hirar, Y.; Konishi, T.; Kanakugi, T.; et al. (2004). High aspect ratio grating fabrication by imprint lithography, *Proceedings of SPIE*, 5515, 187-194
Hiroshima, H, & Komuro, M. (2007). Control of bubble defects in UV nanoimprint, *Japanese Journal of Applied Physics*, 46, 9B, 6391-6394
Hiroshima, H. (2008). Nanoimprint with thin and uniform residual layer for various pattern densities, *Microelectronic Engineering*, 86, 4-6, 611-614
Hong, H.; Lee, J. & Lee, H. (2007). Fabrication of 50 nm patterned nickel stamp with hot embossing and electroforming process, *Microelectronic Engineering*, 84, 5-8, 977-979
Houle, F.; Guyer, E.; Miller, D.; et al. (2007). Adhesion between template materials and UV-cured nanoimprint resists, *J. Vac. Sci. Technol. B*, 25, 4, 1179-1185
Hussain, M.; Labelle, E.; Sassman, B.; et al. (2007). Deposition thickness based high-throughput nano-imprint template, *Microelectronic Engineering*, 84, 594-598
Ishii, Y. & Taniguchi, J. (2007). Fabrication of three-dimensional nanoimprint mold using inorganic resist in low accelerating voltage electron beam lithography, *Microelectronic Engineering*, 84, 912-915
Jeff Myron, L.; Gershtein, L.; Gottlieb, G.; et al. (2005). Advanced mask metrology enabling characterization of imprint lithography templates, *Proc. of SPIE*, 5752, pp. pt.1
Jeong, J.; Choi, Y.; Shin, Y.; et al. (2002). Flow behavior at the embossing stage of nanoimprint lithography, *Fibers and Polymers*, 3, 3, 113-119
Jeong, J.; Kim, K.; Sim, Y.; et al. (2005). A step-and-repeat UV-nanoimprint lithography process using an elementwise patterned stamp, *Microelectronic Engineering*, 82, 180-188
Jiang, L. & Feldman, M. (2005). Accurate alignment technique for nanoimprint lithography, *Proc. of SPIE*, 5752, pp.429-437

Jones, R.; Soles, C.; Lin, E.; et al. (2006). Pattern fidelity in nanoimprinted films using critical dimension small angle X-ray scattering, *Journal of Microlithography, Microfabrication, and Microsystems*, 5, 1, 013001

J-FIL, http://www.molecularimprints.com/Technology/technology2.html, 2009

Jun, G.; Wu, W.; Ganapathiappan, S.; et al. (2005) Issues on nanoimprint lithography with a single-layer resist structure, *Appl. Phys. A.*, 81, 1331–1335

Jun, Y.; Wu, W.; Ganapathiappan, S.; et al. (2005). Issues on nanoimprint lithography with a single-layer resist structure, *Appl. Phys. A*, 81, 1331–1335

Kao, P.; Chua, S.; Zhan, C.; et al. (2005). Fabrication of the patterned flexible OLEDs using a combined roller imprinting and photolithography method, *Proceedings of 2005 5th IEEE Conference on Nanotechnology*, Nagoya, Japan

Kehagias, N.; Chansin, G.; Reboud, V.; et al. (2006). Submicron three-dimensional structures fabricated by reverse contact UV nanoimprint lithography, *J. Vac. Sci. Technol. B*, 24, 3002-3005

Kettle, J.; Coppo, P.; Lalev, G.; et al. (2008). Development and validation of functional imprint material for the step and flash imprint lithography process, *Microelectronic Engineering*, 85,850–852

Key Technologies, http://www.obducat.com/Default.aspx?ID=187, 2009

Khang, D. & Lee, H. (2008). On the Mechanism of low-pressure imprint lithography: capillarity vs viscous flow, *Langmuir*, 24, 5459-5463

Kim, W.; Choi, D. & Bae, B. (2006). Ultraviolet-nanoimprint of 40 nm scale patterns using functionality modified fluorinated hybrid materials, *Nanotechnology*, 17, 13, 3319-3324

Koo, N.; Bender, M.; Plachetka, U.; et al. (2007). Improved mold fabrication for the definition of high quality nanopatterns by soft UV-nanoimprint lithography using diluted PDMS material, *Microelectronic Engineering*, 84, 904-908

Kurihara, M.; Abe, M.; Suzuki, K.; et al. (2007). 3D structural templates for UV-NIL fabricated with gray-scale lithography, *Microelectronic Engineering*, 84, 999-1002

Lan, H.; Ding, Y.; Liu, H.; et al. (2007). Review of the wafer stage for nanoimprint lithography, *Microelectronic Engineering*, 84,4, 684-688

Lan, H.; Ding, Y.; Liu H.; et al. (2009). Mold deformation in soft UV-nanoimprint lithography, *Science in China Series E: Technological Sciences*, 52, 2, 294-302

Lan, S.; Lee, H.; Ni, J.; et al. (2008). Survey on roller-type nanoimprint lithography (RNIL) process, *International Conference on Smart Manufacturing Application*, Korea, 371-376

Lee, J.; Park, S.; Choi, K.; et al. (2008). Nano-scale patterning using the roll typed UV-nanoimprint lithography tool, *Microelectronic Engineering*, 85, 861–865

Lee, H. & Jung, G. (2004). UV curing nanoimprint lithography for uniform layers and minimized residual layers, *Japanese Journal of Applied Physics*, 43, 12, 8369–8373

Lee, M.; Lee, N.; Lim, J.; et al. (2006). Antiadhesion surface treatments of molds for high-resolution unconventional lithography, *Adv. Mater.* 18, 3115–3119

Li, M.; Chen, L.; Zhang W.; et al. (2003). Pattern transfer fidelity of nanoimprint lithography on six-inch wafers, *Nanotechnology*, 14, 33–36.

Li, N.; Wu, W.& Chou, Y. (2006). Sub-20-nm alignment in nanoimprint lithography using Moire´ fringe, *Nano Lett.*, 6,11,2625-2629

Liang, X.; Hua, F.; Fu, Z.; et al. (2007). Air bubble formation and dissolution in dispensing nanoimprint lithography, *Nanotechnology*, 18, 025303

Macintyre, D. & Thoms, S. (2005). A study of resist flow during nanoimprint lithography, *Microel ectronic Engineering*, 78–79, 670–675.
Maltabes, J.; Mackay, R. & Cottle, R. (2005). An integrated approach to the analysis of imprint vs. optical lithography, or why this is not just a mask discussion, *Proc. of SPIE*, 5992, pp.59924A
Maltabes, J. & Mackay, R. (2006). Current overview of commercially available imprint templates and directions for future development, *Microelectronic Engineering*, 83, 933–935
MII, *http://www.molecularimprints.com/index.html*, 2009
Miller, M.; Schmid, G.; Doyle, G.; et al. (2007). Template replication for full wafer imprint lithography, *Microelectronic Engineering*, 84, 885–890
Muhlberger, M.; Bergmair, I.; Schwinger, W.; et al. (2007). A Moire' method for high accuracy alignment in nanoimprint lithography, Microelectronic Engineering, 84, 925–927
Nanoimprint lithography, *Wikipedia*, http://en.wikipedia.org/wiki/Nanoimprint_lithography, 2009
Obducat , *http://www.obducat.com/*, 2009
Pan, C. & Cheng, R. (2008). Development of self-built multilayer aligner for imprint process, *Journal of Materials Processing Technology*, 204,502–507
Park, S.; Schift, H.; Padeste, C.; et al. (2004). Anti-adhesive layers on nickel stamps for nanoimprint lithography, *Microelectronic Engineering*, 73–74,196–201
Pelzer, R.; Gourgon, C.; Landis, S.; et al. (2005). Nanoimprint lithography-full wafer replication of nanometer features, *Proc. of SPIE*, 5650, 256-259
Pfeiffer, K.; Fink, M.; Ahrens, G.; et al. (2002). Polymer stamps for nanoimprinting, *Microelectronic Engineering*, 61-62, 393-398
Piaszenski, G.; Barth, U.; Rudzinski, A.; et al. (2007). 3D structures for UV-NIL template fabrication with grayscale e-beam lithography, *Microelectronic Engineering*, 84, 945-948.
Plachetka, U.; Bender, M.; Fuchs, A..; et al. (2004). Wafer scale patterning by soft UV-nanoimprint lithography, *Microelectronic Engineering*, 73-74, 167–171
Plachetka, U.; Kristensen, A.; Scheerlinck,S.; et al. (2008). Fabrication of photonic components by nanoimprint technology within ePIXnet, *Microelectronic Engineering*, 85, 886–889
Resnick, D.; Schmid, G.; Miller, M.; et al. (2007). Step and flash imprint lithography template fabrication for emerging market applications, *Proc. of SPIE*, 6607, 6 pp.6070T
Reuther, F. (2005). Advanced polymers and resists - A key to the development of nanoimprint lithography, *Journal of Photopolymer Science and Technology*, 18, 4, 525-530
Scheer, H & Schulz, H. (2001). A contribution to the flow behaviour of thin polymer films during hot embossing lithography, *Microelectronic Engineering*, 56, 311–332
Schift, H. (2008). Nanoimprint lithography: An old story in modern times? A review, *J. Vac. Sci. Technol. B.*, 26, 2, 458-480
SCIL, http://www.suss.com/applications/nano_imprint_lithography/scil, 2009
Shao, J.; Liu, H.; Ding, Y.; et al. (2008). Alignment measurement method for imprint lithography using moiré fringe pattern, *Optical Engineering*, 47, 1, 113604

Shao, J.; Ding, Y.; Liu, H.; et al. (2009). Strategy for a loading force induced overlay position shift in step imprint lithography, *Journal of Engineering Manufacture*, 223, 1, 9-17

Steward, M. & Willson, C. (2005). Imprint materials for nanoscale devices, *MRS bulletin*, 30, 947-951.

Suh, D.; Rhee, J. & Lee, H. (2004). Bilayer reversal imprint lithography direct metal–polymer transfer, *Nanotechnology*, 15, 1103–1107

SUSS, http://www.suss.com/index.php, 2009

Tan, H.; Kong, L.; Li, M.; et al. (2004). Current status of nanonex nanoimprint solutions, *Proceedings of the SPIE*, 5374, pp.213-221

Taniguchi, J.; Koga, K.; Kogo, Y.; et al. (2006). Rapid and three-dimensional nanoimprint template fabrication technology using focused ion beam lithography, *Microelectronic Engineering*, 83, 940–943

Toshiba, Toshiba brings track recording technology to prototype of 120GB hard disk drive, http://www.huliq.com/33259/toshiba-brings-track-recording-technology-to-prototype-of-120gb-hard-disk-drive, 2009

Voisin, P.; Zelsmann, M.; Gourgon, C.; et al. (2007). High-resolution fused silica mold fabrication for UV-nanoimprint, *Microelectronic Engineering*, 84, 916-920

Wu, C.; Shen, Y.; Chuang, S.; et al. (2007). Anti-adhesive effects of diverse self-assembled monolayers in nanoimprint lithography, *Sensors and Actuators A*, 139, 145–151

Wu, W.; Tong, W.; Bartman, J.; et al. (2008). Sub-10 nm Nanoimprint lithography by wafer bowing. *Nano Lett.*, 8 , 11, 3865–3869

Yokoo, A & Namatsu, H.. *https://www.ntt-review.jp/archive/ntttechnical.php?contents=ntr200808sp3.html*, 2009

Yoneda,I.; Mikami S.; Ota T.; et al. (2008). Study of nanoimprint lithography for applications toward 22-nm node CMOS device, *Proc. of the SPIE*, 6921, 2, pp.692104

Yoshida, K.; Kojima, K.; Abe, M.; et al. (2008). Optimization of measuring conditions for templates of UV nano imprint lithography, *Proc. of SPIE*, 7028, pp.70281V-1

Yoshida, Y.; Amano, T.; Sasaki, S.; et al. (2006). Dry etch technology development for NIL template, *Proc. of SPIE*, 6283, pp.62833G

Youn, S.; Ogiwara, M.; Goto, H.; et al. (2008). Prototype development of a roller imprint system and its application to large area polymer replication for a microstructured optical device. *Journal of materials processing technology*, 202, 76-85

Yusa, S.; Hiraka, T.; Kobiki, A.; et al. (2007). Progress of Nil template making, *Proc. of SPIE*, 6607, pp.66073H

Zero residual layer nanoimprint lithography, http://techtransfer.universityofcalifornia.edu/NCD/19159.html, 2009

24

Nanoimprint Lithography

Thomas Glinsner and Gerald Kreindl
EVGroup E. Thallner GmbH,
Austria

1. Introduction

Imprinting technology is an ancient technique for the reproduction of writings on appropriate supports. Since 1990´s, one of the imprinting techniques, i.e., injection moulding has been used for compact disk (CD) production. More recently, the semiconductor industry is interested in imprint related techniques because of the mass production requirement of future microelectronic circuits with a possible critical dimension down to a few nanometers. At this deep nanometer scale, traditional photolithography is supposed to rule out because of the optical diffraction or material limitations. In fact, the actual minimum feature size in an integrated circuit (IC) is already less than 50 nm and the actual manufacturing systems are already extremely sophisticated and expensive. The semiconductor industry has always been looking for alternative patterning methods in order to follow Moore´s law, which has been formulated to predict the evolution of the technology nodes. Now, extreme UV lithography (EUV), 193 nm immersion lithography, mask less lithography (MLL) techniques and nanoimprint lithography (NIL) are considered as candidates for the so called *Next Generation Lithography* (NGL) at 32 nm and 22 nm nodes. In parallel, imprint technology has been promoted by a large scientific community and non-IC industry segments including high-density storage, optoelectronics, telecommunication as well as biochips or micro total analysis systems.
The history of imprint technology as lithography method for pattern replication can be traced back to 1970's but the most significant progress has been made by the research group of S. Chou in the 1990's. Since then, it has become a popular technique with a rapidly growing interest from both research and industrial sides and a variety of new approaches have been proposed along the mainstream scientific advances. Indeed, the initial proposal of Chou et al. has been made for the mass production of high density magnetic storage media and it has been used to demonstrate the feasibility of all kind of fine structure patterning at a nanometer scale resolution which is now called nanoimprint lithography.
Nanoimprint lithography is based on surface structuring with a template consisting of topographic patterns. After imprinting, the patterns have to be transferred in order to obtain different functionalities. As lithography method, nanoimprint is fully compatible to the standard micro-fabrication techniques, including different transfer processes such as etching, lift-off, selective re-growth or diffusion. In most cases, a thin layer of resist is deposited on the substrate and then imprinted, resulting in a thickness contrast. This may need a further treatment by reactive ion etching (RIE) to remove the residual resist layer and to transfer the features into the used substrate material. Therefore, the process control of

both imprinting and etching are important in order to obtain a suitable resist profile without any residual layer. For some particular applications, the resist layer can be replaced by functional materials or omitted for an imprint directly into the substrate.

Currently, two main types of nanoimprint methods are existing, i.e., hot embossing and UV-based nanoimprint lithography (UV-NIL). An overview of the most important nanoimprinting techniques and its related stamp materials is shown in figure 1. Thermal processing has been demonstrated by S. Chou et al., which consists of deforming a thermoplastic polymer layer such as a polymer (e.g. PMMA, COC, PC) spin coated on a substrate or a polymer substrate. After pre-bake and heating up above the glass transition temperature of the polymer, the imprint is performed by applying a contact force on the stack consisting of template and substrate. Then, the template-substrate assembly is cooled down before the template release. Alternatively, the UV-based technique is based on the use of photo-curable resist, which can be easily deformed at room temperature and cured by UV-light exposure. Thus the main difference between the two techniques are the material types of both template and resist, i.e., transparent templates and photosensitive resists for UV-NIL, but in some cases non-transparent templates and thermoplastic resists for thermal-NIL.

For both hot embossing and UV-based processes, the template fabrication is one of the most critical issues because of its high cost. This depends on the required minimum feature size, the density of the features and the patterned area. In general, electron beam lithography and reactive ion etching are used to pattern silicon or quartz glass wafers. High resolution patterning over a large wafer area is time consuming but ideally, once the template has been produced, it can be used for many times. In practice, both silicon and quartz glass templates can be damaged due to process imperfection.

Compared to hot embossing, UV-NIL is advantageous when multi-level patterning is needed due to the fact, that high precision alignment can be done much easier with processes performed at room temperature as there are no thermal gradients generated. Basically, the cycle time of UV-NIL is shorter than in hot embossing due to the absence of heating and cooling cycles (heating-up to the imprinting temperature and cooling down to the de-embossing temperature). In addition, both pattern placement accuracy and overlay alignment accuracy of UV-NIL are also inherently higher than in hot embossing processes, because of reduced size variations of templates and wafer materials due to the thermal cycles. It is known that with an increase of the wafer size, the wafer flatness decreases so that a high pressure has to be applied to assure a good imprinting uniformity. Typically, the maximum imprinting area is limited to about 25 mm × 25 mm for UV-NIL processes, considering the stiffness and the thickness variations of quartz glass templates and silicon wafers. In particular cases, the step-and-repeat strategy can be applied to replicate the same small sized template many times over the whole wafer area. However, this requires a more sophisticated imprint tool and can not be applied to the general case of large wafer pattern replication. On the other hand hot embossing is superior over UV-NIL if polymer materials need to be imprinted directly.

Thus, it is highly desirable to have a low cost solution for large area replications, with less risk of template damage but improved process latitude. One possibility is to use soft templates, which are referred to soft working stamps. The use of soft materials should allow for a more conformable imprinting over a large wafer area. Accordingly, soft and hard UV-NIL refers to the material hardness of the template material, hard for a quartz glass template and soft for a soft layer attached to a glass back plane (soft UV-NIL). Another advantage is that such soft working stamps can be easily obtained by casting and curing a liquid

precursor of a soft stamp material onto a silicon master. One of the most widely used materials is polydimethylsiloxane (PDMS). The working stamp fabrication is carried out in the same way as the stamp fabrication for micro-contact printing applications. PDMS stamps have a good optical transparency to the wavelengths used for curing of the UV-NIL resists (between 350 nm and 450 nm) and a low surface energy which ensures an easy separation from the substrate after the UV-NIL process. The drawback of PDMS as a soft stamp material for UV-NIL is that due to the low Young's module of PDMS extremely high resolution pattern in the sub-50 nm range might not be achievable and the local pattern deformation is more critical than in hard UV-NIL. Other materials like perfluorpolyethers (PFPE´s) can be used instead of PDMS for resolving features in the sub-50 nm range.
In this chapter, we are presenting a general process consideration based on the used materials and equipment concepts. The two most important nanoimprint lithography techniques hot embossing and UV-NIL are described in detail with application examples.
This present book chapter focuses on the principles and technologies of hot embossing and UV-NIL structuring techniques; micro contact printing (μ-CP), also considered as soft lithography, is only mentioned in the introduction and is not considered in subsequent paragraphs. Application pictures are added in various sub-chapters to substantiate the described processes.

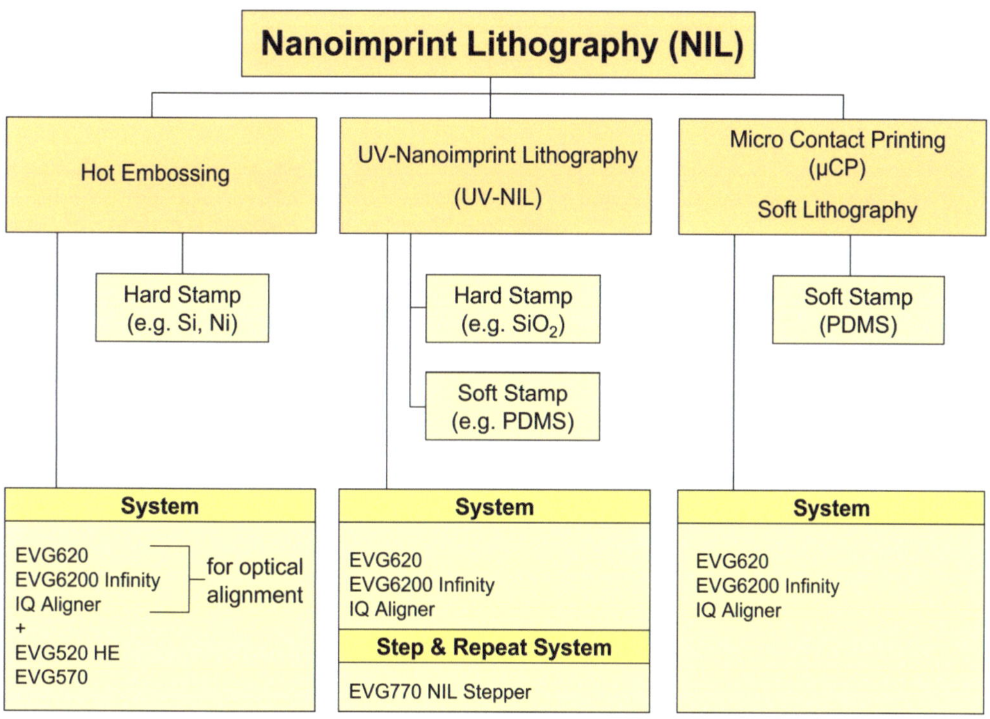

Fig. 1. Main nanoimprint lithography techniques with corresponding stamp materials

2. Nanoimprinting processes

The relevant process parameters used for a nanoimprint lithography process strongly depend on the applied nanoimprint technique. On the one hand hot embossing requires high temperatures and high contact forces, whereas in UV-based NIL processes the imprinted monomer or oligomer is cured by UV-light exposure. A basic process flow of a nanoimprinting process independent on the used technique is demonstrated in figure 2. In case of a hot embossing process using a polymer substrate to be imprinted, the structures are directly transferred into the polymer without any resist and pattern transfer process. A resist is either spin coated or droplet dispensed on a substrate. The template is getting in contact with the resist and distributing the resist evenly in case of droplet dispense. The features on the template are structuring the resist surface in a way that the resist is flowing into the cavities of the topography of the template. Curing of the resist is performed by either cooling down below the glass transition temperature of the resist in case of an HE process or by UV-light exposure in an UV-NIL process. The residual resist layer remaining underneath the elevated features of the template has to be removed in a dry reactive ion etching (DRIE) process in order to transfer the pattern into the substrate by either proceeding with the DRIE process used for the resist removal or by applying a different DRIE process. The alignment keys represented on the template and the substrate surface are used if an aligned imprinting process is required. The structuring process is finished by removing the remaining resist from the substrate.

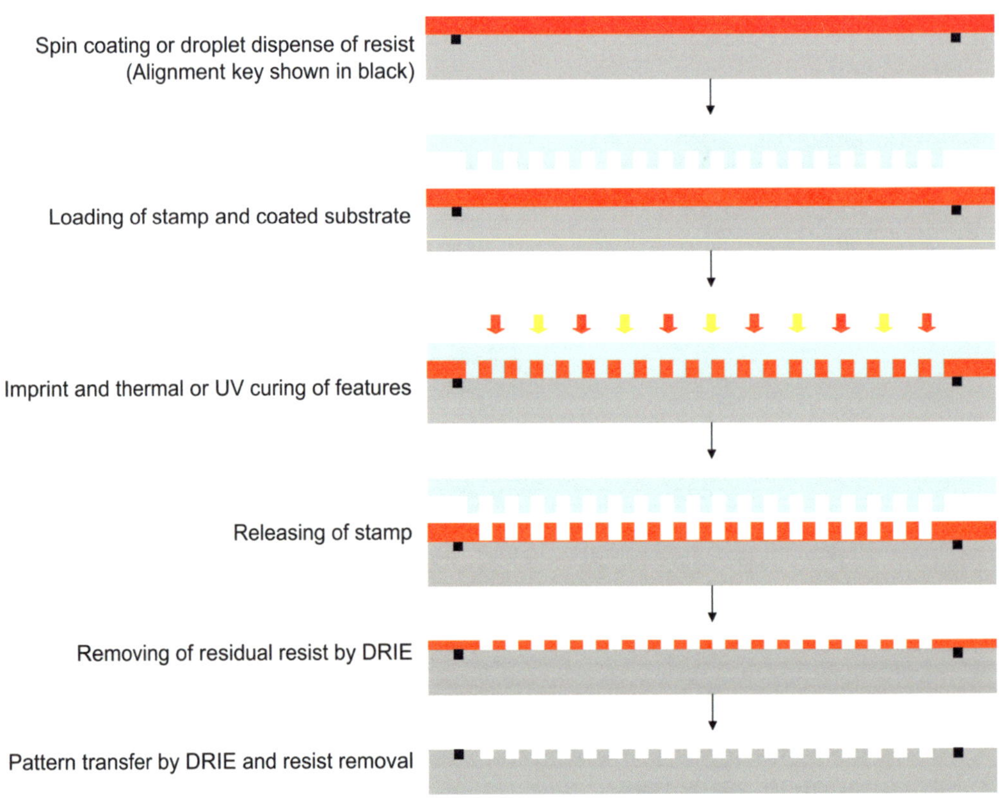

Fig. 2. Graphical representation of an UV-NIL process including pattern transfer

2.1 Hot embossing

In hot embossing processes a polymer sheet or a spin-on polymer is heated above its glass transition temperature and imprinted by applying high contact forces. The stamp material is chosen depending on the required feature sizes and the materials involved.
A typical process flow for hot embossing is depicted in figure 3.

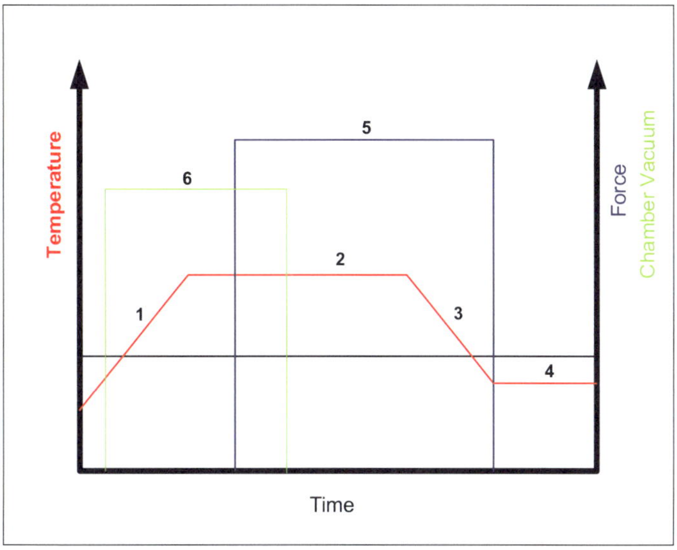

Fig. 3. Hot embossing process flow

The temperatures of top and bottom side heaters are ramped-up above the glass transition temperature of the polymer to be imprinted. In general the imprint temperature is about 20-50°C higher than the glass transition temperature (Tg) of the polymer. Only thermoplastic polymers can be imprinted as they can be dissolved in appropriate solvents for spin coating and are deformable upon the application of contact forces. During the heating time, the hot embossing chamber is evacuated to about 5 mbar. After reaching the imprint temperature of the polymer the contact force is applied. Its amount is dependent on various parameters such as the stamp area, the type of polymer and the feature geometry. The contact force remains applied until the temperature of the heaters reach the de-embossing temperature. This is the temperature level, which allows a reliable and residue-free separation of the stamp from the polymer. The total cycle time for a hot embossing process is strongly dependent on the heating and cooling capabilities of the equipment and is in the order of 3 to 20 minutes. In table 1 a summary of polymers used for hot embossing including the glass transition temperature is listed.
The fastest possible hot embossing processes are isothermal processes. In this scenario the imprint temperature and the de-embossing temperature of the substrate are the same. The heaters can remain at the same temperature level all the time. The resist is fluid enough to guarantee a very fast imprint as soon as it gets in contact with the template and the contact force is applied. De-embossing is performed outside of the chamber in this case. Cycle times of 2 minutes have been demonstrated with high resolution features down to 50 nm on a 200 mm Si substrate by using a EVG520HE hot embossing equipment.
The following stamp materials are widely used for hot embossing processes:

Polymer	Short name	Glass transition temperature [T_g]
Polymer substrates		
Cyclo olefine copolymer	COC	60 – 180°C
Polystyrene	PS	100°C
Polymethylmethacrylate	PMMA	100°C
Polycarbonate	PC	145°C
Spin-on polymers		
mr-I-7000	-	60°C
mr-I-8000	-	115°C
mr-I-9000E	-	35°C
mr-I T85	-	85°C

Table 1. Polymers used for imprinting and its glass transition temperature

- Silicon
- Fused quartz glass
- Nickel
- Soft working stamps

Table 2 lists commonly used stamp materials for hot embossing including the fabrication process, the type of hot embossing process, the thermal expansion of the stamp material as well as its required surface treatment prior to imprinting. The corresponding thermal expansion coefficient plays a role in case of optical alignment is required. In order to achieve optimized alignment conditions, the thermal expansion of template and substrate should be matched.

Stamp material for hot embossing	Fabrication process	Hot embossing process type	Thermal expansion [10^{-6} K^{-1}]	Surface treatment
Silicon	e-beam, optical lithography,	Spin-on layer	2.6	Silane
Fused quartz glass	e-beam	Spin-on layer	0.6	Silane
Nickel	Optical lithography, electroplating	Polymer sheet	13	Not required
Soft working stamps	Replication from a Si or SiO_2 master	Spin-on layer, polymer sheet	310 for PDMS 340 for MD-500 PFPE	Not required

Table 2. Stamp materials used for hot embossing and its characteristics

Soft working stamps can be fabricated from Ni masters, glass masters or directly from resist masters after lithography (e.g. e-beam, optical, laser). In figure 4 the basic process flow of the soft working stamp fabrication process is demonstrated. The liquid pre-cursor of the polymer is dispensed onto the master and squeezed between the master and a glass backplane. Curing of the polymer is achieved by UV-exposure. The soft working stamp attached to the glass backplane is released from the master and represents the reversible image of the master design. The resist master is treated with an anti-sticking layer (fluorinated chemical) in order to ensure a residual-free separation of the polymer from the master. Ni masters are usually not treated as they exhibit good release properties from the master.

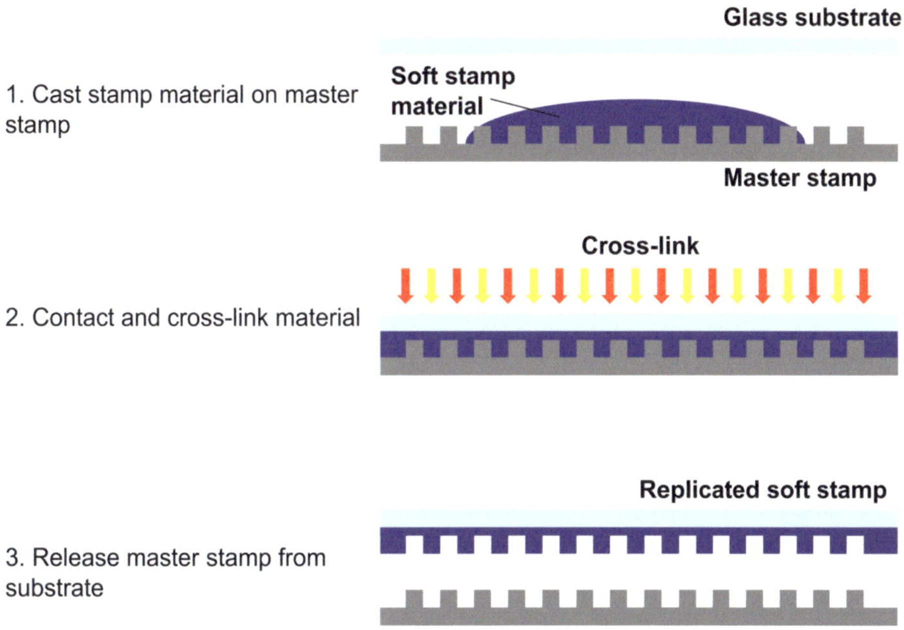

Fig. 4. Basic process of soft working stamp fabrication

In figure 5 and 6 scanning electron microscopy (SEM) pictures of imprinted features are shown replicated by using soft working stamps. In the first case spin-on polymer layers are imprinted whereas in the later case the structures are directly replicated into a polymer.

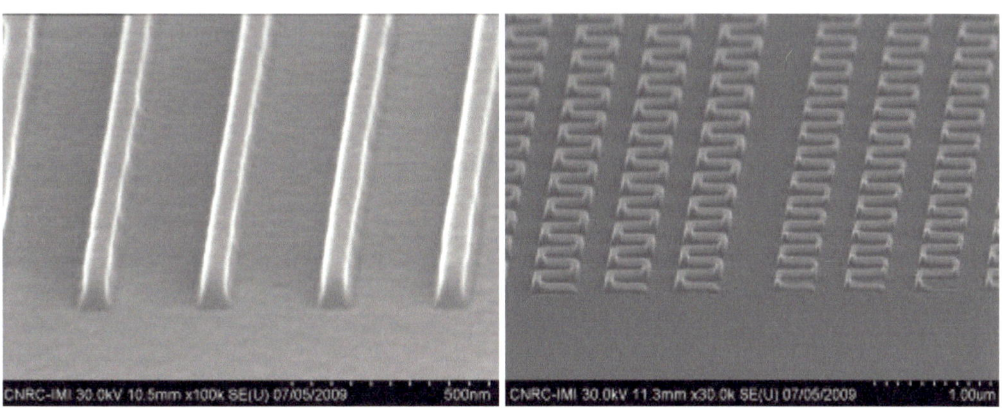

Fig. 5. SEM images of imprinted features by hot embossing utilizing soft working stamps; left: 50 nm lines, 100 nm high, right: 50 nm and 100 nm meander structures, 100 nm high

Fig. 6. SEM image of 200 μm wide hot embossed micro fluidic channels utilizing working stamp substrate top area

2.2 UV-nanoimprint lithography

In UV-based nanoimprint lithography a transparent template with nanostructures on its surface is used to deform a thin resist film or an active material deposited on a substrate followed by a hardening step. The film is cured by photo chemical cross-linking before the stamp is released. When polymer chains are linked together by cross-links, they lose some of their ability to move as individual polymer chains. For example, a liquid polymer (where the chains are freely flowing) can be turned into a "solid" or "gel" by cross-linking the chains together. In polymer chemistry, when a synthetic polymer is cross-linked, it usually means that the entire bulk of the polymer has been exposed to the cross-linking method. The resulting modification of mechanical properties depends strongly on the crosslink density. Low cross-linked densities raise the viscosities of polymers. Intermediate cross-linked densities transform gummy polymers into materials that have elastomeric properties and potentially high strengths. Such materials are usually used as stamp materials. Very high cross-linked densities can cause materials to become very rigid or glassy. Cross-links can be formed by chemical reactions that are initiated by radiation. For example, mixing a non polymerized or partially polymerized resin with specific chemicals called cross-linking reagents results in a chemical reaction that forms cross-links. In hard UV-NIL rigid transparent templates like quartz glass contain the structures whereas in soft UV-NIL polymeric replicas from master templates are used for the imprinting process. Table 3 shows the basic data for both patterning techniques in terms of maximum imprint area and alignment capabilities. Hard- and soft UV-NIL stamps can be re-used many times, in case of hard UV-NIL there is only the need of re-applying the surface release agent called anti sticking layer (ASL), whereas soft stamps are materials with integrated release properties. Soft stamp polymers are cheap and disposable materials, they can be replicated multiple times from the master.

UV-based imprint lithography can be carried out using different types of imprinting machines: single step, full-field imprinting and step-and-repeat.

Single step imprinting machines, like the one shown in figure 7, structure an entire wafer (up to 200 mm) or small areas (called a die) at one time. A step-and-repeat tool (Fig. 8) imprints one die in a hard or soft UV-NIL approach of a wafer at a time and then moves to a new area of the wafer. The process is repeated until the entire wafer is imprinted. One advantage of the step-and-repeat method is that it is easier to achieve a higher alignment accuracy in a smaller area than in a larger area. A second advantage, and probably the most significant for many applications, is that it allows the use of very small hard stamps to create a large imprint

Fig. 7. Single step semi-automated UV-NIL system

Fig. 8. Step-and-repeat semi-automated UV-NIL system

area. It is also possible to use such structured areas as large area template for single step imprinting applications.
NIL has passed the barrier from the laboratory scale to industrial production in various applications such as wafer level camera modules (WLC), optical gratings or LED's. These applications are described in more detail in the following paragraphs.

	Soft Stamps	*Hard Stamps*
Imprinted area	Up to 300 mm	Up to 25 mm x 25 mm*
Resist application	Spin coating, inkjet dispense	Spin coating, inkjet dispense
Alignment	Accuracy limited due to distortion of stamp (~ 1 μm)	High precision alignment possible (< 50 nm)

* in single step imprinting applications

Table 3. Comparison of soft and hard stamps based on different process parameters

2.2.1 Hard UV-Nanoimprint lithography

In hard UV-based nanoimprint lithography processes quartz glass templates are used for replication. Due to the thickness variation of substrates and the rigid templates in hard UV-NIL, the area which can be patterned with one imprint step is limited to about 25 mm x 25 mm. Therefore, in order to pattern large substrates up to 300 mm in diameter, a "step-and-repeat" process has to be applied. In soft UV-NIL processes elastomeric stamps are used as they are able to compensate for any surface roughness and curvature of substrates to be imprinted, offering thereby the possibility to pattern at wafer scale with only one imprint step. Hard UV-NIL is preferred over soft UV-NIL for applications demanding high alignment accuracy. Commercially available resist materials have been used such as Amonil from Amo, UVCur21 from micro resist technology and PAK-01 from Toyo Gosei. All these materials can be spin coated on substrates prior to imprinting.
The adhesion of the resists to the wafer surface has to be increased while the adhesion of the cured resist to the stamp surface has to be as low as possible. In order to get a very low surface energy on the quartz glass template surface an anti-sticking layer (ASL) has to be applied. Silane based ASL´s are covalently bonded onto SiO_2 surfaces as shown in figure 9. It has been reported in literature that the anti-sticking layer can be applied in gaseous phase in a vacuum chamber by exclusion of moisture or by simply immersing the template in a

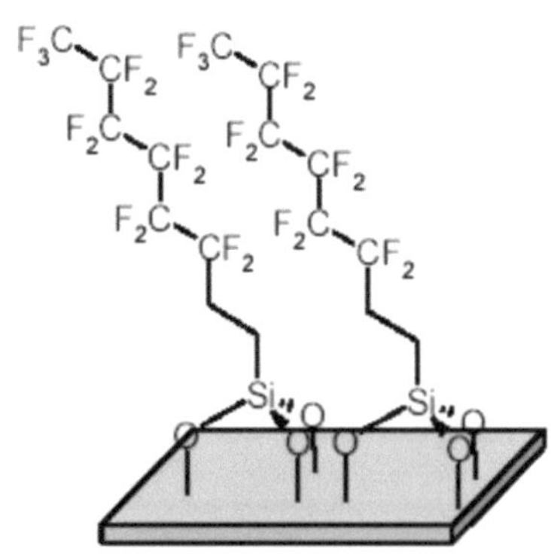

Fig. 9. Fluorosilanized silicon oxide surface

liquid ASL and rinse it with solvent. The so called molecular vapor deposition (MVD), proposed by Applied Microstructures, is a room temperature deposition process and takes place in vacuum. Specific chemicals can be deposited on template surfaces such as perfluorodecyltrichlorsilane (FDTS) to serve as hydrophobic coating for the subsequent nanoimprinting process. It was found out empirically that ASL´s deposited from gaseous phase are preferred over liquid deposited films for features sizes smaller than 50 nm.
In figure 10 an atomic force microscopy (AFM) picture of imprinted sub-µm lines and space features are shown. Figures 11 and 12 demonstrate SEM pictures of imprinted features down to 90 nm resolution using 25 mm x 25 mm quartz glass templates.

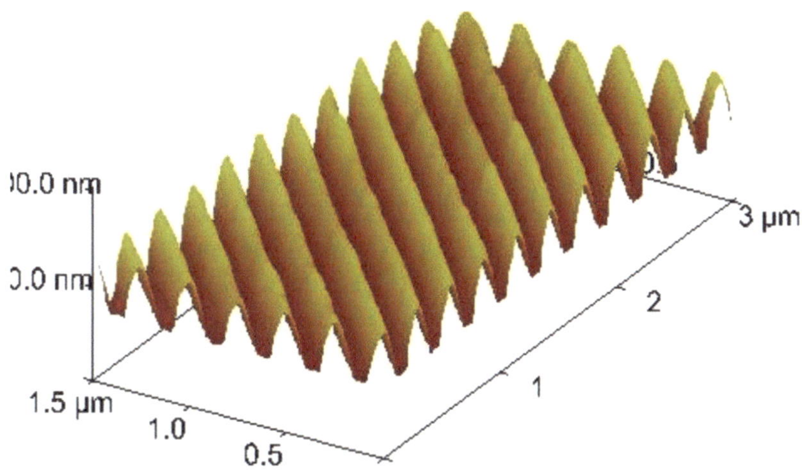

Fig. 10. Three-dimensional AFM image of an imprinted lines and space structure

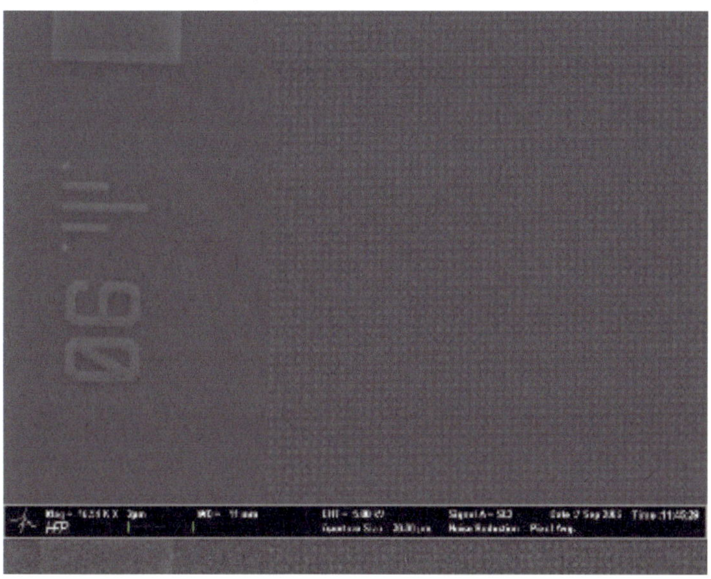

Fig. 11. Surface topography investigation - SEM picture of 90 nm imprinted dots (top view)

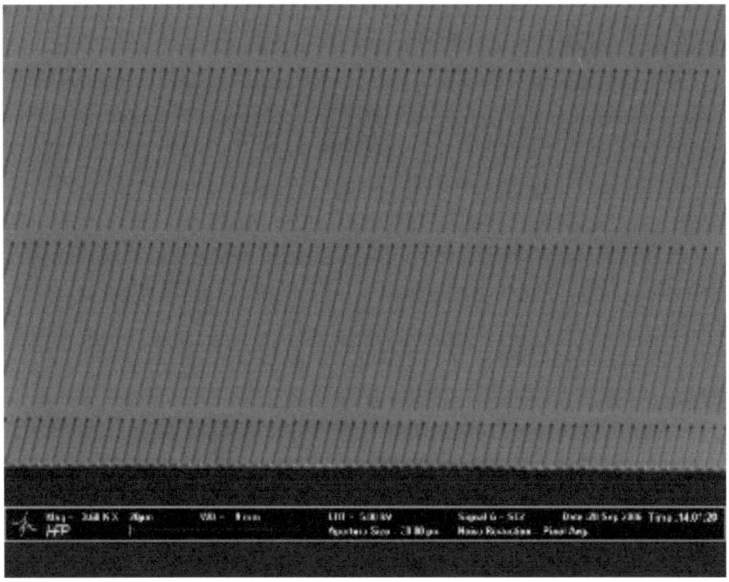

Fig. 12. Cross section investigation - SEM picture of imprinted lines and space structure (chuck tilted by 80°)

The residual layer and its distribution across the substrate surface is an important characteristic for the subsequent pattern transfer process. It needs to be as thin as possible, usually thinner than 50 nm is requested, and its uniformity should be ± 10% from its average value. Figure 13 shows a cross section SEM picture demonstrating a residual layer thickness of about 50 nm.

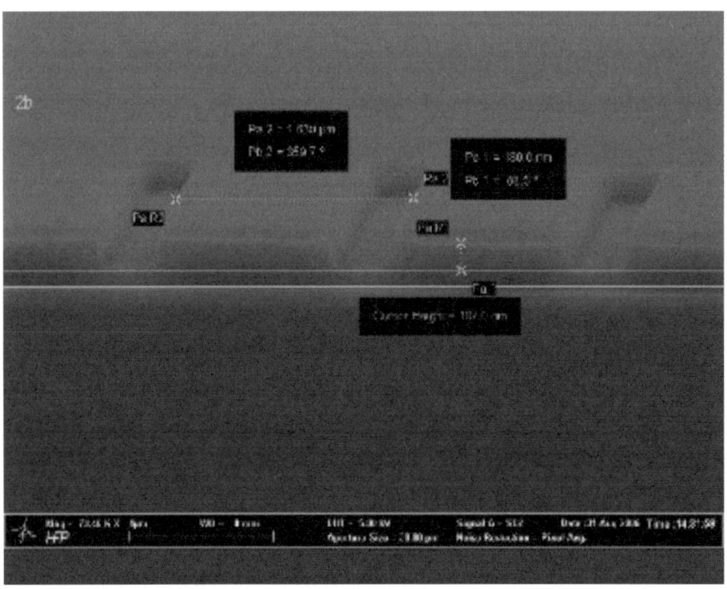

Fig. 13. Close-up of cross sectional SEM picture of imprinted lines and space structure for residual resist measurement (chuck tilted by 80°)

Stacking of lines and space structures has been performed by imprinting of features onto SiO_2 and pattern transfer into SiO_2 with subsequent filling of the etched features with Si. This approach results in 3D photonic crystal woodpile structures exhibiting a full photonic band gap by keeping the required alignment accuracy and the quality of the imprinted layers. In figure 14 all five layers of the Si/SiO_2 woodpile structure are shown. The lines and space structures of the first, third and fifth layers are facing towards the observer. The first and fifth layer are above each other and the third layer is shifted by half of the period according to the design rules for the woodpile structure. This is another example where nanoimprint lithography could play a role in the future as manufacturing technique.

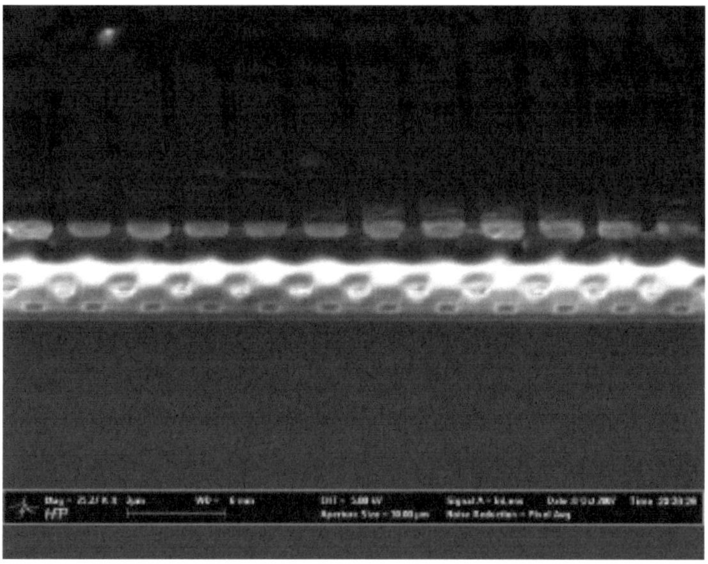

Fig. 14. SEM picture of 5 layers of the Si/SiO_2 woodpile structure

2.2.2 Soft UV-Nanoimprint lithography

Templates for nanoimprint lithography are also called stamps or moulds. In case of soft UV-NIL, soft working stamps can be conveniently reproduced by copying from another template which is called master. Silicon masters for soft UV-NIL template replication are fabricated by electron beam lithography, EUV immersion lithography or other high resolution patterning techniques, followed by reactive ion etching and surface coating with a release agent like mentioned in chapter 2.2.1. Both high resolution lithography and reactive ion etching techniques for silicon processing have been well developed, allowing manufacturing of high resolution (better than 50 nm) and high quality masters. However, the cost of high resolution masters increases rapidly with the wafer size and resolution. It is still a big challenge to cover a large area of a wafer because of throughput limitations. The surface treatment of the master with a release agent is still one of the important research topics, although a large number of recipes or processing protocols have been proposed. Fortunately, soft stamp polymers are materials with low surface energy. They can be easily peeled off after curing from a silicon master, which has been simply coated with FDTS by molecular vapor phase deposition. Now, several companies can provide high resolution silicon masters with feature sizes down to < 20 nm at reasonable prices for standard features such as arrays of gratings, checkerboards, dots and holes. Soft working stamps can be

obtained by curing the polymer on a master. It may also be necessary to introduce a carrier for the flexible stamp in order to facilitate handling of the soft working stamp in an imprinting system. Such a carrier can either be integrated during the curing process of the template material or thermally bonded after a plasma surface treatment of both glass and back surface of the soft stamp polymer template. The soft template bonded onto a carrier is detached from the master manually after curing. For more advanced fabrication, they can be obtained with a dedicated high accuracy molding unit, to apply highly uniform and repeatable contact forces to ensure that the residual layer is kept thin. This is crucial to avoid distortions during the imprinting process, especially for high resolution features. Finally, UV-curable soft materials are cross-linked via UV-exposure in the UV-imprinting system.
The stiffness of the fabricated working stamps can be adjusted by changing the mixture ratio of the base liquid and the curing agent or by changing the molecule end groups in case of UV curable polymers. For high resolution patterning, it is better to use a thin layer of hard polymer bonded on to a softer block. Such a double layer can be obtained by first spin coating the hard layer on the master and then attaching a soft block. After thermal bonding, the soft block can easily be peeled off together with the hard layer. Alternatively, other types of "hard" materials such as PMMA and PFPE can be used for the top layer fabrication, but both attaching and peeling processes are much more delicate. In most cases, a stamp made of a single layer polymer and glass back-plane will be sufficient for the pattern replication of feature sizes between 50 nm and a few micrometers, but the bi-layer template configuration ensures both higher resolution and higher mechanical stability.
The created soft template usually resembles a negative counterpart from the used master. Nowadays there is also the possibility to replicate positive and negative counterparts from one master design by using different stamp polymers. In contrast to the fabrication of quartz glass templates for hard UV-NIL, where the reproduction by copying is much more difficult, a silicon master for soft UV-NIL can be used for copying of many soft working stamps. Soft working stamps have a relative low surface energy which ensures an easier separation from the substrate after UV-NIL. In addition, because of the inherent properties of the soft materials (elastomers), the risk of mechanical damage of the master is also largely reduced. Therefore, the template cost as well as the processing cost of soft UV-NIL is significantly lower than in other nanoimprint techniques.
Soft UV-NIL can be performed in a very simple manner in R&D environments by running the process manually. However, for sophisticated device fabrication and for achieving reproducible results, dedicated systems are necessary with a better imprinting performance and highly controllable process parameters. In principle, a substrate holder, a template holder and a mechanical system that brings the coated substrate and the soft working stamp into contact and a UV-exposure unit are needed. The flexible and forgiving nature of the soft working stamps compensates for the waviness of the substrate and the working stamp itself in order to get full-field highly homogeneous patterns. However, a wedge compensation process has to be performed to ensure that the working stamp and the substrate are parallel before the imprinting process. In addition to that, the imprint force and exposure dose must be controlled over the whole imprinted area and, a suitable detachment unit must be implemented for highly reproducible process results. Finally, an optical alignment is generally required for multilevel device applications.
With the rapid growing interest in nanoimprint techniques, commercial systems are now available from several companies. In particular, *EV Group* has developed dedicated systems for soft UV-NIL.

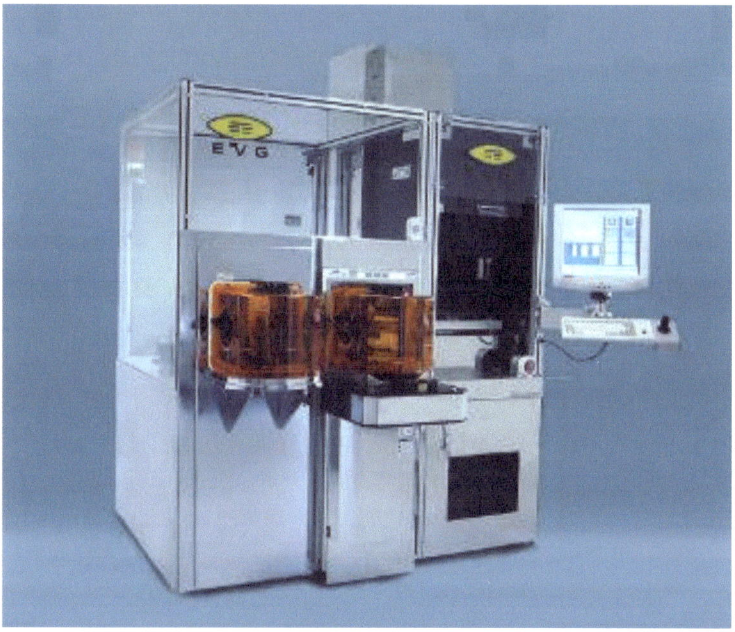

Fig. 15. IQ aligner for large area soft UV NIL (up to 300 mm)

The EVG620 system mentioned in chapter 2.2.1. is able to perform optical lithography as well as hard and soft UV-NIL for R&D applications for substrate sizes up to 150 mm. The IQ aligner like shown in figure 15 is a dedicated system for large area soft UV-NIL applications up to 300 mm substrate size. The available equipment spectrum reaches from semi-automated configurations for research environments to fully automated systems including automated detachment of stamp and substrate. Such systems are already in use for high volume production.

The general imprinting process of soft UV-NIL can be described as follows:

A UV-curable resist layer is coated onto a substrate. For coating of the resists several methods like spin coating, spray coating and droplet or puddle dispense can be used. The applied coating technique is heavily dependent if a pattern transfer process has to be carried out, which requires thin residual layers in the range of below 50 nm. Such a requirement can be fulfilled by either spin coating of thin layers to be imprinted or discrete droplet dispense of small volumes depending on the feature size and feature density. If the residual layer does not play any role, such as the polymer layer is part of the functional device, the resist can be thicker. In this case the resist can be applied by puddle dispense.

Then, a soft working stamp is brought into contact with the UV-curable resist layer. After applying the contact force on the stack consisting of template, resist, and substrate, the resist is cured by UV-exposure through the transparent working stamp. Finally, the working stamp is separated from the imprinted substrate, resulting in an imprinted resist layer, which is the negative image of any topography preserved on the stamp surface. Due to the double-negative process (master/soft working stamp and soft working stamp/imprint) the imprinted pattern have the same tone as the master.

Figures 16 and 17 show AFM measurements of 50 nm boxes on the master and its replica fabricated by nanoimprint lithography. The two structures show a nearly perfect match in pattern fidelity. There is only a variation of < 5% in horizontal and vertical dimensions.

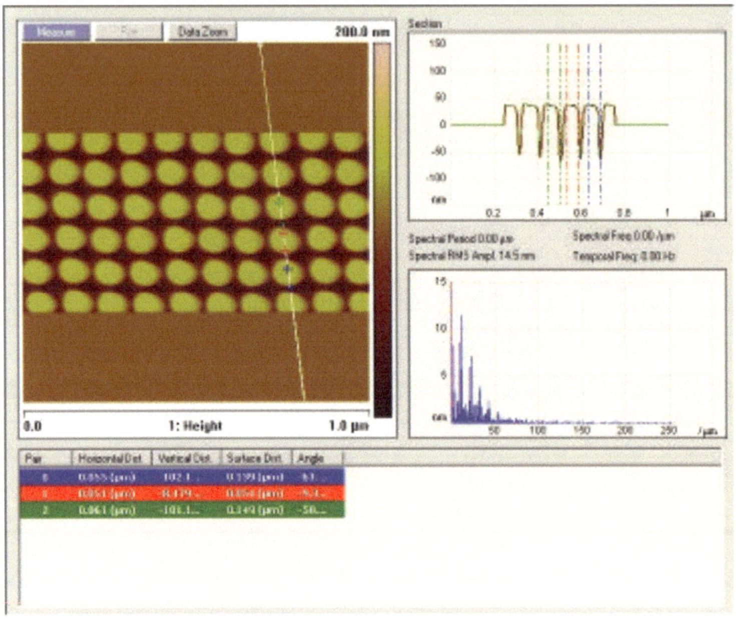

Fig. 16. Master AFM image of 50 nm dots, 100 nm in height

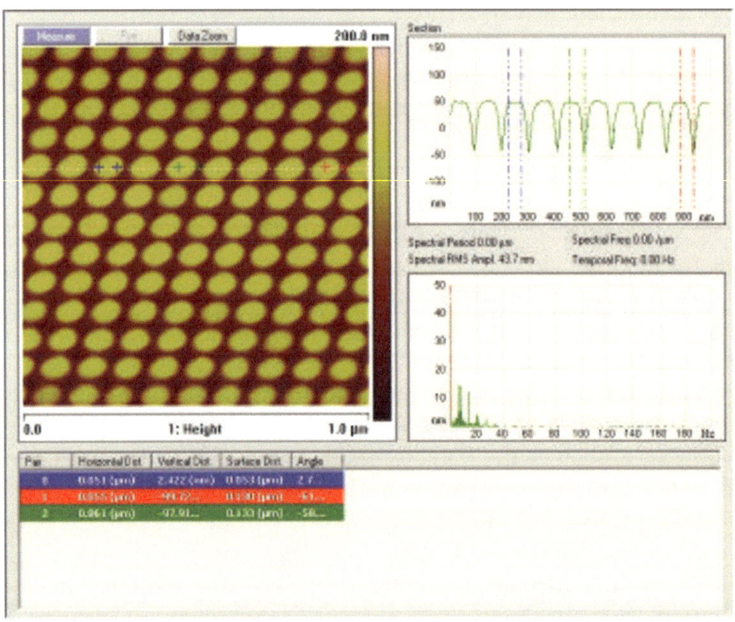

Fig. 17. Corresponding AFM image of 50 nm replicated dots, 100 nm in height

Besides the double-negative process (master/soft working stamp and soft working stamp/imprint) there is the possibility, by tuning the stamp polymer, to replicate positive and negative counterpart working stamps from one master design. In this case a second working stamp imprinted from a first working stamp is used to perform the final imprint as shown in Figure 18.

Fig. 18. SEM Images of 50 nm boxes in a 3 step counterpart imprint

To obtain a high quality soft UV-imprint, the material parameters as well as the equipment capabilities need to be optimized. Ideally, resists for soft UV-NIL should exhibit properties like low viscosity, high UV sensitivity and good etch resistivity. Suitable UV-curable resist materials may be organic materials or inorganic/organic composites. Low viscosity materials tend to fill up pattern on the soft working stamp surface faster and are capable to replicate higher resolutions, whereas higher viscosity resists may be used for thick layer fabrication, where no pattern transfer is needed. Finally, high UV sensitivity is required for high speed replication and a good etch selectivity is needed for a high quality pattern transfer. Several companies such as AMO (Germany), Micro resist Technology (Germany), and Toyo Gosei (Japan) have developed resists for commercial uses and most of them allow reaching a high resolution with reasonable exposing time and etching sensitivity to standard reactive ion etching processes.

The choice of substrates for soft UV-NIL strongly depends on the application. In most cases semiconductor materials such as Si wafers (with or without SiO_2 layer) are in use. These substrates are sometimes covered with metals, other semiconductors or ceramics, but the process can also be carried out on polymer substrates depending on the adhesion properties of the material combinations.

The resolution of a lithographic process is of major importance. In case of soft UV-NIL, the resolution is dependent on both materials (soft polymer stamp, UV-NIL resist) and the performance of the imprinting processes. The resolution achievable is definitely better than 50 nm. Figure 19 illustrates an AFM picture of 35 nm half pitch master structures fabricated by extreme-UV interference lithography (EUV-IL) with the corresponding replicated polymeric stamp in figure 20.

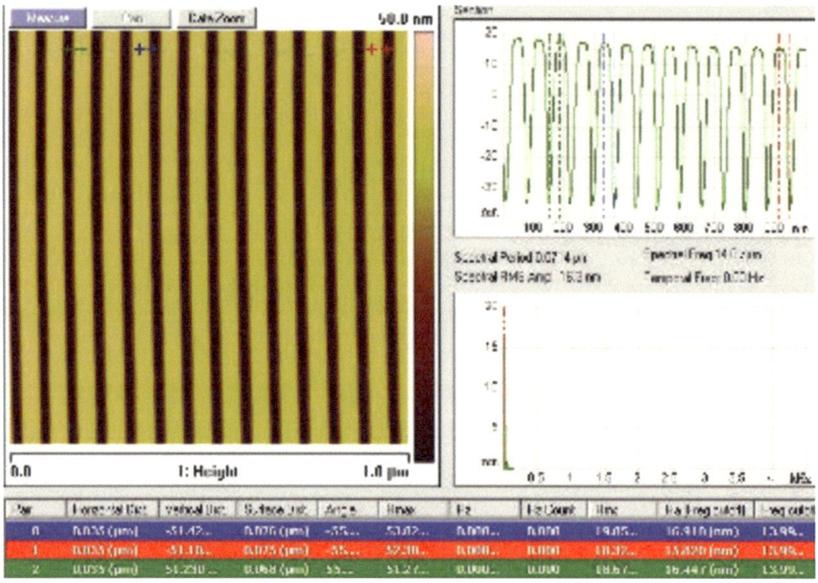

Fig. 19. AFM image of 35 nm half pitch EUV-IL master template

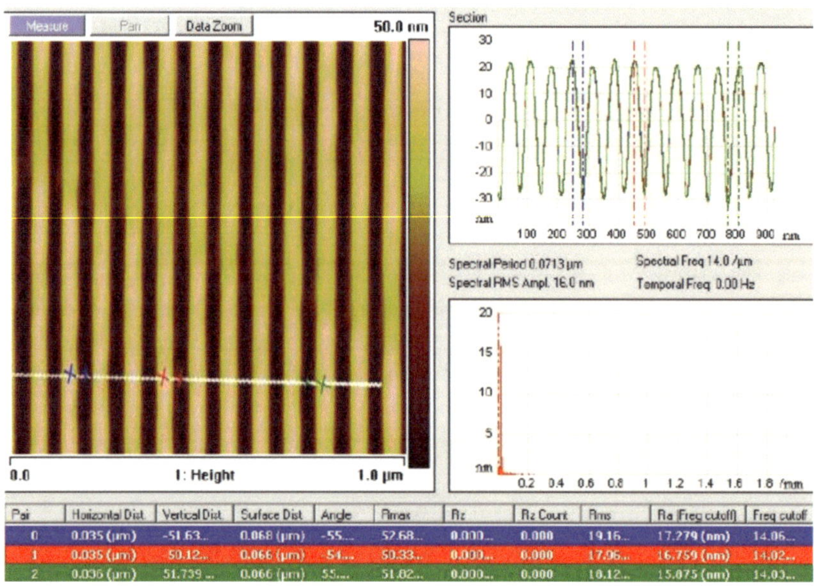

Fig. 20. AFM image of corresponding 35 nm half pitch polymeric working stamp

The unique capability of the soft working stamp to compensate for thickness variations of a given substrate makes it possible to imprint over larger area with high uniformity. Therefore, this technology is an ideal candidate for high throughout and high resolution applications. One key parameter to demonstrate the uniformity of an imprint process is the homogeneity of the residual layer thickness. Other assessment parameters are pattern placement accuracy, overlay accuracy, process latitude, defect density, production rate and cost of ownership of a soft UV-NIL process. However, these are depending on the targeted applications and are different for each application as the variety of possible high volume applications are very broad. Soft UV-NIL is best suited for applications where large areas of continuous pattern need to be fabricated without any stitching error. The ability to pattern over large areas in one imprint step can, for sensor application, be used to provide substrates a "larger" surface by introducing a third dimension on a given area. The process can be performed at room temperature by applying small imprint forces, thus it does not impose additional stress on brittle or heat sensitive substrates.

Due to the flexibility of the template curvatures of the substrate may be neglected, yet for the same reason applications that need high definition alignment in the sub-micron range are hard to realize. Up to now the process has found its way into scientific research domains and production lines in micro-optics (Figure 21) and into prototyping for nanophotonics and biotechnology. It is expected to have additional impact in fabrication of memory devices and in display technology.

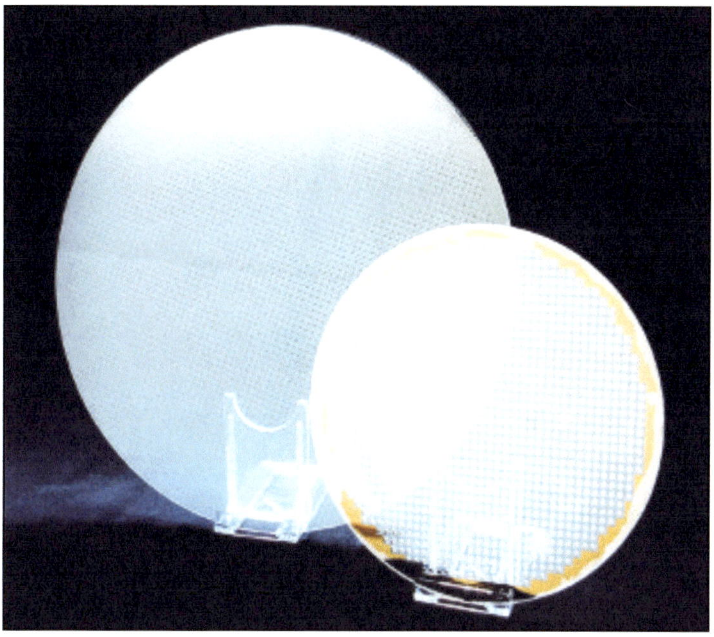

Fig. 21. 300 mm soft stamp replicated lens wafer and 200 mm stacked wafer level camera module

All imprinting techniques are based on material (polymer) transportation from the recessed area to the non recessed area or template cavities. It is obvious that, no matter how large the pressure and how hard the template material, imprinting is easier if only local or short distance material transportation is involved. Large distance resist flow or large feature size patterning may introduce large working stamp deformation. Therefore, soft UV-NIL is also best suited for applications where only small size features are replicated. For the device fabrication where both small and large size features have to be replicated, a mix-and-match technique can be applied, i.e., only high resolution features (contact holes, grating, dot and hole matrices, etc.) are patterned by soft UV-NIL and all large features (cavities, channels, pads, etc.) by standard optical lithography methods. Being still highly parallel, cost effective,

and fully compatible to the requirement of multilevel device manufacturing, such an approach takes the advantage of soft UV-NIL but avoid its inherent drawback.
Probably the most promising high volume application is the fabrication of next generation high density storage medias with a planned capacity in the range of 1 Tb/in^2. Since the working stamps can be cost-effectively copied from a master and a hard disc media consists of homogenously distributed high resolution features (tracks for disc track recording as shown in figure 22 or dots for bit patterned media), soft UV-NIL can be an attractive candidate for the large scale mass production.

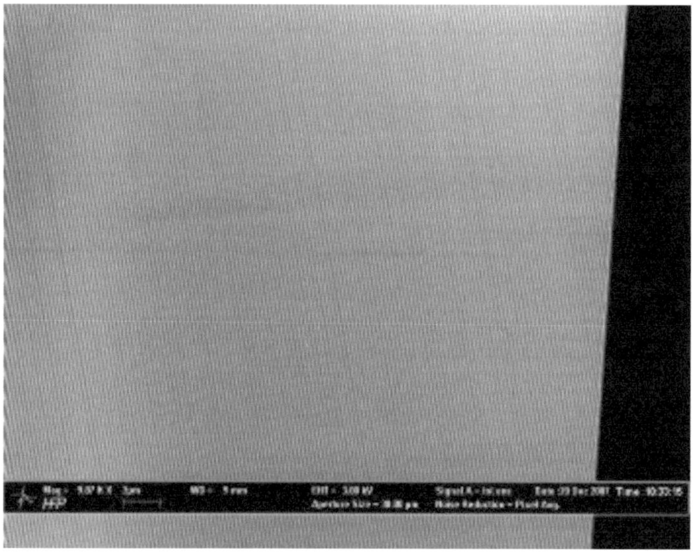

Fig. 22. SEM measurement of 50 nm half pitch imprinted data track features utilizing soft stamp imprint lithography

Fig. 23. SEM image of 350 nm photonic crystal holes, structure depth 600 nm utilizing positive and negative counterpart working stamps

Finally, the soft UV-NIL technique has been used to demonstrate device integration capability. One example is the fabrication of LED devices with photonic crystals. Figure 23 demonstrates a SEM image of photonic crystals replicated with soft UV NIL.

3. Conclusion

The most significant progress in nanoimprint lithography has been made within the last 10 years. Various imprinting modes have been developed and some of them already reached a mature state for research and development environments. Tremendeous progress was seen in building up the infrastructure for NIL. Templates and related fabrication methods, resists and nanoimprinting equipment are commercially available nowadays. This was supported by the foundation of consortia supporting to enhance the infrastructure in NIL. A good example is NILCom, which was founded in 2004 and which aligns activities around the fabrication process for commercial devices.

In addition, a number of potentially high volume future applications have been demonstrated, showing expected performances not only for high resolution patterning, but also for functional device integration. Imprinted micro lenses on up to 200 mm wafers are already in use in wafer-level optics for digital cameras. There are several nanoimprinting process modes available. Depending on the type of imprint material and available equipment either hot embossing or UV-NIL processes can be selected. Both types of processes can be performed with either hard template materials (e.g. Silicon, Quartz glass or Nickel) or soft working stamp materials. All of the described techniques have shown resolution capabilities of better than 50 nm. Hot embossing processes are the preferred process for structuring of polymer substrates directly. For devices with tight requirements on the overlay alignment accuracy hard UV-NIL is preferred. Thermal expansion differences of involved materials do not play any role as this process is performed at room temperature. Comparing to hard UV-NIL processes, soft UV-NIL is advantageous because of the possibility of patterning large area at once. Another benefit of soft UV-NIL is the fact that soft working stamps can be easily replicated from a master and each of them can be used for a large number of pattern replications. The risk of damaging the master due to mechanical means is largely reduced due to the soft properties of the working stamp materials. Considering the available infrastructure and the huge market potential, it can be assumed that a range of industrial applications will be realized by applying nanoimprint lithography processes.

4. Acknowledgement

The authors acknowledge the support from the company EV Group for writing this chapter.

5. References

[1] *International Technology Roadmap for Semiconductor Industry Association*, edition 2006 (http://public.itrs.net/).

[2] International Conference on Nanoimprint and Nanoprint Technology, since 2002.

[3] Other international conferences and proceeding such as MNE, EIBPN, MMC, etc.

[4] *Alternative Lithography*, C.M. Sotomayor (Ed.), Kluwer Academic/Plenum (2003) p 249.

[5] Mamora Kondo, et.al., Fine Pattern Fabrication by Molded Mask, Academic Meeting of Japan Society of Applied Physics, Oct. 1976.

[6] S.Y. Chou, P. R. Krauss and P.J. Renstrom, Appl. Phys.Lett. 67, 3114 (1995).
[7] S. Y. Chou, P. R. Krauss, P. J. Renstrom, Nanoimprint Lithography, J. Vac. Sci. Technol. B 14, 4129 (1996).
[8] P. R. Krauss, S. Y. Chou, Appl. Phys. Lett. 71, 3174 (1997).
[9] J. Haisma, M. Verheijen, K. Van dean Heuvel and J. van den Berg, J.Vac Sci. Technol. B14, 4124 (1996).
[10] Y. Chen, F. Carcenac, C. Ecoffet, D.J. Lougnot, H. Launois, Microelectron. Eng. 46, 69-72 (1999)
[11] M. Colburn, S. Johnson, M. Stewart, S. Damle, B. Jin, T. Bailey, M. Wedlake, T. Michaelson, S. V. Sreenivasan, J. Ekerdt, C. G. Willson. Proceedings SPIE 3676, 379 (1999).
[12] M. Bender, M. Otto, B. Hadam, B. Vratzov, B. Spangenberg, and H. Kurz, Microelectron. Eng. 53, 233 (2000).
[13] Y. Xia, G.M. Whitesides, Angew. Chem., Int. Ed. 37, 550 (1998).
[14] J. L. White "Principles of Polymer Engineering Rheology", John Wiley. &. Sons, New York, 1990.
[15] H. Schift, NaPa Emerging Nanopatterning Methods, Library of Processes, first edition (2008), ISBN 978-3-00-024396-7
[16] http://www.microresist.de
[17] M. Colburn, S. Johnson, M. Stewart, S. Damle, T. Bailey, B. Choi, M. Wedlake, T. Michaelson, S.V. Sreenivasan, J.G. Ekerdt and C.G. Willson. Proc. SPIE 3676(I): 379 (1999).
[18] S. Zaitsev, Report for the European ESPRIT Project No. 29097 "SPINUP", 1999 (unpublished).
[19] http://www.nilt.com; http://www.ntt-at.com
[20] N. Koo, U. Plachetka, M. Otto, J. Bolten, J. Jeong, E. Lee and H. Kurz, Nanotechnology 19, 225304 (2008).
[21] N. Koo, M. Bender, U. Plachetka, A. Fuchs, T. Wahlbrink, J. Bolten, H. Kurz, Microelectron. Eng. 84, 904 (2007).
[22] U. Plachetka, M. Bender, A. Fuchs, T. Wahlbrink, T. Glinsner and H. Kurz, Microelectron. Eng. 83, 944 (2006).
[23] E. Roy, Y. Kanamori, M. Belotti, Y. Chen, Microelectron. Eng. 78-79 (2005) 689
[24] http://www.amo.de; http://www.microresist.de; www.toyogosei.co.jp
[25] Y. Chen, E. Roy, Y. Kanamori, M. Belloti, D. Decanini. Proceedings SPIE 5645, 283 (2005).
[26] J. Shi et al, Microelectron. Eng. 83, 1664 (2006); Microelectron. Eng. 84, 1724 (2007).
[27] U. Plachetka, M. Bender, A. Fuchs, B. Vratzov, H. Kurz, Microelectron. Eng. 73-74, 167 (2004).
[28] J. Shi, A.P. Fang, L. Malaquin, J.L. Viovy, A. Pépin, D. Decanini and Y. Chen, Appl. Phys. Lett. 91, 153114 (2007).
[29] U. Plachetka, N. Koo, T. Wahlbrink, J. Bolten, M. Waldow, T. Plötzing, M. Först, and H. Kurz, IEEE PTL, 20, 7, April 1, 2008
[30] T. Glinsner, P. Hangweier, H. Luesebrink, P. Dorsey, A. Homola, D. Wachenschwanz, Solid State Technology, 51-54, March 2005.
[31] Low Temperature Nanolayer of Metal Oxides by MVD, B.Kobrin, N Dangaria, J. Chinn Applied Microstructures, 1020 Rincon Circle, San Jose, CA 95131 USA
[32] H. H. Solak Nnolithography with coherent extreme ultraviolet light, J. Phys. D – Appl. Phys. 39, (2006) R171-R188
[33] T. Glinsner, Fabrication of 3D-photonic crystals by nanoimprint lithography, Ph.D. Theses, 2007.
[34] http://nilcom.org

25

Effect of Applying Ultrasonic Vibration in Hot Embossing and Nanoimprint

Harutaka Mekaru
National Institute of Advanced Industrial Science and Technology (AIST)
Japan

1. Introduction

Nanoimprint lithography (NIL) is a technology where fine structures on a mold (or template) are transferred onto a substrate coated with thermoplastic or with ultraviolet (UV) curing resins by making contact with the substrate while being heated or exposed to UV lights. Recently, NIL has been applied in semiconductor manufacturing to print fine features of circuits on LSI chips and memories at reduced manufacturing cost. Current nanoimprint technology can be classified as thermal nanoimprint and UV nanoimprint.

In thermal nanoimprint, a mold is heated to above the glass transition temperature (T_g) of a thermoplastic while keeping the mold pattern pressed against the thermoplastic by applying a loading force, as shown in Fig. 1(a). After keeping the mold and the thermoplastic in that position for a while, they are then cooled down to below the $T_{g,}$ and the mold is then released from the solidified thermoplastic. In this technique, depending on thermal deformation, there is a likelihood of deterioration of the positional accuracy and the shape of the pattern. Moreover, the total processing time in thermal NIL also becomes long.

On the other hand, in the case of UV nanoimprint, a template made of material with quartz - like UV-transparency, is brought into contact with a substrate coated with a UV-curing resist. The step is then followed by UV irradiation of the UV-curing resist through the template, as shown in Fig. 1(b).

We are developing a nanoimprint technology for the replication of patterns that employs ultrasonic vibration instead of thermal cycling or UV radiation. In this technique, by maintaining a pressure between mold patterns and thermoplastic, a certain amount of heat is generated at their interface by inducing ultrasonic vibration where the patterns are transformed thermally as shown in Fig. 1(c). In thermal nanoimprint, the molding material is heated to above its T_g, followed by a time consuming process of thermal cycling using electric or oil heating. The use of ultrasonic vibration for the generation of heat had also been proposed in thermal nanoimprinting. However, during the heating, a large amount of energy is lost into the mold material exposing it to mechanical stress. In the present work, the mold is mounted on an ultrasonic generator where the vibration is impressed in a direction of pressure applied on the thermoplastic. Here the mold patterns are pushed and pulled very rapidly within the surface of the thermoplastic. Therefore, the temperature of the mold hardly changes from the room temperature. Here, an assisting effect of ultrasonic vibration in hot embossing and in thermal nanoimprint is shown in a time series which also describes the processes of ultrasonic nanoimprinting.

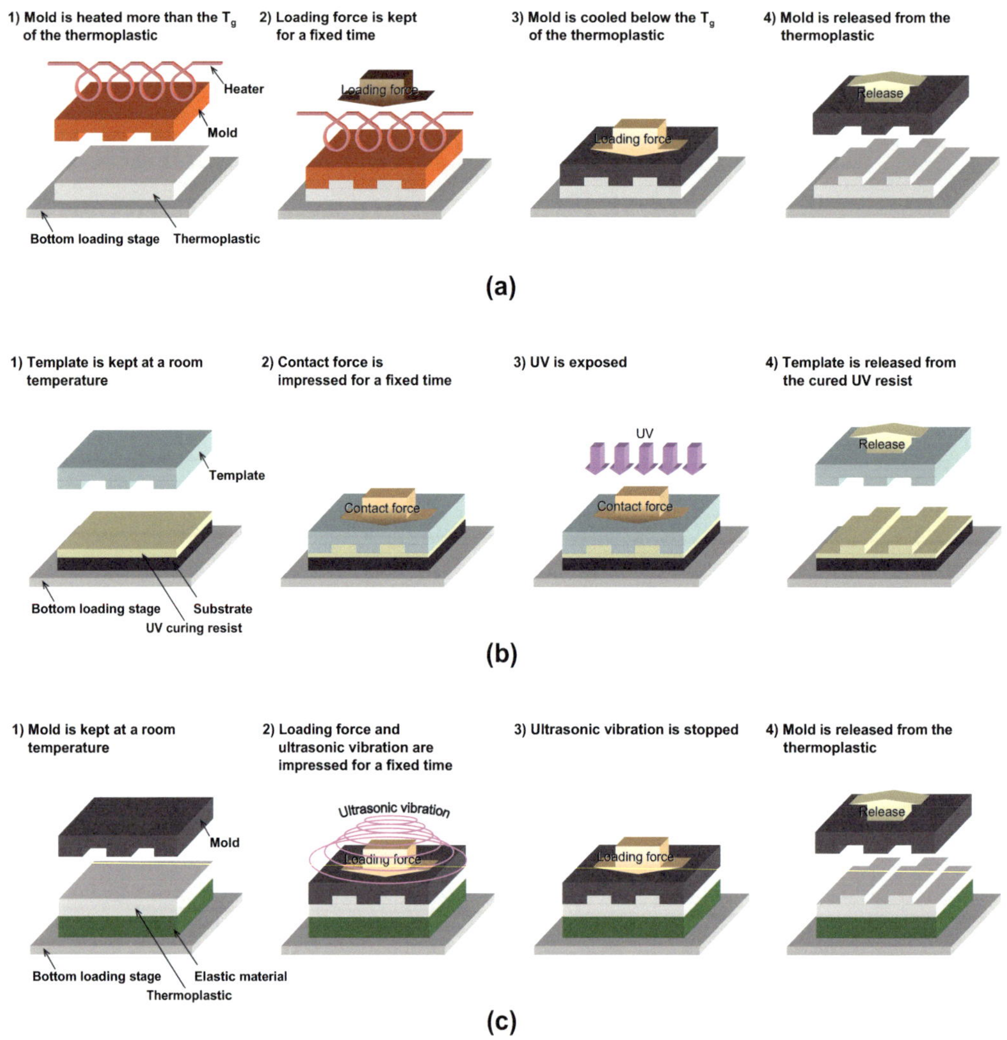

Fig. 1. Concept of process flows in (a) thermal nanoimprint, (b) UV nanoimprint, and (c) ultrasonic nanoimprint.

2. Hot embossing assisted by ultrasonic vibration

2.1 In case of impressing ultrasonic vibration with high amplitude

The introduction of ultrasonic vibration into replication technologies has led to the development of a molding technique for achieving high-aspect ratios (aspect ratio = pattern depth/width of the pattern). As a technique for fabricating the microstructure of the high-aspect ratio, LIGA (Lithographie, Galvanoformung and Abformung) process is widely known. The LIGA process is a total processing technology that combines X-ray lithography, with electroformimg, and molding. X-ray from a synchrotron with its high permeability and directivity is irradiated on a thick resist to form structures with high-aspect ratios. The resist structure is then transformed into a metallic mold processed by electroforming. Electroformed metallic structures are used as mold patterns to mass produce high-aspect-ratio replicated structures in thermoplastic such as in hot-embossing. However, because of tapped air/gas in the cavity of mold pattern the softened molding materials cannot fully enter into it, resulting in the formation of defective imprinted patterns with reduced aspect ratios. Although this problem has been addressed by hot embossing in vacuum but because of the accompanying decompression/purge operation and heating/cooling cycle in an insulated environment of vacuum, the process takes more time. Then, the author thought of assisting the flow of softened material into the mold pattern by applying ultrasonic vibration in hot embossing at room atmosphere.

A new vacuum hot embossing system was developed at University of Hyogo. The maximum pattern size that could be embossed was 50 x 50 mm^2, at a maximum available temperature of 400 °C at which most thermoplastics and sealing glasses are embossed. A servomotor with a maximum output of 50 kN was used in this system. As a result, positional accuracy and pressing speed of the loading stage can be precisely controlled.

Fig. 2. (a) Photograph of an ultrasonic hot embossing system, and (b) bottom loading stage based on an ultrasonic horn.

In this research, a piezoelectric actuator was built into a vacuum hot embossing system as an ultrasonic vibration generator, and the effect of assistance by the ultrasonic vibration in hot embossing was verified experimentally. Figure 2 shows a setup of an ultrasonic vibration generator installed in the hot embossing system. The bottom loading stage installed in the heater of the vacuum hot embossing system was detached, and a longitudinal 15 kHz ultrasonic vibration generator USV-900Z15S (Ultrasonic Engineering Co., Ltd.) with 16 ± 2 μm amplitude and 900 W output was installed.

The metallic mold for the experiment was made by Si dry etching and Ni electroforming. The pattern was in shape of a hollow pyramid with a cut-out apex. There were five kinds of pattern entrances differing in lengths ranging from 100 to 540 μm. All patterns had the same depths of 260 μm and inclined sidewalls with curved surfaces. Figure 3 shows a photograph and details of the pattern size of the Ni mold measured with a three-dimensional (3D) laser microscope VK-9700 (Keyence Corp.). A polycarbonate (PC) was selected for the molding material. The T_g of PC is 144 °C.

(a)

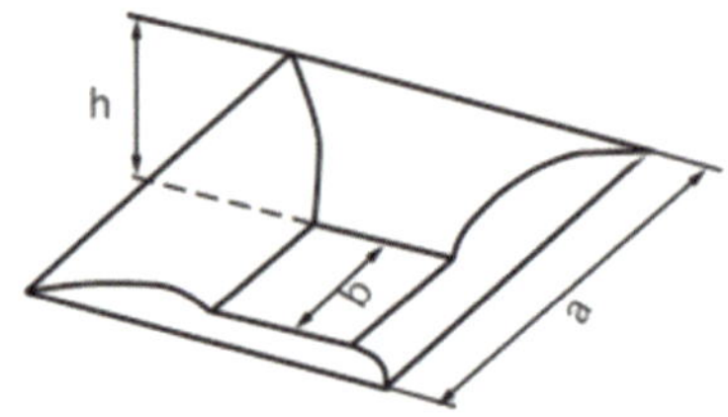

Pattern No.	Width of top; a (μm)	Width of bottom; b (μm)	Height h (μm)
I	1275	534	257
II	1127	429	260
III	968	320	259
IV	830	209	261
V	719	97	260

(b)

Fig. 3. Photograph of electroformed -Ni mold and details of measured pattern size.

The best molding conditions in the vacuum hot embossing were: mold temperature T_h = 180 °C, contact force F = 0.5 kN, and contact time t = 60 s. Figure 4(a) shows a scanning-electron-microscope (SEM) image of the embossed pattern under these conditions. The Roman numerals I through V in this figure correspond to the size of the Ni mold pattern in Fig. 3, where the numeral I being of the largest and V of the smallest size. In the ultrasonic hot embossing, the conditions where the contact time could be shortened to t = 95 s, were T_h = 180 °C, and F = 1.75 kN; the conditions where the contact force could be reduced to F = 1.0 kN, were T_h = 180 °C, and t = 150 s; and the conditions for high reporoducibility were T_h = 180 °C, F = 1.5 kN, and t = 150 s. Figure 4(b) shows the SEM image of the embossed pattern under these conditions of high reproducibility. The resin completely fills to the edge part of the mold pattern, and it can be confirmed that the molding accuracy of the ultrasonic hot embossing reached to the same level as in the vacuum hot embossing. On the other hand, if the ultrasonic vibration were not applied when other molding parameters were same as Fig. 4(b), it would not be possible to mold at all, as shown in Fig. 4(c). Even signs of the molded pattern could not be observed from patterns III to V by the SEM observation. As a result, the assistance of the ultrasonic vibration influencing the molding accuracy became apparent. However, there were one set of conditions e.g., T_h = 200 °C, F = 5.0 kN, and t = 1,800 s (30 min) where molding could be carried out, although barely, without the use of ultrasonic vibration as shown in Fig. 4(d).

Moreover, it turned out that when the contact force was too large in the ultrasonic hot embossing, the overload disturbed the spread of the ultrasonic vibration, and the molding accuracy worsened. Figure 4(e) shows the SEM image of the embossed pattern when the contact force F was set to 4.0 kN, while other parameters being same as in Fig. 4(b). Based on these results, molding conditions giving excellent molding accuracies are plotted in Fig. 5. Here the contact force F (kN) and contact time t (s) are plotted along the horizontal and vertical axes. Molding conditions in the atmospheric hot embossing were plotted in the upper right of the figure; and molding conditions in the vacuum hot embossing are plotted in the lower left of the figure. This figure shows that molding conditions of both methods are quite separate from each other. However, both the contact force and contact time could be greatly reduced by adding ultrasonic vibration even in the room atmosphere. In fact, by employing ultrasonic vibration the contact force and contact time have been reduced to 1/3

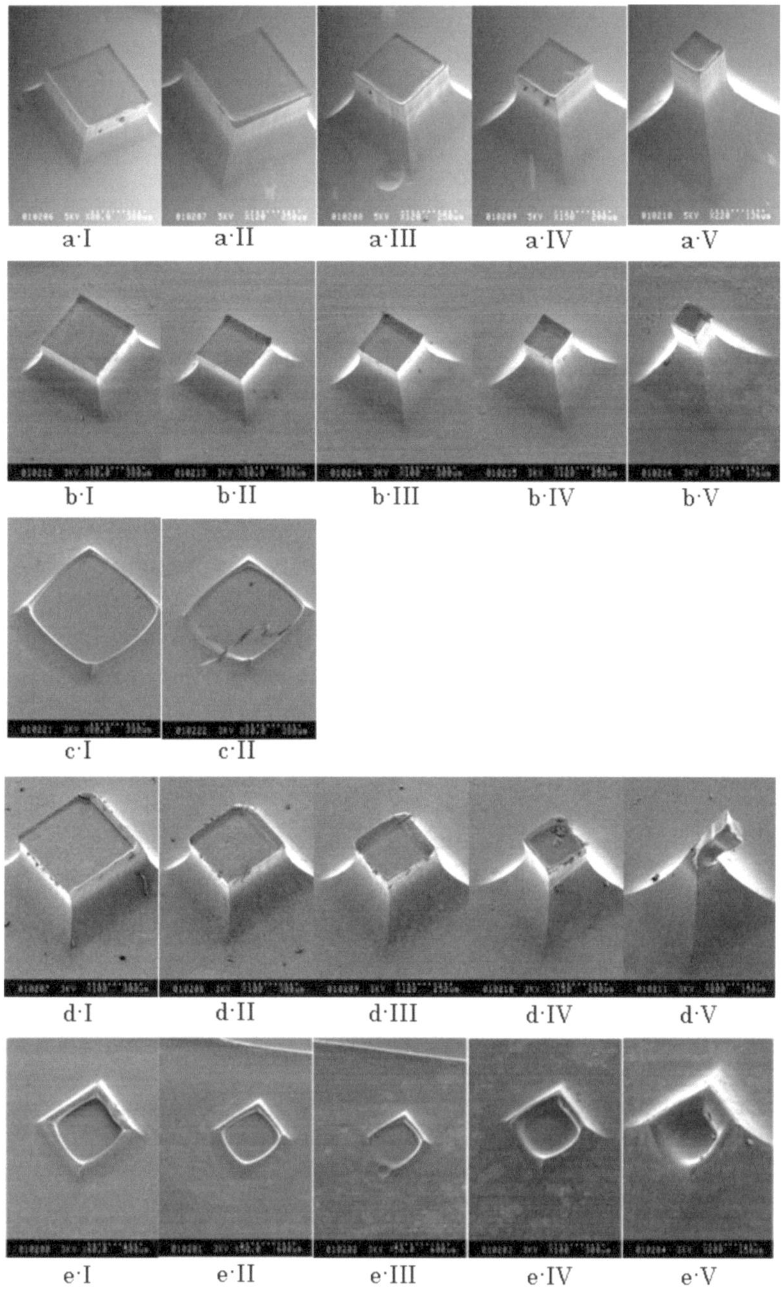

Fig. 4. SEM image of molded pattern (pattern size: I - V) under various conditions: (a) Vacuum hot embossing (T_h =180 °C, F =0.5 kN, t =60 s); (b) Ultrasonic hot embossing (T_h =180 °C, F =1.0 kN, t =150 s); (c) Molded patterns in the same condition as in "b" without ultrasonic vibration (T_h =180 °C, F =1.0 kN, t =150 s, without ultrasonic vibration); (d) Atmosphere hot embossing (T_h =200 °C, F =5.0 kN,t =1,800 s); (e) Ultrasonic hot embossing by overload (T_h =180 °C, F =4.0 kN, t =150 s)

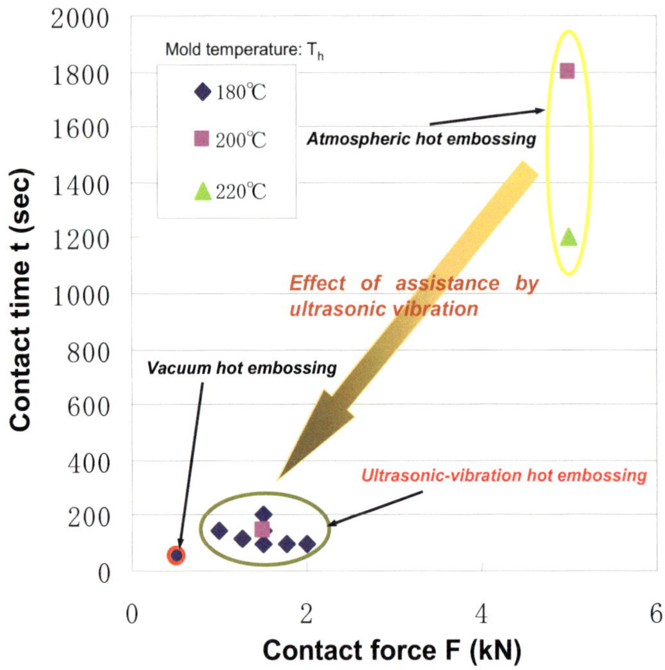

Fig. 5. Effect of assistance by ultrasonic vibration in hot embossing.

and 1/12 of their initial values. These results are surprising. It was possible to bring the contact force and time close to the optimum conditions of vacuum hot embossing, although it was molded in the room atmosphere. Thus, it succeeded in molding by combining ultrasonic vibration and hot embossing with a short contact time and a low contact force.

2.2 Incase of impressing ultrasonic vibration with low amplitude

Later, before the author's new assignment at the National Institute of Advanced Industrial Science and Technology (AIST), another ultrasonic nanoimprint system had also been successful in controlling the flaking from the mold by impressing ultrasonic vibration of the maximum amplitude of 1.8 μm during the de-molding process by a joint development program with Scivax Corporation. The author modified the software of the ultrasonic nanoimprint system where ultrasonic vibration could be applied even to the molding process. By impressing ultrasonic vibration at a maximum amplitude of 1.8 μm the author verified the molding accuracies of patterns ranging in size from 100 μm^2 to 1.2 mm^2.

Figure 6 shows a photograph of the ultrasonic nanoimprint system and its cross-sectional view around its loading stages. A metallic mold of a maximum size of 30 mm square was installed in the upper loading stage. A hand-drum-type horn with 1200 W output power and 19 kHz resonance frequency with 1.8 μm maximum amplitude was installed in the upper loading stage of the system where a longitudinal wave of ultrasonic vibration can be generated. Moreover, a ceramic heater and a circulation cooling system with a thermal medium oil were installed between the upper loading stage and the horn. The metallic mold can be heated up to 200 °C. A molding material sheet was held at the bottom loading stage by a vacuum chuck. The maximum patterning area on the sheet is designed to be 100 x 200 mm² by using a step-and-stamp method. The stage has X/Y translational and θ_z rotational capability, and has an alignment accuracy of ±1 μm with 1 μm/pixel image resolution charge-coupled-device (CCD) cameras. Using a cartridge heater, the molding material sheet can be heated up to 150 °C. A contact force up to 4.9 kN can be applied with a servo motor.

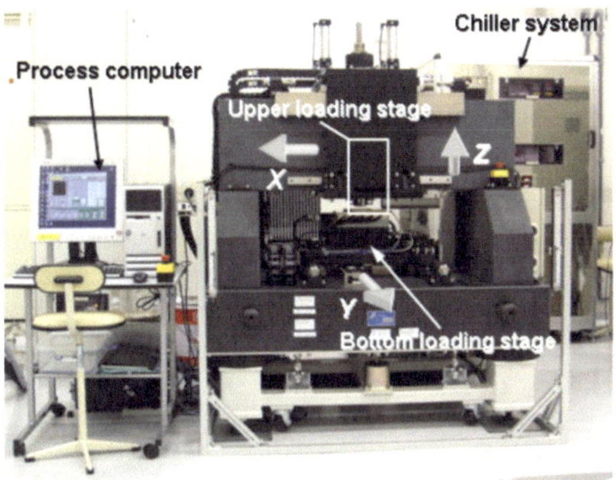

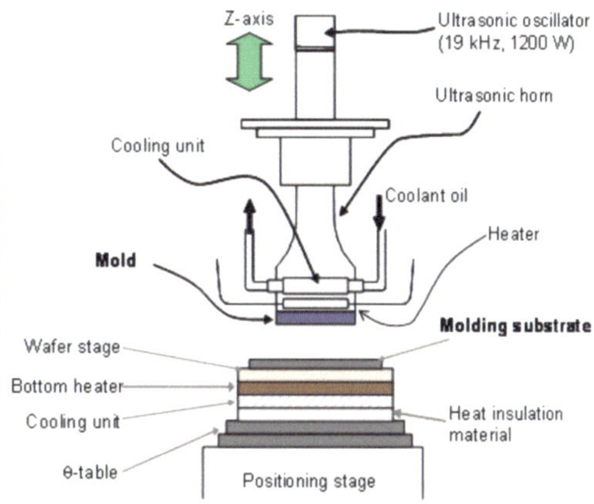

Fig. 6. (a) Photograph of ultrasonic nanoimprint lithography system, and (b) a cross-sectional view of the system around loading stage that includes a piezoelectric actuator.

Based upon the previous experimental result from hot embossing the auther chose the heating temperature as 180 °C and varied the contact force from 1.0 to 2.5 kN. Here also, the contact time and the cooling temperature for each setting were kept at 5 min and 130 °C. Figure 7 shows experimental results from each contact force using the same electroformed-Ni mold as shown in Fig. 3. A total of 40 embossed patterns was obtained in one single molding experiment because there were eight impressions for each of the five kinds. The shapes of all these embossed patterns were evaluated by an optical microscope. With this information, the molding rate was calculated for each pattern size. The molding rate was defined as the ratio of the successful pattern to all 8 patterns that were impressed on PC. For instance, when certain size pattern successfully embossed all 8 impressions then the molding rate would be 8/8 =1. When four impressions successfully embossed then the molding rate would be 4/8 = 0.5. When none embossed, then the molding rate would be 0/8 = 0. For all pattern sizes, the molding rate in the absence of ultrasonic vibration is shown by the left bar chart. When ultrasonic vibration was applied, the molding rate is shown by the adjoining right bar chart. A dramatic change was observed in the molding rate using ultrasonic vibration when the contact force was lowered. In the contact force of 1.0 kN, when ultrasonic vibration was not impressed, the molding rate in pattern size I–V was 0. On the other hand, the molding rate rose up to 0.2 or more when ultrasonic vibration was impressed. As for the molding rate of pattern V, it was found to be low among all contact forces. There are two possible explanations for this. One is that the shape of pattern V was a quadrangular pyramid where the center became a little thinner as shown in Fig. 4. This shape could have been easily damaged during the de-molding process.

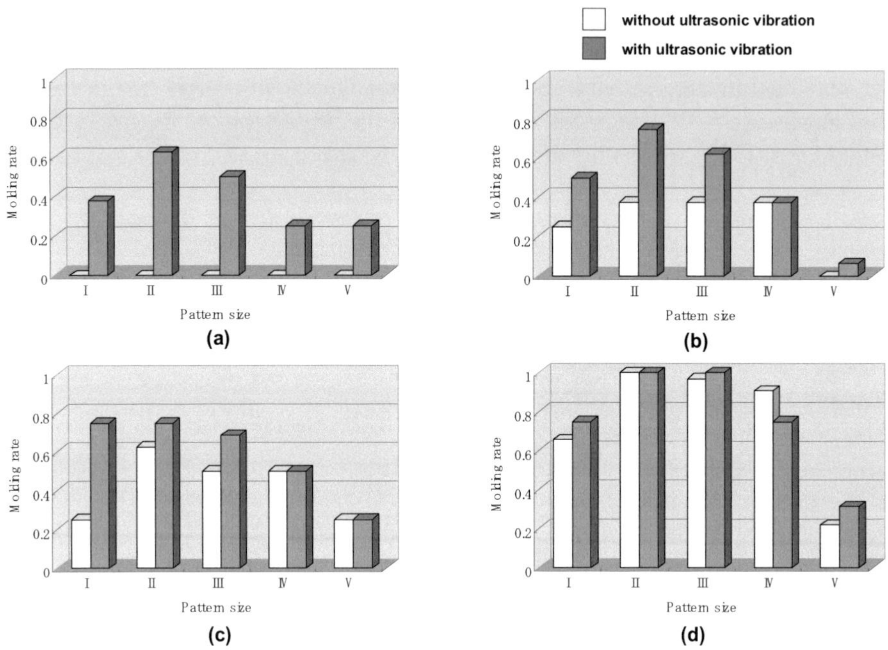

Fig. 7. Relationship between contact force and molding rate. The contact force was: (a) 1.0 kN, (b) 1.5 kN, (c) 2.0 kN and 2.5 kN. The left side shows the usual hot embossing results and the right side shows ultrasonic hot embossing results in the adjoining bar chart. The heating temperature of electroformed-Ni mold, the cooling temperature, the contact time and molding material were 180 °C, 130 °C, 5 min, and PC, respectively.

The other explanation is that the patterns III, IV, and V were located in the central part of the electroformed-Ni mold. For the concave mold, it is necessary to fill the concave pattern by moving the softened molding material from the surroundings of the mold pattern to its central part. Normally, in comparison to the edge of the mold, its center part is not readily molded. Figure 8(a) shows a photograph of an embossed pattern at a contact force of 1.0 kN

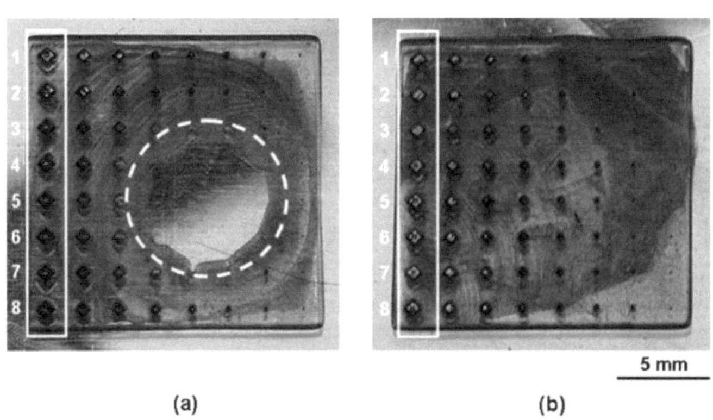

Fig. 8. Photograph of all embossed patterns when the contact force was 1.0 kN: (a) without ultrasonic vibration, and (b) with ultrasonic vibration. An inside of the white dotted circle is a part of flow shortage of PC. White and solid frames are shown as an array of pattern I.

when the ultrasonic vibration was not impressed. It shows that, PC did not reach the center of the mold pattern marked by a white dotted circle. However, by impressing the ultrasonic vibration the PC could be filled to the entire mold pattern as Fig. 8(b).

An examination of individual embossed patterns with an optical microscope showed defective molding caused by residual gas where ultrasonic vibration was not impressed. When ultrasonic vibration was impressed, this bubble defect was diminished or completely disappeared. This information led to a great improvement in the molding rate. Optical microscope photographs of the individual embossed patterns from the pattern I are shown in Fig. 9. The impressions 1 through 8 inside the columns defined by the solid white lines in Figs. 8(a) and 8(b) are shown in two rows in Fig. 9. In the top row, the absence of ultrasonic vibration resulted in zero molding rate. Whereas in the bottom row the presence of ultrasonic vibration successfully imprinted patterns No. 1 and 2 with molding rate of 0.25.

	1	2	3	4	5	6	7	8	Molding rate
Without ultrasonic vibration	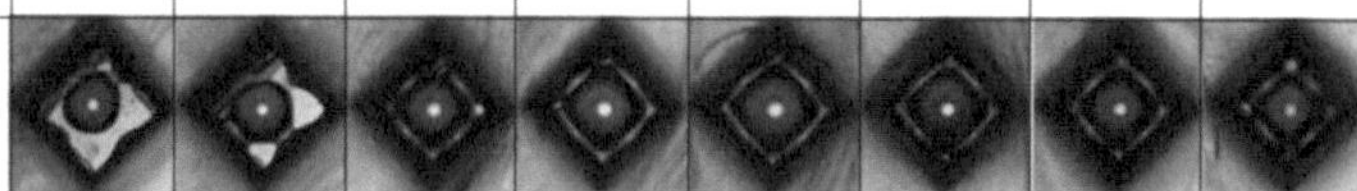								0 / 8 = 0
	↓	↓	↓	↓	↓	↓	↓	↓	
With ultrasonic vibration	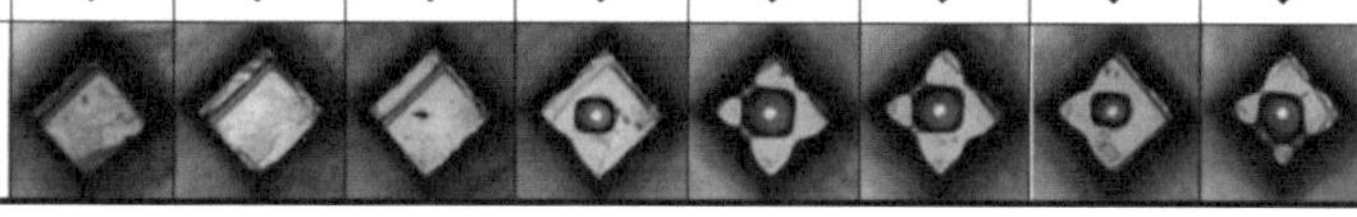								2 / 8 = 0.25

Fig. 9. Optical micrographs of individual embossed patterns of the pattern I in case of the contact force 1.0 kN.

3. Thermal nanoimprint assisted by ultrasonic vibration

3.1 Step-and-stamp type ultrasonic nanoimprint system

In micropatterning experiments of PC by uisng the ultrasonic nanoimprint system, the optimized mold heating temperature was 180 °C. As a part of our experiment the above technique was then extended to thermal nanoimprint where the experiment was executed using the same conditions (heating temperature: 180 °C, contact force: 100 N, and contact time: 10 s). The amplitude of ultrasonic vibration could be changed in ten steps (L1 - H5).

In nanoimprint experiments a concave Si mold was used where features of its patterns were defined by depth = 3 μm, length = 1.8 μm, and linewidths = 500, 750 nm, and 1 μm. The Si mold was fabricated by micro-electro-mechanical-system (MEMS) processing technologies including an e-beam lithography and a reactive-ion-etching (RIE). The Si mold size was 10 mm square. SEM images of thermal-imprinted patterns with linewidth of 500, 750 nm, and 1 μm at 180 °C are shown in the second row of Fig. 10. The heights of imprinted patterns were measured with a 3D optical profiler NewView 5000 (Zygo Corp.). The PC pattern with a maximum aspect ratio of 5.56 (=2.78 μm/500 nm) could be imprinted. However, the thickness of the PC sheet that was originally 0.5 mm had thinned down to 0.27 mm. So as not to consueme the PC sheet to much, the mold heating temperature was reduced to 150 °C; this happens to be the almost same as that of the T_g of PC. The third row in Fig. 10 shows SEM images of imprinted patterns with linewidths of 500, 750 nm, and 1 μm at the heating temperature of 160 °C. A large difference appeared in the molding accuracy when the mold heating temperature was 180 °C. The maximum height of the imprinted pattern was 1.18 μm in the case where the linewidth was 1 μm, and the highest aspect ratio was 1.06 (= 0.53 μm/500 nm). In comparison to the case when the mold heating temperature was 180 °C, the highest aspect ratio had decreased to one-fifth of its original value.

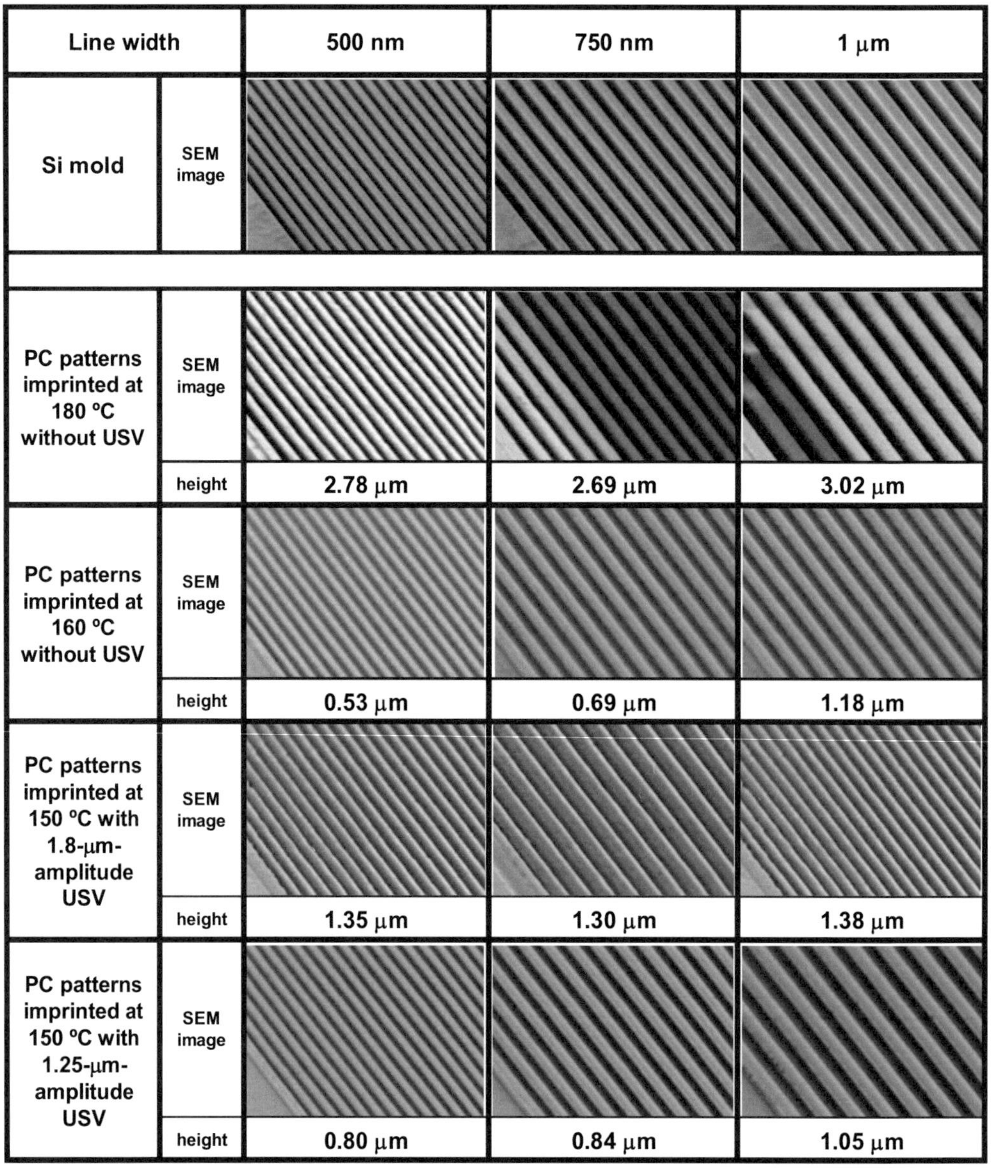

Fig. 10. SEM images of nanopatterns in Si mold and PC imprinted by thermal nanoimprint without/with ultrasonic vibration. "USV" means ultrasonic vibration in the figure.

3.2 Timing of impressing ultrasonic vibration

In the next step, the parameters for ultrasonic nanoimprint were fixed as the following: mold heating temperature = 150 °C, contact force = 100 N, and contact time = 10 s. Figure 11 shows the relationship between the process time and the contact force. The timings I-V in Fig. 11 indicates the starting points of the ultrasonic vibration impression for all ultrasonic vibrations. The first impression marked by I denotes that the timing of ultrasonic vibration began when the contact force was 0 N. In other words, it was the case where the ultrasonic vibration impression began before the Si mold even touched the PC sheet. The markers II, III, IV, and V are the timing points when the ultrasonic vibration impression began after the contact force reached the points 25, 50, 75, and 100 N, respectively. All ultrasonic vibrations were stopped according to the timing VI after being in contact for 10 s. The speed of the press and the release was 1.0 μm/s. Fifty seconds were required from the time Si mold made its initial contact with the PC sheet (contact force = 0 N) to the time when the contact force reached its maximum value of 100 N.

This phenomenon can be clearly confirmed in Fig. 12 (a) where the measurement results of the heights of the imprinted pattern are shown. Especially, the application of ultrasonic vibration according to the early timing seemed to improve the molding accuracy. Heights of the imprinted pattern of all linewidths showed a rise when the impression of ultrasonic vibration began with timings I (contact force = 0 N) and II (contact force = 25 N), as compared with the case where ultrasonic vibration was not applied. Even when the impression of ultrasonic vibration began with other timings, any substantial change was not observed in the pattern with linewidths of 750 nm and 1 μm, though the height of the imprinted pattern with 500 nm linewidth did show a rise. Moreover, when the impression of ultrasonic vibration began on the timing I, the ultrasonic vibration had been continuously impressing for 60 s. Although the contact time was extended to 60 s without applying ultrasonic vibration as shown by the dotted lines in Fig. 11, a noticeable increase in the height of the imprinted pattern was not observed.

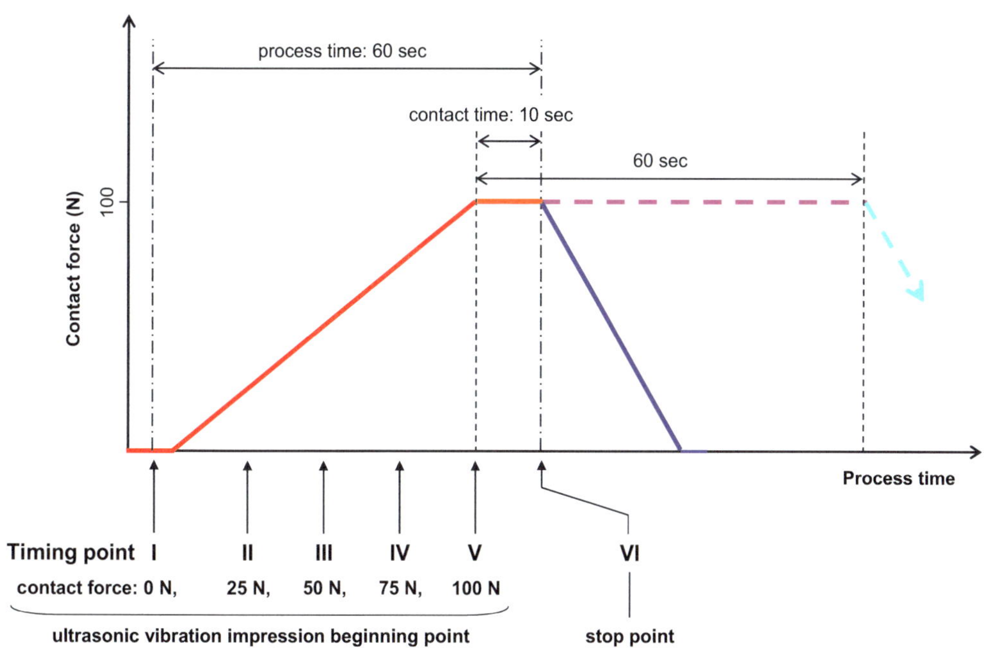

Fig. 11. Relationship between the timing points of applying ultrasonic vibration and contact force.

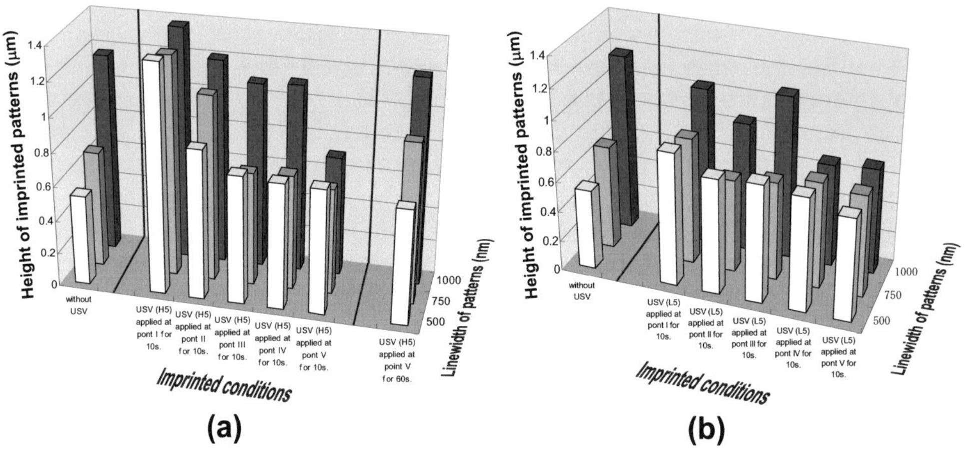

Fig. 12. Height of PC imprinted patterns in ultrasonic nanoimprint: (a) with 1.8-μm amplitude, and (b) with 1.25-μm amplitude. "USV" means ultrasonic vibration in the figure.

When ultrasonic vibration with 1.25 μm maximum amplitude (imput vlotage level = L5) was initiated at a timing I, the molding accuracy of the imprinted patterns was almost same as in the third row, but was less in comparison to that in the fourth row. This phenomenon can be confirmed by the measurement results of the height of the imprinted pattern shown in Fig. 12(b). No dramatic increase in the height of the imprinted pattern by the difference of timing in which the impression of ultrasonic vibration began could be observed in comparison to the case of the maximum amplitude of 1.8 μm (input voltage level = H5). In the case of the linewidth of 500 nm, the height of patterns with impressing ultrasonic vibration increased in comparison to the case where ultrasonic vibration was not impressed. In the case of the linewidth of 750 nm, it was confirmed that the height of the pattern increased where the impressing of ultrasonic vibration started at the timing I (contact force = 0 N). Such a phenomenon shows that the flowing of the thermoplastic cannot be assisted enough by ultrasonic vibration with low amplitude.

3.3 Decrease in mold heating temperature by assistance of ultrasonic vibration

Typically, the optimal heating temperature in thermal nanoimprint lithography using PC is 180 °C, but when ultrasonic vibration is applied, it becomes possible to mold nanopatterns at a lower temperature. As a result, the process time of a thermal cycle could be cut into half; thus improving the throughput greatly. In this experiment, a convex Si mold was fabricated by MEMS processing technologies as shown in Fig. 13. The pattern includes 36 cross-marks, and each cross-mark exists inside of a 500-μm^2 square frame. The shape of the cross-mark is in the form of a cross where two rectangles (400 x 10 μm^2) cross each other. Four crossmarks, enclosed in a frame of white dotted lines in Fig. 12(a), were designed as alignment markers for step-and-stamp nanoimprinting.

In the alignment process, an image of the crossmarks is captured by a 1 μm/pixel CCD camera and is then used by the computer to automatically adjust the position of the stage. The cross-mark was composed of a nanopattern whose size was below the resolution of the CCD camera. Therefore, the entire cross-mark appeared fuzzy, and darker than its surrounding. Figure 13(b) shows the arrangement of figures inside the frame of the dotted white dotted lines in Fig. 13(a). These cross-marks arranged in the frame were composed of four types of nanopatterns. The nanopatterns on the upper left, upper right, lower left, and lower right were labeled as A, B, C, and D, respectively. Pattern A was a line and space pattern with 750 nm linewidth and 1.5 μm pitch, pattern B was a line and space pattern with a 500 nm linewidth and 1 μm pitch, pattern C was a dot pattern of 750 nm^2, and pattern D was a dot pattern of 500 nm^2.

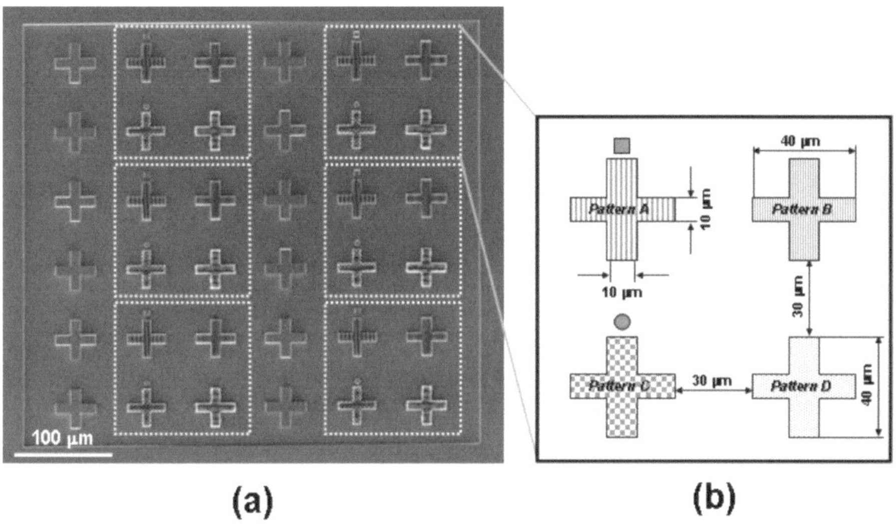

Fig. 13. (a) SEM image of Si mold pattern, and (b) schematic of cross-marks in white square from the left figure.

In the previous work, the molding conditions that could shorten the process time of one cycle in thermal nanoimprint were investigated. As a result, it was possible to imprint in a shorter period of time when the heating temperature, contact force, contact time, cooling temperature and molding and de-molding speeds were 180 °C, 100 N, 10 s, 130 °C and 1.0 μm/s, respectively. After the contact force reached 100 N, the contact force was kept during 10 s and then the cooling process was started. Figure 14(a) shows an optical microscopy photograph of the PC pattern imprinted under these conditions.

The shape of the imprinted pattern was primarily evaluated by optical microscopy considering the fact that in actual alignment it is the optical image that the CCD camera looks at. The cross-marks composed of the four types of nanopatterns can be clearly seen in Fig. 14(a). On the other hand, for the cross-mark with no nanopattern, only the edge part appeared to be somewhat blackish. Next, the cooling temperature, contact force, and contact time were fixed at 130 °C, 100 N, and 10 s, and thermal nanoimprint was carried out with the heating temperatures of the Si mold set at three different low values, i.e., 170, 165, and 160 °C. Optical microscopy photographs of each PC pattern when the heating temperatures were 170, 165, and 160 °C are shown in Figs. 14(b)–14(d), respectively. Similar images were observed for the pattern molded with the heating temperatures of 180 and 170 °C. The contrast between the PC pattern and the surrounding became less pronounced at a heating temperature of 165 °C. When the heating temperature was 160 °C, the cross-mark composed of the nanopattern could not be resolved.

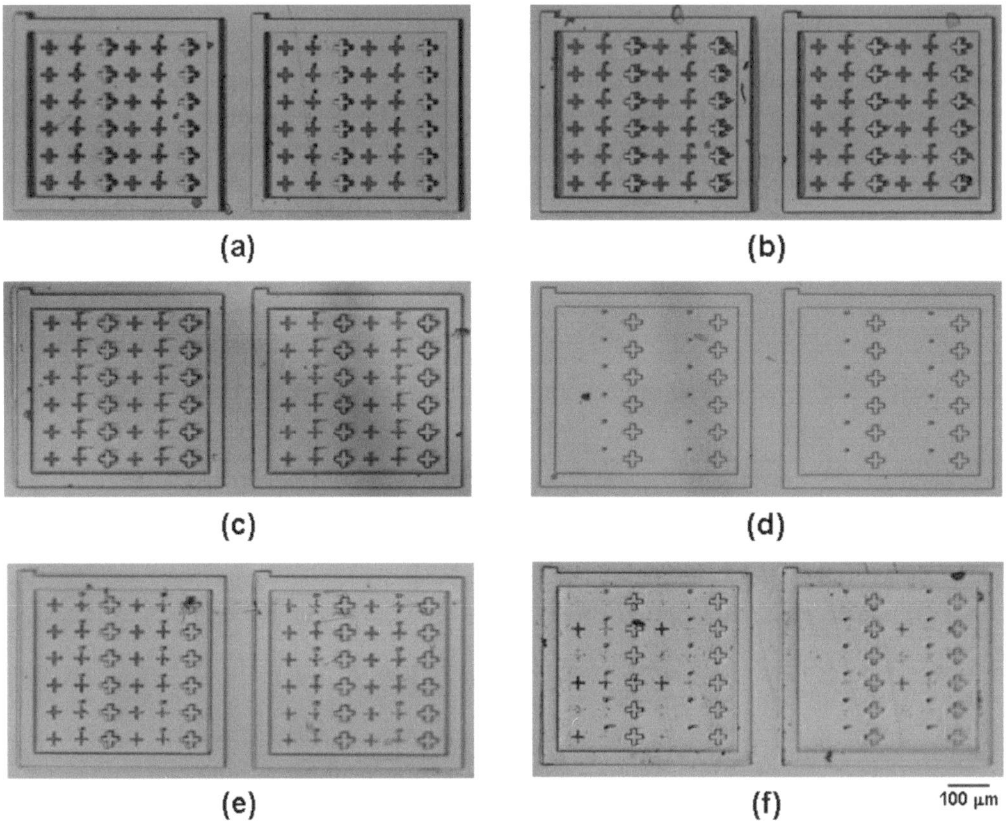

Fig. 14. Optical micrographs of PC imprinted patterns using thermal nanoimprint without ultrasonic vibration. The heating temperatures were: (a) 180, (b) 170, (c) 165, and (d) 160 °C. Optical micrographs of PC imprinted patterns using thermal nanoimprint with ultrasonic vibration. Ultrasonic vibration impression was started when the contact force reached to (e) 20 N and (f) 100N. The contact force and contact time were 100 N and 10 s, respectively.

In the next experiment, the contact time and contact force were 10 s and 100 N, and ultrasonic vibration with a maximum amplitude of 1.8 μm was simultaneously impressed during the contact time. Figures 14(e) and 14(f) show an optical microscopy photograph of the PC pattern from a separate experiment in which ultrasonic vibration impression began when the contact force reached 20, and 100 N, respectively. All the cross-mark images can be confirmed in Fig. 14(e). The contrast of the pattern and the circumference part becomes overwhelmingly clear when compared with the experimental result of thermal nanoimprint at 160 °C. Some cross-marks can be observed in Fig. 14(f). The cross-mark composed of the dotted pattern of 750 nm squares could be observed with high contrast. It turned out that nanoimprint became feasible by impressing ultrasonic vibration below the molding boundary condition of thermal nanoimprint. In addition, it was confirmed that the timing of the initiation of the ultrasonic vibration impression played an important part here, and that the use of ultrasonic vibration was an effective way to nanoimprint. And moreover, the molding accuracy improved even further when a contact force was applied.

4. Ultrasonic nanoimprint

4.1 Exchange of a piezoelectric actuator to magnetostriction actuator in ultrasonic nanoimprint system

The step-and-stamp type ultrasonic nanoimprint system was originally designed and built so that ultrasonic vibration could be impressed in the de-molding process. However, the available power for ultrasonic vibration was insufficient hence the author substituted magnetostriction actuator for piezoelectric actuator in the ultrasonic generator that provided relatively high amplitude. This led to the development of ultrasonic nanoimprint technology capable of imprinting from mold at room temperature that required neither heating nor UV radiation.

An assembly drawing and a photograph around the loading stages of the ultrasonic nanoimprint system are shown in Figs 15(a) and 15(b). Loading stages are composed of an upper loading stage that fixes a mold and a bottom loading stage that fixes a molding material. A parallel adjustment mechanism using spherical sliding was connected to the upper loading stage. A magnetostriction actuator AU-010N300-MS1 (ETREMA Product Inc.) was built into the upper loading stage, and the longitudinal wave of ultrasonic vibration with the frequencies of DC - 20 kHz and the amplitudes of 0 - 5 μm can be generated.

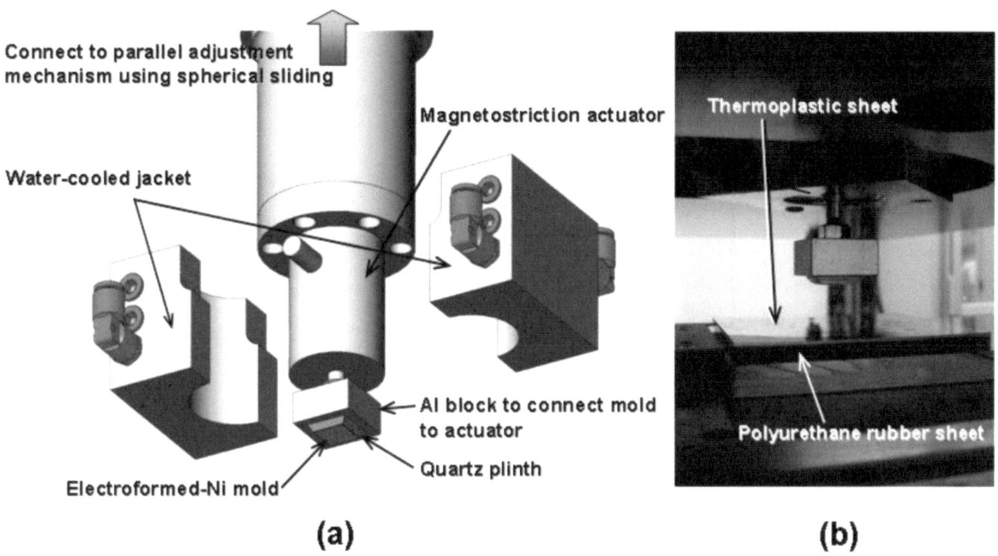

Fig. 15. (a) Assembly drawing of the upper loading stage, and (b) photograph of the upper and bottom loading stages in an ultrasonic nanoimprint system.

The magnetostriction actuator generates heat during its operation. To protect the magnetostriction actuator from heating and to prevent the heat from propagating into the other parts of the mold, the magnetostriction actuator was covered by a water-cooled jacket to cool it down to 20 °C. An Al block with a surface area of 30 mm^2 was fixed on the head of the magnetostriction actuator. The relationship between the frequency and amplitude of ultrasonic vibration had been investigated before the imprint experiments. The amplitude of ultrasonic vibration increased up to 10 μm for the resonance frequency of 8 kHz. At frequencies above 10 kHz the magnetostriction actuator hardly could oscillate. Then by adjusting the input current within a frequency range of DC - 10 kHz it was improved where it could run continuously with an amplitude of 3 μm. When the input current rose up to 4 A, the oscillation of the ultrasonic vibration stopped; it was caused by an overload at the frequency 10 kHz. A mold was bonded on the Al block with a negative-tone UV resist PAK-01 (Toyo Gosei Co., Ltd.) through a quartz plinth. The ditch for a vacuum chuck was processed on the surface of the bottom loading stage, and the molding material of a thickness 0.5 mm was fixed through an elastic material of 3 mm thickness. Clear-through-holes were processed at the same locations as in the ditch of the vacuum chuck in the elastic material.

Figure 16 shows a 15 x 15 mm^2 electroformed-Ni mold (Scivax Corp.) used for this experiment. As for the mold pattern, a basic unit of 2 mm^2 shown in the right half of Fig. 16 was arranged in a 5 x 5 matrix. Convex line/space (L/S) patterns and convex/concave circle dot patterns were put into this basic unit. There are eight sizes of patterns between 0.5 and 10 μm, with pattern height/depth being 1 μm. To ease the release of the mold, a self-assembled monolayer (SAM) of a release agent was formed on the surface of the electroformed-Ni mold. For this purpose, Optool HD-2101TH (Daikin Industries Ltd.) was spread on the electroformed-Ni mold by a nanoimcoater NIC-0703 type N (Eintesla, Inc.), followed by a 30 min of drying that established a SAM on the surface of the mold material.

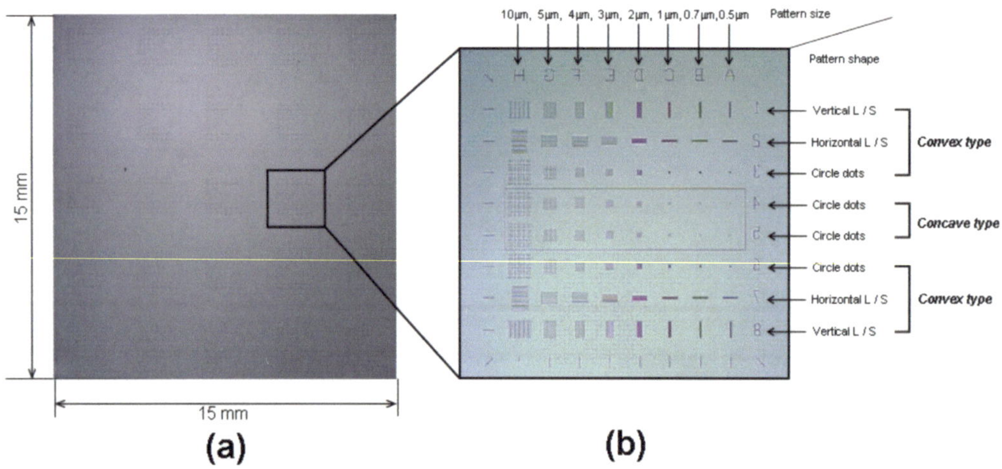

Fig. 16. Photographs of electroformed-Ni mold and mold patterns.

4.2 Optimization of molding conditions for PET in ultrasonic nanoimprint

The molding material selected was polyethylene terephthalate (PET, T_g = 75 °C) because among the engineering plastics its T_g is comparatively low. The PET sheet was 100 mm square in area and 0.5 mm in thickness. As a preliminary experiment in nanoimprinting, the imprinted pattern depth was investigated under conditions where pressure was applied, but no ultrasonic vibration was involved. The depth of the pattern was measured using a five-line confocal microscope Optekics S130 (Lasertec Corp.). The result of this experiment is shown in Figs. 17(a) - 17(c). It was possible to imprint on PET using only the contact force if the imprinted pattern depth were less than 0.5 μm. When the contact force was set at 50, 100, 250, 500, 750, or 1000 N, the expectation that the imprinted pattern depth would increase slightly with increasing force turned out to be the case only where the pattern width was 1 μm or less. Moreover, in the experiment the contact time was also varied and was set to 10, 20, 30, 60, or 90 s. The imprinted pattern depth became deeper accordingly with the contact force when the contact force was 50 and 100 N. However, when the contact force became 250 N or more, any further variation in contact time showed no effect on the pattern depth. The contact force might cause a shift from elastic deformation to plastic deformation between 100 and 250 N on the surface of PET, although it is considerably lower than the compressive strength of bulk PET (110 - 123 MPa). The next experiment addresses the depth of the imprinted pattern when ultrasonic vibration was also applied, while using the same imprint conditions as in the previous case. The results are shown in Figs. 17(d)-17(f). It is now well understood that applying ultrasonic vibration overall deepens the imprinted pattern further. In general, it has also been confirmed that if the contact time became long, the imprinted pattern depth would become deeper. More frictional heat develops when the impressed time of ultrasonic vibration is prolonged. The molding accuracy seemed to improve in ultrasonic nanoimprint because the transformation by contact force was combined with the softening of the plastic by frictional heat. However, the best conditions for the contact force depend on the width of the pattern. The optimized contact force, when the width of the pattern was 500 and 700 nm, was 250 N; and when the width of the pattern was 1 μm, the optimized force was 500 N. Specifically, it succeeded in complete transcriptions of the 1-μm-wide imprinted pattern to a depth of 1 μm, where the optimized conditions were a contact force of 500 N and a contact time of 60 s. We believe that, because ultrasonic vibration was held down while the contact force was growing too much, the effect of the assistance of ultrasonic vibration was impeded.

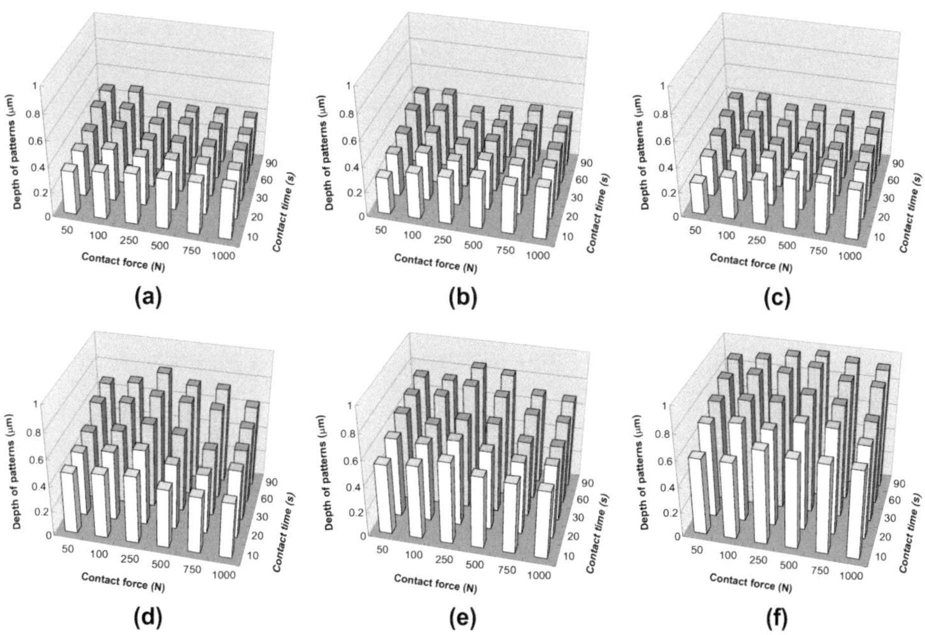

Fig. 17. Relationship between contact force or contact time and imprinted pattern depth without ultrasonic vibration for linewidths of (a) 500 nm, (b) 700 nm, and (c) 1 μm. Relationship between contact force or contact time and imprinted pattern depth in ultrasonic nanoimprint for linewidths of (e) 500 nm, (f) 700 nm, and (g) 1 μm.

In the next step, the author investigated the frequency/amplitude of ultrasonic vibration dependency of the molding accuracy in the ultrasonic nanoimprint by the magnetostriction actuator with a changeable frequency. Because displacement of the magnetostriction material is larger than the piezoelectric materials, the magnetostriction actuator doesn't need a resonance horn like the piezoelectric actuator. Therefore, the magnetostriction actuator can change the frequency of ultrasonic vibration. Then, a bipolar DC power supply BP4610 (NF Corp.) was connected with the magnetostriction actuator. When the frequency was increased from 0 to 10 kHz in five steps the depths of the imprinted patterns also became deeper, regardless of their linewidths. Especially at the frequencies of 5 kHz and higher, the imprinted depths increased at a greater rate as shown in Fig. 18(a). Next, while keeping the frequency fixed at 10 kHz, the amplitude was increased from 0 to 4 μm in nine steps as shown in Fig. 18(b). The figure shows the relationship between the depth of the imprinted pattern and the amplitude of the ultrasonic vibration. The depths of the imprinted patterns became deeper with increasing amplitudes regardless of patterns' linewidths. A 1-μm-width pattern was imprinted fully when the amplitude was 3 μm.

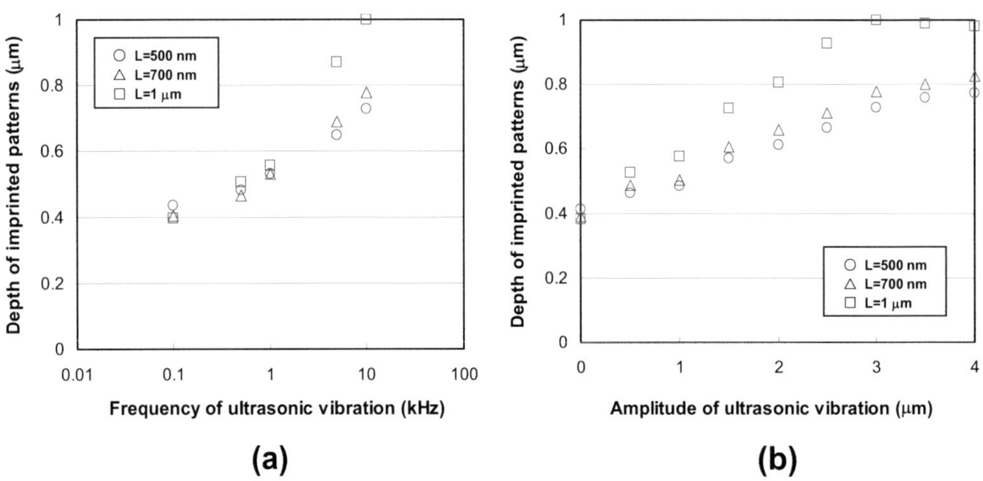

Fig. 18. (a) Relationship between frequency of ultrasonic vibration and depth of imprinted pattern with different pattern widths in ultrasonic nanoimprint, and (b) relationship between amplitude of ultrasonic vibration and depth of imprinted pattern with different pattern widths in ultrasonic nanoimprint.

4.3 Ultrasonic nanoimprint on engineering plastics

Before applying the optimized imprint conditions for PET to other engineering plastics, the author investigated the T_g and the hardness of various engineering plastics and examined them as an index for comparison. There are five kinds of engineering plastics; those, besides PET, are acrylonitrile butadiene styrene (ABS) resin, polymethyl methacrylate (PMMA), cyclo olefine polymer (COP), polycarbonate (PC), and polyimide (PI) as shown in Table 1.

Figure 19 shows optical photomicrographs of the center unit in the imprinted pattern. In all engineering plastics, a concave mold pattern which reverses to a convex imprinted pattern, could not be transferred. However, the transfer of a convex mold pattern, which reverses to a concave imprinted pattern, could be confirmed, although there was a difference in the

Materials	PET	ABS	PMMA	COP	PC	PI
Glass transition temperature (°C)	69	105	114	138	144	336
Indented depth (nm) when loading force was 10 mN	2342	2603	1656	4731	1891	1598

Table 1.Estimated T_g and measured indented depths of PET, ABS, PMMA, COP, PC, and PI.

contrast. In convex patterns the structure that bulges from the mold surface pushes the molding material to the outside of the pattern region. In concave patterns, it is necessary to draw the molding material from the outside of the pattern region toward the inside of the region to fill in the cavities with the molding material. When imprinting where the ambient is room atmosphere, it is difficult to imprint with the concave patterns because the air trapped inside the cavity is likely to cause distortion of the fillings of the molding material. Therefore, a convex pattern generally can be easily imprinted in comparison to that of concave patterns. The vertical stripes on the surface of ABS and PI, as shown in Figs. 19 (b) and 19(f), are caused during the operation when the sheets are polished down to a thickness of 0.5 mm by the plastic supplier.

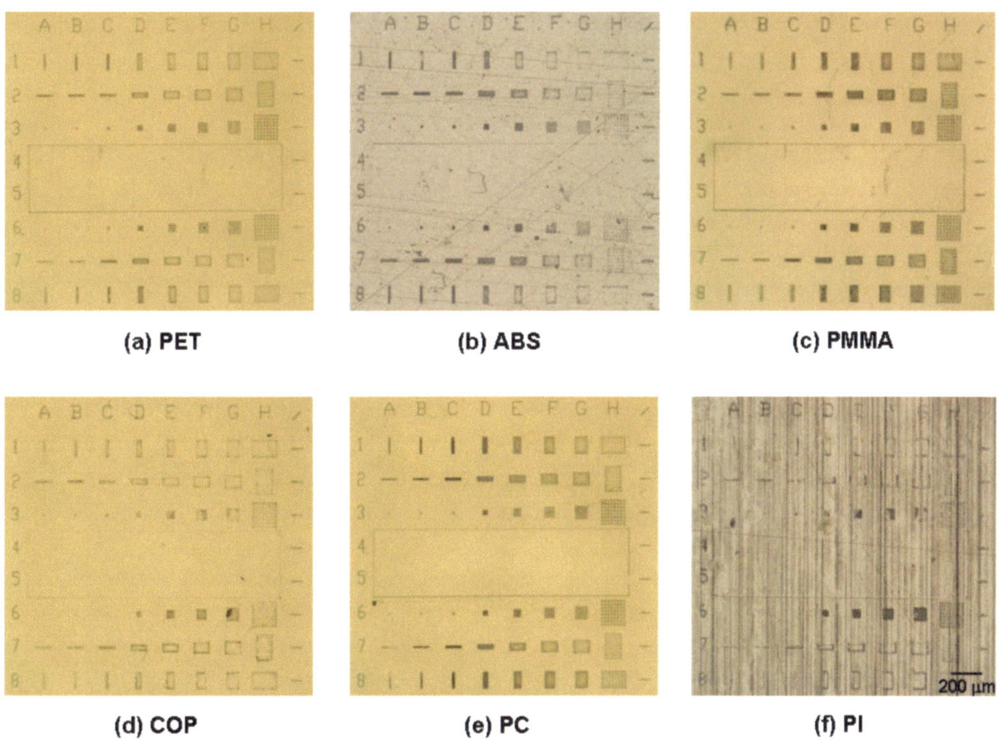

Fig. 19. Optical micrographs of imprinted patterns on (a) PET, (b) ABS, (c) PMMA, (d) COP, (e) PC, and (f) PI.

After observation by an optical microscope, a 20-nm-thick film of Au was sputter deposited on the surface of the imprinted pattern. The result of observing L/S patterns with the

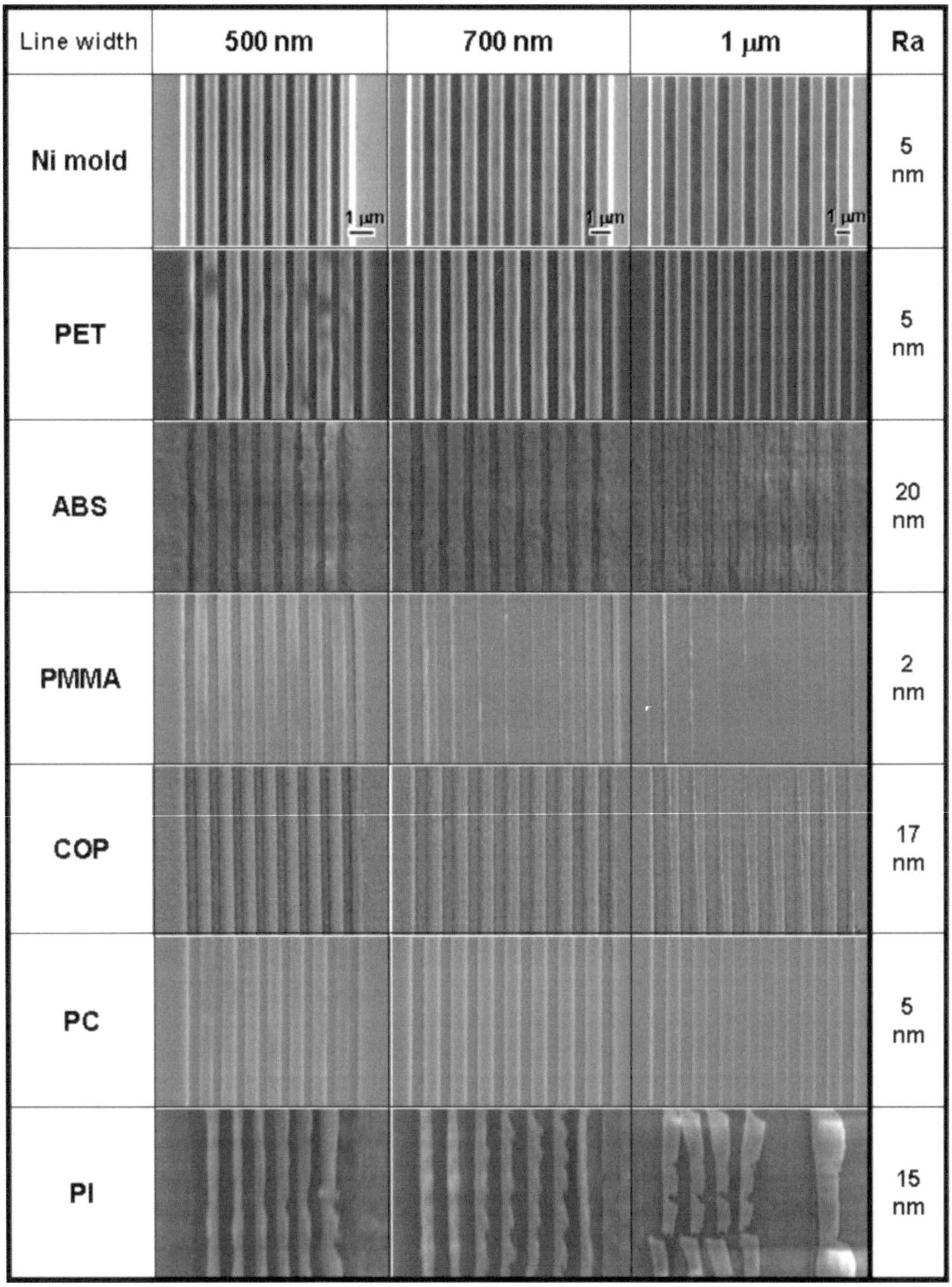

Fig. 20. SEM images and average surface roughness of L/S patterns imprinted on PET, ABS, PMMA, COP, PC, and PI.

linewidths of 500 nm, 700 nm, and 1 μm by SEM is shown in Fig. 20. The upper line in Fig. 20 shows mold patterns of the electroformed-Ni mold. In the SEM observation, we checked the shape of the edge and the surface roughness of the imprinted pattern. The average surface-roughness (R_a) of 1-μm-width imprinted patterns was measured by the 3-D optical profiler as shown in Fig. 20. In PET, the edge is sharp and the surface is also smooth. In the imprinted pattern with the linewidth of 500 nm, the part where the contrast was weak, was observed for the incomplete filling. In the imprinted pattern of all lincwidths on ABS, the edge became rounded, and the shape of the melted plastic solidified within a short time without relaxing. In PMMA, although the edge of the imprinted pattern was sharp, the surface was very smooth. In the imprinted pattern on COP, the edge became duller, and the surface was rough.

In PC, the sharpness in the edge was that of the middle of PMMA and COP, and the surface was as smooth as that of PET. The imprinted pattern on PI was quite different from that of the other engineering plastics. A phenomenon where a surface layer rolled up from the PI substrate was observed.

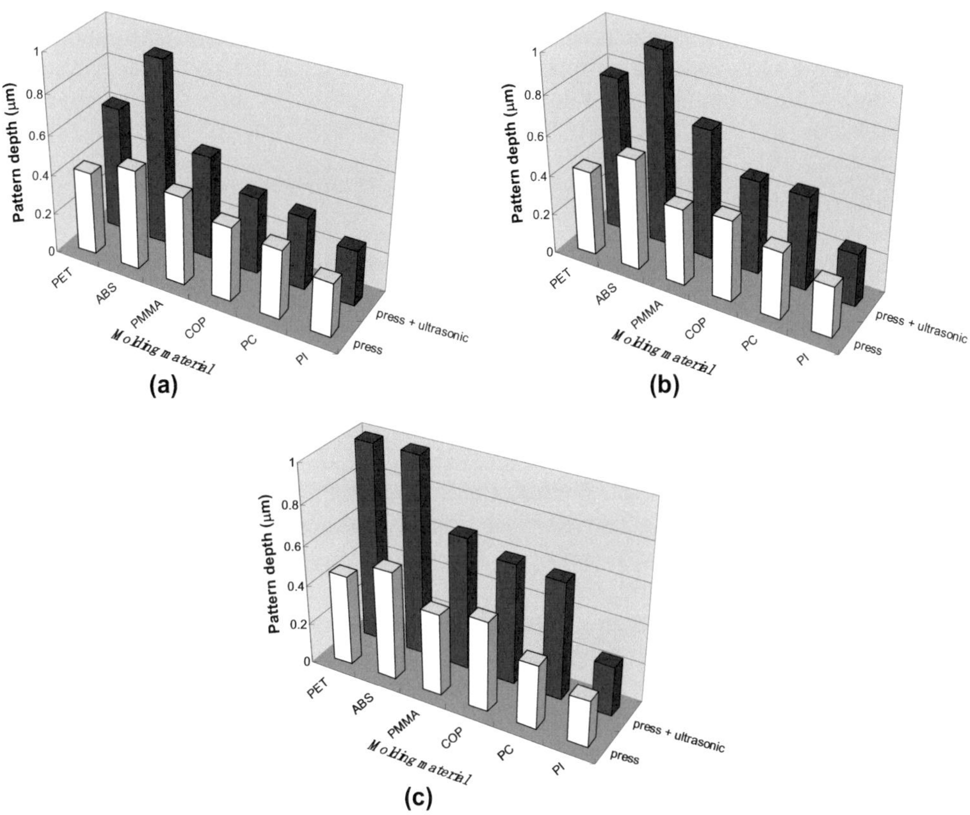

Fig. 21. Relationship between T_g of engineering plastics and depth of imprinted pattern in widths of (a) 500 nm, (b) 700 nm, and (c) 1 μm.

The depths of these imprinted patterns in the widths of 500 nm, 700 nm, and 1 μm were measured using the confocal microscope, and the data are shown in bar chart form in Fig. 21 arranged from the material with lowest T_g. Even where ultrasonic vibration was not applied, the mold pattern was transferred by the plastic deformation. In PET, ABS, and COP, the pattern was imprinted comparatively. The result concerning these materials corresponds to the fact that the maximum depth was largest in Table 1. On the other hand, lower T_g of an engineering plastic resulted in deeper patterns when ultrasonic vibration was applied. In all engineering plastics, the mold pattern was imprinted quite deep when the pattern width became large. This feature was clearly observed in PET and it succeeded to imprint 1 μm pattern depths in PET and ABS. The difference between the depths of the imprinted pattern from the cases where the ultrasonic vibrations were "applied" and "not applied" seemed to diminish with the rise in T_g. Especially, there was hardly any difference in depth of the imprinted pattern on PI with the T_g of more than 300 °C. Black histograms show the depths where the initial plastic deformation was accompanied by thermal deformation of ultrasonic nanoimprint. The difference between these white and black histograms becomes significant in comparison to the cases of materials with high T_g values. Naturally, it is easier to deform thermally an engineering plastic with low T_g values. ABS was processed to a maximum depth at an initial pressure, and it was also processed to a maximum depth in the case of ultrasonic nanoimprint. Therefore, it is concluded that impressing the ultrasonic vibration also deforms the pattern thermally after the pattern has already undergone its initial plastic deformation. In the ultrasonic nanoimprint, the molding material seemed to imprint quite deep by thermal deformation generated by the frictional heat after the transformed plasticity by the contact force in the initial process.

4.4 Ultrasonic nanoimprint on spin-on-glass (SOG)

The ultrasonic nanoimprint technology was also employed for patterning on the spin-on-glass (SOG) coated substrate. Accuglass 512B (Honeywell Co., Ltd.), an organic SOG and known as methyl siloxane was used for experiments on ultrasonic nanoimprint. The chemical formulae of Accuglass 512B is $CH_3Si_2O_{3.5}$, with 14 % content of carbon. A 400-μm-thick and 100-mm-diameter Si wafer was prepared by polishing both its sides by CMP (chemical-mechanical-***polishing)***. A 10-nm-thick Ti layer was sputter-deposited as an interlayer to improve the adhesion between the SOG layer and the Si substrate. In general, SOGs require a two-step thermal-treatment that involves baking at low, and then at a high temperature. We baked the SOG/Ti/Si substrate at 150 °C for 10 s using a hot plate as a 1st annealing after SOG spin-coating. The substrate was then allowed to naturally cool down to room temperature, and then was heated up to 450 °C by a rapid thermal processor AS-One 100 (Annealsys) for 1 h as its 2nd annealing. And then the ultrasonic nanoimprints was verified on the SOG coated substrates. Molding conditions and an electroformed-Ni mold were the same as in the case of the ultrasonic nanoimprint experiments on engineering plastics in the above paragraph. Figure 22 shows the result of transferring line/space patterns with the linewidths of 500 nm, 700 nm, and 1 μm on SOG coated Ti/Si substrates under these conditions. We succeeded in ultrasonic nanoimprinting on all substrates regardless of the linewidths of the mold pattern. Though the edges were relatively sharp, fine structures like detailed cracks could be observed on the bottom surface of the line/space patterns in the SOG/Ti/Si substrate after the 1st annealing. On the other hand, the edge was very shaggy, although the bottom surface of the imprinted pattern was relatively smooth in SOG/Ti/Si substrate that was annealed at 450 °C for 1 h. The surface of the SOG layer before hard baking was damaged, and the part where mold patterns came into contact was rubbed and crushed by ultrasonic vibration. When the molding material is thermoplastics, the temperature of the molding material locally seems to rise to close to the T_g of thermoplastics because of the frictional heat generated by ultrasonic vibration.

Therefore, the bottom surface of concave imprinted patterns, crushed by mold patterns, solidified to a relatively smooth surface after being softened by ultrasonic vibration. However, because our ultrasonic nanoimprint system did not have power strong enough where the annealed SOG could be softened again, the surface of the crushed SOG layer remained rough. On the other hand, a burr appeared in the imprinted pattern like the ones seen in metal structures when cut with a tool bit or milled with an endmill, because the surface of SOG after hard baking had become relatively hard. SOG in the part where mold patterns came in contact was flattened, and grinding swarms of SOG that were pushed away from the center were piled up as a burr on the sidewall of the imprinted pattern. Thus, the processing principle of ultrasonic nanoimprint on SOG seemed to be quite different in the cases of before, and after hard baking. Even if each SOG coated substrate were imprinted without applying ultrasonic vibration under the same contact force and contact time, imprinted patterns could not have been observed.

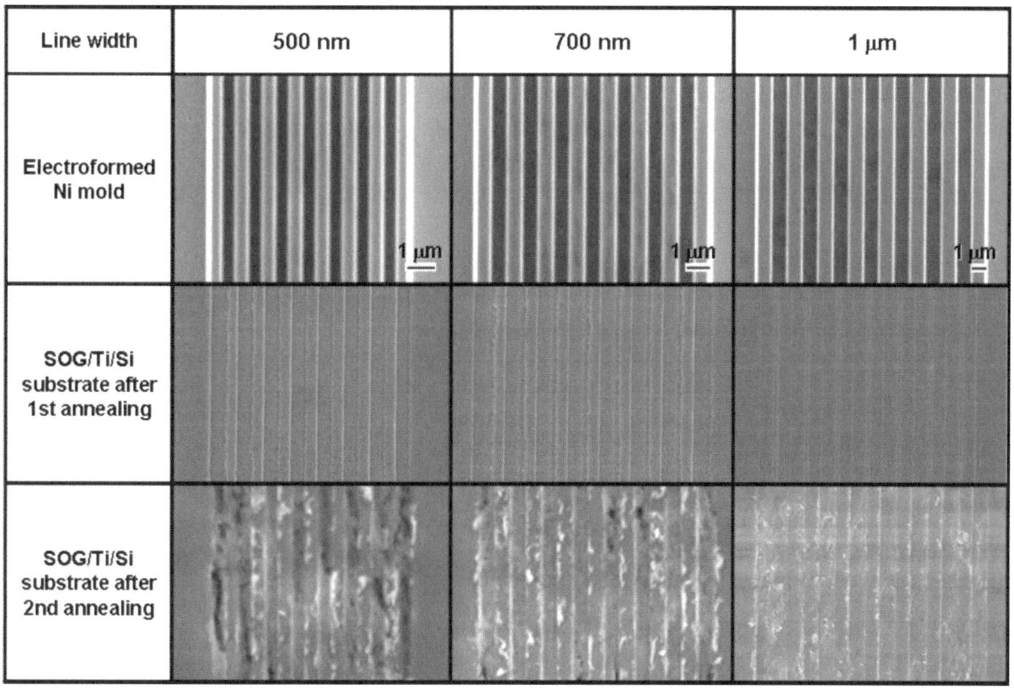

Fig. 22. SEM images of line/space patterns in electroformed-Ni mold and imprinted SOG substrate before and after the 2nd annealing.

5. Conclusion and future work

An ultrasonic hot embossing was proposed as a new and precise replication technology where the effect of assistance of ultrasonic vibration in hot embossing was experimentally verified. It was found that the contact time and contact force decreased greatly by applying ultrasonic vibration in the molding process. It means that molding in room atmosphere can be carried out under conditions similar to that of vacuum hot embossing.

In the marginal cases of hot embossing, the molding rate improved significantly by impressing ultrasonic vibration. Especially, when the contact force was low, the assisting

effect of ultrasonic vibration was quite separate. This phenomenon means that under strong contact force the progression of ultrasonic vibration can be hindered.
Moreover, in order to manage and manipulate the timing at which ultrasonic vibration can be impressed, a system to control the generation of ultrasonic vibration was developed. As for the start time of the impression of ultrasonic vibration set at an earlier stage in the molding process, the assisting effect of ultrasonic vibration was significant.
The effective assistance of ultrasonic vibration in hot embossing was applied to thermal nanoimprinting. Nanopatterns were imprinted on a PC sheet at a temperature close to the T_g using a thermal nanoimprint system with a built-in ultrasonic vibration unit. And the effect of ultrasonic vibration on thermal nanoimprint was experimentally confirmed. The lowest mold heating temperature in thermal nanoimprint on PC was 165 °C. However, the transfer of all the nanopatterns with the widths of 500 and 750 nm succeeded at a heating temperature of 160 °C using ultrasonic nanoimprint. Thus, it was experimentally shown that nanoimprinting at a low temperature can be done when accompanied by ultrasonic vibration, which can also improve the thruput greatly.
As for the ultrasonic nanoimprint system, its ultrasonic generator was remodeled by substituting a magnetostriction actuator for its piezoelctric actuator. And in one of the results, the imprinted pattern depth at room temperature and without the use of ultrasonic vibration, was found to be as deep as 0.5 mm. However, in the case of ultrasonic nanoimprinting with a contact force of 500 N and contact time of 60 s, the depth of the 1-μm-wide pattern reached 1 μm. Thus it showed that ultrasonic nanoimprint with the magnetostriction actuator can be used even with PET for a process time of only 1 min. The possibility of greatly improved throughput was experimentally shown by a push-and-pull ultrasonic nanoimprint technology at room temperature.
In the next step, the optimum conditions for ultrasonic nanoimprint on PET were investigated by using the magnetostriction actuator with changeable frequency and amplitude. After that, ultrasonic nanoimprint was applied to six kinds of engineering plastics, and their practicality was confirmed. The depth of imprinted pattern was found to decrease as the T_g of the plastic rose. The assisting effect of ultrasonic vibration was found to be inversely proportional to the values of T_g. Moreover, the author developed the ultrasonic nanoimprint technology that enabled room temperature-nanoimprinting of engineering plastics and SOG material.
These experimental results prove that ultrasonic vibration with high frequencies and large amplitudes generate a large amount of frictional heat that can induce drastic thermal-deformation. Therefore, it can be said that ultrasonic nanoimprint is a processing technology that combines plastic deformation by the initial contact force and thermal deformation by ultrasonic vibration. Furthermore, the imprinting limit of our ultrasonic nanoimprint system was also determined. Based on these experimental results, the author is planning to develop a new ultrasonic nanoimprint system capable of generating ultrasonic vibration with higher frequencies and larger amplitudes.

6. References

Becker E W; Ehrfeld W; Hagmann P; Maner A & Münchmeyer D. (1986). Fabrication of microstructures with high aspect ratios and great structural heights by synchrotron radiation lithography, galvanoforming, and plastic moulding (LIGA process), *Microelectron. Eng.* 4, 35-42.

Chou S Y; Krauss P R & Renstrom P J. (1995). Imprint of sub-25 nm vias and trenches in polymers, *Appl. Phys. Lett.* 67 3114-3116.

Colburn M; Johnson S C; Stewart M D; Damle S; Bailey T C; Choi B; Wedlake M; Michaelson T; Sreenivasan S V; Ekerdt J & Wilson C G. (1999). Step and flash imprint lithography: a new approach to high-resolution pattering, *Proc. SPIE* 3676 379-390.

Haisma J; Verheijen M; van den Heuvel K & van den Berg J. (1996). Mold-assisted nanolithography: A process for reliable pattern replication, *J. Vac. Sci. Technol.* B 14 4124-4128.

Honeywell International Inc. (2008). Interconnect Dielectrics, Application Note. Publication No. GC0110302Rev2. 3.

Kishi H; Yoshioka H; Jianguo Y; Sumiyoshi N; Goto H; Murakoshi Y & Maeda R. (2003). Thermal Imprinting Stepper with Ultrasonic Vibration Mechanism and Rapid Temperature Control system, *Proc. second International Conference on Nanoimprint Nanoprint Technology* B6.

Lin C & Chen R. (2006). Ultrasonic nanoimprint lithography: a new approach to nano patterning, *J. Microlith. Microfab. Microsyst.* 5 011003.

Liu S & Dung Y. (2005). Hot Embossing Precise Structure Onto Plastic Plate by Ultrasonic Vibration, *Polym. Eng. Sci.* 45 915-925.

Maeda R & Ashida K. (2004). Mechanical lithography - State of the Art and the Future, *J. Jpn. Soc. Precision Engineering* 70 1219-1222 [in Japanese].

Maeda R; Goto H; Awazu K; Mekaru H & Takahashi M. (2005). Other nanoimprint technologies, In: *Scinece & Technology, Essay on Nanoimprint*, Maeda R, (Ed.), 95-107, The Nikkan Kogyo Shimbun, Ltd., ISBN4-526-05574-3, Tokyo [in Japanese].

Mekaru H; Goto H & Takahashi M. (2006). Development of ultrasonic micro hot embossing technology, *Microelectron. Eng.* 84 1282-1287.

Mekaru H; Hattori T; Nakamura O & Maruyama O. (2004). Assisted processing of hot embossing by ultrasonic vibration, *Ultrasonic Technol.* 16 66-69 [in Japanese].

Mekaru H; Nakamura O; Maruyama O & Hattori T. (2004). Development of Precision Transfer Technology of Atmospheric Hot Embossing Replication by Ultrasonic Vibration, J. Soc. Plant Engineers Japan 16 8-14 [in Japanese].

Mekaru H; Nakamura O; Maruyama O; Maeda R & Hattori T. (2007). Development of precise transfer technology of atmospheric hot embossing by ultrasonic vibration, *Microsyst. Technol.* 13 385-391.

Mekaru H; Noguchi T; Goto H & Takahashi M. (2007). Nanoimprint Lithography Combined with Ultrasonic Vibration on Polycarbonate, *Jpn. J. Appl. Phys.* 46 6355-6362.

Mekaru H & Takahashi M. (2008). Ultrasonic Nanoimprint on Poly(ethylene terephthalate) at Room Temperature, *Jpn. J. Appl. Phys.* 47 5178-5184.

Mekaru H & Takahashi H. (2009). Ultrasonic nanoimprint in engineering plastics, *J. Vac. Sci. Technol.* A 27 785-792.

Mekaru H & Takahashi H. (2009). Frequency and amplitude dependencies of molding accuracy in ultrasonic nanoimprint technology, *J. Micromech. Microeng.* 19 125026.

Mekaru H & Takahashi H. (2009). Ultrasonic nanoimprint in Spin-On-Glass (SOG) Coated on Si Substrate, *Jpn. J. Appl. Phys. submitted.*
Motoji T. (2008). Technological trends of fluorine release agents for precise molds, *Tribology* 249 54-56 [in Japanese].
Schift H. (2008). Nanoimprint lithography: An old story in modern time?. A review, *J. Vac. Sci. Technol.* B 26, 458-480.

26

Molecular Dynamics Study on Mold and Pattern Breakages in Nanoimprint Lithography

Masaaki Yasuda, Kazuhiro Tada and Yoshihiko Hirai
Osaka Prefecture University
Japan

1. Introduction

Nanoimprint lithography (NIL) is one of the promising technologies for the fabrication of nanostructures at low cost (Chou et al., 1995) (Chou et al., 1996). In NIL, understanding the deformation behaviour of polymer during imprinting processes is an essential issue for high-speed and uniformed fabrication. Since numerical simulations can be efficient approaches for this issue, several studies using continuum mechanics are performed (Hirai et al., 2001) (Hirai et al., 2004) (Song et al., 2008). Continuum mechanics successfully predict the material deformation in submicron scale. However, as the pattern size becomes smaller than several tens of nanometers, continuum mechanics fails to analyze the material behaviour. Single-nanometre resolution has experimentally been demonstrated in NIL (Hua et al., 2004) (Hua et al., 2006). For the exact analysis of the material deformation in nanoscale system, the behaviour of atoms or molecules should be considered.

Molecular dynamics (MD) simulation is a useful tool to study the deformation mechanism of the materials in atomic scale. Several MD studies on NIL process are reported. Kang et al. propose a MD simulation model of a NIL process imprinting an α-quartz stamp into an amorphous poly-(methylmethacrylate) film (Kang et al., 2007). In their study, the distributions of density and stress in the polymer film are calculated for the detail analysis of deformation behaviour. The qualitative agreement between the MD simulation and the experimental data for the density variation of patterned polymer is reported (Woo et al., 2007). Mold geometry effect on springback phenomenon in NIL process is also studied with the MD simulation (Yang et al., 2009).

For metal direct imprinting, more MD studies are performed. Process parameters such as stamp taper angle, imprint depth, temperature and punch velocity are investigated for copper imprinting (Hsu et al., 2004) (Hsu et al., 2005). The mechanism of the atomic-scale friction is studied for aluminium imprinting (Hsieh & Sung, 2007). The metal film thickness effect on pattern formation is also studied (Cheng et al., 2007). Agreement between MD simulation and experimental results is reported for temperature effects on gold imprinting (Hsiung et al., 2009). MD simulation of nanoimprint for alloys is demonstrated (Fang et al., 2007). In order to save computational time, a multi-scale simulation for nanoimprint process that mixes the atomistic and continuum approaches is proposed (Wu & Lin, 2008). Recently, MD simulation of roller nanoimprint process is performed (Wu et al., 2009).

For an exact understanding of the material deformation mechanism during NIL processes, a comprehensive analysis involving three factors should be conducted: mold deformation, processed material deformation and the interaction between them. However, all the MD studies mentioned above treat the mold as a rigid body. Only the deformations of the processed materials are discussed in the studies.

Here, we review our studies on mold and pattern breakages using MD simulation. Firstly, we study the fractures of the independent silicon (Si) mold. The dependences of the Si mold breakages on the crystalline orientation and the defect structures are investigated. Secondly, the fracture mechanism of the pattern is discussed by the MD analysis of the pressure acting on the mold during glass NIL process. Finally, we introduce our resent approaches to investigate the mold deformations in the processed materials during NIL process.

2. Simulation of mold breakages

2.1 Simulation model

As the first step to understand the deformation of the mold during NIL processes, we study the breakages of the independent mold (Tada et al., 2008). Figure 1 shows a configuration of the calculation. The monocrystalline Si mold, which has line and space structures, is pressed onto the rigid substrate without processed materials such as polymers and glasses. The dashed element in Fig. 1 is defined as a unit cell and periodic boundary conditions are applied in the x- and z-directions. The width of the line and space of the mold, the height of the mold pattern and the thickness of the mold base are 3, 9 and 4 nm, respectively. We investigate the crystalline Si molds having two crystalline orientations. Here, we describe the mold which has {110} top surface and {100} front surface as {110}/{100} mold.

The MD simulation is performed using the Tersoff potential (Tersoff, 1988a) (Tersoff, 1988b) to study the mold breakages. The top surface of the mold is moved at a constant velocity of 5 m/s. The initial temperature of the mold is set to be 300 K. Newton's equation of motion is solved using the Verlet algorithm with a time step of 1 fs.

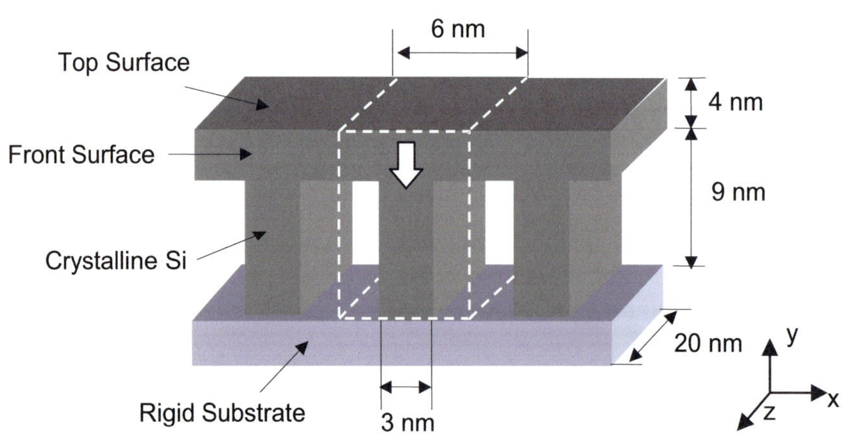

Fig. 1. Configuration of MD simulation for mold breakage analysis. The monocrystalline Si mold is pressed onto the rigid substrate.

2.2 Crystalline orientation dependence on mold breakage

The stress-strain characteristics for {110}/{110} and {110}/{100} molds calculated by the MD simulation are shown in Fig. 2. The compressive strain is calculated as the rate of decrease in the height of the mold. The compressive stress is calculated as the sum of all the atomic forces in the mold top divided by the area of the top surface. The stress increases with an increase in the strain. Yield stresses are 5.5 and 6.7 GPa for {110}/{110} and {110}/{100} molds, respectively.

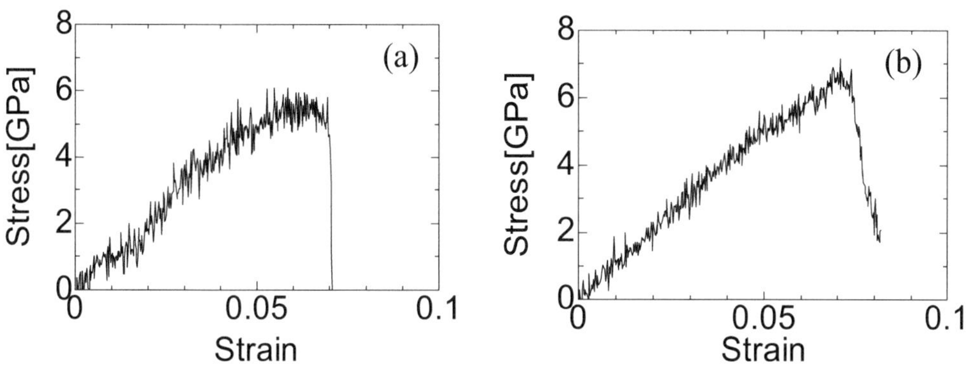

Fig. 2. The stress-strain characteristics for (a) {110}/{110} and (b) {110}/{100} molds.

Figure 3 shows the cross-sectional views at the breaking point for {110}/{110} and {110}/{100} molds. Since the surface energy of {111} crystalline plane is the smallest among those of the other planes, {111} planes could easily slip by shear stress. In {110}/{110} mold, the {111} planes run parallel to the z-direction. This mold could easily fracture along a {111} plane because the width of the mold pattern along the slipping direction is small. In {110}/{100} mold, the {111} planes run parallel to the x-direction. Since the length of the mold pattern along the slipping direction becomes large, the slipping along a {111} plane hardly occurs. Therefore, no slip along a specific crystalline plane is observed in this mold. As a result, {110}/{100} mold exhibits larger strength than {110}/{110} mold. The strength of the mold is strongly associated with the configurations of {111} planes in the mold.

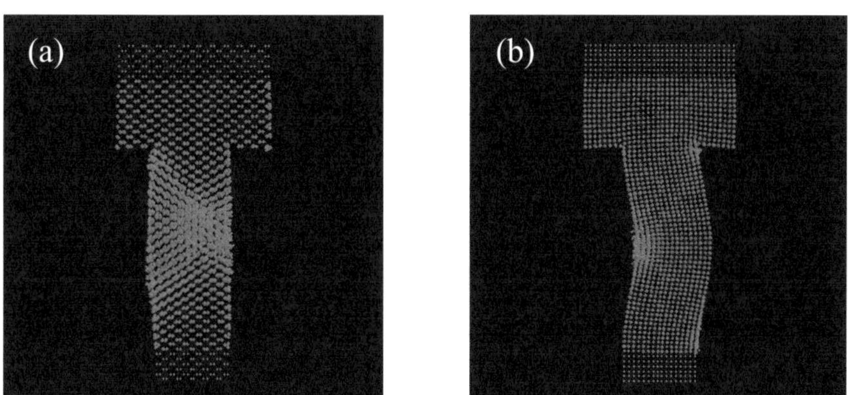

Fig. 3. Cross-sectional views at breaking point for (a) {110}/{110} and (b) {110}/{100} molds.

For comparison, the continuous mechanics simulation using the finite element method is performed. We use the commercially available software MARC distributed by MSC Software. Figure 4 shows the cross-sectional views of the stress distribution calculated by MARC. The aspect ratio of the mold pattern is same as that of MD simulation. Applied pressure is 500 MPa. The compression, shear and von Mises stresses are shown. The maximal stresses are observed at the bottom edge of the mold patterns.

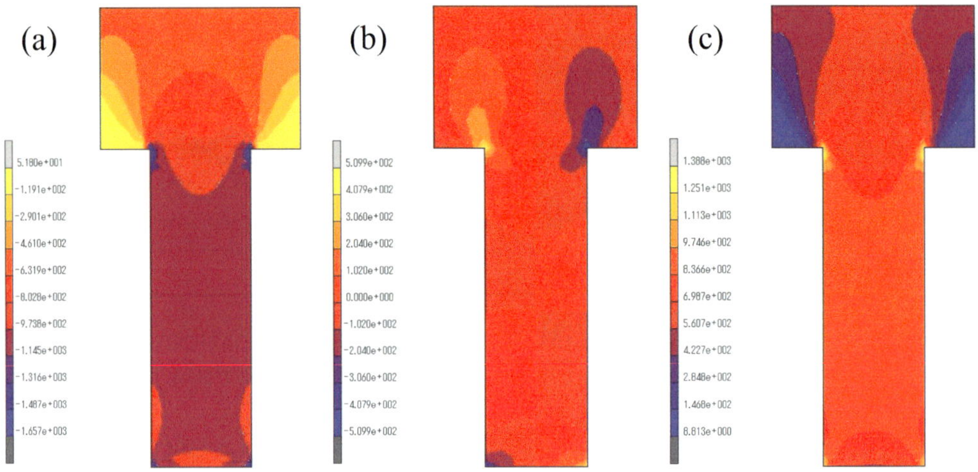

Fig. 4. Stress distributions calculated by the continuum mechanics simulation. (a) compression stress σ_{yy}, (b) shear stress σ_{xy} and (c) von Mises stress.

Figure 5 shows the scanning electron micrograph of the broken pieces of the mold left on the processed material after imprinting. The fracture cross sections of the broken pieces show same crystalline plane orientation. The breakage owing to the slipping of the specific atomic plane is explained by our MD simulation. It is not expected only from the stress distributions obtained by the continuum mechanics simulation.

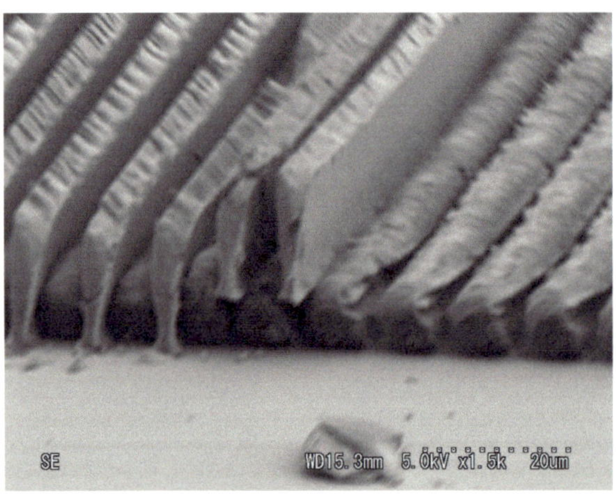

Fig. 5. Scanning electron micrograph of the broken pieces of the mold.

2.3 Breakages of defective mold

The defects caused by the repeated imprinting deteriorate the strength of the mold. Figure 6 shows the cross-sectional views of the stress distributions in {110}/{110} molds with a notch type defect structure during pressing. The mold model has a notch at the center of the mold wall. The depth of the notch is 0.7 nm. The stress is concentrated around the notch defect during pressing as shown in Fig. 6 (a). Finally the mold fractures along {111} plane containing the notch as shown in Fig. 6 (b). The notch acts as a trigger of the crucial mold fracture. Figure 7 shows the stress distribution for {110}/{100} mold. The stress also concentrated around the notch as shown in Fig. 7 (a). The slip along a specific crystalline plane is not observed. However, the notch defect promotes the mold fracture by buckling as shown in Fig. 7 (b).

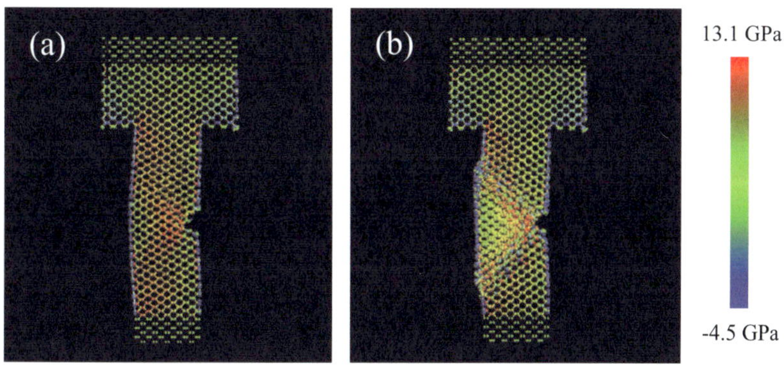

Fig. 6. The stress distributions in {110}/{110} mold with a notch type defect structure (a) before and (b) after mold fracture.

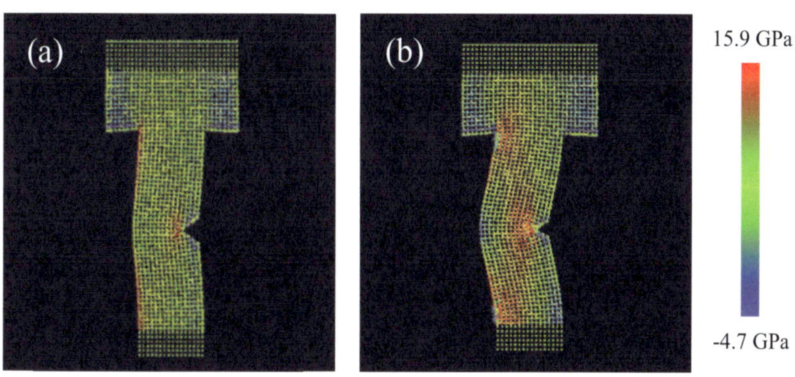

Fig. 7. The stress distributions in {110}/{100} mold with a notch type defect structure (a) before and (b) after mold fracture.

The stress-strain characteristics for {110}/{110} and {110}/{100} molds with a notch type defect structure are shown in Fig. 8. Yield stresses are 4.3 and 4.2 GPa for {110}/{110} and {110}/{100} molds, respectively. These values are smaller than those of the defect-free molds shown in Section 2.2. The strength deterioration of the mold due to the notch defect is more significant for {110}/{100} mold than {110}/{110} mold (Tada et al., 2009a).

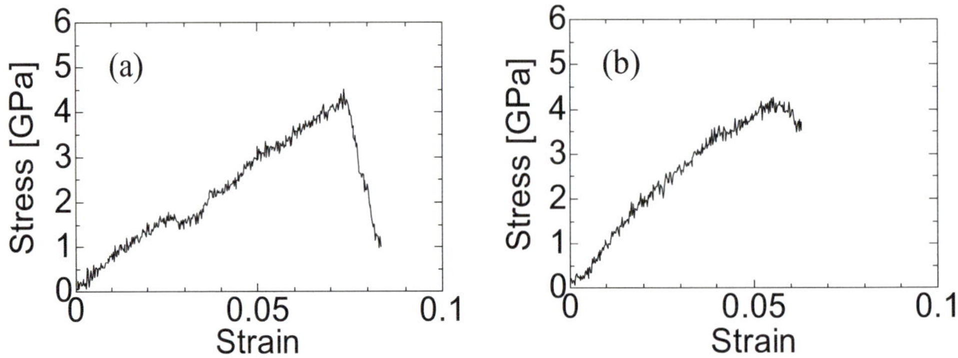

Fig. 8. Stress-strain characteristics for (a) {110}/{110} and (b) {110}/{100} molds with a notch type defect structure.

We also investigate the strength deterioration induced by the vacancy defects. 1.4-nm-diameter spherical vacancies are randomly introduced in the crystalline Si mold. Figure 9 shows the stress distributions in the {110}/{110} mold containing 7 vacancy defects during pressing. Circles in Fig. 9 (a) indicate the positions of the vacancy defects. Before the mold fracture, the stresses around the vacancies are relatively small, because the vacancies act as buffers against compression stress. The fractures along multiple crystalline planes are observed as shown in Fig. 9 (b). Several points where vacancies existed are included in the fracture planes.

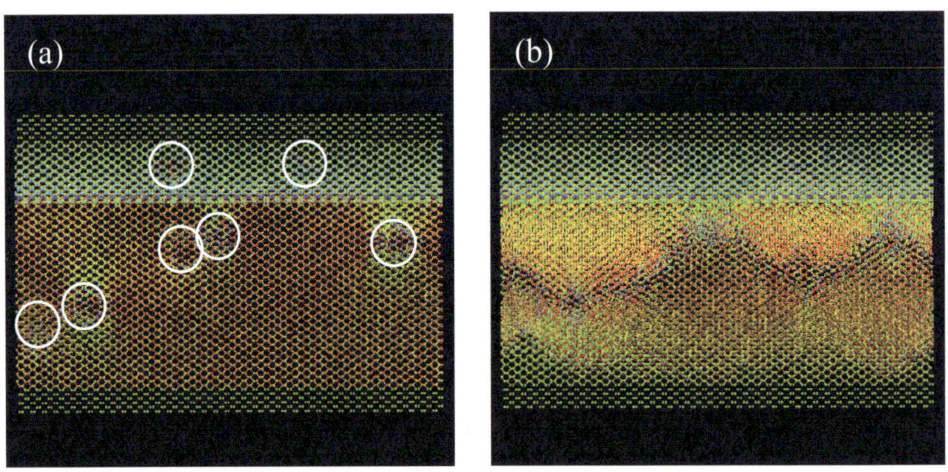

Fig. 9. Stress distributions in y-z plane of the {110}/{110} mold containing 7 vacancy defects (a) before and (b) after mold fracture.

The stress-strain characteristic for {110}/{110} mold containing 7 vacancy defects during pressing is shown in Fig. 10. The stress decreases more gradually after fracture than that of defect-free mold. This gradual decrease of the stress is attributed to the fractures along multiple crystalline planes. Yield stress is 5.3 GPa. The strength deterioration due to the vacancy defects is smaller than that due to the notch defects. The vacancy defects do not act as a trigger of the crucial mold fracture. These results indicate that the surface defects on the mold sidewall such as notches become more serious causes of the mold breakage than the defects originally contained in the mold material, such as vacancies.

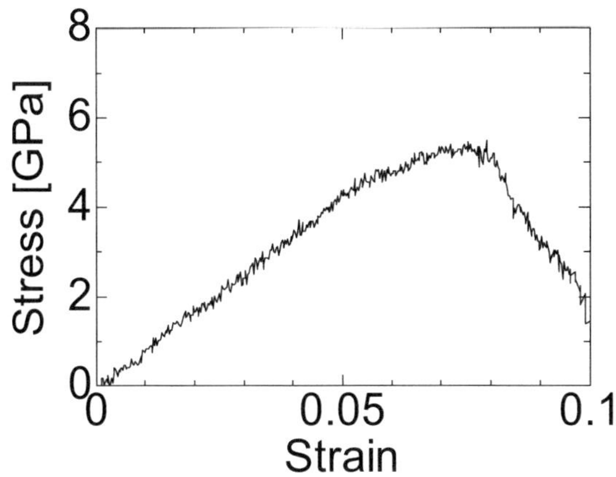

Fig. 10. Stress-strain characteristics for {110}/{110} mold containing 7 vacancy defects.

3. Simulation of pattern breakages

3.1 Simulation model

In this section, we discuss the pattern breakages in NIL. While polymer materials are often used as resist materials in NIL, inorganic glasses are also promising materials for optical devises (Hirai et al., 2003) (Okinaka et al., 2006) (Akita et al., 2007). However, it is more difficult to fabricate nanostructures on glass materials than on polymer materials because of the fragility of glass. Here, we investigate the glass deformation in NIL with a MD simulation.

Figure 11 shows a schematic diagram of the calculation system. The monocrystalline Si mold, which has line and space structures, is pressed onto the SiO_2 glass film. Si mold is treated as a rigid body. We define the dashed element in Fig. 11 as a unit cell. Periodic boundary conditions are applied in the x- and z-directions. 5 nm width and 2 nm thickness are considered as a unit cell in the x- and z-directions, respectively. The height of the mold pattern and the thickness of the glass film are 2.5 and 5 nm, respectively. 1-nm-thick bottom region in SiO_2 glass film is assumed to be rigid as a substrate.

Melt-quench method is used to form the initial structures of the SiO_2 glass by the MD simulation (Delaye et al., 1997). The SiO_2 crystal is melted and rapidly quenched from 8500 to 2500 K at a rate of 5×10^{15} K/s and from 2500 to 300 K at 5×10^{14} K/s to become SiO_2 glass.

Born-Mayer-Huggins potential is adopted to describe the interactions between atoms in the SiO_2 glass (Delaye et al., 1997). Morse potential is used to simulate the interaction between a Si mold and SiO_2 glass (Takada et al., 2004). The mold is pressed and released from glass at a constant velocity of 50 m/s. The temperature of the glass is maintained at 1500 K during the mold pressing using the velocity-scaling method. After filling the cavity of the mold with glass, the system is cooled to 300 K. Finally, the mold is released from the glass.

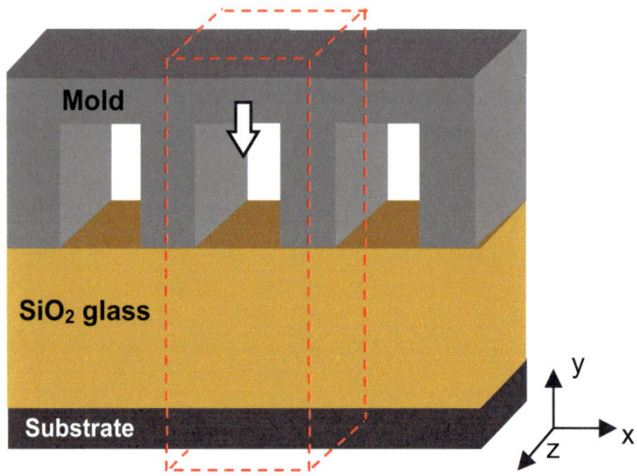

Fig. 11. Schematic diagram of the calculation system. The rigid Si mold, which has line and space structures, is pressed onto the SiO_2 glass film.

3.2 Glass nanoimprint process

Figure 12 shows the variation of the pressure acting on the mold versus mold position obtained by the simulation (Tada et al., 2009b). The cross-sectional views of each step are also shown. The pressure acting on the mold is calculated as the sum of all the atomic forces on the mold divided by the area of the top surface. The position where the mold contacts the glass is 0 nm. Initially, the pressure increases linearly as the mold is pressed to the glass [region (a) to (b)]. In this region, elastic deformation is dominant. In the region where the pressure increases nonlinearly, the plastic flow of the glass to the cavity of the mold is observed [region (b)]. Because of the highly viscous flow above the glass transition temperature, the glass does not fracture in this plastic flow. After filling the cavity with glass, the system is relaxed by resting the mold [region (c)]. During the cooling process the pressure decreases due to the stress relaxation. Then, the mold is released from the glass. Elastic recovery is observed in the first stage of releasing [region (c) to (d)]. The pressure acting on the mold shows a negative value since the tension force acts in the glass by adhesion between the glass and the mold [region (d) to (f)]. Finally, the mold is exfoliated from the mold. The pressure acting on the mold largely disappears [region (e)]. A fluctuation in pressure is observed until the mold is completely released from the glass [region (e) to (f)]. This fluctuation in pressure shows the stick slip phenomena between the side wall of the mold and the glass. The height of the glass after the mold separation [region (f)] is larger than that at the mold holding process [region (c)]. The springback phenomenon is observed. After the mold releasing, the segmentalization is observed at the surface region of the glass.

From our simulation, it is found that the pressure to fill the cavity depends on the glass thickness. If the cavity width and depth are constant, the filling pressure increases as the glass thickness decreases. Because the glass is fixed to the substrate, the mobility of the glass near the substrate is low. Therefore, the thinner glass is hardly deformed. Filling pressure also depends on the cavity width. If the cavity depth and the glass thickness are constant, the pressure increases as the cavity width decreases. These results are consistent with that obtained by the continuous mechanics (Hirai et al., 2004).

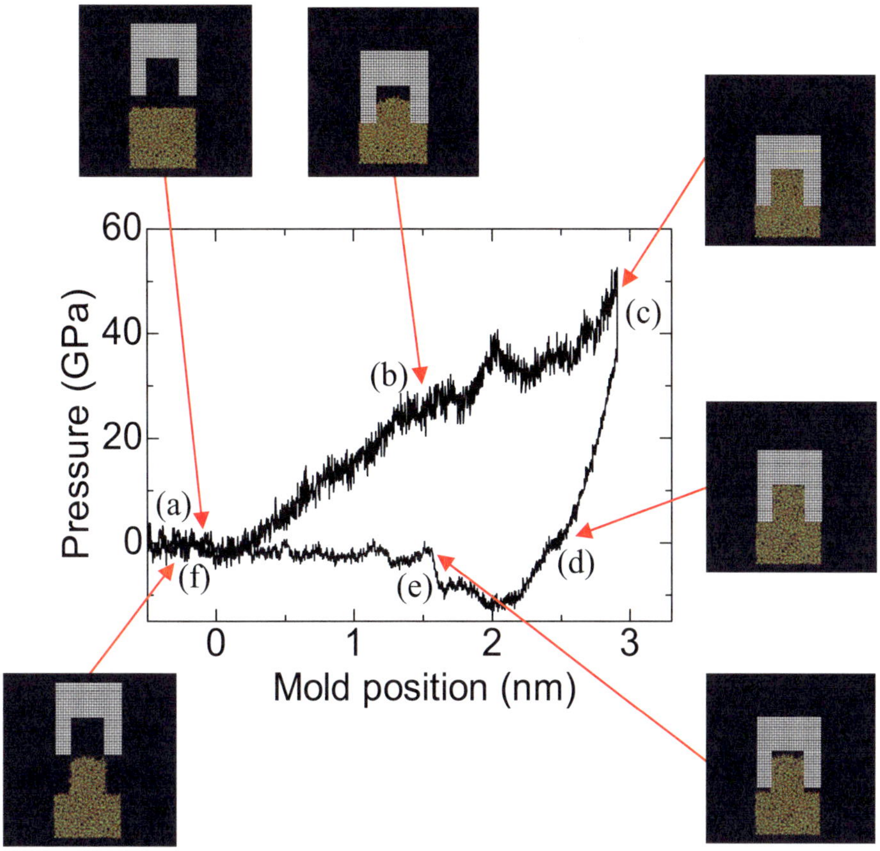

Fig. 12. The variation of the pressure acting on the mold versus mold position.

Kang et al. divided the polymer into three regions to analyze the adhesion and friction forces between the mold and the polymer in detail (Kang et al., 2007). In order to perform the same analysis, we divide the glass into three regions as shown in Fig. 13 (a). The variations of the pressure acting on the mold by regions 1, 2 and 3 versus mold position are shown in Figs. 13 (b), (c) and (d), respectively. During the mold pressing, the pressure from the region 1 mainly acts on the mold. Large compressive stress is concentrated under the protruding portion of the mold. The pressure from the region 2 begins to act at the mold position of 15 nm. This indicates that the glass does not flow into mold cavity soon after the mold pressing starts. At that point, a shear stress arises in the glass. This shear stress reflects the pressure increase in the region 3. During the mold releasing, the pressure acting on the mold is contributed by the regions 1 and 2. The contribution by the region 1 is the adhesion force between the top surface of the mold and the glass. The contribution by the region 2 is the friction force between the side wall of the mold and the glass. A fluctuation in pressure is observed only in the region 2. This indicates that the fluctuation in total pressure shown in Fig. 12 is attributed to the stick slip phenomena between the side wall of the mold and the glass. The contribution by the region 3 is smaller than those by regions 1 and 2 during whole NIL process. The result shown in Fig. 13 is similar to that reported for polymer imprinting (Kang et al., 2007).

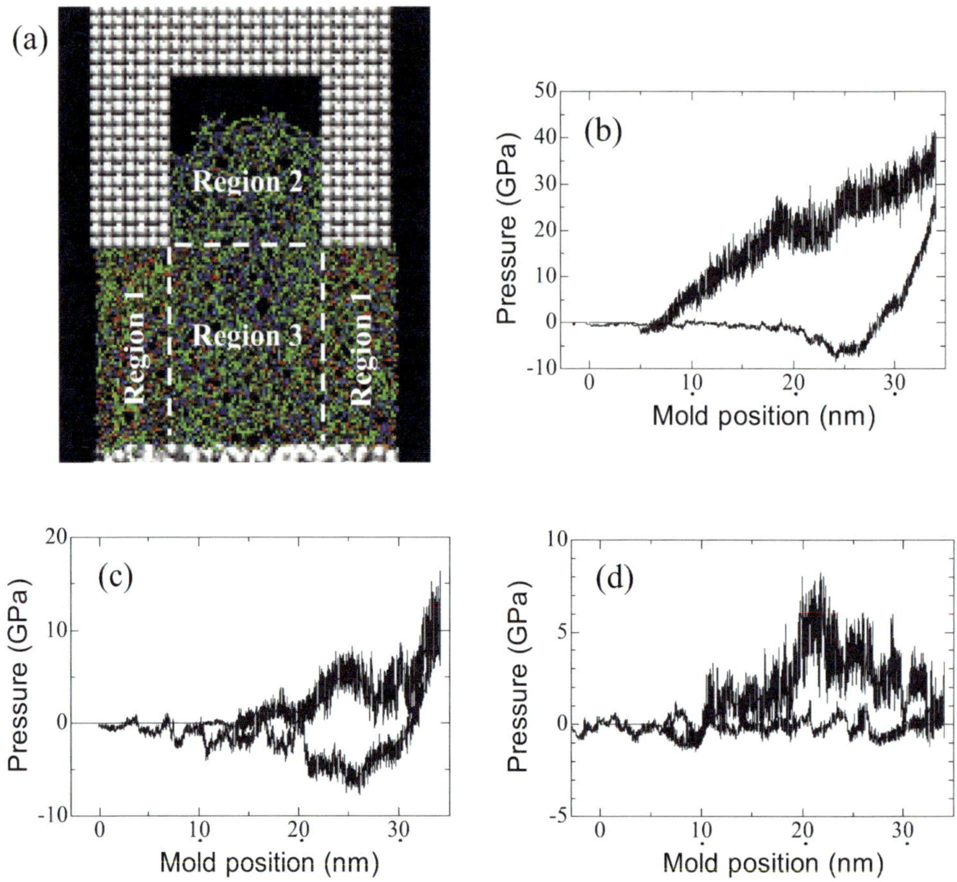

Fig. 13. The variations of the pressure acting on the mold by (a) divided glass regions. The pressures by (b) region 1, (c) region 2 and (d) region 3 are separately analyzed.

3.3 Glass pattern breakages

As discussed in Section 3.2, the friction force appears between the side wall of the mold and the glass during the mold releasing. This friction force induces stretching of the glass. Since the friction force between the side wall of the mold and the glass depends on the contact area, the maximum friction force becomes large for the high aspect ratio pattern (Kang et al., 2007). If the depth of the mold cavity is constant, the maximum tensile stress in the glass increases with the decrease in the cavity width (Tada et al., 2009c). Figure 14 shows the cross-sectional views of the glass pattern after the mold releasing obtained by the MD simulation. The depth of the mold cavity is 2.5 nm. When the cavity width is 2 nm, the pattern is successfully transferred in the glass as shown in Fig. 14 (a). However, for the cavity width of 1 nm, narrow line grass pattern fractures during the mold releasing as shown in Fig. 14 (b). The tension stress induced by the friction force is concentrated at the narrow grass pattern. This is because the minimum line width for the successful pattern transfer depends on the mold geometry. From the MD study, the ultimate resolution in the SiO_2 glass NIL was estimated to be 0.4 nm (Tada et al., 2009c).

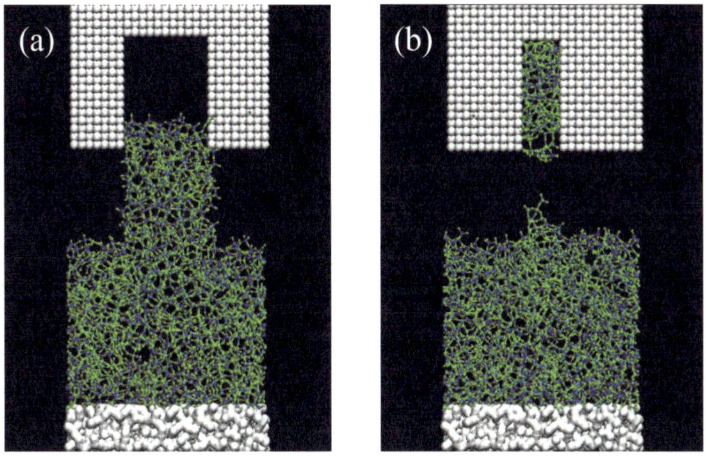

Fig. 14. The cross-sectional views of the glass pattern after the mold releasing. The cavity widths are (a) 2 and (b) 1 nm.

4. Mold deformation in NIL process

Finally, we study the mold deformation in NIL process with the MD simulation. In Sections 2 and 3, either the mold or the processed material is treated as a rigid body. In this section, deformations of both the mold and the processed material are investigated simultaneously.

The configuration of the simulation is almost same as that shown in Fig. 11. However, the Si mold is not treated as a rigid body. The motion of atoms in the mold is also calculated using the Tersoff potential (Tersoff, 1988a) (Tersoff, 1988b). In this simulation, the processed material is the pseudo-glass. For the calculation of the pseudo-glass deformation, we use the Born-Mayer-Huggins potential for SiO_2 glass (Delaye et al., 1997). The potential parameter for pseudo-glass is changed to make Young's modulus smaller (6 GPa) for the NIL process using Si mold. In order to calculate the interaction between a Si mold and SiO_2 glass Lennard-Jones potential is used.

The top surface of the mold is pressed onto the pseudo-glass at a constant speed of 50 m/s. Periodic boundary conditions are applied in the x- and z- directions. The size of a unit cell in the mold structure is 2.7 nm wide, 5.0 nm high and 2.0 nm deep. The thickness of the mold basement is 1.5 nm. The few upper and bottom atomic layers are fixed. The initial temperature of the simulation is 300 K.

Figure 15 shows the cross-sectional views of the stress distributions in {110}/{100} mold during pressing. The maximum stress is concentrated on the sidewall of the Si mold until whole the protruding portion of the mold is buried in the glass as shown in Figs. 15 (a) and (b). Further pressing raises the stress inside the mold as shown in Fig. 15 (c).

Figure 16 shows the cross-sectional views of the stress distributions in {110}/{110} mold during pressing. At the beginning of the pressing process, the relatively large stress is observed at a contact area and the side wall of the mold as shown in Fig. 16 (a). In this case, the mold begins to bend in process of pressing as shown in Fig. 16 (b). The maximum stress concentrates around the bottom edge of the mold with the progress of bending as shown in Fig. 16 (c). Because of the periodic boundary condition in x-direstion, the head of the mold appears from the opposite side of the structure in Fig. 16 (c).

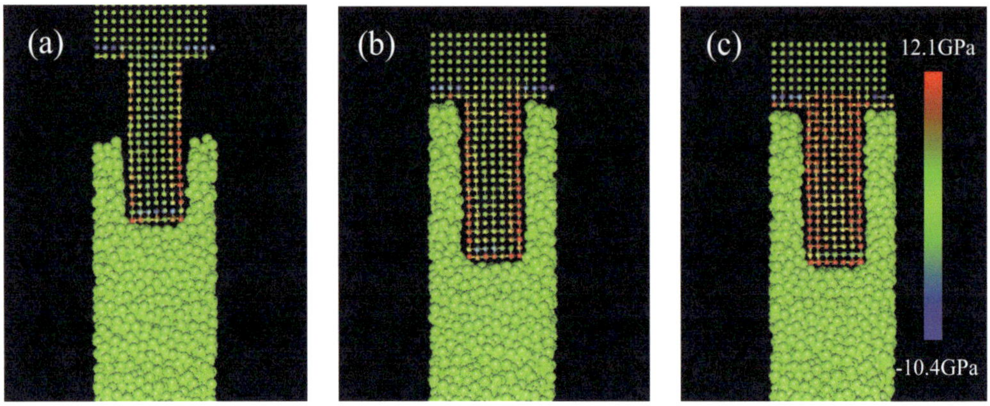

Fig. 15. The cross-sectional views of the stress distributions in {110}/{100} mold during pseudo-glass imprinting.

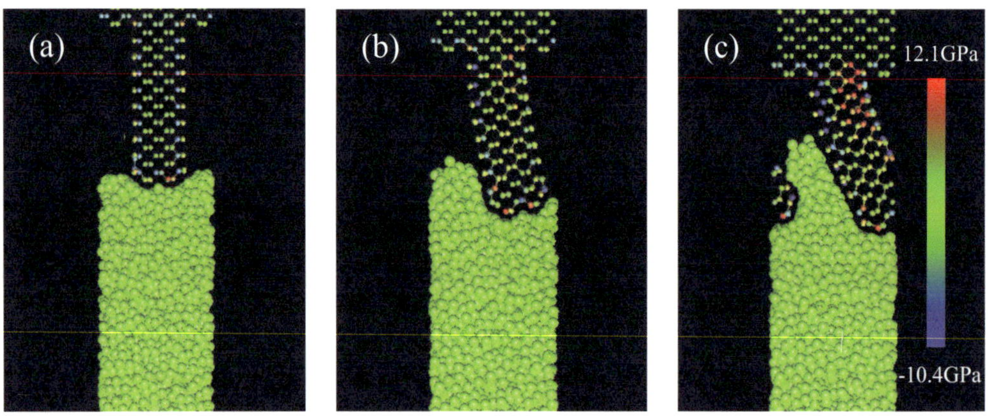

Fig. 16. The cross-sectional views of the stress distributions in {110}/{110} mold during pseudo-glass imprinting.

5. Conclusion

The MD simulations are performed to investigate the mechanism of the mold and the pattern breakages. From the analysis of the independent mold, the dependences of the mold strength on the crystal orientation and the defect structure are revealed. The analysis of the pressure acting on the mold indicates that the frictions between the mold sidewall and the processed material induce the pattern stretching and breakages. The simulation of the mold deformation during NIL process reveals that the excessive pressing or the bending of the mold induces the abnormal stress in the mold, which is the possible cause of the mold breakage.

The breakages of the mold and the pattern largely depend on the surface condition of the materials. Our MD simulations do not consider the anti-sticking treatment and surface geometry such as surface roughness. These are the important future subjects for the MD study of NIL process.

6. Acknowledgement

This work was partially supported by the New Energy and Industrial Technology Development Organization (NEDO) of Japan.

7. References

Akita, Y.; Watanabe, T.; Hara, W.; Matsuda, A. & Yoshimoto, M. (2007). Fine pattern fabrication on glass surface by imprint lithography, *Jpn. J. Appl. Phys.*, Vol. 46, Nos. 12-16, (April 2007) pp. L342-L344

Cheng, M. –C.; Hsiung, H. –Y.; Lu, Y. –T. & Sung, C. –K. (2007b). The effect of metal-film thickness on pattern formation by using direct imprint, *Jpn. J. Appl. Phys.,* Vol. 46, No. 9B, (September 2007) pp. 6382-6386

Chou, S. Y.; Krauss, P. R. & Renstrom, P. J. (1995). Imprint of sub-25 nm vias and trenches in polymers, *Appl. Phys. Lett.,* Vol. 67, No. 21, (November 1995) pp. 3114-3116

Chou, S. Y.; Krauss, P. R. & Renstrom, P. J. (1996). Imprint lithography with 25-nanometer resolution, *Science,* Vol. 272, No. 5258, (April 1996) pp. 85-87

Delaye, J. M.; Louis-Achille, V. & Ghaleb, D. (1997). Modelling oxide glasses with Born–Mayer–Huggins potentials: Effect of composition on structural changes, *J. Non-Cryst. Solids,* Vol. 210, Nos. 2-3, (March 1997) pp. 232-242

Fang, T. -H.; Wu, C. -D. & Chang, W. -J. (2007). Molecular dynamics analysis of nanoimprinted Cu–Ni alloys, *Appl. Surf. Sci.,* Vol. 253, No. 16 (June 2007) pp. 6963-6968

Hsieh, C. -W. & Sung, C. -K. (2007). Atomic-scale friction in direct imprinting process: Molecular dynamics simulation, *Jpn. J. Appl. Phys.,* Vol. 46, No. 9B, (September 2007) pp. 6387-6390

Hsiung, H. Y.; Chen, H. Y. & Sun, C. K. (2009). Temperature effects on formation of metallic patterns in direct nanoimprint technique - Molecular dynamics simulation and experiment, *J. Mater. Process. Tech.,* Vol. 209, No. 9, (May 2009) pp. 4223-4626

Hsu, Q. –C.; Wu, C. –D. & Fang, T. –H. (2004). Deformation mechanism and punch taper effects on nanoimprint process by molecular dynamics, *Jpn. J. Appl. Phys.,* Vol. 43, No. 11A, (November 2004) pp. 7665-7669

Hsu, Q. –C.; Wu, C. –D. & Fang, T. –H. (2005). Studies on nanoimprint process parameters of copper by molecular dynamics analysis, *Comput. Mater. Sci.,* Vol. 34, No. 4, (December 2005) pp. 314-322

Hirai, Y.; Kanakugi, K.; Yamaguchi, T.; Yao, K.; Kitagawa, S. & Tanaka, Y. (2003). Fine pattern fabrication on glass surface by imprint lithography, *Microelectron. Eng.,* Vols. 67-68, (June 2003) pp. 237-244

Hirai, Y.; Fujiwara, M.; Okuno, T.; Tanaka, Y.; Endo, M.; Irie, S.; Nakagawa, K. & Sasago, M. (2004). Study of the resist deformation in nanoimprint lithography, *J. Vac. Sci. Technol. B,* Vol. 19, No. 6, (Novemver 2001) pp. 2811-2815

Hirai, Y.; Konishi, T.; Yoshikawa, T. & Yoshida, S. (2004). Simulation and experimental study of polymer deformation in nanoimprint lithography, *J. Vac. Sci. Technol. B,* Vol. 22, No. 6, (Novemver 2004) pp. 3288-3293

Hua, F.; Sun, Y.; Gaur, A.; Meitl, M. A.; Bilhaut, L.; Rotkina, L.; Wang, J.; Geil, P.; Shim, M.; Rogers, J. A. & Shim, A. (2004). Polymer imprint lithography with molecular-scale resolution, *Nano Lett.,* Vol. 4, No. 12, (December 2004) pp. 2467-2471

Hua, F.; Gaur, A.; Sun, Y.; Word, M.; Jin, N.; Adesida, I.; Shim, M.; Shim, A. & Rogers, J. A. (2006). Processing dependent behaviour of soft imprint lithography on the 1-10-nm scale, *IEEE Trans. Nanotechnol.*, Vol. 5, No. 3, (May 2006) pp. 301-308

Kang, J. -H.; Kim, K. –S. & Kim, K. –W. (2007). Molecular dynamics study of pattern transfer in nanoimprint lithography, *Tribol. Lett.*, Vol. 25, No. 2, (February 2007) pp. 93-102

MARC : http://www.mscsoftware.com

Okinaka, M.; Tsukagoshi, K. & Aoyagi, Y. (2006). Direct nanoimprint of inorganic-organic hybrid glass, *J. Vac. Sci. Technol. B*, Vol. 24, No. 3, (May 2006) pp. 1402-1404

Song, Z.; Choi, J.; You, B. H.; Lee, J. & Park, S. (2008). Simulation study on stress and deformation of polymeric patterns during the demolding process in thermal imprint lithography, *J. Vac. Sci. Technol. B*, Vol. 26, No. 2, (March 2008) pp. 598-605

Tada, K.; Yasuda, M.; Kimoto, Y.; Kawata, H. & Hirai, Y. (2008). Molecular dynamics study of yield stress of Si mold for nanoimprint lithography, *Jpn. J. Appl. Phys.*, Vol. 47, No. 4, (April 2008) pp. 2320-2323

Tada, K.; Yasuda, M.; Fujii, N.; Kawata, H. & Hirai, Y. (2009a). Molecular dynamics study on mold fracture by nano scale defects in nanoimprint lithography, *Proceedings of 25th European Mask and Lithography Conference*, 32, Dresden, Germany, January 2009

Tada, K.; Kimoto, Y.; Yasuda, M.; Kawata, H. & Hirai, Y. (2009b). Molecular dynamics study of nanoimprint lithography for glass materials, *Jpn. J. Appl. Phys.*, Vol. 48, No. 6, (June 2009) 06FH13

Tada, K.; Yasuda, M.; Kimoto, Y.; Kawata, H. & Hirai, Y. (2009c). Molecular dynamics study on resolution in nanoimprint lithography for glass material, *Mater. Res. Soc. Symp. Proc.*, Vol. 1179, BB06-11, San Francisco, USA, April 2009

Takada, A.; Richet, P.; Catlow, C. R. A. & Price, G. D. (2004). Molecular dynamics simulations of vitreous silica structures, *J. Non-Cryst. Solids*, Vols. 345-346, (October 2004) pp. 224-229

Tersoff, J. (1988a). New empirical approach for the structure and energy of covalent systems, *Phys. Rev. B*, Vol. 37, No. 12, (April 1998) pp. 6991-7000

Tersoff, J. (1988b). Empirical interatomic potential for silicon with improved elastic properties, *Phys. Rev. B*, Vol. 38, No. 14, (November 1998) pp. 9902-9905

Woo, Y. S.; Kim, J. K.; Lee, D. E.; Suh, K. Y. & Lee, W. I. (2007). Density variation of nanoscale patterns in thermal nanoimprint lithography, *Appl. Phys. Lett.*, Vol. 91, No. 25, (December 2007) 253111

Wu, C. -D. & Lin, J. –F. (2008). Multiscale particle dynamics in nanoimprint process, *Appl. Phys. A*, Vol. 91, No. 2 (May 2008) pp. 273-279

Wu, C. -D.; Lin, J. –F. & Fang, T. –H. (2009). Molecular dynamics simulations of the roller nanoimprint process: Adhesion and other mechanical characteristics, *Nanoscale Res. Lett.*, Vol. 4, No. 8 (August 2009) pp. 913-920

Yang, S.; Yu, S. & Cho, M. (2009). Molecular dynamics study to identify mold geometry effect on the pattern transfer in thermal nanoimprint lithography, *Jpn. J. Appl. Phys.*, Vol. 48, No. 6, (June 2009) 06FH03

Three Dimensional Nanoimprint Lithography using Inorganic Electron Beam Resist

Jun Taniguchi and Noriyuki Unno
Tokyo University of Science
Japan

1. Introduction

With the advancement in the information technology the need for increasingly complex three-dimensional (3D) structure is eminent. Gray scale lithography can produce 3D structure but it cannot match the resolution of electron beam lithography (EBL). Today the EBL is primarily used for mask making and for making nanoimprint lithography (NIL) molds (Chou *et al.*, 1995), but these structures do not require features with varying depths. Hence EB systems under their current mode of operation cannot build 3D nanostructures. Typically, EB systems are operated at high accelerating voltage (>50 kV) with beam diameter capable of writing at nano-scale (Ishii & Matsuda, 1992). However, high kV EB because of its poor interaction with resist causes poor resist sensitivity resulting in low throughput. It is also difficult to control depth because dose change affects EB blur resulting in feature width spread during develop. This problem can be resolved at low kV operation where increased EB/resist interaction improves resist sensitivity (Olkhovets & Craighead, 1999). Inorganic resist is transparent, so it can be directly applied to an optical surface. Since the transparency of this type of resist extends to ultraviolet (UV) light, it can also be used for making UV-NIL mold. Using the resist we fabricated 3D mold by controlling the acceleration voltage, and then we replicated the mold with nano-order patterns by using UV-NIL. Usually, fabrication of binary optics element involving repeated overlays and dry etch results in poor resolution and is expensive. However, control of acceleration voltage electron beam lithography (CAV-EBL) can fabricate binary optics element by using only EB, and thus makes the process cost effective. Hence fabrication of 3D NIL mold using CAV-EBL is described.

Furthermore, the method of improving the resolution and the contrast using post exposure bake (PEB) with inorganic resist is also described. Although no chemically amplified material is contained in our inorganic resist, PEB was characteristically applied in our process. PEB process causes the anneal effect for inorganic resist and the proximity effect is suppressed, resulting in high contrast.

2. Three dimensional nanoimprint lithography

2.1 Experimental apparatus and procedures

At first, Accuglass-512B, a Spin-on-Glass (SOG) composed mainly of siloxane with 14% organic content whose interlayer dielectrics were developed by Honeywell Co., was used for inorganic resist. The structure is shown in Fig. 1. A buffered hydrofluoric acid (BHF) solution (50%, HF 25 cm^3/l, and 40% NH_4F 30 cm^3/l) was used for developer. The calculated concentration of [HF] was 0.43 mol/l and that of [HF_2^-] was 0.29 mol/l. ERA-8800FE (ELIONIX Co.) was used for EB system with several pA of beam current and about 10 nm beam diameter. The fabrication process of 3D mold involved three steps (Fig. 2). At first, SOG was spin-coated on a Si substrate and cured at 425 ^oC for 1 h resulting in a 450 nm film. Then the sample was written with EB with uniform doses at different acceleration voltages. In the case of the CAV-EBL, the developed depth was controlled by varying the voltage to change the electron range where low and high voltages gave shallow and deep pits. Then the EB-exposed SOG was developed out with BHF solution in 60 s but it did not

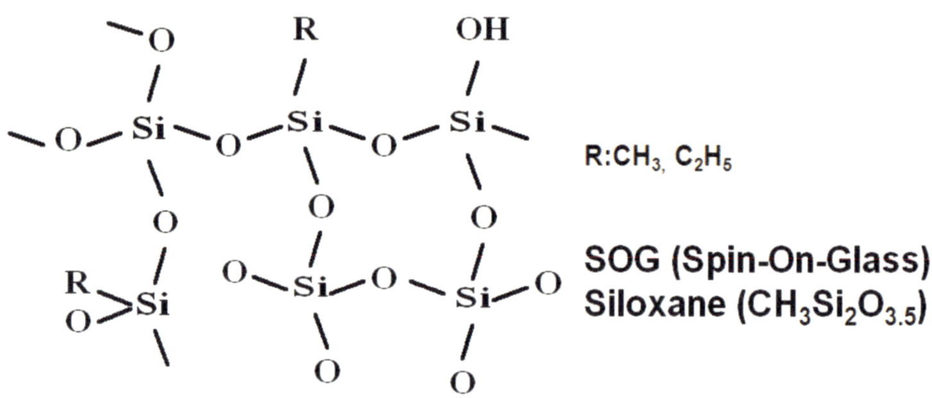

Fig. 1. The structure of spin-on-glass inorganic resist.

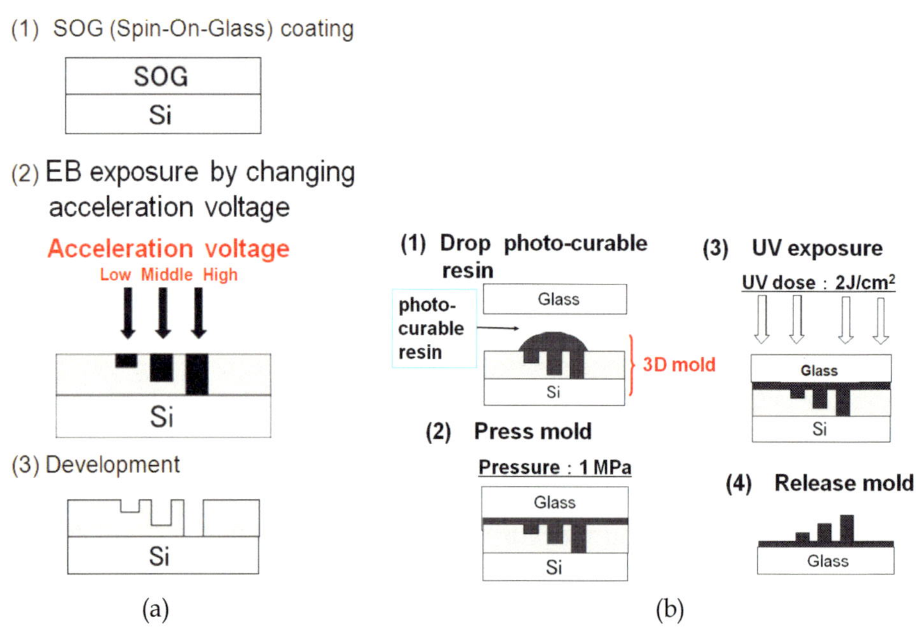

Fig. 2. Process for (a)fabrication of 3D mold with CAV-EBL and (b)UV-NIL.

etch EB-unexposed SOG within 60 s which was helpful in fabricating molds with different depth gradations. For this experiment we used a house-made UV-NIL machine equipped with a UV light source SP-6 (USIO Co.). The process was as follows: at first, a fabricated 3D mold was coated with anti-sticking layer Optool DSX (DAIKIN INDUSTRIES, LTD). Then, a UV photo-curable resin PAK01 (made by Toyo Gosei Co., Ltd.) was dispensed onto a quartz substrate. Next, the mold was pressed against the resin film on the substrate with 1 MPa for 60 s. The photo-curable resin was then exposed to a 2 J/cm^2 dose of UV light through the quartz substrate. The mold was then retracted leaving behind a 3D replica of its pattern. To make the precise measurement of depth dependence on acceleration voltage change, we used a step profilometer (KLA-Tencor alpha-step 500). To enable the profilometer stylus to reach to the bottom of the patterns, we set the drawing pattern width to 10 μm. The mold and replicated pattern was observed with atomic force microscope (SII 100 SPA-400), and scanning electron microscope (SEM, ELIONIX, ERA-8800FE) was used to examine the pattern. The cross-sections of lines were observed with SEM by tilting the specimen to 75°.

2.2 3D UV-NIL

To evaluate the EB dose effect in the CAV-EBL, the acceleration voltage was varied from 1 to 4 kV by 30 V increments. In this case, each accelerating voltage was exposed at 200 and 500 μC/cm^2 dose. The designed line-width for this evaluation was 10 μm. Fig. 3 shows the relationship between acceleration voltage and developed depth. This figure mean that the effect of EB dose on the developed depths was quite noticeable at higher accelerating voltage. In contrast, in low acceleration voltage region (< 2kV) there was no effect of dose change on developed-depth. Thus, in order to control the develop depths, controlling the dose is important. Furthermore, a 5 nm depth control is possible using CAV-EBL and spin on glass as an EB resist.

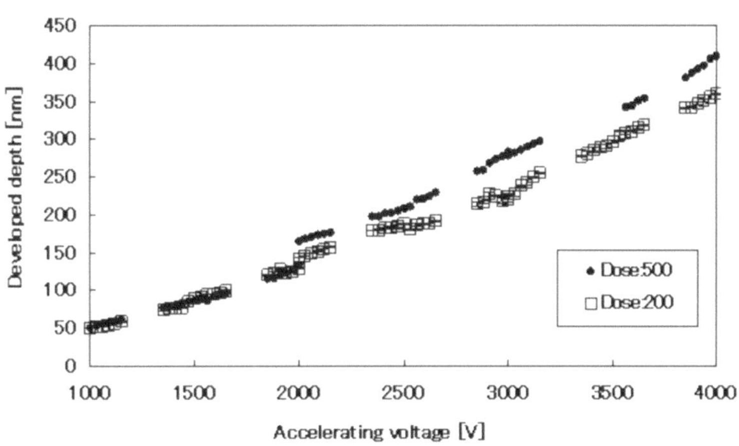

Fig. 3. The relationship between acceleration voltage and developed depth.

Next, we fabricated a blade-shaped mold for binary optics by superposition of different depth areas using stepping motors on SEM stage. However, these motors had limited accuracy of around 500 nm, so superposition of areas width needed to be wider than 500 nm and hence a designed line-width of 2 μm was chosen. Control of depths was carried out by changing the accelerating voltage where seven stairs were formed by using 2, 2.5, 3, 3.5, 4,

4.5 and 5 kV accelerating voltage. For each stair the EB dose was set at 250 μC/cm^2. Fig. 4 shows AFM images of 3D mold where seven stairs were obtained.

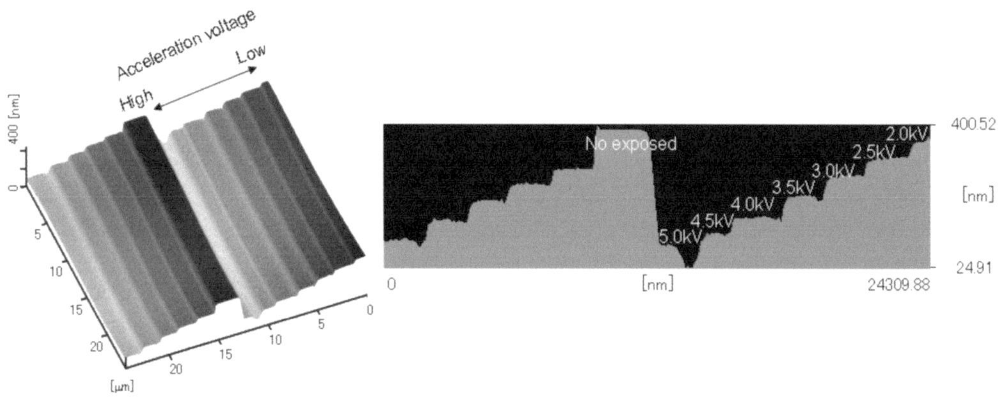

Fig. 4. The AFM images of 3D mold where seven stairs.

Using this mold, UV-NIL was carried out. Fig. 5 shows replicated patterns on photo-curable resin, and table I shows developed-depths of mold and heights of UV-NIL patterns. The mold depths were corresponded to the replicated heights. Hence, volume production of binary optics elements is possible by using CAV-EBL and UV-NIL. The developed depths in table I also agree with the values in Fig. 3.

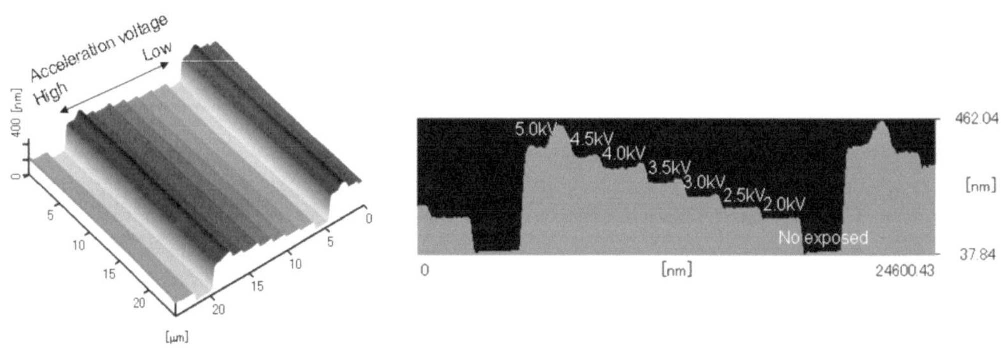

Fig. 5. The AFM images of replicated patterns on photo-curable resin.

Fig. 6 shows a sub-100 nm 3D nanoimprint mold and the UV-NIL pattern on photo-curable resin that we obtained, similar to a blade-shaped mold. The transfer pressure was 3.0 MPa. This result means that the pattern depth can be modulated by using CAV-EBL even at the sub-100 nm scale. The replicated heights were shorter than mold depths, because the photo-curable resin was shrunk by curing at UV exposure. However, the mold line-widths and the replicated pattern line-widths matched, and the replicated pattern heights were also modulated. Thus, the realization of sub-100 nm 3D mold fabrication was possible using CAV-EBL.

Accelerating Voltage [kV]	Depth of stair in 3D mold [nm]	Height of replicated pattern [nm]
2	100.2	106.9
2.5	136.4	143.6
3	173.1	178.4
3.5	220.7	221.9
4	261.6	265.9
4.5	305.3	304.2
5	328.2	325.8

Table I. 3D mold depths and 3D replicated height.

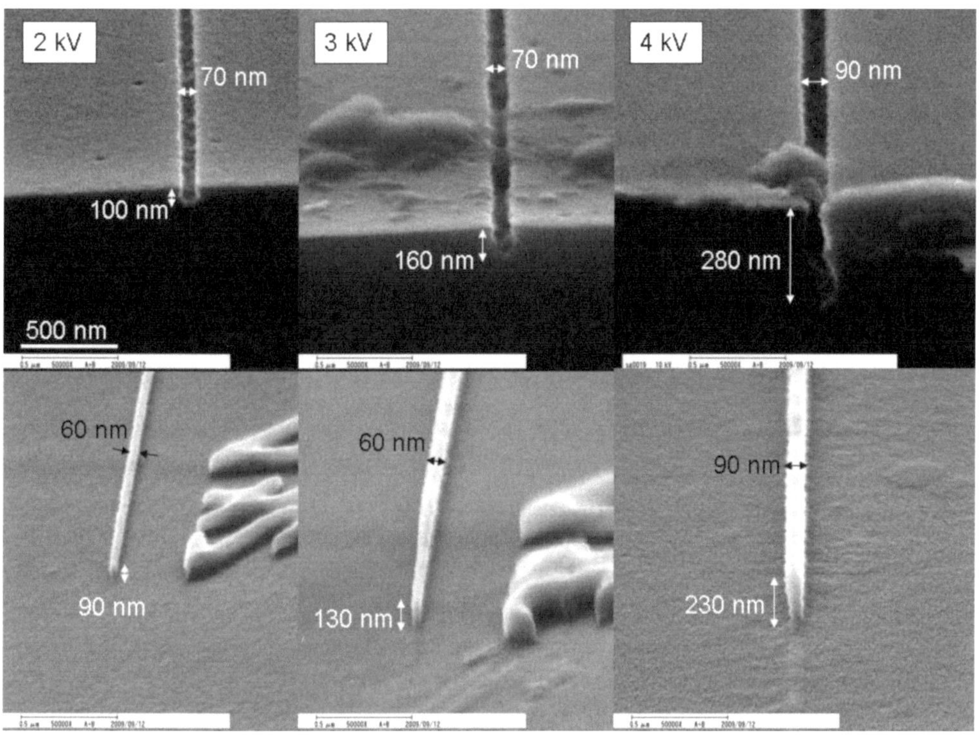

Fig. 6. The fabricated sub-100 nm 3D nanoimprint mold and the UV-NIL pattern

3. Improving the inorganic resist

We used a Spin-On-Glass (SOG) material, Accuglass 512B (Honeywell Co. Ltd.), for positive EB resist that exhibits low sensitivity and LER. On the other hand, negative EB inorganic resist has high sensitivity and resolution (Namatsu, 2001). Hence, positive EB inorganic resists with high sensitivity and resolution, and low line edge roughness (LER) are now required for low kV EBL. Therefore, an inorganic resist known as Nano Imprint Mold Oxide-Positive tone 0701 (NIMO-P0701) made by TOKYO OHKA KOGYO CO., LTD, was developed for improved sensitivity and was employed to investigate the characteristics of EB exposures.

NIMO-P0701 was mainly composed of siloxane and Photo Base Generator (PBG). NIMO-P0701 was spin-coated (pre; 300 rpm (3 s), main; 3000 rpm (10 s)) on silicon substrates. The NIMO-P0701 film was pre-baked at 300 °C (90 s), which is the decomposition temperature of PBG, resulting in an approximately 300 nm film thickness. PBG totally changes Si-O-R into Si-O at low pre-bake temperature of 300 °C and accretive to NIMO-P0701 to achieve high sensitivity and delineate pattern at lower pre-bake temperature. The resist films were then exposed using a Scanning Electron Microscope (SEM; ERA-8800FE, ELIONIX), which was customized for the task.

At first, to investigate the mechanism of lithography for NIMO-P0701, Fourier transform infrared (FT-IR) spectroscopy was employed. Fig. 7 shows FT-IR spectra of exposed area of NIMO-P0701 before, and after develop. After develop shows decrease in Peaks of Si-CH_3 and Si-O indicating the area of NIMO-P0701 developed out.

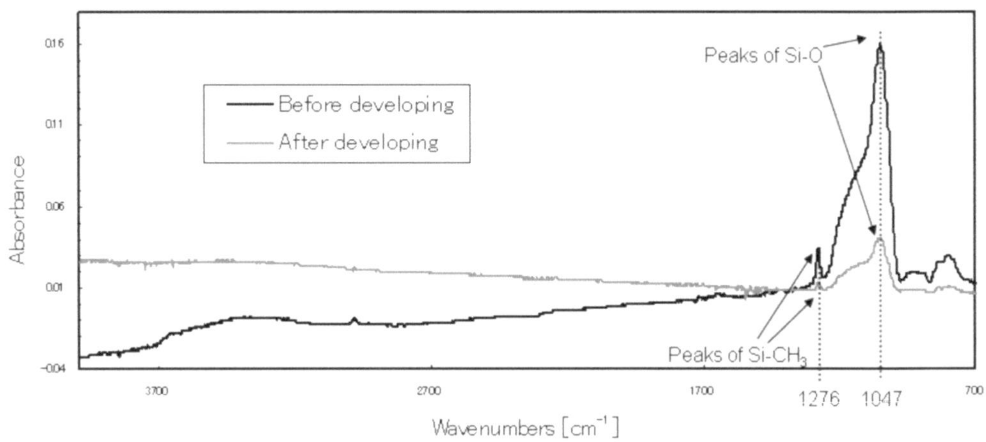

Fig. 7. FT-IR spectra of exposed area of NIMO-P0701 before, and after develop.

Fig. 8 shows FT-IR spectra of exposed and un-exposed areas of NIMO-P0701 after develop. Here the peaks of Si-CH_3 and Si-O in exposed area are significantly lower than the corresponding peaks in the un-exposed area indicating that the exposed area was developed out. The develop mechanism is explained by the conversion of Si-CH_3 into Si-OH when exposed to EB where, unlike Si-CH_3, the newly formed Si-OH is readily soluble in BHF. This behavior results in the appearance of high contrasts.

Next, a developer, buffered hydrofluoric acid (BHF), was optimized. BHF was composed of HF and NH_4F. Mixing ratios and concentrations of these components were varied to optimize the development process. Initially, both resist were developed by 2.4 % concentration BHF (HF:NH_4F=1:1). This concentration has been used for Accuglass 512B. But developing characteristic of NIMO-P0701 turned out to be of low contrast and sensitivity. And hence, the developer for NIMOP0701 had to be optimized. We found that a 2.5% concentration BHF (HF:NH_4F=7:3) was optimum developer for NIMO-P0701. Fig. 9 shows the sensitivity curves of NIMO-P0701 and Accuglass 512B with 4 kV EB. Here NIMO-P0701, developed by 2.5% concentration BHF (HF: NH_4F=7:3), shows higher sensitivity and contrast compared to that of Accuglass 512B.

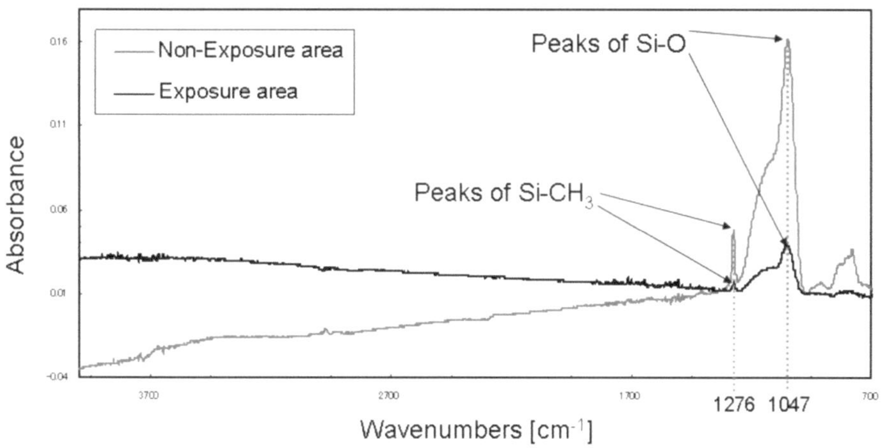

Fig. 8. FT-IR spectra of exposed and un-exposed areas of NIMO-P0701 after develop.

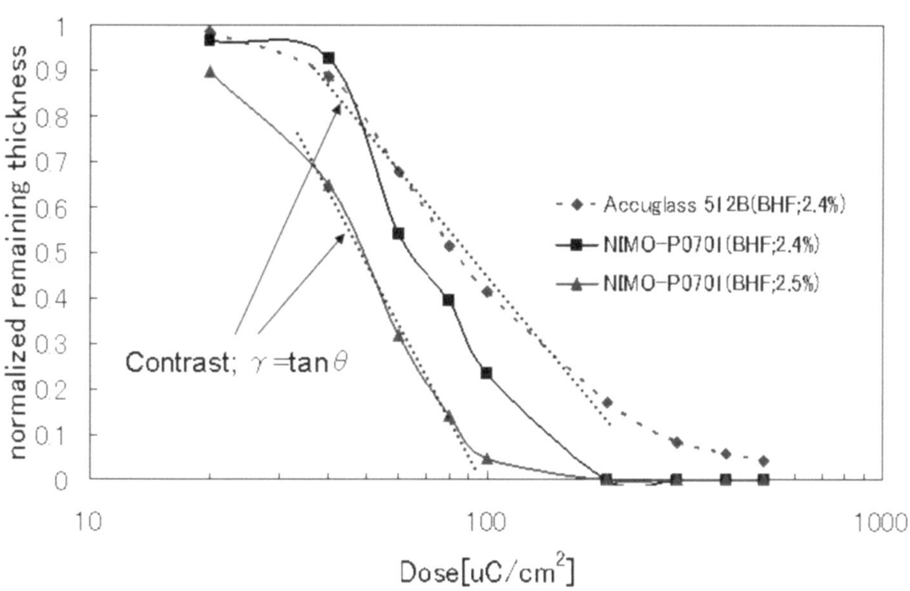

Fig. 9. The sensitivity curves of NIMO-P0701 and Accuglass 512B with 4 kV EB.

Furthermore, the LER was calculated by root-mean-square of line edges which were observed with SEM (Fig. 10). The designed line width is 36nm and space width is 234nm. The condition of EB was 200 $\mu C/cm^2$ at 4 kV. Because of high contrast resulting in high resolution, NIMO-P0701 could be delineated into fine pattern followed by LER measurements.

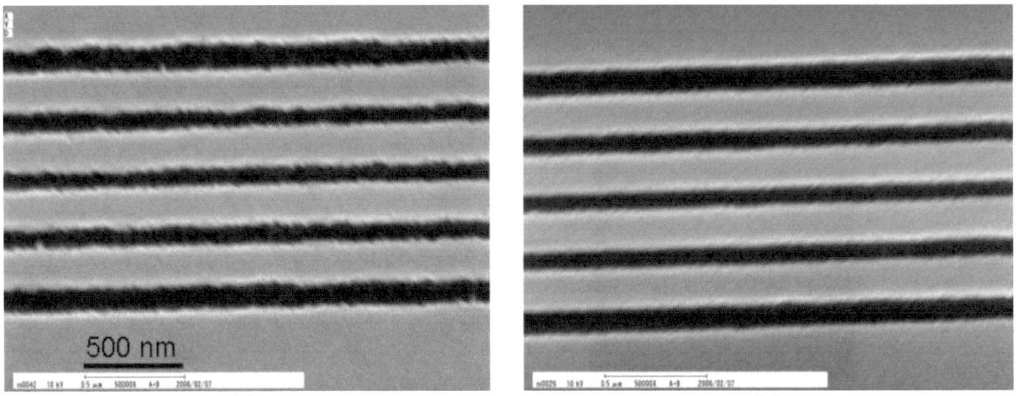

(a) (b)

Fig. 10. Comparison of LER between (a) Accuglass 512B and (b) NIMO-P0701.

The LER for Accuglass 512B was 6.77 nm, whereas the LER for NIMO-P0701 was 5.78 nm, indicating that NIMO-P0701 is better than Accuglass 512B as regards to LER. Since NIMO-P0701 has exhibited high sensitivity and low LER, it was used to image very fine patterns as shown Fig. 11.

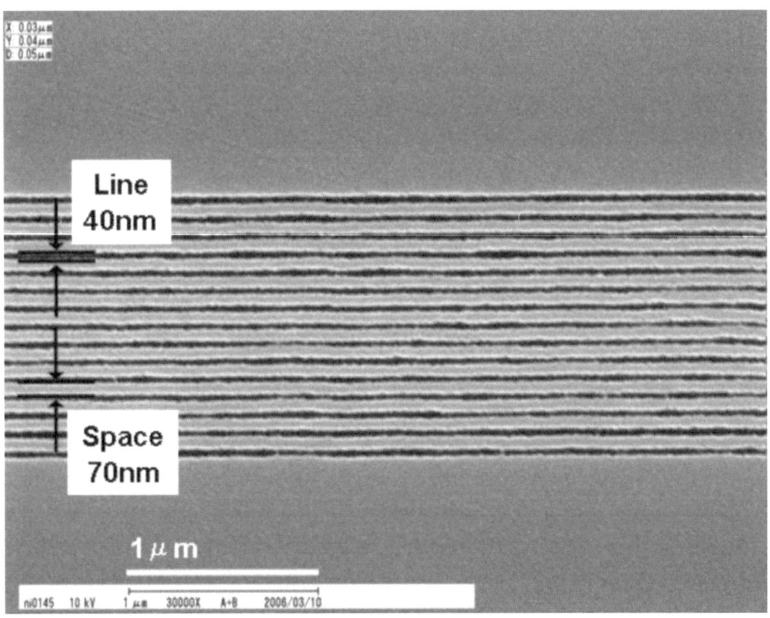

Fig. 11. The fine L&S pattern at 4 kV with NIMO-P0701.

4. Post exposure bake with the inorganic resist

A line and space (L&S) patterns, which line-width was 40 nm and space-width was 70nm, was obtained using low-acceleration-voltage EBL system with inorganic resist, NIMO-P0701 in section 3. However, the smaller space-width was not obtained with 4 kV EB because of the proximity effect. Therefore, post-exposure-bake (PEB) process was employed. Typically,

PEB was only employed for increasing the sensitivity of the resist, which has a chemically amplified material (Ocola, 2003). However, there are no chemically amplified material in NIMO-P0701, but PEB process caused the anneal effect for SOG and the proximity effect was suppressed. The experimental procedure was the same as section 2.1, except for PEB process. PEB was carried out just after EB exposure using hot plate at 200, 300, 425 °C for 5 minutes in air.

Fig. 12 shows the relationship between the anneal effect and PEB temperature. The acceleration voltage and EB dose were constant in 4 kV and 300 μC/cm^2, respectively. Without PEB, the line-width was 140 nm and the higher PEB temperature tended to reduce the developed line-width.

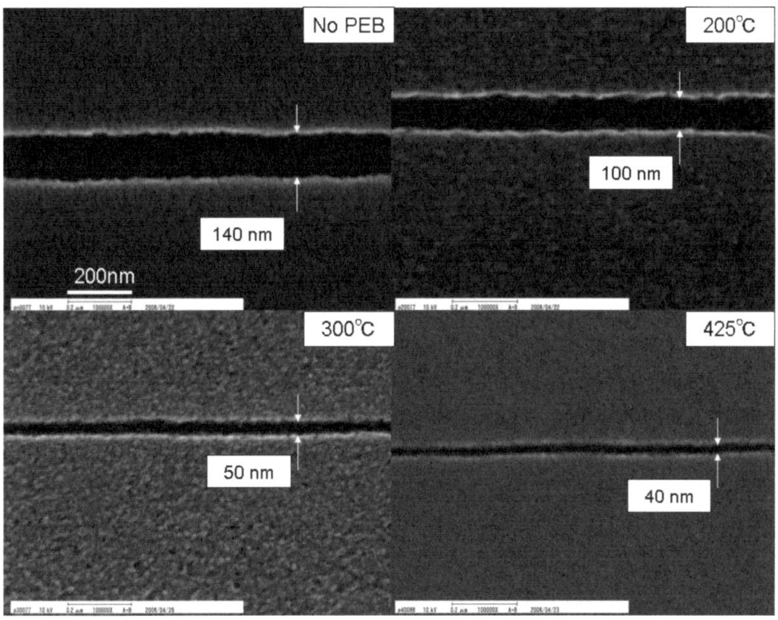

Fig. 12. The relationship between the anneal effect and PEB temperature.

Development mechanism and PEB effect is described as follows: NIMO-P0701 has chemical structures which are Si-OH, Si-CH_3 and Si-O. The main structure of NIMO-P0701 is SiO_2 and SiO_2 reacts with HF to form H_2O and H_2SiF_6, which is dissolves well into H_2O. Thus, the wetting ability between the resist surface and water dominates the etching rate of the resist. The etching rates by BHF of these chemical bonds is as follows: Si-OH > Si-CH_3 >> Si-O. On the formed resist surface, Si-CH_3 is the main structure. EB exposure accelerates the mechanism where Si-CH_3 reacts with an electron to form Si-OH. With PEB in air, Si-CH_3 changes to Si-O at the EB-unexposed area (Ahner *et al.*, 2007). Therefore, the etching rate at the un-exposed area with PEB is smaller than without PEB, result in the developed line-width reduction.

Fig. 13 shows the obtained patterns after development with or without PEB. The designed line-width is 27 nm and the space-width is 108 nm (line and space pattern, L&S). The acceleration voltage and EB dose was 2 kV and 80 μC/cm^2. Without PEB, no L&S pattern was obtained because of the proximity effect. With PEB at 300 °C, in contrast, a 100 nm L&S pattern was obtained.

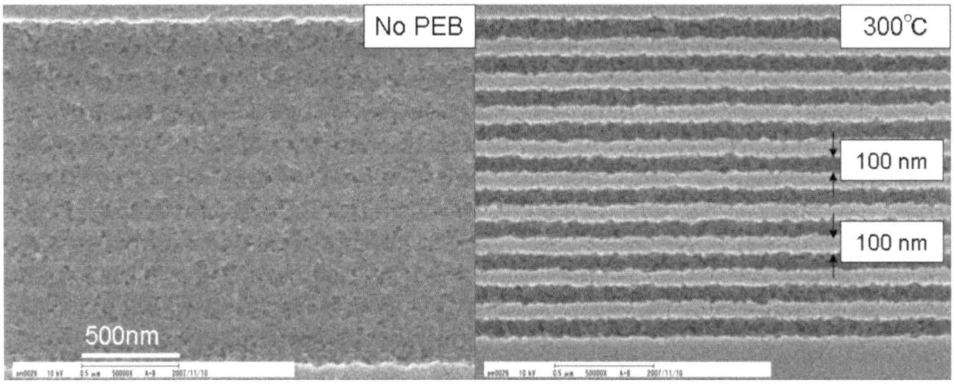

(a) (b)

Fig. 13. The obtained patterns after development (a) with or (b) without PEB.

Fig. 14 shows the cross-sections of developed L&S pattern with or without PEB at 300 °C, whose designed line-width and space-width were 45 nm and 270 nm, respectively. The acceleration voltage and EB dose were 4 kV and 200 μC/cm². The reverse-taper shape, which is undesirable for NIL process, was observed without PEB. However, this reverse-taper shape was prevented by PEB. These results mean that the use of PEB with an inorganic resist helps suppress the proximity effect and causes an annealing effect.

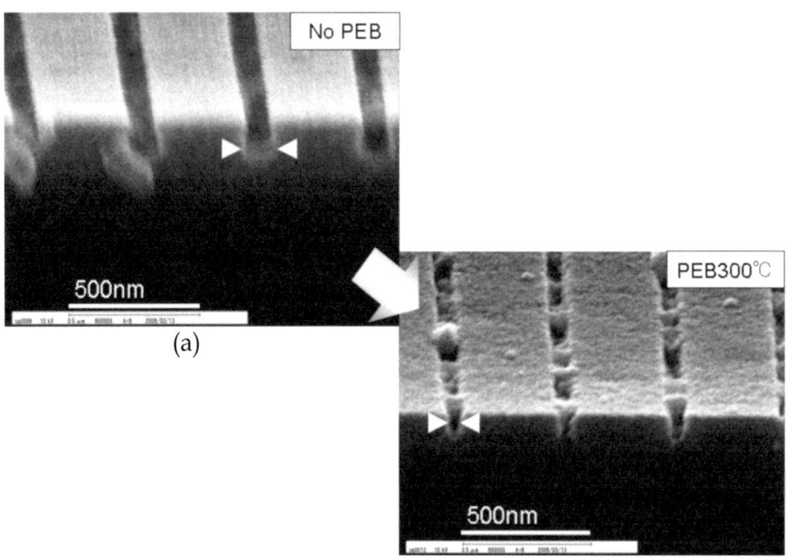

(a)

(b)

Fig. 14. the cross-sections of developed L&S pattern (a) with or (b) without PEB at 300 °C.

In addition, PEB is very effective in fabricating high-aspect ratio pattern in inorganic resist. In this case, thick NIMO-P0701 film (1 μm) on silicon wafer was exposed and developed to achieve the required pattern, because PEB process for the SOG suppress the proximity effect and increase the contrast. NIMO-P0701 was then written with EB. The designed line width and space width were 135 nm and 324 nm, respectively. EBL conditions were EB dose of 2500 μC/cm^2 at acceleration voltage of 30kV. Development temperature was 20 °C and development time was 1.5 min. PEB conditions were temperature of 425 °C and time of 5 min in air.

Fig. 15(a) shows the SEM image of the obtained pattern "without" PEB. This L&S pattern line, space, and depth were was 470 nm, 190 nm and 550 nm, respectively. So aspect ratio "without" PEB in the mold was 1.16. And developed resist surface was rough. In contrast, Fig. 15(b) shows the SEM image of the obtained pattern "with" PEB. This L&S pattern line, space, and depth were was 390 nm, 250 nm and 1000 nm, respectively. So aspect ratio in the case of "with" PEB in the mold was 2.59. Here the developed resist surface was smooth. An aspect ratio of "with" PEB mold was higher than that of "without" PEB, thus, influence of PEB is effective even where pattern size is of the order of submicron. Furthermore, PEB makes the surface smooth, and makes the structure vertically more anisotropic. These are some of the advantages with this type of mold.

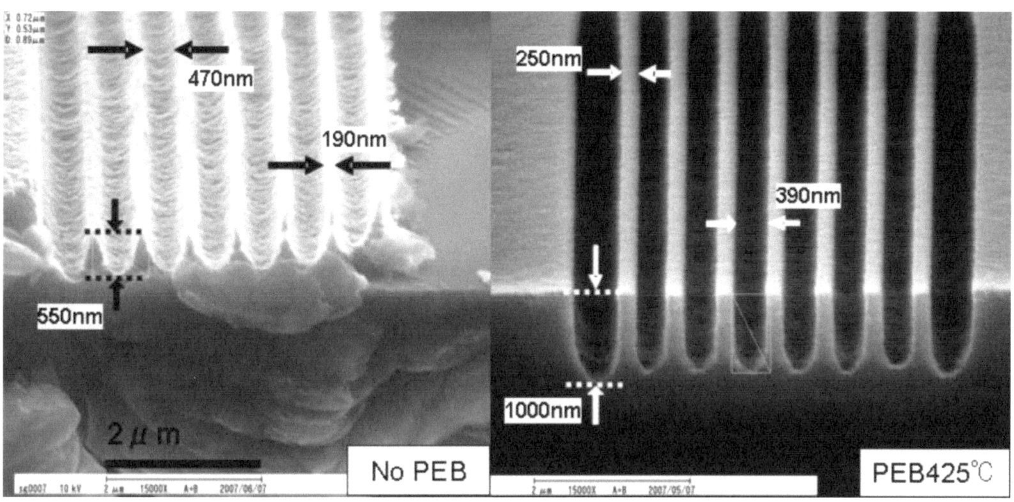

(a) (b)

Fig. 15. The SEM image of the obtained pattern (a) without or (b) with PEB.

Using the this mold with PEB, UV-NIL was carried out, similar to section 2.1. The replicated pattern from this operation is shown in Fig. 16. The obtained L&S pattern's line, space, and depth were was 280 nm, 240 nm and 940 nm, respectively. So aspect ratio of UV-NIL was 3.39. The replicated height and width were slightly smaller than mold depth and width because of volume shrinkage of photo-curable polymer by UV irradiation. As a result, using the developed EB resist layer and UV-NIL, sub-micron L&S mold were fabricated. This method does not involve dry etching which has been known to be a major cause for pattern degradation, and is simpler than conventional LIGA and deep dry etching techniques (Ueno *et al.*, 1997).

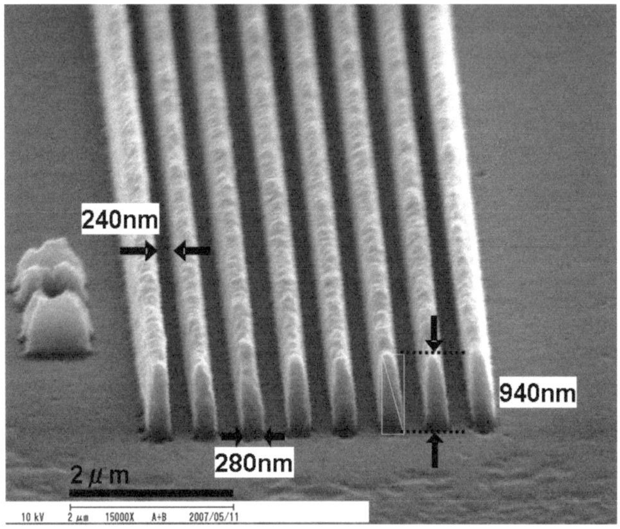

Fig. 16. The replicated pattern using high aspect ratio L&S mold.

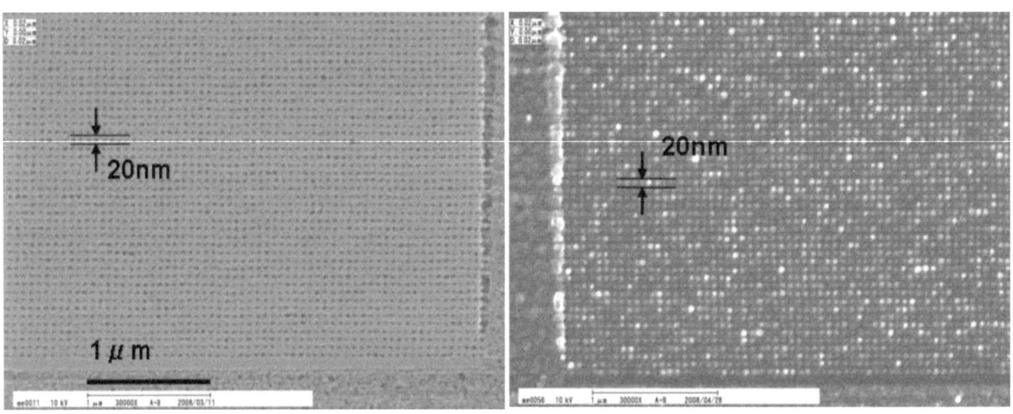

(a) (b)

Fig. 17. The SEM image of (a) 20 nm dots array pattern and (b) its replicated pattern.

Finally, nanodot array pattern was fabricated using PEB with NIMO-P0701. Nanodot array patterns will be used in patterned media and plasmonic devices of the next generation. Nanometer-scale pattern fabrication has already been applied to quantum devices optical (Fujita *et al.*, 1997). The patterned media can have a higher density of dot pattern than existing hard disk drives as a result of their ability to store magnetic information vertically. The storage density of over 1 Tb/in^2 have already been developed (Hosaka *et al.*, 2006). When these nano-meter scale structures fabricate, high acceleration voltage EBL system is normally employed as noted previously in introduction. Using our PEB process with

NIMO-P0701, however, a nanodot array pattern with 20-nm-diameter dots was obtained by 4 kV EBL (Fig. 17). This low acceleration voltage EBL system is quite inexpensive, compared to high kV one. Although the designed diameter of the nanodots was 30 nm, the areas exposed to the electron beam were shrunk to 20 nm as a result of the anneal effect of PEB at 200 °C. Moreover, the EB dose was 60 μC/cm^2 and this value is very faster than high kV EBL.

5. Conclusions

The 3D resolution of a CAV-EBL using inorganic resist has been examined. We found that the CAV-EBL can be employed to fabricate 3D NIL mold which may be in the order of just a sub-100 nm width, and can be employed to control a depth with a precision of 5 nm by simply changing the accelerating voltage by 30 V. Therefore, CAV-EBL is a suitable technique for the fabrication of binary optics element and NIL molds. Furthermore, using these molds, 3D replication patterns were obtained with UV-NIL. Thus, volume production of high value-added binary optics elements is possible using CAV-EBL and UV-NIL.
In addition, inorganic resist and post exposure bake process have been developed. These techniques help suppressing the proximity effect because of annealing effect. Consequently, developed inorganic resist (NIMO-P0701) with PEB process can fabricate high resolution and high aspect ratio NIL molds.

6. Acknowledgements

The authors thank to Mr. K. Ishikawa at TOKYO OHKA KOGYO CO., LTD, for NIMO-P0701 resist development.

7. References

Ahner, N.; Schulz, S. E.; Blaschta, F. & Rennau, M. (2008). Optical, electrical and structural properties of spin-on MSQ low-k dielectrics over a wide temperature range. *Microelectron. Eng.*, 85. pp. 2111–2113.

Chou, S. Y.; Krauss, P. R. & Renstrom, P. J. (1995). Imprint of sub-25 nm vias and trenches in polymers. *Appl. Phys. Lett.*, 67. pp. 3114-3116.

Fujita, J.; Ohnishi, Y.; Manako, S.; Ochiai, Y.; Nomura, E.; Sakamoto, T. & Matsui, S. (1997). Calixarene Electron Beam Resist for Nano-Lithography. *J. Appl. Phys.*, 36. pp. 7769-7772.

Hosaka, S.; Sano, H.; Shirai M. & Sone, H. (2006). Nanosilicon dot arrays with a bit pitch and a track pitch of 25 nm formed by electron-beam drawing and reactive ion etching for 1 Tbit/in^2 storage. *Appl. Phys. Lett.*, 89. pp. 223131.

Ishii, K. & Matuda, T. (1992). Sub-100-nm-Scale Patterning Using a Low-Energy Electron Beam. *Jpn. J. Appl. Phys.* 31.pp. L744-L746

Namatsu, H. (2001). Supercritical resist drying for isolated nanoline formation. *Jpn. Vac. Sci. Technol.*, B 19. pp.2709-2712.

Olkhovets, A. & Craighead, H. G. (1999). Low voltage electron beam lithography in PMMA. *J. Vac. Sci. Technol. B*, 17. pp. 1366-1370.

Ocola, L. E. (2003). Soluble site density model for negative and positive chemically amplified resists. *J. Vac. Sci. Technol. B*, 21. pp. 156-161.

Ueno, H.; Hosaka, M.; Zhang, Y.; Tabata, O.; Konishi, S. & Sugiyama, S. (1997). Study on fabrication of high aspect ratio microparts using the LIGAprocess. *Proceedings of International Symposium on Micromechatronics and Human Science*, pp. 49-54.

28

Three-Dimensional Patterning using Ultraviolet Nanoimprint Lithography

Maan M. Alkaisi and Khairudin Mohamed
*MacDiarmid Institute for Advanced Materials and Nanotechnology,
Department of Electrical and Computer Engineering,
University of Canterbury, Christchurch 8014,
New Zealand*

1. Introduction

Although an extensive number of publications have been reported on nanoimprint lithography (NIL) techniques, the ability of NIL for three-dimensional (3-D) patterning has not been fully addressed in terms of the mold fabrication and imprint processes. Developing technologies for patterning 3-D and multilevel features are important because they eliminate multiple steps and complex interlevel alignments in the fabrication process of nanoscale devices and structures.

The semiconductor industries through the International Technology Roadmap for Semiconductor (ITRS) organization have identified NIL (ITRS 2003; ITRS 2008), especially ultraviolet curable nanoimprint lithography (UV-NIL) as a strong candidate for the next generation lithography (NGL) technology for nodes down to 5 nm.

Three dimensional NIL (3D-NIL) will have a variety of practical applications including generating patterns for MEMS and NEMS devices, on-chip optics, antireflection structures and in biochip reactors.

This chapter explores the 3-D patterning capability using an UV-NIL technique. The 3-D features and multilevel mold design, fabrication, and imprint processes have been studied and analysed and outcome will be presented and discussed.

In the UV-NIL technique, a transparent mold with micro/nanostructure patterns on its surface is allowed to be printed and replicated on UV curable polymer without the need of high applied pressure or temperature. UV-NIL has the potential to fabricate micro/nanostructures with high resolution, high reproducibility, low cost, high throughput and in addition is capable of 3-D patterning. Its resolution is not limited by diffraction as in optical lithography or electron scattering as in electron beam lithography (EBL) and there is no expense that scales with reducing dimensions as in other nanolithography technologies.

The UV-NIL technique is especially useful in avoiding processes that require high pressure and high temperature cycles. Mold pattern writing, pattern transfer and imprint are the three major areas that are described here. The master molds or dies are made using a high-resolution but low-throughput EBL technique. The pattern transfer process for making the mold is achieved using a reactive ion etching (RIE) technique and subsequent imprint lithography is employed for the replication of the micro/nanostructures onto the daughter molds.

The 3-D pattern profiles are created on a negative photoresist using a Raith 150 EBL tool in a single step variable e-beam dose controlled exposure. Resist contrast curves are obtained with a negative photoresist (ma-N2403) and a single pass line (SPL) pattern writing scheme is employed to create the 3-D pattern profiles. Specially designed test patterns are used to demonstrate the capability of UV-NIL in replicating the 3-D and multilevel structures.

Surface charging, electron back scattering and proximity effects during e-beam exposure are the main factors influencing pattern writing on insulating materials such as quartz. Multiple approaches for suppressing the charging, electrons backscattering and proximity effects will be presented in section 3.2. Monte Carlo simulation using CASINO software is employed to investigate the effects of accelerating voltages on electron trajectories during the e-beam exposure.

As a substrate material that has numerous applications at micro/nano scale, quartz is the most common choice for many optical device applications. It is widely used as a UV transparent imprint mold. 3-D pattern transfer onto quartz substrate will be described in section 3.3. The optimization process to achieve a very low surface roughness of below 5 nm suitable for UV exposure and faithful pattern transfer will be presented in section 3.4. In this section an indepth explanation of the underlying science of etching hard and insulating materials such as quartz will be discussed.

2. Research background

In general, 3-D micro/nanostructures, including 3-D NIL molds, can be fabricated using various techniques such as multistep lithography , maskless gray-scale lithography (Totsu et al. 2006), FIB writing (Taniguchi et al. 2006), proton beam writing (Simcic et al. 2005) , x-ray lithography (Romanato et al. 2003); (Romanato et al. 2004) , or combination of EBL and dry etching. These techniques are high cost, low throughputs and some suitable only for relatively large features and not for nanoscale applications. Another method of fabricating 3-D structures is by the surface texturing technique using either dry plasma RIE process (Kumaravelu et al. 2004) or wet chemical etching. A 3-D sub-wavelength surface texturing can also be fabricated using interferometric lithography patterning followed by RIE (Chiu et al. 2006) but these are used for periodic patterns only. As indicated 3-D patterns with structures down to sub-wavelength scale can be fabricated using a number of advanced lithography techniques, however, their reproducibility in mass production, low cost environment is the major limitation.

In EBL, patterns are normally created in a graphic or layout design software such as L-Edit. The patterns may be grouped in multiple clusters called design layer based on their feature sizes, electron dose and acceleration voltage. For 3-D patterning, there are two approaches for defining the patterns: variable acceleration voltage and variable exposure dosage. The variable acceleration voltage technique (Ishii & Taniguchi 2007) uses a specific voltage assigned to a specific pattern design layer where the limitation is set on the maximum number of the possible assigned design layers. When a positive resist such as PMMA is exposed to the e-beam, the depth of the e-beam penetration is determined by the e-beam accelerating voltage as illustrated in Figure 1(a). In this method, for example, the depths of the developed 3-D pattern are controlled by the magnitude of the uniform accelerating voltage assigned to each pattern layer. After pattern development, the depths of the trenches formed are a factor of the uniform accelerating voltage assigned to each design layer.

On the other hand, the variable dosage technique employs an e-beam exposure approach with path by path routing using a single pass line (SPL) scheme, with a specifically assigned dosage for each path as illustrated in Figure 1(b). In this method everything are designed in a single design layer. The 3-D profiles can be achieved on negative resist by bringing the SPL paths close (but not less than e-beam spot size) and parallel to each other. There is no limitations on the number of paths and all paths are designed in a single design layer for any 3-D pattern design (Mohamed et al. 2007).

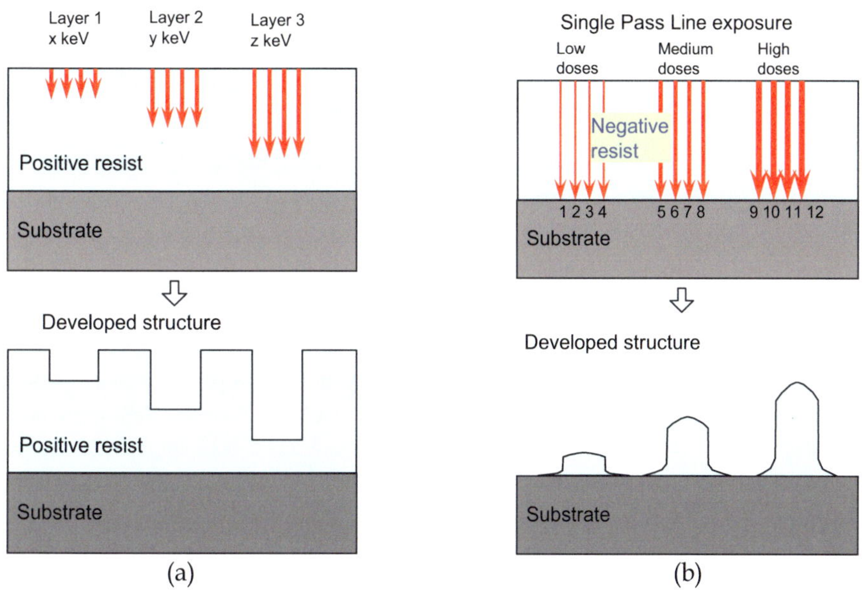

Fig. 1. The two different methods of 3-D patterning using the EBL technique, (a) variable accelerating voltages and (b) variable e-beam dosages.

For 3-D patterning, negative resist such as ma-N2403 from Microresist Technology GmbH can be used for this purpose. Previous work (Konijn et al. 2005) has shown some encouraging results for 3-D and multilevel structures on other resist system. An experiment was designed to study the relationship between the e-beam dosages and the effects of electrons backscattering and proximity effects onto the developed ma-N2403 resist profile. A single pass line (SPL), double pass lines (DPL) and quadruple pass lines (QPL) schemes with various e-beam dosages range were attempted. For parallel multiple line exposure, the spacing between the lines' exposures were about 20 nm to 30 nm. In a group of uniform dosages for each exposure lines, it was observed that the middle of the developed resist structures received higher exposure dosages because of the electrons' backscattering and proximity effects. Figure 2(a) shows an AFM trace of resist cross sections for a range of uniform dosages applied to a QPL exposure scheme. It shows that the remaining resist thickness after development is a factor of the dosages applied.

Figure 2(b) illustrates the contrast curve of the negative resist ma-N2403 when exposed with SPL, DPL and QPL exposure schemes. The SPL, DPL and QPL curves represent an exposure conditions with zero, medium and high proximity effects respectively. There were large differences in the structure height between SPL and DPL writing schemes. For no proximity effect as in SPL, the structure heights were very low. These data have been used for estimating the required dosage factors in designing a 3-D pattern.

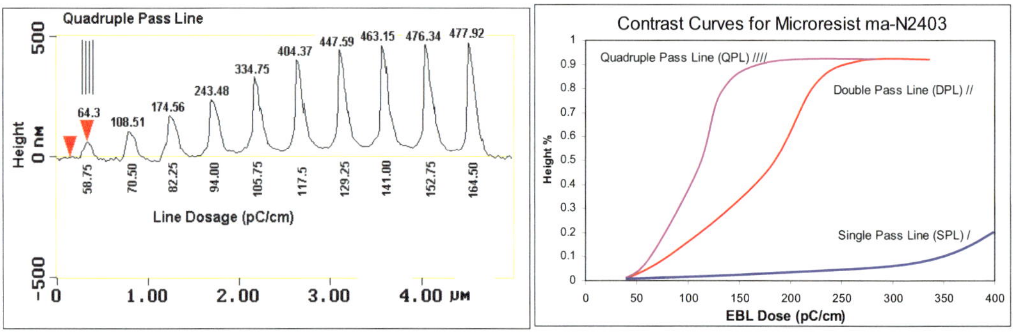

(a) (b)

Fig. 2. (a) AFM traces of the developed ma-N2403 after being exposed with a range of e-beam dosages using QPL scheme and (b) the contrast curves of negative resist ma-N2403 for SPL, DPL and QPL exposure schemes.

3. Mold fabrication

The three-dimensional mold fabrication involves writing a 3-D profile on negative photoresist and then directly transfers the pattern using single-step RIE process. The creation of the 3-D geometrical shapes such as pyramid, hemisphere, cone and other complex shapes will be described in this section. These 3-D structures have many applications especially in optical devices.

A subtractive pattern transfer method (RIE) was employed because the 3-D patterns have to be directly transferred onto the mold substrate in a single-step RIE process to simplify the mold making process. A negative photoresist is employed as a 3-D imaging and masking layer for the EBL and RIE process repectively.

New fabrication processes have been developed by the present work in order to suppress the surface charging effects during the EBL 3-D pattern writing on insulating substrates. Since the 3-D mold fabrication requires direct pattern transfer from mask to substrate, depositing a thin metal coating on the substrate-resist interface as a charge dissipation/grounding layer was not possible. Depositing a thin metal layer on top of the negative resist surface as a charge dissipation/grounding layer was also ruled out. Acidic solution used to remove the thin metal layer prior to pattern development process caused polymer crosslinking on the negative resist surface (Thackeray et al. 1989); Shaw et al., 1997). As a result, no subsequent development process would be possible. For these reasons, the use of conductive polymers was investigated as will be illustrated in the following section 3.2.2.

3.1 Mold design and preparation

The 3-D pattern layout was designed using Raith-150 EBL pattern layout software where the single pass line (SPL) writing scheme is utilised. Figure 3(a) shows an example of a writing scheme for a hemispherical shape. Every SPL path/route was assigned with an individual dosage which was defined through the optimisation process. The spacing between lines is about 20 nm to 30 nm. The developed resist profile is the combined result of proximity effects, back scattering and e-beam dosages. To achieve the desired profile, a trial and error method was employed to optimise the e-beam dosages after a number of experimental

cycles to satisfy the profiles. Figure 3(b) shows another example of a writing scheme for a pyramid profile with a similar method.
The negative tone photoresist from Microresist Technology, ma-N2403 was spun coated on a cleaned quartz substrate at 3000 rpm for 30 seconds and pre-baked in an oven at a temperature of 95 °C for 30 minutes to achieve a 600 nm layer thickness.

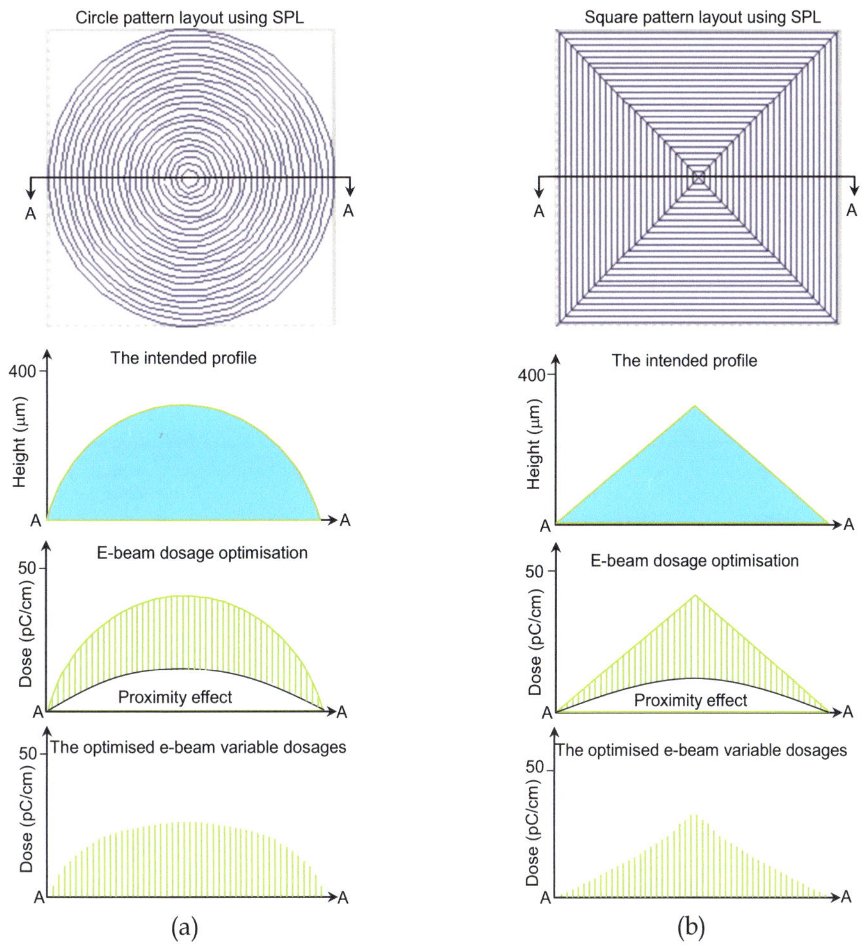

Fig. 3. The SPL writing scheme for 3-D patterns. (a)hemisphere and (b) square base pyramid.

3.2 Fabrication techniques

As mentioned earlier, surface charging is one of the major challenges in patterning on insulating substrate such as quartz. This is even more complicated when the required pattern is in 3-D. Surface charging trapped near the insulating surfaces causes great difficulties in focusing the e-beam onto the resist surface and the unbalanced surface potential of the resist deflects the beam and causes severe pattern distortion (Joo et al. 2006).
Two fabrication approaches were developed to minimise this problem. The first approach is applying e-beam at low accelerating voltages (low keV). The use of low keV leads to reduced beam interaction volume. The penetration depth of the electrons into insulating

materials/substrates is much less than higher keV electrons. Based on this approach a 3-D fabrication method called called Critical Energy (CE) or Critical Accelerating Voltage method was developed and followed (Mohamed et al. 2007). In the CE method as illustrated in Figure 4(a), a variable controlled e-beam dosages at an optimised low accelerating voltage is employed in 3-D pattern writing on a negative resist using EBL. The developed pattern is then used as a masking layer for a single-step RIE pattern transfer process onto quartz substrate.

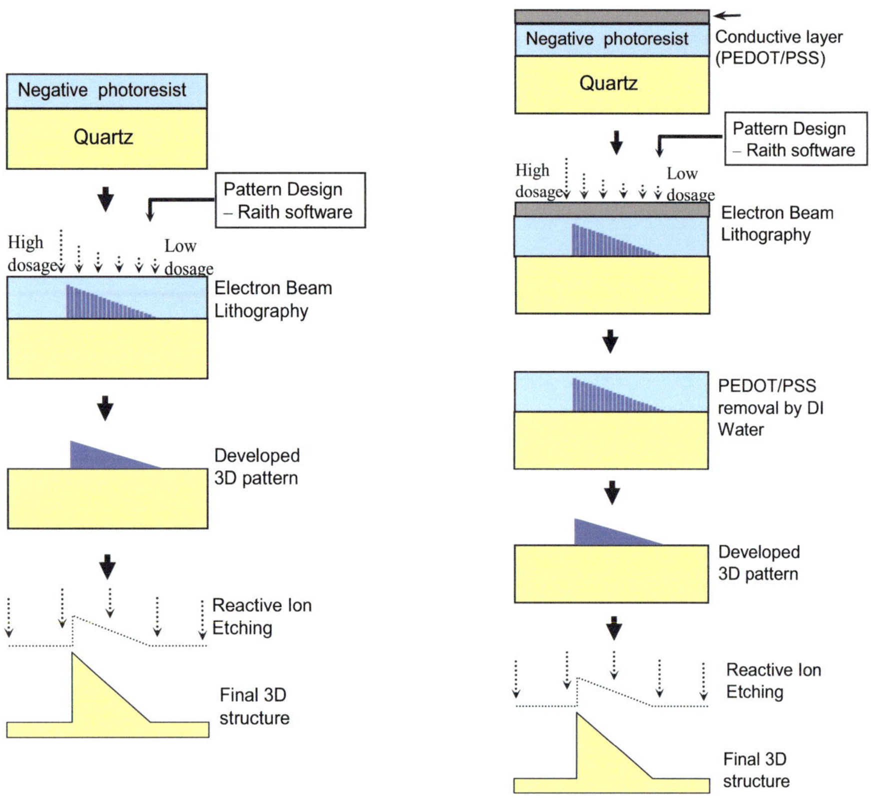

Fig. 4. Flow diagram for the fabrication process employed for 3-D structures defined on insulating substrates using (a) the critical energy method and (b) the high acceleration energy in conjunction with conductive polymer coating method.

The second approach is an application of high accelerating voltages (high energy/keV) in conjunction with the use of top conductive coatings on the resist layer. The use of high keV will produce a greater penetration depth and reduce the number of back scattered electrons which will result in a more confined beam size. Based on this approach, a Top Conductive layer method (TC) was developed and adopted in this work. In the TC method as illustrated in Figure 4(b), a water soluble conductive polymer, poly(3,4-Ethylendioxythiophene) /poly(suyrenesulfonate) or PEDOT/PSS is spun coated on the negative resist as a charge dissipation layer prior to the 3-D pattern writing. A high acceleration voltage that provides a more confined e-beam and less backscattering and proximity effects is used in exposing the 3-D pattern on a negative resist. PEDOT/PSS layer was removed by rinsing with de-ionised water (DIW) prior to the development process. Finally the developed pattern is used as a masking layer in a single-step RIE pattern transfer process.

3.2.1 The critical energy or critical accelerating voltage

Many insulating materials often have certain charge levels where they reach equilibrium state when exposed with an e-beam. This charge equilibrium condition is reached when the charge into the sample material equals to the charge out of the sample. The charge out of the sample will be from secondary electrons (SE), backscattering electrons (BSE), Auger electrons, x-ray and whatever current is absorbed and then transmitted through the sample to the ground. This is referred to as a state of unity (Frey 2007). It can be described as in Figure 5 where the critical energy assumes a neutral charge at points E1 and E2 as suggested by J. Joo and co-workers (Joo et al. 2006). Polarity of charging is dependent on the e-beam configuration.

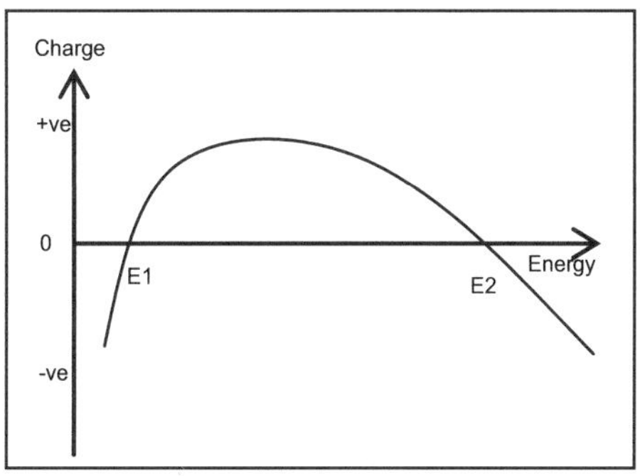

Fig. 5. The schematic diagram of the e-beam critical energy (Joo et al. 2006).

To estimate the range of suitable low energy for writing a 3-D pattern on nonconductive substrates, Monte Carlo simulation using CASINO software (CASINO Software 2006) was employed. Figure 5 illustrates the 200-electron trajectories simulation of electron beams with accelerating voltages of 4.5, 5.5, 6.5 and 20 keV. In these simulations a 600 nm thick ma-N2403 resist layer coated on a quartz substrate was assumed to be exposed by an e-beam.

The volume of electrons interaction increases when the e-beam accelerating voltage increases. At accelerating voltage of 4.5 keV as illustrated in Figure 6(a), the penetration depth of electron trajectories is far from reaching the quartz substrate surface and fully expose the resist. The volume of crosslinked negative resist after the e-beam exposure is therefore present only at the top thin layer of the exposed resist. This leads to a layer of unexposed resist close to the quartz surface and will be washed away during the developing process at a later stage.

Further simulations were carried out to find a suitable/critical accelerating voltage that will crosslink resist all the way down to the quartz surface and neutrally charge the resist. The observation from the simulation shows that high propotion of the electron trajectories settled at about 85% of the longest vertical distance. For example, as illustrated in Figure 6(b), the penetration depth of 5.5 keV accelerating voltage was estimated at approximately 550 nm which is less than the resist thickness (600 nm). At this condition, the resist surface could be negatively charged.

Simulation at an accelerating voltage of 6.5 keV as illustrated in Figure 6(c), the penetration depth is estimated at approximately 660 nm which is larger than the resist thickness of

600nm. At this condition, the resist surface could be positively charged. Based on this analysis, the critical e-beam accelerating voltage for a 600 nm thickness of ma-N2403 resist on quartz substrate is in the range of 5.5 to 6.5 keV. This method can be used as an estimation in finding the critical e-beam accelerating voltage that assumes a neutral charge for certain resist thicknesses. Accordingly, an accelerating voltage of 6.25 keV is suitable for a 600 nm resist thickness exposure. Similar approach can be followed for other resist system.

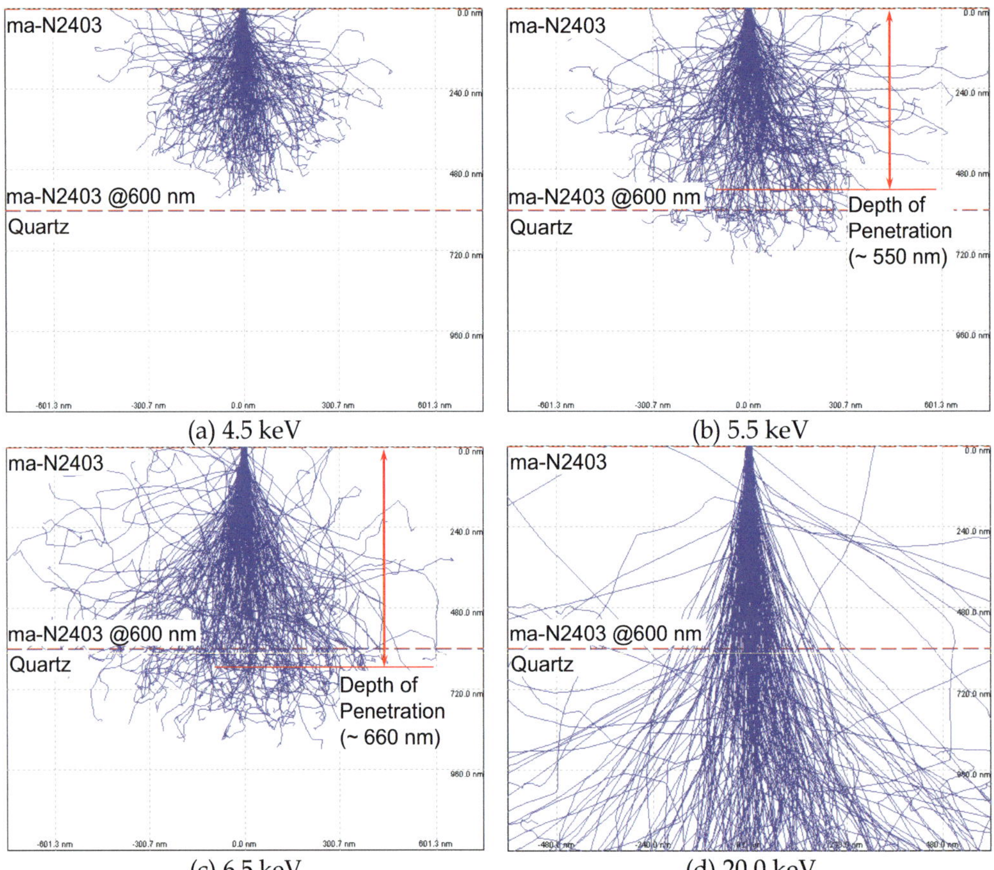

Fig. 6. Electron trajectories of various e-beam acceleration voltages across ma-N2403 resist on quartz substrate by employing Monte Carlo simulation using CASINO software.

Figure 6(d) shows the simulation of electrons trajectories when a high accelerating voltage of 20 keV is used for exposure. A very deep penetration, more confined trajectories, less electrons backscattering and less proximity effects are observed. This is useful for patterning smaller and higher density patterns which will be discussed in the next section.

Charging comes in several species often exhibiting several characteristics, either positive or negative charging. When the sample is charged positively, the image will appear dark and when there is a negative charge, the image will appear bright (Frey 2007).

Further experiments were carried out on a range of ma-N2403 resist layer thicknesses from 350 nm to 800 nm using a Raith-150 EBL tool. The actual critical energy values were determined by gradually increasing the accelerating voltage from the lowest values as

estimated by Monte Carlo simulation to the optimised level. To find an optimized or critical accelerating voltage that leaves no charging marks on the zoomed resist image under scanning electron microscopy (SEM) conditions, images obtained of resist between 200X and 500X magnifications were then zoomed in and out for each accelerating voltage increments.
Figure 7 shows the critical acceleration voltage with respect to the resist thicknesses penetration that was estimated by the Monte Carlo simulations as compared to the experimental results. There were no significant differences between the two and they can be used as a guideline for e-beam exposure at any resist thicknesses.

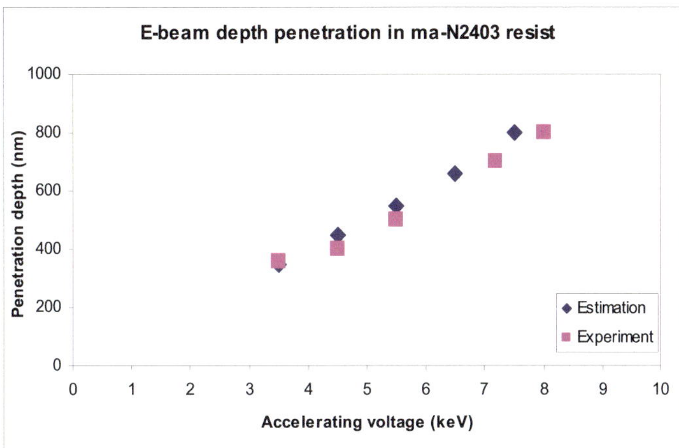

Fig. 7. A comparison of e-beam depth penetration between estimation using CASINO software and experiments, showing penetration depth increases with acceleration voltage.

As explained earlier, the e-beam voltage acceleration of 6.25 keV has been found to be the optimum to realise the minimum charging effect and therefore used for the 3-D patterning on the 600 nm thick ma-N2403 resist. The SPL pattern writing scheme was utilized for 3-D patterning with a line dosage ranging from 0.6 to 42.0 pC/cm. Examples of SPL pattern writing scheme for 3-D structures have been explained in section 3.1. After the e-beam exposure, the samples were developed in Microresist ma-D532 developer for 15 seconds at a temperature of 20 ^{o}C.
Figure 8(a) shows an AFM image of the developed 3-D ring structure test pattern defined on an ma-N2403 resist. This test pattern has three 3-D shaped structures. The first is the centre doom shaped lens 3 microns in diameter and 300 nm on its highest point. The second is the middle 3-D ring, which is 5 microns in inner diameter and 9 microns in outer diameter with a ramp down profile from the highest point of 300 nm at its inner diameter and zero at its outer diameter. The outmost 3-D ring structure which is 11 microns in inner diameter and 13 microns in outer diameter with a ramp down profile from the highest point of 300 nm at its inner diameter and zero at its outer diameter. This test pattern has a potential in micro Fresnel lens fabrication and other optical device applications.
Figure 8(b) shows an AFM image of the developed 3-D pyramid structures with a multilevel array test pattern defined on an ma-N2403 resist after development process. Each outer pyramid has a one micron square base and is 300 nm in height. There are four similar pyramids in the center but at an elevated level of 300 nm. The pyramid array test pattern represent what could be used as light trapping surface texturing structures for solar cells

applications. Multilevel structures are important in eliminating multistep processes and multilevel alignments reducing process complexity and lowering cost.

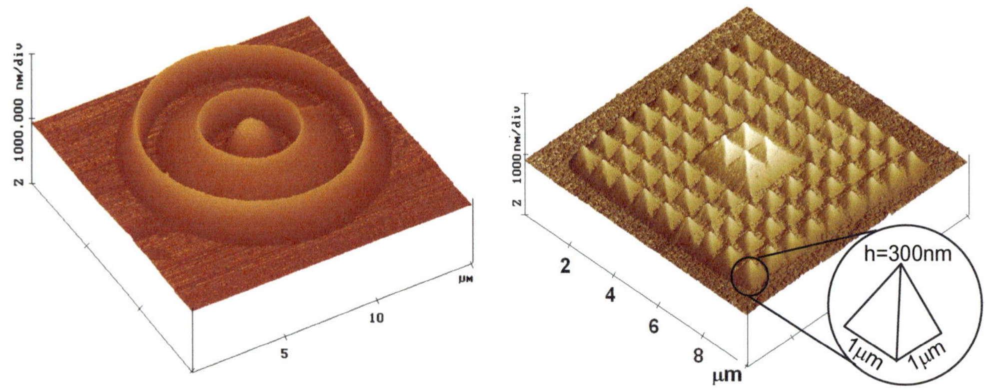

(a) (b)

Fig. 8. AFM images of the developed 3-D structures defined on ma-N2403 resist using critical energy technique (a) 3-D ring structure and (b) 3-D multilevel pyramid array.

3.2.2 Top conductive dissipation layer using PEDOT/PSS

Conducting polymers, particularly the soluble derivatives, are other potential alternatives as a charge dissipators layer for EBL pattern writing on insulating materials. We will describe in this section using a water soluble conductive polymer, PEDOT/PSS as the top conductive coating on the negative resist layer. PEDOT/PSS was spun coated on top of the ma-N2403 resist at 5000 rpm for one minute, and baked on a hot plate at a temperature of 90 ^{o}C for two minutes to achieve a thickness of about 30 nm (Mohamed et al. 2009).

The effectiveness of this material in supressing charging was verified through the exposure experiments. Figure 9 shows SEM images of quartz substrates exposed with an e-beam at a high acceleration voltage of 20 keV and with 20 µm aperture. The bottom half is an illustration of the electron trajectories penetrating the resist layer. Figure 9(a) shows an SEM image of quartz substrate coated with negative photoresist only, without the conductive polymer layer; these display the characteristic bright area of built up negative charges. No objects can be resolved for imaging or writing under the influence of surface charging. Figure 9(b) shows an SEM image of a substrate coated with negative photoresist anda top conductive coating PEDOT/PSS after it has been exposed with e-beam. No charging effects were observed at the resist surface as is evidently shown by the confined 20 nm dot where surface charging is significantly suppressed.

The surface charging build up phenomena could be explained by refering to Figure 9(c), which illustrates the schematic diagram of the electrons path that leads to secondary electron emission and built up charges by the trapped electrons. In comparison, Figure 9(d) illustrates the schematic diagram of the electrons path with the existence of a PEDOT/PSS conductive layer. Most trapped and built up electron charges are grounded through the PEDOT/PSS conductive layer.

This 3-D fabrication method was developed to overcome the limitation of the previous CE method where the creation of smaller feature below 500 nm have been hindered by electrons back scattering and proximity effects. In 3-D patterning on negative resist, the developed resist height at a particular spot is determined by the magnitude of resist crosslinking at that

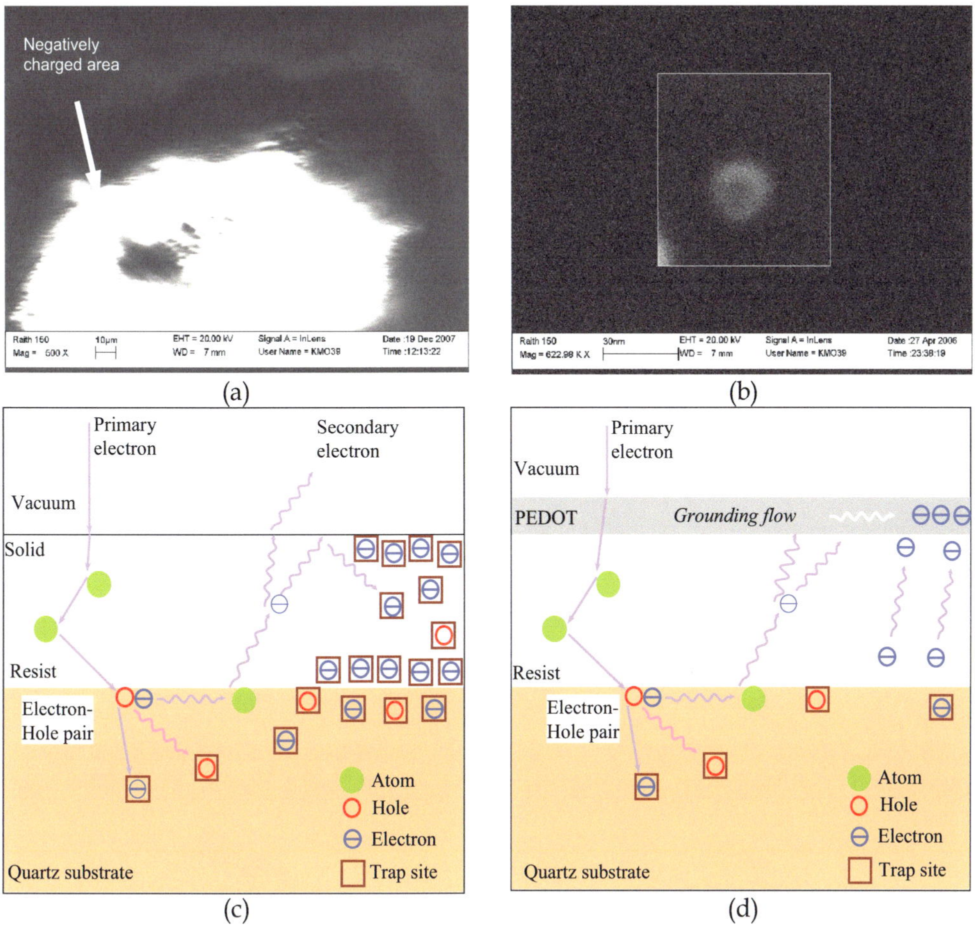

Fig. 9. A comparison between resist surfaces without and with a PEDOT/PSS conductive polymer layer. At the top, SEM micrograph of resist surface (a) without charge dissipator layer showing bright area (negatively charged) and (b) with charge dissipator layer of PEDOT/PSS. At the bottom, schematic diagrams showing possible electron trajectories for resist exposed to e-beam. (c) electron trajectories without the charge dissipator layer and (d) with the charge dissipator layer.

location due to direct e-beam exposure plus the proximity effects from adjacent exposures. At lower voltage acceleration of below 10 keV, the spread of the e-beam is quite large (up to 3 μm) as referred to Figure 5. Hence, the developed resist height at a particular spot is heavily affected by multiple electrons' backscattering and proximity effects caused by line exposures near to the spot. Exposing a line with low dosage does not necessarily produce a lower resist height if the surrounding lines are exposed with high dosages. Manual optimisation of line dosages in order to achieve certain resist height in 3-D patterning requires a long iteration experiment process owing to multiple electrons' scattering and proximity effects from surrounding exposures.

In the following method, a higher voltage acceleration of 20 keV has been employed as compared to 6.5 keV used in previous method that followed the CE method. The single pass line (SPL) pattern writing scheme was utilized for 3-D patterning with line dosages starting at 0.6 pC/cm up to 42.0 pC/cm. Figure 10(a) shows an example of a 3-D pattern writing scheme for a 300 x 300 nm^2 base pyramid. It was created using e-beam writing on the negative resist by routing the multiple SPL parallel paths with various dosages assigned for each SPL (path 1 to path 6). The spaces between each SPL route ranged from 20 nm to 30 nm. Figure 10(b) shows an AFM image of the developed 3-D pyramid test pattern on a ma-N2403 resist. Each pyramid has a 300 x 300 nm^2 base and is 300 nm in height.

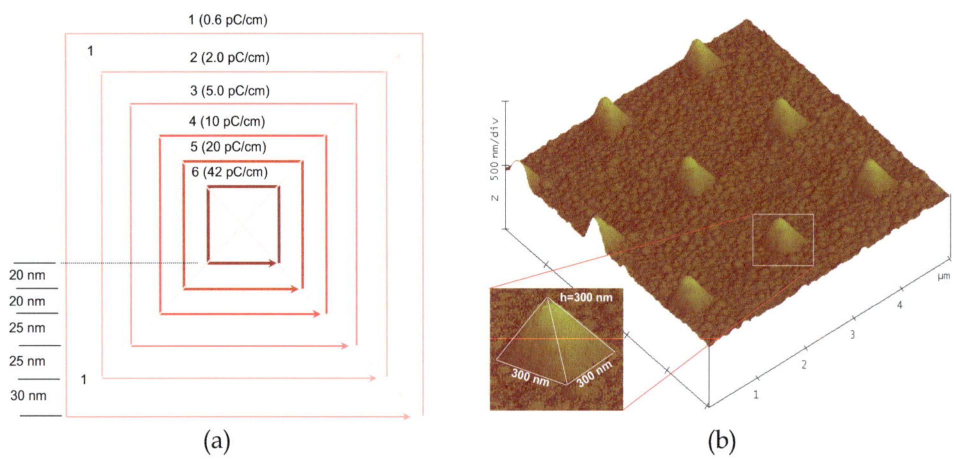

Fig. 10. EBL writing scheme for 3-D patterns (a) SPL routing for writing a 300 x 300 nm^2 pyramid and (b) AFM image of the exposed and developed 300 x 300 nm^2 pyramid array.

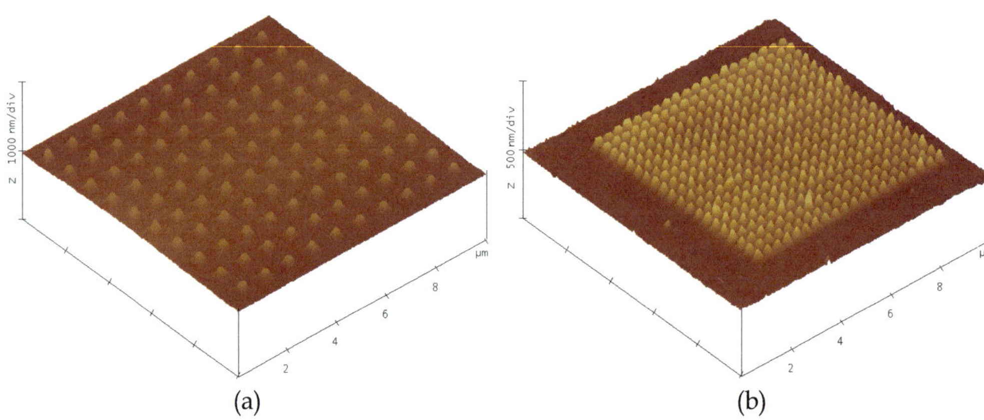

Fig. 11. Maximum pattern density achieved using the two pattern writing methods (a) critical energy method and (b) top conductive coating method, where much higher densities are evident.

The advantage of exposing an e-beam at a higher acceleration voltage is the minimisation of electrons' scattering and lowering of proximity effects on the surrounding exposed region as illustrated by confined electron trajectories as shown in Figure 6(d). This results in a better control over the developed profiles, less time required for the dosage optimization process, and the achievement of higher pattern density of 3-D patterns is possible. Figure 11(a) shows an AFM image of an array of a 500 x 500 nm^2 pyramid structure with 800 nm pitches. These are the maximum density pitches achieved using the previous pattern writing method where low accelerating voltage of 6.25 keV is used. In this example the electrons backscattering and proximity effects have limited the density 3-D structures. Figure 11(b) shows an AFM image of an array of cone structure (300 nm base diameter) with 400 nm pitches which were achieved using a TC pattern writing method. A much higher density 3-D pattern is possible with the proximity effects suppression using a combination of higher accelerating voltages and top conductive coating. This high accelerating voltage allows deeper penetration, more confined beam, and less electrons backscattering and proximity effects.

3.3 Pattern transfer

The developed 3-D patterns have been directly transferred into the quartz substrate by a single-step RIE with suitable selectivity of resist-to-substrate by optimising the etching process parameter. CHF_3/Ar gases were chosen for quartz etching based on their moderate F/C ratio which is important in the 3-D pattern transfer.

RIE Parameter	Setting/measured
Gas	CHF_3/Ar
Flow rate	9.0/6.25 sccm
Pressure	< 6 mTorr
Temperature	295 K
RF Power	125 W on 100 mm electrode
Bias voltage	-344 V
Selectivity resist-to-substrate	1:2
Etch rate	10 nm/min on quartz

Table 1. The optimized CHF_3/Ar RIE parameter for 3-D pattern transfer onto quartz substrate.

We will describe here how to achieve a smooth linear gradient or curved 3-D surface profiles of the fabricated mold with a surface roughness of less than 5 nm after the RIE pattern transfer process. This is an essential requirement for micro optical devices where the general guidelines required the surface roughness have an upper limit of 5 nm.

It is a challenging task to create a relatively tall 3-D structure using the EBL 3-D patterning technique owing to the electrons back scattering and proximity effects even with charge suppression approaches. Hence, an upper limit of 600 nm is imposed on the 3-D structures in this case. This aspect ratio has been achieved through appropriate resist to substrate selectivity. The RIE pattern transfer parameters were optimised to yield 1:2 resist-to-substrate selectivity ratio which is suitable for transferring the 3-D pattern on quartz. Table 1 shows the final optimized RIE parameter employed to achieve a smooth surface finish with surface roughness of less than 5 nm and 1:2 selectivity.

3.4 Etching analysis

Plasma discharges involves quite complex mechanisms and can be difficult to interpret or even understand in some cases. In fluorine–containing plasma for example, surface reaction, etching, and polymerization can occur at the same time. The domination of certain reactions is dependent on the gas feed, the operating parameters and the chemical nature of the polymer/substrate, electrode and geometry (Chan et al. 1996). In a quartz etching process using CHF_3 etchant, the free fluorine radicals, F* are created by the plasma discharge and the etching chemistries can be described as follows;

$$e^- + CHF_3 \rightarrow CHF_2^+ + F^* + \text{radicals} + 2\,e^- \quad (1)$$

$$SiO_2 + xF^* \rightarrow SiF_x + O_2 \quad (2)$$

$$CF_x \text{ radicals} + 2O_2 \rightarrow CO + CO_2 + COF_2 \quad (3)$$

F* is the reactive fluorine atom and SiO_2 is the quartz substrate. RIE plasma generates reactive fluorine atom F* from the supplied CHF_3 gas (Eqn 1). The etching of quartz consumes the F atom to form a SiF_x (Eqn 2). One of the radicals, the CF_x radicals, tend to deposit polymer films on all surfaces under certain condition, but the oxygen liberated in the etching of quartz reacts with CF_x radicals to form volatile CO, CO_2, and COF_2 (Eqn 3). In 3-D pattern transfer, the presence of resist which is polymer based acting as the 3-D masking layer, will increase the C content. The backbone of resist polymer consists of $C_xH_yO_z$ elements which react with plasma gases, creating the volatiles of these elements. This process decreases the F/C ratio, causing domination of polymerization reactions, in other word deposition instead of etching. Figure 12 illustrates the 3-D etching chemistry analysis. The reactive ions bombarding the 3-D surface in a vertical downwards direction, irrespective of the resist pattern shape of profile.

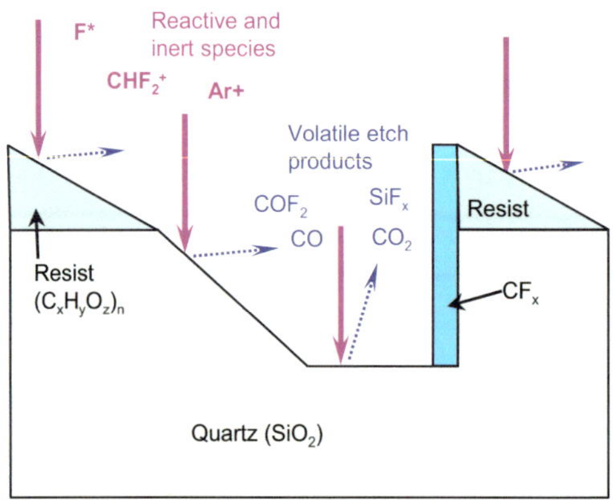

Fig. 12. The schematic of 3-D etching chemistry, showing the fluorinated reactive gases and the by-products during etching quartz.

Let us consider a plasma etching on an inclined quartz surface at an angle $\boldsymbol{\theta}$ where the plasma bombardments are most likely to occur in a vertical downwards direction as illustrated in Figure 13(a). Ion bombardments on the ramped surface will bounce/backscatter in a downhill direction. The bounced ions from the inclined substrate surface collided with the incoming ions from the top causing another bouncing down the slope.. Cumulatively, the plasma concentration will increase towards the downhill direction. Higher plasma concentration induced a higher etch rate (about 10 nm/min) as illustrated in Figure 13(b). The steeper the surface slope is the higher and the narrower the plasma concentration at the bottom of the slope at location D. Moreover, during resist etching, the ion/resist ratio at the thin resist side caused a higher etching rate on the thin resist layer as compared to the thick resist layer. As for the 3-D masking layer, the different resist masking thicknesses will cause a varied etching rate on every location.

Backscattered and bounced ions towards the vertical or sloped walls will collid with the vertically incident ions creating a higher concentration of ion bombardment at the bottom of the structures, thus causing a trenching effect as illustrated in Figure 13(c). At location C, the trenching effects result in a wider channel as compared to the narrow trench observed at location D. Location D represents the trenching effect of an inclined surface when the inclined angle $\boldsymbol{\theta}$ is approaching 90°.

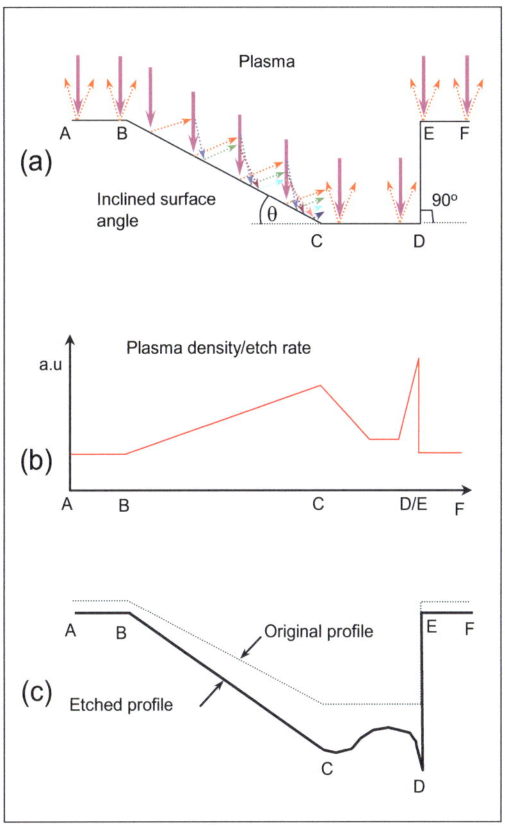

Fig. 13. The 3-D etching mechanism showing the trenching effects at the bottom of the structure.

Figure 14(a) shows an SEM image of the centre of the 3-D ring shape on quartz after the CHF_3/Ar RIE process. It shows the final 3-D ring etched profile with the lens shape in the middle and the vertical sidewall profile for outer ring. The trenching effects formed around the 3-D lens and at the bottom of the sloped 3-D ring structure are observed. Figure 14(b) shows the AFM cross sectional traces of the 3-D ring mold structure with trenching effects. Due to the deep trench profiles it is not possible to trace perfectly the actual topography owing to the limitation of the AFM tip itself.

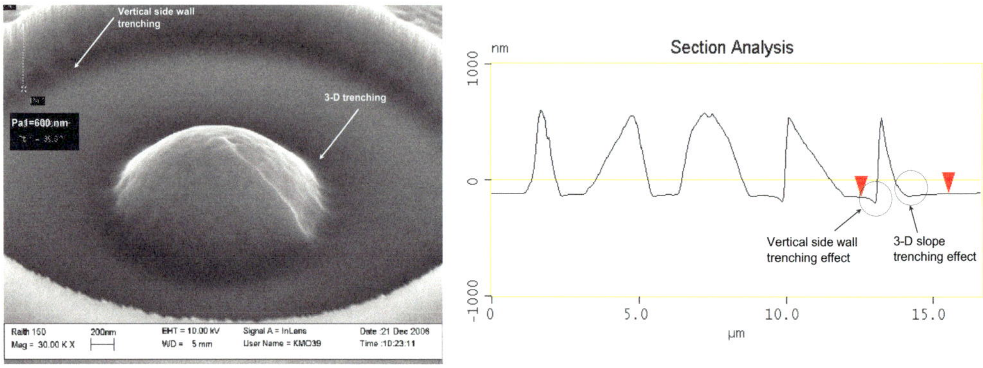

(a) (b)

Fig. 14. Trenching effects (a) SEM image of the central dome shaped lens and the 3-D ring etched profile and (b) AFM traces of cross section into the structure.

In etching hard materials such as quartz where polymerization reactions dominate, high plasma density is essential to increase the etching rate, especially for a deep etching in a 3-D pattern transfer process. High plasma density can be achieved by increasing the RF power and/or reducing the electrode surface area (loading effect). The maximum RF power that can be set on an Oxford Plasma 80plus RIE is 200 Watts. In this experiment, a NiCr coated electrode was clamped with a two-inch (50 mm diameter) PTFE electrode holder. Hence, the maximum plasma density is set to the highest setting of 10.2 W/cm^2 with this arrangement. In this high density etching process where polymerization reactions dominate, some of the etched materials may drop/deposite back on the sample forming a nanomasking island. This will form nanopillars as illustrated in Figure 16 and will be discussed later.

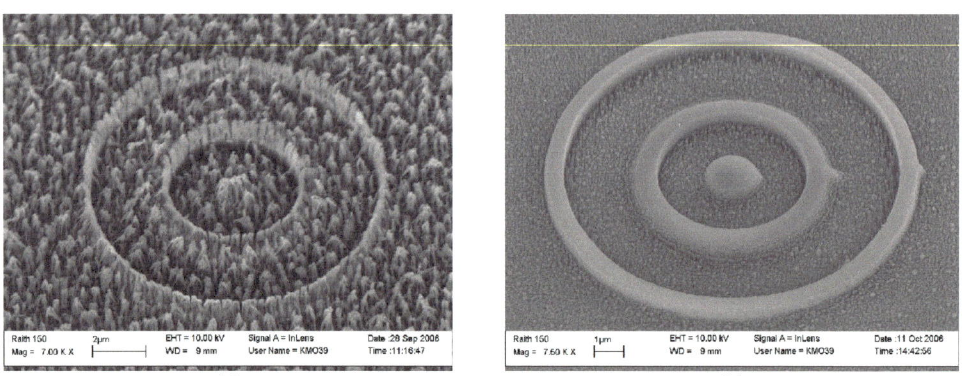

(a) (b)

Fig. 15. SEM images of the etched 3-D structure on quartz substrate. (a) and (b)

The highly dense nanopillar structures that formed as a result of the plasma etching process will finally collapsed after high density plasma bombardment over a long period of time. Figure 15(a) shows an SEM image of a 3-D ring structure when etched with the high density plasma setting of 10.2 W/cm^2 for 40 minutes. The result shows a very rough surface and a vertical sidewall passivation layer can be clearly observed. It can be concluded that increasing the plasma and/or power densities above certain levels to achieve high etching rates would not achieve a smooth etched surface as desiredfor the 3-D patterns used for optical devices or for UVNIL molds.

In the following experiment, the plasma density was reduced to a normal condition by changing the electrode area (by changing the size of the clamp ring) to the 100 mm diameter electrode holder which gives a plasma density of 2.5 Watts/cm^2 with RF power of 200 W. It is found that the surface roughness is improving by lowering the RF power density as shown in Figure 15(b).

Since this etching process was dominated by polymerization reactions, CF_x radicals tended to be deposited anywhere on the etched surface. At an operating pressure of 20 to 30 mTorr, the mean free path of the plasma species is very short (less than the distance between the sample and the RIE anode of about 120 mm) and etching species are in close proximity to each other causing continuous collisions. In this saturated condition, the species produced as a result of ion bombardment of the resist mask material and substrate surface will reach a level at which the vacuum pump is unable to extract away these by-products. The lack of vacuum pumping capacity will lead to dropping of the etching products and particles back onto the substrate surface causing nanomasking effects as illustrated in Figure 16(a) and (b). From the above experiments, it was found that the etching process was limited by the vacuum pumping capacity in extracting the etching by-products during the etching process at this condition. The micro/nanomasking effects in conjunction with polymerisation under certain CHF3 plasma conditions have resulted in the formation of the nanopillars shown in Figure 16. These nanopillars are 30-50 nm in diameter and about 500 nm tall.

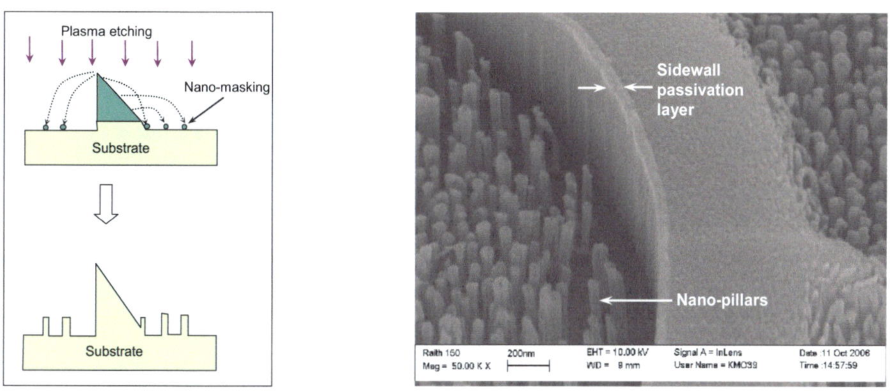

(a) (b)

Fig. 16. (a) Illustration of the possibility of how the pillars are formed through nanomasking effect (b) SEM image of quartz showing the nanopillars and sidewall passivation layer.

To achieve a smooth surface finish, an etching process with a moderate plasma density and a higher physical than chemical reaction should be followed. The optimum etching process should be then set at below the maximum capacity of the vacuum pumping. In that sense, a very low operating pressure is required so that the etching products are extracted immediately before they are allowed to drop on the sample surface. This process is similar to the physical etching mechanism. At a very low operating pressure condition below 20 mTorr, the mean free path is increased, which minimizes the collision between atoms and ions. To achieve low operating pressure, the gas feeds have to be reduced. As a result, the products of the chemical reaction are largely minimized and the etching is mostly carried out by the physical bombardment. This condition was achieved by reducing the incoming flow rates of the etching gases which results in a lower operating pressure but also lower etching rate.

From the above experimental analysis, it was found that the operating pressure has a direct effect on the surface roughness. Further experiment has demonstrated the relationship between the surface roughness and etching pressure as illustrated in Figure 17. The AFM was used to measure the surface roughness of the etched samples. A surface roughness of about 85 nm is measured when the etching pressure is 30 mTorr. At 25 mTorr processing pressure, the measured surface roughness was much lower than actual because of the limitation of the AFM in tracing the nanopillars effectively. A lowest surface roughness $R_{q(rms)}$ of 1.429 nm was measured at etching pressure of below 6 mTorr.

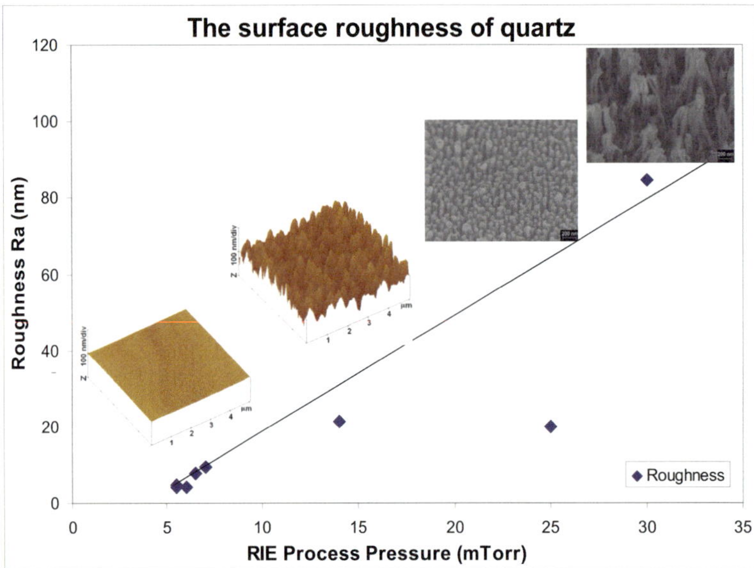

Fig. 17. The surface roughness versus the RIE operating pressure. The upper two SEM images represent high pressure process with high level of roughness, while the bottom AFM images showing very smooth surface obtained by low RIE process pressure.

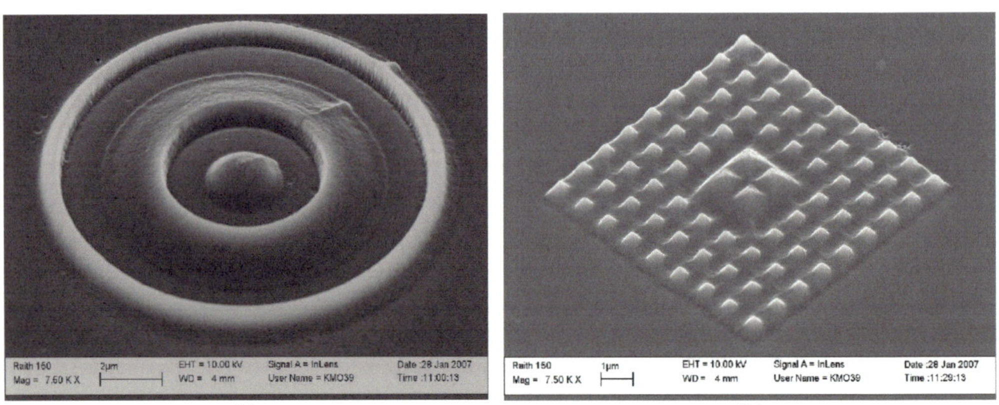

(a) (b)

Fig. 18. SEM images of the etched 3-D structures on quartz substrate (a) 3-D ring and (b) 3-D multilevel pyramid array. The 3-D structures shown are used as the master molds in the subsequent imprint process.

Figure 18(a) shows an SEM image of the 3-D ring structure on a quartz master mold substrate with the highest point on the pattern of 600nm. Figure 18(b) shows an SEM image of 3-D pyramid structures with multilevel features on quartz substrate. These structures were used as the test pattern for the master mold in the 3-D nanoimprint lithography.

4. Imprints

Imprint is primarily a physical deformation process where a mold with micro/nanostructures on its surface is utilized to deform a thin resist film (or an active material) deposited on a substrate (Chou 2003). The resist can be either a thermal plastic or UV curable material. It is a direct pattern transfer technology where the resolution is independent on light wavelength or electron scattering. A wide range of substrates and mold materials are possible. Quartz, SiO2, Si3N4, and Ni are typical molds used in NIL. NIL is a low cost, high throughput technique suitable for patterning micro and nanostructure. Another unique feature of NIL technology is that it is a direct 3-D patterning technology, which is not achievable in other lithography technologies.

There has been a strong demand imposed by semiconductor industries for an innovative lithography technique, where nano-scale resolution, high throughput and cost-effectiveness are fused together, and the NIL technologies can satisfy these criteria.

In the imprint experiments, a simple vacuum operated manual imprint tool was attached onto the UV exposure stage of the Karl Suss Mask Aligner system. An UV illumination of 365 nm wavelength with an intensity of about 0.6 W/cm^2 was utilized for resist curing while imprinting. In imprinting, achieving the thinnest possible residual resist layer is an essential requirement to ensure a successful pattern transfer. This requires optimization of pattern density, resist viscosity, resist thickness and imprint pressure. Hence, the study of the behaviour of resist reflow during and after imprint is important in NIL technologies (Macintyre & Thoms 2005; Mohamed et al. 2006) .

In this section, imprint problems such as sticking, air bubbles, and the uniformity of the residual resist layer will be discussed. For sticking issues, a number of options were followed in order to modify the mold substrate surface properties such that surface chemistry for each material used was considered to minimize the sticking problems. Surface treatment of polymer using plasma to modify the surface energy is a common technique and has been studied extensively with wide range of suggestions and solutions (Chan et al. 1996). Minimization of defect density and contamination are also important aspect in imprinting processes. This is considered one of the limitations of current NIL technique.

De-molding is the process by which the mold is separated from the imprinted polymer structure by a vertical movement of the mold. This is a crucial step in the imprint process and extra care must be taken to minimize failures at this stage. Distortion or damaging of the imprinted structure during this movement can happen as a result of different effects such as adhesion of resist at the surface, friction due to surface roughness of sidewalls and trapping of the polymer due to negative slope of cavity sidewalls. De-molding without resist damage is one of the key elements for successful nanopatterning using NIL technology. Some demolding examples by separating the mold and substrate using an air knife has been demonstrated by S. Merino and co-workers (Merino et al. 2007).

In 3-D imprinting, the pattern on the mold has to be the inverted shape of the intended 3-D structure. The mold making process of the inverted shape of the intended 3-D structure can be quite difficult. Hence, an imprint technique was used to replicate the inverse shape of the mold from the master mold. This process is similar to a casting or moulding process widely used by the manufacturing industry. It allows the master mold structures to be fabricated on any material and later replicated on the intended materials.

4.1 Imprint preparation

In the 3-D imprint, an Ormocomp resist from Microresist Technology was used as the imaging layer for the first imprint step and a mr-UVCur06 resist from Microresist Technology was used in the second imprint step. Ormocomp has excellent properties for UV transmission which is crucial as a transparent mold, while mr-UVCur06 has low viscosity which is essential to achieve a minimum and uniform residual resist thickness in Imprint 2 step. Therefore two steps imprint is usually followed.

The required resist thickness for Imprint 1 is dependent on the height of the master mold structure and the pattern density. The resist should not be too thick compared to features on the mold to avoid the shrinkage stress during the UV curing process because of a high stress is normally present at the edge of the 3-D patterns. A 400 nm resist thickness is found appropriate in replicating the 3-D master mold. To achieve this thin layer, a 30% Ormocomp diluted in Ormothin solution was spun coated on a quartz substrate at a spinning speed of 3000 rpm for one minute and pre baked on a hot plate at 80°C for two minutes (Mohamed et al. 2008).

For imprint 2, the required initial resist thickness is more critical as it determine achieving a minimum residual resist thickness. A 400 nm thick of mr-UVCur06 resist was found appropriate to yield a residual resist thickness below 100 nm. The resist was spun coated at 2800 rpm for 45 seconds and pre baked on a hot plate at a temperature of 80 °C for two minutes to achieve a 400 nm resist thickness.

Prior to the UV resist coating, quartz substrate was cleaned and surface treated using oxygen plasma to promote better adhesion at the interface between quartz and resist surface. It is important ensuring that surfaces in contact during the imprint process have low surface energies to minimize the sticking problems. Surface wetting experiments were carried out to identify the wetability of the surfaces involved in the UV-NIL imprint process. Figure 19 shows the contact angle of different materials in a normal condition and after

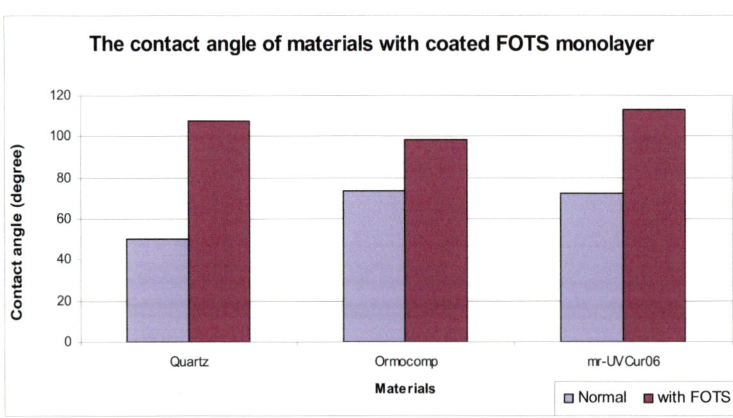

Fig. 19. The contact angle measurement of the imprint materials in normal condition and after being coated with a FOTS anti adhesive monolayer.

being coated with a 1H, 1H,2H,2H-Perfluoroocty-trichlorosilane (FOTS) anti-adhesive monolayer. Another method of minimising the sticking problem is by surface treatment using plasma. A contact angle up to 80° can be achieved on quartz or silicon substrates when exposed to SF_6:CHF_3 1:1 plasma (Sun et al. 2008). The resist surface can also be treated using fluoride containing plasma such as CHF_3 (Chan et al. 1996) in order to increase the contact angle. Table 2 shows the RIE recipes for the surface treatments of quartz and resist.

	Quartz surface	Resist surface
Gases	SF_6 / CHF_3	CHF_3
Flow rate (sccm)	15.0/15.0	30.0
Pressure (mTorr)	28	30
Temperature (K)	295	295
RF Power (W)	50	100
Time (s)	120	30
Max. contact angle	Up to 80°	Up to 85°

Table 2. The RIE parameter for surface treatment employed in minimising the sticking problem and in achieving the desired contact angles.

4.2 Imprint processes

The manual vacuum operated imprint tool was used in this imprint experiment to create a vacuum environment in order to reduce the air bubbles trapped in between the mold and the resist during the imprint process.

Replicating the 3-D structure of the master mold involved two imprint steps. The first imprint (Imprint 1) replicates the inverted shape of the master mold while the second imprint (Imprint 2) replicates the inverted shape of the first imprint (Imprint 1) which became the positive replica of the original master mold.

Two 3-D test master molds were used in this work, the first was the 3-D rings structures (Figure 17(a)) and the other was the 3-D multilevel pyramid structures (Figure 17(b)). A very thin monolayer of FOTS was used as an anti sticking layer on the quartz mold surface. It was deposited on a mold surface by a natural convection method in a petri dish at room temperature.

In this work, the filling of the 3-D cavities with the resist and the required initial resist thickness were investigated by a trial and error method but guided by simulation of resist reflow behaviour (Mohamed 2005). The 100 % cavity filling in the 3-D mold, the required imprint pressure and minimum residual resist thickness were optimized accordingly.

4.2.1 Imprint 1

The purpose of the first imprint was to replicate the inverted shape of the master mold onto a UV curable resist. Figure 20 illustrates the schematic diagram of the Imprint 1 process step. The 3-D mold was placed face up at the bottom of the imprint tool to ease the release of trapped air bubbles from the mold structures. The 400 nm thick Ormocomp resist coated on the quartz substrate with the face down was manually aligned on top of the 3-D master mold. The vacuum pressure of the mask aligner MA-6 which was set at 4 mbar was then activated and followed by UV flood exposure at a room temperature. Under 365 nm UV illumination, Ormocomp resist requires 300 mJ/cm^2 energy for curing, which is about an 8 minute exposure time using 0.6 mW/cm^2 illumination intensity.

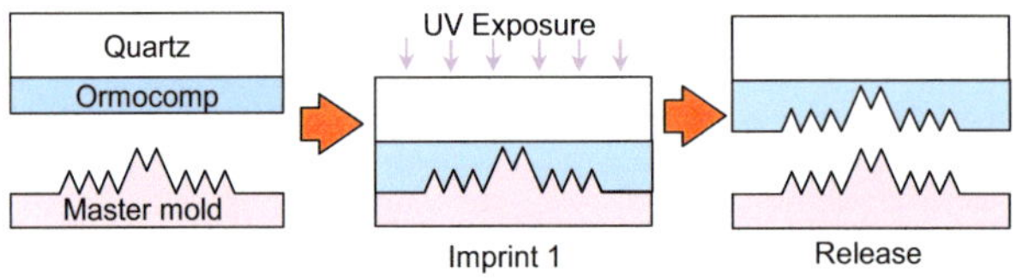

Fig. 20. The process flow of the first imprint step (Imprint 1), showing the master mold and quartz substrate coated resist exposed to Uv light and then separated.

Figure 21(a) shows a close-up AFM image of the associated imprint results of the 3-D ring mold that was made in section 3.3 and 3.4. The master mold profile was faithfully replicated. Figure 21(b) shows the AFM image of the imprint results of the 3-D multilevel molds. In AFM, a STING tip from Mikromash was used to trace the profiles of the 3-D imprint cavities.

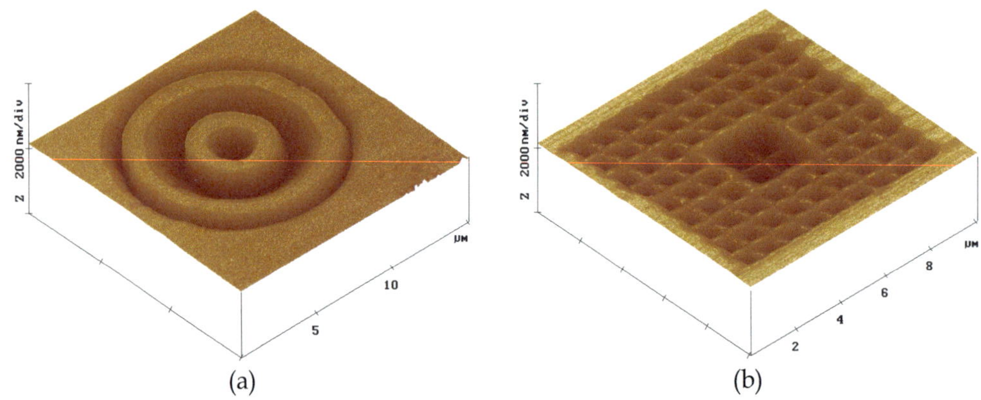

Fig. 21. The AFM images of the replication of 3-D structures (a) 3-D ring and (b) multilevel pyramid, on Ormocomp resist after Imprint 1 process.

The soft mold requires a material which is durable and robustfor a successful repetitive imprint process, it should be solvent resistant and can withstand a long UV exposure. Therefore, the Ormocomp resist was hard baked at 185 ^{o}C for 2 hours to improve its hardness and enhance its solvent resistance properties.

4.2.2 Imprint 2

The purpose of the second imprint is to replicate the inverted shape of the master mold onto another UV curable resist layer in order to produce the replica of the original 3-D structure. Figure 22 shows the schematic diagram of Imprint 2 process steps.

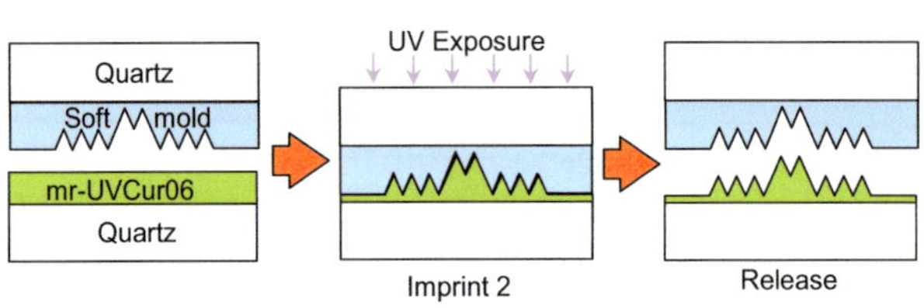

Fig. 22. The process flow diagram of the second imprint step (Imprint 2) to produce the positive replica of original pattern.

The relationship between the resist reflow and cavity filling versus the initial resist thickness is important to understand. To achieve a minimum residual resist thickness, a specially formulated mr-UVCur06 resist which has low viscosity was used. With proper optimisation, a 10 nm thick residual resist layer could be achieved. For Imprint 2 process, the 365 nm UV source was used to expose the samples for 2 minutes at room temperature. A post imprint bake at a temperature of 120 °C on a hot plate for 5 minutes to make it harder as masking layer against RIE was carried out after the sample was released from the soft mold. Imprint 2 results were imaged using the AFM technique as it does not require deposition of anti-charginglayer as in SEM. Figure 23(a) shows AFM image of the positive replica of the imprinted 3-D ring structure. Figure 23(b)} shows the AFM image of the imprinted 3-D multilevel structure on the mr-UVCur06 resist. It can be observed that the pattern has lost some of its original sharpness during the de-molding process.

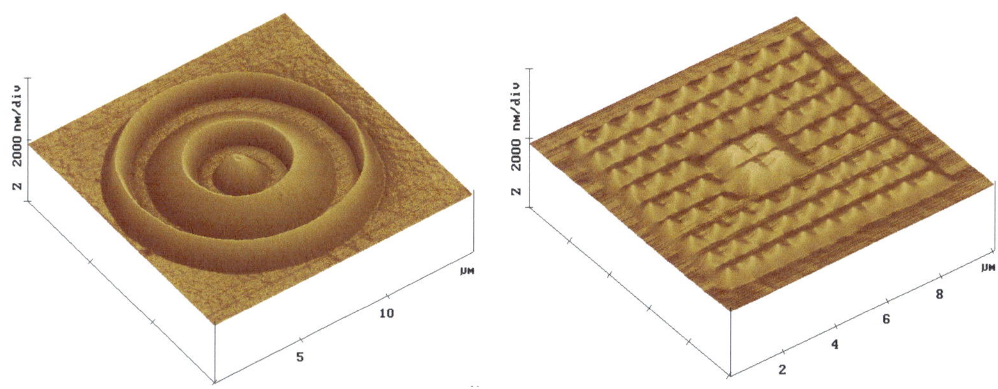

(a) (b)

Fig. 23. The AFM images of the replication of 3-D structures (a) 3-D ring and (b) multilevel pyramid, on mr-UVCur06 resist after the Imprint 2 process. The pattern in (b) has lost some of its sharpness due to de-molding.

4.3 The 3-D imprint analysis

To test the reproducibility of patterns and mold resilience against repetitive use, the imprint process was performed for 35 cycles. The results showed that by using anti-adhesive layer of FOTS coated on the quartz mold surface in Imprint 1 process, faithful replications was achieved without any deterioration of mold or resist pattern. On the other hand it was found that only 5 imprint cycles were successful without the FOTS. In the Imprint 2 process, up to 26 cycles were achieved with the FOTS coating on the soft mold (Ormocomp), however, only 3 imprint cycles were successful without FOTS coating. Imprint failures are mostly due to sticking of residual resist on the mold or as a result of particles contamination present on the mold or on the substrate.

One of the shortcomings of the Ormocomp resist is that it becomes brittle after a long UV exposure. The material brittleness combined with stress cycles in the imprint process make the Ormocomp soft mold easily to crack after about 5 imprint cycles. Micro-cracks on the soft mold surface will limit the repetition of the imprint cycles. A 5 nm of sputtered TiO_2 layer was coated on the soft mold surface as a permanent anti-sticking layer as well as to improve the hardness of the mold surface. The mold surface cleanliness will also be improved resulting in increased 15 repetitive imprint cycles. Cracks in the soft mold were

observed only after 15 imprint cycles, which caused the resist to stick to the mold surface. Figure 24(a) shows the sticking problem of the imprinted 3-D ring structure on the Ormocomp resist during the Imprint 1 process. Figure 24(b) shows a similar sticking problem of the imprinted 3-D multilevel structures on the Ormocomp resist.
The imprint pressure must always be at a low level (below 100 mbar) to avoid such sticking issues, soft mold breakage as well as maintaining the cleanliness.
In imprint 2, the problem of a low viscosity (14 mPa s) of the mr-UVCur06 resist has resulted in resist penetration into mold cracks and microcavities causing sticking issues and mold breakages at high stress points. Other than using a release agent as an anti-sticking layer, the sticking problem can also be minimized by resist surface treatment to change the surface properties.

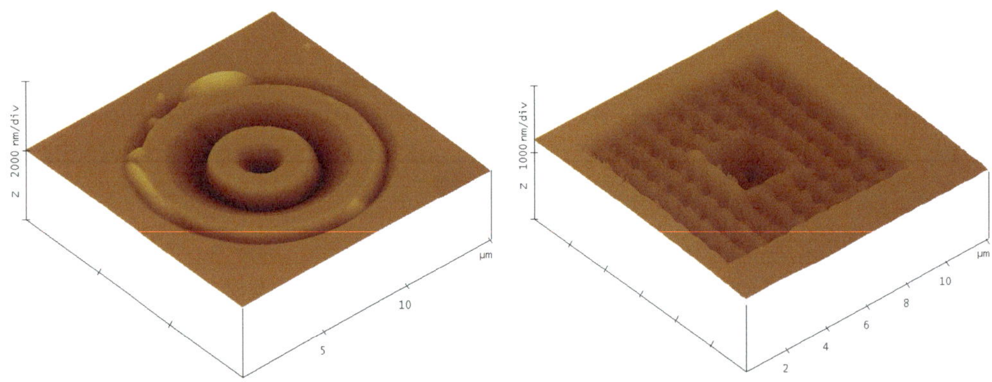

(a) (b)

Fig. 24. Examples of sticking problem of 3-D imprinted structure using Ormocomp resist, (a) sticking on the outer ring of 3-D ring structure and (b) sticking on the multilevel structure.

5. Conclusions

Direct three-dimensional patterning is demonstrated using UV cured nanoimprint lithography (UVNIL). The 3-D mold profiles were created on the ma-N2403 negative tone photoresist using the Raith-150 EBL tool with variable dose controlled exposure. A variable e-beam dose was used to obtain a resist contrast curve to determine the gradient of various 3-D structures. For a 600 nm layer thickness of ma-N2403 resist, the e-beam critical energy was set at 6.25 keV. To suppress charging effects on insulating substrate such as quartz, a conductive polymer layer is used. A higher acceleration voltage of 20 keV was used for the conductive polymer (PEDOT/PSS) top coating is employed. The variable EBL exposure with line dosages ranging from 0.6 pC/cm to 42.0 pC/cm were used for 3-D resist contrast approach. A linear resist contrast profile was obtained with a negative tone photoresist and subsequently was utilized as the 3-D masking layer. The 3-D pattern was transferred onto the quartz mold substrate by an optimized single-step reactive ion etching (RIE) process with selectivity resist-to-substrate of 1:2 using fluorinated plasma CHF_3/Ar.
The main challenging issues in EBL writing on insulating substrates are surface charging, electrons backscattering and proximity effects. The use of critical acceleration voltage and high accelerating voltage in conjunction with conductive polymer PEDOT/PSS have

suppressed the surface charging effects in the fabrication of 3-D structures, where feature size down to 300 nm has been demonstrated using this approach.
The high surface roughness caused by nanomasking and polymerisation effects in the 3-D etching process is the combined result of plasma bombardment/collision products, etched masking particles and dominating polymerization reactions at a high operating pressure. By working at low RIE operating pressure a smooth equartz surface is produced. A surface roughness of below 2 nm was achieved when the RIE process pressure was lower than 6 mTorr. In 3-D imprinting two imprint steps are involved. A 400 nm thick Orcmocomp and mr-UVCur06 resists were used in Imprint 1 and Imprint 2 processes respectively. A 3-D ring with a height of 600 nm and a multilevel pyramid array with each pyramid dimension of one micron square base and 600 nm in height were replicated. Complex three-dimensional and multilevel nanostructures can be reproduced faithfully using UVNIL, this demonstrate the potential of NIL in nanoscale device fabrication.

6. References

CASINO Software. (2006). "Monte Carlo Simulation ".

Chan, C. M., Ko, T. M., and Hiraoka, H. (1996). "Polymer surface modification by plasmas and photons." *Surface Science Reports*, 24(1-2), 1-54.

Chiu, W. L., Alkaisi, M. M., Kumaravelu, G., Blaikie, R. J., Reeves, R. J., and Bittar, A. (2006). "Sub-wavelength Texturing for Solar Cells using Interferometric Lithography." *Advances in Science and Technology*, 51, 115-120.

Chou, S. Y. (2003). "Nanoimprint Lithography." Alternative Lithography: Unleashing the Potentials of Nanotechnology, C. M. Sotomayor Torres, ed., Kluwer Academic/Plenum Publishers.

Frey, M. D. (2007). "Low kV Scanning Electron Microscopy." Scanning Microscopy for Nanotechnology, W. Zhou and Z. L. Wang, eds., Springer.

Ishii, Y., and Taniguchi, J. (2007). "Fabrication of three-dimensional nanoimprint mold using inorganic resist in low accelerating voltage electron beam lithography." *Microelectronic Engineering*, 84(5-8), 912-915.

ITRS. (2003). "ITRS 2003 Edition." International Technology Roadmap for Semiconductors.

ITRS. (2008). "ITRS 2008 Edition." International Technology Roadmap for Semiconductors.

Joo, J., Chow, B. Y., and Jacobson, J. M. (2006). "Nanoscale Patterning on Insulating Substrates by Critical Energy Electron Beam Lithography." *Nano Lett.*, 6(9), 2021-2025.

Konijn, M., Alkaisi, M. M., and Blaikie, R. J. (2005). "Nanoimprint lithography of sub-100 nm 3D structures." *Microelectronic Engineering*, 78-79, 653-658.

Kumaravelu, G., Alkaisi, M. M., Bittar, A., Macdonald, D., and Zhao, J. (2004). "Damage studies in dry etched textured silicon surfaces." *Current Applied Physics*, 4(2-4), 108-110.

Macintyre, D. S., and Thoms, S. (2005). "A study of resist flow during nanoimprint lithography." *Microelectronic Engineering*, 78-79, 670-675.

Merino, S., Schift, H., Retolaza, A., and Haatainen, T. (2007). "The use of automatic demolding in nanoimprint lithography processes." *Microelectronic Engineering*, 84(5-8), 958-962.

Mohamed, K. (2005). "Resist Deformation in Nanoimprint Lithography (NIL)," Master Degree, University of Canterbury, Christchurch.

Mohamed, K., Alkaisi, M. M., and Blaikie, R. J. (2007). "Fabrication of three dimensional structures for an UV curable nanoimprint lithography mold using variable dose control with critical-energy electron beam exposure." *Journal of Vacuum Science & Technology B: Microelectronics and Nanometer Structures*, 25, 2357-2360.

Mohamed, K., Alkaisi, M. M., and Blaikie, R. J. (2008). "The replication of three dimensional structures using UV curable nanoimprint lithography." *Journal of Vacuum Science & Technology B: Microelectronics and Nanometer Structures*, 26, 2500-2503.

Mohamed, K., Alkaisi, M. M., and Blaikie, R. J. (2009). "Surface charging suppression using PEDOT/PSS in the fabrication of three dimensional structures on a quartz substrate." *Microelectronic Engineering*, 85.

Mohamed, K., Alkaisi, M. M., and Smaill, J. (2006). "Resist deformation at low temperature in nanoimprint lithography." *Current Applied Physics*, 6(3), 486-490.

Romanato, F., Businaro, L., Vaccari, L., Cabrini, S., Candeloro, P., De Vittorio, M., Passaseo, A., Todaro, M. T., Cingolani, R., Cattaruzza, E., Galli, M., Andreani, C., and Di Fabrizio, E. (2003). "Fabrication of 3D metallic photonic crystals by X-ray lithography." *Microelectronic Engineering*, 67-68, 479-486.

Romanato, F., Tormen, M., Businaro, L., Vaccari, L., Stomeo, T., Passaseo, A., and Di Fabrizio, E. (2004). "X-ray lithography for 3D microfluidic applications." *Microelectronic Engineering*, 73-74, 870-875.

Simcic, J., Pelicon, P., Rupnik, Z., Mihelic, M., Razpet, A., Jenko, D., and Macek, M. (2005). "3D micromachining of SU-8 polymer with proton microbeam." *Nuclear Instruments and Methods in Physics Research Section B: Beam Interactions with Materials and Atoms*, 241(1-4), 479-485.

Sun, H., Liu, J., Gu, P., and Chen, D. (2008). "Anti-sticking treatment for a nanoimprint stamp." *Applied Surface Science*, 254(10), 2955-2959.

Taniguchi, J., Koga, K., Kogo, Y., and Miyamoto, I. (2006). "Rapid and three-dimensional nanoimprint template fabrication technology using focused ion beam lithography." *Microelectronic Engineering*, 83(4-9), 940-943.

Thackeray, J. W., Orsula, G. W., Canistro, D., and Berry, A. K. (1989). "Evaluation of Deep UV ANR Photoresist for 248.4 nm Excimer Laser Photolithography." *Journal of Photopolymer Science and Technology*, 2(3), 429-443.

Totsu, K., Fujishiro, K., Tanaka, S., and Esashi, M. (2006). "Fabrication of three-dimensional microstructure using maskless gray-scale lithography." *Sensors and Actuators A: Physical*, 130-131, 387-392.

Plasmonic Lithography and Nano Patterning

29

Metal Particle-Surface System for Plasmonic Lithography

V. M. Murukeshan, K. V. Sreekanth and Jeun Kee Chua
School of Mechanical and Aerospace Engineering, Nanyang Technological University
50 Nanyang Avenue,
Singapore 639798

1. Introduction

Optical (Photo) lithography has played a significant role in almost every aspects of modern micro-fabrication technology in the recent years. It has initiated transistor revolution in electronics and optical component developments in photonics. Advances in this field have allowed scientists to improve the resolution of the conventional photolithographic techniques, which is restricted by the diffraction limit [Okazaki, 1991]. To overcome this problem and to reduce the critical dimension, several solutions were introduced. New research suggests that we may be able to develop new low cost photolithographic technique beyond the diffraction limit. The minimum critical dimension (half-pitch resolution) achievable by photolithography (Optical projection lithography) is given by $CD_{half-pitch} = k_1\lambda / NA$, where λ is the incident source wavelength, NA is the numerical aperture of projection optics of the system and k_1 is a constant value as a indication of the effectiveness of the wavefront engineering techniques. To reduce the half-pitch resolution, critical dimension equation demands either to decrease the wavelength of illumination light source or to increase the numerical aperture of projection system. Or in short, fabrication of sub-100nm features generally imposes the requirement of shorter wavelength laser sources.
In this context, the critical challenges that hinder the resolution enhancement approaches are (i) lack of availability of suitable ultra-short wavelength lasers, and (ii) the unavailability of suitable optics and materials such as photoresist for use at suitable wavelengths. Recently, techniques like extreme ultraviolet lithography (EUV) [Gwyn et al., 1998] and X-ray lithography [Silverman, 1998] have been proposed for nanofabrication overcoming the diffraction limit. Here, the illumination wavelength is reduced to the extreme UV (smaller wavelength) to get smaller features. Another reported technique is the immersion lithography in which numerical aperture of the imaging system is increased by inserting high index fluids (prism or liquid) between last optical component and wafer surface [Wu, et al., 2007]. But this technique is either limited by air absorption or availability of high index fluids. Approaches such as electron-beam lithography can also be used to overcome diffraction limit, but these are serial process and cannot be used for high throughput [Chen et al., 2005]. Imprint lithography is another option to improve the resolution beyond the diffraction limit [McAlpine et al., 2003]. Nanometer scale features are possible by stamping a

template on a thin polymer film and it can also generate sub-50nm features by integrating laser beam with AFM, NSOM and transparent particles. The main disadvantages of this technique are: (i) the leveling of the imprint template and the substrate during the printing process, which determine the uniformity of the imprint results, and (ii) slow process speed, which limits their applications in industry. The laser interference lithography (LIL) can be used to fabricate high speed and large area period nanostructures [Prodan et al. 2004]. The basic principle is the interference of coherent light from a laser source to form a horizontal standing wave pattern in the far field, which can be recorded on the photoresist.

Recently near-field lithography techniques have been proposed to overcome the diffraction limit for nanofabrication. One of the emerging areas of research is the scanning probe lithography in which the Scanning Tunneling Microscope (STM) or Atomic Force Microscope (AFM) can be used to pattern nanometer scale features, by the introduction of laser beam in to a gap between an AFM or STM tip and substrate surface with tip scanning over the surface [Jersch, 1997]. But they have stringent limitations with respect to certain materials and effectiveness applies only for certain ambient conditions. Evanescent wave lithography (EWL) is one of the near field interference lithography technique to achieve nano-scale feature at low cost [Blaikie & McNab, 2001; Chua et al., 2007]. It can create a shorter wavelength intensity pattern in the near field of diffraction grating or prism when two resonantly enhanced, evanescently decaying wave superimposed. It provides good resolution, but is limited by low contrast and short exposure depth. These problems can be subdued to a great extent by surface plasmon resonance phenomena due to their characteristics of enhanced transmission in the near field [Ebbesen et al., 1999].

Plasmonic lithography is an emerging area of near field photolithography techniques by which nano resolution features can be fabricated beyond the diffraction limit at low cost [Srituravanich et al., 2004]. Surface plasmon polaritons are electromagnetic waves that propagate along the surface of a metal [Raether, 1988]. Surface plasmon resonances in metallic films are of interest for a variety of applications due to the large enhancement of the evanescent field at the metal/dielectric interface. Hence plasmonic lithography has achieved much progress in the last decade, because it provides us a novel method of nanofabrication beyond the diffraction limit. It can provide high resolution, high density, and strong transmission optical lithography, which can be used to fabricate periodic structures for potential applications such as biosensing, photonic crystals, and high density patterned magnetic storage. Many research groups have already demonstrated that sub-100nm resolution nano structures can be fabricated using plasmonic lithography techniques.

The surface plasmon interference nanoscale lithography based on Kretschmann-Raether attenuated total reflection (ATR) geometry has been proposed numerically [Guo et al. 2006; Lim et al. 2008]. Moreover, a near field interference pattern can be formed by using metallic mask configuration that can generate surface plasmon for periodic structure fabrication [Shao & Chen, 2005; Luo & Ishihara, 2004; Liu, 2009]. In all the above mentioned works, surface plasmon can make a certain pattern on the photoresist layer when the incident p-polarized light passes through a prism or thin metallic mask. However, most of these reported techniques demands the fabrication of fine period mask grating and found to be not cost effective. The recent thrust in this challenging area focuses on exploring novel concepts and configurations to meet the sub-30nm nodes forecasted for the next decade and beyond.

In this chapter, the focus will be on a new plasmonic lithography concept for high resolution nanolithography based on the excitation of gap modes in a metal particle-surface system.

The principle, the excitation of gap modes in a metal particle-surface system and excitation of surface plasmon polaritons mediated by gap modes are illustrated and analyzed numerically from a lithography point of view. In Sect.2, the characteristics of gap modes are discussed on the basis of electromagnetic theory of a metal particle placed near to a metal surface. The concept of gap modes excited plasmonic lithography configuration has been presented in Sect.3 after giving a brief overview on conventional plasmonic lithographic configurations. A detailed analysis on the variation of electric field distribution with various parameters is numerically illustrated in Sect.4. To compute the positional development rates of photoresist domain in response to the normalized intensity profile, a modified cellular automata model is employed. In sect.5, the theoretical analysis of proposed models, followed by resist profile cross section obtained through this proposed concept is discussed. The chapter concludes in Sect.6 with a discussion on the future direction of the proposed concept and related research challenges.

2. Gap modes in metal particle-surface system

At the nanoscale, mainly electric oscillations at optical frequency contribute the optical fields, but the magnetic field component doses not contribute significantly due to weak field component. The existences of localized optical modes on dimensions much smaller than the optical wavelength are responsible for such fields to concentrate and support the nanostructured materials [Stockman, 2008]. These energy concentrating modes are called surface plasmons (SPs) and it is well known that a system should contain both negative and positive dielectric permittivities to support surface plasmons. The shape of the metal nanoparticle and metal surface thickness is an important factor for the surface plasmon resonance. A thin metal surface is associated with surface plasmon polariton (SPP) modes, which are coupled modes of photons and plasmons [Reather, 1988]. Since the SPP modes are nonradiative electromagnetic modes, to excite them the incoming beam has to match its momentum to that of the plasmons. It is possible by passing the incident photons through a bulk dielectric layer to increase the wavevector component and achieve the resonance at the given wavelength. But a fine metal particle is associated with localized surface plasmon (LSP) modes, which are collective oscillations of the conduction electrons in a metal nanoparticle [Kreibig & Vollmer, 1995]. The LSP modes can be excited directly by incident photons since they are radiative electromagnetic modes.

What happens when a system consisting of a fine metal particle placed near to a metal surface? An electromagnetic interaction between LSP modes associated with metal nanoparticle and SPP modes associated with metal surface is possible. This interaction plays an important role to enhance the light emission from metal-insulator-metal tunnel junction, mediated by metal nanoparticles [McCathy & Lambe, 1978; Adams & Hansma, 1979]. Due to this electromagnetic interaction there exist new types of localized electromagnetic normal modes, called gap modes in the space between the nanoparticles and the surface [Rendell et al. 1978; Rendell and Scalapino, 1981]. Figure 1 represents an isolated metal sphere (Al) of dielectric function $\varepsilon(\omega)$ embedded in surrounding medium (SiO_2) of dielectric function ε_m, is placed close to a metal surface (Ag). The retardation effects of electromagnetic fields can be neglected when the radius (R) of the sphere is small and resonant frequencies corresponding to the excitation of LSP are expressed by $\varepsilon(\omega) = -\varepsilon_m(l+1)/l$, where $l = 1,2,....$ is an integer [Boardman, 1982]. If the sphere is much smaller than the wavelength of the incident light, the dipole mode with $l = 1$ is mainly excited.

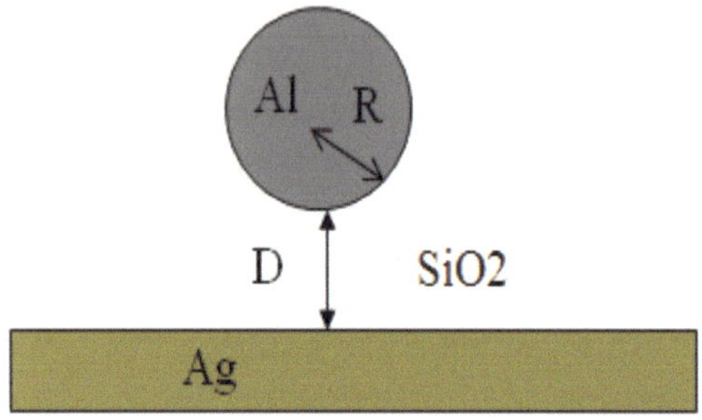

Fig. 1. Schematic diagram of the metal particle-surface system: Al nanosphere of radius R is placed at a distance D from the Ag surface.

Theoretically, this concept can be explained by approximating the LSP modes as a dipole and considering its interaction with image dipole induced inside the metal surface. This dipole-dipole interaction may greatly modify the LSP mode and hence the resonance frequency and field distribution. In other way, new electromagnetic modes are expected to appear by change in symmetry of the system. It means the spherical symmetry of the isolated sphere translate to cylindrical symmetry for a sphere-surface system. These modes also correspond to polarization modes parallel and perpendicular to the symmetry axis [Hayashi, 2001]. When the particle-surface distance is sufficiently small ($D/R<1$), this system can support a series of gap modes and the electric field becomes more and more localized at the gap between the particle and the surface. When the gap mode is excited, the intensity of the electric field is enhanced relative to that of the excitation field and these modes are believed to play an important role in the light emission process. It is reported that the maximum enhancement factor is larger than that achieved with an isolated particle (LSP excitation) or a surface alone (SPP excitation) system [Hayashi, 2001].

The metal particle-surface system supposed to find variety of potential applications in near field optics, although the roles played by the gap modes have not yet been fully explored. One of the promising applications in scanning tunneling microscope (STM) in which the tunneling current are excited by gap modes [Johansson et al., 1990]. In STM, SPP modes in a metallic surface are excited by ATR method. To obtain images with high lateral resolution, the intensity of the reflected and scattered light can be enhanced by placing a sharpened metallic tip very close to a surface. Theoretical treatments of this problem is already reported, in which tip is often modulated by a sphere [Madrazo et al., 1996]. The direct evidence of the existence of gap modes is experimentally demonstrated by Hayashi's group [Hayashi, 2001]. They performed a systematic absorption measurement on Ag island particles placed above an Al surface and realized strong localization and strong enhancement of electromagnetic field under the conditions of resonant excitation of gap modes.

3. Plasmonic lithography configurations

The two well known methods proposed to excite surface plasmons on a thin film and subsequent realization of plasmonic lithography are based on configurations using prism coupling (Kretschmann) and grating coupling (Metal grating mask).

3.1 Kretschmann configuration

Kretschmann (Prism based) configuration is a well known method used to excite surface plasmon polariton, performed with the evanescent field generated by ATR principle and thereby enabling SP interference. Figure 2 represents the SP interference lithography technique using Kretschmann configuration, in which the upper layer is a high refractive index isosceles triangle prism. The coated thin metal layer is at the bottom surface of the prism, which is in contact with the photoresist layer on a substrate.

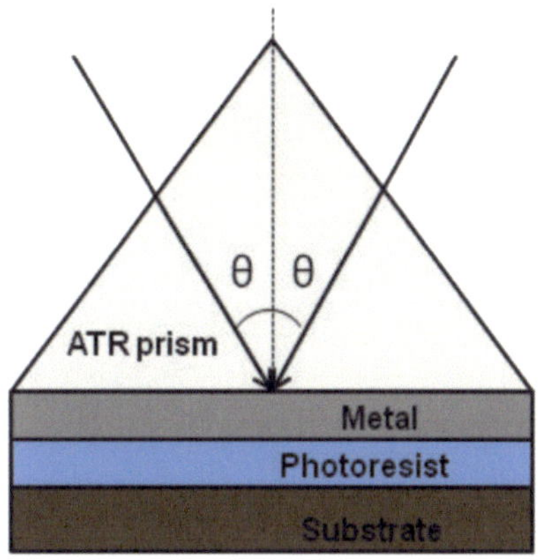

Fig. 2. Schematic diagram of Kretschmann configuration

3.2 Metal grating mask based configuration

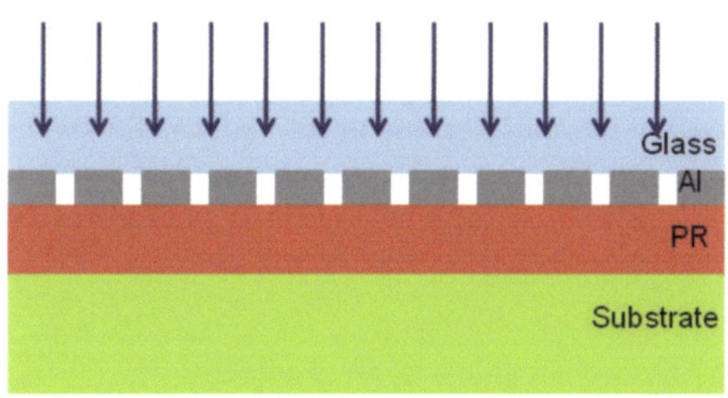

Fig. 3. Metal grating mask based configuration

Metal mask grating based configuration is a commonly used technique for plasmonic lithography. As distinct from a Kretschmann scheme, the mask grating based scheme is much more compact. In this configuration, the period of the grating can be several times greater than the period of the expected interference pattern and interference of various diffraction orders generate the SP interference pattern on the photoresist layer. The optical near field of metallic mask can produce fine features with subwavelength scale resolution. The schematic of plasmonic lithography configuration using metal mask is shown in Fig.3. It consists of metal mask, which can be fabricated on a thin quartz glass by electron-beam lithography and lift off process. And mask is brought into intimate contact with a photoresist coated on a silica substrate. Light is incident normally from the top and light tunnels through the mask via SPP and reradiates in to the photoresist.

3.3 Gap modes excited configuration

Schematic diagram in Fig. 4 represents the plasmonic lithography configuration based on the excitation of gap modes in metal particle-surface system. It consists of a high refractive index thin dielectric layer kept below the incident prism medium, which is in contact with periodic Al metal nanospheres. Period of Al nanosphere array is taken as 24 nm. These Al nanospheres are embedded in a lower refractive index surrounding medium (SiO_2). A thin Ag metal film is coated at distance D from the Al nanospheres in the surrounding medium and in contact with photoresist layer on a silica substrate. The diameter of Al nanospheres is taken as 20nm and thickness of Ag film is 10 nm.

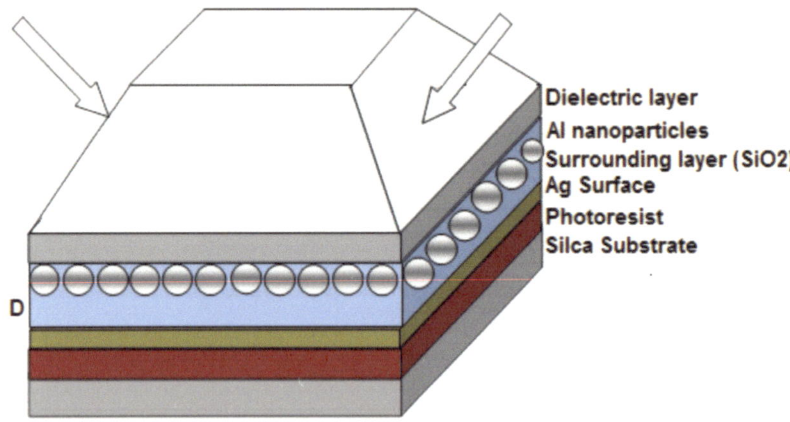

Fig. 4. Schematic diagram of gap modes excited plasmonic lithography configuration (Adapted from Ref. [V. M. Murukeshan and K. V. Sreekanth, Opt. Lett. 34, 845-847 (2009)])

The basic idea of the technique is that the excited gap modes above the metal surface can greatly enhance the surface plasmons on the metal surface and superposition of two excited SP modes generates the interference patterns on the photoresist layer. By applying the appropriate boundary condition at the metal surface/photoresist interface, the dispersion relation for surface plasmon polariton is described as [Raether, 1988],

$$k_{SP} = k_x = \frac{\omega}{c}\sqrt{\frac{\varepsilon_m \varepsilon_d}{\varepsilon_m + \varepsilon_d}} \tag{1}$$

where ω the excitation frequency, c is the speed of light in vacuum, ε_m is the complex dielectric constants of wavelength dependent metal and ε_d is the dielectric constants of photoresist layer. The wavevector of surface plasmon wave becomes significantly larger than that of the incident light wavevector, when the real parts of ε_m approaches $-\varepsilon_d$. At this condition, the wavelength of the excited SPs get normally shorter compared to the wavelength of illumination light. Furthermore the modes responsible for the field distribution could be weakly coupled gap modes of single particle-film systems.

4. Electric field distribution

The finite difference time domain (FDTD) method is used to predict the light intensity distribution on the photoresist layer. The simulation region is terminated with perfectly matched layer (PML) boundary condition on the boundaries perpendicular to the propagation direction of the light and Bloch boundary condition on the other boundaries. The refractive index of the upper prism, dielectric layer and photoresist layer are 1.745 (NLK8 glass), 1.939 (NLAF31A glass) and 1.53 (AZ9200 from AZ Electronic Material) respectively. The dielectric layer thickness is assumed to be 90nm. The wavelength of p-polarized light is taken as 427 nm (as shown in Fig. 5) and incident resonant angle is 56° at this wavelength. The complex dielectric constants of aluminium and silver at this wavelength are -26.728+5.8i and -5.082+0.7232i respectively [Palik, 1985].

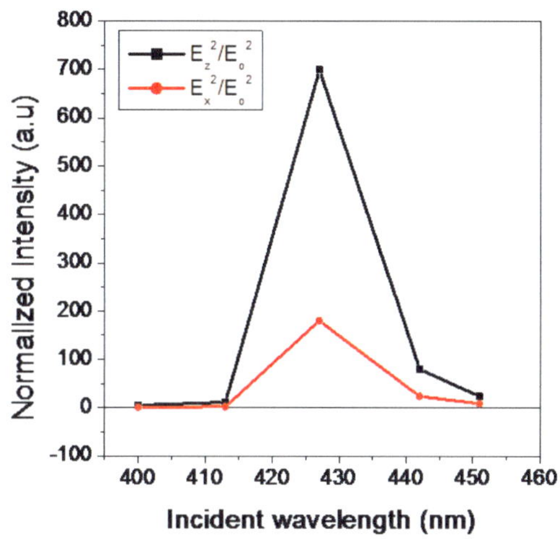

Fig. 5. Variation of normalized intensity with incident wavelength.

The normalized intensities measured along metal surface/photoresist interface as a function of incident wavelengths is shown in Fig. 5, where incident angle is assumed to be 56°. It is evident from the figure that intensity distribution is possible for a band of wavelength and maximum intensity is obtained when incident resonant wavelength is 427 nm. It is also evident that there is a large intensity variation when incident wavelength varies from 413 nm to 442 nm. It could be due to large gap modes excitation at that wavelength range and peak at 427 nm. When two p-polarized 427 nm illumination light beam incident at the dielectric layer/nanosphere interface, the field intensity distribution developed on the photoresist surface is shown in Fig.6. Figures 6(a) and 6(b) respectively shows the field distribution of E_x and E_z components, when particle-surface distance D=0 nm. And Figures 6(c) and 6(d) respectively represents that of conventional prism based configuration (shown in Fig.2) in which high refractive index prism with an Ag metal film coated onto the bottom side and in contact with photoresist layer on a silica substrate. Here the thickness of the Ag film is taken as 20 nm and remaining parameters are same as the proposed configuration. It should be noted that Ag film thickness of 10 nm for proposed configuration and 20 nm for the prism layer configurations are the required respective Ag film thickness for obtaining maximum intensity.

Figure 7 represents the normalized intensity variation, ($|E|^2 / |E_0|^2$) where E_0^2 is the incident intensity) with decay direction. Here Z-axis is considered as the decay direction. The

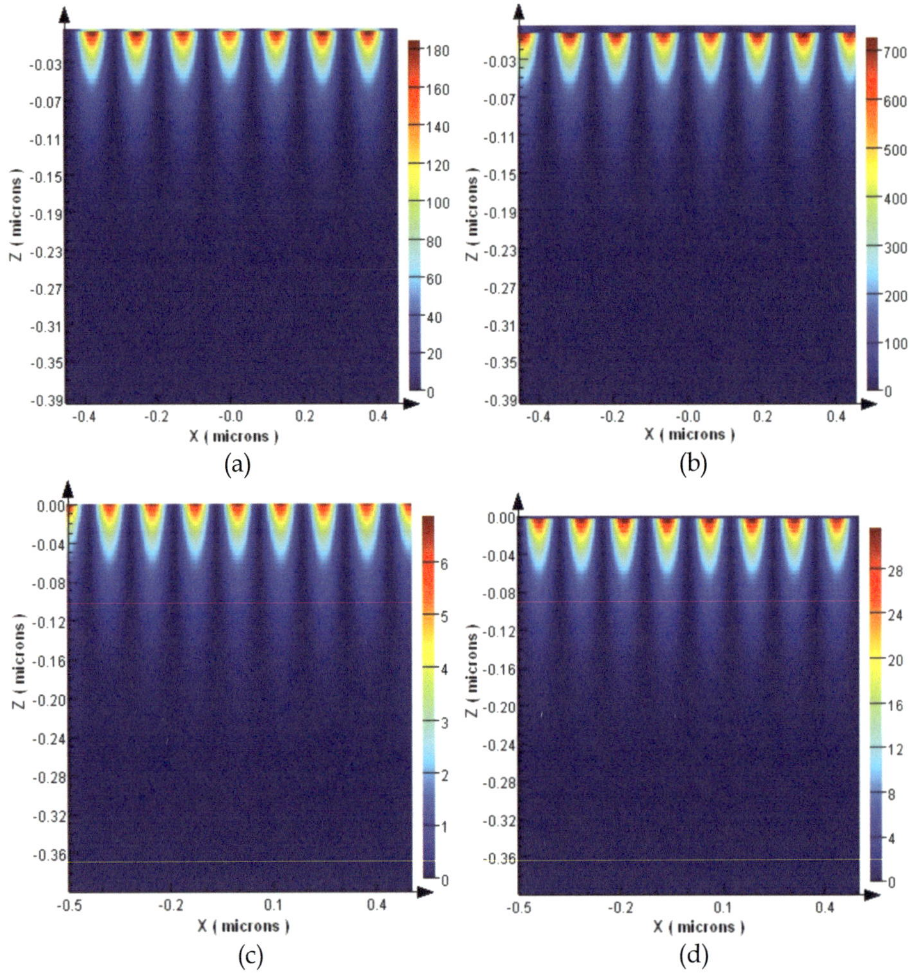

Fig. 6. Electric field distribution of the interference pattern generated by the field components on the photoresist layer (a) E_x (b) E_z of proposed configuration and (c) E_x, (d) E_z of prism based configuration. (Adapted from Ref. [V. M. Murukeshan and K. V. Sreekanth Opt. Lett. 34, 845-847 (2009)])

exposure fields E_z and E_x represents the longitudinal and transverse components respectively. The E_z component is much stronger than the E_x component and these components have $\pi / 2$ phase difference in both configurations. The electric field distribution show that interference patterns with approximate periodicity of 120 nm is formed at the interface for both configurations. From the Fig. 6 and Fig.7, it is evident that the proposed configuration gives high electric field distribution compared to conventional prism based configuration. The intensity transmission calculated at metal surface/ photoresist interface is 4% for conventional prism based configuration compared to that of the proposed configuration. Therefore the maximum enhancement factor is much larger for this configuration compared to that achieved with conventional prism based configuration. From the Fig. 7, it is clear that the exposure depth above 350 nm is achieved for the proposed scheme, but prism based configuration gives around 250 nm. The fringe visibility or contrast can be expressed as $V = (I_{max} - I_{min}) / (I_{max} - I_{min})$, where $I_{max} = E_z^2$ and $I_{min} = E_x^2$. The intensity contrast calculated is around 0.90 for both configurations within the exposure depth of 200 nm, which is sufficiently high for the realization of interference patterns and much higher than the exposure threshold of common negative optical resists.

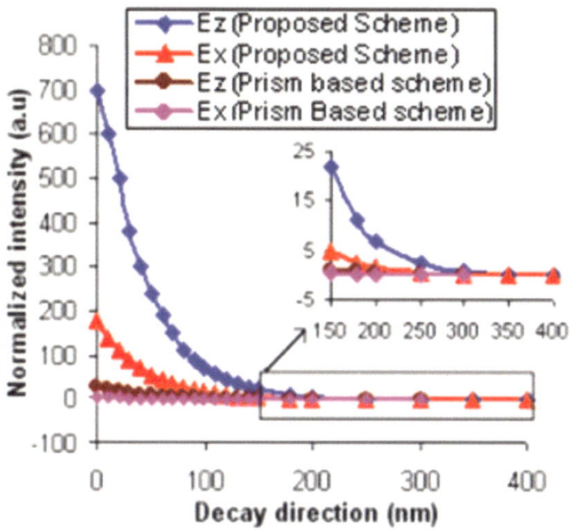

Fig. 7. Normalized intensity variation with decay direction. (Adapted from Ref. [V. M. Murukeshan and K. V. Sreekanth, Opt. Lett. 34, 845-847 (2009)])

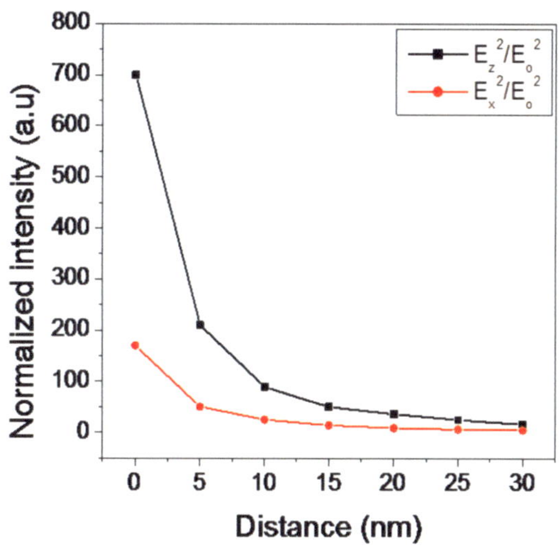

Fig. 8. Normalized intensity variation as a function of metal particle-surface distance (D). (Adapted from Ref. [V. M. Murukeshan and K. V. Sreekanth, Opt. Lett. 34, 845-847 (2009)])

The metal particle-surface distance (D) is an important factor for the gap mode excitation. Therefore the variations of normalized intensities with D need to be analyzed. Figure 8 shows the variation of normalized intensities of E_x and E_z component calculated along metal surface/photoresist interface as a function of metal particle-surface distance D. The distance D is varied from 0 to 30 nm and interference patterns are observed in all cases. From the figure it is clear that when D=0 nm, maximum intensity is obtained and both intensity component decreases when distance increases. It is also evident from the figure that there is a large intensity variation when D varies from 0 to 5 nm. That is when D decreases; the field becomes more and more localized at the gap between the particle and the surface. The localization and excitation of electromagnetic normal modes are caused by resonant excitation of electromagnetic normal modes of the particle-surface system. It correlates well with theory.

5. Resist dissolution process

A modified cellular automata (CA) scheme for simulating resist dissolution during the development process is employed to compute the positional development rates of the photoresist domain in response to the normalized intensity profile (value of 1 mW/cm^2 for the unit normalized intensity value is adopted).

5.1 Theoretical model

Various forms of CA numerical techniques have been reported for simulating the photoresist development process in both the 2D and 3D spatial domain, showing good agreement between the simulated and experimental results [Karafyllidis, et al 2000; Scheckler, et al 1993]. During the development process, the developing reagent dissolves the exposed photoresist in an isotropic manner. In addition to the proven level of accuracy, the CA technique is straightforward in implementation as it allows the user to employ regular and uniform spatial and temporal discretization in modeling the physical system.

A modified three dimensional CA algorithm that follows closely, to the one proposed by Karafyllidis [Karafyllidis, 1999] is employed here. The exposed photoresist domain is divided into identical cubic CA cells, each with edge length a. Unlike previous works, the algorithm implemented here considers all the neighboring cells in the computation sub-domain when solving for the local state of the concerned cell. The local state of each CA cell is defined as the ratio of its etched volume to its total volume and the respective computation sub-domain is illustrated in Fig. 9. Based on Fig. 9, the amount of each cubic cell set at the position (i, j, k), dissolved at time instant t by its local state, $C_{i,j,k}^{t}$ this is defined as,

$$C_{i,j,k}^{t} = h_{diss}/a \tag{2}$$

where h_{diss} is the height dissolved in the development process. Based on the definition, the initial value for $C_{i,j,k}^{t}$ is assigned zero and when the computed value for which exceeds one, a value of one would be assigned. We further introduce the differential cell state terms, dC_{adj}, dC_{edg} and dC_{vtx}. These represent the change of the central cell state(cell for which the state to be solved) due to the flow of the developer from adjacent, edge and vertex cells positions at the new time instance, after one time step of dt from the time instance t_0.

An adjacent neighboring cell shares one face with the central cell, an edge neighboring cell shares an edge with the central cell and a vertex neighboring cell shares a corner vertex with the central cell. As such, the central cell state at the new time instance is defined here as,

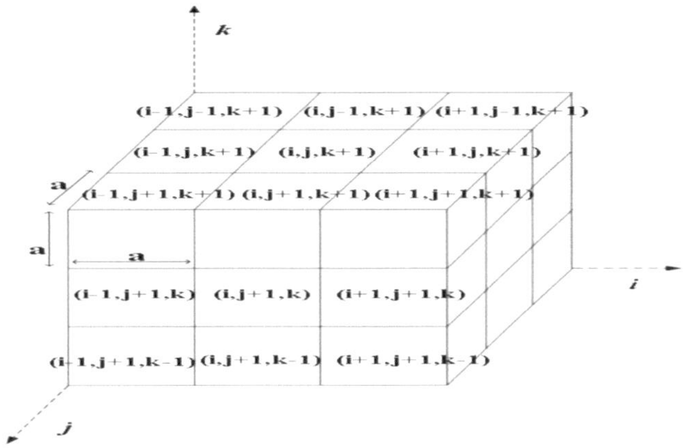

Fig. 9. The neighbourhood of the (i, j, k) cell. (Adapted from Ref. [Murukeshan et al, Opt. Eng. 47, 129001 (2008)])

$$C_{i,j,k}^{t_o+dt} = C_{i,j,k}^{t_o} + dC_{adj}^{dt} + dC_{edg}^{dt} + dC_{vtx}^{dt} \tag{3}$$

Assuming that the dissolution rate of the central cell to be $R_{i,j,k}$, terms would be defined as follow,

$$dC_{adj}^{dt} = \frac{\left(\left(C_{i+1,j,k}^{t_o} + C_{i-1,j,k}^{t_o} + C_{i,j+1,k}^{t_o} + C_{i,j-1,k}^{t_o}\right) + B_{i,j,k+1}^{t_o} + B_{i,j,k-1}^{t_o}\right) R_{i,j,k} dt}{a} \tag{4}$$

$$\begin{aligned} dC_{edg}^{dt} = \gamma_{edg} \Big(& \left(C_{i+1,j+1,k}^{t_o} + C_{i-1,j+1,k}^{t_o} + C_{i+1,j-1,k}^{t_o} + C_{i+1,j-1,k}^{t_o}\right) \ldots \\ & + B_{i+1,j,k+1}^{t_o} + B_{i-1,j,k+1}^{t_o} + B_{i,j+1,k+1}^{t_o} + B_{i,j-1,k+1}^{t_o} \cdots \\ & + B_{i+1,j,k-1}^{t_o} + B_{i-1,j,k-1}^{t_o} + B_{i,j+1,k-1}^{t_o} + B_{i,j-1,k-1}^{t_o} \Big) \frac{2R_{i,j,k}^2 t_o dt}{a^2} \end{aligned} \tag{5}$$

$$\begin{aligned} dC_{vtx}^{dt} = \gamma_{vtx} \Big(& B_{i+1,j+1,k+1}^{t_o} + B_{i-1,j+1,k+1}^{t_o} + B_{i+1,j-1,k+1}^{t_o} + B_{i+1,j-1,k+1}^{t_o} \cdots \\ & + B_{i+1,j+1,k-1}^{t_o} + B_{i-1,j+1,k-1}^{t_o} + B_{i+1,j-1,k-1}^{t_o} + B_{i+1,j-1,k-1}^{t_o} \Big) \frac{3\sqrt{3}}{8a^3} R_{i,j,k}^3 t_o^2 dt \end{aligned} \tag{6}$$

where $B_{I,J,K}^{t_o}$ is defined as,

$$B_{I,J,K}^{t_o} = \begin{cases} 1 & C_{I,J,K}^{t_o} = 1 \\ 0 & \text{otherwise} \end{cases} \tag{7}$$

where I may represents i - 1, i, or i+1 and likewise for J and K. As shown in Eq. (4) to Eq. (6), the binary switch function is implemented at those cell positions with $K = k - 1$ or $K = k + 1$. The hypothesis for the choice of implementation is that the central cell would not be exposed to developer from the higher or lower cell positions until the resist elements occupying those positions are completely dissolved. The time step, dt is defined as, $a/4R_{max}$ where R_{max} refers to largest elemental development rate value in the simulation domain. In this work, the terms γ_{edg} and γ_{vtx} are parameters used to model the indirect flow of developer from the edge cell positions and from the vertex cell positions. The best values for γ_{edg} and γ_{vtx} would be shown and discussed in the later section.

5.2 Homogenous development rate test

The modified 3D CA model is evaluated using a homogeneous etch-rate distribution function [Karafyllidis, 1999] to determine the values of γ_{edg} and γ_{vtx} in Eq (5) and Eq (6) and their effectiveness to describe the effects from the edge and vertex cells. The cells are assigned the same development rate value. At the beginning of the simulation, all the cell states are assigned value of zero except for the one at the centre of the top surface. For the CA model proposed here, the optimum γ_{edg} and γ_{vtx} values would be searched so that the generated profile has the closest conformity to that of a hemispherical crater. For this purpose, we adopt a statistical approach that is explained with the aid of Fig. 10.

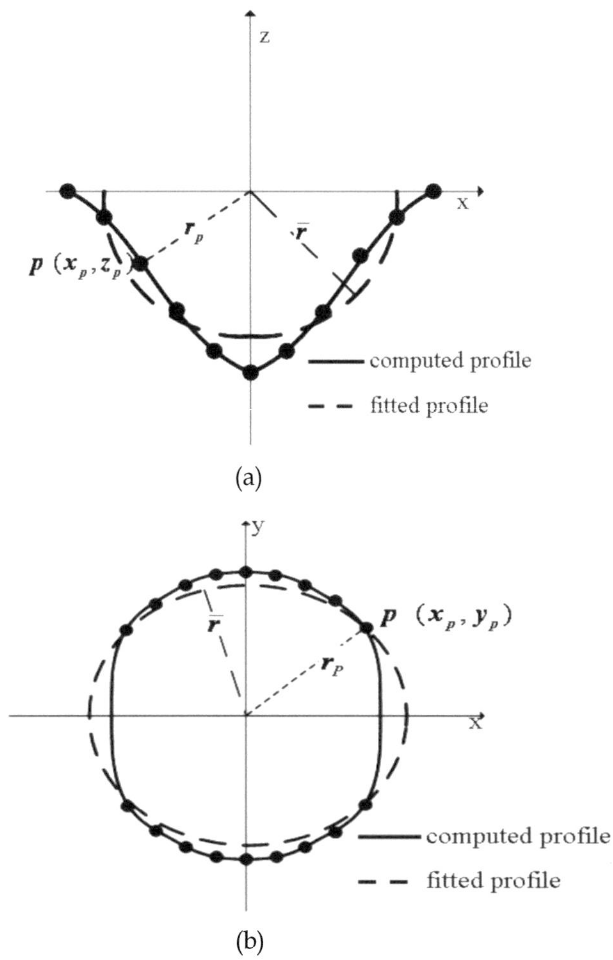

Fig. 10. Computed and fitted profiles for (a) a cross section and (b) the top of the generated hemispherical crater in the homogeneous development rate test. (Adapted from Ref. [Murukeshan et al, Opt. Eng. 47, 129001 (2008)])

Each computed boundary point p is quantified by its distance from the origin given by $r_p = (x_p^2 + y_p^2 + z_p^2)^{1/2}$. The fitted profile is generated by assuming a semicircle for the cross-section profile and a circle for the top profile, as shown in Fig. 10(a) and 10(b), respectively. The radius of $\bar{r}$ is given by $\bar{r} = \sum_{p=1}^{N} r_p \Big/ N$, where N refers to the number of computed profile by the standard deviation σ of the computed r_p values from the $\bar{r}$ value of the fitted perfect hemisphere. A smaller σ value would infer that the computed profile has a higher conformity to a spherical profile. A range of values for γ_{edg} and γ_{vtx} are investigated to identify the combination that generate the smallest σ value. The values for γ_{edg} and γ_{vtx} are varied from 0.1 to 1.0 with coarse step size of 0.1 and with the same total number of time steps. The corresponding results are given in Fig. 11.

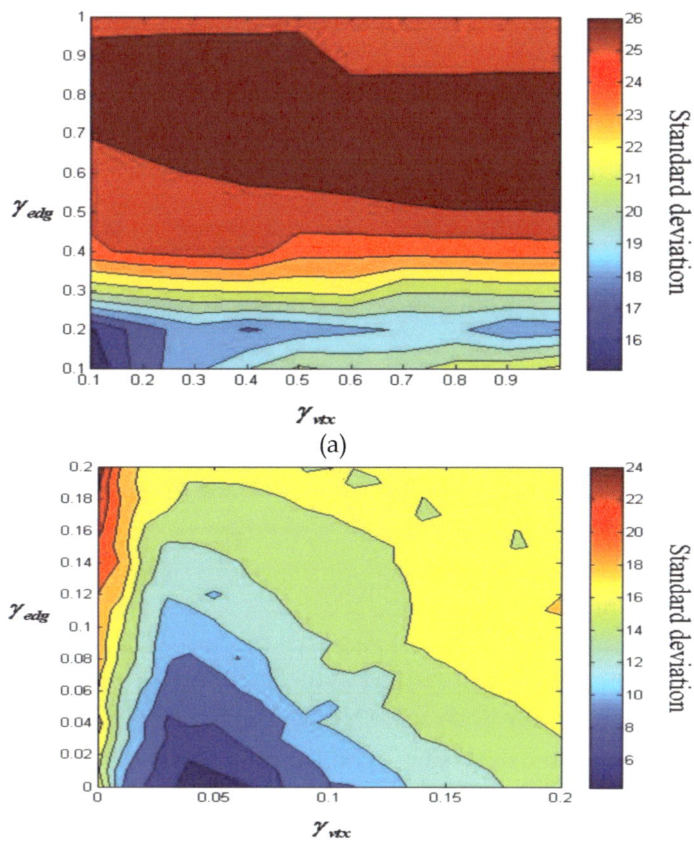

Fig. 11. Contour distribution of standard deviation with respect to coupled variation γ_{edg} and γ_{vtx}, (a) coarse (b) fine variations of γ_{edg} and γ_{vtx} values. (Adapted from Ref. [Murukeshan et al, Opt. Eng. 47, 129001 (2008)])

After identifying the possible combinations of γ_{edg} and γ_{vtx} values, the analysis is repeated, using a smaller range of values and a smaller step size of 0.001. The results in Fig. 11(b) suggest that the best γ_{edg} and γ_{vtx} values are 0 and 0.04, respectively, yielding a standard deviation value of 4.29 nm. The CA modal is simplified by employing the obtained γ_{edg} and

γ_{vtx} values shown in Eqs. (5) & (6). The results for the homogeneous development rate, shown in Fig. 12, are obtained with the optimum γ_{edg} and γ_{vtx} values.

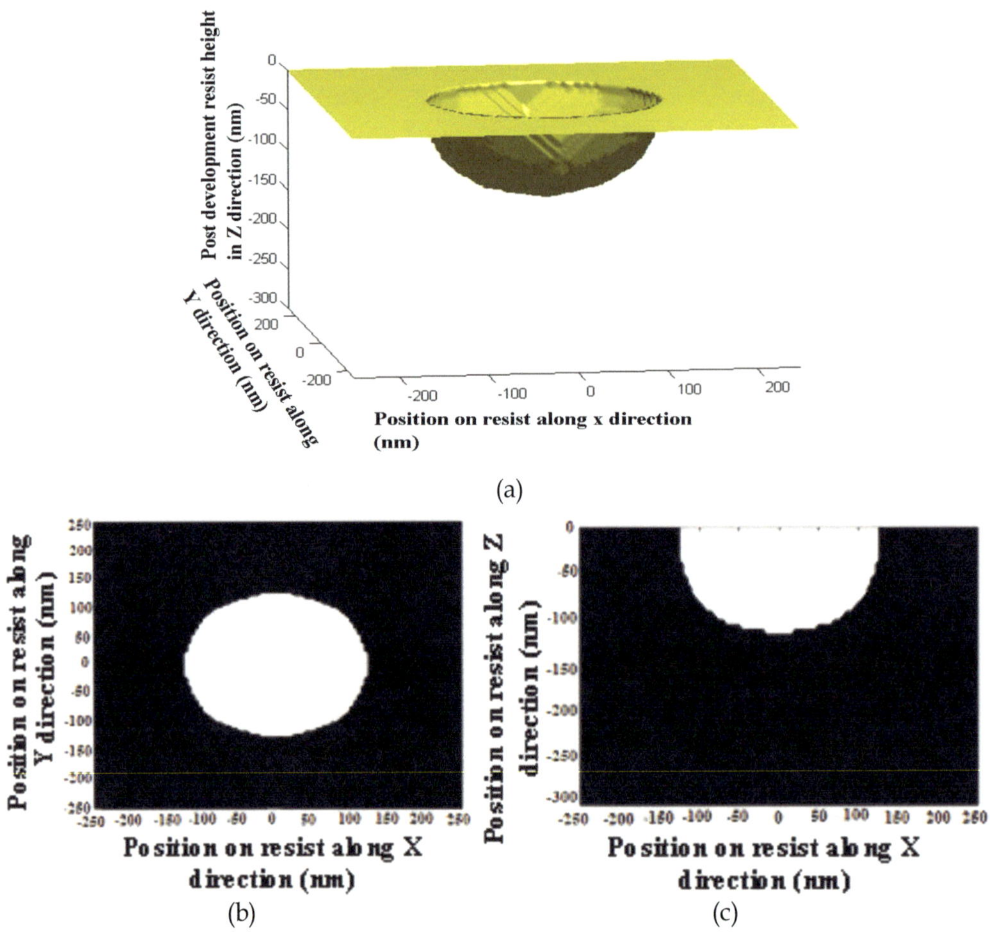

Fig. 12. (a) Three dimensional perspective of the crater generated by the CA model. (b) Computed profile on the x-y plane at z=0 (c) Computed profile on the x-z plane through y=0. (Adapted from Ref. [Murukeshan et al, Opt. Eng. 47, 129001 (2008)])

The generated crater shows a reasonably good hemispherical profile. This result suggests that the proposed CA modal is reasonably accurate in simulating the isotropic nature of the resist-etching mechanism during the resist development process. Besides CA, another widely used numerical scheme for simulating resist development is the fast marching level set method formulated by Sethian and Adalsteinsson [Sethian & Adalsteinsson, 1997]. This method is efficient in handling a large computation domain and accurate in simulating resist topologies with complex geometrics, as in the case of lithography simulation for semiconductor chip fabrication.

5.3 Resist features

The modified CA modal is adopted in order to obtain the resultant resist features, as discussed in the previous sect. 5.2

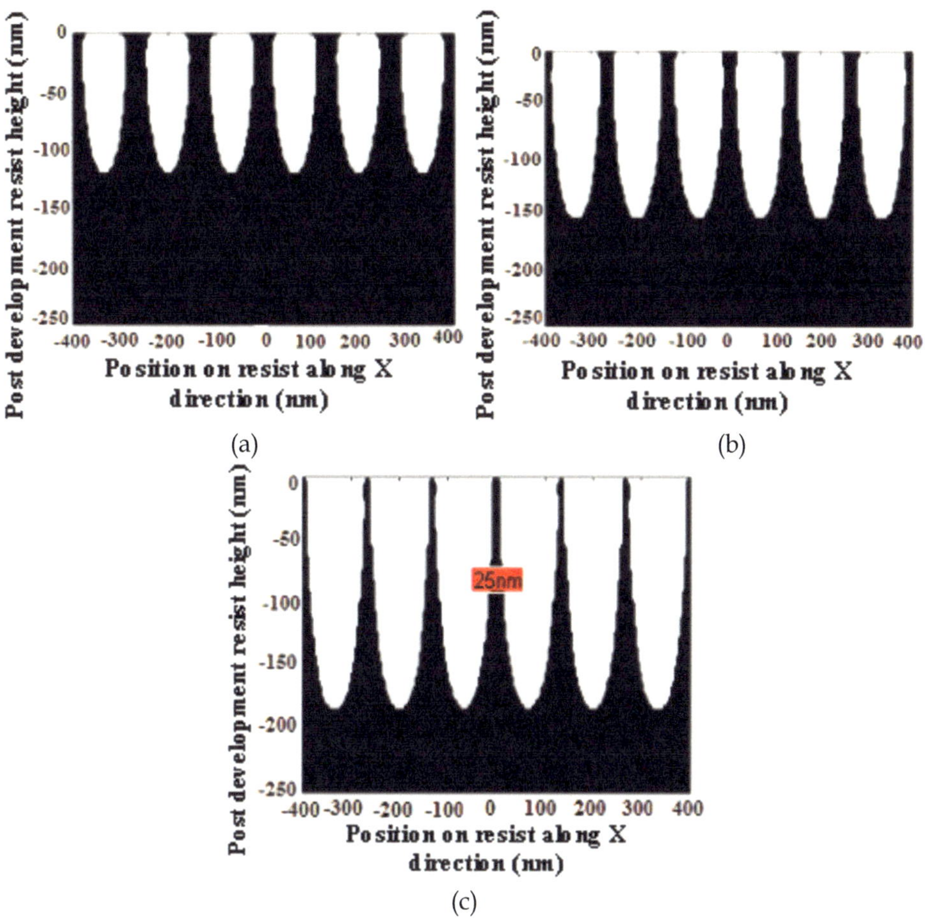

Fig. 13. 2D resist profile cross section at exposure times: (a) =50 s, (b) =100 s and (c) =150 s. (Adapted from Ref. [V. M. Murukeshan and K. V. Sreekanth, Opt. Lett. 34, 845-847 (2009)])

Figures 13 and 14 respectively show 2D and 3D resist cross-section profiles obtained on the photoresist layer by employing the proposed concepts and configuration at different exposure times. The zero value along the vertical axis corresponds to the metal/photoresist interface. The obtained result shows that with increase in exposure time, the line width of the pattern decreases and exposure depth increases. Fig.13 (c) shows that at 150 s exposure duration, the obtained line width, periodicity and exposure depth are around 25 nm, 120 nm and 180nm respectively. Therefore approximate resolution of 25 nm is achievable by this configuration, which is generally not possible with conventional prism based configuration. Also, the advantage of this configuration is that it can provide much larger enhanced fields to give shorter wavelengths of surface plasmons compared to that achieved with an isolated metal particle or a metal surface alone configuration.

6. Conclusion

A recently proposed novel plasmonic lithographic concept and methodology based on the excitation of gap modes in a metal particle-surface system is discussed in this chapter. The

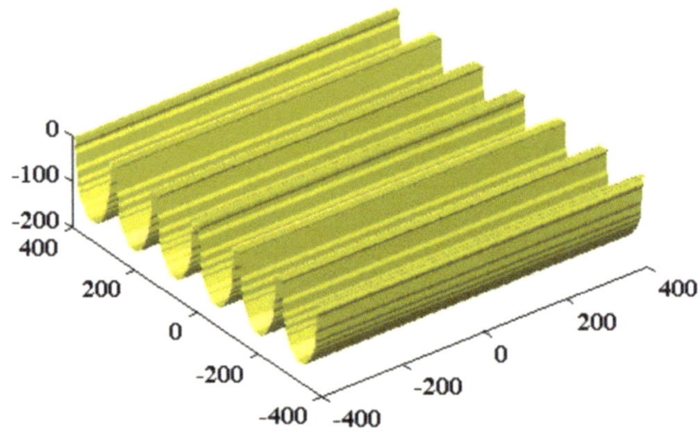

(a)

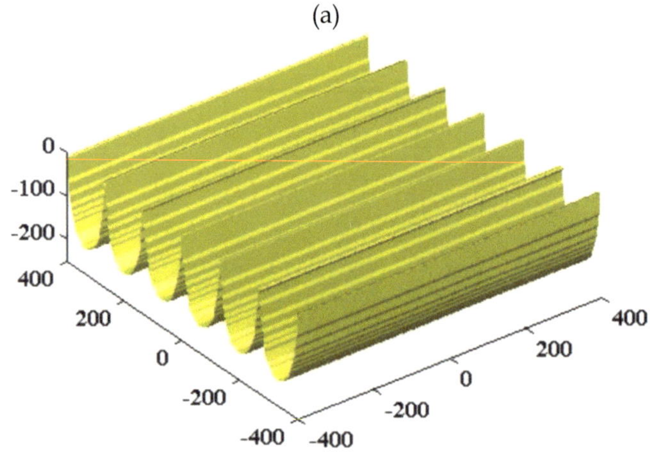

(b)

Fig. 14. 3D resist profile cross section at exposure times: (a) =100 s, (b) =150 s

proposed approach is compared with conventional configurations and illustrated numerically that the exposure depth of the pattern achieved with this configuration is much higher than that achieved with a conventional prism based configuration. The simulation result also shows that this configuration can provide strong enhanced field to give shorter wavelengths of surface plasmons to fabricate sub-25 nm size periodic structures. In order to simulate the resist removal process during the post exposure development stage, a modified CA algorithm was proposed and explained. The theoretical analysis of CA model and resist profile cross section obtained through this proposed configuration is also presented. It is expected that this lithography concept can achieve high resolution, good exposure depth and good contrast to fabricate one-dimensional periodic nanostructures for various applications including biosensors, photonic crystals, and waveguides.

It should be noted that though a dielectric sphere/dielectric surface based configuration contribute gap modes, the degree of localization and enhancement is smaller than that for a metal sphere-surface system. The experimental evidence of the proposed configuration also need to be investigated further. In a practical scenario, to fabricate high aspect ratio nanoscale features, certain requirements must be met. First, the laser output beam should have good pointing stability. Second, a high contrast of the interference fringes must be achieved by keeping the intensity between the incident beams equal, followed by the appropriate control of their polarization state. Third, the mechanical strength of the photoresist must be high enough to withstand the capillary force exerted by fluid between the features. The capillary force is directly proportional to the aspect ratio of the features. These above mentioned potential research challenges augur well for realizing high resolution , high aspect ratio feature fabrication in the near future.

7. Acknowledgment

The authors acknowledge the financial support received through ARC 3/08 and AcRF.

8. References

Adams, A.; Hansma, P.K. (1981). Light emission from small metal particles and thin metal films excited by tunneling electrons. *Phys. Rev. B,* 23, 8 3597-3601, 1550-235X.

Blaikie, R. J.; McNab, S. J.(2001). Evanescent interferometric lithography. *Appl. Opt.*, 40, 10, 1692-1698, 1539-4522.

Boardman, A.D. (1982). Electromagnetic surface modes. John Wiley & Sons Ltd, 0471100773, Chichester.

Chen, K.S.;Lin, I.K.; Ko, F. H. (2005). Fabrication of 3D polymer microstructures using electron beam lithography and nanoimprinting technologies. J. Micromech. Microeng., 15, 10, 1894-1903, 1361-6439.

Chua, J. K.; Murukeshan, V. M.; Tan, S. K.; Lin, Q. Y. (2007). Four beams evanescent waves interference lithography for patterning of two dimensional features. Opt. Exp., 15, 6, 3437-3451, 1094-4087.

Ebbesen, T. W.; Lezec, H. J.; Ghaemi, H. F.; Thio, T.; Wolff, P. A. (1991). Extraordinary optical transmission through sub-wavelength hole arrays. *Nature* (London), 391, 667-669, 1476-4687.

Gwyn, C. W.; Stulen, R.; Sweeney, D.; Attwood, D. (1998). Extreme ultraviolet lithography. *J. Vac. Sci. Technol.* B, 16, 3142-3149, 1520-8567.

Guo, X.; Du, J. ; Guo, Y.; Yao, J.(2006). Large-area surface- plasmon polariton interference lithography. *Opt. Lett,* 31,17, 2613-2615, 1539-4794.

Hayashi, S. (2001). Spectroscopy of gap modes in metal particle-surface systems, In: *Near-field optics and surface plasmon polaritons,* Kawata, S. 71-95, Springer. 978-3-540-41502-2, Berlin Heidelberg.

Jersch.J. (1997). Nanostructuring with laser radiation in the near field of a tip from a scanning forcen microscope, *Appl. Phys. A,* 64, 29-32, 1432-0630.

Johansson, P.; Monreal, R.; Appel, P. (1990). Photon emission from STM-concepts, *Phys. Rev. B,* 42, 9210, 1550-235X.

Karafyllidis, I. (1999). A three-dimensional photoresist etching simulator for TCAD, *Modeling Simul. Mater. Sci. Eng.* 7, 157-168, 1361-651X

Karafyllidis, I; Hagouel, P. I.; Thanailakis, A.; Neureuther, A. R. (2000). An efficient photoresist development simulator based on cellular automata with experimental verification. *IEEE Trans. Semicond. Manuf.* 13, 61-75, 0894-6507.

Kreibig, U.; Vollmer, M. (1995).*Optical properties of metal clusters,* Springer, 0387578366, Berlin/ Heidelberg.

Lim, Y; Kim, S; Kim, H; Jung, J; Lee, B (2008) Interference of surface plasmon waves and plasmon coupled waveguide modes for the patterning of thin film. *IEEE J. of. Quant. Elec.* 44, 305-311, 0018-9197.

Liu, Z.; Wang, Y.; Yao, J.; Lee, H.; Srituravanich, W.; Zhang, X. (2009) Broad band two dimensional manipulation of surface plasmons. *Nano Lett.*, 9, 1, 462-466, 1530-6992.

Luo, X.; Ishihara, T. (2004). Surface plasmon resonant interference nanolithography technique, *Appl. Phys. Lett.*, 84, 23, 4780-4782, 1077-3118.

Madrazo, A; Neito-Vesperinas, M. (1996). Exact calculation of Maxwell equations for a tip-metallic interface configuration: Application to atomic resolution by photon emission. *Phys. Rev. B*, 53, 3654-3657, 1550-235X.

McAlpine, M. C. ; Friedman, R. S; Lieber, C. M. (2003). Nanoimprint Lithography for Hybrid Plastic Electronics. *Nano Lett.*, 3, 443-445, 1530-6992.

McCarthy, S. L.; Lambe, J.(1978). LEIT effect in metal-insulator-semiconductor tunnel junctions. *Appl. Phys. Lett.* 33,10, 858-860, 1077-3118.

Murukeshan, V.M; Chua, J. K.; Tan, S. K.; Lin, Q. Y. (2008) Modeling of subwavelength resist grating features fabricated by evanescent waves interference. Opt. Engi., 47, 12, 129001, 1560-2303.

Murukeshan, V.M.; Sreekanth, K. V. (2009). Excitation of gap modes in a metal particle-surface for sub-30nm plasmonic lithography. *Opt. Lett.*, 34, 845-847, 1539-4794.

Okazaki, S. (1991). Resolution limits of optical lithography. *J. Vac. Sci. Technol.* B, 9, 2829-2833, 1520-8567.

Palik, E. D. (1985). *Handbook of Optical Constants of Solids*. Academic Press, 0125444222, Orlando.

Prodan, L.; Euser, T. G; Wolferen, V.; Beigang, R.; Kuipers.L. (2004) Large-area two-dimensional silicon photonic crystals for infrared light fabricated with laser interference lithography. *Nanotechnology*, 5, 639-642, 1361-6528.

Raether, H. (1988). *Surface plasmons on smooth and Rough Surfaces and on Gratings.* Springer, 978-3-540-17363-2, Berlin Heidelberg.

Rendell, R.W.; Scalapino, D. J.; Muhlschlegel, B. (1978). Role of local plasmon modes in light emission from small-particle tunnel Junctions, *Phys. Rev. Lett.*, 41, 25, 1746-1750, 1079-7114.

Rendell, R.W.; Scalapino, D. J. (1981). Surface plasmons confined by microstructures on tunnel junctions. *Phys. Rev. B*, 24, 6, 3276-3294, 1550-235X.

Scheckler, E. W.; Tam, N. N.; Pfau, A. K.; Neureuther, A. R. (1993). An efficient volume removal algorithm for practical three dimensional lithography simulation with experimental verification", *IEEE Trans. Comput-Aided Des.* 12, 1345, 0278-0070.

Sethian, J. A.; Adalsteinsson. (1997). An overview of level set methods for etching, deposition and lithography development. *IEEE Trans. Semicond. Manuf.*, 10, 167-184, 0894-6507.

Shao, D. B.; Chen, S. C. (2005). Surface plasmon assisted nanoscale photolithography by polarized light, *Appl. Phys. Lett*, 86, 253107 (1-3), 1077-3118.

Silverman, J. P. (1998). Challenges and progress in x-ray lithography. *J. Vac. Sci. Technol.* B, 16, 3137-3141, 1520-8567.

Srituravanich, W.; Fang, N.; Sun, C.; Luo, Q.; Zhang, X. (2004). Plasmonic nanolithography. *Nano Lett.*, 4, 6, 1085-1088, 1530-6992

Stockman, M. (2008). Spasers explained, *nature photonics*, 2 , 327-329, 1749-4893.

Wu, Q; Feke, G. D.; Grober, R.D.; Ghislain, L. P. (2007). Realization of numerical aperture 2.0 using a gallium phosphide solid immersion lens. *Appl. Phys. Lett.*, 75, 4064-4067, 1077-3118.

30

Nanosphere Lithography for Nitride Semiconductors

Wai Yuen Fu and Hoi Wai Choi
The University of Hong Kong
Hong Kong

1. Introduction

Nanolithography has been widely studied in recent years, considering its wide-ranging applications in electronics and photonics. In the ideal case, lithography using short-wavelength electromagnetic wave through a pre-defined mask is probably the most desirable solution. However, as technology progresses further, traditional photolithography is becoming exceedingly complex and incurring higher cost associated with light source as the wavelength of light goes shorter. Therefore, other nanolithography methods become strong competitors excelling with lower equipment cost and simpler fabrication procedures. In particular, nanosphere lithography as a self-assembly bottom-up approach for producing periodic array of spherical particles is simple and inexpensive as compared to other lithographic methods.

Originally developed by Deckman et al. for defining a large area lithographic mask (Deckman & Dunsmuir, 1982), nanosphere lithography was further optimized by Hultenn et al. for applications involving surface-enhanced Raman spectroscopy (Haynes & Van Duyne, 2003; Hulteen & Vanduyne, 1995). Nanospheres that are used in lithography are commercially available in solution form, and through a simple coating procedure with optimized conditions, ordered hexagonal array of nanospheres can be naturally assembled as the nano-particles achieve equilibrium in an initially disordered system, i.e., a solution of nanospheres. Although only hexagonal array can be formed by this lithographic technique, the hexagonal pattern is useful for a range of applications, for instance, in nanophotonics where ordered hexagonal patterns can be used for the fabrication of photonic crystal (Han et al., 2005; Su et al., 2008) and also in plasmonics (Jensen et al., 2000; Malinsky et al., 2001; Stewart et al., 2006). With rivals like nanoimprint lithography (Byeon et al., 2007; Kim et al., 2007) and electron beam (e-beam) lithography (Berrier et al., 2004; David et al., 2006; Noda et al., 2007), the low cost and high through-put (albeit being less precise) nature of nanosphere lithography is especially suited for photonic applications.

Gallium nitride (GaN) semiconductors and devices thus serve well as a platform for nanosphere lithography. In recent years, GaN has emerged as the most successful semiconductor material in optoelectronics, delivering a range of commercialized products which has since revolutionized our lives, including violet laser diodes for ultra-high capacity optical storage and white light LEDs for solid-state lighting (Humphreys, 2008; Schubert & Kim, 2005). While GaN technology is becoming mature, there is still plenty of room for further development. For instance, one major issue with GaN devices is the limited efficiency of light extraction. To tackle this problem, texturing of the surface in a regular manner has been adopted at the micrometer and nanometer scales (David et al., 2008; Han et al., 2006; Ng et al., 2008; Noda & Fujita, 2009).

Therefore in this chapter, we will demonstrate the potential applications of nanosphere lithography including photonic crystals and surface plasmons. Several topics will be outlined in this chapter, including the various techniques of coating nanosphere arrays, together with a detailed analysis of their structural, morphological and optical characteristics. The possibilities of applying nanosphere lithography into practical GaN devices and beyond will be discussed.

2. Process of self-assembled nanosphere lithography

There have been intense efforts targeted at the formation of nanosphere into regular patterns. Although lots of variants of the process have since been developed, including the vertical deposition method (Gil et al., 2007; Jiang et al., 1999; Ye et al., 2001; Zhou & Zhao, 2004), spin coating (Mihi et al., 2006; Ogi et al., 2007) and even merely dipping (Choi et al., 2009; Im et al., 2003), all these fundamentally rely on manipulation of the forces applied amongst nanospheres. Thus, most of these methods require precise control of ambient conditions such as temperature and pressure, together with the viscosity and evaporation rate of the fluid. These conditions could be varied in order to control the number of layers and degree of order of the assembled pattern, as well as the extent of coverage across the sample. It was found that the vertical deposition method yields a better quality of ordered nanosphere array (Wong et al., 2003) compared with others, therefore in this chapter we will concentrate on its discussion.

2.1 Vertical deposition

Known for its good orderliness and simplicity, the vertical deposition method has been adopted in many research studies employing nanosphere lithography (Chiappini et al., 2007; Kuai et al., 2004). A common and general setup for vertical deposition consist of an apparatus for maintaining a constant ambient temperature such as a dry-bath, or a basic oven, that is capable of providing a stable stream of air or gas, and a vial for holding the nanosphere in solution form with the sample to be coated lying upon the sidewall of the vial. A schematic diagram of a typical setup is illustrated in Figure 1. The solution typically consist of nanospheres suspended, and a water-alcohol mixture that controls the evaporation rate (Shimmin et al., 2006). Surfactants are sometimes used together to reduce the viscosity. As the nanosphere solution dries up, the nanosphere will be self-assembled into hexagonal close-packed pattern, adhering to the sample. Unfortunately, cracks often exist in the resultant nanosphere coating, regardless of the degree of order. The reason for the persistent formation of cracks is identified to be caused by the movement of "water line" as the water level changes during evaporation of solution. Also, since the concentration increases as the solution dries up, the thickness of the nanosphere coating, or the number of nanosphere layers, will usually be thicker towards the lower end of the sample than that at the upper end.

The problem of uneven layer distribution along the sample can be resolved by a modified vertical deposition method, namely the flow-controlled vertical deposition method. Instead of waiting for the solution to be evaporated, this modified method proactively scans the

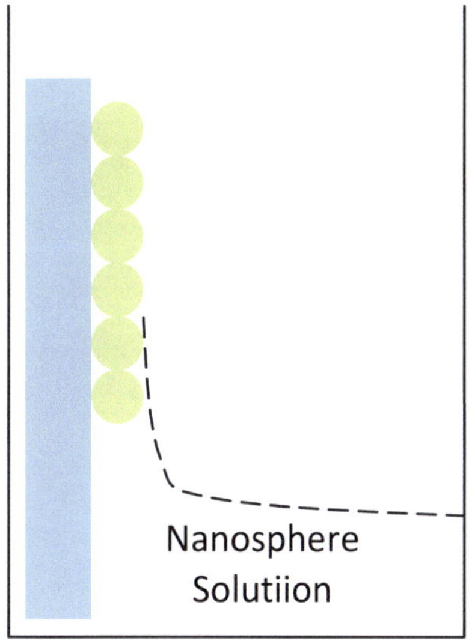

Fig. 1. Schematic diagram of the setup for vertical deposition.

water level along the entire sample by adjusting the relative position between sample and vial, while keeping the concentration changes to a minimum. This allows a much shorter processing time which also minimize possible sedimentation of nanospheres that will aggravate the unevenness during the deposition process. Thus, this method is ideal for maintaining a more even thickness of coating across the sample. For addressing the cracking problem, there have been attempts to use a chemical method in conjunction with vertical deposition method. L. Wang et al. have reported crack-free colloidal crystal by means of hydrolysis of a silica precursor during vertical deposition (Wang & Zhao, 2007). Silica species formed acts like a glue among nanospheres to avoid crack formation during evaporation. Other crack-free approach includes self-assembly at air/water/air interface which will form a free standing nanosphere films, although this is not an approach suitable for lithography.

3. Characterization of nanosphere coating

3.1 Topography of nanosphere coating

After deposition, a hexagonal close-packed nanosphere pattern can be observed under scanning electron microscope (SEM), as shown in Figure 2, if the coating is successful. Unless using very specific approaches to eliminate cracks as suggested in the last section, different kinds of defects, including but not limited to cracks and point defects, will be formed. As mentioned in the previous section, cracks formed through vertical deposition are most likely due to the movement of "water line" across the sample. However, there are also many other causes that could lead to a defect. For example, the vast amount of nanosphere that a solution contains cannot possibly be of identical dimensions. A single apparently larger or smaller nanosphere amongst many in an array is enough to give rise to point defects. Apart from dimensional non-uniformity, irregular nanosphere may also exist, that is, some of the nanosphere may not be perfectly spherical. Point defects will also be resulted from the protruding portions of irregular nanospheres. Due to the limitations in nanosphere synthesis and deposition, defects among nanosphere arrays are inevitable. Nevertheless, with a low defect density, nanosphere lithography is adequate for many applications unless extremely high accuracy is required.

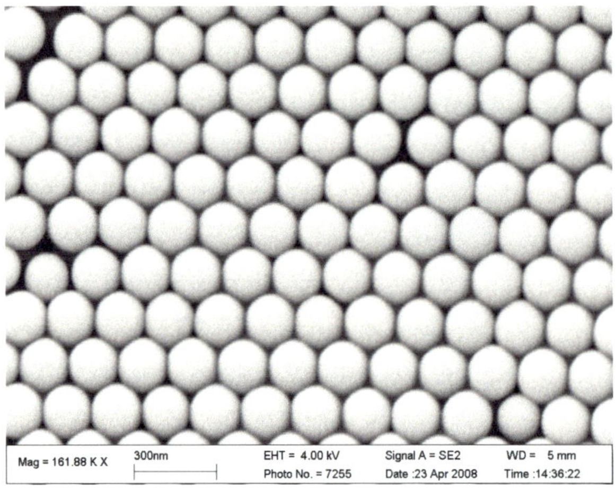

Fig. 2. SEM image of a monolayer of nanospheres, with occasional defects caused by non-uniformity in dimension of nanospheres.

One distinct advantage of nanosphere lithography is that nanospheres that self-assemble into a hexagonal close-packed pattern can produce features of nano- or even sub-nano scales, provided that the required dimensions of nanosphere are available. The voids between nanospheres are at least one order of magnitude smaller than the diameter of nanosphere itself; that is, for a 10 nm-diameter nanosphere, the void can easily be smaller than 1 nm.

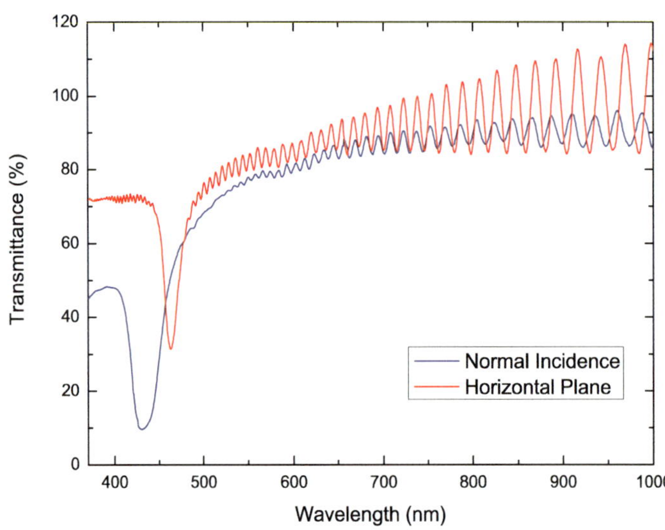

Fig. 3. Transmission spectrum of hexagonal close-packed 192-nm-diameter nanosphere multi-layer array coated on a GaN wafer measured at 2 different angles with reference to an as-grown GaN sample. (with permission for reproduction from Institute of Physics Publishing)

3.2 Optical properties of nanosphere coating

The optical properties of a nanosphere coating are pretty obvious by observing it at different angles under broadband illumination - different colors of light are reflected at different angles, acting like a grating. In fact, the nanosphere coating is a 2-dimensional photonic crystal itself. In the frequency domain there is in fact a gap in the transmission spectrum, an example of which is illustrated in Figure 3. The existence of the gap indicates that certain wavelength of light is being reflected or scattered away. The spectral position of the gap varies according to the material's refractive index, the size of nanosphere, orientation of coating and the number of layers. Typically the gap exists at a position at 2-3 times the diameter of nanosphere spectrally.

4. Nanosphere coating applications

4.1 Nanopillar photonic crystal LED

Besides using the nanospheres themselves directly, a patterned nanosphere layer can act as a good medium for further processing. Amongst different kinds of materials, silica-made nanosphere is especially suitable for etching process, since the good etch resistivity of silica renders the patterned monolayer of nanosphere a naturally good hard mask.

One of the most popular 'secondary' nanostructures derived from nanosphere lithography is the nanopillar array (Cheung et al., 2006; Li et al., 2007; Ng et al., 2008). After coating a monolayer of nanospheres onto a sample, the circular hexagonal close-packed pattern can be easily transferred to the sample by a subsequent dry etch. Figure 4 shows an example of the resultant nanopillar structure. An array of nanopillar is an effective approach to improve light extraction when fabricated on a GaN LED. Due to the high refractive index of GaN, LEDs without any light extraction strategy adopted will suffer from low light extraction efficiency, as a large fraction of emitted light is trapped within the GaN epilayer, propagating as guiding modes and subsequently re-absorbed. A nanopillar array integrated into the surface of an LEDs will have a surface-texturing effect. In one way, the nanopillars increase the total surface area of LED, thus increasing the probability for light to escape, and thereby decreasing photon re-absorption in GaN. At the same time, they reduce the effective refractive index such that total internal reflection is suppressed to a certain extent.

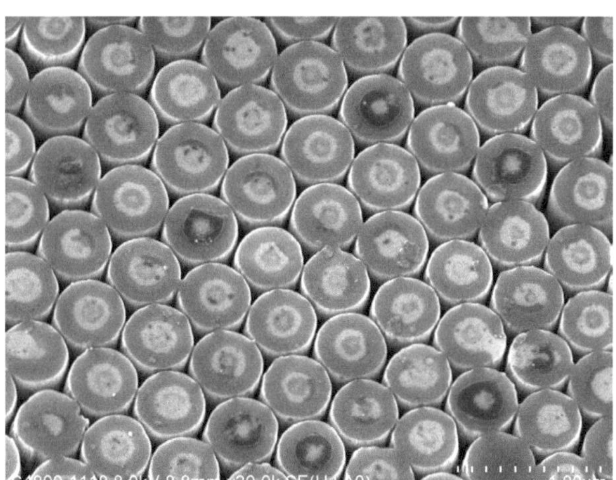

Fig. 4. SEM Image of a nanopillar array with diameter of 500 nm fabricated on GaN wafer.

In the research studies by W. N. Ng et al., a monolayer of silica nanospheres of 500 nm diameter, coated by spin coating method, served as a lithographic mask. Inductively coupled plasma (ICP) etching using a gas chemistry of Cl_2 and BCl_3 at flow rates of 20 sccm and 10 sccm respectively was carried out for a duration of 120 s. The silica nanosphere residual was removed by sonification to expose the nanopillar array beneath. The resulting nanopillar sample contributed to a two-fold enhancement of photoluminescence intensity in the normal direction. In their later studies, the authors successfully demonstrated light extraction enhancement from an electroluminescent LED with an integrated nano-pillar array. Significant improvement to the optical output was achieved compared to an unpatterned LED, as shown in the plot of *L-I* characteristics in Figure 5.

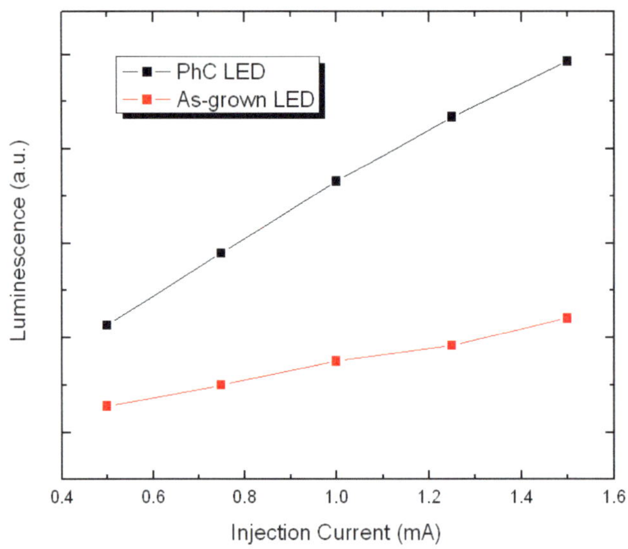

Fig. 5. *L-I* characteristics of a nanopillar photonic crystal LED as compared with an un-patterned LED. (with permission for reproduction from Institute of Physics Publishing)

To further extend the applications of nanospheres to GaN materials, an isotropic etching model with a monolayer of nanosphere has been reported by W. Y. Fu et al. Hemiellipsoid structures were been fabricated, as illustrated in Figure 6, using nanosphere coating mask with a modified recipe from the study in the previous paragraph, using a gas chemistry comprising 12 sccm of Cl_2 and 9 sccm for CHF_3. Compared to the etch recipe adopted by W. N. Ng et al., this modified recipe makes use of CHF_3 that is particularly suited for etching of silica; as a result the dimensions of silica nanosphere in the array will slowly shrink as etching progresses. The etch rates of GaN and SiO_2 can be controlled independently, allowing more freedom in the control of desired shape of the resultant arrays. This is useful for the design of different photonic dispersion characteristics that can be exhibited by photonic crystal structures. For example, a photonic bandgap based on an array of hemiellipsoids can be designed for light extraction purposes. From the angular PL emission plot as shown in Figure 7 (Fu et al., 2009), an obvious overall enhancement to the PL intensity is observed. By comparing the integrated PL intensity between an as-grown with the hemiellipsoid array an enhancement factor of over 3 times can be deduced.

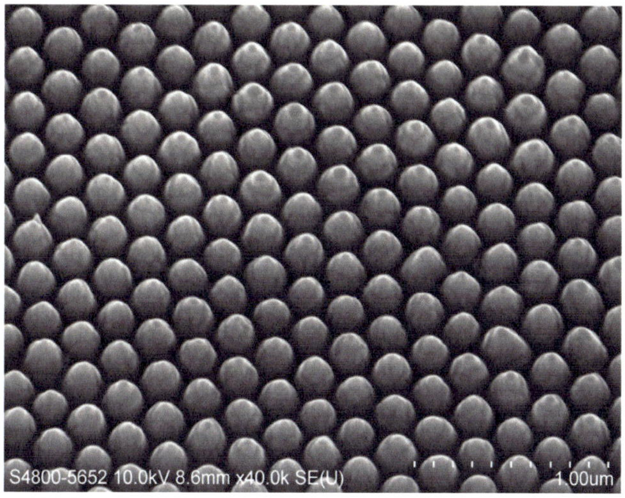

Fig. 6. SEM image of a hemiellipsoid array fabricated by nanosphere lithography. (with permission for reproduction from American Institute of Physics)

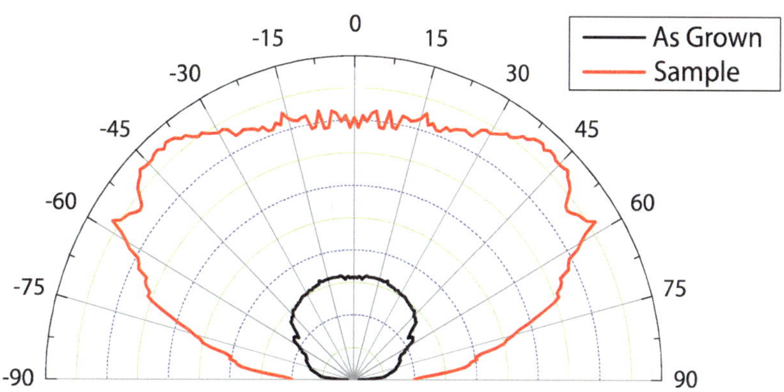

Fig. 7. Angular PL emission pattern from an LED fabricated with a self-assembled hemi-ellipsoid on top for superior light extraction. (with permission for reproduction from American Institute of Physics)

4.2 Fluorescent nanospheres coated white LED

The white-light LED has become a widely available commercial product with numerous applications including solid state lighting, liquid crystal display (LCD) backlighting and signaling etc. There are two common approaches to achieve white light emission from an LED, either by spectral down conversion using YAG phosphor-coated blue LEDs (Pan et al., 2004; Schubert & Kim, 2005) or the mixing of discrete LEDs with the primary colors red, green and blue (Hui et al., 2009; Humphreys, 2008; Muthu et al., 2002; Steigerwald et al., 2002). Due to the technical difficulties of color mixing with three discrete LEDs without bulky optics, phosphor-coated white LED tends to be more widely adopted despite the large Stokes shift losses. Under this circumstance, fluorescent nanospheres coated white LED has been suggested as an alternative with some advantage over phosphor-coated white LED. A schematic diagram of such design of LED device is shown in Figure 8.

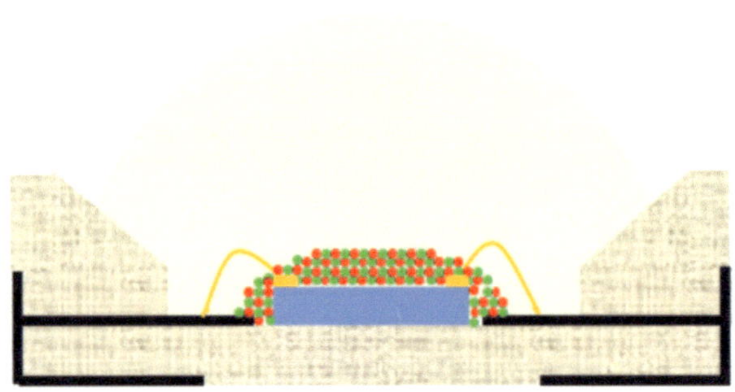

Fig. 8. Schematic diagram of a fluorescent nanosphere coated LED device. (with permission for reproduction from Institute of Physics Publishing)

Nanospheres with fluorescent dyes incorporated internally have been demonstrated to act as an alternative to the conventional approach of coating phosphor on blue LED for white light emission. Commercially available in dimensions ranging from tens of nanometers to micrometers, polystyrene fluorescent-dyed nanospheres have been widely used in the chemical and biological areas (Bhalgat et al., 1998; Matsuya et al., 2003; Seo & Lee, 2004). The swelling and unswelling of the polymeric spheres during preparation ensures that the pores are able to entrap the fluorescent dyes physically, resulting in an enhanced dye photostability, with predicted lifetimes exceeding 36 months. Simply by adjusting the ratio between excited light emission from green- and red- colored nanospheres with respect to the blue light emission intensity from InGaN LED, white light emission can be achieved and the

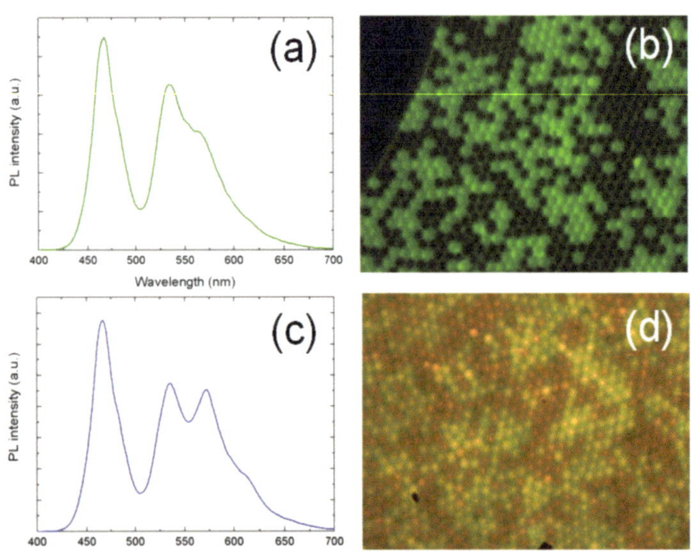

Fig. 9. (a) PL spectrum of the green microspheres, (b) optical image of fluorescence emission from green microspheres (with blue and red light filtered off). (c) PL spectrum of the mixture of green and red microspheres, and (d) optical image of fluorescence emission from green, yellow and red microspheres (with blue light filtered off). (with permission for reproduction from Institute of Physics Publishing)

color temperature can be easily adjusted, although the color conversion is still subjected to Stokes shift loss. However, the hexagonal close-packed patterned and spherical-shaped nature of nanospheres contributes to imrpvoed mixing of color both spatially and angularly, with a slightly better advantage on excitation of emission of other colors compared with conventional phosphors.

Figure 9 shows PL spectra and optical images of fluorescent nanosphere on GaN LEDs with blue emission as in the studies by K.N. Hui et al. (Hui et al., 2008). With different colors of fluorescent nanospheres, including green, yellow and red, emission spectrum of LEDs can be tuned to different shades of white light.

4.3 Surface plasmon

Induced by electromagnetic wave, when collective oscillations of free electrons, i.e., plasmons, are confined to surfaces and couples strongly with light, they are known as surface plasmon. This usually occurs at the interface between a metal (of negative dielectric constant) and a dielectric material (of positive dielectric constant). These surface plasmons are conducive to an important phenomenon, the localized surface plasmon resonance, which occurs when metallic nanoparticle is excited by light when in contact with dielectric material. This phenomenon is exhibited by a strong peak in the extinction spectrum. With GaN devices, localized surface plasmon resonance has great potential for improving light extraction efficiency of LEDs since it can significantly enhance the electric field strength near the surface of nanoparticle (Ross & Lee, 2008; Sundaramurthy et al., 2005).

Typically, when light emitted from multi-quantum wells of LED interacts with a metallic layer, a surface plasmon mode will be formed. This surface plasmon mode increases the density of states and spontaneous emission rate (Gianordoli et al., 2000; Hecker et al., 1999; Vuckovic et al., 2000), and thus helps to improve coupling from the multi-quantum wells (Neogi et al., 2002). There have been reports that successfully demonstrated enhancement in InGaN quantum well emission by means of enhancement of Purcell factor with surface plasmons making use of a metal film on top of an LED (Okamoto et al., 2004). However, light coupled to surface plasmon modes is trapped at the metal/dielectric interface despite the enhancement of internal quantum efficiency. To deal with this problem, localized surface plasmon resonance had been investigated, whereby a periodic metallic array, or plasmonic crystal, is adopted instead of a flat metal film to assist with extracting light from the surface plasmon polaritons (Cesario et al., 2007).

Since the resonance wavelength depends strongly on the size and shape of metallic nanoparticles, and that resonance at visible range often requires a feature size of less than 100 nm, nanosphere lithography is an effective method for the patterning of metallic nanoparticles, particularly because the requirement on orderliness is lower. By controlling the dimensions of the nanospheres, the dimensions of the resultant metallic nanostructures can be adjusted, thereby affecting the spectral position of resonance. With monolayer of nanosphere serving as a mask, deposition of desired metal, typically Ag, Al or Au, can be carried out to form an inverted metallic circular array (Jensen, Malinsky, Haynes & Van Duyne, 2000; Malinsky, Kelly, Schatz & Van Duyne, 2001), providing the metal can penetrate the void regions of the nanosphere array. Figure 10 shows an SEM image illustrating the resultant metallic nanoparticles after removing the nanosphere residual. According to the PL spectra of InGaN/GaN MQWs coated with an array of metallic nanoparticles using this nanosphere lithography method as reported by G.Y. Mak et al., it was demonstrated that the metallic nanoparticles are indeed able to improve light extraction without changes of emission characteristics as demonstrated by Figure 11 (Mak et al., 2009).

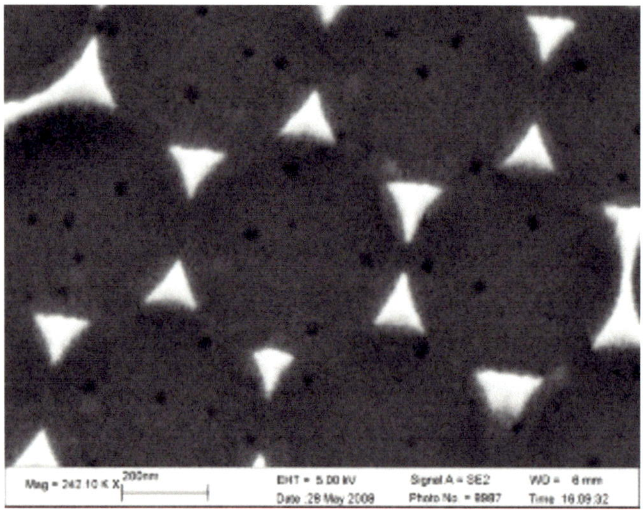

Fig. 10. SEM image showing Au nanoparticles formed by deposition through a nanosphere mask. (with permission for reproduction from WILEY-VCH Verlag GmbH & Co.)

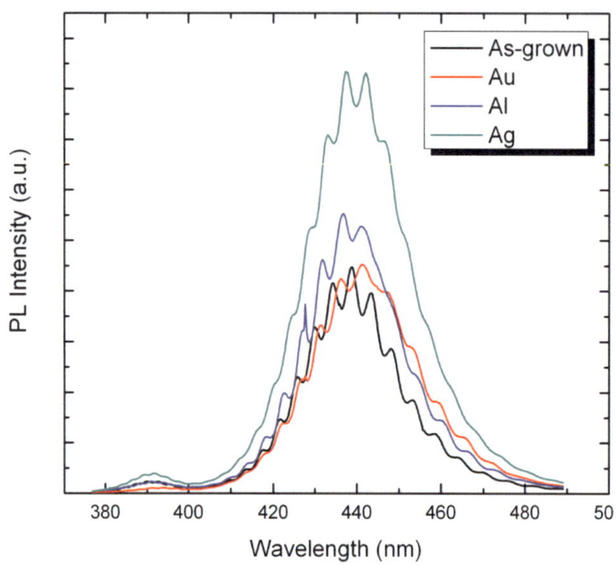

Fig. 11. PL spectra of LED with deposited nanoparticles of Au, Al and Ag. (with permission for reproduction from WILEY-VCH Verlag GmbH & Co.)

5. Summary

An inexpensive and simple approach as compared to other nanolithography methods, nanosphere lithography is able to play an important role in defining periodic-arrayed lithographic mask for photonic applications. Although more optimization works needs to be done on thickness control, the availability of crack-free deposition method made it feasible for mass production with high yield. A few photonic applications have been addressed in this chapter. Fluorescent nanospheres provide an alternative way to produce white LED for solid state lighting with improved color mixing. Nanopillars derived from a monolayer of hexagonal close-packed nanospheres helps to texturize the surface of an LED in order to improve light extraction efficiency. Last but not least, localized surface plasmon resonance not only can improve light extraction, but also has potential sub-wavelength applications including optical energy transport and near field scanning optical microscopy.

6. References

Berrier, A.; Mulot, M.; Swillo, M.; Qiu, M.; Thylen, L.; Talneau, A. & Anand, S. (2004). Negative refraction at infrared wavelengths in a two-dimensional photonic crystal. *Physical Review Letters*, 93, 7, 073902, 0031-9007

Bhalgat, M. K.; Haugland, R. P.; Pollack, J. S.; Swan, S. & Haugland, R. P. (1998). Green- and red-fluorescent nanospheres for the detection of cell surface receptors by flow cytometry. *Journal of Immunological Methods*, 219, 1-2, 57-68, 0022-1759

Byeon, K. J.; Hwang, S. Y. & Lee, H. (2007). Fabrication of two-dimensional photonic crystal patterns on GaN-based light-emitting diodes using thermally curable monomer-based nanoimprint lithography. *Applied Physics Letters*, 91, 9, 091106, 0003-6951

Cesario, J.; Gonzalez, M. U.; Cheylan, S.; Barnes, W. L.; Enoch, S. & Quidant, R. (2007). Coupling localized and extended plasmons to improve the light extraction through metal films. *Optics Express*, 15, 17, 10533-10539, 1094-4087

Cheung, C. L.; Nikolic, R. J.; Reinhardt, C. E. & Wang, T. F. (2006). Fabrication of nanopillars by nanosphere lithography. *Nanotechnology*, 17, 5, 1339-1343, 0957-4484

Chiappini, A.; Armellini, C.; Chiasera, A.; Ferrari, M.; Jestin, Y.; Mattarelli, M.; Montagna, M.; Moser, E.; Conti, G. N.; Pelli, S.; Righini, G. C.; Gonclves, M. C. & Almeida, R. M. (2007). Design of photonic structures by sol-gel-derived silica nanospheres. *Journal of Non-Crystalline Solids*, 353, 5-7, 674-678,

Choi, H. K.; Kim, M. H.; Im, S. H. & Park, O. O. (2009). Fabrication of Ordered Nanostructured Arrays Using Poly(dimethylsiloxane) Replica Molds Based on Three-Dimensional Colloidal Crystals. *Advanced Functional Materials*, 19, 10, 1594-1600,

David, A.; Fujii, T.; Sharma, R.; McGroddy, K.; Nakamura, S.; DenBaars, S. P.; Hu, E. L.; Weisbuch, C. & Benisty, H. (2006). Photonic-crystal GaN light-emitting diodes with tailored guided modes distribution. *Applied Physics Letters*, 88, 6, 061124, 0003-6951

David, A.; Moran, B.; McGroddy, K.; Matioli, E.; Hu, E. L.; DenBaars, S. P.; Nakamura, S. & Weisbuchb, C. (2008). GaN/InGaN light emitting diodes with embedded photonic crystal obtained by lateral epitaxial overgrowth. *Applied Physics Letters*, 92, 11, 113514,

Deckman, H. W. & Dunsmuir, J. H. (1982). Natural Lithography. *Applied Physics Letters*, 41, 4, 377-379, 0003-6951

Fu, W. Y.; Wong, K. K. Y. & Choi, H. W. (2009). Close-packed hemiellipsoid arrays: A photonic band gap structure patterned by nanosphere lithography. *Applied Physics Letters*, 95, 13, 133125, 0003-6951

Gianordoli, S.; Hainberger, R.; Kock, A.; Finger, N.; Gornik, E.; Hanke, C. & Korte, L. (2000). Optimization of the emission characteristics of light emitting diodes by surface plasmons and surface waveguide modes. *Applied Physics Letters*, 77, 15, 2295-2297, 0003-6951

Gil, A.; Vaupel, M.; Guitian, F. & Mobius, D. (2007). Stress-free production and effective medium model of colloidal crystals. *Journal of Materials Chemistry*, 17, 23, 2434-2439, 0959-9428

Han, D.-S.; Kim, J.-Y.; Na, S.-I.; Kim, S.-H.; Lee, K.-D.; Kim, B. & Park, S.-J. (2006). Improvement of light extraction efficiency of flip-chip light-emitting diode by texturing the bottom side surface of sapphire substrate. *IEEE Photonics Technology Letters*, 18, 13, 1406-1408, 1041-1135

Han, S.; Hao, Z. B.; Wang, J. & Luo, Y. (2005). Controllable two-dimensional photonic crystal patterns fabricated by nanosphere lithography. *Journal of Vacuum Science & Technology B*, 23, 4, 1585-1588, 1071-1023

Haynes, C. L. & Van Duyne, R. P. (2003). Plasmon-sampled surface-enhanced Raman excitation spectroscopy. *Journal of Physical Chemistry B*, 107, 30, 7426-7433, 1520-6106

Hecker, N. E.; Hopfel, R. A.; Sawaki, N.; Maier, T. & Strasser, G. (1999). Surface plasmon-enhanced photoluminescence from a single quantum well. *Applied Physics Letters*, 75, 11, 1577-1579, 0003-6951

Hui, K. N.; Fu, W. Y.; Ng, W. N.; Leung, C. H.; Lai, P. T.; Wong, K. K. Y. & Choi, H. W. (2008). Polychromatic light-emitting diodes with a fluorescent nanosphere opal coating. *Nanotechnology*, 19, 35, 355203, 0957-4484

Hui, K. N.; Wang, X. H.; Li, Z. L.; Lai, P. T. & Choi, H. W. (2009). Design of vertically-stacked polychromatic light-emitting diodes. *Optics Express*, 17, 12, 9873-9878, 1094-4087

Hulteen, J. C. & Vanduyne, R. P. (1995). Nanosphere Lithography - a Materials General Fabrication Process for Periodic Particle Array Surfaces. *Journal of Vacuum Science & Technology A-Vacuum Surfaces and Films*, 13, 3, 1553-1558, 0734-2101

Humphreys, C. J. (2008). Solid-state lighting. *MRS Bulletin*, 33, 4, 459-470, 0883-7694

Im, S. H.; Kim, M. H. & Park, O. O. (2003). Thickness control of colloidal crystals with a substrate dipped at a tilted angle into a colloidal suspension. *Chemistry of Materials*, 15, 9, 1797-1802,

Jensen, T. R.; Malinsky, M. D.; Haynes, C. L. & Van Duyne, R. P. (2000). Nanosphere lithography: Tunable localized surface plasmon resonance spectra of silver nanoparticles. *Journal of Physical Chemistry B*, 104, 45, 10549-10556, 1089-5647

Jiang, P.; Bertone, J. F.; Hwang, K. S. & Colvin, V. L. (1999). Single-crystal colloidal multilayers of controlled thickness. *Chemistry of Materials*, 11, 8, 2132-2140, 0897-4756

Kim, S. H.; Lee, K. D.; Kim, J. Y.; Kwon, M. K. & Park, S. J. (2007). Fabrication of photonic crystal structures on light emitting diodes by nanoimprint lithography. *Nanotechnology*, 18, 5, 055306, 0957-4484

Kuai, S. L.; Hu, X. F.; Hache, A. & Truong, V. V. (2004). High-quality colloidal photonic crystals obtained by optimizing growth parameters in a vertical deposition technique. *Journal of Crystal Growth*, 267, 1-2, 317-324,

Li, W.; Xu, L.; Zhao, W. M.; Sun, P.; Huang, X. F. & Chen, K. J. (2007). Fabrication of large-scale periodic silicon nanopillar arrays for 2D nanomold using modified nanosphere lithography. *Applied Surface Science*, 253, 22, 9035-9038, 0169-4332

Mak, G. Y.; Fu, W. Y.; Lam, E. Y. & Choi, H. W. (2009). Metallic nanoparticle array on GaN by microsphere lithography. *physica status solidi (c)*, 6, S2, S654-S657, 1862-6351

Malinsky, M. D.; Kelly, K. L.; Schatz, G. C. & Van Duyne, R. P. (2001). Nanosphere lithography: Effect of substrate on the localized surface plasmon resonance spectrum of silver nanoparticles. *Journal of Physical Chemistry B*, 105, 12, 2343-2350, 1089-5647

Matsuya, T.; Tashiro, S.; Hoshino, N.; Shibata, N.; Nagasaki, Y. & Kataoka, K. (2003). A core-shell-type fluorescent nanosphere possessing reactive poly(ethylene glycol) tethered chains on the surface for zeptomole detection of protein in time-resolved fluorometric immunoassay. *Analytical Chemistry*, 75, 22, 6124-6132, 0003-2700

Mihi, A.; Ocana, M. & Miguez, H. (2006). Oriented colloidal-crystal thin films by spin-coating microspheres dispersed in volatile media. *Advanced Materials*, 18, 17, 2244-2249, 0935-9648

Muthu, S.; Schuurmans, F. J. P. & Pashley, M. D. (2002). Red, green, and blue LEDs for white light illumination. *IEEE Journal of Selected Topics in Quantum Electronics*, 8, 2, 333-338, 1077-260X

Neogi, A.; Lee, C. W.; Everitt, H. O.; Kuroda, T.; Tackeuchi, A. & Yablonovitch, E. (2002). Enhancement of spontaneous recombination rate in a quantum well by resonant surface plasmon coupling. *Physical Review B*, 66, 15, 153305, 1098-0121

Ng, W. N.; Leung, C. H.; Lai, P. T. & Choi, H. W. (2008). Nanostructuring GaN using microsphere lithography. *Journal of Vacuum Science & Technology B*, 26, 1, 76-79, 1071-1023

(2008). Photonic crystal light-emitting diodes fabricated by microsphere lithography. *Nanotechnology*, 19, 25, 255302, 0957-4484

Noda, S. & Fujita, M. (2009). Light-emitting diodes: Photonic crystal efficiency boost. *Nature Photonics*, 3, 3, 129-130, 1749-4885

Noda, S.; Fujita, M. & Asano, T. (2007). Spontaneous-emission control by photonic crystals and nanocavities. *Nature Photonics*, 1, 8, 449-458,

Ogi, T.; Modesto-Lopez, L. B.; Iskandar, F. & Okuyama, K. (2007). Fabrication of a large area monolayer of silica particles on a sapphire substrate by a spin coating method. *Colloids and Surfaces a-Physicochemical and Engineering Aspects*, 297, 1-3, 71-78, 0927-7757

Okamoto, K.; Niki, I.; Shvartser, A.; Narukawa, Y.; Mukai, T. & Scherer, A. (2004). Surface-plasmon-enhanced light emitters based on InGaN quantum wells. *Nature Materials*, 3, 9, 601-605, 1476-1122

Pan, Y. X.; Wu, M. M. & Su, Q. (2004). Tailored photoluminescence of YAG : Ce phosphor through various methods. *Journal of Physics and Chemistry of Solids*, 65, 5, 845-850, 0022-3697

Ross, B. M. & Lee, L. P. (2008). Plasmon tuning and local field enhancement maximization of the nanocrescent. *Nanotechnology*, 19, 27, 275201, 0957-4484

Schubert, E. F. & Kim, J. K. (2005). Solid-state light sources getting smart. *Science*, 308, 5726, 1274-1278, 0036-8075

Seo, J. & Lee, L. P. (2004). Disposable integrated microfluidics with self-aligned planar microlenses. *Sensors and Actuators B-Chemical*, 99, 2-3, 615-622, 0925-4005

Shimmin, R. G.; DiMauro, A. J. & Braun, P. V. (2006). Slow vertical deposition of colloidal crystals: A Langmuir-Blodgett process? *Langmuir*, 22, 15, 6507-6513, 0743-7463

Steigerwald, D. A.; Bhat, J. C.; Collins, D.; Fletcher, R. M.; Holcomb, M. O.; Ludowise, M. J.; Martin, P. S. & Rudaz, S. L. (2002). Illumination with solid state lighting technology. *IEEE Journal of Selected Topics in Quantum Electronics*, 8, 2, 310-320, 1077-260X

Stewart, M. E.; Mack, N. H.; Malyarchuk, V.; Soares, J. A. N. T.; Lee, T. W.; Gray, S. K.; Nuzzo, R. G. & Rogers, J. A. (2006). Quantitative multispectral biosensing and 1D imaging using quasi-3D plasmonic crystals. *Proceedings of the National Academy of Sciences of the United States of America*, 103, 46, 17143-17148, 0027-8424

Su, Y. K.; Chen, J. J.; Lin, C. L.; Chen, S. M.; Li, W. L. & Kao, C. C. (2008). GaN-Based Light-Emitting Diodes Grown on Photonic Crystal-Patterned Sapphire Substrates by Nanosphere Lithography. *Japanese Journal of Applied Physics*, 47, 8, 6706-6708, 0021-4922

Sundaramurthy, A.; Crozier, K. B.; Kino, G. S.; Fromm, D. P.; Schuck, P. J. & Moerner, W. E. (2005). Field enhancement and gap-dependent resonance in a system of two opposing tip-to-tip Au nanotriangles. *Physical Review B*, 72, 16, 165409, 1098-0121

Vuckovic, J.; Loncar, M. & Scherer, A. (2000). Surface plasmon enhanced light-emitting diode. *IEEE Journal of Quantum Electronics*, 36, 10, 1131-1144, 0018-9197

Wang, L. & Zhao, X. S. (2007). Fabrication of Crack-Free Colloidal Crystals Using a Modified Vertical Deposition Method. *Journal of Physical Chemistry C*, 111, 24, 8538-8542, 1932-7447

Wong, S.; Kitaev, V. & Ozin, G. A. (2003). Colloidal crystal films: Advances in universality and perfection. *Journal of the American Chemical Society*, 125, 50, 15589-15598, 0002-7863

Ye, Y. H.; LeBlanc, F.; Hache, A. & Truong, V. V. (2001). Self-assembling three-dimensional colloidal photonic crystal structure with high crystalline quality. *Applied Physics Letters*, 78, 1, 52-54, 0003-6951

Zhou, Z. C. & Zhao, X. S. (2004). Flow-controlled vertical deposition method for the fabrication of photonic crystals. *Langmuir*, 20, 4, 1524-1526, 0743-7463

31

Micro- and Nanopatterning of Surfaces Employing Self Assembly of Nanoparticles and Its Application in Biotechnology and Biomedical Engineering

Claus Burkhardt, Kai Fuchsberger, Wilfried Nisch and Martin Stelzle
NMI Naturwissenschaftliches und Medizinisches Institut an der Universität Tübingen
Germany

1. Introduction

State of the art microelectronics fabrication achieves both high dimensional resolution as well as precision in positioning of structures by using photolithography at relatively high cost / area. There exist, however, numerous applications which only require micro- or nanostructures of well defined dimension without the need for precise positioning. In addition, some of those applications such as filtration membranes or antireflective coatings call for low cost / area, which is not achievable using photolithography. Also, patterning of non-planar surfaces is not feasible by this technology. Materials other than silicon or glass, particularly polymers often are incompatible with solvents and processes employed in conventional micro fabrication.

Diverse and commercially important applications are currently being pursued such as filtration membranes (Fuchsberger, Burkhardt et al.), biosensors (Musil, Jeggle et al. 1995; Neugebauer, Müller et al. 2006; Lohmüller, Müller et al. 2008), optical filters based on plasmon resonance of metallic nanoparticles (Jensen, Duval et al. 1999; Jensen, Schatz et al. 1999; Jensen, Malinsky et al. 2000; Traci R. Jensen 2000; Malinsky, Kelly et al. 2001), model catalysts (Gustavsson, Fredriksson et al. 2004), antireflective coatings (Lohmüller, Helgert et al. 2008; Min, Jiang et al. 2008; Xu, Lu et al. 2008), and model surfaces to study cell / substrate interaction on well defined surface topologies (Hanarp, Sutherland et al. 1999; Dalby, Berry et al. 2004).

Consequently, self assembly of nanoparticles as a means to create nanostructures on surfaces has been investigated by numerous research groups and great effort has been dedicated to devising reproducible deposition techniques as well as to elucidating the physico-chemical effects involved (Denkov, Velev et al. 1992; Adamczyk, Siwek et al. 1994; Kralchevsky & Denkov 2001; Goedel 2003; Hanarp, Sutherland et al. 2003)

Polymers, metals or even liquids may be used as substrates. Chemical functionalization of substrates may be applied to aid in the attraction and assembly of particles.

Electrostatic, van der Waals, capillaric, and steric interactions control particle adhesion and density. Capillary forces (Denkov, Velev et al. 1992; Adamczyk, Siwek et al. 1994; Kralchevsky & Denkov 2001) occurring during drying tend to cause aggregation of particles.

Electrostatic forces (Serizawa, Takeshita et al. 1998; Snyder, Yake et al. 2005) depend on the charge density of particles and surface as well as on the ionic strength of the particle suspension. Van der Waals forces (Hiemenz & Rajagopalan 1997) may cause formation of stable particle micro aggregates in the suspension or on the surface which may induce defects when the monolayer / aggregate structure is transferred to the substrate.

Upon the application of a suspension of these particles or immersion of the substrate into a particle suspension a mono- or multilayer of particles forms on a surface. Adsorption from a particle suspension onto a surface or spin coating procedures have been reported as well. Parameters to achieve homogeneous and defect free particle mono- or multilayers have been elaborated. Yet, although a large number of reports on nanosphere lithography (NSL) may be found in the scientific literature, only very few of them actually demonstrate large scale defect free patterning and are able to quantify the degree of perfection. In most cases, SEM images show a small section of a substrate which - while illustrating the basic principles of the procedure - would be too small to serve any practical purpose. Therefore, particular emphasis will be placed on such technologies holding promise for robust, low defect patterning of large surface areas by NSL.

Once arranged on the surface, particles have been used in a variety of ways to produce nanostructures.

Firstly, particles may serve as shadow masks during the deposition of an etching mask (Musil, Jeggle et al. 1995; Haginoya, Ishibashi et al. 1997; Burmeister, Schäfle et al. 1998; Choi, Yu et al. 2003; Choi, Jang et al. 2004; Lohmüller, Helgert et al. 2008; Lohmüller, Müller et al. 2008). After removal of the particle layers a porous etching mask is obtained having micro- or nanopores at the prior location of the particles. Subsequently, the resulting pattern can be transferred into the bulk material by plasma or wet etching processes. The choice of the material employed for the etching mask depends on the etching process. For example, silicon nitride, titanium or other so-called hard masks will withstand oxygen plasma as is employed to etch polymer substrates. On the other hand, a metallic etching mask may be used to protect silicon oxide surfaces during CF_4 plasma etching. For some applications etching processes need to achieve a considerable aspect ratio. For example nanoporous membranes need to provide sufficient thickness and stability to be of practical use (cf. chapter 6).

Secondly, the particle layer itself may be employed as mask during the deposition of metals to generate arrays of nanoparticles with diverse types of shapes (Hulteen & van Duyne 1995; Burmeister, Schäfle et al. 1998; Hulteen, Treichel et al. 1999; Jensen, Schatz et al. 1999; Jensen, Malinsky et al. 2000; Traci R. Jensen 2000; Haynes & van Duyne 2001; He, Yao et al. 2001; Bullen & Garrett 2002; Gustavsson, Fredriksson et al. 2004; Kosiorek, Kandulski et al. 2004; Ormonde, Hicks et al. 2004; Shemaiah M. Weekes 2004; Yang, Jang et al. 2006). Size and shape of these nanoparticles govern their optical properties. In order to obtain tailor-made particle shapes, complex fabrication schemes comprising the deposition of several layers of nanospheres, etching in plasma, and metallization steps have been devised (Yang, Jang et al. 2006).

After a discussion of general principles of NSL, three application examples will be discussed in detail:

Firstly, optical effects achievable by NSL, secondly, a nanoporous electrode system enabling a considerable enhancement of sensitivity in a electrochemical detection scheme by redox cycling, and thirdly the generation of polymer membranes for ultra filtration.

Finally, technological details of a procedure enabling truly large scale patterning of a polymer membrane by NSL will be discussed and potential future applications of NSL are anticipated.

2. Self assembly of particle monolayers on surfaces

The fundamental challenge for NSL consists in the assembly of mono- or multilayers of particles at fairly even and controllable distribution on a planar or non-planar surface consisting of polymer, oxide or metallic materials. In addition, many applications require large areas and low defect concentration.

In order to devise procedures to achieve this, an in depth understanding of the forces driving the self-assembly process is critical and in fact has been studied over the past 2 decades (Denkov, Velev et al. 1992; Adamczyk, Siwek et al. 1994; Hanarp, Sutherland et al. 2003). Fig. 1 schematically depicts the forces involved.

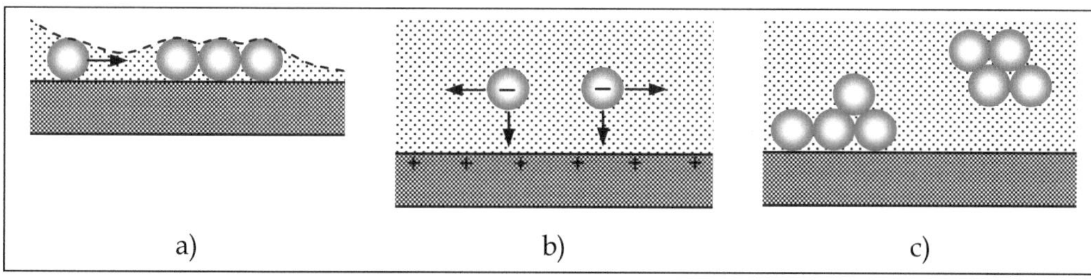

Fig. 1. Forces involved in the formation of particle layers on surfaces: a) when the solvent layer thickness becomes smaller than the particle diameter, capillary forces arise pulling particles together in effect forming 2D crystals. b) Electrostatic interaction results in repulsion between like charged particles and their attraction to an oppositely charged surface. c) Van der Waals forces occur at very low distances and play a role in the formation and stabilization of particle aggregates and 2D crystals

Firstly, long-range electrostatic interaction is observed as a consequence of charged surfaces of both the particles and the substrate. It results in the repulsion of like charged particles and the attraction of particles to a surface bearing a charge density with opposite sign. Charge density of both particles and substrate as well as ionic strength of the buffer solution in which particles are immersed are critical parameters controlling strength and range of electrostatic interaction. Both are reduced by mobile ions present in the buffer solution which will screen the surface charge of particles and substrate. For example, polystyrene particles exhibit negative surface charge due to sulfonate groups stemming from polymer synthesis. Hanarp et.al. (Hanarp, Sutherland et al. 2003) and Semmler et.al. (Semmler, Mann et al. 1998) demonstrated how particle coverage may be controlled by variation of salt concentration. In effect, repulsion of like charged particles is reduced or suppressed at all by addition of mobile ions. While the kinetics of adsorption depends on particle concentration the surface coverage ultimately attained upon saturation seems to be a function of only the ionic strength. If interparticle distance falls below a critical limit short range van der Waals interactions come into play. Then, particles will attract each other and tend to form aggregates. In case of aggregation, interparticle distance is given by the particle size itself. These mechanisms have been studied intensely in colloid physics and have been used to devise stable colloid formulations or to enable formation of colloidal crystals, respectively

(Hiemenz & Rajagopalan 1997). The influence of the surface charge of the substrate on particle assembly has also been studied. Serizawa et.al. investigated both kinetics of and coverage obtained from adsorption of polystyrene particles on substrates coated with ultra thin polymer layers (Serizawa, Takeshita et al. 1998; Serizawa, Kamimura et al. 2000). Surface charge density thereof was modulated by variation of ionic strength of the buffer used during the deposition of polyelectrolyte multilayers. Interestingly, adsorption proceeded faster and resulted in higher coverage onto polyelectrolyte multilayers that had been prepared using higher salt concentration in the buffer. As expected, no particle adsorption was observed on surfaces bearing like signed charge as the particles. Surface charge may also be modulated by variation of the pH if an appropriate substrate material is used. This was shown by Antelmi et.al. (Antelmi & Spalla 1999) with sapphire substrates and by Semmler et.al. with mica substrates (Semmler, Mann et al. 1998). At the point of zero surface charge density (PZC) no particle adsorption was observed.

Capillary forces (Fig.1a) act on particles as a result of the surface tension between the solvent and the particle material (Denkov, Velev et al. 1992; Kralchevsky & Denkov 2001). After application of a droplet of a particle suspension onto a substrate (Denkov, Velev et al. 1992; Burmeister, Schäfle et al. 1997; Burmeister, Badowsky et al. 1999), during the controlled withdrawal of a sample from a particle suspension (Choi, Jang et al. 2004; Prevo & Velev 2004), during spin coating (Denis, Hanarp et al. 2002; Choi, Jang et al. 2004; Jiang & McFarland 2004) or after adsorption of particles to a substrate in a particle suspension (Semmler, Mann et al. 1998; Serizawa, Takeshita et al. 1998; Antelmi & Spalla 1999; Serizawa, Kamimura et al. 2000; Hanarp, Sutherland et al. 2003) the evaporation of the solvent plays a critical role with respect to the structure of the particle layer. As the thickness of the solvent layer becomes comparable to the particle diameter, attractive forces arise pulling particles together to form 2D crystals. Depending on the application pursued this may or may not be desirable. For example densely packed 2D crystals have been employed as masks during a metallization process leading to a regular arrangement of separated metal particles (Hulteen & van Duyne 1995; Hulteen, Treichel et al. 1999; Jensen, Malinsky et al. 2000; Haynes & van Duyne 2001; Bullen & Garrett 2002). Hard core repulsion between particles results in maximum packing density with a lattice constant corresponding to the particle diameter. On the other hand, well separated pores are for example required to provide for sufficient mechanical stability of a filter membrane. To this end, a homogeneous distribution of separated particles has been achieved by freeze drying (Lohmüller, Müller et al. 2008). After preparation of a monolayer of homogeneously distributed particles, while the thickness of the solvent layer is still much larger than the particle diameter so that electrostatic interactions between particles and substrate are predominant, the substrate is cooled below freezing point of the solvent, followed by sublimation of the frozen solvent in a vacuum chamber. Hanarp et.al. explored heat treatment of polystyrene particles and co-adsorption of smaller silica nanoparticles in order to prevent aggregation of polystyrene particles during the drying process (Hanarp, Sutherland et al. 2003). Also, the choice of a solvent with lower surface tension or the addition of surfactant may aid in the reduction of capillary forces. Also, swelling of particles in an appropriate solvent during the deposition process or etching of a 2D particle crystal in oxygen plasma will produce monolayers of separated particles as will be described in detail below.

Repulsive interaction between nanoparticles can be achieved through polymer chain interaction, i.e. stochastic processes. For this purpose metallic nanoparticles were functionalized with hydrogel polymers such as polyethylene glycol. Thus, monolayers of homogeneously distributed metallic particles were obtained (Lohmüller, Helgert et al. 2008).

3. Preparation of particle mono- and multilayers

A large variety of substrates has been used in NSL. However, general principles apply when large scale high quality particle layers shall be produced.

Often, polymeric particles such as polystyrene particles which are intrinsically negatively charged, silica particles or even metallic nanoparticles bearing charges due to specific surface functionalization are employed in NSL. In order to induce particle adhesion, particles and substrate surface have to bear opposite charges. In case of the substrate this has been achieved by choice of an appropriate material such as an oxide or mica (Semmler, Mann et al. 1998; Antelmi & Spalla 1999) or by deposition of synthetic or biological polyelectrolyte mono- or multilayer coatings (Serizawa, Takeshita et al. 1998; Serizawa, Kamimura et al. 2000; Lohmüller, Müller et al. 2008), which can be deposited under well controlled conditions yielding reproducible results. The charge density of such a polymer layer can be tuned within a certain range by the choice of the salt concentration of the polymer solution during deposition. As has been shown by Serizawa et al., this will affect final particle density as well as adhesion kinetics (Serizawa, Takeshita et al. 1998; Serizawa, Kamimura et al. 2000). Other synthetic and biological polymers such as polyethylene-imine and bovine serum albumin, respectively, which provide for a positive charge density at neutral pH have been used to coat substrates in order to promote electrostatic interaction with negatively charged polystyrene particles (Lohmüller, Müller et al. 2008). Polyimide membranes were successfully modified using diethylene triamine (DETA) to introduce positive charges and provide for favourable wetting properties (Santoso, Albrecht et al. 2003).

Spreading of suspensions of functionalized particles at the air / water or water / oil interface has been used to generate 2D particle assemblies (Aveyard, Clint et al. 2000; Xu & Goedel 2003; Cayre & Paunov 2004; Xu, Yan et al. 2005). Gödel et.al. used functionalized silica particles in a suspension containing polyisoprene (Xu & Goedel 2002). After generation of a monolayer consisting of hydrophobized silica particles and polyisoprene at the air water interface, cross linking of the polymer layer was induced by UV illumination. Silica particles were dissolved using HF and polymer membranes with a regular arrangement of pores were obtained.

As has been pointed out earlier, capillary forces in combination with evaporation play a dominant role in the arrangement of particles if samples are allowed to dry following the addition of a particle suspension. Special emphasis therefore has been put on the design of procedures and instrumentation for controlled deposition of particles.

According to the model proposed by Denkov et.al. (Denkov, Velev et al. 1992) capillary forces and convective flow contribute to the assembly process. Due to evaporation the solvent layer eventually becomes thinner than the particle diameter giving rise to capillary forces pulling particles into close contact with each other in effect forming 2D crystal while at the same time convective flow transports particles from areas exhibiting thicker solvent layers towards the 2D crystal.

Based on this model, refined procedures using heating stages, evaporation chambers to control evaporation have been devised (Denkov, Velev et al. 1992; Burmeister, Schäfle et al. 1997; Burmeister, Badowsky et al. 1999; Zhen Yuan 2006) which, however, proof to be quite

time consuming. In a refined procedure Prevo et.al. employ a meniscus between the substrate and a moving plate to generate a controllable evaporation zone (Prevo & Velev 2004).
Finally, spin coating has been used in an attempt to provide a fabrication method which would both be compatible to conventional micro fabrication technology and enable generation of large scale 2D and 3D colloid crystals (Denis, Hanarp et al. 2002; Jiang & McFarland 2004). However, data reported generally show defects such as dislocations, voids and aggregates.

4. Properties of particle monolayers

Figure 2 shows micrographs obtained by scanning electron microscopy of typical results obtained by NSL and is intended to visualize commonly observed defect structures. 2D crystals are always imperfect in that dislocations and unoccupied lattice positions are present. Rarely, long range order exceeds more than ~100 µm (Fig.2a). While electrostatic interaction between nanospheres and surface usually result in a homogeneous distribution of well separated particles, upon drying of the surface capillary forces draw particles together to form small 2D aggregates and voids (Fig.2b). As Fig.2b shows compression may be strong enough to even cause deformation of particles (arrow). A third type of defect observed in NSL is 3D particle aggregates of different sizes (Fig.2c). Presumably, these aggregates are already present in the suspension employed in NSL and adsorb to the substrate surface along with individual nanospheres by electrostatic interaction.

5. Applications of surfaces patterned using NSL

5.1 Optical applications: surface plasmon resonance and optical filters

Haynes et.al. used NSL to produce 2D crystals of polystyrene particles which were used as shadow masks during the deposition of Ag on mica substrates (Haynes & van Duyne 2001). Regular 2D patterns of discrete Ag nanoparticles of various shapes were fabricated and investigated with respect to their optical properties. Metallic nanoparticles give rise to localized surface plasmon resonance (LSPR). This phenomenon has two consequences with potentially interesting applications. Firstly, light in a certain spectral region is absorbed providing optical filter properties. Secondly, electric field strength in close vicinity of the particle is enhanced. This is useful in sensor applications where molecules in close contact with these particles can be probed with superior sensitivity by surface enhanced Raman spectroscopy (SERS) to obtain chemical information about individual molecules (Kneipp, Wang et al. 1997; Nie & Emory 1997). Through an appropriate choice of the particle diameter used for NSL and deposition parameters a variety of shapes of metallic nanoparticles was demonstrated. In particular, by introduction of a tilt angle between the surface normal and the direction of evaporation, asymmetric particle shapes were obtained. While nanospheres showed extinction wavelength around 300 nm, strongly asymmetric nanotriangles exhibited absorptions in the mid-IR range, demonstrating the ability to tune the optical properties of nanoparticles by variation of their shape. In addition to devising fabrication procedures Haynes et al. studied the effect of nanoparticle material, size, shape, interparticle spacing, external dielectric medium, thin film overlayer, substrate dielectric constant and molecular adsorbates on the optical properties of nanoparticle arrays. Finally, the sensitivity towards thin adsorbate layers is demonstrated by measurement of the

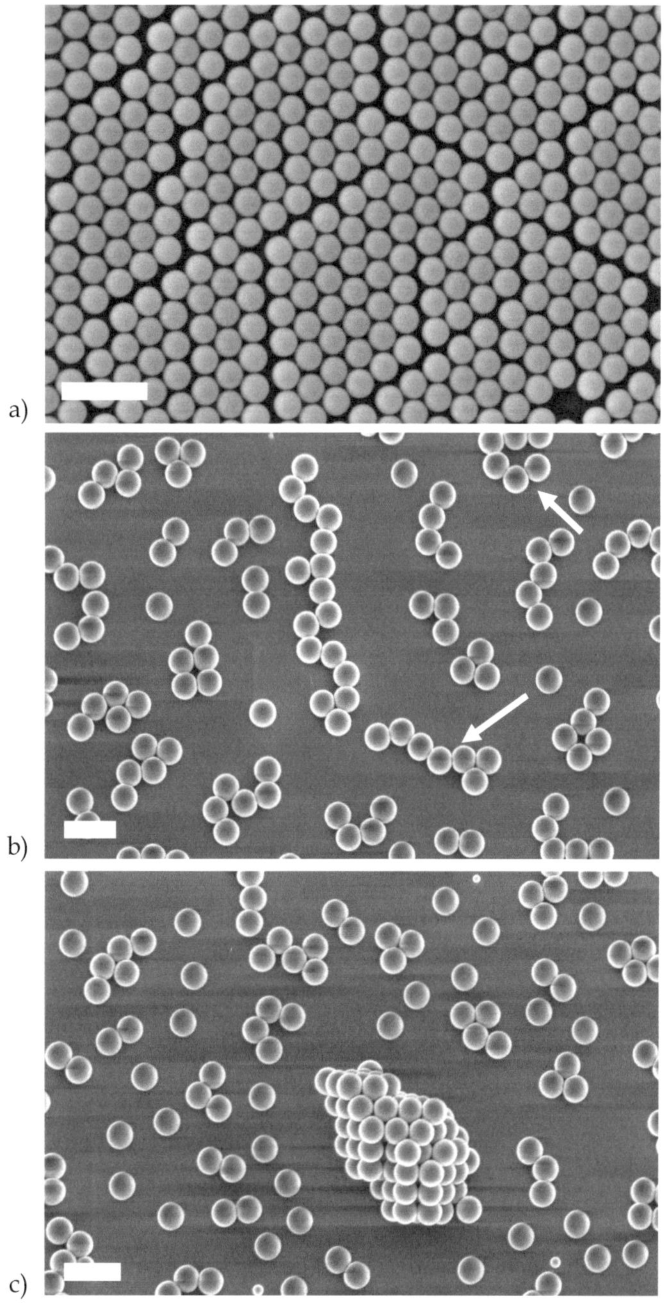

Fig. 2. typical morphology of particle monolayer and commonly observed defect structures: a) cracks, dislocations and voids in 2D nanosphere crystal. b) small 2D aggregates separated by voids, generated by capillary forces during the drying process. c) 3D nanosphere aggregates, presumably already present in the suspension used for NSL. Bar corresponds to 1μm.

adsorption of polylysine to Ag nanoparticles that had been functionalized with a carboxy-terminated alkane thiol (McFarland & Van Duyne 2003). A comprehensive review of this particular application was published by Haes et.al. (Haes & Van Duyne 2004; Stuart, Haes et al. 2004).

In addition, van Duyne et al. published a number of studies exploring NSL as a means to fabricate arrays of metallic nanoparticle of different shapes and their relationship to their optical properties which were mostly measured by UV-VIS spectroscopy (Hulteen & van Duyne 1995; Jensen, Duval et al. 1999; Jensen, Schatz et al. 1999; Jensen, Malinsky et al. 2000; Traci R. Jensen 2000; Malinsky, Kelly et al. 2001; Ormonde, Hicks et al. 2004) and related their results to theoretical considerations (Jensen, Duval et al. 1999; Jensen, Schatz et al. 1999; Haynes & van Duyne 2001).

NSL also has been employed to provide a low cost large scale etching mask on glass and silicon substrates to produce nanoporous surfaces which exhibit reduced reflectivity in the UV and visible spectrum by utilizing the moth eye effect (Lohmüller, Helgert et al. 2008), (Xu, Lu et al. 2008). A reduction of reflectivity results in an improved efficiency of solar cells. Also, optical throughput of UV optics can be improved in this way.

5.2 Sensors

NSL already has found significant applications in sensor devices. Van Duyne and co-workers demonstrated biosensors based on surface enhanced Raman spectroscopy on metallic nanoparticles deposited through a mask fabricated by NSL (Haynes & van Duyne 2001; Haes & Van Duyne 2004; Stuart, Haes et al. 2004). Upon adsorption or binding of molecules to the nanoparticles, a spectral shift in the absorption maximum is observed and chemical signatures of molecules may be observed in the spectrum obtained by surface enhanced Raman spectroscopy.

Porous metal layers have been proposed as transducers for electrochemical biosensors as they allow for the deposition of the biological recognition elements in close proximity to the sensing element. Musil et al. used NSL, deposited a thin metal layer by evaporation or plasma enhanced chemical vapour deposition (PECVD) and subsequently removed the particles to obtain a sensor element having a nanoporous, yet laterally connective metal layer (Musil, Jeggle et al. 1995). Though not demonstrated in this particular publication, it is easily conceivable how molecules may be selectively either bound to the surface exposed in the pores employing silane chemistry or to the gold electrode using thiol groups.

Electrochemical sensors are widely employed for immunosensing applications (Marquette & Blum 2006; Wang, Xu et al. 2008). In order to enhance sensitivity and/or signal strength, so-called redox cycling may be employed (Aoki 1989; Niwa, Xu et al. 1993). In this particular detection scheme, redox molecules generated by the enzyme label of the immuno assay may be oxidized and reduced several times giving rise to an increased current signal. However, this requires very closely spaced electrodes in order to obtain a reasonably high number of cycles (Shim, Rust et al. 2008). While this may be achieved with interdigitated electrode arrays (IDE) using conventional photolithography or e-beam patterning, the costs associated with using these technologies are incompatible with the requirements of the diagnostic market.

As a cost effective alternative, nanoporous metal - insulator - metal electrode systems (Fig.3) have been developed and were shown to exhibit superior redox-cycling properties (Neugebauer, Müller et al. 2006; Lohmüller, Müller et al. 2008). Amplification ratios of up to

100 fold have been demonstrated. Deposition of a first gold layer creates the bottom electrodes and their connecting leads. Shadow masks may be used to pattern the relatively coarse structures of the sensor pads and the connecting leads. Subsequently, an insulation layer of 100 - 200 nm thickness is deposited and NSL using polystyrene particles with a diameter ranging from 200 nm to 1 μm is carried out. A second gold layer is deposited onto this substrate. After removal of the particles the porous top electrode is used as an etching mask during a plasma etching process to expose the bottom electrode in the area of the pores (Fig. 3). In this way, a large number of electrically connected electrochemical nano-

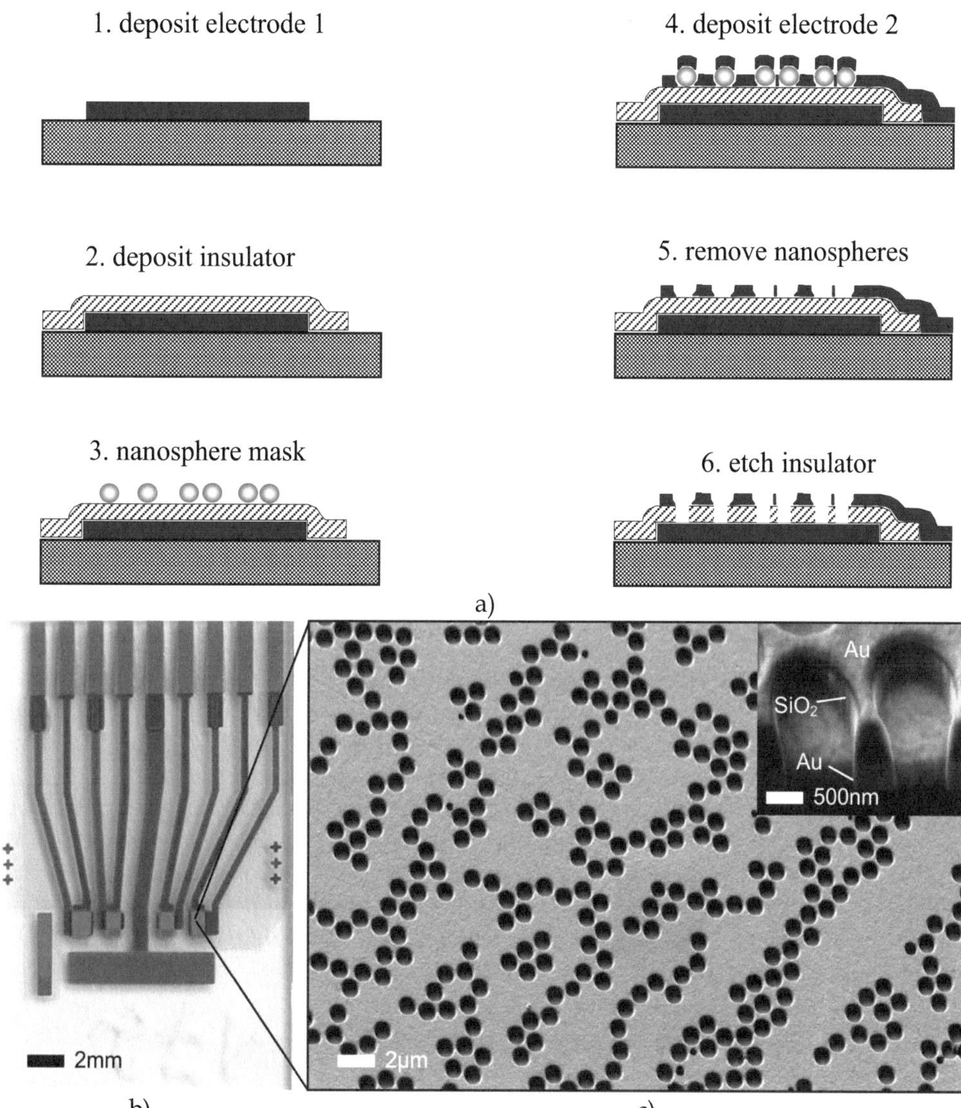

Fig. 3. NanoBioPore sensors: a) fabrication process, b) sensor with four sensor pads, c) SEM image of sensor surface showing pores and cross-sectional view of the porous metal - insulator - metal electrode system (insert).

cells are created and upon application of oxidizing and reducing potentials to top and bottom electrodes, respectively, a large signal due to repeated oxidation and reduction of molecules in the nanopores is observed. Here, the thickness of the insulation layer defines the electrode spacing. As it is quite easy to produce insulation layers with a thickness as small as 100 - 200 nm, highly efficient redox - cycling devices may be fabricated at relatively low cost when compared to IDE sensors.
While homogeneous pore distribution in principle is not very important in this application, it was observed that 2D aggregates may contribute to short circuits, when small electrode portions are destabilized by under-etching during the pore development (Lohmüller, Müller et al. 2008). Also, back sputtering of metal from the bottom electrode to the pore walls due to prolonged pore etching was discussed as another possible cause for short circuits between top and bottom electrodes.

5.3 Filter membranes

Ultra filtration membranes are used in such diverse applications as for filtration of fermenter suspensions to remove virus or bacterial contamination or to filter solvents in micro fabrication technology. Conventional ultra filtration membranes are manufactured by phase inversion technology. A polymer is dissolved in an organic solvent which evaporates as a thin film is continuously drawn on an air/water interface. Porosity, thickness and effective pore diameter depend on process parameters and require careful control of parameters such as temperature, humidity, and drawing speed.
Usually, thickness is on the order of 100 - 200 µm both to provide mechanical stability as well as to obtain the required filtration properties through the joint effect of the micro- or nanoporous polymer mesh making up the membrane. However, particularly for very small pore sizes, fluidic resistance of these membranes become exceedingly high thus limiting flux and ultimately the productivity of the filtration process (DiLeo, Allegrezza et al. 1992; van Reis & Zydney 2001).
In contrast, a membrane manufactured by a combination of NSL and plasma etching technology should exhibit much larger flux and possibly a narrower pore size distribution. The requirements are quite demanding, though: the reduction ratio for virus filtration for example should be $>10^6$. This requires an extremely low defect density in NSL. In particular, aggregate deposition or formation has to be avoided by all means. At the same time continuous reproducible and reliable NSL on very large surface areas is critical.
The Nanohybrid project employed polyimide as membrane substrate, a polymer which even at a thickness of only 5 µm provides sufficient mechanical stability to be useful as filtration membrane. At this thickness combined with a porosity on the order of 20 – 30 % theoretical calculations predict an increase of flux by up to 100 fold when compared to conventional membranes. Results obtained by an optimized fabrication procedure based on NSL are presented in chapter 6.

5.4 Applications of NSL in biology and biotechnology

NSL has been used to generate surfaces to study the effect of well defined nano-topologies and chemistries on cellular functions such as adhesion (Dalby, Berry et al. 2004; Dalby, Riehle et al. 2004; Graeter, Huang et al. 2007; Huang, Graeter et al. 2009), proliferation, endocytosis (Dalby, Berry et al. 2004), differentiation (Steinberg, Schulz et al. 2006), apoptosis (Ranzinger, Krippner-Heidenreich et al. 2009) and gene regulation. In particular,

the group of Spatz et.al. showed the deposition of polymer-functionalized gold colloids on substrates followed by an etching step to remove polymer ligands. This results in regular arrays of nano-sized anchoring sites for biochemical functionalization via thiol chemistry. Control of spacing, order and biochemical function immobilized onto gold nanospots allowed for in depth studies of cellular functions (Spatz, Mössmer et al. 2000; Roos, Ulmer et al. 2005; Steinberg, Schulz et al. 2006; Graeter, Huang et al. 2007; Huang, Graeter et al. 2009; Ranzinger, Krippner-Heidenreich et al. 2009).
The fabrication of nanopatterned films of protein by NSL yielding ring shaped structures has been demonstrated (Garno, Amro et al. 2002; Li, Henry et al. 2006). In particular it was shown that proteins deposited in this manner retained the capability to bind specifically to immunoglobulin molecules (Garno, Amro et al. 2002). Results obtained from this research are expected to result in novel biomaterials provided with tailor made surface topologies and chemical composition to induce specific tissue responses.

6. Origin of defects in NSL ultra low defect & large scale patterning for filter membranes

As pointed out in Chapter 4, typical defect structures found in NSL involve crystal defects in 2D crystals as well as 2D and 3D aggregates where a homogeneously array of separated particles was intended.
A robust fabrication process involving NSL to produce nano-porous filtration membranes was developed. Defect densities on the order of 10^{-5} or smaller were achieved. Particular care was taken to devise and apply such process steps that would be scalable and

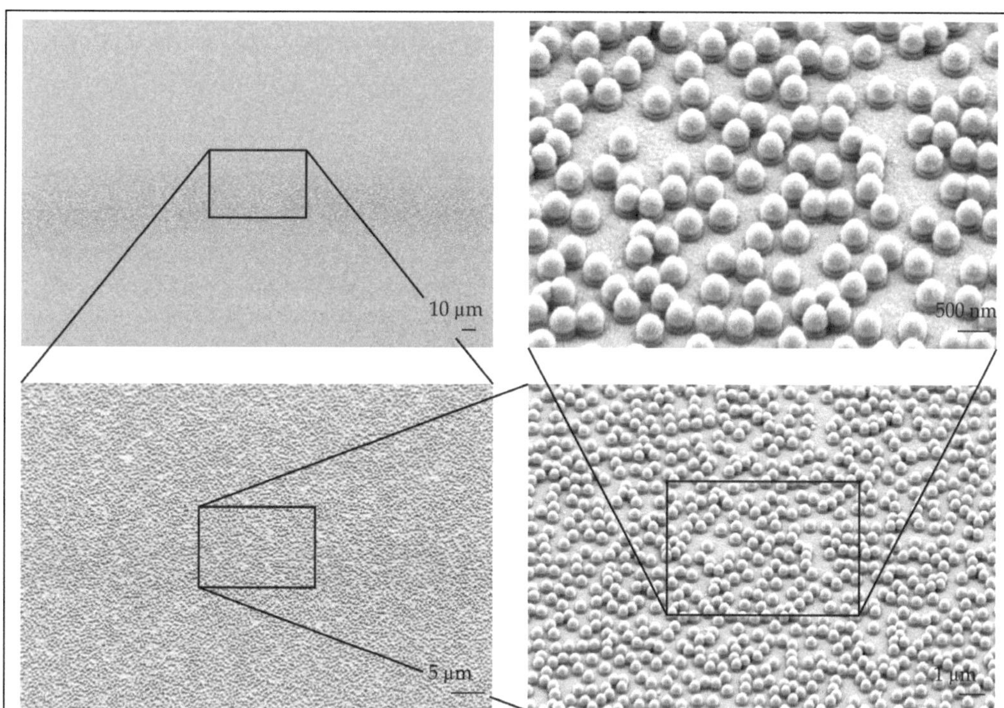

Fig. 4. Large area ultra low defect NSL on polyimide membranes

compatible with state of the art production technologies. Details of this process will be published elsewhere (Fuchsberger, Burkhardt et al.).

Briefly, polyimide was used as substrate because of its favourable mechanical and chemical properties. Even at 5 μm, polyimide membranes exhibit sufficient mechanical stability to produce filter membranes. After deposition of nanospheres (Fig.4), an etching mask was deposited, nanospheres were removed and the nanostructure was transferred into the polyimide membrane by plasma etching in an oxygen / argon plasma.

In conclusion, this method opens up the possibility to employ NSL to pattern pores with a diameter of less than 100 nm at very high aspect ratios which was previously not possible (Fig.5).

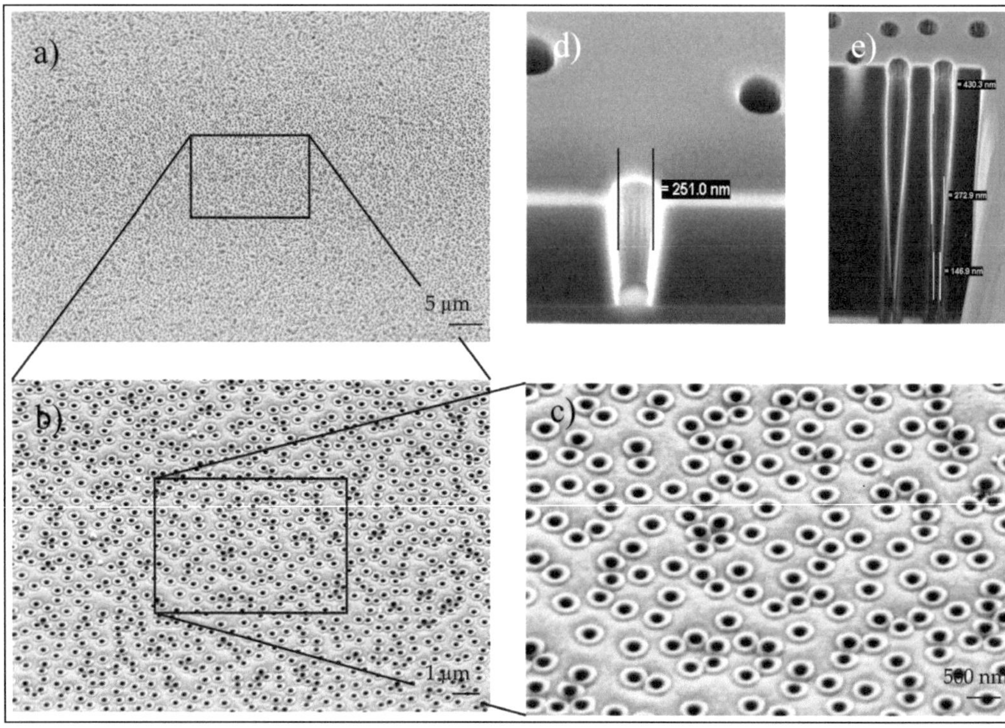

Fig. 5. Membrane comprising well separated nanopores (a-c). Cross-section of a porous membrane with a thickness = 600nm (d) and 6μm (e), respectively, demonstrates high aspect ratios of up to ≈15 with a small degree of conicity.

7. Conclusion

NSL has been studied intensely as it provides a cost effective means to generate well defined nanostructures where positional order is not required. Numerous applications have been proposed and investigated in optics, sensors and membrane technology. However, in spite of broad knowledge about forces governing the self-assembly of nanospheres on surfaces, widespread application of NSL has been severely hampered by typical defect structures, namely dislocations and voids in 2D crystals as well as 2D and 3D aggregates where a homogeneous monolayer of well separated nanospheres would be required. Very few if any patterning procedures are currently available which actually have been demonstrated on large (>cm^2) areas.

To this end, a comprehensive process scheme was devised comprising all necessary steps from preparation of surfaces and pure particle suspensions, application of nanospheres to surfaces up to pattern transfer into polymeric membranes. Defect densities below 10^{-5} have been demonstrated and nanoporous filter membranes exhibiting pores with an aspect ratio of >15 were fabricated.

We believe that these developments open the door for more widespread application of NSL in industrial fabrication.

8. Acknowledgements

Helpful discussions with Volkmar Thom, Sartorius Stedim, are acknowledged. Funding for this research was in part obtained from the German ministry for education and research through grant 13N9122.

9. References

Adamczyk, Z., B. Siwek, et al. (1994). "Kinetics of Localized Adsorption of Colloid Particles." Adv.Colloid and Interface Sci. 48: 151-280.

Antelmi, D. A. and O. Spalla (1999). "Adsorption of Nanolatex Particles to Mineral Surfaces of Variable Surface Charge." Langmuir 15: 7478-7489.

Aoki, K. (1989). "Theory of the steady-state current of a redox couple at interdigitated array electrodes of which pairs are insulated electrically by steps." J.Electroanal.Chem. 270(1-2): 35-41.

Aveyard, R., J. H. Clint, et al. (2000). "Compression and Structure of Monolayers of Charged Latex Particles at Air/Water and Octane/Water Interfaces." Langmuir 16: 1969-1979.

Bullen, H. A. and S. J. Garrett (2002). "TiO2 Nanoparticle Arrays Prepared Using a Nanosphere Lithography Technique." Nanoletters 2(7): 739-745.

Burmeister, F., W. Badowsky, et al. (1999). "Colloid monolayer lithography - A flexible approach for nanostructuring of surfaces." Appl.Surf.Sci. 144-145: 461-466.

Burmeister, F., C. Schäfle, et al. (1998). "From Mesoscopic to Nanoscopic Surface Structures: Lithography with Colloid Monolayers." Adv.Mater. 10(6): 495-497.

Burmeister, F., C. Schäfle, et al. (1997). "Colloid Monolayers as Versatile Lithographic Mask." Langmuir 13: 2983-2987.

Cayre, O. J. and V. N. Paunov (2004). "Fabrication of microlens arrays by gel trapping of self-assembled particle monolayers at the decane–water interface." J. Mater. Chem. 14: 3300 - 3302.

Choi, D.-G., S. G. Jang, et al. (2004). "Two-Dimensional Polymer Nanopattern by Using Particle-Assisted Soft Lithography." Chem.Mater. 16: 3410-3413.

Choi, D.-G., H. K. Yu, et al. (2003). "Colloidal LIthographic Nanopatterning via Reactive Ion Etching." J.Am.Chem.Soc. 126: 7019-7025.

Dalby, M. J., C. C. Berry, et al. (2004). "Attempted endocytosis of nano-environment produced by colloidal lithography by human fibroblasts." Experimental Cell Research 295: 387-394.

Dalby, M. J., M. O. Riehle, et al. (2004). "Changes in fibroblast morphology in response to nano-columns produced by colloidal lithography." Biomaterials 25: 5415-5422.

Denis, F. A., P. Hanarp, et al. (2002). "Fabrication of Nanostructured Polymer Surfaces Using Colloidal Lithography and Spin-Coating." Nanoletters 2(12): 1419-1425.

Denkov, N. D., O. D. Velev, et al. (1992). "Mechanism of Formation of Two-Dimensional Crystals from Latex Particles on Substrates." Langmuir 8: 3183-319.

DiLeo, A. J., A. E. Allegrezza, et al. (1992). "High Resolution Removal of Virus from Protein Solutions Using a Membrane of Unique Structure." Bio/Technology 10: 182-188.

Fuchsberger, K., C. Burkhardt, et al. "manuscript in preparation."

Garno, J. C., N. A. Amro, et al. (2002). "Production of Periodic Arrays of Protein Nanostructures Using Particle Lithography." Langmuir 18: 8186-8192.

Goedel, W. A. (2003). "A simple theory of particle-assisted wetting." Europhysics Letters 62(4): 607-613.

Graeter, S. V., J. Huang, et al. (2007). "Mimicking Cellular Environments by Nanostructured Soft Interfaces." Nano Letters 7(5): 1413-1418.

Gustavsson, M., H. Fredriksson, et al. (2004). "Nanostructured platinum-on-carbon model electrocatalysts prepared by colloidal lithography." J.Electroanal.Chem. 568: 371-377.

Haes, A. J. and R. P. Van Duyne (2004). "A unified view of propagating and localized surface plasmon resonance biosensors." Analytical and Bioanalytical Chemistry 379(7-8): 920-930.

Haginoya, C., M. Ishibashi, et al. (1997). "Nanostructure array fabrication with a size-controllable natural lithography." Appl.Phys.Lett. 71(20): 2934-2936.

Hanarp, P., D. Sutherland, et al. (1999). "Nanostructured model biomaterial surfaces prepared by colloidal lithography." NanoStructured Materials 12: 429-432.

Hanarp, P., D. S. Sutherland, et al. (2003). "Control of nanoparticle film structure for colloidal lithography." Colloids and Surfaces A 214: 23-36.

Haynes, C. L. and R. P. van Duyne (2001). "Nanophere Lithography: A Versatile Nanofabrication Tool for Studies of Size-Dependent Nanoparticle Optics." J.Phys.Chem. B 105: 5599-5611.

He, S., J. Yao, et al. (2001). "Formation of Silver Nanoparticles and Self-Assembled Two-Dimensional Ordered Superlattice." Langmuir 17: 1571-1575.

Hiemenz, P. C. and R. Rajagopalan (1997). Principles of Colloid and Surface Chemistry, Marcel Dekker.

Huang, J., S. V. Graeter, et al. (2009). "Impact of Order and Disorder in RGD Nanopatterns on Cell Adhesion." Nano Letters 9(3): 1111-1116.

Hulteen, J. C., D. A. Treichel, et al. (1999). "Nanosphere Lithography: Size-Tunable Silver Nanoparticles and Surface Cluster Arrays." J.Phys.Chem. B. 103: 3854-3863.

Hulteen, J. C. and R. P. van Duyne (1995). "Nanosphere Lithography: A materials general fabrication process for periodic particle array surfaces." J.Vac.Sci.Technol. A 13(3): 1553.

Jensen, T. R., M. L. Duval, et al. (1999). "Nanosphere Lithography: Effect of the External Dielectric Medium on the Surface Plasmon Resonance Spectrum of a Periodic Array of Silver Nanoparticles." J.Phys.Chem. B. 103: 9846-9853.

Jensen, T. R., M. D. Malinsky, et al. (2000). "Nanosphere Lithography: Tunable Localized Surface Plasmon Resonance Spectra of Silver Nanoparticles." J.Phys.Chem. B 104: 10549-10556.

Jensen, T. R., G. C. Schatz, et al. (1999). "Nanosphere Lithography: Surface Plasmon Resonance Spectrum of a Periodic Array of Silver Nanoparticles by Ultraviolet -

Visible Extinction Spectroscopy and Electrodynamic Modeling." J.Phys.Chem. B. 103: 2394-2401.

Jiang, P. and M. J. McFarland (2004). "Large-Scale Fabrication of Wafer-Size Colloidal Crystals, Macroporous Polymers and Nanocomposites by Spin-Coating." J.Am.Chem.Soc. 126: 13778-13786.

Kneipp, K., Y. Wang, et al. (1997). "Single Molecule Detection Using Surface-Enhanced Raman Scattering (SERS)." Phys. Rev. Lett. 78(9): 1667-1670.

Kosiorek, A., W. Kandulski, et al. (2004). "Shadow Nanosphere Lithography: Simulation and Experiment." Nanoletters 4(7): 1359-1363.

Kralchevsky, P. A. and N. D. Denkov (2001). "Capillary forces and structuring in layers of colloidal particles." Curr.Opinion in Colloid & Interface Sci. 6: 383-401.

Li, J.-R., G. C. Henry, et al. (2006). "Fabrication of nanopatterned films of bovine serum albumin and staphylococcal protein A using latex particle lithography." The Analyst 131: 244-250.

Lohmüller, T., M. Helgert, et al. (2008). "Biomimetic Interfaces for High-Performance Optics in the Deep-UV Light Range." Nano Lett. 8(5): 1429-1433.

Lohmüller, T., U. Müller, et al. (2008). "Nano-spaced electrode systems by colloidal lithography for sensitive electrochemical detection: fabrication and properties." J.Micromech.Microeng. 18: 115011.

Malinsky, M. D., K. L. Kelly, et al. (2001). "Nanosphere Lithography: Effect of Substrate on the Localized Surface Plasmon Resonance Spectrum of Silver Nanoparticles." J.Phys.Chem. B. 105: 2343-2350.

Marquette, C. A. and L. J. Blum (2006). "State of the art and recent advances in immunoanalytical systems." Biosens Bioelectron 21(8): 1424-33.

McFarland, A. D. and R. P. Van Duyne (2003). "Single Silver Nanoparticles as Real-Time Optical Sensors with Zeptomole Sensitivity." Nano Lett. 3: 1057-1062.

Min, W.-L., P. Jiang, et al. (2008). "Large-Scale assembly of colloidal nanoparticles and fabrication of periodic subwavelength structures." Nanotechnology 19: 475604.

Musil, C. R., D. Jeggle, et al. (1995). "Nanostructuring of gold electrodes for immunosensing applications." J.Vac.Sci.Technol.B 13(6): 2781-2786.

Neugebauer, S., U. Müller, et al. (2006). "Characterization of Nanopore Electrode Structures as Basis for Amplified Electrochemical Assays." Electroanalysis 18(19-20): 1929-1936.

Nie, S. and S. R. Emory (1997). "Probing Single Molecules and Single Nanoparticles by Surface-Enhanced Raman Scattering." Science 275(5303): 1102-1106.

Niwa, O., Y. Xu, et al. (1993). "Small-volume voltammetric detection of 4-aminophenol with interdigitated array electrodes and its application to electrochemical enzyme immunoassay." Anal Chem 65(11): 1559-63.

Ormonde, A. D., E. C. Hicks, et al. (2004). "Nanosphere lithography: fabrication of large-area Ag nanoparticle arrays by convective self-assembly and their characterization by scanning UV-visible extinction spectroscopy." Langmuir 20(16): 6927-31.

Prevo, B. G. and O. D. Velev (2004). "Controlled, rapid deposition of structured coatings from micro- and nanoparticle suspensions." Langmuir 20(6): 2099-107.

Ranzinger, J., A. Krippner-Heidenreich, et al. (2009). "Nanoscale Arrangement of Apoptotic Ligands Reveals a Demand for a Minimal Lateral Distance for Efficient Death Receptor Activation." Nano Letters.

Roos, W., J. Ulmer, et al. (2005). "Microtubule Gliding and Cross-Linked Microtubule Networks on Micropillar Interfaces." Nano Letters 5(12): 2630-2634.

Santoso, F., W. Albrecht, et al. (2003). "A novel technique for preparation of aminated polyimide membranes with microfiltration characteristics." J.Membrane Science 223(1-2): 171-185.

Semmler, M., E. K. Mann, et al. (1998). "Diffusional Deposition of Charged Latex Particles on Water-Solid Interfaces at Low Ionic Strength." Langmuir 14: 5127-5132.

Serizawa, T., S. Kamimura, et al. (2000). "Electrostatic adsorption of polystyrene particles with different surface charges onto the surface of an ultrathin polymer film." Colloids and Surfaces A: Physicochemical and Engineering Aspects 164: 237-246.

Serizawa, T., H. Takeshita, et al. (1998). "Electrostatic Adsorption of Polystyrene Nanospheres onto the Surface of an Ultrathin Polymer Film prepared by Using an Alternate Adsorption Technique." Langmuir 14: 4088-4094.

Shemaiah M. Weekes, F. Y. O., and William A. Murray (2004). "Fabrication of Large-Area Ferromagnetic Arrays Using Etched Nanosphere Lithography." Langmuir 20(25): 11208–11212. .

Shim, J. S., M. J. Rust, et al. (2008). "Interdigitated Array Electrodes with Nano Gaps Using Optical Lithography and Controlled Undercut Method." Nanotechnology 2008, Arlington, TX, IEEE.

Snyder, C. E., A. M. Yake, et al. (2005). "Nanoscale Functionalization and Site-Specific Assembly of Colloids by Particle Lithography." Langmuir 21: 4813-4815.

Spatz, J. P., S. Mössmer, et al. (2000). "Ordered Deposition of Inorganic Clusters from Micellar Block Copolymer Films." Langmuir 16(2): 407.

Steinberg, T., S. Schulz, et al. (2006). "Early Keratinocyte Differentiation on Micropillar Interfaces." Nano Letters 7(2): 287-294.

Stuart, D. A., A. Haes, et al. (2004). Refractive index sensitive, plasmon resonant scatterin, and surface enhanced Ramn scattering nanoparticles and arrays as biological sensing platforms. Plasmonics in Biology and Medicine, SPIE.

Traci R. Jensen, M. D. M., Christy L. Haynes, and Richard P. Van Duyne (2000). "Nanosphere Lithography: Tunable Localized Surface Plasmon Resonance Spectra of Silver Nanoparticles " J. Phys. Chem. B 104(45): 10549 -10556.

van Reis, R. and A. Zydney (2001). "Membrane separations in biotechnology." Current Opinion in Biotechnology 12: 208-211.

Wang, Y., H. Xu, et al. (2008). "Electrochemical Sensors for Clinic Analysis." Sensors 8: 2043-2081.

Xu, H. and W. A. Goedel (2002). "Polymer-Silica Hybrid Monolayers as Precursors for Ultrathin Free-Standing Porous Membranes." Langmuir 18: 2363-2367.

Xu, H. and W. A. Goedel (2003). "From Particle-Assisted Wetting to Thin Free-Standing Porous Membranes." Angewandte Chemie International Ed. 42: 4694-4696.

Xu, H., N. Lu, et al. (2008). "Biomimetic Antireflective Si Nanopillar Arrays." Small 4(11): 1972-1975.

Xu, H., F. Yan, et al. (2005). "Particle-assisted wetting." Journal of Physics: Condensed Matter 17(9): S465.

Yang, S.-M., S. G. Jang, et al. (2006). "Nanomachining by Colloidal Lithography." Small 2(4): 458-475.

Zhen Yuan, D. B. B., Plamen Atanassov and Hongyou Fan (2006). "Convective self-assembly to deposit supported ultra-thin mesoporous silica films." J. Mater. Chem. 16: 4637 - 4641.

32

Strategies for High Resolution Patterning of Conducting Polymers

Lin Jiang and Lifeng Chi
Physikalisches Institut Westfälische Wilhelms-Universität Münster & Center for Nanotechnology (CeNTech), 48149 Münster Germany

1. Introduction

The discovery of conducting polymers by MacDiarmid, Shirakawa and Heeger in 1977 has generated enormous interest in π-conjugated macromolecules.[1] Since then, their unique properties such as the metal-insulator transition induced by doping-undoping process, controllable optoelectronic property through molecular design and attractive mechanical properties and processing advantages have stimulated many progresses in researches and technically relevant applications. They have been extensively studied allowing the demonstration of different functional devices including organic light emitting diodes (OLED),[2] electrochromic devices,[3] field effect transistors[4] and integrated circuits,[5] batteries,[6] photodetectors[7] and sensors.[8] In microelectronic-related device fabrication, conducting polymers need to be patterned into micro- and nanostructures. For example, conducting polymer microstructures are used as interconnects and as source, drain and gate electrodes in all-polymer integrated circuits[5] or as red, green, blue (RGB) pixels in multi-color OLED displays.[9] In contrast to the striking achievements in conducting polymer microelectronic devices, the application of conducting polymer nanostructures in nanoelectronic devices is still under development due to the current limitation of nanofabrication. However, their applications in other less demanding areas such as optics, sensors and biotechnology have been demonstrated. For example, Boroumand et al. demonstrated a series of electrically driven nanoscale light sources, which might be used in near-field optical communication system and storage devices.[10] Roman et al. gave strong evidence that nanopatterning of conducting polymer was capable of improving the performance of photodiodes.[11] Matterson demonstrated the increasing efficiency and controllable light output from structured LEDs.[12] Recently, several groups reported the use of conducting polymer nanostructures as chemical or biological sensors, which show higher sensitivity than conventional sensors.[13] For instance, Craighead and co-workers have reported a polyaniline nanowire chemical sensor more sensitive than the traditional filmbased sensors for the detection of ammonium gas. [14] Huang et al. reported a nanofiber film sensor that responds much faster than the conventional film sensor to HCl gas. [15] Apart from the detection of chemical vapors, improved sensing behaviour is also observed when detecting biological molecules, such as glucose[16] and biotin. [17]

There are basically two approaches to fabricating nanoscale conducting polymeric structures: the synthesis and post-assembly method[18] or the microfabrication method[19-22] (photolithography, soft lithography, etc.). In the former approach, sub-100 nm structures can be readily obtained by either template synthesis[23] or electrospinning. [24] However, manipulation and positioning of the synthesized nanoscale objects with respect to microelectrodes is rather difficult and not precise. In the latter case, conducting polymers can be precisely patterned by a variety of microlithographic methods based on different principles. For example, soft lithography[[19] can be used to fabricate conducting-polymer patterns by means of selective deposition, embossing, and so on; photo- or electron (e)-beaminitiated polymerization or degradation[20] can define patterns on conducting polymer films; dip-pen lithography[21] is capable of patterning conducting polymers, based on a principle of ink transfer from the tip to the substrate; electrochemical lithography[22] represents another series of methods of obtaining patterned conducting polymers, which can be obtained by scanning electrochemical microscopy (SECM), electrochemical dip-pen lithography, or scanning tunneling microscopy (STM). Although these microlithographic methods yield conducting- polymer micro/nanostructures, the methods which are capable of integrating high resolution with high throughput are still highly demanded.
In this review article, we intent to summarize our recent works in high resolution patterning of conducting polymers. Three different strategies will be included, together with the demonstration of advanced functions, e.g. in gas sensing and anisotropic conducting performance.

2. Surfactant assisted fabrication of polypyrrole nanowires between microelectrodes

Since conductive/semiconductive polymers are promising candidates for future molecular electronics, opto-electronics, and electro-optics, the reproducible fabrication of these materials on pre-defined positions, e.g. between microelectrodes, attracts increased scientific and technological attentions. To achieve this goal, one of the key issues has to be addressed: the improvement of the adhesion of the conducting polymers to the substrate without reducing the conductivity, especially with insulating surfaces. We used firstly a strategy by employing both surfactants and surface coupling reagents. E-beam as well as AFM lithography was used to create defined surface patterns. The successful construction of PPy structures (20-120 nm in thickness and down to 200 nm in width) was achieved.
It has been demonstrated that polypyrrole synthesized in the presence of surfactants is superior to that polymerized in the presence of small inorganic anions because anionic surfactants induce structural regularity into the polymers[23] These amphiphilic molecules inside the polypyrrole films will also help to improve the adhesion of polypyrrole to the substrate. [23a] Another way to immobilize polypyrrole is to treat the substrate with a surface-reactive reagent bearing pyrrole, for example, N-(3-trimethoxysilylpropyl) pyrrole (abbreviated as py-silane). Polymerization in the presence of a self-assembled monolayer of this reagent yields a polypyrrole film that is more adherent to the substrate. [24] We found that only the employment of both surfactant and py-silane would result in satisfying submicrometer structures, with the surfactant helping to wet the substrate and py-silane helping to immobilize the polypyrrole film.
First, we utilized an e-beam or tip-scratching to define channels in the resist of poly(methyl methacrylate) (PMMA) on a silicon substrate. Second, we modified the patterned wafer with

pysilane by placing the wafer into py-silane vapor at 140 mbar for 20 min. Third, we covered this wafer with a layer of oxidant solution, which was prepared by dissolving iron chloride (0.5 g) and sodium dodecyl sulfate (SDS; 0.125 g) in purified water (2.5 mL) by either spin-coating or immersion. Then, the wafer was exposed to pyrrole vapor to induce the polymerization. Finally, the resist and the polypyrrole on top of it were removed while the polypyrrole directly in contact with the silicon substrate remained.
Since this method offers the opportunity to grow polypyrrole wires in situ at a desired position, submicrometer wires can thus be introduced between two microelectrodes. We first fabricated gold electrodes by standard e-beam lithographic patterning. The PMMA was spin-coated onto this wafer. Using AFM-tip scratching, the coated PMMA between the electrodes was selectively removed to form a channel. The polypyrrole wire and microelectrode contacts can be generated with different dimensions within this channel by post-polypyrrole growth and using a lift-off process. For example, a polypyrrole wire (500 nm wide, 2.5 mm long,and 60 nm thick) that bridges two electrodes is shown in Figure 1a. In order to confirm the contact between the polypyrrole wire and the electrodes, current and voltage studies were carried out. As a control, we also studied the junction before polypyrrole growth, as shown in Figure 1b (dashed line), which indicates that it has no conductance. After polypyrrole growth, a linear dependence of the current on the applied potential was observed (Figure 1b, solid line), thus confirming the ohmic contact. The conductivity of this wire was calculated to be 0.8–1.2 Scm^{-1}, which is comparable with that of the bulk polymer. Polypyrrole wires as small as 250 nm in width and 120 nm in thickness, which grew in the channel defined by e-beam lithography, can also be fabricated.

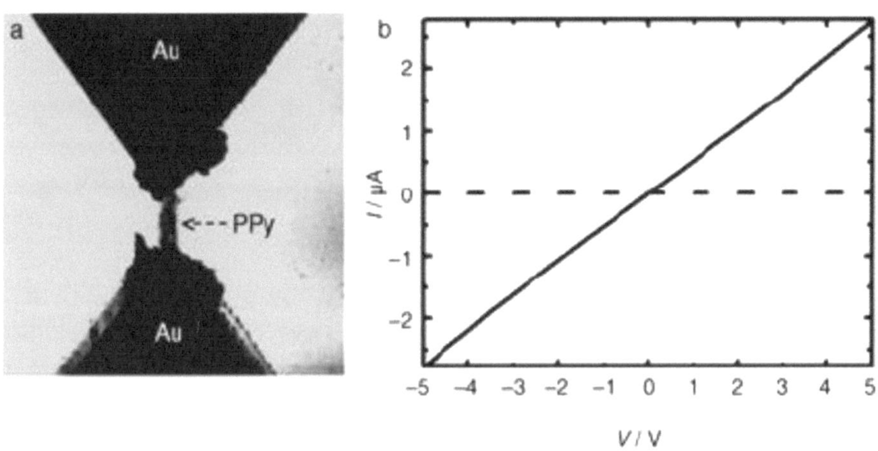

Fig. 1. a) AFM image of a 500-nm-wide polypyrrole wire grown between Au electrodes (image=10-10 mm^2); b) I–V characteristics of this junction before (b) and after (c) polypyrrole wire growth.

In order to clarify the functions of both SDS and pysilane during the construction process, we performed two control experiments. First, fabrication was carried out in the absence of py-silane. It was found that SDSdoped polypyrrole had only limited adhesion to the substrate, as evidenced by the fact that only structures on the micrometer scale, rather than with features of several hundred nanometers could be fabricated. We attributed this result to the weak adhesion based on the physical adsorption between SDS-doped polypyrrole and the substrate when the contact area was in the submicrometer scale. Second, we employed only py-silane. There is no doubt that py-silane will immobilize the polypyrrole onto the substrate through covalent bonds. However,submicrometer wire formation was not observed because the oxidant solution without SDS will not wet the substrate. Hence, based on the above observations, successful fabrication requires the combination of both SDS and the pyrrole-bearing silane.

3. High resolution and high throughout patterning based on a random copolymer strategy

Due to a lack of control over the film thickness during deposition the introduced above is not suitable for high-resolution (< 100 nm) fabrication. We further developed a copolymer strategy to control the thickness and the adhesive and electrical properties of conducting polymers by incorporating a surface-active monomer into the main chains. The method is suitable for the fabrication of devices with a sub-100 nm resolution, and allows the device exclusively based on conducting polymers. Combining with nanoimprinting lithography (NIP), high resolution of conductive polymers with high throughout become then possible.
Figure 2 depicts the fabrication process, which, in accordance with the standard lift-off process, includes three steps. In the first step, a pattern is defined on the photoresist using e-beam lithography or NIP. Second, a copolymer film thinner than the resist layer is deposited by oxidizing pyrrole (or aniline) and Py-silane with iron chloride. Finally, the resist is lifted off in acetone by sonication.

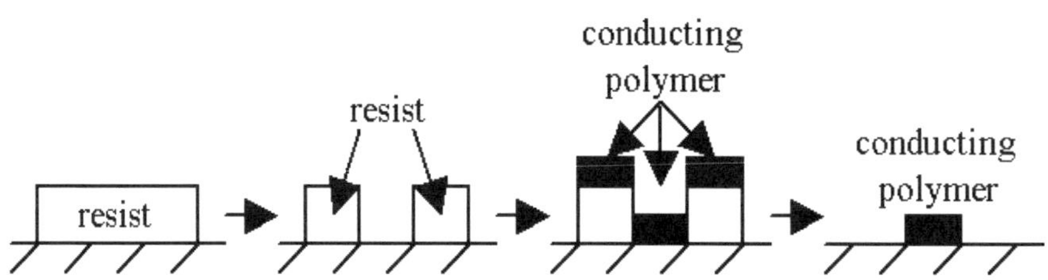

Fig. 2. Schematic representation of the fabrication process.

In order to realize successful fabrication of conducting-polymer structures by the lift-off process and to facilitate their further application as sensors, conductivity and adhesion, as well as their combination, are the essential parameters that should be taken into account. Here we employed the two-point measurement method and the adhesiontape test[19c] to estimate the electrical and adhesive properties of the copolymer film, respectively. Taking polypyrrole as an example, the introduction of py-silane has a significant influence on the properties of polypyrrole. Notably, copolymers that contain py-silane (9 vol.-%) can successfully withstand the adhesion-tape test whereas pure polypyrrole cannot. Moreover, upon increasing the fraction of py-silane, electrical conductivity and film thickness of the copolymer decrease. [25] The film thickness can be controlled by the deposition time, [19c] thus it can be kept thinner than the resist layer, thereby facilitating the lift-off process. In order to obtain a strongly adherent film that retained as high conductivity as possible, we carried out all depositions at a pyrrole to py-silane volume ratio of 91:9 unless stated otherwise. Under these conditions, the actual copolymer composition is estimated to be roughly 97:3 (pyrrole to py-silane), [25] based on X-ray photoelectron spectroscopy (XPS) measurements.

Construction of sub-100 nm conductingpolymer structures on insulating surfaces is normally hardly achievable by microlithographic techniques, except for a few examples.[20c] The strategy presented here offers efficient fabrication of conducting-polymer structures with a resolution of less than 100 nm. For example, we were able to fabricate an 80 nm wide, 11 μm long polypyrrole nanowire on a silicon substrate with a 100 nm thermally oxidized SiO2 layer (Figure 3a). The wire structure was initially defined on the resist by e-beam lithography.

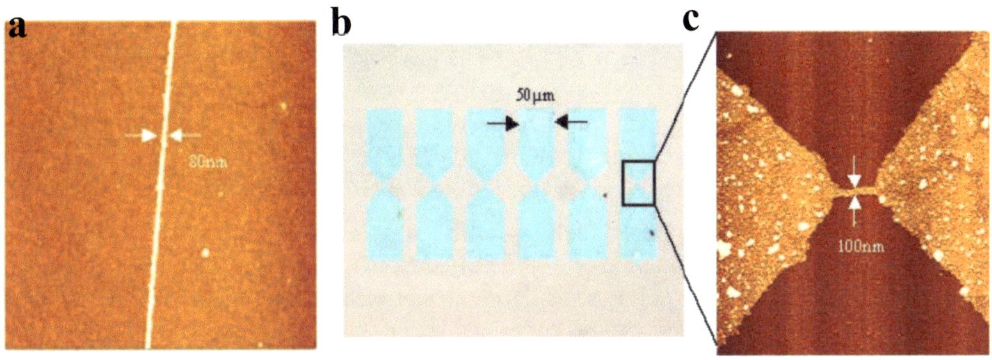

Fig. 3. a) Atomic force microscopy (AFM) image of an 80 nm wide polypyrrole wire (image size: 11 μm×11 μm). b) Optical images of six pairs of polypyrrole microelectrode shaped structures (size: 475 μm× 390 μm). c) Enlarged AFM image of the area inside the black box of Figure 2b, showing one 100 nm wide polypyrrole wire that bridges the ends of two polypyrrole microelectrodes (size: 9 μm×10 μm).

Precise manipulation and placement of the nanostructures at desired locations and their integration with larger-scale systems has long been a challenge and proposes a serious hindrance for the further development of nanoelectronics. [18] Exploiting advantages of the copolymer strategy, our nanowires can be integrated in between the conducting-polymer microelectrodes in one deposition step, facilitating the fabrication of nanosensors exclusively based on polymers. Figure 3b shows an optical image of six pairs of polypyrrole microelectrodes on a silicon substrate with a 100 nm SiO_2 layer (structures defined by e-beam lithography). The polypyrrole nanowires in between the microelectrodes are not visible at this magnification. The overall electrodes consist of homogeneous, closely packed, globular-shaped polypyrrole structures (see for example Figure. 3c, electrode part). All six pairs of electrodes were successfully connected with polypyrrole nanowires, proving the reliability of this strategy. In Figure 3c one of the six junctions, consisting of one single polypyrrole nanowire (100 nm wide, 1.25 μm long) bridging the ends of two polypyrrole microelectrodes, is shown.

By combining the copolymer strategy presented above with NIP, we successfully fabricated high density, large area conducting polymer nanostructures with a resolution of 100 nm (see Figure 4a). Each field of polypyrrole nanowires can extend over a large area, as seen in the light diffraction images captured by a CCD camera shown in Figure 4b. Position 1, 2, 3, 4, 5 and 6 correspond to 100, 150, 250, 350, 320 and 220 nm wide polypyrrole nanowires separated by 150, 200, 300, 400, 380 and 280 nm, respectively. The uniformity of each color reflects a good uniformity of each nanowire array.

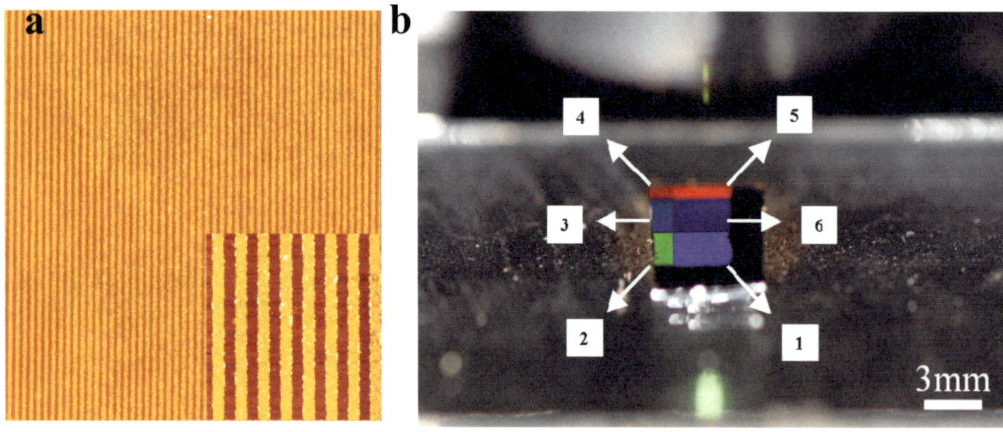

Fig. 4. a) 150 nm wide polypyrrole nanowires with a separation of 150 nm (size: 30 ×30 μm²). Inset: enlarged AFM image showing one 5 ×5 μm² area; b) CCD image showing the light diffraction from high density polypyrrole nanowires fabricated by NIL and a lift off process. (1) 100 nm wide polypyrrole nanowires separated by 150 nm. (2) 150 nm wide polypyrrole nanowires separated by 200 nm. (3) 250 nm wide polypyrrole nanowires separated by 300 nm. (4) 350 nm wide polypyrrole nanowires separated by 400 nm. (5) 320 nm wide polypyrrole nanowires separated by 380 nm. (6) 220 nm wide polypyrrole nanowires separated by 280 nm.

To demonstrate the functionality of this structure, a current-versus-potential (*I–V*) study was carried out. A linear dependence of the current on the applied potential was observed (Figure 5a), confirming the ohmic behavior of the material. We further evaluated the sensing performance of this device by exposing it to a 240 ppm NH3 stream. The detection of NH3 in air is of interest from an environmental point of view because of the high toxicity of this gas. The sensing principle is based on the fact that the electron-donating molecule (in this case NH3) can reduce the charge-carrier concentration of polypyrrole (a p-type conducting polymer or electron acceptor) and decrease its conductivity.[26] The device shown in Figure 5c exhibits a reversible response upon the addition of NH3 with a sensitivity (defined as $R/R0 - 1$, where $R0$ is the initial resistance and R is the resistance after exposure to NH3) of

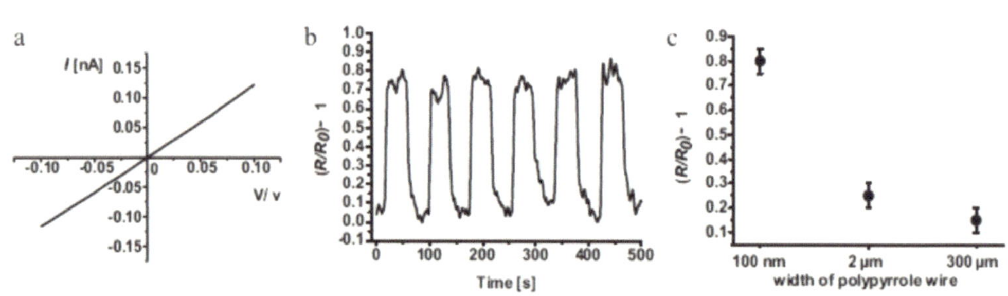

Fig. 5. a) *I–V* characteristics of a 100 nm nanodevice consisting exclusively of polypyrrole as shown in Figure 3c. b) Real-time response (R/Ro – 1 = sensitivity, see text) of this 100 nm nanowire sensor (Fig. 3c) to a 240 ppm NH3 stream (40 s on/40 s off). c) Sensitivity dependence on the width of the wire between the electrodes (at constant thickness).

0.8 (Figure 5b), which is of the same order of magnitude as reported for polypyrrole-based sensors. [27] This value can be further improved by the introduction of other anionic dopants.[27] We have compared the sensitivity of this nanosensor consisting exclusively of polypyrrole with a polypyrrole-nanowire sensor with metal electrode contacts and observed no difference, indicating that the nanowire is the key element that determines the sensitivity. By increasing the width of the wire from the nanometer regime to the micrometer (e.g.,2μm in width) or even into bulk, a decrease in sensitivity was observed (Figure 5c), proving the superiority of the nanoscale device. This is most likely due to the high surface area to volume ratio of the nanowire. A conductivity enhancement trend was also observed with decreased wire width (data not shown here), which is in agreement with results reported in literature.

With the large area conductive polymer structures based on the NIP patterns, the electronic connection became quite straightforward: They may be easily addressed electrically by connection to suitably sized metal electrodes. Moreover, since the residual layer is not

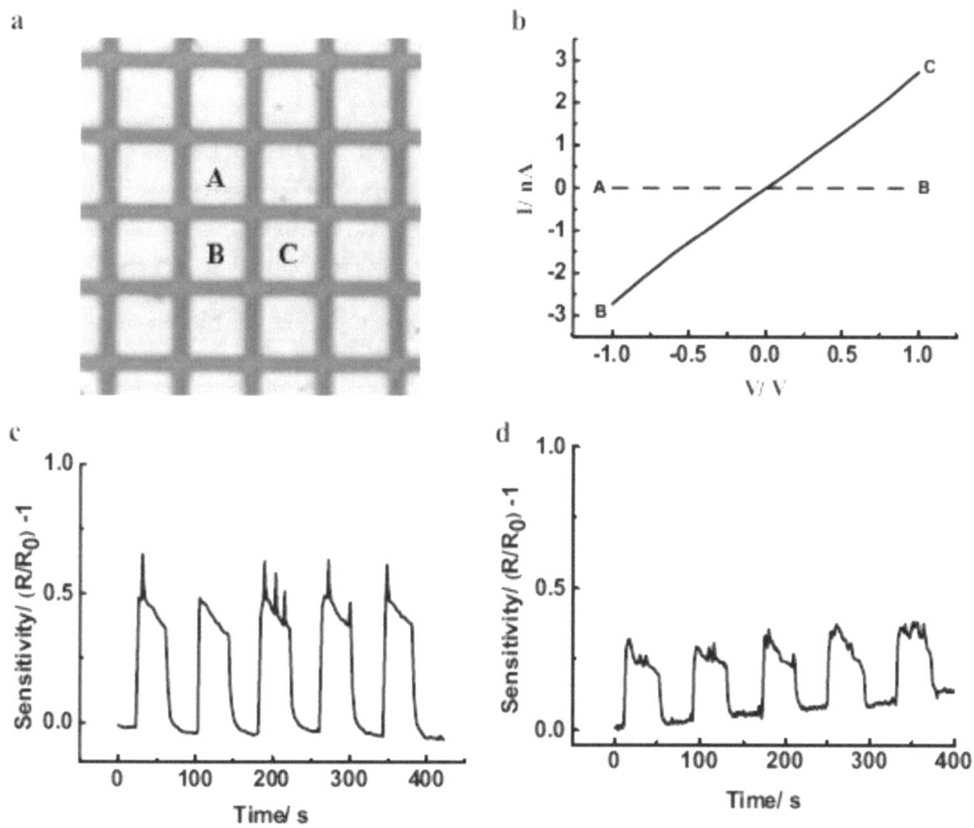

Fig. 6. a) Optical image of the gold pads evaporated on the nanowire arrays (image size: 330 μm× 330 μm). b) Resistance parallel (BC solid line) and perpendicular (AB dashed line) to the nanowire direction (BC direction). The measured anisotropy ratio is 1.6 × 103. c) Real-time response of a sensor consisting of 300 nm nanowire arrays to a 240 ppm NH3 stream (40 s on/40 s off). d) Real-time response of a sensor consisting of 5 μm wire arrays to a 240 ppm NH3 stream (40 s on/40 s off).

present, this should guarantee high resistance anisotropy in the parallel and perpendicular directions. Taking into account the sensing properties of conducting polymers, these electrically addressable conducting polymer nanowires could be directly used as nanosensors. For nanosensor applications, the high density nanowires offer a high error tolerance in the device performance if one or several nanowire fabrication failures occur.
In order to demonstrate the above advantages, gold pads were deposited on the nanowire arrays by shadow mask evaporation (Figure 6a). Figure 6b shows the typical current-voltage curves measured from the 300 nm wide nanowire arrays. As expected, the resistance in the parallel and perpendicular directions shows great anisotropy. Resistance values of 370 MΩ and 600 GΩ were obtained parallel and perpendicular to the nanowire direction, respectively, resulting in a value of 1.6 × 103 for the anisotropy ratio. The resistance varied from 200 to 400 MΩ for different wire arrays. These isolated electrically addressable wire arrays could be directly used as nanowire sensors. A demonstration of their sensing potential is shown in Figure 6c and d. Note that the electrical addressing of the nanowire array can also be realized by connecting them with silver glue and so on, because the nanowire array can extend over a larger area. Given the advantages of NIL, we thus developed a facile method to fabricate conducting-polymer nanosensors at low cost.

4. Scaling down from micrometer patterns to nanometer structures

In the combination of NIL and copolymer strategy introduced above, the conducting polymer (CP) layer must be thinner than the resist layer for a successful lift-off.[28] It restricts the construction of CPs with arbitrary thickness, which sometimes limits the application of the resulting polymer patterns. Furthermore, in NIL, the imprinted structure is usually a negative copy of the stamp. Thus, a high resolution stamp is required for creating a high resolution pattern, which leads to an expensive and time-consuming fabrication process. Herein, we demonstrate an approach to reduce the lateral size of CP structures using an isotropic plasma etching (IPE). Using this method, we are able to improve the printed pattern resolution without the thickness limit of the CP layer, and the CP can be protected from the etching process.
The fabrication process is illustrated in Figure 7. First, the polyaniline (PANI) layer was deposited onto a silicon dioxide (or quartz) substrate. Then a 400-nm poly(methyl methacrylate) (PMMA) film was spin-coated on the PANI film. Subsequently, the stamp was pressed into the PMMA film heated above Tg, followed by peeling away from the substrate after cooling down the system. Figure 8a is an atomic force microscopy (AFM) image of the line patterns on the stamp with gradually changing channel width, which ranges from 1.4mm to 2.6mm. The stamp pattern can be completely transferred into thePMMAfilm by the above printing processes as presented in Figure 8b. Anisotropic plasma etching (APE) is usually used for faithful pattern transferring rather than IPE because IPE will also etch off the profiles of protrusions at the same time as the residual layer. In our case, we expect to scale down the lateral dimension of the CP pattern when removing the residual PMMA and the PANI covered by the residual PMMA using IPE. The width correlation of the produced PANI line and the imprinted PMMA line is illustrated in Figure 9. It is evidently demonstrated that by using a stamp features larger than 1 μm, one can scale down till 250 nm. No obviously changes in UV-Vis spectrum indicated that PANI remained intact after by the imprinting and etching process.

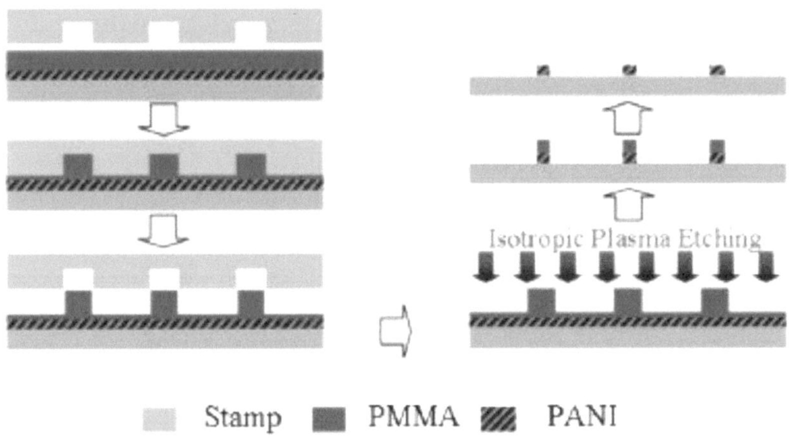

Fig. 7. Schematic illustration of the fabricating process with NIL.

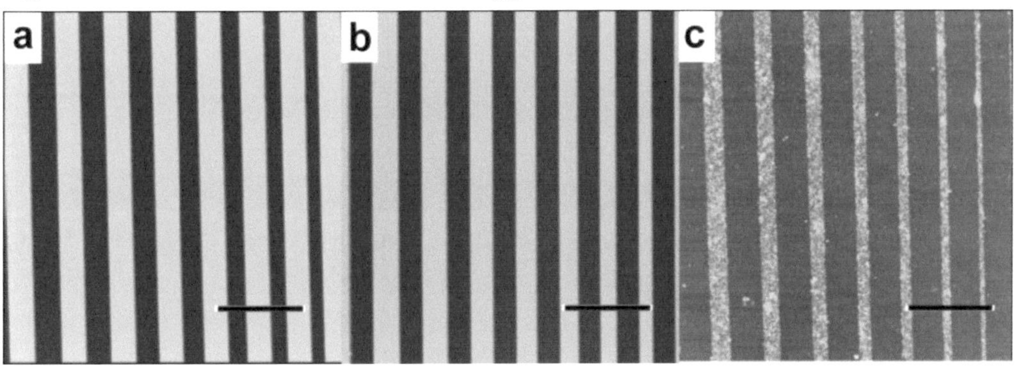

Fig. 8. AFM images of (a) the silicon stamp, (b) an imprinted PMMA structure using the stamp shown in (a), and (c) fabricated PANI structures with NIL (Scale bar¼7.5mm).

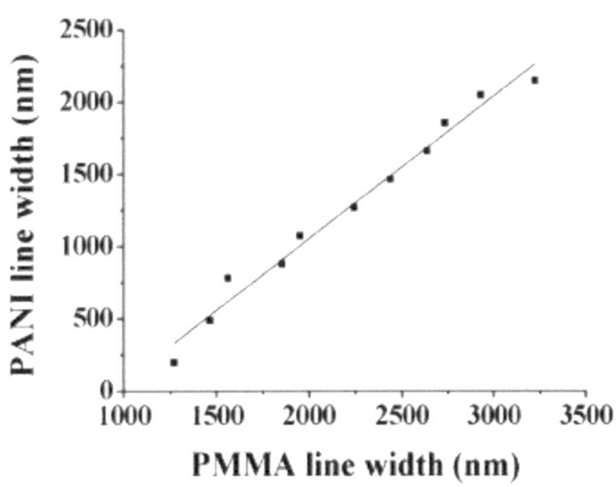

Fig. 9. The width correlation of the produced PANI line and the imprinted PMMA line.

The charged PANI structure is an ideal candidate for selective absorption of the negatively charged particles. As shown in Figure 10, Silver nanoparticles (SNPs) were selectively adsorbed on the patterned PANI. Previous investigation shows that electrostatic interaction is the main interaction between the sodium citrate-covered SNPs and the substrate, [29,30] and that the suitable pH of the particle solution is between 2 and 4. [31-33] The SNP patterns were obtained by keeping the PANI patterned substrate in the SNP aqueous solution (pH 3) for several hours. The result demonstrates that the PANI pattern is of high selectivity for SNP adsorption.

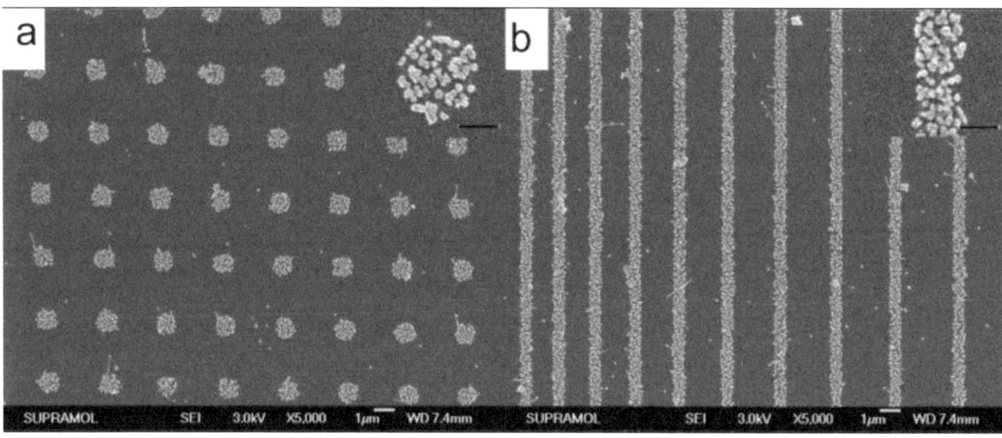

Fig. 10. SEM images of SNPs adsorbed onto the PANI dots (a) and lines (b). Scale bar=1 mm. Scale bar of the insets=500 nm.

5. Summary and outlook

We introduce here three different strategies to fabricate ordered conductive polymer structures with high special resolution and high throughput. The advanced functions in e.g. sensitive gas sensing are demonstrated. The methods presented above should be general to pattern other functional polymers. However, even we are able to achieve the resolution till 80 nm, the reproducibility below 100 nm remains a problem. Improvement of the presented methods are on going. For instance, for copolymer strategy, by using warm bathing the silane can be hydrolyzed adequately. Only few amount of py-silane (0.08 vol.-%) is needed for adhesion of copolymer film thus results in a significantly increase of the electrical conductivity and a decrease of the polymer film thickness. By such an improvement, the smallest feature size of ppy wire can be reduced to 50nm with well reproducibility. The investigation on conductivity of polymer wires well below 100 nm will be a very interesting topic both for theoretical study and practice use.

6. References:

[1] for reviews, see: a) H. Shirakawa, *Angew. Chem. Int. Ed.* 2001, *40*, 2574. b) A. G. MacDiarmid, *Angew. Chem. Int. Ed.* 2001, *40*, 2581. c) A. J. Heeger, *Angew. Chem. Int. Ed.* 2001, *40*, 2591.

[2] J. H. Burroughes, D. D. C. Bradley, A. R. Brown, R. N. Marks, K. Mackay, R. H. Friend, P. L. Burns, A. B. Holmes, *Nature* 1990, *347*, 539-541.
[3] a) G. Sonmez, C. K. F. Shen, Y. Rubin, F. Wudl, *Angew. Chem. Int. Ed.* 2004, *43*, 1498. b) G. Sonmez, H. B. Sonmez, C. K. F. Shen, F. Wudl, *Adv. Mater.* 2004, *16*, 1905.
[4] a) F. Garnier, R. Hajlaoui, A. Yassar, P. Srivastava, *Science*, 1994, *265*, 1684-1686. b) Z. Bao, Y. Feng, A. Dodabalapur, V. R. Raju, A. J. Lovinger, *Chem. Mater.* 1997, *9*, 1299-1301.
[5] C. J. Drury, C. M. J. Mutsaers, C. M. Hart, M. Matters, D. M. de Leeuw, *Appl. Phys. Lett.* 1998, *73*, 108-110.
[6] C. Arbizzani, M. Mastragostino, S. Panero, P. Prosperi, B, Scrosati, *Synth. Met.* 1989, *28*, C663.
[7] C. H. Lee, G. Yu, C. Zhang, A. J. Heeger, *Appl. Phys. Lett.* 1994, *64*, 1540.
[8] C. G. J. Koopal, B. Ruiter, R. J. M. Nolte, *J. Chem. Soc. Chem. Commun.* 1991, 1691.
[9] C. D. Mueller, A. Falcou, N. Reckefuss, M. Rojahn, V. Wiederhirn, P. Rudati, H. Frohne, O. Nuyken, H. Becker, K. Meerholz, *Nature* 2003, *421*, 829.
[10] F. A. Boroumand, P. W. Fry, D. G. Lidzey, *Nano lett.* 2005, *5*, 67.
[11] L. S. Roman, O. Inganas, T. Granlund, T. Nyberg, M. Svensson, M. R. Andersson, J. C. Hummelen, *Adv. Mater.* 2000, *12*, 189.
[12] B. J. Matterson, J. M. Lupton, A. F. Safonov, M. G. Salt, W. L. Barns, I. D. W. Samuel, *Adv. Mater.* 2001, *13*, 123.
[13] a) M. H. Yun, N. V. Myung, R. P. Vasquez, C. Lee, E. Menke, R. M. Penner, *Nano Lett.* 2004, *4*, 419. b) K. Ramanathan, M. A. Bangar, M. H. Yun, W. Chen, A. Mulchandani, N. V. Myung, *Nano Lett.* 2004, *4*, 1237. c) K. Ramanathan, M. A. Bangar, M. H. Yun, W. Chen, N. V. Myung, A. Mulchandani, *J. Am. Chem. Soc.* 2005, *127*, 496. d) X. Y. Wang, Y. G. Kim, C. Drew, B. C. Ku, H. Kumar, L. A. Samuelson, *Nano Lett.* 2004, *4*, 331. e) H. Q. Liu, J. Kameoka, D. A. Czaplewski, H. G. Craighead, *Nano lett.* 2004, *4*, 671. f) J. X. Huang, S. Virji, B. H. Weiller, R. B. Kaner, *J. Am. Chem. Soc.* 2003, *125*, 314. g) Y. F. Ma, J. M. Zhang, G. J. Zhang, H. X. He, *J. Am. Chem. Soc.* 2004, *126*, 7097. h) E. S. Forzani, H. Q, Zhang, L. A. Nagahara, I. Amlani, R. Tsui, N. J. Tao, *Nano Lett.* 2004, *4*, 1785-1788.
[14] H. Q. Liu, J. Kameoka, D. A. Czaplewski, H. G. Craighead, *NanoLett.* 2004, *4*, 671.
[15] J. X. Huang, S. Virji, B. H. Weiller, R. B. Kaner, *J. Am. Chem. Soc.*2003, *125*, 314.
[16] E. S. Forzani, H. Q. Zhang, L. A. Nagahara, I. Amlani, R. Tsui, N. J.Tao, *Nano Lett.* 2004, *4*, 1785.
[17] K. Ramanathan, M. A. Bangar, M. H. Yun, W. Chen, N. V. Myung,A. Mulchandani, *J. Am. Chem. Soc.* 2005, *127*, 496.
[18] See, for example: a) Y. Huang, X. F. Duan, Q. Q. Wei, C. M. Lieber,*Science* 2001, *291*, 630. b) P. A. Smith, C. D. Nordquist, T. N. Jackson,T. S. Mayer, B. R. Martin, J. Mbindyo, T. E. Mallouk, *Appl. Phys. Lett.* 2000, *77*, 1399.
[19] a) Y. N. Xia, G. M. Whitesides, *Angew. Chem. Int. Ed.* 1998, *37*, 550.b) C. B. Gorman, H. A. Biebuyck, G. M. Whitesides, *Chem. Mater.*1995, *7*, 526. c) Z. Y. Huang, P. C. Wang, A. G. MacDiarmid, Y. N.Xia, G. M. Whitesides, *Langmuir* 1997, *13*, 6480. d) K. M. Vaeth,R. J. Jackman, A. J. Black, G. M. Whitesides, K. F. Jensen, *Langmuir*2000, *16*, 8495.
[20] a) J. Y. Bai, C. M. Snively, W. N. Delgass, J. Lauterbach, *Adv. Mater.*2002, *14*, 1546. b) F. Shi, B. Dong, D. L. Qiu, J. Q. Sun, T. Wu,X. Zhang, *Adv. Mater.* 2002, *14*, 805. c) S. H. M. Persson, P. Dyreklev,O. Inganas, *Adv. Mater.* 1996, *8*, 405.

[21] a) J. H. Lim, C. A. Mirkin, *Adv. Mater.* 2002, *14*, 1474. b) M. Su,M. Aslam, L. Fu, N. Q. Wu, V. P. Dravid, *Appl. Phys. Lett.* 2004, *84*,4200.
[22] a) A. J. Bard, F. R. F. Fan, J. Kwak, O. Lev, *Anal. Chem.* 1989, *61*, 132. b) B. W. Maynor, S. F. Filocamo, M. W. Grinstaff, J. Liu, *J. Am. Chem. Soc.* 2002, *124*, 522. c) R. Yang, D. F. Evans, W. A. Hendrickson, *Langmuir* 1995, *11*, 211.
[23] a) E. L. Kupila, J. Kankare, Synth. Met. 1995, 74, 241–249; b) Y. Kudoh, Synth. Met. 1996, 79, 17–22; c) J. P. Pouget, Z. Oblakowski, Y. Nogami, P. A. Albouy, M. Laridjani, E. J. Oh, Y. Min, A. G. MacDiarmid, J. Tsukamoto, T. Ishiguro, A. J. Epstein, Synth. Met. 1994, 65, 131–140; d) A. J. Epstein, J. Joo, R. S. Kohlman, G. Du, A. G. MacDiarmid, E. J. Oh, Y. Min, J. Tsukamoto, H. Kaneko, J. P. Pouget, Synth. Met. 1994, 65, 149–157; e) J. P. Pouget, C. H. Hsu, A. G. MacDiarmid, A. J. Epstein, Synth. Met. 1995, 69, 119–120.
[24] a) R. A. Simon, A. J. Ricco, M. S. Wrighton, J. Am. Chem. Soc. 1982, 104, 2031–2034; b) R. J. Willicut, R. L. McCarley, J. Am. Chem. Soc. 1994, 116, 10823–10824; c) C. O. Noble, R. L. McCarley, J. Am. Chem. Soc. 2000, 122, 6518–6519.
[25] a) J. R. Reynolds, P. A. Poropatic, R. L. Toyooka, *Macromolecules* 1987, *20*, 958. b) D. Stanke, M. L. Hallensleben, L. Toppare, *Synth. Met.* 1995, *72*, 89. c) D. Stanke, M. L. Hallensleben, L. Toppare, *Synth. Met.* 1995, *72*, 167. d) D. Stanke, M. L. Hallensleben, L. Toppare, *Synth. Met.* 1995, *73*, 261.
[26] a) J. P. Blanc, N. Derouiche, A. El Hadri, J. P. Germain, C. Maleysson, H. Robert, *Sens. Actuators, B* 1990, *1*, 130. b) J. Janata, M. Josowicz, D. M. DeVaney, *Anal. Chem.* 1994, *66*, 207R.
[27] a) H. Nagase, K. Wakabayashi, T. Imanaka, *Sens. Actuators B* 1993, *14*, 596. b) M. Brie, R. Turcu, C. Neamtu, S. Pruneanu, *Sens. Actuators, B* 1996, *37*, 119.
[28] S. Y. Chou, P. R. Krauss, P. J. Renstrom, Appl. Phys. Lett. 1995, 67, 3114.
[29] L. Y. Wang, X. H. Ji, X. T. Zhang, Y. B. Bai, T. J. Li, Z. Z. Zhi, X. G. Kong, Y. C. Liu, Chem. J. Chinese U. 2002, 23, 2169.
[30] T. Yonezawa, S. Onoue, T. Kunitake, Adv. Mater. 1998, 10, 414.
[31] D. W. Thompson, I. R. Collins, J. Colloid Interface Sci. 1992, 152, 197.
[32] J. C. Chiang, A. G. MacDiarmid, Synth. Met. 1986, 13, 193.
[33] A. G. Macdiarmid, J. C. Chiang, A. F. Richter, A. J. Epstein, Synth. Met. 1987, 18, 285.